HUMAN GROWTH HORMONE

HUMAN GROWTH HORMONE

Edited by
Salvatore Raiti
National Hormone and Pituitary Program
Baltimore, Maryland
and
Robert A. Tolman
National Institute of Arthritis, Diabetes and Digestive and Kidney Diseases
Bethesda, Maryland

PLENUM MEDICAL BOOK COMPANY
New York and London

Library of Congress Cataloging in Publication Data

Human growth hormone.

Includes bibliographies and index.
1. Growth disorders. 2. Somatotropin. 3. Somatostatin. 4. Somatomedin. I. Raiti, Salvatore, 1934– . II. Tolman, Robert A. [DNLM: 1. Growth Disorders. 2. Somatotropin. WK 515 H918]
RB140.H86 1986 618.92′47 86-3288
ISBN 0-306-42053-8

233 Spring Street, New York, N.Y. 10013

Plenum Medical Book Company is an imprint of Plenum Publishing Corporation

Printed in the United States of America

Preface

It has been ten years since the National Hormone and Pituitary Program (then called the National Pituitary Agency) sponsored a symposium on human growth hormone (hGH). Numerous advances have occurred during this period. This book does not attempt to summarize past achievements. Rather, it deals with the contemporary issues in hGH research. A discussion of the present state of the art, of necessity, includes a review of the past.

Some of the topics herein discussed include the following:

1. *Growth hormone releasing factor (GRF).* In 1973, the growth hormone inhibitory factor (somatostatin) had recently been discovered. The search for a releasing factor in humans led to its discovery not in the pituitary but in a pancreatic tumor that secreted growth hormone. The advances are discussed in this book. The current hope is that GRF will eventually become an effective therapeutic agent for idiopathic hypopituitarism in childhood and adolescence.

2. *Biosynthesis of hGH by recombinant DNA technology.* Current advances are discussed. Although hGH is not yet an approved drug, it will eventually become one. This will broaden our horizons in terms of hGH effectiveness in disorders other than hypopituitary dwarfism. The current experience with this type of hGH in both the United States and Europe is reviewed by several authors.

3. *Clinical studies.* Much remains to be learned. The quality and quantity of hGH have improved in the last 7 years. The newer national data show that the older recommendation of doses of 0.1 IU/kg needs to be upgraded to 0.08 IU/kg of the new and better hGH. Much more work is needed to answer such questions as maximal effective dose and frequency of administration (daily versus twice a week versus three times a week). Subcutaneous administration of hGH has now been shown to be safe and is preferred by most patients. The long-term psychological impact of hGH therapy is discussed.

The preliminary results of hGH therapy for Turner's syndrome and for short normal children are discussed. Data are updated on such disorders as "normal variant" short stature, bioinactive, and GH-resistant syndromes.

4. *Chemistry; Receptors.* These topics are discussed and updated; including coverage of hGH action on adipocytes and binding to receptors.

5. *Somatomedin C.* This was a rapidly expanding field of research at the time of the first symposium in 1973. Ten years later, much more has been accomplished. Somatomedin C has been synthesized and its binding proteins, inhibitors, and receptors have been identified. Its mode of action on cell growth and on cartilage is discussed. Recent data on its regulation and clinical application are presented.

Overall, this book includes comprehensive data on many aspects of human growth hormone. The participants found the symposium to be of great value, as shown by both their comments and their persevering attendance at all the sessions. It is hoped that the reader, too, will find this book to be of great value.

Salvatore Raiti
Robert A. Tolman

Baltimore, Maryland
Bethesda, Maryland

Contents

Introduction I

The National Hormone and Pituitary Program

Achievements and Current Goals

Salvatore Raiti

Following the discovery of human growth hormone (hGH) in 1956 and proof of its effectiveness as a therapeutic agent for hypopituitarism in 1958, there developed an urgent need to obtain human pituitary glands at autopsy. Individual parents sought liaison with cooperating pathologists. Investigators tried to establish their own collection programs to meet the needs of their patients. By the early 1960s, it became evident that there was urgent need for a single national program which would organize the collection of all human pituitaries, arrange for the extraction and purification of hGH, and then distribute this to individual programs throughout the United States.

In 1963, this program, which was then called the National Pituitary Agency, was established through the joint efforts of the College of American Pathologists and the National Institute of Arthritis, Diabetes and Digestive and Kidney Diseases. The first part-time director was Dr. Robert Blizzard, and the program was located at the Johns Hopkins Hospital, Baltimore. In 1967, the program was transferred to the University of Maryland School of Medicine under the directorship of the late Dr. M. S. Grossman. In 1970, I was appointed the first full-time director and have continued in that capacity since that time.

1. Historical Perspective (Table 1)

1.1. hGH

The original goal of the program was to provide hGH, and this remains as one of the very important goals. The activities have expanded considerably

---

Salvatore Raiti • The National Hormone and Pituitary Program, University of Maryland School of Medicine, Baltimore, Maryland 21201.

Table 1 National Hormone and Pituitary Program Achievements

1963	Established by NIADDK and CAP
1965	TSH
1966	LH
1967	FSH
1970	FSH and LH for clinical studies
1972–74	PRL
1973	First hGH symposium
1975	Neurophysins
1978–79	β-Lipotropin
1982	PRL for clinical studies
1983	Second hGH symposium

as techniques for isolating other pituitary hormones were developed. hGH will be discussed again later.

1.2. RIA

The diagnosis of hypopituitarism in 1963 was more complex. Many used metabolic balance studies looking at hGH effects on nitrogen retention and other such parameters. The bioassay for sulfation factor (now called somatomedin C) was also used but was expensive and tedious. The radioimmunoassay (RIA) for hGH was developed in 1963 and began to find widespread use after 1965. Even this was tedious at first, since the technique required paper chromatography to separate the bound from the free fraction for every sample. It became a more readily available investigative tool when the double-antibody technique for separation of bound from free fractions became widely accepted. This development of RIA led to the need to prepare highly purified hGH for iodination, and this was the beginning of a new goal for this program, namely, to provide highly purified pituitary hormones for research, using RIA as the investigative tool. It also led to the need for production and distribution of highly specific antisera for each of these pituitary hormones.

1.3. hTSH

By 1964, there was a pressing need to provide human thyroid-stimulating hormone (hTSH) for use in RIA, which was being developed. It was decided to "sacrifice" about 10,000 acetone glands for this project. I say "sacrifice" because the objective was to obtain as much hTSH as possible, recognizing that little or no hGH would be recoverable from the side fractions. Such decisions have never again been needed. In 1965, Parlow *et al.*[1] described a method for preextracting the glycoprotein fraction without losing any of the GH fraction. This has become the standardized technique.

1.4. hFSH and hLH

Until the mid-1960s, it was thought that the human had a single glycoprotein hormone that had both follicle-stimulating (FSH) and luteinizing (LH) activities. No one had been able to identify two separate hormones. With the availability of more human pituitary glycoprotein fraction, first LH was isolated and then FSH. The first immediate needs for these hormones were to provide highly purified hFSH and hLH, which could be used for iodination for RIA. Several investigators were selected to receive the precious little of each of these hormones to try to raise antisera for use in RIA. As a consequence of these actions, by the late 1960s, these hormones (hTSH, hLH, and hFSH) and antisera became available to investigators. This led to numerous publications in basic physiology and clinical research through use of these assays. This also led to a decline in the need for bioassays of these hormones so that, today, only one or several laboratories remain equipped to run such bioassays.

As more and more LH and FSH were produced, the next urgent goal was to provide each of these hormones in purified form for use in clinical studies. This was achieved by 1971.

1.5. hPRL

Until 1971, it was believed that the human did not produce prolactin (PRL) since no one could find it in human tissue or plasma. Indeed, a lecture delivered at the NIH at about that time, strongly argued the case that hGH had both growth and prolactin activities and that it was the single such hormone in the human. Very shortly afterward, hPRL was isolated from human pituitary tissue. It could only be recovered from pituitaries that had been preserved frozen and was only recoverable in microgram quantities. Such a discovery led to the urgent need to isolate more hPRL for use in RIA and also to produce a specific antiserum for this technology. These RIA materials were first distributed through the program in 1974.[2] This led to an explosion in publications related to hPRL research. Even today the hPRL materials are the ones most commonly sought from this program.

There has been and there remains an urgent need to provide hPRL for clinical studies. Dr. A. F. Parlow, who has been responsible for isolating all human pituitary hormones since around 1977, has succeeded in producing sizable quantities of hPRL from human pituitaries. About 1 g of hPRL with a biological activity of about 30–40 IU/mg is now being produced each year. The first 4 g (representing 4 years of collections of pituitaries) has been vialed and is ready to be distributed for clinical research.

1.6. hACTH

This hormone can be and has been synthesized. However, quantities have been recovered from the human pituitary. A specific antiserum was developed. Both are distributed through this program for research.

1.7. *First hGH Symposium*[3]

This was sponsored by NIADDK, and more than 450 investigators attended from over 20 countries, indicating the degree of interest in hGH. A publication of almost 1000 pages resulted.[3] This has remained one of the classic books used by investigators. It summarized most of the new advances in hGH research up to that time.

1.8. *Neurophysins*

These are found in the posterior pituitary gland. Dr. A. G. Robinson recovered, for the program, both estrogen-stimulated and nicotine-stimulated neurophysins and produced specific antisera for each. These have been distributed for RIA studies through this program since that time.

1.9. *β-Liptropin (βLPH)*

Both the purified hormone and the antisera produced by Dr. Parlow have been distributed through this program since 1978. These are used for RIA studies. Larger quantities of βLPH are being prepared for eventual use in clinical studies.

1.10. *Newer Hormones and Activities in the Pituitary*

(1) Growth hormone-releasing factor (GHRH) has been recovered from the stalks of the pituitaries collected. It has been used to confirm the structure of GHRH recovered from other sources. (2) There is evidence that human chorionic gonadotropin (hCG) exists in the pituitary. We do not know why it is there and what function it serves. (3) An androgen-stimulating hormone has been identified in the human pituitary. Much more work is needed to produce larger quantities and to further characterize it. (4) and (5) Both bombesin and a calcium-regulating factor have been thought to exist in the pituitary. More work is currently needed to characterize and further identify these. (6) Another peptide has been identified in the human pituitary which, when injected into dogs, produces a definite diabetogenic effect. This peptide appears to share some of the structure of hGH and hPRL. Much more work is needed to confirm and characterize it. (7) A corticotropin-releasing factor has been identified in animals. Work is in progress to recover it from human pituitaries or their stalks. (8) The known hormones account for only about 10% of the weight of the human pituitary. Three other new peptides were recently identified. Their function remains unknown. The remaining 90% of the pituitary is a vast unknown. Tissue structure accounts for some of this weight. But there are surely other activities in the human pituitary yet to be discovered and of vital importance to the better understanding of human physiology and disease. The need to collect all available human pituitaries remains as urgent today as it was in 1963.

Table 2 Hormone Yields from 1000 Frozen Human Pituitaries[a]

Hormone	Activity	Weight
hGH	14,000 IU	6000 mg
LH	1,000,000 IU	200 mg
FSH	100,000 IU	20 mg
TSH	3,000 IU	150 mg
PRL	3,000 IU	75 mg
ACTH	1,250 IU	50 mg

[a] It takes the national pituitary collections from 4 to 5 years to obtain enough hormone to produce a single batch of LH, FSH, TSH, and PRL. Each of the hormones listed is isolated from all pituitaries without loss of any of the others.

2. The Human Pituitary, Recoverable Hormones, Collection Techniques

2.1. Recoverable Hormones from the Pituitary

Table 2 outlines the current yields of the commonly used hormones from "frozen" human pituitaries. hGH is the most abundant and hFSH is the least abundant representing 1/300 of the hGH content by weight, 1/10 of the LH content, and 1/6 of the TSH content. It takes the total national pituitary collections of 4–5 years to produce enough hFSH and hPRL for a batch for clinical research. Neither hPRL nor hFSH nor hLH nor hTSH can be produced synthetically now or in the near future. There is a compelling need to collect all available human pituitaries for many years to come, to provide these hormones for the much-needed clinical research.

2.2. Available Pituitaries

In recent years, there has been a 30% decline in the available human pituitaries. This is a direct consequence of the change in requirements for accreditation of hospitals. Prior to 1978, autopsies were required on a certain percentage of deaths in hospitals in order to maintain accreditation. When this requirement was lifted, the number of autopsies carried out declined precipitously. This action has affected the program.

There are other, more far-reaching effects. The autopsy is the "final court of appeal" in medicine. It identifies all disease processes in the deceased. It is of vital importance to maintain the high standards of medical care. It discourages identifying only one or two causes of death without accounting for other disease processes that coexist (identified or unidentified). It is not replaceable by the considerable advances in technology in biochemical diagnosis. It must remain a very important part of medical care that is carried through to and after death. All disease processes must be accounted for in that individual and be potentially identifiable in all future similar individuals during their living years.

2.3. Changes in Pituitary Collection Techniques

Prior to 1977, the emphasis was on collection of pituitaries stored in acetone. This was cheap and effective. The newer data showed that collection of frozen pituitaries was vastly superior. Almost twice as much hGH can be recovered. PRL cannot be recovered at all from acetone glands. Its only source is from frozen glands. The TSH recovered from acetone glands has shown a biological activity of 1.5–2.0 IU/mg. However, the TSH recovered from frozen glands shows a high biological activity of 15–20 IU/mg. The reason for this discrepancy is not understood at present. It does make the frozen pituitary a more desirable source of TSH. The quality of FSH and LH is better when recovered from frozen glands.

This urgent need for frozen pituitaries led to an immediate change in the method for collecting pituitaries. It now became important to personally collect these frozen pituitaries. This has increased that cost of recovering glands to about $1.50–$2.00/pituitary. This cost includes both the increased manpower needed and transportation. It is minimal in terms of the significant benefits, but there is no doubt that it costs more to recover frozen than acetone pituitaries.

Other methods of transportation have been tried with disastrous results. Volunteers have collected pituitaries and shipped them, but on several occasions they have arrived thawed and smelling. If each hospital were to air-freight their pituitaries to Baltimore, the cost for 20 hospitals in Chicago, for example, would be three times the cost of sending a courier, who can also visit several cities and many more hospitals for the same cost. The personal visit has the added advantage of maintaining good rapport with the diener and pathologist.

3. Pituitary Hormones for Clinical Studies

3.1. Awards of Glycoprotein Hormones (Table 3)

As can be seen, considerable amounts are awarded each year. These are minimal quantities needed because the committee has had to remain highly selective in making such awards because of limited availability. Currently the

Table 3 Human Glycoprotein Hormones Awards

	Clinical studies (IU in thousands)		Other research (mg)		
Year	hFSH	hLH	hFSH	hLH	hTSH
1977	17.0	8.8	51	43	1
1978–79	24.1	785.4	1163	172	157
1979–80	134.8	998.6	1440	928	167
1981	41.8	561.0	24	5	7
1982 (to 5/31)	72.0	836.0			
Total	289.7	3189.8	2578	1148	332

supplies are improving, and with renewed encouragement to investigators to apply for these hormones, the awards are likely to increase significantly in the next few years.

3.2. PRL for Clinical Studies

The first batch has been prepared but has not been distributed pending preliminary studies on dosage, metabolic clearance, absorption, and other factors. Its distribution might be restrictive at first, but this will change as new supplies become available.

3.3. hGH Distribution (Table 4)

When I became the first full-time director of this program in July 1970, only about 170,000 IU of hGH was available for distribution each year. This hGH was being produced by three investigators, the late Dr. M. S. Raben, Dr. A. Wilhelmi, and Dr. B. B. Saxena. They continued these activities each year until about 1977. At that time, Dr. A. F. Parlow was given the responsibility for isolating all the hormones from all the pituitaries collected.

During the early 1970s, most pituitaries were stored in acetone and shipped by mail to this program. In retrospect, this was a cheap way to recover pituitaries. A few "frozen" pituitaries (stored frozen, transported in dry ice, and kept frozen until extraction) were collected for specific research projects. The hGH recovery from acetone glands was 2–4 mg/pituitary, and it had a mean biological activity of about 1 IU/mg. Monomers were not separated from dimers. The clinical significance of these two moieties was not then appreciated. At that time, it was not possible to recover much more hGH from frozen glands than

Table 4 hGH for Clinical Studies

Year	hGH (IU)
1968	105,062
1969	129,575
1970	168,191
1971	181,769
1972	212,190
1973	277,195
1974	243,634
1975	219,821
1976	386,469
1977	533,375
1978	653,620
1979	733,964
1980	707,504
1981	819,749
1982	900,137

from acetone glands, and there was no need, therefore, to go to the added expense of collecting pituitaries of the frozen type.

Our only recourse to obtaining more hGH was to try to recover more pituitaries. This was achieved, and our peak collections rose within a few years by about 25,000 pituitaries per year. Through this mechanism and through further improvements in the isolation of hGH, the amount available for distribution rose, by 1972, to a plateau of about 250,000 IU/year. As a consequence, more new patients were entered into the program and studied. However, restrictions remained. hGH was only available for about half of the diagnosed hypopituitary dwarfs. They were given 8 months' supply for each 12-month period. HGH was stopped when they achieved a height of 5 ft (153 cm).

One of the many outstanding achievements of Dr. Parlow, when he took over the isolation of hormones, was to improve both the quantity and quality of hGH recovered. Only high-monomeric hGH, and of high biological activity of 2.4 IU/mg was produced. The consequence of using only the monomeric form was that the phenomenon of antibody production in treated patients disappeared overnight. Dr. Parlow worked avidly with both types of pituitaries. His recoveries of hGH from acetone pituitaries increased remarkably to 8 IU/pituitary of monomeric hGH, and his recoveries from frozen glands were almost double (14 IU/pituitary). Since I was personally responsible for encouraging Dr. Parlow to work with frozen glands (I had been persuaded by our European colleagues), I can take some pride in watching his accomplishments grow.

As Table 4 shows, the available hGH since 1977 has increased steadily so that we are now producing 500–600% more hGH from much fewer pituitaries than were available to us in 1972. The benefits have been that every hypopituitary child who needs hGH can have it made available to him. A surplus of hGH remained. Therefore, the other restrictions were lifted. HGH was provided for 12 months of the year and until these children achieved heights of 164–166 cm (5 ft 4 in. to 5 ft 6 in.). Since there were still surpluses, the Committee on Growth Hormone Usage decided that further studies could be carried out. These included the reevaluation of dosage and frequency of administration and the use of hGH in non-hGH-deficient dwarfs (short normal children) and in Turner's syndrome. The results of these investigations are reported elsewhere in this book.

However, as described earlier, the number of available pituitaries continued to decline as a consequence of the changes in hospital accreditation requirements and fell to a new plateau. The amount of hGH available also declined to a new plateau of about 750,000–800,000 IU/year. In order to cope with the deficit, it became necessary again to restrict the amount of hGH given to each patient to only 8 months' supply per year.

Currently there is a flurry of activity in state legislatures to have laws permitting the retention of human pituitaries by medical examiners and coroners. This has resulted in a steady increase in available pituitaries. It is hoped that the current restriction on hGH supply to patients can be lifted shortly.

The emphasis on hGH research has been directed toward the hypopituitary dwarf. There is a potential vast need for hGH for other short but growing children. The promised availability of hGH from biosynthetic sources is sure to meet this need. However, the hGH produced from human pituitaries remains the standard against which all others are studied. It is highly purified and very cheap to produce and is likely to remain the standard for years to come. Currently the total cost of hGH production from human pituitaries is about $1.00/IU.

4. *Pituitary Hormones for RIA and Other Studies*

Table 5 outlines the number of ampules of highly purified hormones distributed each year. Each ampule has enough hormone for 3–12 months' work. New supplies ought to be requested every 6–9 months so that the investigator has "fresh" hormone to work with.

Table 5 shows the distribution of hormones of human, rat, ovine, and bovine origin. As can be seen, PRL is the hormone most often requested for all species. FSH is also a hormone very much needed.

Not shown in Table 5 is the distribution of antisera for each of these hormones. An ampule of antiserum generally contains enough material for 10,000–15,000 RIA tubes and, again, is enough for 3–12 months' work. Fresh supplies ought to be requested every 6–12 months.

Currently there is a handling charge of $12.00/ampule of hormone or antiserum, which must be prepaid. The animal hormones used for *biological* studies are also shown in Table 5. The comments made for RIA also apply to the distribution of these. The handling charge also applies.

The list of available materials for distribution is described by Dr. Tolman in Introduction II. These lists are updated twice a year and appear in the January and July issues of *Endocrinology* and *Journal of Clinical Endocrinology and Metabolism* and in several other journals. The updated list is also freely distributed on request to this program.

Table 5 Hormones Distributed for Immunochemical Research in 1981

Hormone	Human (vials)	Rat (vials)	Ovine (g)	Bovine (g)
GH	179	205	15.6	8.4
FSH	340	306	1.8	—
LH	214	327	2.9	0.8
TSH	171	161	0.4	0.2
PRL	448	422	35.5	8.6

5. Current Goals

1. Collect all available human pituitary glands in the United States.
2. Organize preparation and distribution of hGH, FSH, LH, and PRL (and TSH soon) for clinical research.
3. Distribute all pituitary hormones and their antisera of human, rat, ovine, bovine, and porcine origin for research.
4. Encourage research in newer pituitary peptides by making available supplies of side fractions or of whole pituitaries for such research.
5. Upgrade the quality of hormones produced if possible, for example, FSH that is completely devoid of LH.
6. Seek methodology for increasing the sensitivity and specificity of antisera used for RIA and other purposes.
7. Encourage further chemical and structural studies of the glycoprotein hormones so that synthesis can eventually be achieved.
8. Continually seek methodologies that will improve the yields and quality of hormones isolated from the pituitary.
9. Continually encourage pathologists, internists, and others so that more pituitaries can become available.
10. Work toward production of more TSH, PRL, lipotropins, and other hormones so that they can become available for clinical research.

6. Current Activities

1. *Collect* all human pituitary glands.
2. *Clinical research:* Distribute hGH, FSH, LH, and PRL.
3. *Biochemical research:* Distribute quantities of all pituitary hormones.
4. *RIA and similar studies:* Distribute such materials for research.
5. *Animal pituitary hormones:* Distribute rat, ovine, bovine, and porcine hormones and their antisera for biological studies, for RIA, and for other research.

7. Other Sources of Animal Pituitary Hormones

1. The bovine and porcine hormones are produced by contract from the U.S. Department of Agriculture with Dr. Parlow and are given to this program to distribute for research.

2. The hCG and the monkey hormones and antisera are produced by contract from National Institute of Child Health and Human Development and are given to this program to distribute for research.

3. Canine, rabbit, cat, and mouse pituitary hormones are produced and distributed by Dr. Parlow on request to him.

8. General Perspective

This program was conceived in the early 1960s and implemented in 1963. It had a single goal of providing hGH for hypopituitary dwarfs. It soon became possible to isolate the other hormones from the same pituitaries. Each of these were added as the technology became available.

The needs for hormones for RIA became evident by 1964. An additional goal has been to provide such highly purified hormones and also the specific antisera that are needed.

As larger quantities of hormones became available, the goals expanded again to provide these hormones for clinical use.

When the needs for hGH increased as a consequence of easier diagnosis of hypopituitarism, efforts were implemented to increase both the quality and quantity of hGH. This was achieved predominantly by a change to collection of frozen glands. Though this is a more costly procedure, it has resulted in enormous benefits not only in terms of hGH supplies but also as the sole source of hPRL and the greatly improved quality and quantity of TSH, FSH, and LH.

Currently, the program supplies the needs for the hormones for clinical studies within the United States. It provides the hormones and antisera needed for RIA studies for scientists from any country who request them. It provides the animal pituitary hormones and their antisera needed in basic research. It encourages research into new or yet-undiscovered hormones within the pituitary. It continually upgrades the quality and quantity of materials needed as newer technology makes this possible.

Although far from the original concept, it has provided a model on how the needs of thousands of scientists can be met efficiently, cheaply, and with the highest standards. The enormous advances in pituitary hormone research are attributable, in large measure, to the ready availability of these materials.

The National Institute of Arthritis, Diabetes and Digestive and Kidney Diseases, the College of American Pathologist, and the original supporters of this program are to be congratulated for their efforts and foresight. The scientific community throughout the world owes a great deal of gratitude to them and to the program. The better understanding of pituitary physiology and disease is a direct consequence of such actions. Numerous hypopituitary dwarfs and patients with other pituitary disorders are deeply thankful that this program was conceived and implemented 20 years ago.

References

1. A. F. Parlow, A. E. Wilhelmi, L. E. Reichert, Jr. Further studies on the fractionation of human pituitary glands, *Endocrinology* **71:**1125, 1965.
2. M. L. Aubert, R. L. Becker, B. B. Saxena, *et al.*, Report of the National Pituitary Agency. Collaborative study of the radioimmunoassay of human prolactin, *J. Clin. Endocrinol.* **38:**1115, 1974.
3. S. Raiti (ed.), *Advances in Human Growth Hormone Research*, U.S. Dept of Health, Education and Welfare. Government Printing Office, 1974.

Introduction II

The NIADDK Hormone Distribution Program

Robert A. Tolman

By the middle 1950s researchers had established that the human pituitary produced human growth hormone (hGH), which was species specific—thus explaining why animal growth hormone would not work in humans. Because human pituitaries were then the only source of hGH, many individual efforts were mounted around the country to collect human pituitaries at autopsy so that the growth hormone could be extracted and given to children with demonstrated GH deficiency. It soon became obvious that chaos would result if various groups were all trying to collect and compete for the scarce and valuable materials. In order to consolidate the collection of pituitaries, the College of American Pathologists and the National Institute of Arthritis and Metabolic Diseases [now the National Institute of Arthritis, Diabetes and Digestive and Kidney Diseases (NIADDK)] agreed to cooperate on the formation of a centralized collection agency; thus the National Pituitary Agency (NPA) was established. (NIADDK's involvement in this program is possible because the material produced is used for research. Each child who receives GH and is helped to grow must be involved in a research protocol.)

The NPA was first housed at the Johns Hopkins University Medical School with Dr. Robert Blizzard as its director. Glands were collected from pathologists around the country, sent to Baltimore, and then distributed to various investigators who extracted the GH, which was then used for clinical investigation. Other hormones were also distributed as shown in a 1963 announcement from the back cover of *Endocrinology* (Fig. 1). Originally distribution was coordinated through the National Institutes of Health Endocrinology Study Section. Since 1970 it has been handled by NIADDK or the Distribution Contractor.

Robert A. Tolman • Endocrinology Research Program, Division of Diabetes, Endocrinology and Metabolic Diseases, National Institute of Arthritis, Diabetes and Digestive and Kidney Diseases, National Institutes of Health, Bethesda, Maryland 20205.

NATIONAL INSTITUTES OF HEALTH

Research Materials Available to Qualified Investigators

The National Institutes of Health, Public Health Service, through the Endocrinology Study Section, is supporting a program which makes the following materials available free of charge to qualified investigators:

I. *Pituitary Hormones:*

	Vial
Human Pituitary Gonadotrophin (urinary) (NIH-HPG-UE)	25 mg.
Follicle-stimulating hormone (ovine), sterile (for assay)	5 mg.
Follicle-stimulating hormone (ovine), sterile (experiment)	25 mg.
Growth hormone (ovine or bovine), non-sterile	(Bulk)
Luteinizing hormone (ovine or bovine), sterile	10 mg.
Luteinizing hormone (ovine) Reference Standard for bioassay only: 2mg.LH (2 units) + 8 mg. lactose	10 mg.
Prolactin (ovine), sterile	25 mg.
Thyroid-stimulating hormone (ovine or bovine)	10 mg.
Lysine Vasopressin:	
2 or 20-units in acetic acid, or chlorbutanol, sterile	
260 or 2600 units in nitrogen, lyophilized	1 or 10 mg.

II. *Radioactively-labeled Steroids:*

	Specific Activity	Vial
Aldosterone 1,2-H^3	100 mc/mg.	100 μc.
Corticosterone 1,2-H^3	160 mc/mg.	100 μc.
Cortisol 1,2-H^3	50 mc/mg.	100 μc.
Desoxycorticosterone 1,2-H^3	110 mc/mg.	100 μc.
Progesterone 1,2-H^3	100 mc/mg.	25 μc.
Testosterone 1,2-H^3	150 mc/mg.	200 μc.

To obtain supplies of these materials, investigators should send a brief description of the proposed usage and an estimate of the amount of material required. Requests are filled as promptly as possible after review by the Committee on Hormone Distribution.

Requests for materials listed in Group II should be accompanied by the applicant's Atomic Energy Commission registration number governing the use of isotopic materials, and a purchase order certifying that the handling charge of $5.00 per ampule will be paid to Tracerlab, Inc., Waltham, Massachusetts. Institutions requiring clearance for receipt of radioactive materials must comply with safety regulations and indicate clearance with each request.

Requests for supplies in either group should be submitted to

MORRIS M. GRAFF, *Executive Secretary*
Endocrinology Study Section
Division of Research Grants
National Institutes of Health
Bethesda 14, Maryland

Figure 1 An early announcement of the availability of research materials. From the back cover of *Endocrinology*, Volume 72, April 1963.

About 1967, the NPA was transferred to the University of Maryland. More recently the name has been changed to the National Hormone and Pituitary Program (NHPP), which more closely describes its function.

In addition to the growth hormone other anterior pituitary hormones are also extracted and purified so that LH, FSH, PRL TSH, and ACTH are distributed as well as the alpha and beta subunits of TSH, FSH, and LH. Other hormones produced and delivered are rat and ovine pituitary hormones and, in conjunction with the U.S. Department of Agriculture, bovine and porcine pituitary hormones. The National Institute of Child Health and Human

ANNOUNCEMENTS

National Institute of Arthritis, Diabetes and Digestive and Kidney Diseases

Research Materials Available to Qualified Investigators

The research materials listed on the following page are procured and distributed through programs supported by the National Institute of Arthritis, Diabetes and Digestive and Kidney Diseases, by the Center for Population Research of the National Institute of Child Health, and the Agricultural Research Service of the U.S. Department of Agriculture, to qualified investigators, and are distributed only for research purposes, not for therapeutic, diagnostic or commercial purposes.

Supplies of these materials may be obtained by submitting a one page letter IN DUPLICATE to: Doctor Salvatore Raiti, Director, National Hormone and Pituitary Program, Suite 201-9, 210 West Fayette Street, Baltimore, Maryland 21201, U.S.A.

Applications (one page letter) must include: (a) a progress report and/or a recent publication list of results from materials previously awarded, (b) a description of the proposed use of materials, brief but sufficient to justify the kinds and amounts requested, (c) evidence of experience with radioimmunoassay with all new requests for reagents for radioimmunoassay, (d) your correct mailing address including country (if from outside the U.S.A.). (e) Letters from trainees must also be signed by a senior staff member.

Only one vial, or one kit, of the reagents, animal or human for radioimmunoassay, is awarded at a time. Requests for large amounts of materials may be granted, subject to review by the Committee on Hormone Distribution.

NOTE: There is a shipping and handling charge at the rate of *$12.00 per vial* for all hormones or antisera. There is no handling charge for the following: bFSH, pLH, pGH, pPRL, hCG, mFSH, mLH, rABP, and PMSG. No materials for which there is a handling and shipping charge will be shipped *if not prepaid*. Monies must be sent in U.S. dollars. Checks, etc. are to be made out to *National Hormone and Pituitary Program* and must accompany each request. The name of the investigator should appear on all check vouchers and/or purchase orders. Purchase Orders can be accepted *only* from U.S. universities or institutions.

For Investigators within the USA Only

(a) *Biochemical Research*. hTSH, hLH and hFSH are available in highly purified form and in quantities sufficient for biochemical studies.

(b) *Human pituitaries* in very limited numbers can be requested for specific research if the purified pituitary materials are not otherwise available.

(c) *Clinical Research*. Human growth hormone (hGH) (14 IU/vial) and human luteinizing hormone (hLH) (2200 IU/vial) are available.

The application forms for the above and other human pituitary hormones for chemical and other studies may be obtained by writing also to Dr. Raiti.

Figure 2 First page of current announcement on the availability of research materials.

Development also distributes materials through NHPP, e.g., chorionic gonadotropin. In addition, antisera to most of these hormones have been developed and are distributed for use in radioimmunoassay work. Figures 2 and 3 are copies of the two-page announcement published semiannually in several journals, listing the materials available and describing how they may be obtained.

1. *Administration of the Hormone Distribution Program*

The NHPP has always been administered as a contract. However, the production and purification of the hormones was originally accomplished by

ANNOUNCEMENTS

	Human RIA vials			Rat RIA vials				Ovine RIA vials				Bovine RIA vials			Porcine RIA vials		
	H	AS	RP	H	AS	RP	Biol	H	AS	RP	Biol	H	AS	Biol	H	AS	Biol
GH	1	1	1	1	1	1	5 mg	1	1	-	50 mg	1		10 mg	1		10 mg
PRL	1	1	1	1	1	1	5 mg	1	1	-	50 mg	1		10 mg	1		10 mg
FSH	1	1	1	1	1	1	5 mg	1	1	1	0.25 mg	5		3.3 mg	5		10 mg
LH	1	1	1	1	1	1					5 mg	1		5 mg	1		10 mg
TSH	1	1	1	1	1	1	-	1	1			1	1	0.5 mg	1		-
ACTH	1	1	-	-	-	1		-	-			-	-	-	-		-

H, Hormone; AS, antiserum; RP, reference preparation.

Subunits

Human(h), ovine(o), rat(r) kits

oLHα	1
oLHβ	1
hCGα	1
hCGβ	1
hTSHα	1
hTSHβ	1
hFSHα	1
hFSHβ	1
hLHα	1
hLHβ	1
rLHα	Antiserum only
rLHβ	1
rTSHβ	1
hCGα	2–5 mg
hCGβ	2–5 mg

Miscellaneous

1) *Human*	
PL	1 vial
PL antiserum	1 vial
CG	1 vial
CG antiserum	1 vial
β-LPH	1 vial
β-LPH antiserum	1 vial
β-LPH Ref. Prep.	1 vial
Neurophysins	
ESN	1 kit (no Ref. Prep)
NSN	1 kit
Somatomedin-C antiserum[a]	1 vial
Somatomedin-C Ref. Prep.[b]	1 vial
Antialdosterone serum	5 ml
bPTH(RIA)	2 μg
bPTH(Bio)	1 mg
2) *Canine*	
GH	1 kit
PRL	1 kit
3) *Rabbit*	
GH	1 kit
PRL	1 kit
LH	1 kit
FSH	1 kit
TSH	1 kit
4) *Mouse*	
GH(RIA)	0.1 mg
PRL	1 kit
5) *Monkey*	
FSH(RIA) non-human primate	1 kit
Hemagglutination inhibition for rhCG	1 kit
LH(RIA)	1 kit
6) *Rat* Hypothalamic extract	250 ME
Rat ABP[c]	1 kit
7) *Other*	
LH-FSH-RH	100 μg, 10 mg
Relaxin	1 mg
PMSG	2000 IU
Labeled Somatomedin-C for RIA for non-human species (U.S.A. only)	
8) *Antisera for Immunocytochemistry*	
hGH, hPRL, hFSH beta, hLH beta, hTSH beta, and rPRL	1 vial
9) *Insulin, etc.*	
Chicken insulin	100 μg
Guinea pig insulin	80 μg
Porcine desoctapeptide insulin	100 μg

No handling charge for the following: *hFSH, pLH, pGH, pPRL, hCG, mFSH, mLH, rABP,* and *PMSG*

[a] You must have Somatomedin-C for labeling before you can apply for the antiserum.

[b] Two types of reference preparation for Somatomedin-C: (a) lyophilized acromegalic serum, and (b) acid chromatographed acromegalic plasma.

[c] These will be shipped in the *frozen state by air express.* Overseas investigators *must* supply the following additional information for shipping purposes: 1) Telex number, 2) point or city of entry of materials into that country.

Figure 3 Second page of current announcement. Published twice yearly in endocrine journals.

individual investigators working on research grants who performed the extraction as part of their research, or else extracted pituitary glands in order to keep small amounts of the hormone and/or the side fractions from the extraction for their own research needs. Some of the workers who participated in this manner were Dr. Brij Saxena, Dr. Alfred Wilhelmi, and the late Dr. Maurice Rabin. In particular, Dr. Wilhelmi had a grant that was dedicated in large part to the preparation and purification of pituitary hormones.

In 1974, the National Academy of Sciences was requested by NIADDK to review and evaluate the NHPP. Their report recommended that the entire Hormone Distribution Program be run on a contract basis because this allows for greater control of the production and distribution of these very valuable materials. In compliance with this recommendation, the NIADDK now administers three contracts in order to produce and distribute hormones.

The first contract is to the University of Maryland to support the NHPP. Its purpose is to collect and store human pituitaries, send them to the extraction contractor, and vial, store, and distribute the extracted human pituitary hormones used in clinical research. The NHPP also stores and distributes animal hormones and antisera. Dr. Salvatore Raiti, the director, supervises a staff of about 10 persons. A major effort of this program is the collection of frozen human pituitaries, which give a much better yield of all hormones. Currently, the NHPP staff collects between 50,000 and 60,000 human pituitary glands in this country. The number of glands collected has declined somewhat from earlier years, because of the decrease in number of autopsies being done in the United States. The total cost of this operation is about $750,000 per year.

The other two contracts, awarded to Harbor General Hospital in Torrence, California, are for the extraction and purification of hormones. The first contract is for the extraction and purification of human pituitary hormones. The hormones are recovered from pituitaries that are either frozen or acetone-preserved. The second contract is for the production of rat and ovine pituitary hormones. All the hormones are highly purified by Dr. Albert Parlow and his staff. The total cost of these contracts is about $500,000 per year.

2. *Medical Advisory Board and Subcommittees*

The NHPP has established a Medical Advisory Board, consisting of 8–10 members, including two pathologists representing the College of American Pathologists. Two subcommittees have been established, which include some members of the Medical Advisory Board as well as outside members. These are (1) the Growth Hormone Subcommittee, which advises NHPP on GH distribution and reviews protocols, and (2) the Glycoprotein Subcommittee, which advises on distribution of FSH, LH, and TSH, and their purity and methods of assay. In addition, this group will be used to review requests for animal hormones, review research protocols, and assess purity of the various hormones produced. These committees meet at least once a year, usually at the annual meeting of the Endocrine Society, and may meet more often if necessary, depending on the business to be transacted.

3. *Future Plans for NHPP*

The NIADDK staff feels that the NHPP has filled a need in supplying hormones and antisera to the research community. Because new sources of hormones are developing, the emphasis of our work may change in the near future. However, we expect to continue to collect human pituitaries for some time and to make the hormones available for research purposes. At the same time we are looking for new materials to distribute that are potentially useful in research and are not currently available from other sources. To this end, suggestions of what materials might be distributed are always welcome, and the program hopes to continue to add to its inventory. We also anticipate that there will be a NHPP review similar to the 1974 review, a year or two from now.

We do expect to remain in operation for the foreseeable future.

I

HYPOPITUITARISM AND HYPERPITUITARISM

Clinical Studies

1

A Perspective on Growth Hormone and Growth

A Tribute to Maurice Raben

Henry G. Friesen, Heather J. Dean, and Susan Kasper

It is a fitting tribute to recognize Dr. Maurice Raben because his pioneering contributions served as a great catalyst to studies on growth hormone (GH). His research ushered in a quarter-century of progress in understanding growth and GH-related problems. Maury Raben was born in New York and graduated with an M.D. in 1941 from New York University following which he served with distinction in the Armed Forces until 1946. After his residency training he became a research fellow at the New England Center Hospital in 1948 in Dr. E. B. Astwood's laboratory. It was here that his career in endocrine research was launched and where he remained for the rest of his professional life. At the time of his death he was Director of the Cardiovascular Research Program at New England Center Hospital and Professor of Medicine at Tufts.

When Maury Raben joined Astwood, two principal areas of investigation were being pursued in Astwood's laboratory. The first involved the mechanism of action of antithyroid drugs and the second the purification of corticotropin. Raben participated in both studies, but it was undoubtedly the second theme that led to his interest in GH. As early as 1951 he coauthored a paper on "Recovery of Growth Hormone in Purification of Corticotropin."[(1)] In the intervening years between 1951 and 1957, he authored other papers on the separation of pituitary hormones and their metabolic effects. His classic on "Preparation of Growth Hormone from Pituitaries of Man and Monkey" appeared in *Science* in 1957[(2)] along with a companion paper by Beck *et al.*[(3)] on "Metabolic Effects of Human and Monkey Growth Hormone in Man" in 1957. Purification and characterization of proteins in that period in many

Henry G. Friesen, Heather J. Dean, and Susan Kasper • Departments of Physiology and Pediatrics, University of Manitoba, Winnipeg, Manitoba, Canada R3E OW3.

respects was primitive by today's standards. Paper chromatography and paper electrophoresis were ill suited to separate and resolve proteins, but that was the best that was available.

The absolute requirement for human GH to promote growth in man was suggested by studies of Knobil *et al.*,[4] who showed that monkey GH, but not nonprimate GH was effective in monkeys. Several years earlier Wilhelmi had reported that purified fish GH was inactive in rats whereas it promoted nitrogen retention in fish.[5] In retrospect it seems strange that the concept of zoological specificity of GH took so long to be recognized, but one must appreciate that no similar species specificity was observed when insulin or corticotropin derived from nonprimate sources was used in man.

The efficacy of human GH (hGH) in treatment of GH-deficient dwarfs was widely confirmed using a number of hGH preparations. Initially the collection of human pituitaries for extraction of hGH was a fairly informal arrangement, which was largely dependent on the enthusiasm and interest of local investigators. As the need for larger supplies of hGH became evident, a coordinated approach for the collection and extraction of pituitaries was developed under the auspices of the National Institutes of Health (NIH). Thus the National Pituitary Agency was formed. I cannot think of any program along with its successor the National Hormone and Pituitary Program that has been so beneficial and cost-effective to so many investigators in endocrinology. As "foreign" investigators, we know we speak for endocrinologists around the world when we express our appreciation to the NIH for its generosity in making a variety of pituitary hormones available to the entire endocrine community.

That same generous approach was the hallmark of Maury Raben's style. From the outset he purified hGH for all who requested it regardless of country of origin. We all owe a great deal to this thoughtful, generous, and kindly man whose pioneering contributions were instrumental in demonstrating an effective approach to treatment of children of short stature and whose discovery spurred the creation of the National Pituitary Agency. The demonstration of the effectiveness of hGH therapy also was the catalyst for a great deal of research on the regulation of secretion of GH and its mechanism of action in the next two decades. It clearly provided the *raison d'être* for the "gene jockeys" to produce hGH by recombinant DNA techniques. Thus, the significant discovery by Maurice Raben of the effectiveness of hGH launched an era of progress in our understanding of growth.

1. Advances in Research on GH

In reflecting on the advances that have occurred in the quarter-century since Raben's seminal research, we have identified several that are particularly significant (Table 1). One can also identify them by looking at the topics that are discussed in this volume. It it not our intention to review all of these, but we do wish to focus on a few particular aspects that interest us.

Table 1 Milestones in Research on Growth Hormone

Purification of hGH
Treatment of hGH-deficient dwarfs
Neuroendocrine control
RIA
Somatostatin
GHRH
Mechanism of action
Sulfation factor
Somatomedins, IGF1,2
GH receptors
Multigene family
Biosynthetic hGH
Gene transfer
Clinical studies
Therapeutic trials, NPA MRC
Diagnostic criteria
Laron dwarf

2. The Politics of Height and Heightism

With the demonstration that hGH therapy increased the growth rate of GH-deficient dwarfs,[6] there was considerable interest in the possibility that hGH might also be used to treat short, otherwise healthy children. A few limited studies in patients of this kind using hGH were attempted, largely with negative results.[7,8] Enthusiasm for more extensive studies of this kind was tempered in this early period by the recognition that supplies of hGH were limited and that it would be prudent to restrict use of the material to bona fide cases of hGH deficiency. With the imminent availability of recombinant DNA-produced hGH in large amount, there is renewed discussion and enthusiasm for using hGH in short but otherwise healthy children. Interest has been fueled by the demonstration that some children whose secretion of hGH seems normal do respond to hGH therapy.[9,10] These findings raise several questions. (1) How does one identify GH-deficient patients? (2) What is the error factor in making the diagnosis or what is the specificity of the diagnostic criteria? (3) How does one select short, otherwise healthy children who would benefit from hGH therapy? (4) How do we define "benefit"? (5) Why treat short, otherwise healthy children at all? (6) Are we susceptible to the politics of height and heightism?

In an interesting monograph entitled *Too Tall, Too Small* John S. Gillis has examined Society's preoccupation with the importance of height.[11]

> "Society is obsessed with height and our language reflects that obsession. Height bias is very deeply embedded in the English Language. The synonym for height is stature and words such as stature and status are used synonymously. There is a very strong association between height estimates and occupational esteem. The bottom line is that our economic status is influenced by our height. Height can be important in employment, salary and promotional decisions. Heightism, like racism or sexism, may

be an important source of prejudice and discrimination on the employment (or unemployment) scene.

If we accept this analysis and the conclusions about the importance of height in our society, then we begin to understand reasons behind the pressures that are generated to treat short children to enhance their stature. Pediatricians are very cognizant of the forcefulness and persistence of the pressures exerted by parents when they request treatment of their offspring who are short. The analysis by Gillis provides some of the reasons why parents at times are so demanding. Is it possible that those of us involved in treatment of hGH-deficient children too may have been mesmerized, albeit inadvertently, by "stature" and changes in growth rates produced by hGH therapy?

In reviewing the various reports on treatment with hGH of GH-deficient children,[6] it is remarkable that (1) most are concerned with changes in growth rates during the course of treatment; (2) few with final heights achieved, for example, in relation to midparental height[12,13]; (3) none with growth after discontinuing hGH treatment; and (4) none with the level of psychosocial integration into the community of hGH-treated patients. How well do individuals who are treated with GH for many years function as adults in society? What is their educational, employment, and marital status? Has the treatment program improved the prospects of these individuals to lead useful, productive lives or, even worse, unintentionally decreased their chances of doing so by focusing so much attention on their stature that we have unwittingly reinforced their sense of being different and thereby enhanced feelings of dependency? Regrettably and strikingly, none of the studies of hGH therapy that we are aware of have been designed to address these questions, but they are even more germane now that the prospect of treatment of many more short healthy children is at hand.

3. Medical Research Council of Canada Therapeutic Trial of hGH—Analysis of Psychosocial Factors

As a result of some of these concerns, we have tried to obtain answers to some of these questions. We have interviewed graduates of the Medical Research Council (MRC) of Canada Therapeutic Trial of hGH, a collaborative Canadian study of the treatment of hGH-deficient dwarfs that began in 1967. Since that time we have treated 756 patients, with 376 currently receiving hGH treatment.

More than 100 GH-deficient adults, aged 18–40 years, who were treated with at least two 6- or 10-month courses of growth hormone during childhood were interviewed regarding their educational attainment, current employment, and marital status. The education of patients who had completed their formal education was similar to that of their siblings and the general population, except that post-secondary-school education of the GH-deficient patients was more frequently vocational training than university. The rate of unemployment of patients who were in the labor force (i.e., had completed their formal education) was approximately three-fold greater than the rate of unemployment in the general population. Educational achievement did not appear to affect employment status. The distribution of the employed persons in occupational categories and the ratio of full-time to part-time work were similar to those of the general population. Remarkably, of the GH-deficient patients not attending school, almost 80% were living with their parents or relatives, suggesting a higher-than-expected "dependent status." The percentage of GH-deficient patients who were married was more than 50% below normal for each age group in the general population, strongly suggesting psychosexual maladjustment in many patients. There was no association between the educational achievement, occupational or marital status, and the diagnosis, pretreatment stature, or response to hormone therapy.

In summary, from our survey, GH-deficient patients appear to perform satisfactorily in the public school system, but their rates of employment and marriage are greatly reduced compared to control subjects. This outcome appears unrelated to the response to hGH therapy and emphasizes the need to develop strategies that lead to more satisfactory psychosocial integration and productive employment of GH-deficient patients in adult life. It is hoped that our results will stimulate others to examine the educational, vocational and marital status of GH-deficient adults treated with GH during childhood. If the conclusions of our studies are confirmed by others, they emphasize one inadequacy of our present program and impress upon us the need for a more inquiring attitude about the reasons for the difficulties encountered in adult life by patients treated with hGH.

4. Specificity of hGH—Diagnosis and Unanswered Questions regarding Treatment

A second significant gap in our knowledge of hGH deficiency is the problem of diagnosis. What is the "gold standard" upon which the diagnosis must be based? Several years ago it seems we were somewhat more confident than now that the specificity of diagnosis was very high. After all, if someone were diagnosed as being "hGH-deficient," it would dictate years of treatment, a decision not to be taken lightly. The criteria employed in establishing the diagnosis in the MRC Therapeutic Trial of hGH included clinical, radiological, and biochemical criteria widely accepted elsewhere. When conducting our survey of adults who are treated with hGH during childhood, we inquired about their height at the time of the interview. To our considerable surprise, a number of patients have increased in height by several inches in the years since hGH therapy was discontinued. One example of such a patient is summarized in Fig. 1. Do these individuals reflect the degree and extent of errors of diagnosis? Is it possible that some of these patients had transient hGH deficiency and therefore, when first investigated, satisfied the diagnostic criteria for establishing hGH deficiency? In some patients with age/or treatment with hGH therapy, do abnormalities of the regulation of hGH secretion correct themselves?

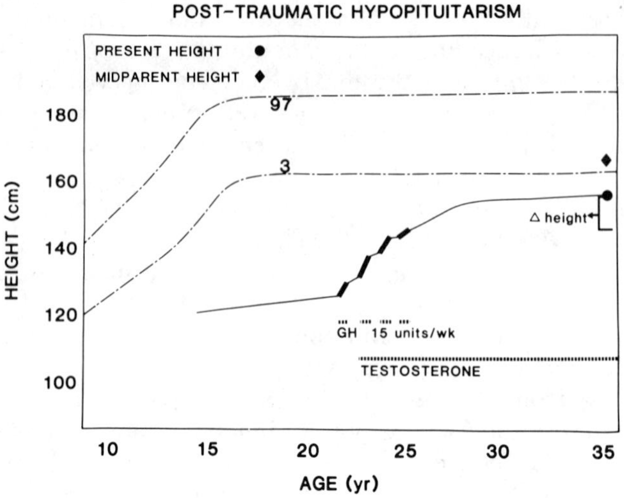

Figure 1 A hypopituitary patient was treated with hGH plus testosterone for several periods, as shown. As indicated, his growth velocity increased on each occasion. Despite no further hGH treatment, he continued to grow ultimately reaching a height that was just below the 3rd percentile. The mechanism leading to continued growth in the absence of hGH therapy raises the possibility that transient hGH deficiency occurs in some patients or that other growth stimulatory factors not detected by RIA may be secreted.

Is this the explanation for the subsequent surprisingly large spontaneous growth that was observed in some of our patients? Or do these patients represent those who were mistakenly diagnosed as hGH-deficient and therefore constitute the "error factor" in diagnosis? If so, they highlight the inadequacy of present criteria for establishing the diagnosis of hGH deficiency.

The confidence with which the diagnosis of hGH deficiency is made is further challenged by the reports of children with bioinactive GH who respond to hGH therapy.[9,14,15] Unanswered questions are posed by the term "bioinactive hGH." What is it and what are adequate criteria for defining it?[16]

Healthy short children are reported to respond at least in the short term with an increased growth velocity.[10] In this group, are we at our peril failing to distinguish between physiological and pharmacological effects of hGH? No one should deny that pharmacological amounts of hGH will cause almost any normal child to increase his growth rate and somatomedin levels, as is the case in children who have gigantism.[17] With the usual doses of hGH, peak hGH levels have been reported to approximate 107 ng/ml,[18] a level frequently reported in gigantism or acromegaly.[19]

5. Novel Growth Factors

Finally, is it possible that growth factors other than 22K hGH and/or somatomedin C are importantly involved in growth? The possibility that this may be the case is suggested by several observations. The estimates of GH in pituitary extracts and serum by bioassay and RIA may be greater by a factor of 20 to 200.[19,20] In a number of craniopharyngioma patients, "catchup" growth occurs in the absence of circulating hGH as measured by RIA.[21] Acromegaloid features occur in patients who exhibit elevated levels of hGH-like activity with normal levels of hGH measured by RIA.[22] There is a poor correlation between clinical features of acromegaly, circulating hGH levels determined by RIA, and somatomedin C concentrations.[23] In hGH gene transfer experiments at least two forms of hGH are expressed, 22K and a second form that is much less reactive in an hGH RIA.[24] Correlation between changes of somatomedin and growth velocity in GH-treated children is variable.[25]

In studies in our laboratory we have identified a novel pituitary protein that stimulates chondrocyte growth in tissue culture.[26] This factor appears to be secreted by human pituitaries both *in vitro* and *in vivo*.[27] Its significance in stimulating body growth is unknown at present.

The possibility that one or other of these "growth factors" may be of importance in the regulation of growth in man must be considered. If this were the case, then present concepts of the hormonal regulation of growth would be incomplete because they failed to include one or more of the putative growth factors that have been suggested by the preceding observations.

In conclusion, in the quarter-century since Dr. Maurice Raben's demonstration of the efficacy of hGH treatment of hGH-deficient dwarfs notable advances in our understanding of growth have occurred. At the same time we recognize areas where our knowledge is incomplete and where opportunities exist to gain further insight into the biology of growth.

ACKNOWLEDGMENTS. The research was supported by the Medical Research Council of Canada, U.S. Public Health Services HD-07843-12, Manitoba Medical Services Foundation, and National Cancer Institute of Canada.

References

1. M. S. Raben and V. W. Westermeyer, Recovery of growth hormone in purification of corticotropin, *Proc. Soc. Exp. Biol. Med.* **78:**550, 1951.
2. M. Raben, Preparation of growth hormone from pituitaries of man and monkey, *Science* **125:**883, 1957.
3. J. C. Beck, E. E. McGarry, I. Dyrenfurth, and E. H. Venning, Metabolic effects of human and monkey growth hormone in man, *Science* **125:**884, 1957.
4. E. Knobil, R. C. Wolf, R. O. Greep, Some physiological effects of primate growth hormone preparations in hypophysectomized rhesus monkeys, *J. Clin. Endocrinol. Metab.* **16:**916, 1956.
5. A. E. Wilhelmi, Comparative biochemistry of growth hormone from ox, sheep, pig, horse and fish pituitaries, in: *Henry Ford International Symposium on Hypophyseal Growth Hormone: Nature and Actions* (R. W. Smith, O. H. Gaebler and C. N. Long eds.), McGraw Hill, New York, pp. 59, 1955.
6. S. D. Frasier, Human pituitary growth hormone (hGH) therapy in growth hormone deficiency, *Endocrine Rev.* **4:**155, 1983.
7. J. M. Tanner, R. H. Whitehouse, P. C. R. Hughes, and F. P. Vance, Effect of human growth

hormone treatment for one to seven years on growth of 100 children with growth hormone deficiency, low birth weight, inherited smallness, Turner's syndrome, and other complaints, *Arch. Dis. Child.* **46:**745, 1971.
8. H. L. Lenko, S. Leisti, and J. Perheentupa, The efficacy of growth hormone in different types of growth failure: An analysis of 101 cases, *Eur. J. Pediatr.* **138:**241, 1982.
9. D. Rudman, M. H. Kutner, R. D. Blackston, R. A. Cushman, R. P. Bain, and J. H. Patterson, Children with normal-variant short stature: Treatment with human growth hormone for six months, *N. Engl. J. Med.* **305:**123, 1981.
10. G. VanVliet, D. M. Styne, S. L. Kaplan, and M. M. Grumbach, Growth hormone treatment for short stature, *N. Engl. J. Med.* **305:**123, 1981.
11. J. S. Gillis, *Too Tall, Too Small,* Institute for Personality and Ability Testing Inc., Champaign, IL, 1983.
12. E. C. Burns, J. M. Tanner, M. S. Preece, and N. Cameron, Final height and pubertal development in 55 children with idiopathic growth hormone deficiency, treated for between 2 and 15 years with human growth hormone, *Eur. J. Pediatr.* **137:**155, 1981.
13. E. Joss, K. Zuppinger, H. P. Schwartz, and H. Raten, Final height of patients with pituitary growth failure and changes in growth variables after long term hormonal therapy, *Pediatr. Res.* **17:**676, 1983.
14. A. A. Kowarski, J. Schneider, E. Ben Galin, V. V. Weldon, and W. H. Daughaday, Growth failure with normal serum RIA-GH and low somatomedin activity: Somatomedin resotration and growth acceleration after exogenous GH, *J. Clin. Endocrinol. Metab.* **47:**461, 1978.
15. T. Frazer, J. R. Gavin, W. H. Daughaday, R. E. Hillman, and V. V. Weldon, Growth hormone-dependent growth failure, *J. Pediatr.* **101:**12, 1982.
16. E. Tokohiro, H. J. Dean, H. G. Friesen, and D. Rudman, Comparative study of serum hGH measurement with Nb_2 lymphoma cell bioassay, IM-9 receptor modulation assay and RIA in children with disorders of growth, *J. Clin. Endocrinol. Metab.* **58:**549, 1984.
17. E. M. Whitehead, S. M. Shalet, D. Davies, B. A. Euach, D. A. Price, and C. G. Beardwell, Pituitary gigantism a disabling condition, *Clin. Endocrinol.* **17:**271, 1982.
18. Frasier S. D., Costa G, Ley S. M., and Kaplan S. A., Plasma growth hormone concentration after a single intramuscular injection of human growth hormone, *Pediatr. Res.* **3:**557, 1969.
19. S. Ellis, M. A. Vodian, and R. E. Grindeland, Studies on bioassayaable growth hormone-like activity of plasma, *Recent Prog. Hormone Res.* **34:**213, 1978.
20. S. M. Russel and G. L. Mayer, Rat anterior pituitaries contain and secrete a low molecular weight (14-15K) protein with growth hormone activity, 65th Annual Meeting Endocrine Society, Williams and Wilkins, San Antonio, Texas, abstr. 917, 1983.
21. H. Bucher, J. Zapf, T. Torresani, A. Prader, E. R. Froesch, and R. Illig, Insulin-like growth factors I and II, prolactin, and insulin in 19 growth hormone-deficient children with excessive, normal, or decreased longitudinal growth after operation for craniopharyngioma, *N. Engl. J. Med.* **309:**1142, 1983.
22. M. W. Ashcroft, D. I. Hartzband, A. J. Van Herle, N. Bersch, and D. W. Golde, A unique growth factor in patients with acromegaloidism, *J. Clin. Endocrinol. Metab.* **57:**272, 1983.
23. J. A. H. Wass, D. R. Clemmons, L. E. Underwood, I. Barrow, G. M. Besser, and J. J. Van Wyk, Changes in circulating somatomedin C levels in bromocriptine-treated acromegaly, *Clin. Endocrinol.* **17:**369, 1982.
24. G. N. Pavlakis, N. Hizuta, P. Gorden, P. Seeburg, and D. H. Hamer, Expression of two human growth hormone genes in monkey cells infected by simian virus 40 recombinants, *Proc. Natl. Acad. Sci. USA* **78:**7398, 1981.
25. H. J. Dean, J. G. Kellett, R. M. Bala, H. J. Guyda, B. Bhaurnich, B. I. Bosner, and H. G. Friesen, The effect of growth hormone treatment on somatomedin levels in growth hormone deficient children, *J. Clin. Endocrinol. Metab.* **55:**1167, 1982.
26. S. Kasper, I. G. Worsley, J. M. Rowe, R. P. C. Shiu, and H. G. Friesen, Chrondrocyte growth factor from human pituitary gland, *J. Biochem.* **257:**5226, 1982.
27. S. Kasper and H. G. Friesen, Human pituitary tissue secretes a potent factor for chondrocyte proliferation, *J. Clin. Endocrinol. Metab.* (in press).

2

Growth Hormone Hypersecretory States

Mark E. Molitch

1. Introduction

Growth hormone (GH) hypersecretion in the human may be due to a GH-secreting tumor, to a GH-releasing-hormone (GHRH) secreting tumor, to disorders of GH feedback regulatory mechanisms, or to functional disturbance of GH regulation in association with systemic disorders, malnutrition, or psychiatric disease (Table 1). It is only in the first two circumstances that the clinical syndrome of acromegaly develops. In this chapter the pathophysiology of these disorders will be reviewed, the primary emphasis being given to the pathogenesis of GH-producing pituitary tumors.

2. GH-Secreting Pituitary Tumors

Marie, in 1866, was the first to associate acromegaly with pituitary dysfunction.[(1)] A tumor has been found in virtually all patients studied with current techniques, who had elevated blood GH levels and had complete sellar explorations. Hyperplasia has been associated primarily with ectopic GHRH secretion.

2.1. Pathology of GH-Secreting Pituitary Tumors

Extensive studies of GH-secreting adenomas have been performed in the last few years using electron microscopy (EM) and hormone-specific immunohistochemistry.[(2–7)] A number of different subtypes of GH-secreting tumors have been described, some of which also secrete prolactin (PRL) and other

Mark E. Molitch • Center for Endocrinology, Metabolism, and Nutrition and Department of Medicine, Northwestern University Medical School, Chicago, Illinois 60611.

Table 1 Syndromes of Growth Hormone Hypersecretion

GH-secreting pituitary tumors
GHRH-secreting tumors
Polyostic fibrous dysplasia
Laron-type dwarfism
Diencephalic syndrome
Systemic disorders
Stress
Malnutrition
Cirrhosis
Renal failure
Psychiatric disorders

hormones. Recently Kovacs, Horvath, and their co-workers[7] have proposed a new classification of GH-secreting tumors, shown in Table 2.

The tumors found on EM to be densely or sparsely granulated comprise the two largest groups, accounting for almost 60% of patients. The number of granules in the cells correlates well with the GH content of the tumors but does not correlate with blood GH levels.[3] However, blood GH levels did correlate with the secretory activity of the cells, as reflected by the size and pleomorphism of the nuclei, the existence of nucleoli, and the development of the Golgi and the rough endoplasmic reticulum.[3] Clinically, the sparsely granulated tumors grow more rapidly and often have extrasellar extension.[7]

PRL levels are commonly elevated in patients with acromegaly. Hyperprolactinemia has been found in 4 of 17 acromegalic patients seen in our institution in the past 10 years in whom PRL was measured before therapy. In most reported series of patients between 20 and 40% of patients are hyperprolacti-

Table 2 Classification of GH-Secreting Pituitary Tumors[a]

Tumor type	Frequency (%)	Immunostaining GH	Immunostaining PRL	Serum levels[b] GH	Serum levels[b] PRL	Clinical features
Densely granulated GH cell	27	+	−	I	N	Acromegaly
Sparsely granulated GH cell	32	+	−	I	N	Acromegaly, invasive tumor
Mixed GH cell and PRL cell	15	+	+	I	N-I	Acromegaly
Acidophil stem cell	12	+	+	N-I	N-I	Galactorrhea, amenorrhea, impotence, invasive tumor
Mammosomatotroph cell	4	+	+	I	N-I	Acromegaly
Plurihormonal	6	+	+	I	N-I	Acromegaly
Hyperplasia	<1	+	±	I	N-I	Acromegaly
Carcinoma	<1	?	?	I	?	Acromegaly

[a] Based on data of Melmed *et al.*[7]
[b] I = increased; N = normal.

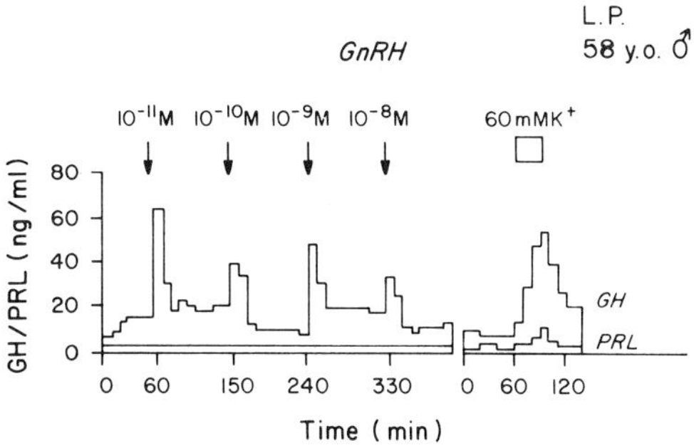

Figure 1 Results of stimulation of the tumor in an *in vitro* capillary perfusion culture system. Note the GH response but absent PRL response to GnRH and the minimal PRL response to depolarizing amounts of potassium (K+) added to the medium.

nemic.[6,8–19] In ours and in most of these series there is no correlation between GH and PRL levels; there appear to be two populations, normoprolactinemic and hyperprolactinemic. The hyperprolactinemic group is made up of two populations, one in which the PRL elevation appears to be due to hypothalamic/stalk dysfunction due to the tumor mass and one in which PRL is secreted together with GH from the pituitary tumor.

Four different tumor types are capable of secreting both hormones (Table 2). The mixed GH cell and PRL cell tumors have two separate populations of cells within the tumor, each making their own hormone. In all cases GH levels are high and the patients have acromegaly, but PRL levels may be normal or elevated.[7] In an acromegalic patient we have recently seen, GH levels were 20–50 ng/ml and PRL levels were always below 15 ng/ml, yet immunohistochemical staining of the pituitary tumor revealed both GH- and PRL-containing cells. Obviously, there is little PRL actually being secreted by this tumor, which is full of PRL. When this tumor was cultured and studied in a capillary perfusion system (Fig. 1), there were obvious GH responses to gonadotropin-releasing hormone but no response of PRL; similar findings occurred with other secretogogues, and even with depolarizing amounts of potassium there was only minimal immunoreactive PRL released, compared to GH. In this patient's tumor, therefore, there is a discrepancy in its ability to release one hormone compared to the other, implying a selective defect in the sequence leading from hormone synthesis to release, which, at present, is unexplained. Thus the measurement of serum PRL levels in acromegalic patients actually underestimates the true number of tumors containing both hormones.

The acidophil stem cell is the common precursor for the mature somatotrophs and lactotrophs (Fig. 2). Acidophil stem cell tumors are often invasive and have one cell type that contains both GH and PRL. However, GH levels are usually normal, and patients have minimal evidence of acromegaly, usually presenting with symptoms of hyperprolactinemia, such as amenorrhea, galactorrhea, or impotence.[7]

Mammosomatotroph cell adenomas are mature, slow-growing tumors in which the same cells secrete both hormones. In these patients GH levels are high, the patients have acromegaly, and PRL levels are usually normal to slightly elevated.[7]

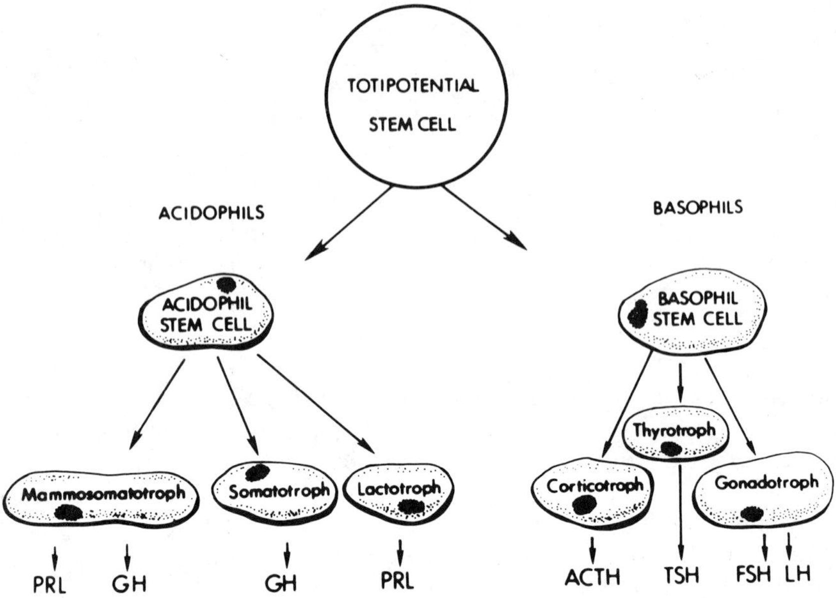

Figure 2 Proposed development on anterior pituitary cell types (reproduced from ref. 7, by copyright permission of The Endocrine Society).

The plurihormonal adenomas secrete GH in association with a variety of hormones, including PRL, TSH, and ACTH from either one cell type or a mixed population of cell types within the tumor. GH cell carcinomas are extremely rare.[7]

Relationship of Pathological Findings to Pathogenesis of Acromegaly

Discussions of the pathogenesis of GH-secreting pituitary tumors[7,20–23] have focused on whether such tumors arise spontaneously within the pituitary or are due to prolonged excessive stimulation by hypothalamic factors, such as GHRH (Fig. 3). Excessive stimulation by GHRH may occur either because of an actual increased GHRH secretion into the hypothalamic—pituitary portal vessels or because of increased sensitivity of somatotroph cells to normal GHRH levels. If tumors are due to excessive stimulation by GHRH, it is likely that a two-stage process is needed, involving both excessive stimulation and neoplastic transformation by a specific cell type in response to such stimulation.[7]

Three findings have emerged from the studies of the pathology of these tumors that are pertinent to the question of the role of the hypothalamus in the pathogenesis of GH-secreting tumors: (1) there are at least six different tumor cell types that are associated with GH hypersecretion; (2) some GH-secreting tumors also secrete PRL, a finding seemingly incompatible with the possibility of the tumor originating from increased GHRH activity, since GHRH

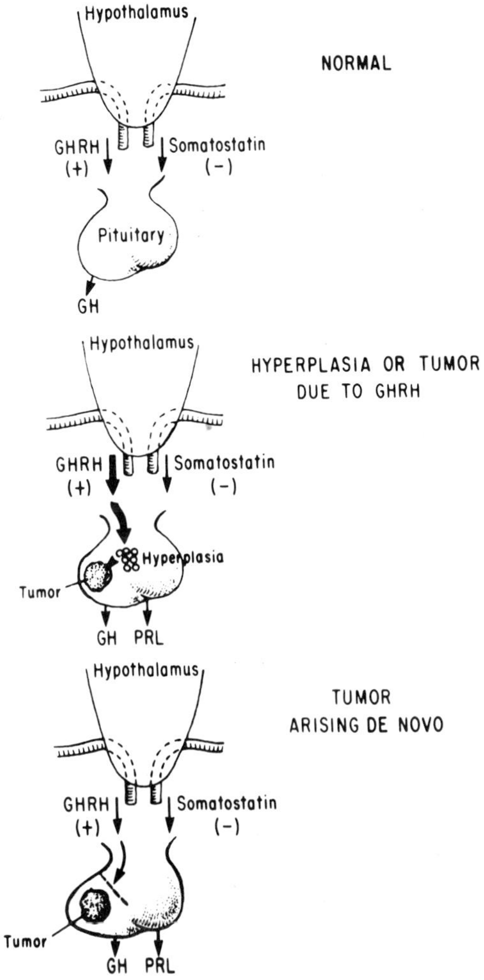

Figure 3 Possible mechanisms leading to the formation of GH-secreting pituitary tumors in acromegaly.

has not been regarded as a PRL-releasing factor; and (3) the pituitary tissue surrounding the tumor does not show hyperplasia,[24,25] as would be expected if the cause of the disease was increased GHRH activity.

Because cell types ranging from the primitive stem cell to the mature somatotroph are found in GH-secreting tumors, it is obvious that neoplastic transformation may occur at any stage of development of these cells, regardless of whether the initiating factor was GHRH or some unknown cellular phenomenon. The finding of different cell types in these tumors, therefore, does not provide evidence for or against a hypothalamic role in the pathogenesis of these tumors.

The fact that some tumors secrete both GH and PRL has been regarded as evidence against the hypothalamic theory of pathogenesis, since GHRH has not been regarded as a PRL-releasing factor.[7] However, we have recently

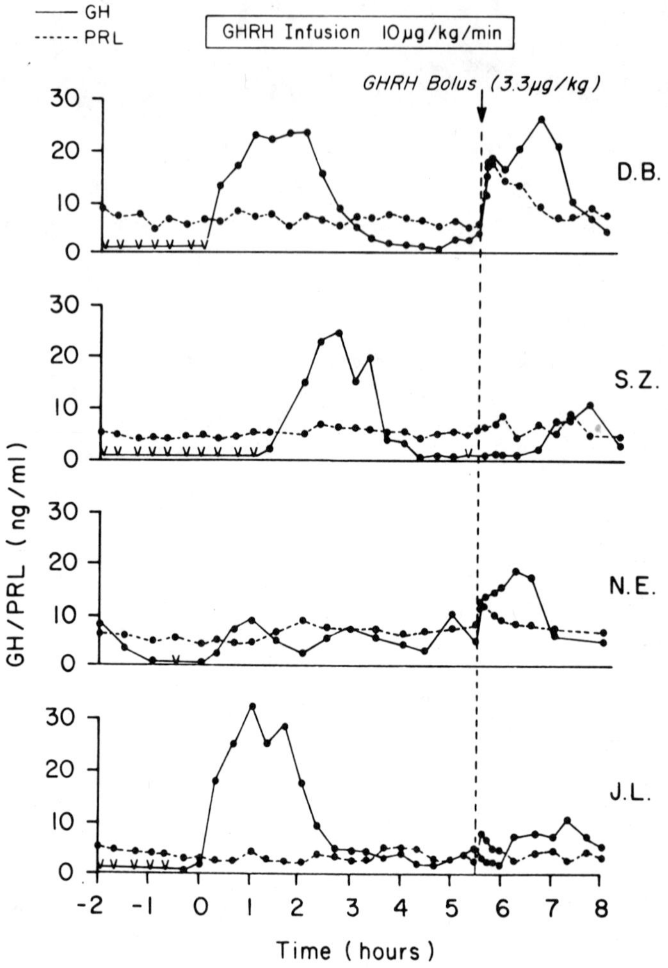

Figure 4 Effects of hGHRH (1–40) intravenous infusion (1.0 μg/kg per min) for 6 hr, with an acute intravenous bolus injection of hGHRH (1–40) (3.3 μg/kg) given at 330 min, on GH and PRL levels in normal subjects.

found that GHRH can stimulate both GH and PRL release in both normal individuals and acromegalic patients. In normal volunteers prolonged (5-hr) hGHRH (because human pancreatic GH-releasing factor appears to be identical to hypothalamic human GH-releasing factor,[26] I have chosen to refer to this material as hGHRH in this paper) infusion appears to sensitize the PRL-secreting cells to the stimulatory action of hGHRH. In Fig. 4, the effects of a prolonged hGHRH (1–40) infusion (6 hr at 1.0 μg/kg per min) are shown. At 330 min a bolus of 3.3 μg/kg was given, and there was a small rise in PRL in each of the four subjects. In these subjects there was also a small PRL response to the acute injection of hGHRH (1–40) without prior infusion.[27] In other

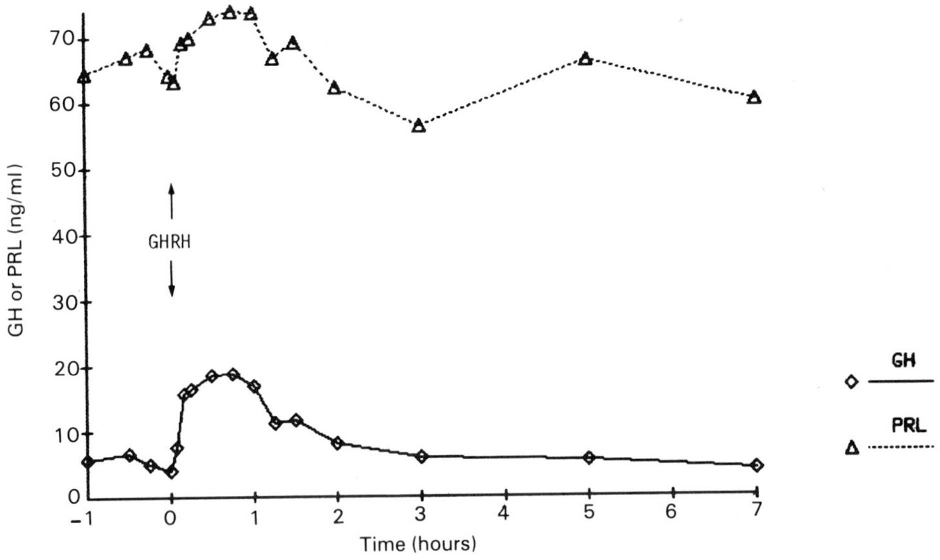

Figure 5 Effects of an acute hGHRH (1–40) intravenous bolus on GH and PRL levels in an acromegalic man.

studies Borges *et al.*[28] have shown that very large single boluses (10 μg/kg) of hGHRH (1–40) could elevate PRL in three of six normal subjects and in all 11 subjects with idiopathic GH deficiency. Whether the PRL release in these subjects is due to PRL release from GHRH-stimulated lactotrophs or somatotrophs is unknown. Although recent studies in the rat show that hpGRF (1–40) stimulates mRNA transcription of pituitary rGH but not rPRL, such studies have not yet been performed on human pituitaries.[29]

Whereas GH has been shown to respond to GHRH in patients with acromegaly *in vivo*[30,31] and *in vitro*,[32,33] we have also found that in acromegalic patients with GH- and PRL-secreting tumors there is a parallel release of GH and PRL in response to hGHRH (1–40) (Fig. 5); no PRL response was seen in acromegalic subjects who were not hyperprolactinemic.[31] Thus, as in normals, PRL has been shown to be released by GHRH in acromegalic patients who have elevated prolactin levels. Webb *et al.*[32] have also recently demonstrated the release of GH and PRL *in vitro* in response to GHRH from an adenoma from a patient with acromegaly and hyperprolactinemia. This stimulation of both GH and PRL, then, may explain the development of bihormonal tumors, depending on the initial cell type that is stimulated, according to the hypothalamic theory of tumor pathogenesis.

The lack of hyperplasia in the normal tissue surrounding the tumor is strong evidence against excessive GHRH secretion in patients with acromegaly. However, if only some cells are hypersensitive to normal levels of GHRH, then only these cells would hypersecrete and proliferate. Alternatively, the increase in somatomedin levels in such patients may feedback negatively at the level of the pituitary,[34] reducing the stimulatory effects of GHRH on the nonneoplastic somatotrophs. Thus, this finding does not absolutely rule out a hypothalamic role.

2.2. GH Regulation in Acromegaly

One argument used in favor of a hypothalamic etiology for acromegaly is that in patients with acromegaly there often appears to be some measure of hypothalamic influence over GH secretion. In the normal individual hyperglycemia suppresses and hypoglycemia stimulates GH secretion. In patients with acromegaly, one of the hallmarks is the nonsuppressibility of GH by glucose, and indeed, this is used as a diagnostic criterion. In our own series, of 18 patients whose GH responses to an oral glucose load were evaluated prior to any therapy (Fig. 6), five had a paradoxical increment (defined as being greater than 50% of baseline values) of GH, 12 had no suppression, and one patient had suppression of GH levels but not to <2 ng/ml; these results are similar to those reported by others.[8,10,12,13,20,21,35–37] Of 11 of our patients whose GH responses to insulin-induced hypoglycemia were measured prior to therapy, eight had an increment in GH values of greater than 50% of basal values (a normal response would be an increment of greater than 100% of basal values)

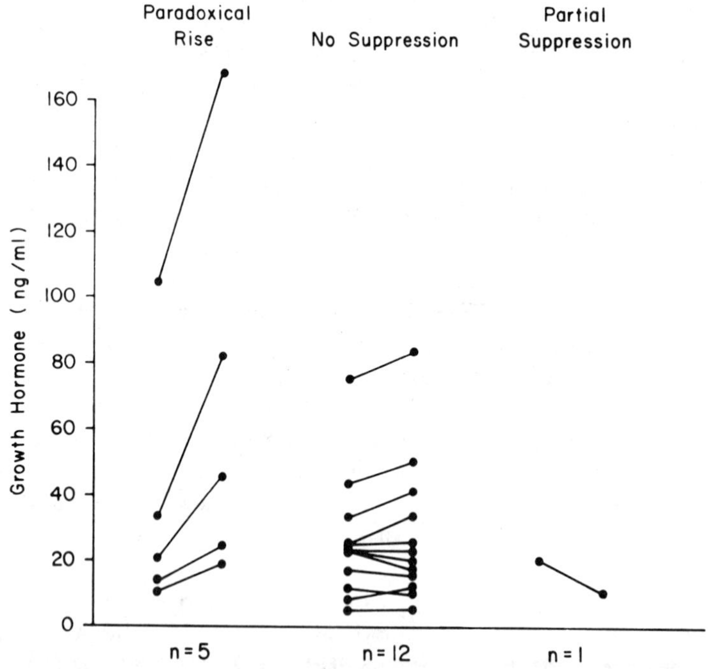

Figure 6 GH responses to hyperglycemia achieved by an oral glucose load in 18 patients with acromegaly prior to therapy.

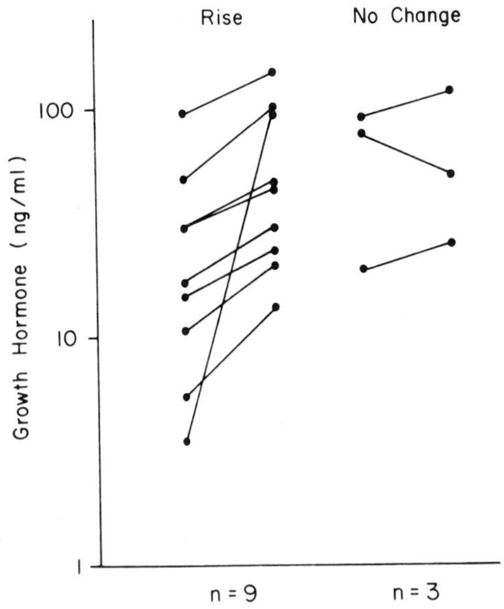

Figure 7 GH responses to insulin-induced hypoglycemia in 11 patients with acromegaly prior to therapy.

while only three had no response to hypoglycemia (Fig. 7). This proportion of patients responding to hypoglycemia is somewhat higher than that reported in the literature[8,12,20,21] and, in fact, is not very different than the proportion of normal subjects who would respond to hypoglycemia with a GH increase.[38,39] Therefore, many patients with acromegaly continue to display some response

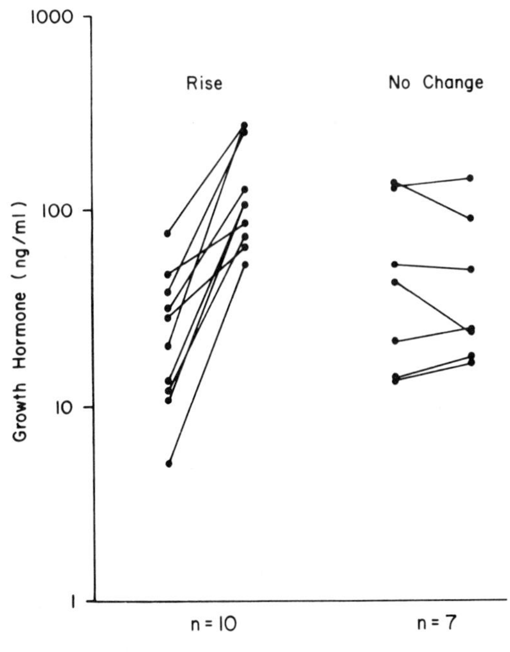

Figure 8 GH responses to TRH (500 μg) in 17 patients with acromegaly prior to therapy.

to changing glucose levels, although the response may be paradoxical. Since it is thought that such responses are mediated through hypothalamic mechanisms, this implies retention of hypothalamic influence over GH secretion.

TRH was initially demonstrated to release GH in the majority of acromegalic patients by Irie and Tsushima.[40] In our own series of 17 patients given TRH prior to therapy (Fig. 8), 10 responded with an increment of GH (>50% increment over basal values) and seven did not have a response; these proportions are similar to those reported by others.[8,10–15,19,35,36,40–42] This effect is mediated by a direct effect at the level of the pituitary.[43] The common explanation given for this response to TRH in patients with acromegaly is that in this abnormal, neoplastic tissue there is an "unmasking" of TRH receptors.[22] An alternative possibility, compatible with a hypothalamic etiology for this tumor, is that prior stimulation with GHRH enables this response. Borges *et al.*[44] have recently shown in *in vitro* studies in the rat that prolonged exposure to hGHRH (1–40) sensitizes the somatotrophs to TRH (Fig. 9). It is likely,

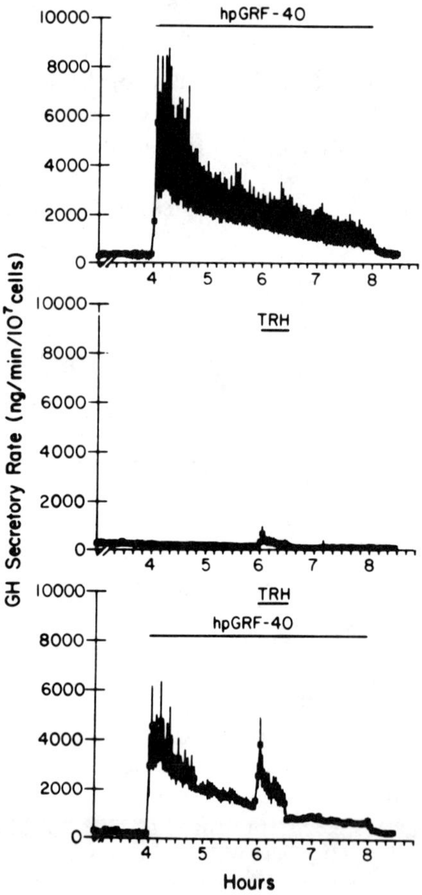

Figure 9 GH response to TRH with and without prior perfusion with hGHRH (1–40) in rat pituitary cells *in vitro* (reproduced from ref. 44, by copyright permission of The Endocrine Society).

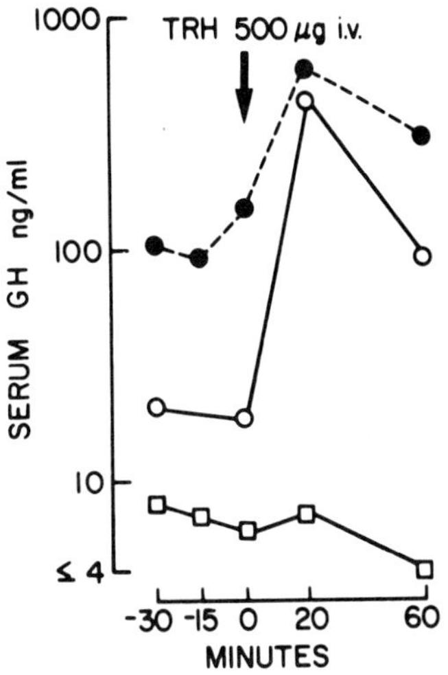

Figure 10 Serum GH levels following the injection of TRH (500 μg) in a patient before and after removal of a GHRH-secreting pancreatic tumor. The filled circles represent the response prior to pituitary surgery, the open circles represent the response 7 weeks after pituitary surgery, and the squares represent the response 17 weeks after resection of a GHRH-secreting pancreatic tumor. Note that the paradoxical response of GH to TRH was abolished after removal of the pancreatic tumor (reproduced from ref. 45, by copyright permission of the American Society for Clinical Investigation).

therefore, that all somatotrophs have TRH receptors, but additional factors, such as increased GHRH activity, are needed to unmask or activate them.

Thus, there remains at least some hypothalamic influence over GH secretion in most patients with acromegaly, although the GH responses may not be the same as those in normal individuals. Importantly, the response to GHRH is retained and even exaggerated in some individuals. Recent evidence suggests that some of these abnormal responses can be explained by postulating prior stimulation of the cells by GHRH, i.e., the response to TRH and the PRL response to GHRH. This formulation is supported by the findings in the patient with ectopic GHRH secretion reported by Thorner *et al.*,[45] in whom the GH response to TRH (Fig. 10) and basal hyperprolactinemia (Fig. 11) were abolished after removal of the GHRH-secreting tumor. Thus, these data derived from experiments with GHRH administration and from the patient with ectopic

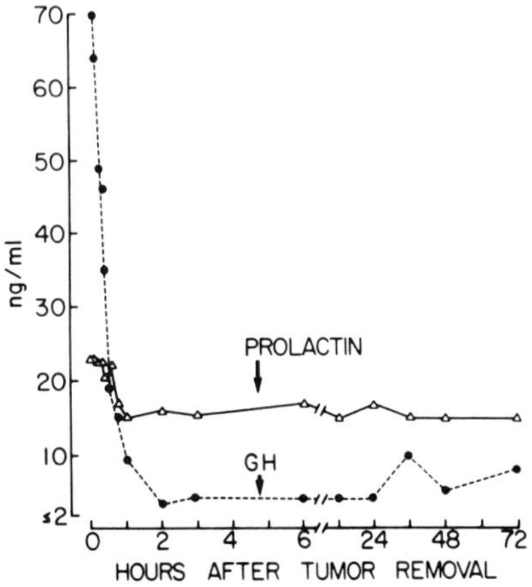

Figure 11 Parallel fall in GH and PRL levels in an acromegalic patient following resection of an hGHRH (1–40) producing pancreatic tumor (reproduced from ref. 45, by copyright permission of the American Society for Clinical Investigation).

GHRH secretion provide new, powerful support for the hypothesis that excess secretion of GHRH by the hypothalamus or hypersensitivity of the pituitary to normal levels of GHRH may play a primary role in the pathogenesis of these tumors.

Presumably such a postulated increase in GHRH would occur only within the hypothalamic–pituitary portal system. Whereas several preliminary reports found increased bioassayable "GRF" in the peripheral blood of patients with acromegaly,[46–48] radioimmunoassay of hGHRH (1–40) in the blood of several acromegalics by Dr. Lin in our laboratories has not shown elevated levels.

An alternative explanation to the hypothalamic etiology is that these tumors, like so many endocrine tumors, remain partially responsive to normal stimuli and may show aberrant responses to a variety of secretogogues.

2.3. GH Secretion in Acromegalic Patients Cured by Surgery

The return of GH secretory dynamics to normal after selective tumor resection, with failure of regrowth of the tumor with prolonged follow-up, has been the strongest evidence against the hypothesis that the hypothalamus plays a primary role in the pathogenesis of these tumors. Figure 12 shows that the lack of suppression by glucose and the response to TRH found preoperatively

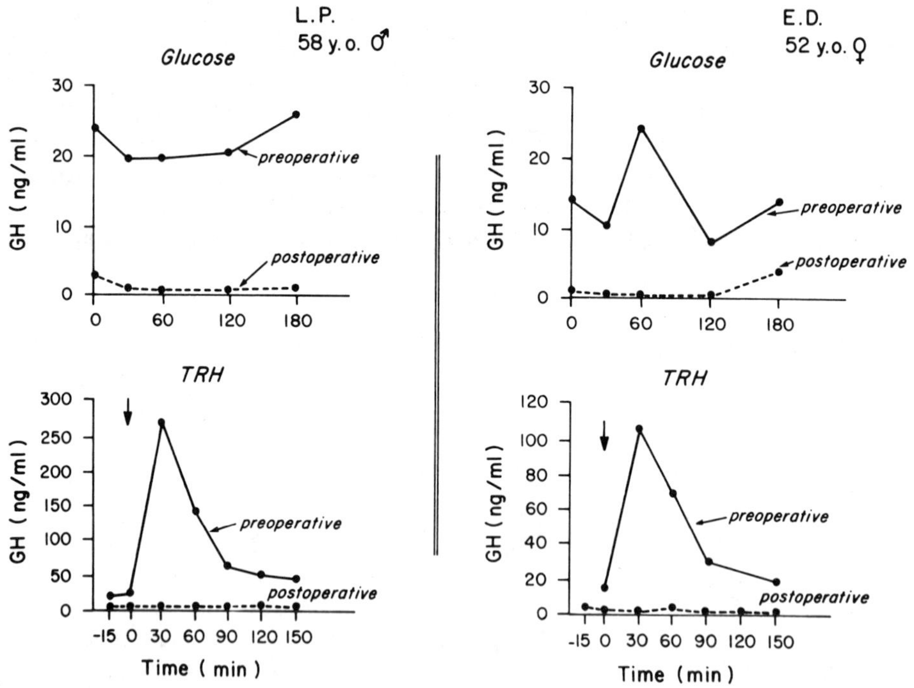

Figure 12 Resolution of abnormal GH secretory dynamics in two patients after transsphenoidal selective resection of GH-secreting pituitary adenomas.

in two of our patients have been reversed by surgery. No recurrence has occurred in these two patients after 2 (L.P.) and 8 (E.D.) years of follow-up. A number of similar patients have been reported by others.[12,13,19,36,49–53] In all these series, however, there were reported a large number of patients with basal GH levels persistently below 5 ng/ml but who had abnormal responses to glucose or TRH. Whether such patients have residual tumor tissue or underlying hypothalamic dysregulation is unknown.

3. Ectopic Hormone Syndromes

3.1. Growth Hormone

GH has been found to be elevated basally and to respond paradoxically to glucose in patients with a number of different types of cancer,[54–58] variably associated with pulmonary osteoarthropathy but not with acromegaly. GH has also been documented to be present in up to 62.5% of cancers in amounts never greater than 51 μg/g of wet tissue weight but not associated with significant serum GH elevations (although there may have been paradoxical elevations of GH in response to glucose administration) or acromegaly.[55,59–63] In the report of Greenberg *et al.*[60] a poorly differentiated lung cancer was demonstrated to synthesize and release immunoreactive GH from cultured cells. In one case of ectopic GH secretion leading to acromegaly, reported by Dabek,[64] a 30-year-old woman with acromegalic features was found to have elevated basal levels of GH with a paradoxical response to glucose. Her resected carcinoid tumor of the lung was found to contain 122 ng/g wet tumor weight of GH and to contain no GHRH-like activity when injected into monkeys. The patient had a marked lowering of GH levels after tumor resection with partial resolution of acromegalic features. This patient did, however, have an enlarged sella turcica on plain skull X rays so that GHRH may have been present in the tumor but was not detected by the bioassay system used. Leveston *et al.*[65] have also reported a patient with acromegaly who had both GH and GHRH activity in a foregut carcinoid tumor. Recently Melmed *et al.*[65a] reported a case of acromegaly in which the ectopic secretion of GH by an islet-cell tumor was very well documented.

3.2. GH-Releasing Hormone

The hormone secreted ectopically that clearly can cause acromegaly is GHRH, and this syndrome is reviewed elsewhere in this volume. Of interest in these patients, and pertinent to the earlier discussion of acromegaly, is the fact that in many of these patients there were paradoxical GH responses to glucose and increases of GH in response to TRH. Both these responses are characteristic of acromegaly, but in these patients the abnormal responses were reversible with resection of the GHRH-producing tumors.[45,64,66–69] Thus, the excess GHRH appeared to be responsible for these abnormal responses rather than a tumor cell that has become altered or "dedifferentiated." In addition, of the three patients who had PRL measurements performed,[45,65,58] two were hyperprolactinemic.[45,65] In the case described by Thorner *et al.*[45] there was a parallel fall of GH and PRL after resection of the GHRH-producing islet cell tumor (see above). It is apparent, therefore, that chronic GHRH excess can produce sustained hypersecretion of both GH and PRL.

A possible variant of the ectopic GHRH syndrome is that of intrasellar neuronal hamartomas found in association with GH-secreting pituitary adenomas causing acromegaly. Six of seven cases reported with intrasellar neuronal hamartomas were found in association with GH-secreting tumors, and it has been hypothesized that in these cases the hamartomas were producing GHRH.[70] Recently the hamartoma reported by Asa *et al.*[70] has been demonstrated by histochemical techniques to contain immunoreactive GHRH.[71]

4. Polyostotic Fibrous Dysplasia

Polyostotic fibrous dysplasia (McCune–Albright syndrome) is a syndrome consisting of fibrous dysplastic areas of bones, patches of cutaneous pigmentation (café au lait spots), and a variety of endocrinological disorders, including acromegaly, Cushing's syndrome, hyperprolactinemia, and hyperthyroidism.[72,73] Because of this large variety of endocrine disorders, Hall and Warrick[74] proposed that the underlying endocrine defect in this syndrome was disordered hypothalamic–pituitary regulation, perhaps an uncontrolled secretion of hypothalamic releasing factors. Nine patients have been reported with polyostotic fibrous dysplasia associated with acromegaly,[75–84] and we are currently studying a tenth, who has a GH level of 95 ng/ml and a PRL level of 935 ng/ml. In all nine patients tested oral glucose loading did not cause a reduction in GH levels to normal, and all five patients tested showed GH release in response to TRH. Of six patients who had PRL levels measured, five had elevated basal levels. Pathological examination of pituitary and hypothalamic tissue from one patient at autopsy was normal, as was hypothalamic tissue. However, in two patients large pituitary tumors were visualized on computerized tomography.

Increased secretion of GHRH could explain the sustained hypersecretion of GH and possibly PRL in these patients, as GHRH has been shown to increase both hormones under some circumstances (see above). Dr. Hong Da Lin in our laboratories, however, has found GHRH levels in the extracted peripheral blood of our patient to be 20.2 pg/ml, which is within the normal range for extracted serum (range 13–163 pg/ml); however, it may be that portal blood levels would be elevated.

5. Laron Dwarfism

In this variety of dwarfism, individuals have markedly retarded skeletal growth and bone maturation and are similar clinically to individuals with GH deficiency. However, radioimmunoassayable GH levels are elevated in these patients, and their growth defect is due to their inability to generate somatomedins in response to their own or exogenous GH.[85]

6. Diencephalic Syndrome

Children with this syndrome develop poor weight gain and lack subcutaneous fat despite good caloric intake while initially maintaining a normal to advanced gain in height and bone age. Subsequently, growth becomes retarded. In over 90% of cases there is a tumor in the hypothalamus, usually anterior and midline.[86–87] In the few patients tested, basal GH levels have been reported to be elevated, and some patients have had paradoxical increments in GH levels in response to an oral glucose load.[86,87]

There are several possible explanations for the GH hypersecretion in these patients: (1) the hypothalamic lesions may be irritative, causing release of GHRH; (2) the hypothalamic lesions may interrupt inhibitory somatostatinergic fibers passing from the preoptic/anterior hypothalamic areas to the median eminence; or (3) the emaciation of these individuals interferes in some way with somatomedin generation, causing decreased somatomedin negative feedback. Somatomedin levels have not been reported in any children with this syndrome so that it is not possible to distinguish between these possibilities. Increased thyroid function has not been found, so that it is unlikely that alteration of somatostatinergic pathways is involved.

7. Systemic Disease Associated with GH Hypersecretion (Table 3)

GH hypersecretion is one of the hormonal responses to stress. With prolonged stress basal GH levels may either remain mildly elevated or return

Table 3 GH Hypersecretion Associated with Systemic Disorders[a]

TSH levels	Basal GH levels	GH response to TRH	Basal TSH levels
Acute physical stress	I		N-D
Chronic physical stress	N	N-I	N-D
Psychiatric disorders			
Acute anxiety	I		
Depression	N	N-I	N-D
Schizophrenia	N	I	N
Anorexia nervosa	I	I	D
Protein–calorie malnutrition	I	I	D
Diabetes mellitus (poorly controlled)	I	I	D
Cirrhosis	I	I	N-D
Chronic renal failure	I	I	N-D

[a] N = normal; I = increased; D = decreased.

to normal. However, even in this latter situation GH may increase in response to thyrotropin-releasing hormone (TRH), a response that does not occur in the normal individual. In patients with such illness there is a suppression of thyrotropin (TSH) secretion, and the TSH response to TRH is blunted.[88] It should be noted, however, that serial studies of GH (and other pituitary hormone) secretory dynamics have not been performed in the same patients who are under prolonged, severe stress. Such studies need to be done.

Basal GH levels are elevated, and GH increases in response to TRH in patients with anorexia nervosa, protein–calorie malnutrition, poorly controlled diabetes mellitus, cirrhosis, and renal failure. GH also increases in response to TRH in many patients with schizophrenia and depression, although their basal GH levels are not usually elevated. In most of these disorders the TSH response to TRH is blunted.[88]

Because of information gained from recent work with GHRH and other studies, it is possible to construct a hypothesis regarding the neuroendocrine mechanisms causing GH hypersecretion in at least some of these disorders. It is likely that the GH hypersecretion is due either to an absolute increase in hypothalamic GHRH secretion or to an increase in activity of secreted GHRH at the somatotroph level rather than to a decrease in somatostatin. Many studies have shown that GHRH can cause a rapid rise in GH levels sufficient to explain the acute rise of GH in response to stress. In most of the disorders causing a chronic elevation of basal GH levels there is also a concomitant decrease in somatomedin levels, and it is possible that the GH elevation is due to lack of negative feedback of somatomedin at the hypothalamic[89,90] and/or pituitary[34] levels (Fig. 13).

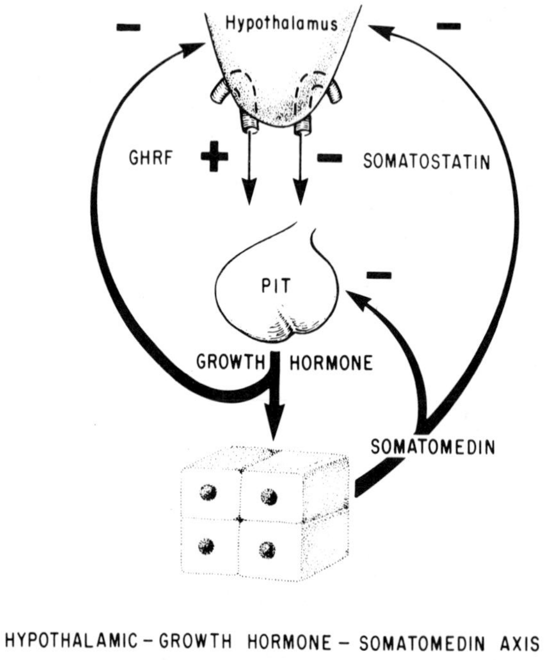

Figure 13 Proposed feedback regulatory mechanisms for GH secretion (PIT - pituitary; GHRF - growth-hormone-releasing factor). Figure courtesy of S. Reichlin.

Chronic stimulation with GHRH may explain the acquired responsivity of the somatotrophs to TRH in these disorders, as discussed previously. The fact that TSH is suppressed in these illnesses suggests that the GH elevation is not due to a decrease in somatostatin, as TSH would be expected to be elevated in such circumstances.

8. *Conclusions*

The role of the hypothalamus in the pathogenesis of GH-secreting tumors remains controversial. Excessive secretion of GHRH as the cause for acromegaly has only been proven for the small number of patients with ectopic GHRH-secreting tumors. Although increased GHRH secretion by the hypothalamus or increased pituitary sensitivity to GHRH may play a primary role in the pathogenesis of the more common GH-secreting pituitary tumor, this theory is not yet proved. Evidence in favor of a hypothalamic etiology includes (1) retention of hypothalamic influence on GH secretion; (2) retention of GH response to GHRH; (3) explanation of the nonsuppressibility by glucose and the abnormal response to TRH by prior exposure to GHRH as shown experimentally and in the patients with ectopic GHRH secretion; and (4) explanation of the occurrence of GH- and PRL-secreting tumors by finding PRL release in response to GHRH. Evidence in favor of such tumors arising *de novo* in the pituitary includes (1) the pituitary tissue surrounding the GH-secreting adenoma does not show hyperplasia, as it should if there were excess GHRH; (2) many patients cured by surgery remain cured for long periods of observation and have a return of secretory dynamics to normal; and (3) many endocrine tumors display partial or aberrant responses to various stimuli, and the occurrence in acromegaly of such responses is not unexpected.

It is quite possible that there is more than one etiology for the pituitary tumors of patients with acromegaly. In some patients some cells may be particularly sensitive to GHRH, causing focal hyperplasia and eventually tumor formation with the rest of the cells appearing normal. Such a tumor may then function semiautonomously. When such a tumor is removed, the rest of the cells do not then respond abnormally to GHRH. Although measurement of GHRH levels in portal blood may yield the answer to this question, the finding of normal or low levels does not rule out a hypothalamic role, since the elevated levels of somatomedins may feed back to the hypothalamus to reduce GHRH secretion once the tumor has developed semiautonomy. Furthermore, if the disorder is due to hypersensitivity of the somatotrophs to GHRH, then portal vessel GHRH levels may be normal or even low.

The recent data obtained with GHRH and extrapolation from data obtained in patients with Laron dwarfism allow speculation as to the mechanism of the GH hypersecretion that may occur for short periods of time in association with stress and a variety of systemic disorders. In most circumstances this appears to be due to disordered hypothalamic regulation, the paradoxical responses to glucose, and the abnormal response to TRH suggesting excessive secretion of or sensitivity to GHRH as the etiology. The lack of TSH elevation in these settings is evidence against there being a decrease in somatostatin. In some of these disorders somatomedin levels are low, suggesting that the GH elevations may be due to decreased somatomedin negative feedback on GHRH secretion or action.

ACKNOWLEDGMENTS. The studies presented were performed when I was at the Tufts University School of Medicine and The New England Medical Center. I wish to thank Drs. James Goldman, Hong Da Lin, Arthur Lee, and Seymour Reichlin for their permission to allow me to present some of their studies and for their discussions which have helped to formulate some of the concepts presented here. I wish to thank Drs. Thorner, Vale, and Rivier for their provision of generous quantities of hGHRH (1–40) for use in these studies. I also wish to thank Ms. Sheila O'Connell-Nielson and Ms. Ann McDermott and the staff of the Clinical Study Unit of the Tufts-New England Medical Center for their help in the care of the patients and subjects whose data were presented, the National Hormone and Pituitary Agency and the University of Maryland for provision of the radioimmunoassay hormones and antibodies used in these studies, and Ms. Sara Gottesman, Ms. June Vincent, and Ms. Leticia Morales for expert technical assistance. These studies were supported in part by USPHS Grant #RR-0054-7 to the Clinical Study Unit of the Tufts-New England Medical Center, NIH Grant #AM20870 (MEM), and NIH Grant #AM16684 (Seymour Reichlin).

References

1. P. Marie, On two cases of acromegaly: Marked hypertrophy of the upper and lower limbs and the head, *Rev. Med.* **6**:297, 1866.
2. E. A. Zimmerman, R. Defendini, and A. G. Frantz, Prolactin and growth hormone in patients with pituitary adenomas: A correlative study of hormone in tumor and plasma by immunoperoxidase technique and radioimmunoassay, *J. Clin. Endocrinol. Metab.* **38**:577, 1974.
3. J. Trouillas, C. Girod, M. Lheritier, B. Claustrat, and M. P. Dubois, Morphological and biochemical relationships in 31 human pituitary adenomas with acromegaly, *Virchows Arch. A. (Pathol. Anat.)* **389**:127, 1980.
4. T. Kameya, M. Tsumuraya, I. Adachi, K. Abe, K. Ichikizaki, S. Toya, and R. Demura, Ultrastructure, immunohistochemistry and hormone release of pituitary adenomas in relation to prolactin production, *Virchows Arch. A. (Pathol. Anat.)* **387**:31, 1980.
5. N. S. Halmi, Occurence of both growth hormone- and prolactin-immunoreactive material in the cells of human somatotropic pituitary adenomas containing mammotropic elements, *Virchows Arch. A. (Pathol. Anat.)* **398**:19, 1982.
6. A. C. Nieuwenhuijzen-Kruseman, G. T. A. M. Bots, F. Roelfsema, M. Frolich, and H. Van Dulken, Immunocytochemical growth hormone and prolactin in pituitary adenomas causing acromegaly and their relationship to basal serum hormone levels and the growth response to thyrotrophin releasing hormone, *Clin. Endocrinol. (Oxford)* **19**:1, 1983.
7. S. Melmed, G. D. Braunstein, E. Horvath, C. Ezrin, and K. Kovacs, Pathophysiology of acromegaly, *Endocrine Rev.* **4**:271, 1983.
8. G. Tolis, L. Kovacs, H. Friesen, and J. B. Martin, Dynamic evaluation of growth hormone (GH) and prolactin (hPRL) secretion in active acromegaly with high and low GH output, *Acta Endocrinol.* **78**:251, 1975.
9. S. Franks, H. S. Jacobs, and J. D. N. Nabarro, Prolactin concentrations in patients with acromegaly: Clinical significance and response to surgery, *Clin. Endocrinol. (Oxford)* **5**:63, 1976.

10. S. W. J. Lamberts, J. G. M. Klijn, G. H. Kwa, and J. C. Birkenhager, The dynamics of growth hormone and prolactin secretion in acromegalic patients with "mixed" pituitary tumors, *Acta Endocrinol.* **90:**198, 1979.
11. P. Moriondo, P. Travaglini, M. Rondena, P. Beck-Peccoz, F. Conti-Puglisi, B. Ambrosi, and G. Faglia, Prolactin secretion in acromegaly, in: *Pituitary Microadenomas* (G. Faglia, M. A. Giovanelli, and R. M. MacLeod, eds.), Academic Press, New York, 1980, pp. 247–255.
12. H. S. G. Tucker, S. R. Grubb, J. P. Wigand, C. O. Watlington, W. G. Blackard, and D. P. Becker, The treatment of acromegaly by transsphenoidal surgery, *Arch. Intern. Med.* **140:**795, 1980.
13. B. M. Arafah, J. S. Brodkey, B. Kaufman, M. Velasco, A. Manni, and O. H. Pearson, Transsphenoidal microsurgery in the treatment of acromegaly and gigantism, *J. Clin. Endocrinol. Metab.* **50:**578, 1980.
14. K. Hanew, M. Kokubun, A. Sasaki, T. Mouri, and K. Yoshinaga, The spectrum of pituitary growth hormone responses to pharmacological stimuli in acromegaly, *J. Clin. Endocrinol. Metab.* **51:**292, 1980.
15. K. Hanew, A. Sasaki, T. Mouri, and K. Yoshinaga, Plasma growth hormone and prolactin responses to TRH and arginine given before and during bromocriptine therapy to patients with acromegaly, *Acta Endocrinol.* **95:**298, 1980.
16. F. De Pablo, R. C. Eastman, J. Roth, and P. Gorden, Plasma prolactin in acromegaly before and after treatment. *J. Clin. Endocrinol. Metab.* **53:**344, 1981
17. F. Camanni, F. Massara, M. Santia, G. M. Molinatti, and E. E. Muller, Impaired prolactin responsiveness to dopamine antagonists in acromegaly, *Metabolism* **31:**1090, 1982.
18. S. Balagura, P. Derome, and G. Guiot, Acromegaly: Analysis of 132 cases treated surgically, *Neurosurgery* **8:**413, 1981.
19. M. Arosio, M. A. Giovanelli, Riva E, C. Nava, B. Ambrosi, and G. Faglia, Clinical use of pre- and postsurgical evaluation of abnormal GH response in acromegaly, *J. Neurosurg.* **59:**402, 1983.
20. A. M. Lawrence, I. D. Goldfine, and L. Kirsteins, Growth hormone dynamics in acromegaly, *J. Clin. Endocrinol. Metab.* **31:**239, 1970.
21. W. H. Daughaday, P. E. Cryer, and L. S. Jacobs, The role of the hypothalamus in the pathogenesis of pituitary tumors, in: *Diagnosis and Treatment of Pituitary Tumors* (P. O. Kohler and G. T. Ross, eds.), American Elsevier, New York, 1973, pp. 26–34.
22. W. H. Daughaday, Pathophysiology of acromegaly, in: *Hormone-Secreting Pituitary Tumors* (J. R. Givens, ed.), Year Book Medical Publishers, Chicago, 1982, pp. 209–223.
23. S. Reichlin, and M. E. Molitch, Neuroendocrine aspect of pituitary tumors in: *Pituitary Hyperfunction: Physiopathology and Clinical Aspects* (F. Camanni, E. E. Müller, eds.), Raven Press, New York, 1984, pp. 47–70.
24. W. Saeger, Die morphlogie der paraadenomatosen adenohypophyse, *Virchows Arch. A. Path. Anat. Histol.* **372:**299, 1977.
25. A. M. Landolt, Biology of pituitary microadenomas, in: *Pituitary Microadenomas* (G. Faglia, M. A. Giovanelli, and R. M. MacLeod, eds.), Academic Press, New York, 1980, pp. 107–122.
26. K. E. Mayo, W. Vale, Rivier, J. M. G. Rosenfeld, and R. M. Evans, Expression-cloning and sequence of a cDNA encoding human growth hormone-releasing factor, *Nature* **306:**86, 1983.
27. J. A. Goldman, M. E. Molitch, M. O. Thorner, W. Vale, J. Rivier, and S. Reichlin, Growth hormone releasing hormone (1-40) is a prolactin releasing factor (submitted for publication).
28. J. L. C. Borges, R. M. Blizzard, M. C. Gelato, R. Furlanetto, A. D. Rogol, W. S. Evans, M. L. Vance, D. L. Kaiser, R. M. MacLeod, G. R. Merriam, D. L. Loriaux, J. Spiess, J. Rivier, W. Vale, and M. O. Thorner, Effects of human pancreatic tumour growth hormone releasing factor on growth hormone and somatomedin C levels in patients with idiopathic growth hormone deficiency, *Lancet* **2:**119, 1983.
29. M. Barinaga, G. Yamonoto, C. Rivier, W. Vale, R. Evans, and M. G. Rosenfeld, Transcriptional regulation of growth hormone gene expression by growth hormone-releasing factor, *Nature* **306:**84, 1983.
30. S. M. Wood, J. L. C. Ch'ng, E. F. Adams, J. D. Webster, G. F. Joplin, K. Mashiter, and S. R. Bloom, Abnormalities of growth hormone release in response to human pancreatic growth

hormone releasing factor (GRF 1-44) in acromegaly and hypopituitarism, *Br. Med. J.* **286:**1687, 1983.
31. M. C. Gelato, G. R. Merriam, M. L. Vance, J. A. Goldman, C. Webb, W. S. Evans, J. Rock, E. H. Oldfield, M. E. Molitch, J. Rivier, W. Vale, S. Reichlin, L. A. Frohman, D. L. Loriaux, and M. D. Thorner, Effects of growth hormone-releasing factor on growth hormone secretion in acromegaly, *J. Clin. Endocrinol. Metab.* **60:**251, 1985.
32. C. B. Webb, J. L. Thominet, and L. A. Frohman, Ectopic growth hormone releasing factor stimulates growth hormone release from human somatotroph adenomas in vitro, *J. Clin. Endocrinol. Metab.* **56:**417, 1983.
33. E. F. Adams, C. L. J. Winslow, and K. Mashiter, Pancreatic growth hormone releasing factor stimulates growth hormone secretion by pituitary cells, *Lancet* **1:**1100, 1983.
34. P. Brazeau, R. Guillemin, N. Ling, J. Van Wyk, and R. Humbel, Inhibition par les somatomédines de la sécrétion de l'hormone de croissance stimulée par le facteur hypothalamique somatocrinine (GRF) ou le peptide de synthèse hpGRF, *C. R. Seances Acad. Sci. (III)* **295:**651, 1982.
35. A. Gomez-Pan, W. M. G. Tunbridge, A. Duns, R. Hall, G. M. Besser, D. H. Coy, A. V. Schally, and A. J. Kastin, Hypothalamic hormone interaction in acromegaly, *Clin. Endocrinol. (Oxford)* **4:**455, 1975.
36. M. E. Leavens, N. A. Samaan, R. H. Jesse Jr, and R. M. Byers, Clinical and endocrinological evaluation of 16 patients treated by transsphenoidal surgery, *J. Neurosurg.* **47:**853, 1977.
37. L. D. Schuster, J. P. Bantle, J. H. Oppenheimer, and E. L. Seljeskog, Acromegaly: Reassessment of the long-term therapeutic effectiveness of transsphenoidal pituitary surgery, *Ann. Intern. Med.* **95:**172, 1981.
38. R. L. Eddy, P. F. Gilliland, J. D. Ibarra Jr, J. F. McMurry Jr, and J. Q. Thompson, Human growth hormone release. Comparison of provocative test procedures, *Am. J. Med.* **56:**179, 1974.
39. T. Lin and J. R. Tucci, Provocative tests of growth-hormone release. A comparison of results with seven stimuli, *Ann. Intern. Med.* **80:**464, 1974.
40. M. Irie and T. Tsushima, Increase of serum growth hormone concentration following thyrotropin-releasing hormone injection in patients with acromegaly or gigantism, *J. Clin. Endocrinol. Metab.* **35:**97, 1972.
41. D. S. Schalch, D. Gonzalez-Barcena, A. J. Kastin, A. V. Schally, and L. A. Lee, Abnormalities in the release of TSH in response to thyrotropin-releasing hormone (TRH) in patients with disorders of the pituitary, hypothalamus and basal ganglia, *J. Clin. Endocrinol. Metab.* **35:**609, 1972.
42. G. Faglia, P. Beck-Pecoz, C. Ferrari, P. Travaglini, B. Ambrosi, and A. Spada, Plasma growth hormone response to thyrotropin-releasing hormone in patients with active acromegaly, *J. Clin. Endocrinol. Metab.* **36:**1259, 1973.
43. M. Ishibashi and T. Yamaji, Effect of thyrotropin-releasing hormone and bromoergocriptine on growth hormone and prolactin secretion in perfused pituitary adenoma tissues of acromegaly, *J. Clin. Endocrinol. Metab.* **47:**1251, 1978.
44. J. L. C. Borges, D. R. Uskavitch, D. L. Kaiser, M. J. Cronin, W. S. Evans, and M. O. Thorner, Human pancreatic growth hormone-releasing factor-40 (hpGRF-40) allows stimulation of GH release by TRH, *Endocrinology* **113:**1519, 1983.
45. M. O. Thorner, R. L. Perryman, M. J. Cronin, A. D. Rogol, M. Draznin, A. Johanson, W. Vale, E. Horvath, and K. Kovacs, Somatotroph hyperplasia. Successful treatment of acromegaly by removal of a pancreatic islet tumor secreting a growth hormone-releasing factor, *J. Clin. Invest.* **70:**965, 1982.
46. T. C. Hagen, A. M. Lawrence, and L. Kirsteins, *In vitro* release of monkey pituitary growth hormone by acromegalic plasma, *J. Clin. Endocrinol. Metab.* **33:**448, 1971.
47. T. C. Hagen, A. M. Lawrence, R. E. Kacherian, and L. Kirsteins, A dialyzable heat stable factor in acromegalic plasma associated with increased in vitro growth hormone release, *Proc. Soc. Exp. Biol. Med.* **141:**536, 1972.
48. T. C. Hagen, A. M. Lawrence, and L. Kirsteins, Preliminary studies of plasma growth hormone releasing activity during medical therapy of acromegaly, *Horm. Metab. Res.* **10:**310, 1978.
49. J. P. Allen, D. M. Cook, M. A. Greer, H. Paxton, and A. Castro, Evidence that acromegaly is not a hypothalamic disease, *Trans. Assoc. Am. Physicians* **86:**272, 1973.

50. K. M. Hoyte and J. B. Martin, Recovery from paradoxical growth hormone responses in acromegaly after transsphenoidal selective adenomectomy, *J. Clin. Endocrinol. Metab.* **41**:656, 1975.
51. Z. J. Wanski, A. G. Robinson, and P. J. Jannetta, Selective total removal of a growth-hormone-secreting adenoma: Evidence that acromegaly is a primary pituitary disease, *Metabolism* **28**:624, 1979.
52. I. Bachelot, S. Halimi, J. de Rougemont, and C. Feuerstein, Hypothalamic and pituitary functions in two cases of cured acromegaly, in: *Pituitary Adenomas. Biology, Physiopathology and Treatment.* Second European Workshop, La Pitie-Salpetriere, Paris (P. J. Derome, C. P. Jedynak, and F. Peillon, eds.), Asclepios Publishers, Paris, 1980, p. 315.
53. D. S. Baskin, J. E. Boggan, and C. B. Wilson, Transsphenoidal microsurgical removal of growth hormone-secreting pituitary adenomas. A review of 137 cases, *J. Neurosurg.* **56**:634, 1982.
54. H. Steiner, O. Dahlback, and J. Waldenstrom, Ectopic growth-hormone production and osteoarthropathy in carcinoma of the bronchus, *Lancet* **1**:783, 1968.
55. M. Sparagana, G. Phillips, C. Hoffman, and L. Kucera, Ectopic growth hormone syndrome associated with lung cancer, *Metabolism* **20**:730, 1971.
56. D. N. Glass, A. S. Russel, and R. Davies, Human growth hormone and lung carcinoma, *Lancet* **1**:683, 1972.
57. G. C. Ennis, D. P. Cameron, and H. G. Burger, On the aetiology of hypertrophic pulmonary osteoarthropathy in bronchogenic carcinoma: Lack of relationship to elevated growth hormone levels, *Aust. NZ J. Med.* **3**:157, 1973.
58. J. M. Feldman, J. W. Plonk, C. H. Bivens, H. E. Lebovitz, and S. Handwerger, Growth hormone and prolactin secretion in the carcinoid syndrome, *Am. J. Med. Sci.* **269**:333, 1975.
59. D. P. Cameron, H. G. Burger, D. M. DeKretzer, K. J. Catt, and J. B. Best, On the presence of immunoreactive growth hormone in a bronchogenic carcinoma, *Australasian Ann. Med.* **18**:143, 1969.
60. P. B. Greenberg, C. Beck, T. J. Martin, and H. G. Burger, Synthesis and release of human growth hormone from lung carcinoma in cell culture, *Lancet* **1**:350, 1972.
61. C. Beck and H. G. Burger, Evidence for the presence of immunoreactive growth hormone in cancers of the lung and stomach, *Cancer* **30**:75, 1972.
62. J. Podmore, B. Wilson, E. A. Cowden, G. H. Beastall, and J. G. Ratcliffe, Multiple hormones in human tumours, in: *Carcino-Embryonic Proteins* (F. G. Lehmann, ed.), Elsevier/North-Holland Biomedical Press, New York, 1979, pp. 457–463.
63. A. Kaganowicz, N. H. Farkouh, A. G. Frantz, and A. U. Blaustein, Ectopic human growth hormone in ovaries and breast cancer, *J. Clin. Endocrinol. Metab.* **48**:5, 1979.
64. J. T. Dabek, Bronchial carcinoid tumour with acromegaly in two patients, *J. Clin. Endocrinol. Metab.*, **38**:329, 1974.
65. S. A. Leveston, D. W. McKeel Jr, P. J. Buckley, K. Deschryver, M. H. Greider, B. M. Jaffe, and W. H. Daughaday, Acromegaly and Cushing's syndrome associated with a foregut carcinoid tumor, *J. Clin. Endocrinol. Metab.* **53**:682, 1981.
65a. S. Melmed, C. Ezrin, K. Kovacs, R. S. Goodman, and L. A. Frohman, Acromegaly due to secretion of growth hormone by an ectopic pancreatic islet-cell tumor, *N. Engl. J. Med.* **312**:9, 1985.
66. P. H. Sonksen, A. B. Ayres, M. Braimbridge, B. Corrin, D. R. Davies, G. M. Jeremiah, S. W. Oaten, C. Lowy, and T. E. T. West, Acromegaly caused by pulmonary carcinoid tumors, *Clin. Endocrinol. (Oxford)* **5**:503, 1976.
67. R. H. Caplan, L. Koob, R. M. Abellera, A. S. Pagliara, K. Kovacs, and R. V. Randall, Cure of acromegaly by operative removal of an islet cell tumor of the pancreas, *Am. J. Med.* **64**:874, 1978.
68. S. M. Shalet, C. G. Beardwell, I. A. MacFarlane, M. L. Ellison, C. M. Norman, L. H. Rees, and M. Hughes, Acromegaly due to production of a growth hormone releasing factor by a bronchial carcinoid tumour, *Clin. Endocrinol. (Oxford)* **10**:61, 1979.
69. M. S. uz Zafar, R. C. Mellinger, G. Fine, M. Szabo, and L. A. Frohman, Acromegaly associated with a bronchial carcinoid tumor: Evidence for ectopic production of growth hormone-releasing activity, *J. Clin. Endocrinol. Metab.* **48**:66, 1979.

70. S. L. Asa, J. M. Bilbao, K. Kovacs, and J. A. Linfoot, Hypothalamic neuronal hamartoma associated with pituitary growth hormone cell adenoma and acromegaly, *Acta Neuropathol.* **52:**231, 1980.
71. S. L. Asa, B. W. Scheithauer, J. M. Bilbao, E. Horvath, N. Ryan, K. Kovacs, R. V. Randall, E. R. J. Laws, W. Singer, J. A. Linfoot, M. O. Thorner, and W. Vale, A case for hypothalamic acromegaly: A clinicopathological study of six patients with hypothalamic gangliocytomas producting growth hormone-releasing factor, *J. Clin. Endocrinol. Metab.* **58:**796, 1984.
72. D. J. McCune and H. Bruch, Osteodystrophia fibrosa: report of a case in which the condition was combined with precocious puberty, pathologic pigmentation of the skin and hyperthyroidism, with a review of the literature, *Am. J. Dis. Child.* **54:**806, 1937.
73. F. Albright, A. M. Butler, A. O. Hampton, and P. Smith, Syndrome characterized by osteitis fibrosa disseminata, areas of pigmentation and endocrine dysfunction, with precocious puberty in females, *N. Engl. J. Med.* **216:**727, 1937.
74. R. Hall and C. Warrick, Hypersecretion of hypothalamic releasing hormones: A possible explanation of the endocrine manifestations of polyostotic fibrous dysplasia (Albright's syndrome), *Lancet* **1:**1313, 1972.
75. M. T. Scurry, J. M. Bicknell, and S. S. Fajans, Polyostotic fibrous dysplasia and acromegaly, *Arch. Intern. Med.* **114:**40, 1964.
76. E. S. Lightner, R. Penny, and S. D. Frasier, Growth hormone excess and sexual precocity in polyostotic fibrous dysplasia (McCune-Albright syndrome): Evidence for abnormal hypothalamic function, *J. Pediatr.* **87:**922, 1975.
77. E. S. Lightner, R. Penny, and S. D. Frasier, Pituitary adenoma in McCune–Albright syndrome: Follow-up information, *J. Pediatr.* **89:**159, 1976.
78. S. K. Joishy and L. B. Morrow, McCune-Albright syndrome associated with a functioning pituitary chromophobe adenoma, *J. Pediatr.* **89:**73, 1976.
79. D. G. B. Powell, Polyostotic fibrous dysplasia with acromegaly (Albright's syndrome). A case report, *S. Afr. Med. J.* **50:**182, 1976.
80. D. Carr, I. K. Mathie, A. R. Manners, and C. Colman, Hyperprolactinaemia in a patient with the McCune–Albright syndrome, *Br. J. Obstet. Gynaecol.* **86:**330, 1979.
81. J. Albin and R. Wu, Abnormal hypothalamic–pituitary function in polyostotic fibrous dysplasia, *Clin. Endocrinol. (Oxford)* **14:**435, 1981.
82. A. Lipson and T.-H. Hsu, The Albright Syndrome associated with acromegaly: Report of a case and review of the literature, *Johns Hopkins Med. J.* **149:**10, 1981.
83. K. F. Chung, J. Alaghband-Zadeh, and A. Guz, Acromegaly and hyperprolactinemia in McCune Albright syndrome, *Am. J. Dis. Child.* **137:**134, 1983.
84. C. Polychronakos, G. Tsoukas, J. R. Cucharme, J. Letarte, and R. Collu, Gigantism and hyperprolactinemia in polyostotic fibrous dysplasia, *J. Endocrinol. Invest.* **5:**323, 1982.
85. W. H. Daughaday, Z. Laron, A. Pertzelan, and J. N. Heins, Defective sulfation factor generation: A possible etiological link in dwarfism, *Trans. Assoc. Am. Physicians* **82:**129, 1969.
86. I. M. Burr, R. K. Slonim, R. K. Danish, N. Gadoth, and I. J. Butler, Diencephalic syndrome revisited, *J. Pediatr.* **88:**439, 1976.
87. S. L. S. Drop, H. J. Guyda, and E. Colle, Inappropriate growth hormone release in the diencephalic syndrome of childhood: Case report and 4 year endocrinological follow-up, *Clin. Endocrinol. (Oxford)* **13:**181, 1980.
88. M. E. Molitch and S. H. Hou, Neuroendocrine alterations in systemic disease, *Clin. Endocrinol. Metab.* **12:**825, 1983.
89. H. Abe, M. E. Molitch, J. J. Van Wyk, and L. E. Underwood, Human growth hormone and somatomedin C suppress the spontaneous release of growth hormone in unanesthetized rats, *Endocrinology* **113:**1319, 1983.
90. G. S. Tannenbaum, H. J. Guyda, and B. I. Posner, Insulinlike growth factors: A role in growth hormone negative feedback and body weight regulation via brain, *Science* **220:**77, 1983.

3

Clinical Features, Tests, and Causes of Growth Hormone Deficiency

Margaret H. MacGillivray

In the early 1960s it was easy to diagnose growth hormone (GH) deficiency. At that time, only children with severe growth failure and the classical features of hypopituitarism were eligible for diagnosis and treatment because supplies of GH were very limited. During the past decade, however, the staff of the National Hormone and Pituitary Program has established an efficient system for collecting pituitaries, especially frozen glands, and the total yield of GH has become sufficient to support a wide variety of clinical and basic science projects. These investigations have demonstrated that GH treatment is beneficial not only to patients with classical GH deficiency (complete or partial) but also to those with idiopathic growth failure who have normal GH responses to provocative stimuli and low somatomedin levels.[1–6] It is unfortunate that the term "normal-variant short stature" has been used on occasion to describe this population because the linear growth rates of these children are clearly abnormal. Whether they represent subjects who produce bioinactive GH or have defective neurosecretory regulation of GH needs to be clarified.[7] Recently, GH therapy has increased the linear growth velocity in another population of children with idiopathic growth failure whose GH responses and somatomedin C levels are normal.[8] These observations may be interpreted in various ways. Subtle forms of GH deficiency may be present in some children whose only features include height below the 3rd percentile, abnormal linear growth rates, and normal body proportions. Alternatively, these patients may benefit from the general anabolic properties of GH, analogous to that seen following treatment with oxandrolone and similar steroids. Finally, Dr. John Crawford has recommended that we focus more attention on the nutritional state of

---

Margaret H. MacGillivray • Department of Pediatrics, State University of New York at Buffalo School of Medicine and The Children's Hospital of Buffalo, Buffalo, New York 14222.

children with idiopathic growth failure because they may need "meat and potatoes" rather than GH treatment.[9]

1. Clinical Features

The clinical features of GH deficiency vary with the etiology, age at onset, and severity of the disorder.[10–14] A majority of the children have the idiopathic form of hypopituitarism, which, by definition, implies GH deficiency in the absence of an identifiable cause. Infants with congenital idiopathic GH deficiency have normal birth length and weight, grow normally for 3–6 months, and thereafter exhibit deceleration of linear growth rates. Their linear growth curves deviate progressively away from the mean. The severity of the growth disorder is not usually appreciated by parents, relatives, and even family physicians until the child is 2–5 years of age at which time it is obvious that a marked height discrepancy exists between the patient and agemates. Growth failure is the primary concern that prompts families to seek medical consultation. After 3 years of age, growth rates in affected children are almost always less than 4–5 cm/year. Characteristically, these patients have normal body proportions for chronological age. Most, but not all, children with GH deficiency are overweight for height and some are frankly obese. Excessive accumulation of fat occurs predominantly in the chest and abdominal areas, and the overlying skin has a grossly ripply or bosselated appearance. The general appearance of the child is always younger than is appropriate for chronological age. In early childhood, the facies are frequently cherubic, infantile, or dolllike, and the voice may be high pitched. Eruption and shedding of primary teeth usually are delayed. The permanent teeth also may appear late and are often crowded and irregularly positioned. Muscle mass is diminished. The hair is thin and nail growth is poor. Skeletal age in the prepubertal hypopituitary patient is always delayed, and the degree of retardation of bone maturation is proportional to that of height age. Head circumference is within the normal range for age, but the sella turcica is often small. Closure of the anterior fontanelle is usually very delayed.

During early infancy and childhood, patients with GH deficiency may experience recurring episodes of fasting hypoglycemia and seizures, which, untreated, may result in mental retardation, a seizure disorder, and even death. Therapy with GH is highly effective. Some of these infants have a concomitant deficiency of ACTH; their treatment includes maintenance doses of hydrocortisone (10–15 mg/M^2 per day in two or three divided doses per os) in addition to GH (0.1 IU/kg, i.m. on Monday, Wednesday, and Friday or given daily if necessary to prevent hypoglycemia). During illnesses, stress doses of hydrocortisone given orally (60 mg/M^2 hydrocortisone in three divided doses) or parenterally (60 mg/M^2, i.m. or i.v. given in two or three divided doses) are required. Intravenous use of glucose-containing electrolyte solutions is frequently necessary when oral intake is curtailed in these children. The tendency to fasting hypoglycemia usually disappears by 5 years of age but may reappear if severe illnesses prevent oral intake for prolonged periods. Apart from the problems of hypoglycemia and seizures during the early childhood years, children with idiopathic GH deficiency are generally healthy and have normal intelligence. Emotional problems are more frequent, however, in this population because patients are perceived to be much younger than their chronological age and are treated accordingly. Excessive teasing and social isolation are common complaints.

GH-deficient boys with micropenis and hypoplastic testes are likely to also have gonadotropin deficiency, which may be partial or complete. Usually, the gonadotropin status cannot be accurately assessed until adolescence. Small doses of androgen should be prescribed during infancy and early childhood to bring the penis size into the normal range for age. During the adolescent and adult years, testicular growth and enhanced virilization can be facilitated in gonadotropin-deficient hypopituitary males by the use of human chorionic gonadotropin in combination with human menopausal gonadotropin.[15] Children with isolated idiopathic GH deficiency who are treated with GH from an early age are likely to have normal onset of pubertal maturation, whereas those who begin treatment late in childhood often experience delayed puberty. Hypopituitarism of organic origin (e.g., tumor, trauma) carries a high risk of multiple pituitary hormone deficiencies.

2. *Tests for GH Deficiency*

The diagnostic tests for GH deficiency are summarized in Table 1.[10–13] The usual indications for evaluation of GH deficiency include (1) pathological short stature in a child with normal body proportions (height >2 SD below the mean; some centers require height be >3 SD below the mean), (2) abnormal growth rate for chronological age (<7 cm/year prior to the age of 3 years; 4–5 cm/year from 3 years to onset of puberty and <5.5–6 cm/year during the pubertal years); (3) delayed skeletal maturation (bone age >2 SD below the mean for chronological age); and (4) special clinical considerations (e.g., microgenitalia, hypoglycemia, history of head trauma, or central nervous system (CNS) tumor, hypoxemia, or intracranial hemorrhage). In addition, tests of pituitary function are necessary when children with suspected acquired hypopituitarism have deceleration or cessation of linear growth even if their height is within the normal range.

Radioimmunoassay of plasma GH levels during physiological or pharmacological stimulation is still the most practical method of detecting GH deficiency. A peak plasma GH level >9 (or 10) ng/ml is interpreted as a normal response. Radioreceptor assays of plasma GH are not widely available, are more difficult to perform, and are not sufficiently well standardized for widespread clinical application.[1,4] Determinations of somatomedin C/IGF1 concentrations prior to and following GH administration are being used with increasing frequency. However, interpretation of these tests is difficult because (1) many conditions other than GH deficiency cause low somatomedin C/IGF1 levels; (2) the

Table 1 Diagnostic Tests for GH Deficiency

Screening tests	Method	Comment
Exercise	Fasting state Use 20 min moderate to vigorous exercise Take sample 20–40 min after start of exercise	Lacks standardization Variable compliance, 70% of normals have peak GH >10 ng/ml
Clonidine[21–24]	Fasting state 25 μg/os Take sample at (60) *90* min	In 90% normals, peak GH >10 ng/ml
Somatomedin C	Random serum sample	Age-dependent normal values
Definitive Tests	**Requirements**	**Comment**
1. Pharmacological		
Insulin	Reg. insulin 0.05–0.1 IU/kg IV 50% fall in glucose necessary Glucose nadir (<45 mg/dl) at 20–30 min Samples taken 0, 20, 40, 60, 90, 120 min	Peak GH at 45–75 min Peak cortisol >15 μg/dl
Arginine	L'Arginine monohydrochloride 5–10% sol., 0.5 g/kg i.v. over 30 min Samples taken 0, 30, 60, 90, 120 min	Negligible side effects
Clonidine	1. 25 μg/os to all subjects Take 0, 30, 60, 90, 120 min	Peak GH level at +60 min or +90 min; no drowsiness or hypotension
	or	
	2. 0.15 mg/M^2 per os	Somnolence, fall in blood pressure
Levodopa	Overnight fast 0.5 g/1.73 M^2 per os Sample at 0, 60, 90, 120 min	Normal GH responses in 74% normal children Nausea, vomiting, 30–40 min after ingestion Peak GH at 45–120 min
Propranolol	Overnight fast 0.75 mg/kg per os 30–60 min before glucagon, insulin, arginine, or exercise	Mainly used to augment primary stimuli
Levodopa–propranolol	Overnight fast L-Dopa, 0.5 g/1.73 M^2 per os plus propranolol, 0.75 mg/kg per os Sample at 0, 30, 60, 90 min	Normal GH response in 95% normal children Do not use if patient prone to hypoglycemia or has asthma or reduced cardiac reserves
Glucagon	0.03 mg/kg i.m. or s.c. (maximum dose = 1 mg)	Peak GH at 120–180 min

(*continued*)

Table 1 (*continued*)

Definitive tests	Requirements	Comment
2. Physiological		
Sleep	Take sample 1 hr after onset deep sleep or Use EEG to document stages 3 & 4 and obtain samples at 20-min intervals for more accurate evaluation of GH secretion	Normal responses in 69% normal children if single sample used
Diurnal GH profile	Sample at 20-min intervals for 24 hr Interpretation based on number and amplitude of GH pulses	Requires special support staff Not a routine procedure Uncomfortable for patients
3. Others		
Somatomedin generation test	Baseline sample for Sm C Give GH 0.1 μg/kg i.m. q.d. ×5 days Take sample for Sm C 12 hr after last injection	Sm C levels should double and post-GH Sm C level should be in normal range for age

methodology used for somatomedin C/IGF1 determinations has not been standardized; (3) GH-dependent growth factors apart from somatomedin C/IGF1 may be essential for normal growth; (4) somatomedin C levels vary with age and there is overlap between the range of somatomedin C levels in very young children and hypopituitary patients; and (5) the short-term response of somatomedin C to growth hormone treatment correlates with bone age[16–19] but not with growth velocity during long-term treatment with growth hormone.[16,17,19,20]

3. *Causes of GH Deficiency or Defective GH Action*

A classification of the causes of GH deficiency and GH resistance is outlined in Table 2. No attempt has been made to distinguish whether the defect in GH secretion is due to disease in the hypothalamus or the pituitary gland. Prior to the discovery of growth hormone releasing factor (GHRF), it was presumed that the lesion was localized in the hypothalamus if patients with GH, thyroid-stimulating hormone (TSH), and/or follicle-stimulating hormone/luteinizing hormone (FSH/LH) deficiencies exhibited a rise in plasma concentrations of TSH and/or FSH/LH following TRH and gonadotropin-releasing hormone stimulation, respectively. The recent isolation and synthesis of GHRF should result in an improved etiological classification of GH deficiency based on the

Table 2 Causes of GH Deficiency or Defective GH Action

- Congenital GH deficiency
  1. Decreased GH secretion
     - Idiopathic
     - Hereditary—autosomal recessive
       —autosomal dominant
     - Embryological defects—aplasia, hypoplasia, ectopia
       —anencephaly, holoprosencephaly
       —septo-optic dysplasia
       —midline facial dysplasia
       —empty sella syndrome
       —miscellaneous syndromes
     - Bioinactive GH
     - Neurosecretory defects
  2. GH resistance—Laron dwarf
     —pygmy
- Acquired GH deficiency
  - Idiopathic
  - Neurosecretory defects
  - CNS tumors—craniopharyngioma
    —optic glioma
    —dysgerminoma
    —hamartoma
  - Trauma—perinatal insult—breech deliveries, hypoxemia, asphyxia, difficult forceps, intracranial hemorrhage, precipitous or prolonged delivery, twin pregnancy
    —child abuse
    —accidental trauma
  - Inflammatory diseases—viral encephalitis
    —bacteria, group B streptococcal meningitis, etc.
    —fungal
    —granuloma—tuberculosis, syphilis, sarcoid, unknown etiology
  - Autoimmunity—lymphocytic hypophysitis
  - Irradiation—CNS radiation for brain tumors, leukemia
  - Vascular lesions—aneurysms, pituitary vessels
    —infarction
  - Hematological disorders—hemochromatosis
    —sickle cell disease
    —thalassemia
  - Histiocytosis
  - Transient defects in (1) GH secretion, or (2) action
    —peripuberty (1)
    —primary hypothyroidism (1, 2)
    —psychosocial stress (1, 2)
    —malnutrition (2)
    —glucocorticoid excess (1?, 2?)
    —drugs

anatomical site involved, i.e., GH deficiency due to primary pituitary disease or secondary to dysfunction in the hypothalamus or higher brain centers.[25–29]

The prevalence of GH deficiency has been estimated to be as low as 1 in 30,000 births or as high as 1 in 4018 in two studies carried out in Great Britain. It is possible that identification of the milder forms of GH deficiency will result in a disease frequency greater than 1 in 4000.[30,31]

4. Congenital GH Deficiency

4.1. Idiopathic Hypopituitarism

The most common type of hypopituitarism is idiopathic in origin. By definition, no organic lesion or etiological factor can be identified. Males are affected more frequently than females. The disorder may involve only GH (isolated GH deficiency) or may be associated with deficiencies of TSH and/or gonadotropins and/or ACTH (idiopathic multiple pituitary deficiencies). Prepubertal children with GH plus TSH or ACTH deficiency are likely to exhibit gonadotropin deficiency (partial or complete) in adolescence. Defective hypothalamic regulation of the pituitary is believed to exist in a majority of individuals with idiopathic hypopituitarism. Onset may be congenital or acquired; the latter mandates thorough and repeated evaluations for possible organic lesions.

4.2. Inherited Forms of GH Deficiency

Both autosomal recessive and autosomal dominant patterns of inheritance have been found in the familial forms of GH deficiency.[32,33]

4.3. Embryological Defects

Aplasia, hypoplasia, or ectopic location of the pituitary may occur as an isolated, congenital defect or in association with anencephaly or holoprosencephaly. The holoprosencephalies comprise a spectrum of embryological abnormalities that interfere with midline cleavage of the forebrain and cause midline dysplasia of the face.[34]

The syndrome of septo-optic dysplasia (De Morsier's syndrome) consists of optic nerve and optic disc hypoplasia with or without abnormalities of the septum pellucidum and corpus callosum. Growth failure and hypopituitarism occur in 60% of cases. Other features include median facial cleft, hyperprolactinemia, and diabetes insipidus. Septooptic dysplasia may represent a mild form of holoprosencephaly.[35–37]

An increased incidence of hypopituitarism has been documented in patients with cleft lip and palate.[38]

4.4. Empty Sella Syndrome

Extension of the subarachnoid space into the sella turcica is responsible for the empty sella syndrome. It is presumed to result from an incompetent sella diaphragm and an increase in cerebrospinal fluid pressure. In the absence of surgery or radiation, the condition is called primary empty sella syndrome. Familial cases are rare. In most patients, pituitary function is normal and the sella turcica is diffusely enlarged. Empty sella syndrome, seen most frequently in obese middle-aged women, can be readily distinguished from pituitary tumor by computed axial tomography or pneumoencephalogram. Empty sella syndrome is rarely recognized during childhood; most of the reported cases have associated cranial defects. One kindred has been described in which empty sella syndrome was transmitted as an autosomal dominant trait in association with Rieger's anomaly of the anterior chamber of the eye.[39–42]

4.5. Bioinactive GH

Affected children present with growth failure and have the following characteristics: height >3 SD below the mean, abnormal growth rate (<4 cm/year), apparent good health, normal body proportions, normal increments of GH by radioimmunoassay measurement after provocative stimuli, abnormally low radioreceptor/radioimmunoassay GH ratios, low somatomedin C concentrations, improved levels of somatomedin C following 3–10 daily injections of GH, and improved linear growth when given GH therapy. The assumption is made that the endogenous GH in these children is biologically inactive but immunologically reactive. Bioinactive GH is thought to be the etiological factor in "normal variant short stature."[1–5]

4.6. Neurosecretory Dysfunction

This entity has been identified in children with heights <1 percentile, growth velocity <4 cm/year, peak GH >10 ng/ml after provocative testing, but low somatomedin C levels and abnormally low diurnal frequency and amplitude of GH pulses.[7] Somatomedin C levels doubled after administration of exogenous GH, 0.05–0.1 U/kg, i.m., daily for 5 days. The number and amplitude of GH secretory pulses during a 24-hr period were significantly below those observed in healthy children but greater than those documented in classical GH deficiency. Treatment with exogenous GH enhanced linear growth in all subjects. The data suggest that these children have defective neuroregulation of GH secretion and suffer from a less severe form of GH deficiency. They may be overlooked because of their normal GH responses to provocative tests. Identification of this disorder is somewhat difficult because it requires blood sampling at 20-min intervals for 24 hr.

4.7. Defective GH Action

The clinical characteristics of children with Laron-type dwarfism are identical to those observed in GH-deficient children. In contrast, however, they exhibit very elevated basal levels of GH and exaggerated GH responses to provocative testing. The endogenous GH in these children is biologically active and binds normally in standard GH receptor assays. Laron dwarfs have low somatomedin C levels, which do not increase after GH therapy. Administration of GH does not enhance growth or stimulate nitrogen retention. Adults rarely exceed 130 cm in height and are presumed to be fertile. This growth disorder results from an abnormal GH receptor. The disease is seen most commonly in the products of consanguineous marriages involving Jewish parents of Middle Eastern extraction. It is transmitted by autosomal recessive inheritance. Affected children may benefit from somatomedin therapy when it becomes available.[43,44]

The pygmy may represents another example of individuals with GH resistance; their IGF1 levels are low and IGF2 levels are normal. Treatment with exogenous GH fails to correct the IGF1 deficiency. Merimee and colleagues postulate that the pygmy has an isolated deficiency of IGF1 due to either a defect in a complex carrier protein system or an abnormal GH receptor.[45,46] Malnutrition is a possible alternative explanation for the IGF1 deficiency.

5. Acquired GH Deficiency

5.1. CNS Tumors

Craniopharyngioma is the most common CNS tumor to cause GH deficiency and multiple other pituitary hormone deficits in childhood. The tumor is believed to rise from embryonic squamous cell rests located at the junction of the adenohypophysis and neurohypophysis. Because of its location, expansion may cause permanent injury of varying severity to the hypothalamus, optic nerves, and pituitary gland. The only complaint may be growth failure. Other clinical features include visual field defects, signs and symptoms of increased intracranial pressure, and multiple pituitary endocrine deficiencies. After surgical extirpation, a majority of children have panhypopituitarism. Destruction of the satiety center may result in uncontrolled hyperphagia and excessive obesity. Normal or excessive growth has been observed postoperatively in some children who have abnormal GH responses and normal somatomedin levels. A recent study suggests that hyperprolactinemia or hyperinsulinism may account for the normal IGF1 levels and sustained growth in these children.[47–50] Other CNS tumors that cause hypopituitarism in childhood are optic gliomas, germinomas, ependymomas, meningiomas, colloid cysts of the 3rd ventricle, and, rarely, chromophobe adenomas.

5.2. Trauma

It appears that perinatal injury to the pituitary and/or hypothalamus may be the cause of hypopituitarism in a significant number of children with presumed congenital idiopathic GH deficiency. A careful review of the birth histories of these children has shown that 50–65% of cases had one or more significant perinatal insults. Birth size was generally normal; the sex ratio was four males to one female. The perinatal risk factors included intrapartum hypoxemia or asphyxia, breech deliveries, difficult forceps deliveries, intracranial hemorrhage, precipitous or prolonged labor, twin pregnancies, and postnatal seizures. Other risk factors include bleeding during pregnancy and toxemia.[51,52] Throughout childhood accidental head trauma or child abuse (battered-child syndrome) may result in hypopituitarism.

5.3. *Inflammatory Diseases*

Bacterial meningitis, viral encephalitis, and fungal infections of the CNS may cause permanent injury to the pituitary and hypothalamus. Giant cell granuloma of the pituitary gland is a rare cause of hypopituitarismm. Its presentation may be indistinguishable from that of pituitary adenoma.[53] The diagnosis is usually made by obtaining pituitary tissue for histological examination or at autopsy. The disease is usually of unknown etiology but has been seen in association with tuberculosis, syphilis, and sarcoidosis.

5.4. *Autoimmunity*

Lymphocytic hypophysitis is considered an autoimmune disorder whose clinical characteristics resemble those of pituitary tumors. Hypopituitarism is frequently present. The disease has been described most frequently in women who are or were recently pregnant.[54–56]

5.5. *Irradiation*

Irradiation for tumors of the head and neck carries a high risk for impaired pituitary function.[57–59] The doses of irradiation usually exceed 4000 rads. GH deficiency is the most common abnormality. TSH and ACTH secretion usually are preserved. Less is known about the gonadotropin status of these children. Primary thyroid dysfunction due to radiation injury of the thyroid gland is not uncommon. Usually the patient appears euthyroid and has an elevated TSH level with normal T_4 and T_3 concentrations. Clinical hypothyroidism with low T_4 and T_3 levels has also been observed in these children.

5.6. *Vascular Lesions*

Aneurysms of the pituitary vessels and infarction of the pituitary gland are extremely rare causes of hypopituitarism in childhood.[60]

5.7. *Hematological Disorders*

Severe, widespread impairment of endocrine glands is frequently observed in patients with thalassemia major who have iron overload. Normal GH responses and low somatomedin levels have been documented.[61–63]

5.8. Histiocytosis

Permanent vasopressin deficiency and diabetes insipidus is the most common CNS complication of disseminated histiocytosis. Many affected children have inadequate GH responses to provocative tests but have satisfactory growth and do not require treatment. Those with growth failure and abnormal GH responses are benefited by GH therapy. Usually TSH, ACTH, FSH, and LH secretion remain normal.[(64)]

5.9. Transient Defects in GH Secretion or Action

Transient or functional hypopituitarism has been documented in peripubertal males. Underwood and colleagues have given this entity the name "lazy pituitary syndrome." Therapy with GH or androgens will accelerate growth velocity. Normal growth hormone responses to provocative stimuli have been observed after the onset of puberty.[(65,66)]

5.10. Psychosocial Stress

Hypopituitarism has been documented in young children with growth failure who come from hostile home environments. Psychosocial stress and nutritional factors are the probable causes of the entity known as "psychosocial dwarfism." These children have delayed speech, are withdrawn, and have abnormal sleep patterns and bizarre eating habits. After these children are placed in foster homes or after the home stresses are resolved, pituitary function returns to normal and they exhibit accelerated growth, improved behavior, and normal dietary habits.[(67–71)]

5.11. Hypothyroidism

Transient GH deficiency due to myxedema of the pituitary gland has been documented in primary hypothyroidism. Treatment with thyroid hormone is associated with catchup growth and normal GH responses to provocative stimuli. Children with secondary or tertiary hypothyroidism should be treated with thyroxine replacement for 3–4 weeks before their pituitary function is evaluated. The results of tests of GH secretion cannot be interpreted accurately if children are evaluated while hypothyroid.

5.12. Malnutrition

Poor nutrition is the most common cause of growth failure in the world. These children have elevated levels of GH and low somatomedin concentrations.[(72)] Improved nutrition results in restoration of normal GH and somatomedin levels.

6. Conclusion

Extraordinary progress has been made during the past 20 years in the treatment of patients with clinical GH deficiency. However, new questions have arisen because of recent studies which report that some children with idiopathic growth failure respond to GH treatment even though their GH responses to traditional pituitary stimulation tests were normal.[1–6,8,73,74] It has been hypothesized that some of these children may have a GH neurosecretory disorder and others may be producing GH that is immunologically reactive but biologically inactive. These observations point out the potential shortcomings of our current diagnostic protocols and suggest that the clinical spectrum will need to be broadened to include children with subtle and atypical GH deficiency. The major area of interest and controversy in the next decade will pertain to the use of pharmacological doses of GH to augment heights of short boys and girls whose linear growth is on one of the lower percentiles for normal children. The long-range safety and efficacy of such treatments are unknown. A strong argument could be made against using pharmacological doses of GH in healthy short children with normal rates of linear growth because of the potential risks of causing diabetes mellitus or antibody-mediated resistance to exogenous and endogenous GH.

References

1. A. A. Kowarski, J. Schneider, E. Ben-Galim, V. V. Weldon, and W. H. Daughaday, Growth failure with normal serum RIA-GH and low somatomedin activity: Somatomedin restoration and growth acceleration after exogenous GH, *J. Clin. Endocrinol. Metab.* **47:**461, 1978.
2. D. Rudman, M. H. Kutner, R. D. Blackston, R. A. Cushman, R. P. Bain, and J. H. Patterson, Children with normal-variant short stature: Treatment with human growth hormone for six months, *N. Engl. J. Med.* **305:**123, 1981.
3. A. Hayek and G. T. Peake, Growth and somatomedin-C responses to growth hormone in dwarfed children, *J. Pediatr.* **99:**868, 1981.
4. T. Frazer, J. R. Gavin, W. H. Daughaday, R. E. Hillman, and V. V. Weldon, Growth hormone-dependent growth failure, *J. Pediatr.* **101:**12, 1982.
5. L. P. Plotnick, C. D. R. Q. L. Van Meter, and A. A. Kowarski, Human growth hormone treatment of children with growth failure and normal growth hormone levels by immunoassay: Lack of correlation with somatomedin generation. *Pediatrics* **71:**324, 1983.
6. G. M. Bright, A. D. Rogol, A. J. Johanson, and R. M. Blizzard, Short stature associated with normal growth hormone and decreased somatomedin-C concentrations: Response to exogenous growth hormone, *Pediatrics* **71:**576, 1983.
7. B. E. Spiliotis, G. August, W. Hung, W. Sonis, and B. B. Bercu, Broadening of the spectrum of growth hormone (GH) secretory abnormalities in short children. The Endocrine Society Programs & Abstracts for 65th Annual Meeting, San Antonio, June 8–10, 1983, p. 172A.
8. G. Van Vliet, D. M. Styne, S. L. Kaplan, and M. M. Grumbach, Growth hormone treatment for short stature, *N. Engl. J. Med.* **309:**1016, 1983.
9. J. D. Crawford, Meat, potatoes, and growth hormone (editorial), *N. Engl. J. Med.* **305:**163, 1981.
10. M. H. MacGillivray and M. L. Voorhess, Disorders of growth and development, in: (P. Felig, J. Baxter, A. E. Broadus, and L. A. Frohman, eds.), *Endocrinology and Metabolism* McGraw-Hill, New York, 1981, p. 1305.

11. L. E. Underwood and J. J. Van Wyk, Hormones in normal and aberrant growth, in: *Textbook of Endocrinology*, 6th ed. (R. H. Williams, ed.), W. B. Saunders, Philadelphia, 1981, p. 1149.
12. W. Hung, G. P. August, and A. M. Glascow, Anterior pituitary gland, in: *Pediatric Endocrinology*, Medical Examination Publishers, New York, 1983.
13. S. A. Kaplan, Growth and growth hormone, in: *Disorders of the Anterior Pituitary*. W. B. Saunders, Philadelphia, 1982.
14. M. A. Preece, Diagnosis and treatment of children with growth hormone deficiency, *Clin. Endocrinol. Metab.* **11:**1, 1982.
15. R. R. Clopper, T. Mazur, M. H. MacGillivray, R. E. Peterson, and M. L. Voorhess, Data on virilization and erotosexual behavior in male hypopituitarism during gonadotropin and androgen treatment, *J. Androl.* **4:**303, 1983.
16. R. G. Rosenfeld, S. F. Kemp, and R. L. Hintz, Constancy of somatomedin response to growth hormone treatment of hypopituitary dwarfism, and lack of correlation with growth rate, *J. Clin. Endocrinol. Metab.* **53:**611, 1981.
17. H. J. Dean, J. G. Kellett, R. M. Bala, *et al.*, The effect of growth hormone treatment on somatomedin levels in growth hormone-deficient children, *J. Clin. Endocrinol. Metab.* **55:**1167, 1982.
18. S. L. Blethen, W. H. Daughaday, and V. V. Weldon, Kinetics of the somatomedin C/insulinlike growth factor I: response to exogenous growth hormone in GH-deficient children. *J. Clin. Endocrinol. Metab.* **54:**986, 1982.
19. D. S. Schalch, S. E. Tollefsen, G. J. Klingensmith, R. W. Gotlin, and M. J. Diehl, Effects of human growth hormone administration on serum somatomedins, somatomedin carrier proteins, and growth rates in children with growth hormone deficiency. *J. Clin. Endocrinol. Metab.* **55:**49, 1982.
20. K. C. Copeland, L. E. Underwood, and J. J. Van Wyk, Induction of immunoreactive somatomedin D in human serum by growth hormone: Dose–response relationships and effect on chromatographic profiles, *J. Clin. Endocrinol. Metab.* **50:**690, 1980.
21. S. Lal, G. Tolis, J. B. Martin, G. M. Brown, and H. Guyda, Effect of clonidine on growth hormone, prolactin, luteinizing hormone, follicle stimulating hormone and thyroid stimulating hormone in the serum of normal man, *J. Clin. Endocrinol. Metab.* **41:**827, 1975.
22. I. Gil Ad, E. Topper, and Z. Laron, Oral clonidine as growth hormone stimulation test, *Lancet* **3:**278, 1979.
23. R. Lanes and E. Hurftado, Oral clonidine—An effective growth hormone-releasing agent in prepubertal subjects, *J. Pediatr.* **100:**710, 1982.
24. R. Lanes, A. Herrera, A. Palacios, and G. Moncada, Decreased secretion of cortisol and ACTH after oral clonidine administration in normal adults, *Metabolism* **32:**568, 1983.
25. R. Guillemin, P. Brazeau, P. Bohlen, F. Esch, N. Ling, and W. Wehrenberg, Growth hormone-releasing factor from a human pancreatic tumor that caused acromegaly, *Science* **218:**585, 1982.
26. J. Rivier, J. Spiess, M. Thorner, and W. Vale, Characterization of a growth hormone-releasing factor from a human pancreatic islet tumour, *Nature* **300:**276, 1982.
27. J. L. C. Borges, R. M. Blizzard, M. C. Gelato, R. Furlanetto, *et al.*, Effects of human pancreatic tumor growth hormone releasing factor on growth hormone and somatomedin C levels in patients with idiopathic growth hormone deficiency, *Lancet,* **11:**119, 1983.
28. M. L. Vance, J. L. C. Borges, D. L. Kaiser, R. Furlanetto, and M. O. Thorner, Growth hormone responses to human pancreatic growth hormone-releasing factor in man. The Endocrine Society Program & Abstracts, 65th annual meeting, San Antonio, June 8–10, 1983, p. 83.
29. S. M. Rosenthal, E. A. Schriock, S. L. Kaplan, R. Guillemin, and M. M. Grumbach, Synthetic human pancreatic growth hormone releasing factor (hpGRF(1-44)-NH2) stimulates growth hormone (GH) secretion in normal men. The Endocrine Society Program & Abstracts, 65th annual meeting, San Antonio, June 8–10, 1983, p. 83.
30. G. V. Vimpani, A. F. Vimpani, G. P. Lidgard, E. H. D. Cameron, and J. W. Farquhar, Prevalence of severe growth hormone deficiency, *Br. Med. J.* **2:**427, 1977.
31. J. M. Parkin, Incidence of growth hormone deficiency, *Arch. Dis. Child.* **49:**905, 1974.
32. D. L. Rimoin, T. J. Merimee, and V. A. McKusick, Growth hormone deficiency in man: An isolated, recessively inherited defect, *Science* **152:**1635, 1966.

33. D. L. Rimoin and R. N. Schimke, *Genetic Disorders of the Endocrine Glands*, C. V. Mosby, St. Louis, 1971.
34. D. W. Smith, *Recognizable Patterns of Human Malformation: Genetic, Embryologic and Clinical aspects*, 2nd ed., W. B. Saunders, Philadelphia, 1976.
35. S. L. Kaplan, M. M. Grumbach, and W. F. Hoyt, A syndrome of hypopituitary dwarfism, hypoplasia of optic nerves and malformation of the prosencephalon, *Pediatr. Res.* **4:**480, 1970.
36. H. Patel, W. J. Tze, J. U. Crichton, *et al.*, Optic nerve hypoplasia with hypopituitarism, *Am. J. Dis. Child.* **129:**175, 1975.
37. C. Stewart, M. Castro-Magana, J. Sherman, M. Angulo, and P. J. Collipp, Septo-optic dysplasia and median cleft face syndrome in a patient with isolated growth hormone deficiency and hyperprolactinemia, *Am. J. Dis. Child.* **137:**484, 1983.
38. D. Rudman, G. T. Davis, J. H. Priest, *et al.*, Prevalence of growth hormone deficiency in children with cleft lip or palate, *J. Pediatr.* **93:**378, 1978.
39. J. P. Berke, L. F. Buxton, and E. Kokmen, The empty sella, *Neurology* (*Minneap.*) **25:**1137, 1975.
40. R. M. Jordan, J. W. Kendall, and C. W. Kerber, The primary empty sella syndrome: Analysis of the clinical characteristics, radiographic features, pituitary function and cerebrospinal fluid adenohypophysial hormone concentrations, *Am. J. Med.* **62:**569, 1977.
41. K. Onur, V. Lala, J. Zimmer, C. S. Juan, and T. W. AvRuskin, The primary empty sella syndrome in a child, *J. Pediatr.* **90:**425, 1977.
42. R. E. Kleinmann, E. L. Kazarian, V. Raptopoulos, and L. E. Braverman, Primary empty sella and Rieger's anomaly of the anterior chamber of the eye, *N. Engl. J. Med.* **304:**90, 1981.
43. Z. Laron, A. Pertzelan, M. Karp, A. Kowaldo, and W. H. Daughaday, Administration of growth hormone to patients with familial dwarfism with high plasma immunoreactive growth hormone: measurement of sulfation factor, metabolic and linear growth responses, *J. Clin. Endocrinol. Metab.* **33:**332, 1971.
44. D. W. Golde, N. Bersch, S. A. Kaplan, D. L. Rimoin, and C. H. Li, Peripheral unresponsiveness to human growth hormone in Laron dwarfism, *N. Engl. J. Med.* **303:**1156, 1980.
45. T. J. Merimee, J. Zapf, and E. R. Froesch, Dwarfism in the pygmy, *N. Engl. J. Med.* **305:**965, 1981.
46. T. J. Merimee, J. Zapf, and E. R. Froesch, Insulin-like growth factors (IGFs) in pygmies and subjects with the pygmy trait: characterization of the metabolic actions of IGF I and IGF II in man, *J. Endocrinol. Metab.* **55:**1081, 1982.
47. M. J. Thomsett, F. A. Conte, S. L. Kaplan, *et al.*, Endocrine and neurologic outcome in childhood craniopharyngioma. Review of effect of treatment in 42 patients, *J. Pediatr.* **97:**728, 1980.
48. F. M. Kenny, H. J. Guyda, J. C. Wright, and H. G. Friesen, Prolactin and somatomedin in hypopituitary patients with "catch-up" growth following operations for craniopharyngioma, *J. Clin. Endocrinol. Metab.* **36:**378, 1973.
49. J. W. Finkelstein, J. Kream, A. Ludan, and L. Hellman, Sulfation factor (somatomedin): An explanation for continued growth in the absence of immunoassayable growth hormone in patients with hypothalamic tumors, *J. Clin. Endocrinol. Metab.* **35:**13, 1972.
50. H. Bucher, J. Zapf, T. Torresani, A. Prader, E. R. Froesch, and R. Illig, Insulin-like growth factors I and II, prolactin, and insulin in 19 growth hormone-deficient children with excessive, normal, or decreased longitudinal growth after operation for craniopharyngioma, *N. Engl. J. Med.* **309:**1142, 1983.
51. R. J. Rona and J. M. Tanner, Aetiology of idiopathic growth hormone deficiency in England and Wales, *Arch. Dis. Child.* **52:**197, 1977.
52. W. H. Craft, L. E. Underwood, and J. J. Van Wyk, High incidence of perinatal insult in children with idiopathic hypopituitarism, *J. Pediatr.* **96:**397, 1980.
53. J. M. Del Pozo, J. E. Roda, J. G. Montoya, J. R. Iglesias, and A. Hurtado, Intrasellar granuloma, *J. Neurosurg.* **53:**717, 1980.
54. G. F. Bottazzo, C. McIntosh, W. Stanford, and M. Preece, Growth hormone cell antibodies and partial growth hormone deficiency in a girl with Turner's syndrome, *Clin. endocrinol.* **12:**1, 1980.

55. C. J. Portocarrero, A. G. Robinson, A. L. Taylor, and I. Klein, Lymphoid hypophysitis, *JAMA* **246:**1811, 1981.
56. S. L. Asa, J. M. Bilbao, K. Kovacs, R. G. Josse, and K. Kreines, Lymphocytic hypophysitis of pregnancy resulting in hypopituitarism: a distinct clinicopathologic entity, *Ann. Intern. Med.* **95:**166, 1981.
57. G. E. Richards, W. M. Wara, M. M. Grumbach, *et al.*, Delayed onset of hypopituitarism: sequelae of therapeutic irradiation of central nervous system, eye and middle ear tumors, *J. Pediatr.* **89:**553, 1976.
58. N. A. Samaan, M. M. Bakdash, J. B. Caderao, *et al.*, Hypopituitarism after external irradiation: Evidence for both hypothalamic and pituitary origin, *Ann. Intern. Med.* **83:**771, 1975.
59. P. K. Duffner, M. E. Cohen, S. W. Anderson, M. L. Voorhess, M. H. MacGillivray, A. Panahon, and M. L. Brecher, Long-term effects of treatment on endocrine function in children with brain tumors, *Ann. Neurol.* **14:**528, 1983.
60. J. D. Russell, P. H. Wise, and H. G. Rischbieth, Vascular malformation of the hypothalamus: A cause of isolated growth hormone deficiency, *Pediatrics* **66:**306, 1980.
61. P. Saenger, E. Schwartz, A. L. Markenson, J. H. Graziano, L. S. Levine, M. I. New, and M. W. Hilgartner, Depressed serum somatomedin activity in B-thalassemia, *J. Pediatr.* **96:**214, 1980.
62. G. Costin, M. D. Kogut, C. B. Hyman, and J. A. Ortega, Endocrine abnormalities in thalassemia major, *Am. J. Dis. Child.* **133:**497, 1979.
63. N. McIntosh, Endocrinopathy in thalassaemia major, *Arch. Dis. Child.* **51:**195, 1976.
64. G. D. Braunstein and P. O. Kohler, Pituitary function in Hand-Schuller-Christian disease. Evidence for deficient growth hormone release in patients with short stature, *N. Engl. J. Med.* **286:**1225, 1972.
65. R. Penny and R. M. Blizzard, The possible influence of puberty on the release of growth hormone in 3 males with apparent growth hormone deficiency, *J. Clin. Endocrinol. Metab.* **34:**82, 1972.
66. M. Gourmelen, M. T. Pham-Huu-Trung, and F. Girard, Transient partial hGH deficiency in prepubertal children with delay of growth, *Pediatr. Res.* **13:**221, 1979.
67. R. G. Patton and L. I. Gardner, Influence of family environment on growth: The syndrome of "maternal deprivation," *Pediatrics* **30:**957, 1962.
68. G. F. Powell, J. A. Brasel, and R. M. Blizzard, Emotional deprivation and growth retardation simulating idiopathic hypopituitarism. I. Clinical evaluation of the syndrome, *N. Engl. J. Med.* **276:**1271, 1967.
69. G. F. Powell, J. A. Brasel, S. Raiti, and R. M. Blizzard, Emotional deprivation and growth retardation simulating idiopathic hypopituitarism. II. Endocrinologic evaluation of the syndrome, *N. Engl. J. Med.* **276:**1279, 1967.
70. I. Krieger and R. C. Mellinger, Pituitary function in the deprivation syndrome, *J. Pediatr.* **79:**216, 1971.
71. C. F. Whitten, Evidence that growth failure from maternal deprivation is secondary to undereating, *JAMA* **209:**1675, 1969.
72. R. L. Hintz, R. Suskind, K. Amatayakul, O. Thanangkul, and R. Olson, Plasma somatomedin and growth hormone values in children with protein-calorie malnutrition, *J. Pediatr.* **92:**153, 1978.
73. J. M. Gertner, M. Genel, S. P. Gianfredi, R. L. Hintz, R. G. Rosenfeld, W. V. Tamborlane, and D. M. Wilson, Prospective clinical trial of human growth hormone in short children without growth hormone deficiency, *J. Pediatr.* **104:**172, 1984.
74. L. E. Underwood, Growth hormone treatment for short children, *J. Pediatr.* **104:**237, 1984.

4

Dose–Response Relationships in Growth Hormone Therapy

S. Douglas Frasier

Although human growth hormone (hGH) has been administered to GH-deficient patients for 25 years,[1] there are few published data on the relationship of dose to response. This chapter will review the dose–response information that is available and will attempt to make recommendations regarding the dosage and dose schedule of hGH to be used during an initial year of treatment in GH-deficient patients.

1. Dose of hGH

The minimal effective dose of hGH in GH deficiency has been incompletely defined. However, two studies bear on this point. Rosenbloom[2] found that 0.4 mg (presumed to be 0.4 IU) hGH daily failed to accelerate linear growth. Frasier *et al.*[3] found that 0.01 IU hGH/kg body weight three times a week (t.i.w.) was ineffective, whereas 0.03 IU/kg t.i.w. produced a significant acceleration of growth in GH-deficient patients.

There are very few studies in which different doses of hGH were compared without a change in dose schedule. Two of those which are available are shown in Table 1.

The British Medical Research Council trial compared a dose of 5 IU twice a week (b.i.w.) with 10 IU b.i.w. As originally reported by Preece *et al.*[4] and subsequently expanded in the review of Milner *et al.*,[5] there was a significantly greater ($p < 0.01$) effect at the higher dose. The U.S. Collaborative Project reported similar findings.[6] Patients with idiopathic GH deficiency who received

S. Douglas Frasier • Department of Pediatrics and Physiology/Biophysics, University of Southern California School of Medicine and the Los Angeles County–University of Southern California Medical Center, Los Angeles, California 90033.

Table 1 *Direct Comparisons of the Effect of Dose of hGH on the Linear Growth Response to an Initial Year of Therapy*

Author	Number of patients	Dose of hGH: IU/dose	Dose of hGH: IU/year	Growth rate[a] (cm/year)
Preece *et al.*[4] and	22	5 b.i.w.	520	7.3 ± 1.4
Milner *et al.*[5]	30	10 b.i.w.	1040	9.9 ± 2.2
Aceto *et al.*[6]	20	2 t.i.w.	312	9.3 ± 2.2
	9	10 t.i.w.	1560	11.9 ± 2.4

[a] Mean ± SD.

10 IU hGH t.i.w. showed a significantly greater response ($p < 0.01$) than comparable patients receiving a more conventional dose of 2 IU t.i.w. The Canadian Medical Research Council (MRC) Study also reported an increasing response to an increasing dose of hGH expressed as IU/kg body weight.[7] It is evident from the British MRC study and the U.S. Collaborative Project data that although increasing the dose leads to a better response, the difference in growth rate is relatively much less than the change in dose.

In order to further define these dose–response relationships we[8] administered varying doses of hGH on the basis of body weight to 93 GH-deficient patients over an initial 12 months of treatment. To be included in this analysis, GH-deficient patients were required to have a pretreatment growth rate of 5.5 cm/year or less. All patients were tested with insulin-induced hypoglycemia,[9] and 60 were also tested with an intravenous infusion of arginine, either as a separate procedure[10] or as part of a sequential arginine–insulin tolerence test.[11]

Sixty-five male patients and 28 female patients met the criteria for inclusion. Sixty-three patients had idiopathic GH deficiency, and in 30 there was an underlying intracranial lesion.

The diagnosis of thyroid-stimulating hormone, adrenocorticotropic hormone, and antidiuretic hormone deficiency was made by standard methods, and patients with deficiencies of these pituitary hormones received appropriate replacement therapy. No patient received gonadotropin or gonadal steroid treatment during hGH administration. Forty patients received GH treatment alone, and 25 were treated with a combination of hGH and thyroid replacement. Seven patients received hGH, thyroid replacement, and glucocorticoid. Seven patients received this combination and antidiuretic hormone, as well. Ten patients received GH, thyroid replacement, and antidiuretic hormone, and four patients received hGH and only antidiuretic hormone.

Study patients received either 30 (n = 27), 60 (n = 38), 80 (n = 12), or 100 (n = 16) milli International Units (mIU) of hGH/kg of pretreatment body weight intramuscularly t.i.w.

hGH was obtained from the National Hormone and Pituitary Program and was administered to patients under protocols approved by the appropriate

Table 2 Response to hGH Therapy

Dose of hGH (mIU/kg t.i.w.)	Number of patients	Absolute growth rate[a] (cm/year)	Increase in growth rate[a] (cm/year)
30	27	5.59 ± 2.30	2.53 ± 2.26
60	38	7.31 ± 1.75	3.97 ± 1.85
80	12	7.22 ± 3.12	3.79 ± 2.52
100	16	8.94 ± 1.19	5.79 ± 1.35

[a] Mean ± SD.

human subjects protection procedures. Treatment was continuous over 1 year, and the dose was not modified during that time.

Differences between means were compared by the Student *t* test. The chi-square statistic and correlation coefficients were calculated using standard methods.[12] The statistical analysis of the log dose–response curve was carried out as described by Bliss[13] and with the expert assistance of Dr. Elliot Landow, Departments of Biomathematics and Pediatrics, University of California, Los Angeles.

The response to therapy, expressed as absolute growth rate while receiving hGH, is shown in Table 2. In each group of patients there was a highly significant difference between the treatment growth rate and the pretreatment growth rate. At each dose of hGH p was less than 0.001. The growth rate of patients receiving 60 mIU hGH/kg t.i.w. was significantly greater ($p < 0.01$) than that of patients receiving 30 mIU hGH/kg. The growth rate of patients receiving 100 mIU hGH/kg t.i.w. was significantly greater than that of patients receiving either 30 mIU hGH/kg ($p < 0.01$) or 60 mIU hGH/kg ($p < 0.001$). Doubling the dose from 30 to 60 mIU hGH/kg t.i.w. increased the response approximately 1.3 times. A 3.3-fold increase in the dose from 30 to 100 mIU hGH/kg t.i.w. increased the response 1.6 times.

Table 2 also shows the response to therapy expressed as the increase between the pretreatment growth rate and the rate of growth while receiving hGH. When expressed in this way, the response to 60 mIU hGH/kg t.i.w. was significantly greater than the response to 30 mIU hGH/kg ($p < 0.01$). The response to 100 mIU hGH/kg t.i.w. was significantly greater than the response to either 30 mIU hGH/kg ($p < 0.01$), 60 mIU hGH/kg ($p < 0.02$), or 80 mIU hGH/kg ($p < 0.001$). Doubling the dose from 30 to 60 mIU hGH/kg increased the response approximately 1.6 times.

The calculated log dose–response equation for the absolute growth rate while receiving hGH over the range of doses used in this study is

$$Y = -3.12 + 5.80 \log X$$

Y is the expected response in cm/year, and $\log X$ is the common logarithm of the dose in mIU hGH/kg given three times a week. The log dose–response curve calculated from this equation is shown in Fig. 1. The slope of this curve

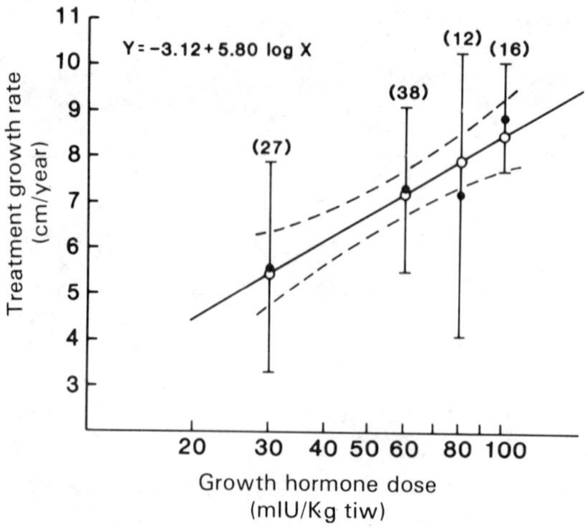

Figure 1 Log dose–response curve of the annual growth rate as a function of hGH dose. The closed circles and vertical bars represent a mean response ±1 SD at each dose. The numbers of patients at each dose are shown in the parentheses. The calculated dose–response equation is shown in the upper left-hand corner. The open circles show the expected mean response at each dose as calculated from the dose–response equation, and the diagonal line is the calculated log dose–response curve. The curved lines encompass the 95% confidence interval for the mean response at each dose.

is significantly different from zero ($p < 0.001$), with a 95% confidence interval of 3.5–8.1. The 95% confidence interval for the mean response is 5.5 ± 0.74 cm/year at 30 mIU hGH/kg t.i.w., 7.2 ± 0.44 cm/year at 60 mIU hGH/kg t.i.w., and 8.5 ± 0.72 cm/year at 100 mIU hGH/kg t.i.w.

The calculated log dose–response equation for the increase in growth rate while receiving hGH over the range of doses used in this study is

$$Y = -6.09 + 5.67 \log X$$

Y is the change in growth rate in cm/year, and $\log X$ is the common logarithm of the dose in mIU hGH/kg given three times a week. The log dose–response curve calculated for this equation is shown in Fig. 2. The slope of the curve calculated from this equation is also significantly different from zero ($p < 0.001$). The 95% confidence interval of the slope is 3.4–7.9. The 95% confidence

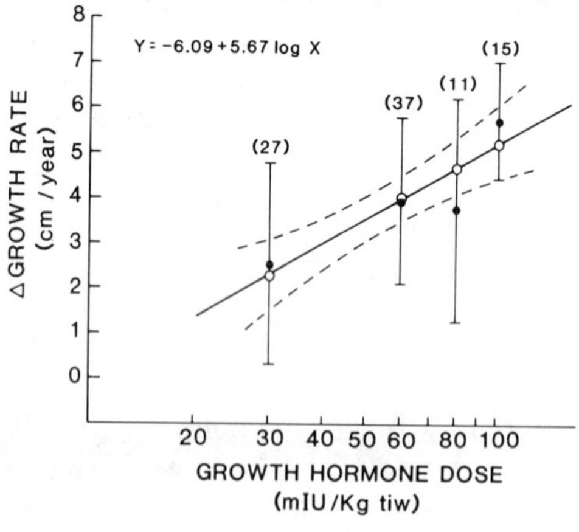

Figure 2 Log dose–response curve of the change in annual growth rate as a function of hGH dose. The closed circles and vertical bars represent a mean response ±1 SD at each dose. The numbers of patients at each dose are shown in the parentheses. The calculated dose–response equation is shown in the upper left-hand corner. The open circles show the expected mean response at each dose as calculated from the dose–response equation, and the diagonal line is the calculated log dose–response curve. The curved lines encompass the 95% confidence interval for the mean response at each dose.

interval for the mean increase in growth rate is 2.3 ± 0.73 cm/year at 30 mIU hGH/kg t.i.w., 4.0 ± 0.44 cm/year at 60 mIU hGH/kg t.i.w., and 5.2 ± 0.72 cm/year at 100 mIU hGH/kg t.i.w.

There are few data with which this analysis may be directly compared. Rudman and co-workers administered three doses of hGH to GH-deficient patients for 10 days and followed their growth response over the next 8 weeks.[14] The doses used were 20, 60, and 200 mIU hGH/kg daily.[15] Although the design used in this short-term study is quite different from ours, it is possible to derive a log dose–response equation and a dose–response curve from the data provided. The curve has a shallow slope similar to that derived from our data. Increasing the dose 3.3 times from 60 to 200 mIU hGH/kg increased the linear growth response only 1.6 times. The validity of this dose–response curve is also supported by the recent study of Nugent *et al.*[16] who found that the administration of 67 mIU hGH/kg t.i.w. resulted in an average linear growth response of 6.0 cm/year.

The effect of doses of hGH in excess of 100 mIU (0.1 IU)/kg thrice a week has not been systematically investigated. Although studies using a fixed dose of hGH have included patients receiving higher doses per kilogram body weight, there are no direct data comparing patients receiving 0.1 IU/kg t.i.w. with those receiving higher doses such as 0.15 or 0.2 IU/kg t.i.w. Such studies would be very helpful, especially now that problems in maintaining the supply of hGH appear to be nearing resolution.

2. *Schedule of GH Administration*

GH has been administered continuously, using a weekly, twice-weekly, or thrice-weekly schedule (Table 3).

The effect of weekly therapy was originally reported by Rosenbloom,[2] who claimed success in treating nine patients for short and variable periods of time. Trygstad's review paper[17] also indicates that weekly therapy was successful, at least in patients weighing less than 25 kg. Rosenbloom has continued to use weekly injections of hGH as initial therapy in GH deficiency and has recently described his successful experience with 2.5 IU hGH once a week in 29 patients.[18] This degree of success has not been duplicated by others. Soyka *et al.*[19] gave 2 mg (2 IU) three times a week and 6 mg (6 IU) once a week to six patients selected on the basis of a good response during a short-term trial of hGH. The response to a single weekly injection was inferior to three-times-weekly administration in two-thirds of the comparisons, and there was no difference in the remaining comparisons. The mean overall extrapolated annual growth rate in this study was 5.0 cm during nine periods of one dose each week compared to 9.5 cm during 29 periods of t.i.w. dosage. The U.S. Collaborative Project also compared 6 IU once a week with 2 IU three times a week.[20] It was concluded that weekly treatment was less effective than treatment divided over the week. However, this conclusion must be regarded as very tentative since weekly hGH administration followed 18 months of t.i.w. admin-

Table 3 Comparisons of Continuous Schedules of hGH Administration on the Linear Growth Response to an Initial Year of Therapy

Author	Number of patients	Dose of hGH		Growth rate[a] (cm/year)
		IU/dose	IU/year	
Rosenbloom *et al.*[18]	29	2.5 weekly	130	7.3 ± 2.4
Preece *et al.*[4] and Milner *et al.*[5]	22	5 b.i.w.	520	7.3 ± 1.4
Tanner *et al.*[24]	59	10 b.i.w.	1040	8.1 ± 3.4
Preece *et al.*[4] and Milner *et al.*[5]	30	10 b.i.w.	1040	9.9 ± 2.2
Milner *et al.*[5][b]	8	10 b.i.w.	1040	9.0 ± 1.4
Root *et al.*[25]	18	2 t.i.w.	312	9.0 ± 3.2
Aceto *et al.*[6]	60	2 t.i.w.	312	7.5 ± 3.9
Milner *et al.*[5][b]	8	5 t.i.w.	780	11.2 ± 2.6
Aceto *et al.*[6]	9	10 t.i.w.	1560	11.9 ± 2.4

[a] Mean ± S.D.
[b] Direct comparison of 10 IU t.i.w. and 5 IU b.i.w. ($p < 0.01$).

istration, and the difference could be explained by the waning effect of therapy over time.

A great deal of information has been accumulated on the effect of twice-weekly hGH administration. This dose schedule results in a significant increase in linear growth as reported in early studies from Switzerland.[21–23] The average growth velocity in the Swiss studies varied from 7.2 to 8.1 cm/year. These studies are not included in Table 3 because the dose was 5 mg hGH/m^2 b.i.w. of hGH preparations that varied in potency between 0.9 and 1.8 IU/mg, and because the total annual dose cannot be calculated from these authors' published data. The response shown by British patients given 5 IU b.i.w.[4,5] is comparable to the Swiss experience. The three series of patients in which 10 IU b.i.w. were used by the British MRC show a narrow range of response with an average annual growth velocity between 8.1 and 9.9 cm/year.[4,5,24]

There is also a great deal of information on the effect of hGH given thrice a week. Root *et al.*,[25] the U.S. Collaborative Project, and the British MRC study have reported results using this schedule. As was the case with twice-weekly dosage, the response was dependent on the quantity of hGH administered. The average growth velocity during the entire year of treatment varied from 7.5 to 11.9 cm.

When Table 3 is scanned, the data suggest the conclusion that the response is greater, in relation to the total annual dose of hGH, when therapy is given three times a week rather than twice. The only study that directly supports this conclusion is the British MRC comparison of eight patients receiving 10 IU b.i.w. with eight receiving 5 IU t.i.w.[5] As shown in Table 3, the lower total dose given three times a week resulted in a significantly greater response ($p < 0.01$) than a higher dose divided into two weekly injections.

The data presented so far relate to the continuous administration of hGH. GH has also been administered by an intermittent schedule during an initial year of treatment. Wright *et al.* treated 11 hypopituitary patients with hGH given in a course of daily injections for 2 weeks followed by no therapy for 2 weeks.[26] The total duration of initial therapy varied from 4 to 10 months, and the daily dose of hGH ranged from 2 to 5 mg of preparations containing approximately 1 IU/mg. The average extrapolated annual growth rate while receiving therapy was 8.1 cm. The total annual dose cannot be calculated but can be estimated to be approximately 350–400 IU. Since these patients were all treated for less than 1 year, and the dose of hGH was variable, the overall effectiveness of this approach cannot be determined.

A different protocol for intermittent hGH therapy has been used by Kirkland *et al.*[27] Patients received hGH daily for no more than 31 days every 3 months. In 13 of the 18 patients treated, the duration of each course of therapy was 15 or 16 days. The average annual dose of hGH was 176 ± 39 (SD) IU/m^2 per year. The growth rate of these 18 patients increased significantly during the initial year of therapy and averaged 5.5 ± 1.2 (SD) cm/year. This compares to average annual growth rates of 6.4 ± 1.4 (SD) cm/year seen in 12 patients who received 126.4 ± 14.9 IU hGH/m^2 per year given continuously as 0.03 IU/kg t.i.w.[3] Thus, continuous therapy appears to be more efficient than this intermittent regime when comparable annual doses of hGH are given during the initial year of treatment.

3. Recommendations

Although the data reviewed in this chapter are subject to different interpretations, I believe that they provide the basis for several recommendations regarding GH therapy during the initial year of treatment of GH deficiency. Although these recommendations are derived from information obtained using pituitary GH, it is likely that they are applicable to GH manufactured by recombinant DNA technology as well.

GH should be administered on the basis of body weight. The initial dose should be 0.08–0.1 IU/kg administered three times a week. Although higher doses may be more effective, the difference in linear growth that would be expected from 0.15 or 0.2 IU/kg t.i.w. would be only a few centimeters. The safety of these higher doses has not yet been evaluated. Therapy should be continuous rather than intermittent whenever possible.

ACKNOWLEDGMENT. The author wishes to acknowledge the expert secretarial assistance of Margaret A. Bloom.

References

1. M. S. Raben, Treatment of a pituitary dwarf with human growth hormone, *J. Clin. Endocrinol. Metab.* **18:**901, 1958.
2. A. L. Rosenbloom, Growth hormone replacement therapy, *JAMA* **198:**364, 1966.
3. S. D. Frasier, T. Aceto, A. B. Hayles, and V. G. Mikity, Collaborative study of the effects of

human growth hormone in growth hormone deficiency: IV. Treatment with low doses of human growth hormone based on body weight, *J. Clin. Endocrinol. Metab.* **44:**22, 1977.
4. M. A. Preece, J. M. Tanner, R. H. Whitehouse, and N. Cameron, Dose dependence of growth response to human growth hormone in growth hormone deficiency, *J. Clin. Endocrinol. Metab.* **42:**477, 1976.
5. R. D. G. Milner, T. Russell-Fraser, C. G. D. Brook, P. M. Cotes, J. W. Farquhar, J. M. Parkin, M. A. Preece, G. J. A. I. Snodgrass, A. S. Mason, J. M. Tanner, and F. P. Vince, Experience with human growth hormone in Great Britain: The report of the MRC working party, *Clin. Endocrinol.* **11:**15, 1979.
6. T. Aceto, S. D. Frasier, A. B. Hayles, H. F. L. Meyer-Bahlburg, M. L. Parker, R. Munschauer, and G. DeChiro, Collaborative study of effects of human growth hormone in growth hormone deficiency. I. First year of therapy, *J. Clin. Endocrinol. Metab.* **35:**483, 1972.
7. H. Guyda, H. Friesen, J. D. Bailey, G. Leboeuf, and J. C. Beck, Medical Research Council of Canada therapeutic trial of human growth hormone: First 5 years of therapy, *Can. Med. Assoc. J.* **112:**1301, 1975.
8. S. D. Frasier, G. Costin, B. M. Lippe, T. Aceto, and P. F. Bunger, A dose–response curve for human growth hormone, *J. Clin. Endocrinol. Metab.* **53:**1213, 1981.
9. S. D. Frasier, The serum growth-hormone response to hypoglycemia in dwarfism, *J. Pediatr.* **71:**625, 1967.
10. M. L. Parker, J. M. Hammond, and W. H. Daughaday, The arginine provocative test: An aid in the diagnosis of hyposomatotropism, *J. Clin. Endocrinol. Metab.* **27:**1129, 1967.
11. R. Penny, R. M. Blizzard, and W. T. Davis, Sequential-arginine and insulin tolerance tests on the same day, *J. Clin. Endocrinol. Metab.* **29:**1499, 1969.
12. T. D. V. Swinscoe, *Statistics at Square One*, British Medical Association, London, 1978.
13. C. I. Bliss, *The Statistics of Bioassay with Special Reference to the Vitamins*, Academic Press, New York, 1952, pp. 452–471.
14. D. Rudman, M. H. Kutner, G. A. Fleming, R. C. Harris, E. E. Kennedy, A. Bethel, and J. H. Priest, Effect of 10-day courses of human growth hormone on height of short children, *J. Clin. Endocrinol. Metab.* **46:**28, 1978.
15. D. Rudman, S. B. Chyatte, J. H. Patterson, F. F. Gerron, I. O'Beirne, J. Barlow, P. Ahmann, A. Jordan, and R. C. Mosteller, Observations on the responsiveness of human subjects to human growth hormone. Effects of endogenous growth hormone deficiency and myotonic dystrophy, *J. Clin. Invest.* **50:**1941, 1971.
16. J. N. Nugent, D. C. Moore, R. H. A. Ruvalcaba, and V. C. Kelley, Dose response curve for human growth hormone and oxandrolone by weight in children with hypopituitarism, *Clin. Res.* **29:**111A, 1981.
17. O. Trygstad, Human growth hormone and hypopituitary growth retardation, *Acta Paediatr. Scand.* **58:**407, 1969.
18. A. L. Rosenbloom, W. J. Riley, J. H. Silverstein, A. D. Garnica, M. L. Netzloff, and F. T. Weber, Low dose single weekly injections of growth hormone: Response during first year of therapy with hypopituitarism, *Pediatrics* **66:**272, 1980.
19. L. F. Soyka, H. H. Bode, J. D. Crawford, and F. J. Flynn, Effectiveness of long-term human growth hormone therapy for short stature in children with growth hormone deficiency, *J. Clin. Endocrinol.* **30:**1, 1970.
20. S. D. Frasier, T. Aceto, and A. B. Hayles, Collaborative study of the effects of human growth hormone in growth hormone deficiency. V. Treatment with growth hormone administered once a week, *J. Clin. Endocrinol. Metab.* **47:**686, 1978.
21. A. Prader, R. Illig, J. Szeky, and H. Wagner, The effect of human growth hormone on hypopituitary dwarfism, *Arch. Dis. Childh.* **39:**535, 1964.
22. A. Prader, M. Zachmann, J. R. Poley, R. Illeg, and J. Szeky, Long-term treatment with human growth hormone (Raben) in small doses. Evaluation of 18 hypopituitary patients, *Helv. Paediatr. Acta* **22:**423, 1967.
23. A. Ferrandez, M. Zachmann, A. Prader, and R. Illeg, Isolated growth hormone deficiency in prepubertal children. Influence of human growth hormone on longitudinal growth, adipose tissue, bone mass and bone maturation, *Helv. Paediatr. Acta* **25:**566, 1970.

24. J. M. Tanner, R. H. Whitehouse, P. C. R. Hughes, and F. P. Vince, Effect of human growth hormone deficiency, low birthweight, inherited smallness, Turner's syndrome and other complaints, *Arch. Dis. Childh.* **46:**745, 1971.
25. A. W. Root, A. M. Bongiovanni, and W. R. Eberlein, Diagnosis and management of growth retardation with special reference to the problem of hypopituitarism, *J. Pediatr.* **78:**737, 1971.
26. J. C. Wright, J. A. Brasel, T. Aceto, J. W. Finkelstein, F. M. Kenny, J. S. Spaulding, and R. M. Blizzard, Studies of human growth hormone (hGH). An attempt to correlate metabolic response during short-term administration with linear growth during prolonged therapy, *Am. J. Med.* **38:**499, 1965.
27. R. T. Kirkland, J. L. Kirkland, L. Librik, and G. W. Clayton, Results of intermittent human growth hormone (hGH) therapy in hypopituitary dwarfism, *J. Clin. Endocrinol. Metab.* **37:**204, 1973.

5

Dose Studies for Human Growth Hormone in Hypopituitarism

Current Recommendations

Salvatore Raiti and the Committee on Growth Hormone Usage (S. L. Kaplan, W. V. Moore, J. S. Parks, M. H. MacGillivray, A. W. Root, A. J. Johanson, L. E. Underwood, S. A. Kaplan, T. P. Foley, and A. E. Wilhelmi)

A number of studies have reported growth responses to different doses of human growth hormone (hGH). Most such reports deal with the first year or first 18 months of therapy in hypopituitarism. Twice-a-week (b.i.w) therapy has also been compared to three-times-a-week (t.i.w) therapy and has been found to be not quite as effective. The report of Aceto *et al.*[1] in 1973 suggested that an effective therapeutic dose could be 0.1 IU/kg body weight given t.i.w., and this has found widespread acceptance. This work was based on a collaborative study carried out on hypopituitary patients in the late 1960s and early 1970s. The hGH used in those studies varied from batch to batch and had mean biopotencies ranging from 0.8 to 1.2 IU/mg. It was about 50% monomeric and included a considerable amount of dimers and polymers. The significance of these other forms of hGH was not appreciated until the mid-1970s.

The technology for production of hGH for clinical use has improved dramatically and has met or exceeded the newer accepted criteria.[2] Since the late 1970s, the hGH has been of uniform character. It has been highly monomeric and has shown a consistently high mean biological potency of 2.4 IU/mg.

Salvatore Raiti and the Committee on Growth Hormone Usage (S. L. Kaplan, W. V. Moore, J. S. Parks, M. H. MacGillivray, A. W. Root, A. J. Johanson, L. E. Underwood, S. A. Kaplan, T. P. Foley, and A. E. Wilhelmi) • The National Hormone and Pituitary Program, University of Maryland School of Medicine, Baltimore, Maryland 21201.

In view of such dramatic improvements, the Committee for Growth Hormone Usage of the National Hormone and Pituitary Program decided that the question of dose should be reevaluated. The studies of Frasier *et al.*[3] suggested that a log–dose growth responsiveness hGH therapy might be expected. The effective dose range appeared to lie between 0.06 and 0.1 IU/kg when given t.i.w. The Committee selected a dose of 0.08 IU/kg t.i.w. for the studies for these reasons. First, it was well within the effective dose range. Second, the 0.1-IU/kg dose used in Aceto's studies probably, in fact, reflected an actual dose of about 0.08 IU/kg of monomeric hGH.

In designing this study, the Committee felt that it would also be worth reexploring the effectiveness of b.i.w. therapy compared to t.i.w. therapy. The patients were to receive the same total amount of hGH each week.

1. Study Plan

Two studies were carried out. The first was on patients with idiopathic hypopituitarism who had not received hGH previously. The second study was carried out in children with organic hypopituitarism. This diagnosis included postoperative craniopharyngioma, postirradiation for cerebral tumors (e.g., medulloblastoma, dysgerminoma, optic glioma), empty sella syndrome, and septooptic dysplasia.

The criteria for acceptance of patients into the studies are shown in Table 1. Patients were assigned to treatment schedules alternatively as they were accepted into the studies. In all studies, the patients received a total of 0.24 IU/kg per week for 2 years.

In the first group, the patients were given doses of 0.08 IU/kg t.i.w. for 6 months, and then 0.12 IU/kg for b.i.w. for 6 months, and then 0.08 IU/kg t.i.w. for 6 months, and then 0.12 IU/kg b.i.w. for the final 6 months of the 2 years. The second group were started on 0.12 IU/kg b.i.w. for 6 months, then given 0.08 IU/Kg t.i.w. for 6 months, then 0.12 IU/kg b.i.w. for 6 months, and then 0.08 IU/Kg t.i.w. for the final 6 months. This same study plan was used for patients with idiopathic hypopituitarism and with the organic hypopituitarism group.

The patients were measured at 2-month intervals for the 1st year and at 3-month intervals for the 2nd year. Reports on the data collected were sent in

Table 1 Criteria for Acceptance

1. Short, 2 SD below mean.
2. Poor growth rate, 4 cm/year or less.
3. Bone age delay by 2 years or more.
4. Euthyroid (normally or on replacement therapy).
5. Normal adrenal–ACTH status.
6. Low GH response to two definitive stimuli (arginine, insulin, L-dopa, i.m. glucagon, or clonidine) with peak responses below 8–10 ng/ml.

Table 2 Growth Responses to Human Growth Hormone, Isolated Deficiency

		Growth rate					
			cm/6 months				
Study period	Number	cm/year	0.08 t.i.w.	0.12 b.i.w.	0.08 t.i.w.	0.12 b.i.w.	cm/year (mean)
Before	69	2.78 ± 1.02					2.78
1st	61		4.49 ± 1.39				8.98
2nd	44			3.29 ± 0.72			6.58
3rd	27				3.18 ± 0.99		6.36
4th	16					3.06 ± 0.89	6.12

every 6 months at which time the dose frequency was changed and the dose was adjusted for the current body weight.

2. Results

2.1. Idiopathic Hypopituitarism

Table 2 give the results to date on the 69 patients whose therapy was started at 0.08 IU/kg t.i.w. and then changed at 6-month intervals. The mean pretreatment growth rate was 2.78 cm/year. In the 1st 6 months, they grew at a mean rate of 4.49 cm/6 months, or 8.98 cm/year. In the 2nd 6 months on b.i.w. therapy the mean growth rate fell to 3.29 cm/6 months, or 6.58 cm/year. During the 3rd 6 months (now on t.i.w. therapy) the growth response was 3.18 cm, or 6.36 cm/year. During the final 6 months of the 2-year study the growth response in the 16 patients (who had completed the study) was 3.06 cm, or 6.12 cm/year.

Table 3 shows the results in the 67 patients who were started on b.i.w. therapy. Their mean pretreatment growth rate was 2.47 cm/year. In the 1st 6 months on 0.12 IU/kg b.i.w., their mean growth response was 9.04 cm/year During the 2nd 6 months on 0.08 IU/kg t.i.w., the mean growth response was 3.99, or 7.98 cm/year. During the 3rd 6 months on 0.12 IU/kg t.i.w., the mean

Table 3 Growth Responses to Human Growth Hormone, Isolated Deficiency

		Growth rate					
			cm/6 months				
Study period	Number	cm/year	0.12 b.i.w.	0.08 t.i.w.	0.12 b.i.w.	0.08 t.i.w.	cm/year (mean)
Before	67	2.47 ± 1.12					2.47
1st	66		4.52 ± 1.50				9.04
2nd	44			3.99 ± 1.46			7.98
3rd	27				3.43 ± 1.27		6.86
4th	12					3.03 ± 0.88	6.06

Table 4 Growth Responses to Human Growth Hormone, Organic Hypopituitarism

		Growth rate					
			cm/6 months				
Study period	Number	cm/year	0.08 t.i.w.	0.12 b.i.w.	0.08 t.i.w.	0.08 b.i.w.	cm/year (mean)
Before	36	2.19 ± 1.17					2.19
1st	22		3.68 ± 1.79				7.36
2nd	15			2.66 ± 1.10			5.32
3rd	4				2.08 ± 1.10		4.16
4th	3					3.63 ± 0.91	7.26

response was 3.43 cm, or 6.86 cm/year. During the last 6 months on 0.08 IU/kg t.i.w., the 12 patients who had completed the study grew at a mean rate of 3.03 cm, or 6.06 cm/year.

2.2. Organic Hypopituitarism

Table 4 shows the responses for the 36 patients who were started at 0.08 IU/kg t.i.w. Their mean pretreatment growth rate was 2.19 cm/year. During the 1st 6 months on 0.08 IU/kg t.i.w., their mean growth response was 3.68 cm, or 7.36 cm/year. During the 2nd 6 months on 0.12 IU/kg b.i.w., their mean growth response was 2.66 cm, or 5.32 cm/year. During the 3rd 6 months on 0.08 IU/kg t.i.w., the mean response for the four patients who had completed the study was 2.08 cm, or 4.16 cm/year. During the final 6 months on 0.12 IU/kg b.i.w., the mean growth response for the three patients was 3.63 cm, or 7.26 cm/year.

Table 5 shows the responses for the 33 patients who were started at 0.12 IU/kg b.i.w. During the 1st 6 months, their mean growth response was 2.9 cm, or 5.8 cm/year. During the 2nd 6 months at 0.08 IU/kg t.i.w. their mean growth response was 3.05 cm, or 6.10 cm/year. During the 3rd 6 months on 0.12 IU/kg b.i.w., their mean growth response was 2.48 cm, or 4.96 cm/year. During

Table 5 Growth Responses to Human Growth Hormone, Organic Hypopituitarism

		Growth rate					
			cm/6 months				
Study period	Number	cm/year	0.12 b.i.w.	0.08 t.i.w.	0.12 b.i.w.	0.08 t.i.w.	cm/year (mean)
Before	33	1.86 ± 1.08					1.86
1st	25		2.90 ± 0.98				5.80
2nd	19			3.05 ± 1.39			6.10
3rd	6				2.48 ± 1.19		4.96
4th	3					3.20 ± 1.37	6.40

Table 6 Overview of Human Growth Hormone Responses[a]

	Growth response (cm/year)		
	Before	1st year	2nd year
Idiopathic			
0.08 → 0.12 etc.	2.78	7.70	6.24
0.12 → 0.08 etc.	2.47	8.52	6.46
Mean	2.62	8.11	6.35
Organic			
0.80 → 0.12 etc.	2.19	6.34	
0.12 → 0.08 etc.	1.86	5.95	
Mean	2.02	6.15	

[a] 0.24 IU/kg per week.

the final 6 months on 0.08 IU/kg t.i.w., the mean growth response for the three patients was 3.2 cm, or 6.4 cm/year.

2.3. Annual Growth Responses for All Patients

Table 6 summarizes the actual growth responses each year.

2.3.1. Idiopathic

In the 1st year, the 136 patients grew at a mean rate of 8.11 cm, and during the 2nd year, they grew at a mean rate of 6.35 cm.

2.3.2. Organic

During the 1st year, the 69 patients grew at a mean rate of 6.15 cm. The data for the 2nd year to date are insufficient to analyze.

3. Discussion

Overall, the study does not show any marked differences in growth responses when t.i.w. therapy is compared to b.i.w. therapy. The data suggest that t.i.w. therapy is marginally better and that the growth response is better sustained.

This crossover design was selected because it was thought that each patient could serve as his or her own control. It would appear that b.i.w. therapy is quite effective at this dose range using the same total weekly dose of 0.24 IU/kg per week. Use of b.i.w. therapy reduces the frequency of administration of hGH and possibly allows for more patient compliance for the difficult patient.

3.1. Idiopathic

Aceto *et al.*[1] reported results for the 1st 18 months using doses of 0.1 IU/kg t.i.w. (=0.3 IU/kg per week). Their best mean growth response for their group of 12 patients was 12.7 cm in 18 months. In the present study using 80% of his dose (=0.24 IU/kg per week), the growth response for the 1st 18 months was 11.94 cm (Table 3) and 10.96 cm (Table 2). Given the crossover design of our study and the fact that 54 patients had completed 18 months (compared to 12 patients in ref. 1, the responses are considered comparable. Frasier *et al.*,[3] using a dose of 0.1 IU/kg t.i.w. in 16 patients, reported a mean growth response of 8.94 cm in the 1st year. This is higher than our mean growth response of 8.11 cm for our 136 patients using the lower dose. In their 12 patients who received 0.08 IU/kg t.i.w. as we did, the mean growth response was 7.22 cm, and this is lower than what we observed using the same dose in 136 patients.

Our patients showed a mean increase in growth rate over the pretreatment growth rate of three- to fourfold in the 1st 6 months and two to threefold in the 2nd, 3rd, and 4th 6-month study periods. In the patients in Table 2, the growth rate remained fairly constant after the 1st 6 months. In the patients in Table 3, there was a slow progressive decline during each 6-month period. The growth response in the final 6 months of the 2-year period was the same for both groups.

3.2. Organic

In the study of Aceto *et al.*,[1] the nine patients with organic hypopituitarism grew at a mean rate of 6.8 cm in 18 months. In our study the mean growth response in many more patients during the first 18 months was 8.42–8.43 cm (Tables 4 and 5). There was considerably better growth, but our sample size was also much larger.

Our patients showed a three- to fourfold increase in growth rate over the pretreatment rate during the 1st 6 months and a two- to threefold increase thereafter, as we observed for the patients with idiopathic hypopituitarism. Our data are also consistent with other data in the literature which show a lower growth response in patients with organic hypopituitarism than in those with idiopathic hypopituitarism.

4. Conclusions

During the 1980s, only highly monomeric and highly purified hGH with high biological activity has been used for the management of hypopituitarism. Our study shows that the recommended dose of 0.08 IU/kg given t.i.w. is as effective as or more effective than the previously recommended dose of 0.1 IU/kg of less purified hGH. This recommendation has been accepted by the Committee and is currently in general use throughout the United States.

References

1. T. Aceto, Jr., S. D. Frasier, A. B. Hayles, H. Meyer-Bahlburg, M. L. Parker, R. Munschauer, and G. DiChiro, Collaborative study of the effects of human growth hormone in growth hormone deficiency. First eighteen months of therapy, in: *Advances in Human Growth Hormone Research* (S. Raiti, ed.), DHEW Publication 74-612, 1974 pp. 695–724.
2. S. Raiti and R. A. Tolman, Isolation and purity of human pituitary hormones, *Endocrinology* **102:**1444, 1978.
3. S. D. Frasier, G. Costin, B. M. Lippe, T. Aceto, Jr., and P. F. Burger, A dose–response curve for human growth hormone, *J. Clin. Endocrinol.* **53:**1213, 1981.

6

Subcutaneous versus Intramuscular Injection of Human Growth Hormone

Wayne V. Moore and Libbie J. Russo

Growth hormone therapy has been classically given by the intramuscular route. Subcutaneous (s.c.) injection of human growth hormone (hGH) has been avoided owing to the effect that this route of administration might have on the development of antibodies to hGH. Underwood *et al.*[1] reported development of growth-attenuating antibodies in several patients who inadvertently received subcutaneous injection of "clinical-grade" hGH. Between 30 and 60% of children treated with hGH prepared by the method of Raben or Wilhelmi are reported to develop antibodies to hGH[2–6] compared to 5% of children receiving hGH prepared by the method of Roos.[7] Even with the higher incidence of antibody formation, growth attenuation during therapy secondary to antibodies was uncommon so that only 5–10% of children develop growth-attenuating antibodies.[2] However, the appearance of growth-attenuating antibodies can seriously jeopardize the treatment of an individual patient.

In the United States, clinical-grade GH preparations were contaminated with significant quantities of high-molecular-weight protein (Fig. 1). Wilhelmi has indicated that the peak with molecular weight corresponding to dimeric hGH, although not completely resolvable into monomer by urea or guanidine HCl, indeed has the amino acid composition of hGH. In contrast, the amino acid composition of the highest-molecular-weight peak does not resemble hGH indicating that this peak represents impurities as well as aggregated and denatured hGH. The critical step in removing this aggregate from hGH prepared by Wilhelmi[8] and Parlow[9] is a terminal DEAE–cellulose chromatography (A-31) or Sephadex G-100 gel filtration (P4-8), respectively.

We have used these different hGH preparations as well as clinical-grade hGH purified by Sephadex gel filtration in our laboratory to study the antibody

Wayne V. Moore and Libbie J. Russo • Section of Pediatric Endocrinology, Department of Pediatrics, University of Kansas Medical Center, Kansas City, Kansas 66103.

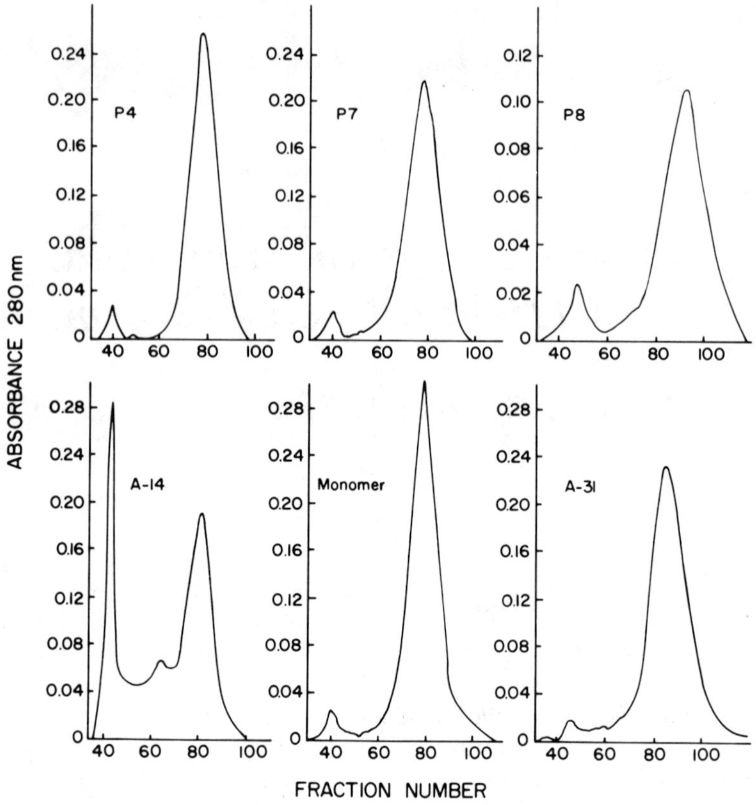

Figure 1 The Sephadex gel filtration profile of various lots of hGH. The growth hormone (4–5 mg) was dissolved in H_2O and applied to a Sephadex G-100 column (1.5 × 90 cm). The protein was eluted with 0.05 M phosphate buffer (pH 7.4) with 0.1% sodium azide. The absorbance of the 1.0 ± 0.1 ml fractions is plotted versus the fraction number. Lots P4, P7, and P8 were the packaged hGH prepared by Dr. Parlow from acetone-preserved and frozen pituitaries. Lot A-14 was packaged clinical grade hGH prepared by Dr. Wilhelmi from acetone-preserved glands. Lot A-31 was packaged hGH prepared and purified by Dr. Wilhelmi by a terminal DEAE–cellulose chromatography. Monomer represents clinical grade hGH (from Dr. Wilhelmi in bulk form) purified by Sephadex gel filtration in our laboratory prior to distribution in solution for treatment of hGH deficiency.

response in patients receiving intramuscular hGH therapy and have observed three distinct responses.[10,11] Approximately 50% of the patients did not develop antibodies to hGH. Approximately 50% of the patients developed antibodies to hGH as determined by a statistically significant increase (>2.5 SD above the mean for control serum) in binding of [^{125}I]hGH by the patient's serum. This occurred regardless of the hGH preparation used in treatment. The antibodies appear by 3–6 months of therapy in most cases (Fig. 2). The presence of antibodies to hGH tended to be transient in patients receiving hGH purified by Sephadex gel filtration while the antibodies persisted and were of

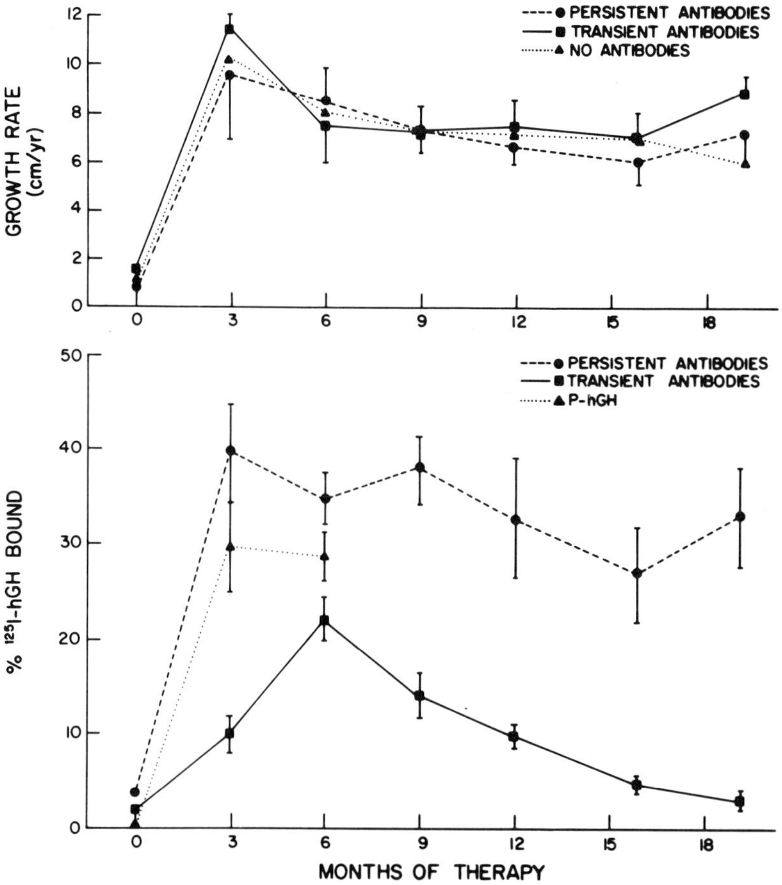

Figure 2 The growth rate and percent [^{125}I]hGH bound by patient plasma at 3-month intervals during hGH therapy given intramuscularly. The growth rate was determined by increment in height at 3-month intervals from the averages of 10 measurements on a clinical statiometer for children with a persistent or transient antibody response to hGH and children who did not develop antibodies to hGH. The percent of [^{125}I]hGH bound by 0.1 ml of the patients' plasma was determined at 3-month intervals by the dextran–charcoal method. The mean ± SD percent of [^{125}I]hGH bound by plasma of patients with persistent or transient antibody response is compared to that of patients receiving hGH prepared by Parlow (P-hGH).

higher capacity in patients receiving the impure preparations of hGH (Fig. 3). Only 2 out of 18 patients have developed persistent antibodies to P series hGH, which is comparable to the antibody formation with Roos hGH. We have not observed growth-attenuating antibodies since switching to the purified preparations of hGH. We conclude that the development of antibodies during hGH therapy is the result of individual susceptibility and the presence of denatured or aggregated hGH or impurities in the preparation.[10,11]

With the preceding conclusions, we compared the growth and antibody

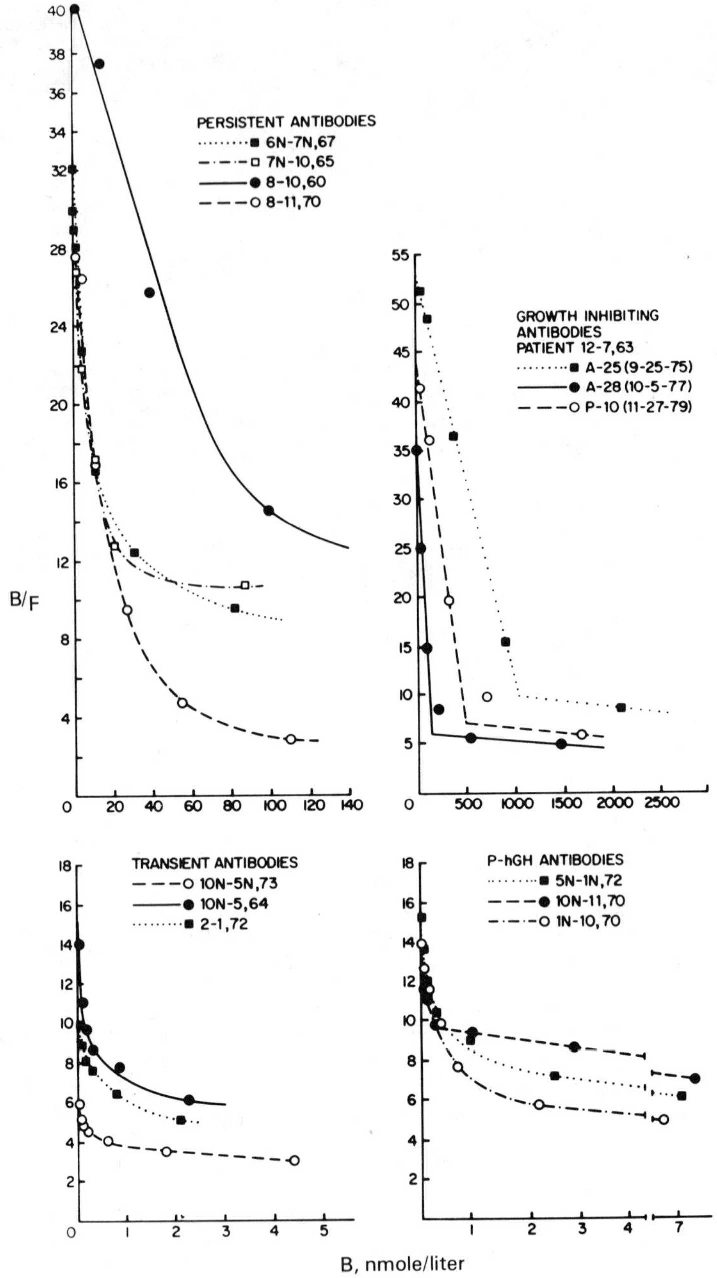

Figure 3 Representative Scatchard plots of [^{125}I]hGH binding by plasma of hGH-deficient patients during hGH therapy. The ratio of the amount bound to the concentration of free hormone (B/F) is plotted versus the amount bound. The plasma sample used from patients with transient antibodies or antibodies to P-hGH was the one yielding maximum binding. The plasma samples used from the patient with growth attenuation were chosen at times of growth attenuation while receiving lot A-25 hGH, during a maximum growth response while receiving lot A-28 hGH, and while receiving lot P-10 hGH.

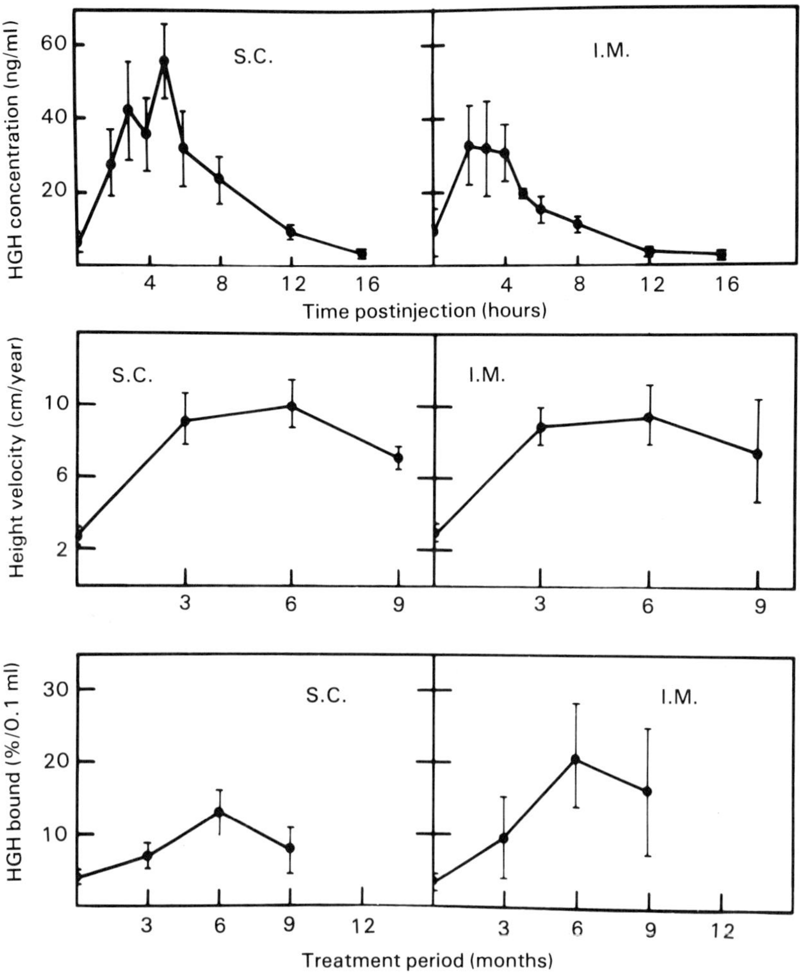

Figure 4 The comparison of the effect of subcutaneous (s.c.) or intramuscular (i.m.) injection of growth hormone on the plasma concentration of hGH (top panel), height velocity (middle panel), and development of antibodies to hGH (bottom panel). The plasma concentration of hGH was determined by radioimmunoassay on plasma obtained following the initial injection of 0.1 IU/kg hGH given subcutaneously (s.c.) or intramuscularly (i.m.). Height velocity was determined at 3-month intervals from the increment in height measured on a clinical statiometer. The percent of 0.1 ng/ml [^{125}I]hGH bound by 0.1 ml of the patients' plasma was determined at 3-month intervals by the dextran–charcoal technique.

response of GH injected s.c. or intramuscularly (i.m.)[(12)] using hGH prepared by Dr. Parlow. The trial of subcutaneous injection of hGH was also prompted by noncompliance in patients secondary to the pain of i.m. injections and a personal communication from Dr. A. Parlow of increased biological response to GH following subcutaneous injection in animals.

Children were identified with GH deficiency by lack of hGH response to

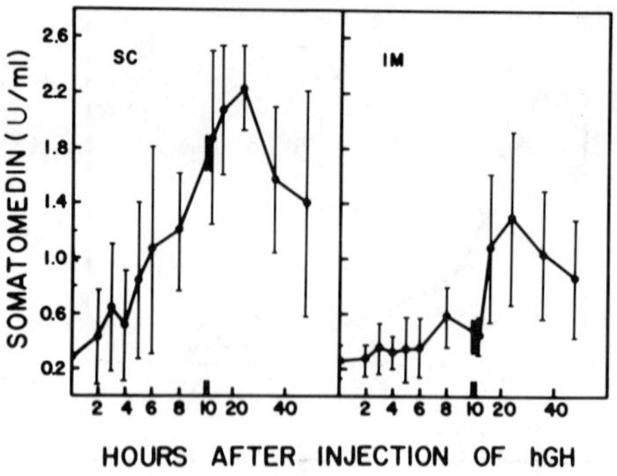

Figure 5 The plasma somatomedin concentration following subcutaneous (s.c.) or intramuscular (i.m.) injection of hGH. The plasma concentration of somatomedin C was determined by radioimmunoassay (Dr. Louis Underwood) at specified time intervals following the initial injection of 0.1 IU/kg hGH. The results represent mean ± SD concentration of somatomedin C in six patients receiving i.m. injections and five patients receiving s.c. injections.

two stimuli of GH secretion (L-dopa and arginine), growth rate of <4 cm/year, and lack of other causes of growth failure. Thyroxine or cortisol deficiencies were treated prior to testing and treatment with hGH.

The plasma peak and integrated concentration of hGH attained after the initial injection of 0.1 IU/kg hGH (2 IU/ml and 4 IU/ml for the i.m. and s.c. group, respectively) did not differ in the s.c. or i.m. group (Fig. 4). The height velocity and antibody response during the first year of therapy also did not differ between the groups (Fig. 4). The difference between the peak somatomedin C concentrations following the initial injection of hGH was not statistically significant (Fig. 5).

The initial 13 patients involved in this study underwent a crossover at 9–12 months of therapy, and all except one preferred the s.c. injection route. Since completing this study, we have switched the approximately 60 patients on our program to s.c. injection and started 30 new patients on s.c. injection. The binding of [^{125}I]hGH by the patients' plasma has been determined at 3- to 4-month intervals in all patients, and we have detected a persistent antibody response in only 1 of 13 patients treated solely with s.c. injections. Several patients have exhibited a decrease in height velocity following several months of s.c. injections, but this resolved following rotation of the injections to areas not involved with mild lipoatrophy resulting from repeated injections into the same site. Since switching to s.c., injections, we have not encountered any noncompliance with the treatment regimen.

In summary, s.c. injection of hGH appears to be a safe, effective, and acceptable mode of treatment of hGH deficiency in children. The ease of administration and patient acceptance have made s.c. injection the preferred route of therapy in our clinic.

ACKNOWLEDGMENT. This work was supported by NIH, NIAMDD Grant AM19544 and Research Career Development Award AM00309 (to W.V.M.).

References

1. L. S. Underwood, S. J. Voina, and J. J. Van Wyk, Restoration of growth by hGH (Roos) in hypopituitary dwarfs immunized by other hGH preparation: Clinical and immunological studies, *J. Clin. Endocrinol. Metab.* **38**:288–297, 1974.

2. S. L. Kaplan, D. Savage, S. Suter, R. Walter, and M. M. Grumbach, Antibodies to hGH arising in patients treated with hGH: Incidence, characteristics and effect of growth, in: *Advances in hGH Research* (S. Raiti, ed.), DHEW #74-612, 1973, pp. 725–744.
3. R. Illig, Growth hormone antibodies in patients treated with different preparations of hGH, *J. Clin. Endocrinol. Metab.* **31:**679–688, 1970.
4. D. S. Frasier, T. Aceto, A. B. Hayles, M. L. Parker, and H. Meyer-Bahlburg, Collaborative study of the effects of hGH in growth hormone deficiency: Development and significance of antibodies to hGH during the first year of therapy, *J. Clin. Endocrinol. Metab.* **38:**14–18, 1974.
5. S. R. Shalkey and J. M. Tanner, Incidence and effects on growth of antibodies to hGH, *Arch. Dis. Child.* **46:**160–166, 1971.
6. W. K. Waldhausal and R. Franz, Development of antibodies to hGH in children of short stature treated with hGH, *Acta Endocrinol.* **68:**345–354, 1971.
7. K. Hall and P. Olin, Sulfation factor activity and growth rate during long-term treatment of patients with pituitary dwarfism with hGH, *Acta Endocrinol.* Kbh **69:**417–433, 1972.
8. J. B. Mills, R. B. Ashworth, A. E. Wilhelmi, and A. S. Hartree, Improved methods for the extraction and purification of hGH, *J. Clin. Endocrinol Metab.* **29:**1456–1459, 1969.
9. A. Parlow and B. Shome, Improvement in the quality and quantity of growth hormone isolated from acetone-preserved human pituitaries for clinical usage, *J. Clin. Endocrinol. Metab.* **43:**229–231, 1976.
10. W. Moore, The role of aggregated hGH in the therapy of hGh deficient children, *J. Clin. Endocrinol. Metab.* **46:**20–27, 1978.
11. W. Moore and P. Leppert, Role of aggregated human growth hormone (hGH) in development of antibodies to hGH, *J. Clin. Endocrinol. Metab.* **51:**691–698, 1980.
12. L. Russo and W. V. Moore, A comparison of subcutaneous and intramuscular administration of hGH in the therapy of growth hormone deficiency, *J. Clin. Endocrinol. Metabol.* **55:**1003–1006, 1982.

7

Psychosocial Impact of Long-Term Growth Hormone Therapy

Robert M. Blizzard, Susan Joyce, Tina Mitchell, Ann J. Johanson, Samuel Libber, Leslie P. Plotnick, and Claude J. Migeon

1. Introduction

Human growth hormone has been administered to children of short stature for over two decades. Some research has focused on aspects of the patients while they were involved in the treatment program, but remarkably little information is available concerning the longer-range impact of the treatment on the patients. This study was designed to assess important areas of the patients' lives as adults, and to use this information in helping physicians be aware of long-term effects of this treatment program.

2. Method

2.1. Subjects

Seventy patients who were growth hormone deficient and who were treated with growth hormone during middle and/or late childhood were contacted to determine their current sociological and psychological status. All patients were participants in long-term growth hormone treatment at either Johns Hopkins University or the University of Virginia; all were Caucasian. Fifty-eight of these

Robert M. Blizzard and Ann J. Johanson • Department of Pediatrics, University of Virginia School of Medicine, Charlottesville, Virginia 22901. ***Susan Joyce and Tina Mitchell*** • Department of Psychology, University of Virginia, Charlottesville, Virginia 22901. ***Samuel Libber, Leslie P. Plotnick, and Claude J. Migeon*** • Department of Pediatrics, Johns Hopkins University School of Medicine, Baltimore, Maryland 21205.

patients and/or their parents responded to the initial questionnaire. There were 44 males and 14 females. The average age was 26 ± 6.6 SD (range, 16–46 years). The diagnoses for these patients are listed in Table 1. There was only one set of siblings.

The average height was 157.5 cm. Of the 44 males, 33 or 75% achieved a height greater than 152.4 cm (5.0 ft). Eleven did not. Of the 14 females, four achieved a height of 147.3 cm (4.10 ft) or greater. Ten did not.

2.2. Measures

2.2.1. *Questionnaire*

Two similar questionnaires (one for the patient, and one for the patient's parents) were created to obtain general descriptive information (e.g., age, sex, height, weight, employment status). In addition, educational and employment histories were gathered on the targeted patient, the patient's siblings, and the parents. Finally, open-ended questions were included to tap overall education achievement, satisfaction with employment, benefits and hindrances of short stature, positive and negative impacts of the growth hormone treatment program on the patient and the family, initial treatment expectations of both the patient and the parents, and (for the parents only) a comparison of the patient's adjustment to that of his or her siblings. Both the patient and the parent(s) were sent the applicable version of the questionnaire.

2.2.2. *Tennessee Self-Concept Scale (TSCS)*

One area of interest in this study was the self-concept of the treated sample, which was compared to a nontreated sample. This scale, created by Fitts,[1] had 100 self-statements which the respondent rated on a five-point Likert-type scale ("completely false" to "completely true"). These items were classified into 10 dimensions of self-concept; identity, self-satisfaction, behavior, physical self, moral–ethical self, personal self, family self, social self, self-criticism, and global self-concept. The reliabilities (internal consistency) of each scale were checked on this sample, and found to be satisfactory (0.81–0.91, mean, 0.87). Twenty-nine patients filled out this form.

2.2.3. *Joyce Measure*

The goal of this measure was twofold: first, to examine perceptions by both the patient and the parent(s) concerning how the patient had changed from childhood to adulthood; second, to explore the congruence between patient and parent perceptions of the patient. Thirty-two adjectives (e.g., "irresponsible," "relaxed," "competitive," "self-sufficient") were generated in conjunction with experts in the field and the research literature to accomplish this. Focusing first on how the patient was as a child, and next, on how the patient is presently, each respondent rated each adjective on a five-point Likert-type scale ["never true of me (him/her)" to "always true of me (him/her)").

Although the 32 items could have been analyzed individually, there are several important reasons why this would have been unwise. First, many of the items were highly correlated with each other, rendering univariate analyses such as F tests invalid. Second, since groups of items were intercorrelated, one must assume that they were measuring the same underlying attribute. Yet a single item may have had a low direct correlation with the underlying attribute owing to extraneous factors. For example, in a spelling test, a child who reads baseball stories may be better able to spell "umpire" than a child who does not read such stories, even though over a *variety* of words the second child is a better speller. Thus, single items may correlate with the attribute of interest but also with other irrelevant attributes (e.g., an interest in baseball). Third, single items have considerable random error—that is, people do not always repeatedly respond to the same item with exactly the same response. However, the tendency of items to relate to extraneous factors usually averages out when correlated items are combined. In addition, random error tends to decrease when multiitem scales[(2)] are utilized, thus producing more stable or reliable statistical results.

Thus, in order to create reliable scales, endorsement frequencies were first examined; six items with extremely low variance (that everyone answered the same way) were discarded from further analyses. These were "relaxed," "compliant," "cooperative," "serious," "kind," and "inconsiderate." Next, using a rational–empirical method of measure development,[(3)] the remaining variables were grouped into two rational scales and their internal consistency checked using Cronbach's alpha.[(4)] Those items with low corrected item–total correlations were discarded. In this manner two scales were created: (1) "maturity," which contained areas such as "calm," "selfish," "responsible," "secure," "angry," and "tense" ($\alpha = 0.92$), and (2) "assertive," which contained such areas as "assertive," "compliant," and "competitive" ($\alpha = 0.80$). Since these two scales were also correlated with each other, a third, more global evaluative scale was created which consisted of both scales. Thus, there were three average-item scale scores for the patient as a child and three for the patient as he or she is presently. In addition, since both the patient and the parent(s) completed this measure, there were scores for patient and parent perceptions of each of these.

2.3. Procedure

After the list of 70 patients was compiled, current addresses were verified by telephone. Five were completely lost to any follow-up because of inability to track down a current address.

Once the address was verified, the initial questionnaire was sent with a return, stamped envelope. One questionnaire was sent to the parents and one to the patient. If no reply had been received within 4 weeks, either another questionnaire was sent or the patient or parent(s) were contacted by phone. Persistence was necessary, as some individuals responded only after the fourth contact. Once the questionnaire was returned, a phone interview was arranged to obtain detailed information and fill in missing data as much as possible. The TSCS and Joyce measures were mailed at that time.

Fifty-eight families returned the initial questionnaire; of these, 12 families had only one person return the questionnaire—seven parents and five patients. In 46 families both questionnaires were returned. Of the initial 70 patients only 12 were not utilized in this study. Six failed to respond. Five were lost to follow-up, and one was not utilized because she was from an alien culture which negated a reasonable comparison with those in the American culture.

With the Joyce measure, at least one member of 28 families returned the measure; in 13 cases, only one member complied (six patients and seven parents).

Twenty-nine patients returned the TSCS questionnaire.

Twenty-seven families completed the five questionnaires.

3. Results

3.1. Diagnoses

The diagnoses versus incidence by sex are listed in Table 1. Eight of the patients had *organic hypopituitarism* as defined by the presence of a tumor mass or a space-occupying intracranial lesion. Interestingly, all eight patients were males.

Nine patients had *idiopathic panhypopituitarism*. Seven of these were females.

Forty-one patients had *idiopathic hypopituitarism* with growth hormone deficiency, but *normal gonadotropin and testosterone or estrogen* production. Thirty-four were males and seven were females. Therefore, of the 50 patients with idiopathic hypopituitarism, 41 had normal gonadotropin production. Unfortunately, it was not possible to satisfactorily classify the patients with idiopathic hypopituitarism with respect to thyroid-stimulating hormone (TSH) and/or ACTH deficiency. A number of these patients had received metyrapone 15 or 20 years previously and were thought to have limited ACTH reserve. A number of these patients also intermittently had low thyroxine determinations a number of years ago while on growth hormone treatment and were believed to have

Table 1 Classification according to Diagnosis and Sex[a]

	Males	Females	Total
Organic hypopituitarism	8	0	8
Idiopathic hypopituitarism	36	14	50
Panhypopituitarism	2	7	9
GH deficiency + normal LH	34	7	41

[a] GH = growth hormone; LH = luteinizing hormone.

Table 2 School Performance as Perceived by the Patients

Average	47%	100%
Above average	41%	
Below average	12%	

limited TSH reserve. Subsequent studies in a number of those initially thought to have ACTH and/or TSH deficiency have been normal, and consequently, we often were not able to delineate which patients may have had isolated growth hormone deficiency and which had growth hormone deficiency plus TSH and/or ACTH deficiency.

The high incidence of males versus females in both organic hypopituitarism and idiopathic hypopituitarism is as previously noted by many authors. The high incidence of idiopathic panhypopituitarism in females as compared to males has not been reported previously to our knowledge.

3.2. Open-Ended Questions

These areas were explored on the initial questionnaire. They were clarified later during phone interviews.

3.2.1. School Performance

The first area evaluated dealt with performance in school (Table 2). Forty-seven percent of the patients rated their performance as average, 41% believed that they were above average, and only 12% rated themselves as being below average.

In Table 3 the evaluation of the short stature in respect to the impact it had on school performance is listed. The parents reported different perceptions than the patients. Fifty-one percent of the parents reported a negative impact of short stature. Only 20% of the patients so reported. A large percentage of the parents (47%) and an even larger percentage of the patients (62%) felt that stature had no impact on school performance.

Although a variety of responses were offered concerning the manner in which short stature affected academic performance, the most frequently cited

Table 3 The Impact of Short Stature on School Performance

	Self-Rating	Parental Rating
Negative impact	20%	51%
Positive impact	18%	2%
No impact	62%	47%

positive effects were "working harder at school as a result of the short stature" (24%) and "receiving special attention for being different" (24%). The most frequently cited negative factor was teasing by peers. The parents noted that the major way that the short stature affected the patients was that they were not able to participate fully in physical education classes or in sports and had problems with dating and peer relationships.

3.2.2. *Employment*

The second area explored dealt with employment and employment satisfaction. Thirty-three of the patients were employed full-time, six were employed part-time, and 19 were unemployed. Six of the latter were students and would not be expected to be employed.

Sixty-eight percent were satisfied with their current status of employment. Twenty-three percent were dissatisfied. Nine percent did not answer.

With respect to employment and the effect of short stature, 57% of the patients stated that short stature was no major factor in their obtaining employment. However, 37% did indicate that it had been a negative factor in their employment status. Two percent stated that it was a positive factor and 4% were unsure.

The parents had very similar reports. The most frequently cited reason for the negative impact of short stature focused on general limitations of size (e.g., "not enough strength," "could not reach things").

3.2.3. *Appearance versus Chronological Age*

The patients were evaluated for their perception of their chronological age versus their apparent age. The patients' and parents' responses generally were in agreement. Fifteen percent believed that the patient looked at least 10 years younger than his/her chronological age. Forty-two percent believed their appearance to be between 5 and 9 years younger than the chronological age. Twenty-eight percent perceived himself/herself to appear 1–4 years younger than their chronological age, and 15% believed that the patient appeared his or her chronological age.

When asked if they were treated as being younger than their chronological age when they were children, 40% of the patients replied in the negative and 60% in the affirmative. Parents, however, seemed to believe differently: 20% of the parents did not note any such occurrences while 80% said that the patient had been treated as if he or she was younger than would be expected for the chronological age. Interestingly, when the patients were asked whether they now were treated younger than their chronological age, 59% said "no," and 41 replied "yes."

3.2.4. *Positive and Negative Effects of Short Stature*

On the questionnaire, one set of questions was established to detect both positive and negative effects of short stature on the patients and their families.

Table 4 The Positive Effects of Short Stature as Perceived by the Patients

1.	Striving for perfection was increased
2.	The patient received immediate recognition as a person because of his size
3.	The patient became more compassionate
4.	Family ties were increased
5.	The siblings of the patients also learned compassion

Although there was great variability in the responses, the most commonly cited *positive* effects of short stature on the patients as perceived by them are listed in Table 4.

The two most commonly reported *negative* effects of short stature on the patient were the lack of relationship with peers of the same age and difficulties with heterosexual social relationships. It was also noted that there were limited options socially for extremely short people. For example, there frequently was the inability to participate in sports and difficulty in finding shoes and clothes that fit.

The two most commonly cited negative effects of short stature on the family were the strain involved in trying to diagnose and treat the problem, and in the sibling jealousy that was aroused by the special attention the patient(s) frequently received.

3.2.5. Expectations and Results of Treatment

The expectations of the patients and parents with respect to height are listed in Table 5.

Some patients (4%) and parents (10%) stated that their expectations were too high.

Table 5 The Initial Expectations of the Parents and Patients to Growth Hormone Treatment

	Patients	Parents
Expected to reach normal height	37%	37%
Expected to grow "some"	41%	47%
Could not recall expectations	18%	6%
Did not have specific idea but expectations were too high in retrospect	4%	10%
	100%	100%

Table 6 The Extent to Which the Patients' and Parents' Expectations Were Met

	Patients	Parents
Completely	64%	50%
Partially	32%	48%
Not at all	4%	2%
	100%	100%

In assessing how well expectations were met, we learned that these expectations were stated to be met by 64% of the patients and by 50% of the parents (Table 6). This was true even though many of these patients were treated late and a number did not achieve a height that would be considered by many to be in the normal range.

A large percentage of patients (36%) and parents (50%) were disappointed in the treatment because their expectations were only partially met or not met at all.

In one set of questions there was focus on both the benefits and the negative aspects of treatment, coupled with recommendations of treatment of others. Overwhelmingly, the greatest benefit of the program was the fact that the patients grew; more specifically, many patients and parents noted that the patient was better able to interact in society by being taller. The worst aspect of the treatment was, again overwhelmingly, the pain of the injections and difficulties of the treatment and traveling to and staying at the treatment site.

Patients and parents were in accord about recommendations of treatment to others. Four percent of the patients said they would not recommend treatment to others; 10% would recommend it with some qualifications (e.g., not for children who just want to grow faster), and 86% of the patients and 84% of the parents definitely would recommend the treatment to patients with growth hormone deficiency.

Recommendations that the patients and parents might make to physicians currently working with short-statured children and adolescents were obtained. The most salient and frequent recommendation from both the patients and the parents was to establish self-help support groups of families and patients with short stature problems. In addition, the need to downplay the defiance of being short was emphasized (e.g., have successful short people talk with the patient, just in case the patient does not attain "normal" height). In addition, the patients complained about the impersonal treatment they often received from physicians (e.g., large numbers of physicians in the room, and speaking in front of the patients as if they were not there). They emphasized the need for prospective patients to be given very realistic expectations about the possibility of growing in response to treatment.

3.2.6. Sibling Comparisons

The area of sibling comparisons was explored only with the parents. The majority (54%) of the parents noted no essential differences between the patient and his (her) siblings of normal stature. Fifteen percent of the patients believed that the patient was better adjusted than the siblings, and 29% believed the siblings were better adjusted than the patient.

3.2.7. Psychosexual Development

Five of the fifty-one patients responding to this question had never dated and 46 had. The average age of first dating was 16.5 years.

Thirty-nine percent reported having had sexual intercourse at least once, 26% never had experienced intercourse, and 35% did not respond to the question. Seventeen percent of the sample did not currently have a sexual partner and 22% reported that they did. Sixty-four percent did not report an age at first intercourse; the mean for those responding was 23 years (range 15–29 years).

Of the 52 responding to inquiries regarding marital status, 35 were single and 17 had been or are married currently. Two of these seventeen are divorced.

Ten of the seventeen couples were parents. Eight couples had children by conception and two by adoption. There were 17 children distributed among these 10 couples.

Upon analysis of the characteristics of those with natural children, the following data were obtained: (1) short stature was a more negative factor ($p < 0.04$) in school performance for this group than for those without children, (2) short stature was a more negative factor ($p < 0.02$) in employment for those in this group than for those without children, (3) the expectations for treatment were less well met for this group than for those without children, and (4) these patients made more negative statements about themselves than did patients who were not parents.

3.2.8. Self-Concept by TSCS

The scores of the TSCS for the sample were compared to the norms which were developed previously. F tests were used. The patients as a group felt more satisfied with their self image in general than did the control population ($p < 0.05$). When this general perception was dissected, it was learned that the patients rated higher in self-satisfaction ($p < 0.05$), personal worth ($p < 0.05$), and social being ($p < 0.05$) than controls, but lower in their evaluation of physical self ($p < 0.03$) and self-criticism ($p < 0.03$).

3.2.9. Perception of the Patient by the Joyce Measure (Table 7)

As seen in Table 7 by reading across the table, the patients and parents evaluated the patients similarly as a child in respect to the maturity of behavior,

Table 7 Perception of the Patient by the Patients and Their Parents Using the Joyce Measure[a]

	Patients	Parents	*p*
As child			
Mature versus immature	3.3 ± .6	3.3 ± .8	NS
Assertive versus passive	3.1 ± .6	3.3 ± .8	NS
Total perception	3.2 ± .5	3.3 ± .7	NS
As adult			
Mature versus immature	3.6 ± .5	3.5 ± .9	NS
Assertive versus passive	3.5 ± .6	3.4 ± .8	NS
Total perception	3.6 ± .5	3.5 ± .8	NS

[a] NS = not significant.

assertiveness versus passiveness, and total perception. The scale scores hovered around the midpoints of the scale (e.g., 3.28 ± 0.65 of a scale range of 1.00–5.00). The patients and parents evaluated each patient similarly as an adult in respect to these parameters.

As seen in Table 7 by reading vertically down each column, the evaluation of the patient "across time" by both the patients and their parents did not change significantly.

These data are indicative that there were no significant changes in how either the patients or their parents viewed the patient from childhood to adulthood. Since all scale scores hovered around the midpoint, the data are indicative that the individuals in the sample do not seem to feel either extremely positive or extremely negative toward themselves.

3.2.10. *Subsample Comparisons*

Data from the group returning only the major questionnaire were compared to the data from those returning all three questionnaires. F tests (with integer and ratio data) and chi-square analyses (with nominal and ordinal data) were used. On approximately 40 variables from the questionnaire, the only significant differences between the two groups was that the male returnees reported an earlier onset of shaving ($p < 0.01$) and those returning the measures were slightly more likely to be employed, although this did not reach significance ($p < 0.07$). Therefore, differences between the group returning the TSCS and Joyce measure and those who did not appear to be a meaningful confound. One also can probably rule out a potential bias in the sample who did not return the initial questionnaire, as only six patients failed to comply.

4. Discussion

Before the findings and implications of this study are discussed, it is important to note some of the research design and statistical caveats. First, since

12 patients could not be contacted or never responded to the initial questionnaire, it is impossible to know what kind of sample bias may exist in the reported data set. However, it can be assumed that the six who were completely lost to follow-up (i.e., no current address could be found) were normally distributed across the sample range. Thus, only six families of the 64 who received the questionnaire refused to cooperate at all. Even if all six were at one extreme of the distribution, the changes in these findings would probably not be dramatic.

Second, in some of the subsample comparisons, there were very small cells. For example, in comparing parents of natural children to those with no children on the TSCS data, one of the cells was made up of only six subjects. Because of this, some of these analyses may be statistically less stable, requiring future replications.

Third, when this sample was compared to "normal-stature" groups, few differences were discovered. However, one must be wary of "accepting the null hypothesis"; just because no differences were found does not necessarily imply that there *were* no differences at all in the groups. Perhaps there were differences in untapped areas; perhaps the instruments utilized were not powerful enough to discriminate. Finally, the research design used was a purely post hoc, correlational design, which means that much of the information was retrospective. Unfortunately, perceptions of the past sometimes may be altered by the passage of time. In addition, no implications of causality can be drawn from *any* correlational design; one can merely report the occurrence or nonoccurrence of relationships between variables without inferring the direction of a causal relationship between them.

However, when these cautions are considered, one can explore some of the broader implications of this study. Four general areas deserve further investigation and thought. One important finding was that patients, in general, reported average and above-average school performance. This finding offers further support to those refuting the lack of academic motivation in short-stature students.(5,6)

Second, correlations between parent and patient reports were quite high. There were two notable exceptions to this. The parents reported that short stature had a more negative impact on the patient in school than was reported by the patients, and the parents reported more often than did the patients that the patient as a child had been treated more in line with his or her "height age" rather than in accord with the chronological age. In all other areas where both the patient and the parent were asked the same questions, there was a high degree of agreement. It is possible that this disagreement was simply an artifact of retrospective data; perhaps the parents *always* remember the past differently than the child does. However, there were no similar discrepancies between parent and child remembrances on the Joyce measure, which contained retrospective data from the same time period. This indicates that something else was occurring. Perhaps the parents were more *observant* about these aspects of specialness than the patients were, or perhaps the parents *interpreted* events in a more negative light than the patient saw them as a child. Although the true interpretation cannot be discerned from this study, there is support in the literature for the latter explanation. Parents (especially mothers) reported early negative views of their short-stature children (e.g., sickly, vulnerable); in addition, there was a tendency to doubt one's parenting competence.[7] This may have resulted in parental misattributions of events as more serious than the child perceived them.

A third interesting area is the comparison of the treated sample to the norms on aspects of self-concept by the TSCS. Not surprisingly, this sample felt less positive about themselves physically. This finding has been reported by others.[8] However, beyond this, it appears that this sample fared rather well compared to others; they seemed to feel better about themselves as worthwhile people in general and as social beings. In addition, there were no other discernible differences on aspects of self-concept between the sample and the normal group. Since others have reported more negative self-concepts,[9] these results are a bit surprising. However, the fact that this group also scored *lower* on self-criticism in general is an important key in interpreting this situation. Typically, to score low on self-criticism is interpreted as evidence of defensiveness against saying negative things about oneself. Indeed, Rotnem *et al.* reported that short-stature populations tend to use denial in dealing with poblems.[9] Denial of problems, coupled with the counterpart of "accentuating the positive," might explain these results. Thus, it is possible that high defensiveness may have resulted in inflated reports of positive self-perception.

An alternative explanation might be that this sample has *learned* to regard themselves more kindly and to exhibit more compassion toward themselves than occurs in unafflicted people. This could occur because of the adversity they have experienced. In fact, there is some support for this perspective in this data set; many parents and patients noted compassion toward self and others as an important positive effect of short stature on the patient.

The fourth area of intriguing findings was the consistently negative reports about themselves from the six people with natural children. Again, there are several alternatives for explaining this finding. It is possible that the patients without children were inflating their self-images as a defense, as discussed previously. Perhaps those who have families, being more settled and secure, no longer feel the need to "kid" themselves or others. Another possibility might be that this subgroup felt negatively about themselves all along, and one in which they attempted to make their lives better has been has been through raising a family. Again, because this is a correlational design, none of these interpretations can be verified. Further research and more longitudinal designs are needed to clarify this.

The research findings having been discussed, certain statements concerning the implications are appropriate. It is apparent that there are problems of adjustment to being short. However, it is quite apparent that many of these patients do very well with respect to employment, scholastic achievement, and social and family adjustment.

It also is apparent that the expectations of the patients and parents often were not met. The same effect has been noted by others in short-term and semi-long-term treatment with growth hormone.[7] Rotnem *et al.* previously reported a discrepancy in the way physicians who supervised the administration of growth hormone differed in their enthusiasm for modest growth from the patients and their patients, who were disappointed that the growth rate and the total growth were not more significant.[7] Physicians must be more realistic in explaining to patients the benefits of treatment. For example, the physician should emphasize that the long-range goal is to "catch up in height," but the short-term goal is to "not fall further behind."

Unequivocally, in retrospect growth hormone treatment has been a positive factor for most of these patients. Growth hormone treatment is here to stay because it obviously is beneficial. Earlier treatment might have minimized the handicaps of being short, and physicians should strive to assist growth in patients with proven growth hormone deficiency at an early age. Regardless, these data demonstrate that growth hormone treatment often can be instituted in late childhood with positive benefits to the patient.

Further studies comparing the results in this sample with samples where the patients were treated earlier will be enlightening and helpful in our approach to the treatment of all types of short stature.

5. Summary

A group of patients who had received growth hormone treatment in late childhood were assessed as adults in areas of school performance, employment, and employment satisfaction, "height age" versus "chronological age," and effects of short stature and growth hormone treatment. In addition, they were compared to siblings and previously developed norms on areas such as overall adjustment and self-concept. The majority reported above-average school performance, satisfaction with employment, a discrepancy between "height age" and chronological age, and an overall strong recommendation for treatment to others (although differences between initial expectations and treatment results were noted). Parents reported no consistent difference in school performance or overall adjustment between the patient and his or her siblings. The patients tended to rate themselves more highly on self-concept measures; however, they also exhibited greater defensiveness about saying negative things about themselves. Finally, implications for treatment, such as cautions about offering realistic treatment expectations, and suggestions for further research were noted.

ACKNOWLEDGMENT. These studies were supported by the Human Growth Foundation and U.S.P.H. General Clinical Research Grant RR-847.

References

1. W. H. Fitts, *Tennessee Self Concept Scale*, Western Psychological Sciences, Los Angeles, 1965.
2. J. C. Nunally, *Psychometric Theory*, McGraw-Hill, New York, 1978.

3. D. N. Jackson, A sequential system for personality scale development, in: *Current Topics in Clinical and Community Psychology* (C. Spielberger, ed.), Academic Press, New York, 1970.
4. L. J. Cronbach, Coefficient alpha and the internal structure of tests, *Psychometrika* **16:**297, 1951.
5. A. L. Rosenbloom, D. W. Smith, and D. G. Loeb, Scholastic performance of short statured children with hypopituitarism, *J. Pediatr.* **69:**1131, 1966.
6. E. Pollitt and J. Money, Studies in the psychology of dwarfism I. Intelligence quotient and school achievement, *J. Pediatr.* **64:**418, 1968.
7. D. Rotnem, D. J. Cohen, R. Hintz, and M. Genel, Psychological sequela of relative "failure" for children receiving human growth hormone replacement, *Am. Acad. Child Psychiatry* **2:**505, 1978.
8. M. Kusalic and C. Tortin, Growth hormone treatment in hypopituitary dwarfs: Longitudinal psychological effects, *Can. Psychiatr. J.* **20:**329, 1975.
9. D. Rotnem, M. Genel, R. L. Hintz, and D. J. Cohen, Personality development in children in growth hormone deficiency, *Am. Acad. Child Psychiatry* **16:**412, 1977.

II

HUMAN GROWTH HORMONE IN OTHER FORMS OF SHORT STATURE

Gene Studies

8

Growth-Stimulating Effects of Human Growth Hormone Therapy in Turner's Syndrome

Preliminary Report

Salvatore Raiti and the Committee on Growth Hormone Usage (S. L. Kaplan, G. P. August, S. A. Kaplan, M. H. MacGillivray, W. V. Moore, J. S. Parks, A. F. Parlow, A. W. Root, and A. E. Wilhelmi)

Prior to 1982 and during the first 25 years of its clinical availability, the use of human growth hormone (hGH) for the treatment of Turner's syndrome was very limited because of inadequate supplies. A review of the published literature is enlightening in that a number of studies did describe growth acceleration during hGH administration to patients with Turner's syndrome. However, in comparison to the growth responses achieved in hypopituitarism, the growth acceleration described in Turner's syndrome was not considered adequate to justify the expenditure of hGH needed. A few investigators reported no significant growth acceleration at all.

In 1964, Soyka *et al.*[1] studied two patients. The first showed no growth acceleration during the 12 months of study using doses as high as 18 mg (presumably 18 IU) per week. The second patient was treated for 13 months with doses of 3 mg t.i.w. and showed a growth acceleration from 3.0 to 5.3 cm/year. This was considered to be a poor response for an investment of 486 mg of precious hGH at that time.

Salvatore Raiti and the Committee on Growth Hormone Usage (S. L. Kaplan, G. P. August, S. A. Kaplan, M. H. MacGillivray, W. V. Moore, J. S. Parks, A. F. Parlow, A. W. Root, and A. E. Wilhelmi) • The National Hormone and Pituitary Program, University of Maryland School of Medicine, Baltimore, Maryland 21201.

In 1965, Hutchings *et al.*[2] reported studies on two patients with gonadal dysgenesis. The 1st patient received 5 mg of hGH daily (35 mg/week) for 96 of 114 days of the study and was then observed again without therapy for another 6 months. The pretreatment growth rate was 3.8 cm/year. During the 3 months of hGH therapy, she grew at 7.5 cm/year. During the posttreatment phase, she grew at a rate of 2.5 cm/year. The 2nd patient grew at a rate of 2.5 cm/year before therapy. She then received hGH in doses of 5 mg/day (=35 mg/week) for 4 months and grew at a rate of 6 cm/year during this period. During the follow-up period her growth rate returned to 2.4 cm/year.

In 1965, Wright *et al.*[3] reported studies of two patients with gonadal dysgenesis. The pretreatment growth rates were 2.54 and 3.8 cm/year, respectively. The 1st was treated with hGH in doses of 5 mg daily (35 mg/week) for 14 out of every 28 days over a 6-month period. Her growth rate increased from 2.54 to 5.08 cm/year during this period. The 2nd patient was treated in a similar manner over a 7-month period. Her growth rate increased from 3.8 to 6.35 cm/year during this period.

Tzagouris,[4] studied one patient using hGH doses of 5 mg twice a week (10 mg/week) for 1 year. Prior to therapy the growth rate was 2.5 cm/year. During the 1 year of treatment, the patient grew 6.0 cm. In the posttreatment year, the patient received estrogen therapy and grew 3.5 cm.

Tanner *et al.*[5] reported results on five patients who were treated with 10 mg of hGH twice a week (=20 mg/week). The pretreatment growth velocity of 2.9 cm/year rose to 3.9 cm during the year of treatment and fell to 1.8 cm during the year after treatment. The acceleration during the threapy year was canceled by the posttreatment deceleration.

Rudman *et al.*[6] compared the growth-stimulating effects of oxandrolone and of hGH. Study periods of only 3 months each were used. In the pretreatment year, the mean growth rate was 2.04 cm/year. With oxandrolone, this increased to 3.24 cm/year. During the hGH study period, the growth rate was 3.94 cm/year. When both hGH and oxandrolone were used for the 3-month study period, the growth rate increased to 7.56 cm/year.

With this background of studies that suggest that hGH either does or does not improve the growth rate of patients with Turner's syndrome, and with the availability of hGH that is highly monomeric with a high biological potency of 2.4 IU/mg (compared to 1.0 IU/mg or less for the hGH used in the earlier studies), it was decided to reevaluate the question of hGH effectiveness as a growth stimulant for Turner's syndrome. The results reported here represent the progress achieved at the end of 12–18 months of a 4 to 5-year continuing project.

1. Patient Selection and Study Design

Fifty-one patients with Turner's syndrome were accepted from pediatric endocrinologists from multiple centers within the United States. The criteria for acceptance into the study are listed in Table 1. One requirement was that

Table 1 *Turner's Syndrome, Criteria for Acceptance*

1.	Short (<1st percentile)
2.	Growth rate (4 cm/year or less)
3.	Euthyroid
4.	No previous therapy

all patients had to have had growth studies for at least 1 year prior to entering the study. All patients had chromosome studies. Table 2 lists the tests that were carried out at the beginning of the study. All the patients had normal GH responses to two of the definitive stimulation tests.

The hGH dose selected was 2.5 times the dose currently accepted as physiological replacement (0.08 IU/kg body weight t.i.w). The dose used was 0.2 IU/kg body weight, given every Monday, Wednesday, and Friday (Table 3). The hGH was given on a 12-months-on, 6-months-off schedule. The patients' heights and weights were measured every 2 months for the 1st year and then at 3-month intervals. At the end of the 12 months of therapy bone ages, thyroid, glucose tolerance, and somatomedin studies were carried out.

2. *Clinical Growth Responses*

All patients were growing at 4 cm or less per year for 1 year or more prior to entering the study. The actual mean growth rate for the group was 3.18 cm/year.

Table 2 *Turner's Syndrome, Tests Required*[a]

1.	hGH—two definitive tests
2.	Glucose—oral GTT
3.	T_4 and TSH
4.	Bone age
5.	Somatomedin generation, 0.2 IU/kg per day × 4 days

[a] hGH = human growth hormone; GTT = glucose tolerance test; TSH = thyroid-stimulating hormone.

Table 3 *Turner's Syndrome, Human Growth Hormone (hGH) Dose and Follow-Up*[a]

1.	hGH, 0.2 IU/kg on Monday, Wednesday, and Friday
2.	Height at 2, 4, 6, 8, 10, and 12 months
3.	Bone age at 6 and 12 months
4.	T_4 and TSH at 6 and 12 months
5.	GTT at 12 months
6.	Somatomedin at 12 months

[a] TSH = thyroid-stimulating hormone; GTT = glucose-tolerance test.

Table 4 Turner's Syndrome

Study period	Number	Growth rates		
		cm/year	cm/6 months	cm/year (mean)
Before	51	3.18 ± 0.79		3.18
1st 6 months	35		3.31 ± 0.94	6.62
2nd 6 months	22		2.69 ± 0.63	5.38
1st 12 months	22			5.87 ± 1.18
6 months off	8		1.45 ± 0.70	2.90

The data for the growth responses are summarized in Table 4. During the 1st 6 months of hGH therapy, the mean growth response was a 6.6 cm/year, that is, a doubling of the pretreatment growth rate. There was a falloff during the 2nd 6 months of therapy with a mean growth rate of 5.38 cm/year (a drop in rate of about 0.5 in./year). The overall mean response for the first 12 months of hGH was 5.87 cm/year. Eight patients have completed a 6-month poststudy observation period. The mean growth response fell to 2.9 cm/year, which is similar to the pretreatment growth rate. These patients are now beginning a 2nd 12 months of hGH therapy again at a dose of 0.2 IU/kg t.i.w.

3. Discussion

The reports of the studies carried out in the 1960s[1–4] suggested that a growth response of 6–7 cm/year might be expected. Admittedly, some of those studies involved hGH administration for only 3 months. The data published in the 1970s[5,6] were at variance with these findings and suggested little or no growth response with growth rates remaining at 3–4 cm/year. The data of Rudman *et al.*, however, suggested that there was potential value in combined therapy using hGH and oxandrolone even though these patients showed little or no response when hGH was used alone.

The present study, using a large sample group for 1 year confirmed the expectations of the earlier studies of the 1960s. The mean overall growth response of almost 6 cm for the year of therapy was almost double the pretreatment growth rate of just over 3 cm/year. As might be expected, the best growth response occurred during the 1st 6 months of therapy. While this is not nearly as good as the growth response of 7–9 cm/year that one expects to see in the 1st year of therapy in idiopathic hypopituitarism, the response still remains highly significant and well worth the treatment used.

It is interesting to note that when hGH therapy was stopped for 6 months, the growth rate fell to the pretreatment level. This is at variance with observations in hypopituitary patients. When their hGH therapy is stopped, they either stop growing or grow at a much slower rate than for the pretreatment period.[7] The probable reason for this observation in Turner's syndrome is that they do continue to produce normal amounts of GH, that this is not irreversibly suppressed, and they are, therefore, able to grow again at the rate that is "normal" for them. The hypopituitary patients, on the other hand, lose most or all of their capacity to produce any GH and cannot grow.

The data collected to date in this study, suggest a beneficial growth-stimulating effect of hGH therapy. The conclusions must be regarded as tentative and viewed with some caution at this time. The overall growth response over 3 or more years of therapy must be reviewed to determine whether these patients are indeed better off after such long-term hGH therapy or whether a deceleration occurs to the extent of eliminating the gains achieved.

4. Postscript

Since the time that the data were first presented, many of the patients have completed the first 6 months of the 2nd year of hGH therapy. Nineteen of twenty-one such patients grew at a rate of 5–6 cm/year. The other two patients grew at a rate of 2.4 and 4.6 cm/year during that period. These preliminary observations suggest that the growth response found in the 1st year of therapy, might continue into the 2nd and later years of hGH treatment.

References

1. L. F. Soyka, A. Ziskind, and J. D. Crawford, Treatment of short stature in children and adolescents with human pituitary growth hormone (Raben), *N. Engl. J. Med.* **271**:754, 1964.
2. J. J. Hutchings, R. F. Escamilla, C. H. Li, and P. F. Forsham, Li human growth hormone administration in gonadal dysgenesis, *Am. J. Dis. Child.* **109**:318, 1965.
3. J. C. Wright, J. A. Brasel, T. Aceto, Jr., J. W. Finkelstein, F. M. Kenny, J. S. Spaulding, and R. M. Blizzard, Studies with human growth hormone (hGH), *Am. J. Med.* **38**:499, 1965.
4. M. Tzagouris, Response to long-term administration of human growth hormone in Turner's syndrome, *JAMA* **210**:2373, 1969.
5. J. M. Tanner, R. H. Whitehouse, P. C. R. Hughes, and F. P. Vince, Effect of human growth hormone treatment for 1 to 7 years on growth of 100 children, with growth hormone deficiency, low birth weight, inherited smallness, Turner's syndrome and other complaints, *Arch. Dis. Child.* **46**:745, 1971.
6. D. Rudman, M. Goldsmith, M. Kutner, and D. Blackston, Effect of growth hormone and oxandrolone singly and together on growth rate in girls with X chromosome abnormalities, *J. Pediatr.* **96**:132, 1980.
7. S. Raiti, E. Trias, L. Levitsky, and M. S. Grossman, Oxandrolone and human growth hormone, *Am. J. Dis. Child.* **126**:597, 1973.

9

Turner's Syndrome and Human Growth Hormone

Biochemical Studies

G. Van Vliet, D. M. Styne, S. L. Kaplan, and M. M. Grumbach

The mechanism responsible for short stature in patients with Turner's syndrome is unknown. Growth retardation in the syndrome of gonadal dysgenesis is often present *in utero*; growth velocity after infancy and early childhood is subnormal, and there is no pubertal growth spurt. Clinical incidence suggests that impaired growth is related to resistance of the growing cartilage to growth factors.[1,2] However, a case report of a patient with Turner's syndrome who developed a growth hormone (GH) producing tumor before epiphyseal fusion which was associated with growth acceleration[3] raises the possibility that pharmacological amounts of GH may override the putative cartilage resistance. By analogy with other syndromes of hormone resistance, one would expect increased concentrations of GH or of somatomedin in the circulation. Yet, GH levels after pharmacological stimulation[4] are normal in these patients; furthermore, plasma somatomedin activity[5] and immunoassayable somatomedin C (SMC)[6] are also normal. Rudman *et al.* reported a significant linear correlation between growth velocity and serum SMC concentration during GH treatment in five girls with Turner's syndrome[6]; the slope of the regression line was more shallow than in short normal children. They interpreted their findings as evidence for cartilage resistance to SMC. In cultured skin fibroblasts from patients with Turner's syndrome, Rosenfeld *et al.* found SMC receptors that were normal in number and affinity and normal enhancement of DNA synthesis and cell replication following SMC stimulation[7]; comparable studies in cartilage cells are not available.

G. Van Vliet, S. L. Kaplan, and M. M. Grumbach • Department of Pediatrics, University of California–San Francisco, San Francisco, California 94143. ***D. M. Styne*** • Department of Pediatrics, University of California–Davis, Davis, California 95616.

In the present study, we measured serum SMC concentrations before and after GH injections in 39 patients with Turner's syndrome in an attempt to elucidate the mechanism responsible for short stature in these patients and to explore possible usefulness of this procedure in selecting patients with gonadal dysgenesis for a trial on GH therapy. The effect of GH on glucose tolerance and thyroid function was also monitored.

1. Patients and Methods

1.1. Patients

Of the 50 patients enrolled in the Collaborative Growth Hormone Treatment Study of Turner's syndrome under the auspices of the National Hormone and Pituitary Program, NIADDK, we received samples for SMC determinations from 39 patients. For each of these patients, the following information was obtained: bone age, height velocity before and during a year of treatment with GH, oral glucose tolerance tests [1.75 g/kg body weight (BW)], and thyroid function tests. The samples for SMC determinations were obtained before and 24 hr after the last of four daily injections of GH 0.2 U/kg BW. The glucose tolerance tests after treatment were in general performed more than 48 hr after the last GH injection. Seven of these thirty-nine patients were followed in the Pediatric Endocrine Clinic of the University of California San Francisco.

1.2. SMC Radioimmunoassay

Serum SMC was measured after acid ethanol extraction of serum(8) as previously described.(9) The tracer used in the assay was either purified(10) or synthetic(11) SMC, obtained from Dr. R. E. Humbel (Zurich) or Dr. C. H. Li (San Francisco), respectively. The antiserum(12) was provided by Dr. L. E. Underwood and J. J. Van Wyk and distributed by the National Hormone and Pituitary Program, NIADDK. The results are expressed in reference to a pool of serum from healthy adults (Ortho #1770-5, Ortho Diagnostics, Raritan, New Jersey) to which a potency of 1 U/ml was assigned. One unit is equivalent to about 200 ng synthetic (Li 4378B) SMC.

1.3. Statistical Analysis

Linear correlation analysis and Student's t test (paired or unpaired) were used. Results are presented as means ± SEM.

2. Results

Table 1 shows the results of the GH stimulation tests before treatment. For all three tests used, the mean peak GH concentrations attained by the

Table 1 GH Response to Pharmacological Stimuli in the Syndrome of Gonadal Dysgenesis[a]

	Arginine	Insulin	L-Dopa
Mean peak GH (ng/ml)	11.3	15.5	15.3
Range (ng/ml)	0.9–18.0	7.6–35.5	6.3–34.5

[a] $n = 39$.

patients with Turner's syndrome were similar to those observed in normal children.[4,13,14] All of the patients had at least two different tests and achieved a peak GH level of 8 ng/ml or more during one test. Figure 1 shows the baseline serum SMC concentrations; these fell within the range that we[9] and others[15] have observed in prepubertal children of normal height. As in normal children,[15] the basal SMC concentrations increased progressively with age ($r = 0.62$, $p < 0.001$). Figure 2 shows the serum SMC concentrations achieved after four daily i.m. injections of 0.2 IU/kg BW of GH: again, these levels correlate positively with bone age ($r = 0.66$, $p < 0.001$).

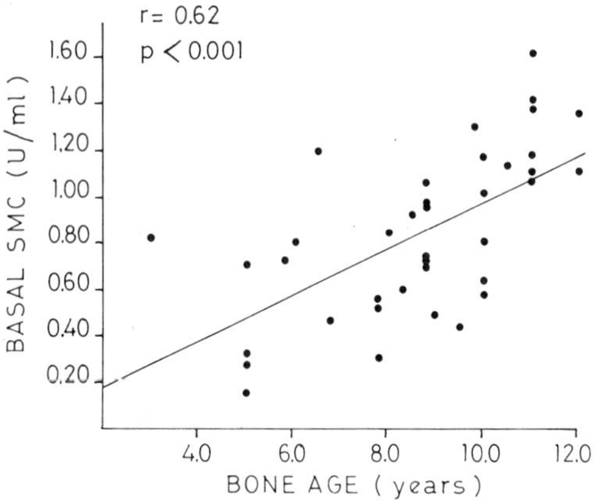

Figure 1 Correlation between basal serum somatomedin C (SMC) in units per milliliter (U/ml) and bone age in years.

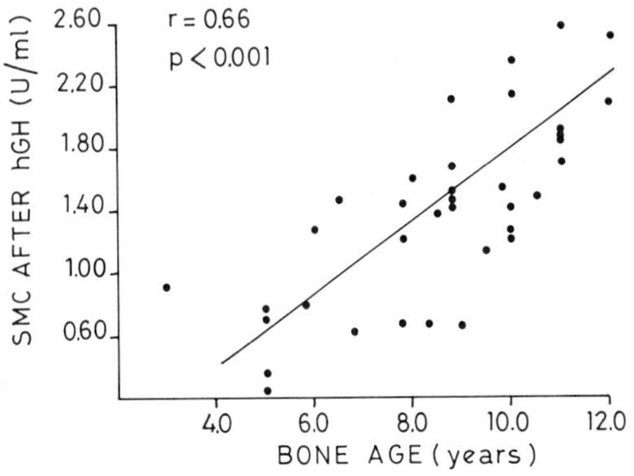

Figure 2 Correlation between serum somatomedin C (SMC) in U/ml after short serum growth hormone administration (see text) and bone age in years.

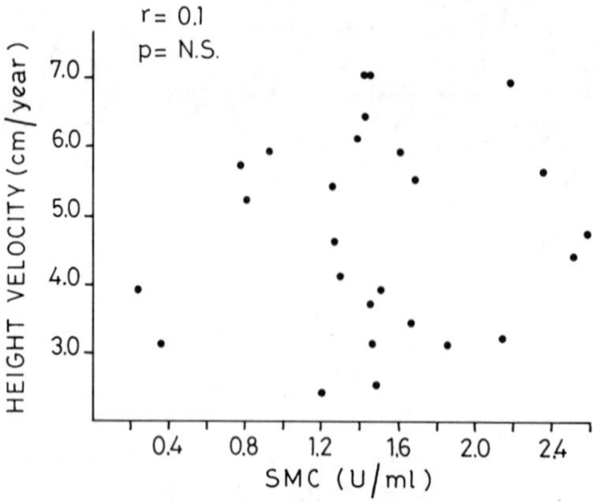

Figure 3 Lack of correlation between height velocity during a year of growth hormone (GH) treatment and serum SMC after short-term GH.

Figure 3 illustrates the lack of correlation between the height velocity observed during the year of treatment with GH and the serum SMC concentration obtained after the four daily injections of GH ($r = 0.1$, p not significant). The results of the oral glucose tolerance tests performed before and after 6 and 12 months of GH therapy are shown in Fig. 4. There was no significant difference in the results. Similarly, the glucose assimilation coefficient was not

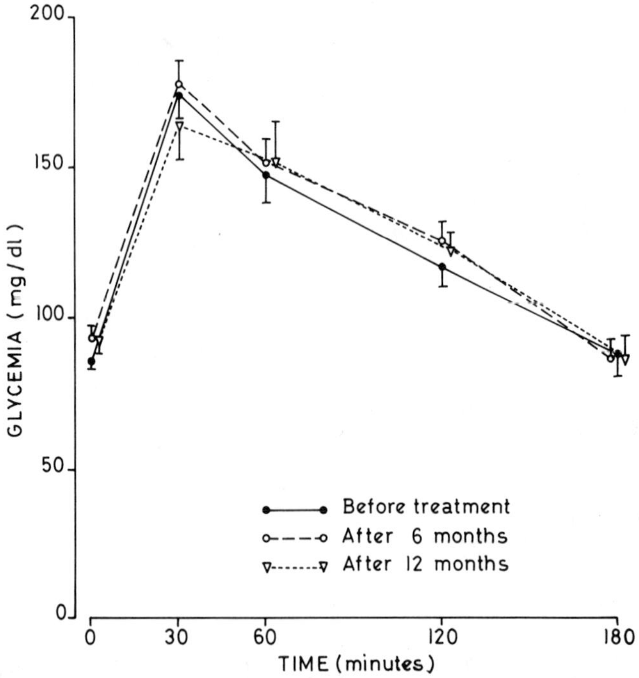

Figure 4 Oral glucose tolerance test before (closed circles) and after 6 (open circles) and 12 months (open triangles) of growth hormone treatment.

Table 2 Human GH Therapy in the Syndrome of Gonadal Dysgenesis: Thyroid Function Tests (Mean ± SEM)[a]

	Before treatment	After 6 months	After 12 months
T_4 (μg/dl)	9.2 ± 0.4	8.8 ± 0.5	8.6 ± 0.6
TSH (μg/ml)	5.1 ± 0.8	4.5 ± 0.4	5.0 ± 1.3
T_3 resin uptake (%)	29 ± 0.8	27 ± 1.6	NA

[a] NA = not available.

altered by GH treatment (1.83 ± 0.09 before treatment versus 1.65 ± 0.30 after 12 months of GH treatment).

The lack of change in thyroid function during GH treatment is illustrated in Table 2. The values for serum thyroxine, serum TSH, and T_3 resin uptake test were not altered by hGH treatment.

3. Discussion

The normal serum GH concentrations after all three pharmacological stimuli (Table 1) confirm our earlier observation[4] of normal stimulated levels of GH after an insulin tolerance test in patients with the syndrome of gonadal dysgenesis. This finding suggests that the association of Turner's syndrome and GH deficiency[16] is rare and does not account for the slow growth rate in the syndrome.

Baseline serum SMC concentrations have been reported to be normal in Turner's syndrome.[6] We found that SMC levels increased with age during childhood in these patients as in normal children (Fig. 1). The positive correlation between the SMC concentration achieved after four injections of GH and bone age (Fig. 2) is also similar to that observed in short normal children and in hypopituitary children. Interestingly, the mean increment in serum SMC attained after four daily injections of 0.2 U/kg BW of GH is similar in girls with Turner's syndrome and in hypopituitary and short normal children[9] (0.66 ± 0.09, 0.64 ± 0.19, and 0.57 ± 0.16 U/ml, respectively). This indicates that there is no abnormality in the generation of circulating SMC in Turner's syndrome although it does not exclude the possibility that circulating SMC may not reflect the tissue concentrations of the peptide. Indeed, it has recently been shown that SMC is synthesized by multiple tissues[17] (possibly including cartilage[18]). This also explains the lack of correlation between circulating SMC after the administration of GH and the growth velocity on treatment (Fig. 4). Such a discrepancy between changes in growth rate and in serum SMC has now been extensively documented during GH treatment of hypopituitary[9,19,20] and short normal children[9,21] and during estrogen treatment of girls with Turner's syndrome.[22]

A discrepancy between abnormal oral glucose tolerance and normal intravenous glucose tolerance has been reported in Turner's syndrome.[23] Our results show that 1 year of GH treatment had no adverse side effects on either oral or intravenous glucose tolerance in these patients, although the dose of human GH was twice that usually used in the treatment of hypopituitarism.[9]

The prevalence of autoimmune thyroiditis in the syndrome of gonadal dysgenesis is increased. Furthermore, development of hypothyroidism affects the response to GH treatment[24] in hypopituitarism. In our patients thyroid function was normal and was not affected by treatment with GH.

In summary, we found normal serum GH and SMC concentrations in a group of patients with Turner's syndrome. Both basal and GH-stimulated serum SMC concentrations correlated positively with bone age but not with the growth response to GH. The mean increment in serum SMC after GH was similar to that observed in hypopituitary and short normal children. Glucose tolerance and thyroid function tests remained normal during a year of treatment with GH.

The cause of growth retardation in Turner's syndrome is unknown, but it is not related to an abnormality in GH-releasing factor–pituitary GH–somatomedin C cascade.

We suspect that the abnormality resides in the response of the chondrocyte to somatomedin and/or other growth factors. The measurement of circulating SMC is not an enlightening approach to unraveling the underlying cellular abnormality.

ACKNOWLEDGMENT. The endocrinologists who provided the patient data described are listed in Chapter 8.

References

1. M. M. Grumbach and F. A. Conte, Disorders of sex differentiation, in: *Textbook of Endocrinology* (R. H. Williams, ed.) W. B. Saunders, Philadelphia, 1981, pp. 423–514.
2. C. G. D. Brook, G. Murset, M. Zachmann, and A. Prader, Growth in children with 45,XO Turner's syndrome, *Arch. Dis. Child.* **49**:789, 1974.
3. C. H. Willemse, A patient suffering from Turner's syndrome and acromegaly, *Acta Endocrinol.* **39**:204, 1962.
4. S. L. Kaplan, C. A. L. Abrams, J. J. Bell, F. A. Conte, and M. M. Grumbach, Growth and growth hormone. I. Changes in serum levels of growth hormone following hypoglycemia in 134 children with growth retardation, *Pediatr. Res.* **2**:43, 1968.
5. P. Saenger, E. Schwartz, E. Wiedemann, L. S. Levine, M. Tsai, and M. I. New, The interaction of growth hormone, somatomedin and estrogen in patients with Turner's syndrome. *Acta Endocrinol.(Kbh)* **81**:9, 1976.
6. D. Rudman, S. D. Moffit, P. M. Fernhoff, W. J. McKenzie, J. M. Kenny and R. P. Bain, The relation between growth velocity and serum somatomedin C concentration, *J. Clin. Endocrinol. Metab.* **52**:622, 1981.
7. R. G. Rosenfeld, L. A. Dollar, R. L. Hintz, and C. A. Conover, Demonstration of normal somatomedin C/insulin like growth factor I binding and action in cultured human fibroblasts from Turner syndrom, *Acta Endocrinol.(Kbh)* **104**:502, 1983.
8. W. H. Daughaday, I. K. Mariz, and S. L. Blethen, Inhibition of access of bound somatomedin to membrane receptors and immunobinding sites: A comparison of radioreceptor and radioim-

munoassay of somatomedin in native and acid ethanol extracted serum, *J. Clin. Endocrinol. Metab.* **51:**781, 1980.
9. G. Van Vliet, D. M. Styne, S. L. Kaplan, and M. M. Grumbach, Growth hormone treatment for short stature, *N. Engl. J. Med.* **309:**1016, 1983.
10. E. Rinderknecht and R. E. Humbel, The aminoacid sequence of human insulin-like growth factor I and its structural homology with proinsulin, *J. Biol. Chem.* **253:**276, 1978.
11. C. H. Li, D. Yamashiro, D. Gospodarowicz, S. L. Kaplan, and G. Van Vliet, Total synthesis of insulinlike growth factor I (somatomedin C), *Proc. Natl. Acad. Sci. USA*, **80:**2216, 1983.
12. R. W. Furlanetto, L. E. Underwood, J. J. Van Wyk, and A. J. D'Ercole, Estimation of somatomedin C levels in normals and patients with pituitary disease by radioimmunoassay, *J. Clin. Invest.* **60:**648, 1977.
13. R. Youlton, S. L. Kaplan, and M. M. Grumbach, Growth and growth hormone. IV. Limitations of the growth hormone response to insulin and arginine and of the immunoreactive insulin response to arginine in the assessment of growth hormone deficiency in children, *Pediatrics* **43:**989, 1983.
14. R. Collu, G. Leboeuf, J. Letarte, and J. R. Ducharme, Stimulation of growth hormone secretion by Levodopa propanolol in children and adolescents, *Pediatrics* **56:**262, 1975.
15. R. M. Bala, J. Lopatka, A. Leung, E. McCoy, and R. G. McArthur, Serum immunoreactive somatomedin levels in normal adults, pregnant women at term, children at various ages and children with constitutionally delayed growth, *J. Clin. Endocrinol. Metab.* **52:**508, 1981.
16. C. G. D. Brook, Growth hormone deficiency in Turner's syndrome, *N. Engl. J. Med.* **298:**1203, 1978.
17. A. J. d'Ercole, G. T. Applewhite, and L. E. Underwood, Evidence that somatomedin is synthesized by multiple tissues in the fetus, *Dev. Biol.* **75:**315, 1980.
18. O. G. P. Isaksson, J. O. Jansson, and I. A. M. Gause, Growth hormone stimulates longitudinal bone growth directly, *Science* **216:**1237, 1983.
19. R. G. Rosenfeld, S. F. Kemp, and R. L. Hintz, Constancy of somatomedin to growth hormone to growth hormone treatment of hypopituitary dwarfism, and lack of correlation with growth rate, *J. Clin. Endocrinol. Metab.* **53:**611, 1981.
20. H. J. Dean, J. G. Kellett, R. M. Bala, J. H. Guyda, B. Bhaumick, B. I. Posner, and H. G. Friesen, The effect of growth hormone treatment on somatomedin levels in growth hormone deficient children, *J. Clin. Endocrinol. Metab.* **55:**1167, 1982.
21. L. P. Plotnick, Q. L. Van Meter, and A. A. Kowarski, Human growth hormone treatment of children with growth failure and normal growth hormone levels by immunoassay: Lack of correlation with somatomedin generation, *Pediatrics* **71:**324, 1983.
22. J. Levine Ross, F. G. Cassorla, M. C. Skerda, I. M. Valk, D. L. Loriaux, and G. B. Cutler, A preliminary study of the effect of estrogen dose on growth in Turner's syndrome, *N. Engl. J. Med.* **309:**1104, 1983.
23. E. Rasio, A. Antaki, and J. Van Campenhout, Diabetes mellitus in gonadal dysgenesis: studies of insulin and growth hormone secretion, *Eur. J. Clin. Invest.* **6:**59, 1976.
24. S. D. Frasier, Human pituitary growth hormone (hGH) therapy in growth hormone deficiency, *Endocr. Rev.* **4:**155, 1983.

10

Short Normal Children in Therapeutic Trial of Human Growth Hormone Responsiveness

Salvatore Raiti and the Committee on Growth Hormone Usage (S. L. Kaplan, G. P. August, S. A. Kaplan, M. H. MacGillivray, W. V. Moore, J. S. Parks, A. F. Parlow, A. W. Root, and A. E. Wilhelmi)

This project was designed to evaluate the effectiveness of human growth hormone (hGH) in children with short stature who are growing at subnormal rates. Although it has previously been believed that such patients do not respond to hGH therapy, more recent data suggest that this might not be correct.

1. Study Plan

An attempt was made to study 100 short children who met the criteria described later. To our surprise, we have found only about 50 such children to date, although we advertised this widely on a national basis.

The criteria for selection are given in Table 1. The age group of 4 up to 10 years was selected in order to avoid problems with expected faster growth rates in the earlier years of life. We also wanted to avoid the pubertal years and interference with interpretation of data by the pubertal growth spurt. Very few patients could be identified in this wide age range of 4–10 years.

A growth rate of 4 cm/year or less was required as determined from height

---

Salvatore Raiti and the Committee on Growth Hormone Usage (S. L. Kaplan, G. P. August, S. A. Kaplan, M. H. MacGillivray, W. V. Moore, J. S. Parks, A. F. Parlow, A. W. Root and A. E. Wilhelmi) • The National Hormone and Pituitary Program, University of Maryland School of Medicine, Baltimore, Maryland 21201.

Table 1 Short Normal, Criteria for Acceptance[a]

1. Age 4–10 years
2. Short (<1st percentile)
3. No puberty
4. Growth <4.0 cm/year
5. Bone age, delay of 1 year or more
6. Skull X-ray, normal
7. Thyroid and adrenal, normal
8. hGH tests, normal response
9. No previous therapy

[a] hGH = human growth hormone.

measurements for the previous 12 months. A minimum bone age delay of 1 year or more was required. Normal endocrine functions in terms of thyroid, adrenal, and growth hormone status were required. Table 2 lists the tests needed prior to entry into the study.

The selected hGH dose was 0.1 IU/kg and was given three times a week (Table 3). The patients received hGH for 6 months and were then off hGH for 6 months so as to serve as their own controls. The patients' heights were measured at 2-month intervals. Skeletal age and thyroid function were measured at 6-month intervals.

After the 1st year, this study plan was repeated. It is planned to continue this into the 3rd and possibly subsequent years. The patients who qualified for a 2nd year were those who showed a significant growth response to hGH in

Table 2 Short Normal, Tests Required[a]

1. hGH, two definitive tests
2. T_4 and TSH
3. Bone age
4. RRA
5. Somatomedin generation, 0.2 IU/kg per day × 4 days

[a] hGH = human growth hormone; TSH = thyroid-stimulating hormone; RRA = radio receptive assay.

Table 3 Small Normal, hGH Dose and Follow-Up[a]

1. hGH, 0.1 IU/kg on Monday, Wednesday, and Friday, 6 months *on*, then 6 months *off*
2. Height at 2, 4, 6, 8, 10, and 12 months
3. Bone age at 6 and 12 months
4. T_4 and TSH at 6 and 12 months

[a] hGH = human growth hormone; TSH = thyroid-stimulating hormone.

the 1st year *and* whose growth rate during the 6 months off therapy dropped to approximate pretreatment rates.

2. Results

Only preliminary results can be discussed at this time. The mean pretreatment growth rate for the 33 patients accepted into the study at this time was 3.48 cm/year. All had a growth rate of 4.0 cm/year or less. Not all patients responded to hGH. The mean growth response of the 21 patients who showed acceleration was 3.26 cm in 6 months, or 6.56 cm/year. In the posttherapy period, 18 patients (to date) showed a mean growth response of 2.98 cm/6 months, or 4.16 cm/year.

The completed data on the 1st year and on subsequent years will be reported at a future time.

3. Discussion

When this study was first advertised, it was anticipated that the requests would be too numerous to handle. It was decided to limit the study to 100 patients.

In practice, it took almost 1 year to collect 33 patients who could meet the criteria selected. These criteria were not very restrictive. The patients sought were those who were short, who were growing at 4 cm/year or less, who had a bone age delay of 1 year or more, and who had normal endocrine functions. The age group was restrictive in that only patients in the prepubertal years were accepted. Every program in the United States that uses hGH was invited to submit such patients. The minimal number of patients collected over a long period of time indicates that there are, in fact, very few patients who meet these criteria, the incidence being far less than for hypopituitarism in the same age group.

Numerous investigators have indicated that they each have many short normal patients in the pubertal age range. If such numbers cannot be found in the prepubertal years, then one must wonder whether they are dealing with a different problem. The most likely diagnosis in such patients in the pubertal years is constitutional delay in sexual development. This is a separate diagnostic group and must be studied separately. Their growth response from pubertal hormones must be dissected from the response that can be directly attributable to hGH therapy.

The results of the present study show that for the age group selected, only about two-thirds showed an accelerated growth response during the first 6 months of therapy. Some of these continued to grow at this accelerated growth rate when hGH was stopped, suggesting a transient growth disorder prior to study. This observation is of importance because if hGH therapy is to be considered, it must be interrupted by periods of 4–6 months off therapy from

Table 4 Small Normal Children[a]

Study period	Number	Growth rates cm/year	cm/6 months	cm/year (mean)
Before	33	3.48 ± 0.70		3.48
1st 6 months on hGH	21		3.26 ± 0.86	6.56
2nd 6 months off hGH	18		2.08 ± 1.09	4.16
3rd 6 months on hGH				
4th 6 months off hGh				

[a] hGH = human growth hormone.

time to time, in order to determine that hGH is still needed. The extra mean growth response for this group was double the pretreatment growth rate, similar to the observations in patients with Turner's syndrome (described in this volume, Chapter 8). The growth response is far short of what is observed in the hypopituitary patient during the first 6 months or year of therapy.

The mean growth response for the 18 patients after hGH was stopped (Table 4) shows a slightly better mean growth than for the pretreatment period.

4. Conclusions

This report must be considered as a preliminary one. It shows that the incidence of short normal children with slow growth rates and a 1-year delay in bone age prior to 10 years of age is extremely small. It shows that about half of such patients can be expected to respond with a doubling of their growth rate. It shows that some of these patients will continue to grow at the new accelerated growth rate once the hGH therapy is stopped. It suggests that the vast majority of patients with so-called "normal variant" short stature and bioinactive forms of circulating GH must be extremely small. It suggests that the so-called "therapeutic trial" of hGH for short stature should be limited to patients who are growing at less than 4.0–4.5 cm/year and who have a bone age delay of 1 year or more.

11

Bioinactive Growth Hormone and Secondary Growth Hormone Deficiency

A. Avinoam Kowarski

It is now well established that a significant number of growth-impaired children respond to treatment with replacement doses of growth hormone (GH).[1–4] These patients, who were formerly thought to have either constitutional delay or familial short stature, are presently included in a new syndrome called the "normal variant short stature" or "GH-dependent growth failure." I am going to present some of our old data and our most recent results which lead us to conclude that we are not dealing with a single syndrome but with a group of distinct diagnostic entities that will require the use of new diagnostic tools.

1. Secondary GH Deficiency

Figure 1 was taken from a paper we published in 1979,[5] which reports our experience with the exercise screening test for GH deficiency. The upper panel depicts the distribution of the response of children who were subsequently found to have no GH deficiency. The lower panel depicts maximum response of a group of children who were subsequently found to be GH-deficient. As can be seen, most of the GH-deficient children were unable to respond to the exercise test with a rise in GH level above 10 ng/ml. However, three of the GH-deficient patients had a normal GH response. These three children were treated with GH despite their normal GH response to exercise. They were diagnosed as suffering from GH deficiency mostly because of their response to the treatment. This result led us to investigate the possibility that stimulating tests for GH do not always reflect the actual secretion of GH. We decided to use a

A. Avinoam Kowarski • Department of Pediatrics, University of Maryland School of Medicine, Baltimore, Maryland 21201.

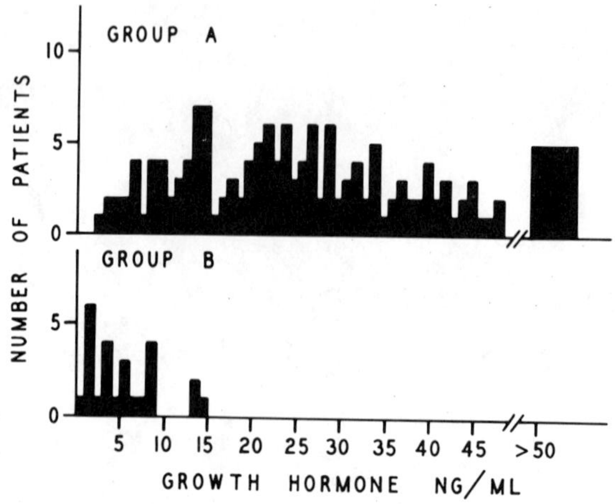

Figure 1 Experience with the exercise screening test for growth hormone (GH) deficiency. The upper panel depicts the distribution of the maximal GH response to exercise of children who have no GH deficiency. The lower panel depicts maximum GH response to exercise of children who are GH-deficient.

new tool, our nonthrombogenic continuous blood withdrawal system, which enabled us to see the entire secretory pattern of GH during a 24-hr period.

Figure 2 describes the 24-hr integrated concentration of GH (ICGH) secretory pattern of two children who were unresponsive to the arginine–insulin stimulation test (AITT). The two panels represent the two opposite extreme results in patients suffering from GH deficiency. Both children were unable to respond to GH stimulation (AITT) and both had a lower-than-normal ICGH. Yet the child depicted in the upper panel never had a peak ICGH above 3 ng/ml, while the ICGH of the child depicted in the lower panel had a peak just above 10 ng/ml.

Figure 3 describes the 24-hr secretory pattern of GH in two of our patients

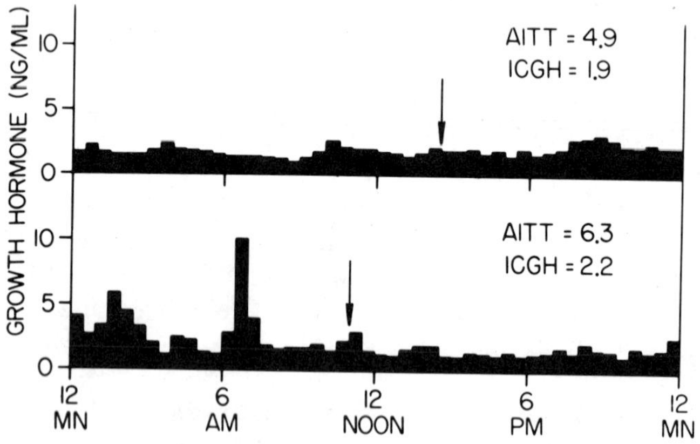

Figure 2 The 24-hr integrated concentration of growth hormone (ICGH) secretory pattern of two children who were unresponsive to the arginine–insulin stimulation test (AITT). The two panels represent the two opposing extremes of the spectrum of the ICGH found in patients suffering from GH deficiency.

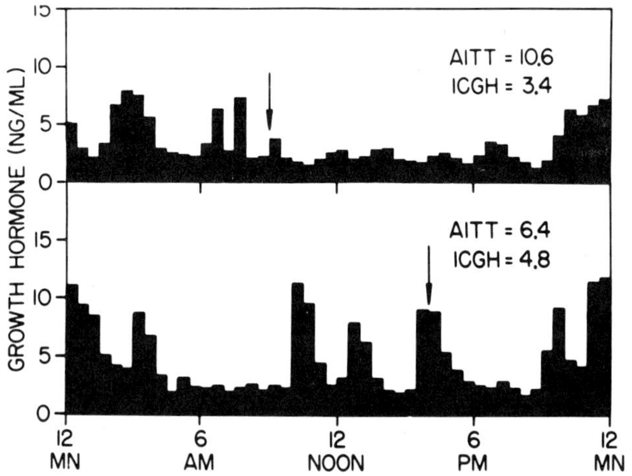

Figure 3 The 24-hr secretory pattern of growth hormone (GH) in two patients who exhibited a discrepancy between the arginine–insulin stimulation test (AITT) and the integrated concentration of growth hormone (ICGH). In the lower panel is the GH secretory pattern of a child who was GH-deficient by the AITT but had a normal ICGH. In the upper panel is a child who had a normal response to AITT, yet had a subnormal ICGH.

who exhibited a discrepancy between the AITT and the ICGH. The lower panel represents the GH secretory pattern of a child who was GH-deficient by the AITT but had a normal ICGH. The upper panel shows the reverse situation, a child who had a normal response to AITT, yet had a subnormal ICGH. These findings convinced us that the results of the stimulation tests do not always reflect the actual secretion of GH. We therefore proceeded to measure the ICGH in a large number of subjects.

Figure 4 depicts unpublished data from our ongoing studies. On the left side the dots represent the ICGH of children of normal height and weight (NS) who came to us for workup of short stature. The center dots represent children who had a normal response to AITT (NGH). The dots on the right side represent the ICGH of 13 GH-deficient children (GHD). As can be seen, the children with classical GH deficiency had lower ICGH, but there were 19 children whose ICGH was similar to that of the children with GH deficiency.

The question we asked ourselves at this point was: which of the two tests is the best predictor of response to GH therapy? In order to answer this question, we decided to treat a few of the children whose 24-hr ICGH was at the hypopituitary range even though the AITT was normal. We obtained permission to treat 10 of these children. At the present time, we are treating eight children. Four of these children, who had a subnormal growth rate, have been on therapy for more than 4 months; three of them have more than doubled their growth rate, and the remaining child has increased his growth rate from 3.2 to 5.4 cm/year.

B. E. Spiliotis, G. August, W. Hung, and B. B. Bercu reported at the 1983

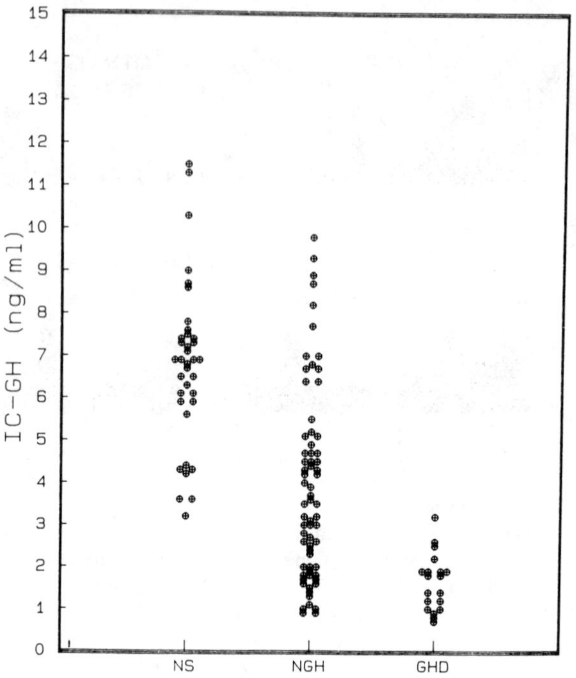

Figure 4 Comparison of individual IC–GH levels in all subjects grouped according to normal stature (NS), short with normal GH response to stimulation (NGH), and short with deficient GH response to stimulation (GHD).

meeting of the Endocrine Society that some patients with short stature have an "intermittent irregularity in GH secretion, resulting in either reduced number and/or amplitude of pulses, not readily identified with provocative tests [and that] hGH treatment will significantly increase the growth rate in most of these patients." Poster presentations at this meeting by Drs. K. Albertsson-Wikland *et al.* and Spiliotis *et al.* also include results that are in agreement with ours.

We conclude that some children have inadequate daily secretion of GH even though their pituitary's ability to release GH in response to stimulation is intact. The growth impairment of these children is therefore due to a secondary defect in the complex regulatory mechanism above the pituitary gland.

2. *Variant Structures of GH in the Peripheral Circulation and the Bioinactive GH Syndromes*

Since the amino acid sequence of GH is genetically determined, we might expect that mutations of the gene for GH would be manifested by secretion of GH with a slightly altered structure. A number of different mutations may actually occur, each manifested by a different alteration of the GH molecule.

Some structural and functional abnormalities in GH may remain undetected by the presently available assay techniques. GH concentration is measured by either a radioimmunoassay or a radioreceptor assay. Both methods depend on the affinity of the GH molecule to specific binding sites. The radioimmunoassay requires binding to an antibody. The radioreceptor assay requires binding to plasma membrane taken from an appropriate GH target tissue. The specificity of the methods depend on the specificity of the binding sites. Any molecule that competes with GH for the binding sites is interpreted by these methods as GH. Consequently, any molecule that alters the affinity of the binding sites to GH will also alter the result of the tests. Since the antibodies currently used in the radioimmunoassay bind at multiple sites on the GH molecule, it is reasonable to expect that some of the mutated GH will be measured by the immunoassay even though its biological activity is reduced or abolished. Since biologically inactive GH may still bind to target tissues, the radioreceptor assay is also fallible.

In 1978 we described, in collaboration with Daughaday's group, two cases of growth failure who responded to GH therapy.(1) These children exhibited all the symptoms of GH deficiency even though their plasma level of GH was normal. The two patients responded to GH supplied by the National Pituitary Agency, but were unresponsive to their own GH. Consequently, we postulated that the GH secreted by the two children was biologically inactive. We also speculated that the biological inactivity is due to a variation in the structure of the GH secreted by these children. Since similar variability is exhibited by many other natural polypeptides, we should expect some structural variability of the GH secreted by all individuals. Indeed, Lewis and his collaborators(7) have found, in the RIA grade of GH, several electrophoretic variants of GH. The regular GH size is 22,000 daltons (22K). One of the variants reported by Lewis was a 20K molecule which is encoded by a separate species of mRNA. The 20K variant of GH resembles the 22K GH except for a missing 2K chain. The 20K variant was found to have growth-promoting bioactivity in rats, as well as a prolactin bioactivity. But the 20K GH variant did not compete with 22K GH in the GH radioreceptor assay.

Dr. Agarwal had raised in our laboratory a large number of monoclonal antibodies to GH. We have selected one of these monoclonal antibodies, which bound to the 22K regular GH but showed no binding to the 20K variant of GH. This antibody has its binding site (epitope) at the area of the regular GH that is missing in the 20K variant. An immunoassay based on this monoclonal antibody would not detect any molecule that does not include this epitope. The established immunoassay for GH which is based on a polyclonal antibody measures the 20K and the 22K variants. The difference between the level of GH as measured by the two methods would represent a fraction of the patient's GH that does not have the specific epitope.

We have compared the GH level by the two methods in a group of children who have been treated with GH. The results of these studies are depicted in Fig. 5. Only in two of the children was the ratio between the level measured by the two methods equal to 1, indicating that all of their GH contained the

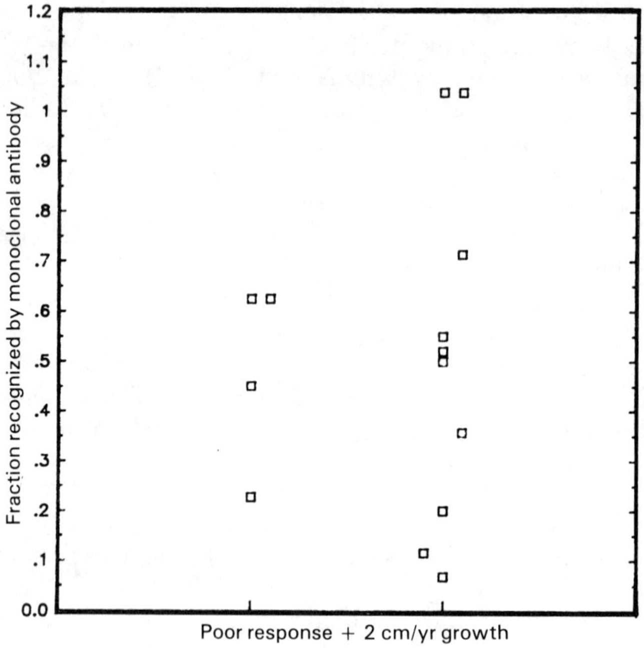

Figure 5 The concentration of growth hormone (GH) in plasma of children who did not respond to GH therapy (left column) and those who responded to GH therapy increasing their growth rate by more than 2 cm/year (right column) was measured by two immunoassays. One of the immunoassays was based on a monoclonal antibody that does not recognize the 20K variant. The second immunoassay was based on a polyclonal antibody that recognizes both the 20K and the regular 22K GH. The difference between the level of GH as measured by the two methods represents a variant structure GH. The results are represented as the fraction of GH recognized by immunoassay based on the monoclonal antibody.

epitope. In the remaining patients, a fraction of the GH did not contain the epitope. Since we could not find a difference between the patients who responded to GH therapy and those who did not, the antibody we have selected would not differentiate between the bioactive and bioinactive GH. It is expected, however, that a different monoclonal antibody may enable us to identify variants of GH that are biologically inactive.

Blethen and Chasalow[8] have carried out a similar study using the Tandem GH kit. The Tandem kit for immunoassay of GH is based on two monoclonal antibodies. They reported that the ratio between the GH measured by the Tandem and the polyclonal immunoassays may serve as a predictor of responsiveness to GH therapy. We have independently carried out a similar study using the same Tandem GH monoclonal immunoassay kit and our regular polyclonal GH immunoassay. We also discovered a large individual variability of the GH ratio in normal adults and an overlap of the GH ratio with the GH ratio in the patients. However, unlike Blethen and Chasalow, we have found no correlation between the Tandem ratio and the responsiveness to GH therapy.

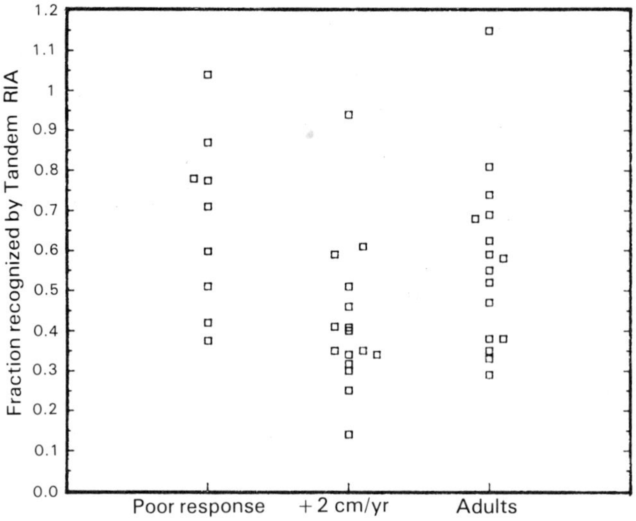

Figure 6 The concentration of growth hormone (GH) in plasma of children who did not respond to GH therapy (left column) and those who responded to GH therapy increasing their growth rate by more than 2 cm/year (center column) and of normal adults (right column) was measured by two immunoassays. One was the Tandem immunoassay, which is based on two monoclonal antibodies that do not recognize the 20K variant. The second immunoassay was based on a polyclonal antibody that recognizes both the 20K and the regular 22K GH. The difference between the level of GH as measured by the two methods represents a variant structure GH. The results are represented as the fraction of GH recognized by immunoassay based on the monoclonal antibody.

Our results are displayed in Fig. 6. Since we have also found that the Tandem assay is insensitive to the 20K variant of GH, these results are in agreement with our previous studies. It should be stressed that the immunoassay based on our monoclonal antibody and the Tandem immunoassay do not measure the same epitope since we have found no significant correlation between the measurement by the two assays on the same plasma samples.

Our studies with monoclonal antibodies indicate that circulating GH is a variable mixture of several structural variants. A significant percentage of the GH secreted by patients and normal adults has an altered structure since it is immunoassayable by the polyclonal antibody but is not measured by two immunoassays based on monoclonal antibodies to an epitope missing in the 20K variant of GH.

3. Conclusion

We have presented data that support the existence of two distinct causes for the GH-dependent growth failure syndrome. (1) Secondary GH deficiency is implied by responsiveness to GH therapy in patients who have a normal response to GH-provocative tests, but low ICGH. (2) Significant structural variability of the circulating GH was revealed through the use of appropriately selected monoclonal antibodies. Although the structural variants of GH that have been described to date are biologically active, their existence lends support to the presence of the putative bioinactive structural variants.

ACKNOWLEDGMENT. The studies presented in this chapter were carried out in collaboration with a large number of investigators. I cannot hope to name all of them but I would like to note specifically the contribution of Drs. Zvi Zadik, Stuart Chalew, Judith McLaughlin, Leslie Plotnick, Peter Lee, Claude Migeon, and Salvatore Raiti. Dr. Sudha Agarwal carried out the studies with the monoclonal antibodies. Dr. U. J. Lewis contributed a sample of 20K variant of GH.

References

1. A. A. Kowarski, J. Schneider, E. Ben-Galim, V. V. Weldon, and W. H. Daughaday, Growth failure with normal serum RIA-GH and low somatomedin activity: Somatomedin restoration and growth, *J. Pediatr.* **47**:461, 1978.
2. D. Rudman, M. H. Kutner, M. A. Goldsmith, J. Kenny, H. Jennings, and R. P. Bain, Further observations on four subgroups of normal variant short stature, *J. Clin. Endocrinol. Metab.* **51**:1378, 1980.
3. T. E. Frazer, J. R. Gavin, W. H. Daughaday, R. E. Hillman, and V. V. Weldon, Growth hormone-dependent growth failure, *J. Pediatr.* **101**:12, 1982.
4. L. Plotnick, Q. L. Van Meter, and A. A. Kowarski, Human growth hormone treatment of children with growth failure and normal growth hormone levels by immunoassay: lack of correlation with somatomedin generation, *Pediatrics* **71**:324, 1983.
5. E. Eisenstein, L. Plotnick, R. Lanes, P. A. Lee, C. J. Migeon, and A. A. Kowarski, Evaluation of growth hormone exercise test in normal and growth hormone deficient children, *Pediatrics* **62**:526, 1978.
6. L. P. Plotnick, P. A. Lee, C. J. Migeon, and A. A. Kowarski, Comparison of physiologic and pharmacologic tests for growth hormone function in children with short stature, *J. Clin. Endocrinol. Metab.* **48**:811, 1979.
7. U. J. Lewis, J. T. Dunn, L. F. Bonewald, B. K. Seavey, and W. P. VanderLann, A naturally occurring structural variant of human growth hormone, *J. Biol. Chem.* **253**:2679, 1978.
8. S. L. Blethen and F. I. Chasalow, Use of a two-site immunoradiometric assay for growth hormone (GH) in identifying children with GH-dependent growth failure, *J. Clin. Endocrinol. Metab.* **57**:1031, 1983.

12

The Hyposomatomedinemic Short Child

Daniel Rudman, Rajender K. Chawla, William P. Heath, Carol J. Berry, and Michael H. Kutner

1. Introduction

Until 1978, the usual way of identifying growth-hormone-deficient (GHD) children in the short population was to measure the serum level of immunoreactive growth hormone (GH) during provocative tests, after exercising or during sleep. An alternative approach became possible in 1978 with the introduction of the radioimmunoassay (RIA) for somatomedin C (SMC).[1] When this assay was applied on a large scale to the population of short children, it became apparent that there were substantially more short children with low SMC levels than there were GHD children.[2] Furthermore, some of the hyposomatomedinemic boys and girls who had normal levels of immunoreactive GH responded with accelerated linear growth to long-term human growth hormone (hGH) treatment.[3–5] It was suggested, therefore, that some of these responsive, non-GHD short children might have a structurally abnormal form of endogenous GH.[3–6]

In light of this previous work, we undertook the present study with the following objectives:

(1) To estimate normal range for plasma SMC in children of normal height, 7–10 years old.

(2) To determine the prevalence of subnormal SMC in the population of short (<3rd percentile) children, 7–10 years old.

(3) In the hyposomatomedinemic short children, aged 7–10, to determine what proportion are GHD and what proportion are "non-GHD."

Daniel Rudman, Rajender K. Chawla, William P. Heath, and Carol J. Berry • Department of Medicine and Clinical Research Facility, Emory University School of Medicine, Atlanta, Georgia 30322. ***Michael H. Kutner*** • Department of Biometry and Clinical Research Facility, Emory University School of Medicine, Altanta, Georgia 30322.

(4) To compare the linear growth responses to hGH treatment of hyposomatomedinemic, GHD short children with hyposomatomedinemic, non-GHD short children.
(5) To compare the RIA profile of the endogenous GH in normal children, with hyposomatomedinemic non-GHD short children, using newly available monoclonal RIA techniques as well as the conventional polyclonal assay.

2. Methods

The study proceeded in three phases.

In Phase 1 we determined the prediction limits of plasma SMC in healthy children, 7–10 years old, of normal stature (5th to 95th percentile for chronological age).

In Phase 2 we measured plasma SMC in 97 children, aged 7–10, who were below the 3rd percentile in height for chronological age. We selected those whose SMC was below the mean for chronological age by at least 2 SD. These children received a clinical workup in order to classify them as GHD or non-GHD. Both groups then received 6-month courses of hGH at standard doses, to determine the linear growth responsiveness to the hormone.

In Phase 3 the immunochemical and size properties of the nocturnal serum GH were compared in healthy children of normal stature and in hyposomatomedinemic, non-GHD children.

2.1. Phase 1

For each yearly interval between 7 and 10 years, 14 or 15 healthy prepubertal children were studied. They were referred by pediatricians in greater Atlanta if they met these criteria: (1) height between the 5th and 95th percentile; (2) body weight between 95 and 115% of ideal; (3) no known chronic medical condition; (4) no drug treatment during the period of study; (5) normal history and physical examination within 6 months before the beginning of the study; and (6) normal hemogram and urinalysis within 6 months before the beginning of the study.

In addition, seven prepubertal GH-deficient children were also studied. Their clinical data are summarized in Table 1. Criteria for diagnosing deficiency of GH and other pituitary hormones have been given before.[4] Deficiencies of ACTH, thyroid-stimulating hormone, or antidiuretic hormone, when present, were treated as described elsewhere.[4] Human GH had not been given for at least 3 months.

Blood for SMC was drawn in EDTA at 1–4 P.M. The blood was spun immediately at room temperature for 10 mins; the plasma was then promptly stored at −20°C and mailed in dry ice within 2 weeks to Nichols Institute, Los Angeles, for analysis by RIA.[1] Intra- and interassay coefficients of variation

Table 1 Clinical Data of the GHD Subjects (Phase 1)

Case	Cause of GHD	Age (yr)	Sex	Ht (cm)	Wt (kg)	Bone age (yr)	Hormone deficiency
1	Idiopathic	7	F	103	20	4	GH
2	Carniopharyngioma	10	M	125	27	7	GH, ACTH, TSH, ADH[a]
3	Idiopathic	8	F	109	18	5	GH, ACTH
4	Idiopathic	7	F	100	21	4	GH
5	Idiopathic	7	M	106	22	4	GH, TSH
6	Idiopathic	8	F	112	24	5	GH, TSH
7	Craniopharyngioma	8	M	115	19	6	GH, ACTH, TSH, ADH

[a] GH = growth hormone; TSH = thyroid-stimulating hormone; ADH = antidiuretic hormone.

Table 2 Clinical Evaluation of the Short Child (Phase 2)

History:	Duration of gestation, weight at birth, parental short stature (one or both parents below the 3rd percentile in height), socioeconomic background
Physical examination:	Full examination including funduscopic exam and visual field testing
Laboratory:	Hemogram, urinalysis, SMA-18, serum Ca^{2+}, T_4, free T_4, 8 A.M. fasting cortisol, metyrapone test, 2-hr postprandial blood glucose
X ray:	Wrist X ray for bone age, skull X ray for sella turcica evaluation
GH evaluation:	Serum GH determination during two provocative tests (dopa and glucagon); serum GH determination during two sleep tests
Buccal smear and karyotype (girls).	

were reported by Nichols to be 6% and 10%, respectively, and cross-reaction with somatomedin A to be 3%.

2.2. *Phase 2*

The subjects in Phase 2 were 97 children, 7–10 years old, referred to this clinic during 1978–1983 by Georgia pediatricians because of height <3rd percentile for chronological age. They had not previously been worked up for cause of short stature. All were prepubertal. Exclusion criteria were birth weight <5.5 lb; previously diagnosed gastrointestinal, pulmonary, renal, or hepatic disease; chronic treatment with steroids or Ritalin; inborn metabolic disorders; previously diagnosed Turner's syndrome; and previously diagnosed endocrine disease. Mental retardation, seizures, and congenital heart disease were not exclusions.

At entry we began to measure the child every 2 weeks on a wall-mounted stadiometer. This continued as long as the child remained in the study. All children had 4 plasma SMC determinations at 6 ± 1 wk intervals. Those with average SMC below the mean for chronological age by at least 2 SD ("hyposomatomedinemic children"), as defined during Phase 1, then entered the next phase of the study by having a clinical evaluation (Table 2). Two or more provocative tests were done with dopa or glucagon[7] (if the peak value in the first two tests averaged >12 ng/ml, no more tests were done; if the first two peaks averaged <12 ng/ml, two more provocative tests were done). The hyposomatomedinemic children then entered 6-month alternating periods of

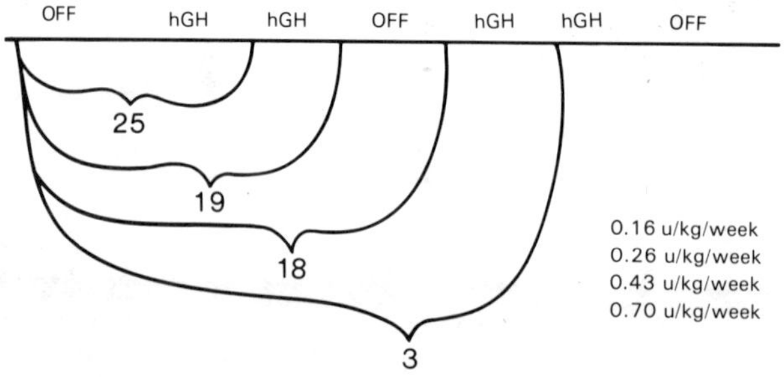

Figure 1 Protocol for hGH treatment in Phase 2. Each period lasted 6 months. The hGH dose during each period was 0.16, 0.26, 0.43, or 0.70 IU/kg per week, selected from a randomized table. The number of children who have completed each segment of the plan at the time of this report is indicated.

Table 3 *Clinical Database Summary of Short Children Studied in Phases 2 and 3*[a]

		GHD	Partial GHD	Non-GHD
Chronological age	Years	9.0 ± 1.0	8.0 ± 0.9	9.1 ± 0.6
Height age/chronological age		0.58 ± .15	0.59 ± 0.10	0.65 ± 0.09
Weight age/chronological age		0.59 ± 0.30	0.57 ± 0.14	0.64 ± 0.13
Height age/weight age		1.18 ± 0.58	1.06 ± 0.12	1.05 ± 0.30
Bone age/chronological age		0.61 ± 0.04	0.66 ± 0.03	0.66 ± 0.04
% 1–3 percentile in height		40%	43%	63%
% <1 percentile in height		60%	57%	37%
Sex	Male/female	10/2	4/3	17/3
Birth weight	kg	3.1 ± 0.2	3.4 ± 0.2	3.0 ± 0.1
IQ		92 ± 5	92 ± 9	86 ± 8
Growth before hGH	Cm/year	3.1 ± 0.4	4.0 ± 1.1	4.0 ± 0.2
SMC (avg. of 4) before hGH	U/ml	0.23 ± 0.03	0.21 ± 0.01	0.30 ± 0.03
Sleep test peak serum GH	ng/ml	2.5 ± 0.7	4.1 ± 1.2	13.8 ± 0.8
Provocative tests peak serum GH				
L-Dopa	ng/ml	1.6 ± 0.2	4.8 ± 1.4	16.3 ± 5.1
Glucagon	ng/ml	1.7 ± 0.3	7.5 ± 0.6	13.1 ± 1.5
Congenital anomalies	+/−	2/10	2/5	4/16
TSH deficiency	+/−	7/5	0/7	0/20
ACTH deficiency	+/−	3/9	0/7	0/20
ADH deficiency	+/−	1/11	0/7	0/20
Sit/Stand height		0.55 ± 0.01	0.58 ± 0.01	0.56 ± .02
Mother's height (inches)		63 ± 1	65 ± 2	62 ± 1
Father's height (inches)		69 ± 2	70 ± 1	69 ± 1
Family history +/−		3/6	0/4	3/14

[a] TSH = thyroid-stimulating hormone; ADH = antidiuretic hormone.

treatment with hGH. During each period, one of four logarithmically equally spaced doses was administered: 0.16, 0.26, 0.43, and 0.70 IU/kg per week, given Monday, Wednesday, and Friday at 10 P.M. The plan of the study is shown in Fig. 1. The order of doses was randomized.

2.3. Phase 3

2.3.1. *Subjects*

The normal group in Phase 3 comprised six boys and 10 girls, 12–18 years old. History and physical examination revealed no evidence of acute or chronic illness.

The hyposomatomedinemic non-GHD short children consisted of the 17 boys and three girls, 7–10 years of age, who had been identified during Phase 2. Their clinical characteristics are summarized in Table 3. Mean serum immunoreactive GH* response to dopa or glucagon was >12 ng/ml, showing normal immunoreactive GH status.

* The conventional, commercially available RIA with polyclonal rabbit antiserum was used to measure immunoreactive GH in the provocative tests. Our technique for carrying out this assay is described under methods, Phase 3.

T_4, free T_4, and cortisol at 8 A.M. were normal. Bone age/chronological age ratio averaged 0.66. In 16 of the 20 short children, history, physical examination, hemogram, urinalysis, and SMA 18 were normal. Two children were mentally retarded and two had polydactyly.

2.3.2. Study Design

The subjects were admitted for 5 days to the Clinical Research Facility. No medications were given. The short children had not received hGH injections during the previous month. Their diet was 2300 cal, 100 g protein, 250 g carbohydrate, and 100 g fat. Activity was unrestricted. Via an indwelling heparin lock inserted at 8–9 P.M., 5 ml heparinized blood was removed on retiring, and at 30, 60, 90, and 120 min after falling asleep. The blood was kept at 4°C until 8 A.M., when it was spun and the serum then stored at −20°C for up to 2 weeks before analysis by RIA.(1) Some sera were also examined by gel filtration.

2.3.3. Materials

The RIA kits were obtained from Kallestad Laboratories ("polyclonal kit") and from Hybritech Laboratories ("biclonal" or immunoradiometric kit). Two mouse monoclonal antibodies (#33.2 and 101.3) to hGH were purchased from Hybritech. According to the manufacturers, the antibodies were directed at different epitopes of the hGH molecule. The precipitating antibody (rabbit antiserum to mouse IgG) was from Calbiochem-Behring. The labeled hGH used in the monoclonal assays was from Kallestad Laboratories. Immunochemical-grade hGH, used as the cold hormone in these assays, was from the National Pituitary Agency (NPA), as were the other hormones—human placental lactogen (hPL), human prolactin (hPRL), human thyroid-stimulating hormone (hTSH), human follicle-stimulating hormone (hFSH), human luteinizing hormone (hLH), and β-lipotropin. For gel filtration, Ultrogel AcA 44 was from LKB, and disposable columns (1 cm × 50 cm) were from BioRad, Richmond, California. All other reagents and chemicals were from Sigma Chemicals Co., St. Louis.

2.3.4. Analytical Methods

2.3.4a. The Polyclonal RIA. The polyclonal RIA kit (Kallestad Laboratories, Inc.) contained hGH calibration standards (NIH immunochemical grade, according to the manufacturers) ranging from 1.0 ng/ml to 16.0 ng/ml, rabbit antiserum to hGH, ^{125}I-labeled hGH, goat antiserum to rabbit gamma globulin, and 0.01 M phosphate buffered saline containing bovine serum albumin. The assay was based on a double-antibody sequential saturation technique. The samples were incubated with hGH antiserum and [^{125}I]hGH to form hGH–antibody complexes. The complexes were then precipitated with the second antibody, and the bound fractions were counted in a gamma counter.

A control serum containing a known amount of reference preparation hGH from the NPA was included in triplicate in each assay. In addition, a displacement curve of hGH-free serum containing varying amounts (1–12 ng/ml) of NPA immunochemical-grade hGH was also included in each assay. The serum curve was always identical with the displacement curve obtained from the material supplied in the kit.

2.3.4b. The Biclonal RIA. This immunoradiometric assay kit (Hybritech, Inc.) comprised plastic beads coated with monoclonal mouse antibody to hGH, ^{125}I-labeled monoclonal mouse antibody to hGH (different from the one coating the beads), calibration standards of 1.0 ng/ml through 50 ng/ml (NIH immunochemical grade, according to the manufacturers), and a buffer wash. The assay was a solid-phase, two-site immunoradiometric assay. Samples bind to the antibody on the bead and to the different [^{125}I]antibody. The bead is then washed to remove all unbound labeled antibody and counted in a gamma counter.

As in Section 2.3.4a, in each assay, a control serum containing a known amount of the reference hGH was included. In addition, a displacement curve in serum, as described previously, was also included. The serum curves were found to be identical to the standard curve.

2.3.4c. Two Monoclonal RIA's. Two RIA's using monoclonal mouse antibodies supplied by Hybritech (#33.2 and 101.3) were developed to measure serum hGH levels. These assays were of the sequential saturation type; their details are described below.

Optimal dilution of the monoclonal antibody. A determination of the optimal antibody dilution for RIA was the first step in development. A stock solution of 1 mg/ml monoclonal antibody was serially diluted from 1 in 100 to 1 in 300,000 in the assay buffer, a 0.01 M phosphate solution of pH 7.4 containing 0.5% bovine serum albumin. Each antibody dilution was tested in triplicate. An aliquot (500 μl) of the assay buffer was mixed with 100 μl of each antibody dilution and 100 μl of [^{125}I]hGH (approximately 12,000 cpm). The tubes were vortexed and incubated for 4 hr at 37°C. Thereafter, 200 μl of the precipitating antibody solution was mixed into each tube of the assay, excluding the total-counts tubes. The precipitating antibody was rabbit antiserum to mouse IgG supplied by Calbiochem-Behring. This was diluted 1 in 5 parts in the assay buffer, and then equally with 12% polyethylene glycol. The total assay was then incubated for 16 hr at 4°C after which the tubes were centrifuged for 30 min at 3000 g. The supernatant was decanted and the residue counted in a gamma counter. The percent absolute binding was plotted against the dilution to determine the dilution of monoclonal antibody that bound about 30% of the labeled hGH.

Displacement curve. Once the optimal antibody dilution was determined, a displacement curve for the antibody using varying amounts of hGH was examined for binding characteristics such as sensitivity and specificity. Immunochemical-grade hGH from NPA was used as the standard. The assay was based on sequential saturation technique where 500 μl of assay buffer, 100 μl diluted antibody, and 100 μl of sample, control, or standard (0, 3, 6, 12, 25, 50, and 100 ng/ml) were incubated for 4 hr at 37°C. The standards were prepared in hGH-free serum. Nonspecific binding was determined by spiking duplicate tubes in the assay with 10 μg hGH in place of sample or standard in order to complex only cold hormone. After the initial 4-hr incubation, 100 μl of [^{125}I]hGH was mixed into all tubes and incubated for 2 hr at 37°C. The precipitating antibody solution (200 μl) was then mixed into all tubes excluding the total-counts tubes. The entire assay was incubated for 16 hr at 4°C and then spun, drained, and counted in order to calculate the percent of bound [^{125}I]hGH in each tube. A log–logit plot was established to correlate percent binding and the amount of hormone in the standards. The concentrations of the samples were then interpolated from the standard displacement line. As in polyclonal and biclonal assays, a control serum containing a known amount of the reference hGH was included in each assay. In addition to tests of sensitivity, and inter- and intraassay variability, the antibody was tested for specificity to hGH. Cross-reactivity was measured by testing the ability of ACTH, hPRL, hTSH, hLH, hFSH, and β-lipotropin to displace the labeled hGH.

2.3.4d. Gel Filtration Analyses. The chromatography was carried out on a 1 cm × 50 cm Ultrogel AcA 44 column. The eluting buffer used was 0.023 M ammonium carbonate at pH 8.6 with 0.1% albumin and 0.02% sodium azide.

The column and buffer were stored at 4°C and were run at 25°C. One ml aliquot of serum was taken from a sleep test sample and was loaded on the column. The flow rate was 1 ml/1.5 min, and 1-ml fractions were collected. The fractions were separately lyophilized, reconstituted in 250 μl of water, and assayed in the polyclonal and biclonal RIA's.

3. Results

3.1. Phase 1

3.1.1. Variability Studies

Three normal prepubertal 10-year-old girls were recruited in order to estimate four sources of variability associated with measuring plasma SMC samples. A single plasma sample for each girl was split into three aliquots and sent to Nichols Institute, in a single-blind fashion. Plasma samples were also collected on 3 consecutive days and at 6-week intervals over 4 consecutive months. This allowed us to estimate intrassay, day-to-day, month-to-month, and intersubject variability.

3.1.1a. Intraassay Variability. A one-way analysis of variance random model was used to assess intraassay variability. Data in Table 4 were analyzed to estimate intraassay variability. The statistical model for the jth plasma SMC sample from subject i, denoted by Y_{ij}, is represented as

$$Y_{ij} = \mu + s_i + e_{ij} \quad i = 1, 2, 3 \quad j = 1, 2, 3$$

Table 4 Plasma SMC Values for Three 10-Year-Old Normal Girls

	SmC (units/ml)					
Subject #	Day 1	Day 2	Day 3	Day 45	Day 90	Day 135
#1	0.69	0.74	0.80	0.40	0.94	0.56
	0.68					
	0.72					
#2	1.15	1.32	1.00	0.85	1.43	0.90
	1.18					
	1.26					
#3	0.85	0.75	0.70			
	0.85			1.14	0.67	0.60
	0.80					

where μ is a constant component common to all subjects; s_i is the random subject effect which is independently normally distributed with mean zero and variance $\sigma_s{}^2$ (between subject variability); e_{ij} is the random replicate effect for each subject which is independently normally distributed with mean zero and variance $\sigma_e{}^2$ (intraassay variability). Table 5 summarizes the analysis of variance where the individual variances $\sigma_s{}^2$ and $\sigma_e{}^2$ and the variance of an observation

Table 5 Estimation of Sources of Variability for Plasma SMC (Phase 1): Analysis of Variance Table for Estimating the Intraassay Variability and Related Estimates

Source of variation	Sums of squares	Degrees of freedom	Mean squares	Expected mean squares
Between subjects	0.4007	2	0.20035	$3o_s^2 + o_e^2$
Within assay	0.0090	6	0.0015	o_e^2
Total	0.4097	8		

Estimates

$o_e^2 = 0.0015$ (intraassay variance component)

$$o_s^2 = \frac{0.20035 - 0.0015}{3} = 0.06628 \text{ (between-subject variance component)}$$

$o^2\,(Y_{ij}) = o_s^2 + o_e^2 = 0.06778$ (estimated variance of an observation)

Proportion of total variability of the Y_{ij} which is accounted for by the variability of the subjects is estimated as

$$\frac{o_s^2}{o_s^2 + o_e^2} \times 100 = 97.8\%$$

$\bar{Y}.. = 0.909$ (common mean of all subjects)

$$\text{CV (coefficient of variation): } \frac{o_e}{\bar{Y}..} \times 100 = 4.3\%$$

Table 6 Analysis of Variance Table for Estimating the Day-to-Day Variability and Related Estimates

Source of variation	Sources of squares	Degrees of freedom	Mean squares	Expected mean squares
Between subjects	0.32349	2	0.161745	$3\sigma_s^2 + \sigma_d^2$
Between days	0.00690	6	0.0115	σ_d^2
Total	0.33039	8		

Estimates

$$\sigma_d^2 = 0.0115 \text{ (day-to-day variance component)}$$

$$\sigma_s^2 = \frac{0.161745 - 0.0115}{3} = 0.05008 \text{ (between-subject variance component)}$$

$$\sigma^2 (Y_{ij}) = \sigma_s^2 + \sigma_d^2 = 0.06158 \text{ (estimated variance of an observation)}$$

Proportion of total variability of the Y_{ij} which is accounted for by the variability of the subjects is estimated as

$$\frac{\sigma_s^2}{\sigma_s^2 + \sigma_d^2} \times 100 = 81.3\%$$

$$\bar{Y}.. = 0.889 \text{ (common mean of all subjects)}$$

$$\text{CV (coefficient of variation) } \frac{\sigma_e}{-\bar{Y}..} \times 100 = 12.1\%$$

Y_{ij}, given by $\sigma^2 (Y_{ij}) = \sigma_s^2 + \sigma_e^2$, are each estimated. The intraassay variability was estimated to be 0.0015 and the estimated coefficient of variation was 4.3% (Table 5).

3.1.1b. Day-to-Day Variability. An analysis of variance analogous to that for estimating intraassay variability was performed for the SMC samples taken on 3 successive days in Table 4. These results are summarized in Table 5B. Note the significant eightfold increase of the day-to-day variability over the intraassay variability (0.0115 versus 0.0015). The estimated coefficient of variation increased to 11.9% (Table 6).

3.1.1c. Month-to-Month Variability. An analysis of variance analogous to that for estimating intraassay variability was performed for the SMC samples taken every 6 weeks in Table 4. These results are also summarized in Table 7. Note the significant sixfold increase of the month-to-month variability over the day-to-day variability (0.0627 versus 0.0115). The estimated coefficient of variation increased markedly to 30.1%.

3.1.1d. Intersubject Variability. In each of the above analyses, we also esti-

Table 7 Analysis of Variance Table for Estimating the Month-to-Month Variability and Related Estimates

Source of variation	Sums of squares	Degrees of freedom	Mean squares	Expected mean squares
Between subjects	0.29195	2	0.145975	$4\sigma_s^2 + \sigma_m^2$
Between months	0.564475	9	0.06272	σ_m^2
Total	0.856425	11		

Estimates

$$\sigma_m^2 = 0.06272 \text{ (month-to-month variance component)}$$

$$\sigma_s^2 = \frac{0.145975 - 0.06272}{4} = 0.02081 \text{ (between-subject variance component)}$$

$$\sigma^2 (Y_{ij}) = \sigma_s^2 + \sigma_m^2 = 0.08353 \text{ (estimated variance of an observation)}$$

Proportion of total variability of the Y_{ij} which is accounted for by the variability of the subject is estimated as

$$\frac{\sigma_s^2}{\sigma_s^2 + \sigma_m^2} \times 100 = 24.9\%$$

$$\bar{Y}.. = 0.8325 \text{ (common mean of all subjects)}$$

$$\text{CV (coefficient of variation)} \frac{\sigma_m}{-\bar{Y}..} \times 100 = 30.1\%$$

mated the intersubject variability. The estimated variances were 0.066, 0.050, and 0.021. Note that the intersubject estimated variances were generally of the same magnitude as or smaller than the month-to-month variance. The substantial amount of month-to-month variability in plasma SMC samples suggested that single determinations in children were of very limited value in estimating tolerance limits for normal children.

3.1.2. Estimating Tolerance Intervals (Normal Values)

The lower limit of the 95% tolerance limits based on single plasma SMC samples were always below the sensitivity of the assay (<0.10 unit/ml) (Table 8). In view of the substantial variability in SMC determinations at 6-week intervals, we investigated the feasibility of estimating tolerance limits for normal children based on the average of several monthly plasma SMC samples.

Separate analyses were performed for each sex and age group based on the average of one, two, three, or four determinations of plasma SMC samples obtained at 6-week intervals. Each analysis was performed by calculating the mean for each child based on one, two, three, or four determinations and then

Table 8 *Means, Standard Deviations, and Tolerance Intervals by Age and Sex Based on One, Two, Three, and Four Determinations of Plasma SMC Levels (Phase 1)*[a]

		Number of determinations										
		1		2		3		4				
									Tolerance intervals			
Age	Number of hildren	Mean ± SD	95% Tolerance interval	Mean ± SD	95% Tolerance interval	Mean ± SD	95% Tolerance interval	Mean ± SD	50%	80%	95%	98%
Boys												
6.5–7.4	15	0.451 ± 0.218	(0, 0.934)	0.494 ± 0.200	(0.051, 0.937)	0.491 ± 0.161	(0.134, 0.848)	0.495 ±0.125	(0.406, 0.584)	(0.321, 0.669)	(0.218, 0.772)	(0.156, 0.834)
7.5–8.4	15	0.565 ± 0.300	(0, 1.230)	0.544 ± 0.241	(0.010, 1.078)	0.577 ± 0.184	(0.149, 0.965)	0.536 ± 0.124	(0.447, 0.625)	(0.364, 0.708)	(0.259, 0.813)	(0.200, 0.872)
8.5–9.4	15	0.583 ± 0.272	(0, 1.191)	0.578 ± 0.199	(0.133, 1.023)	0.578 ± 0.145	(0.263, 0.911)	0.543 ± 0.132	(0.449, 0.637)	(0.359, 0.727)	(0.248, 0.838)	(0.181, 0.905)
9.5–10.4	15	0.551 ± 0.255	(0, 1.116)	0.614 ± 0.219	(0.129, 1.099)	0.630 ± 0.163	(0.269, 0.991)	0.607 ± 0.135	(0.510, 0.704)	(0.419, 0.795)	(0.308, 0.906)	(0.241, 0.973)
Girls												
6.5–7.4	15	0.585 ± 0.246	(0, 1.130)	0.574 ± 0.165	(0.209, 0.939)	0.589 ± 0.164	(0.226, 0.952)	0.605 ± 0.144	(0.502, 0.708)	(0.405, 0.805)	(0.280, 0.924)	(0.215, 0.995)
7.5–8.4	15	0.673 ± 0.304	(0, 1.346)	0.659 ± 0.204	(0.207, 1.111)	0.664 ± 0.160	(0.310, 1.018)	0.659 ± 0.145	(0.555, 0.763)	(0.458, 0.860)	(0.338, 0.980)	(0.266, 1.052)
8.5–9.4	14	0.748 ± 0.340	(0, 1.501)	0.788 ± 0.267	(0.197, 1.397)	0.798 ± 0.234	(0.280, 1.316)	0.821 ± 0.247	(0.644, 0.998)	(0.473, 1.166)	(0.274, 1.368)	(0.143, 1.499)
9.5–10.4	15	1.036 ± 0.588	(0, 2.351)	1.054 ± 0.388	(0.186, 1.922)	1.046 ± 0.328	(0.313, 1.799)	1.056 ± 0.329	(0.821, 1.291)	(0.599, 1.513)	(0.322, 1.792)	(0.164, 1.948)

[a] Tolerance limits for the mean based on one, two, three, or four determinations were obtained by taking the mean ± t (1 − /2; n − 1)(SD) 1 + 1/n, where t is the appropriate percentile of Student's t distribution and n is the number of children. For example, the 95% tolerance interval for 7-year-old boys based on four determinations is 0.495 ± t (0.975; 14) (0.125) 1 + 1/15 = 0.495 ± (2.145) (0.125) 1 + 1/15 = 0.495 ± 0.277.

Table 9 Plasma SMC Values at 6-Week Intervals for 15 7-Year-Old Normal Boys (Phase 1)

Subject #	SMC	Units/ml			Mean	Z Score[a]
1	0.69	0.23	0.81	0.50	0.5575	0.502
2	0.64	0.55	0.30	0.20	0.4225	−0.578
3	0.36	0.27	0.65	0.80	0.5200	0.202
4	0.21	0.32	0.84	0.60	0.4925	−0.018
5	0.27	0.80	0.60	0.40	0.5175	−1.82
6	0.10	0.30	0.19	0.45	0.2600	−1.878
7	0.50	0.25	0.23	0.47	0.3625	−1.058
8	0.51	0.83	0.20	0.23	0.4425	−0.418
9	0.60	0.83	0.94	0.22	0.6475	1.222
10	0.25	0.49	0.33	0.88	0.4875	−0.058
11	0.30	0.63	0.82	0.71	0.6150	0.962
12	0.64	0.96	0.26	0.51	0.5925	0.782
13	0.85	0.60	0.19	0.90	0.6350	1.122
14	0.60	0.80	0.63	0.40	0.6075	0.902
15	0.24	0.20	0.30	0.31	0.2625	−1.858

Mean of means = .4948
Standard deviation of means (SD) = .125

[a] $Z = \frac{\text{Mean} - \text{mean of means}}{\text{SD}}$

computing the mean and standard deviation of these means for each group. This is demonstrated in Table 9 for 7-year-old boys, based on four determinations at 6-week intervals. The results for each sex and age group are summarized in Table 8 and Fig. 2. Note the decrease in the width of the tolerance intervals as the number of determinations increases.

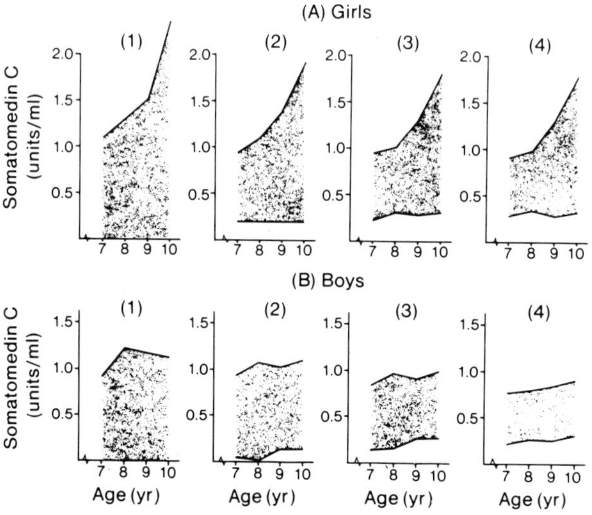

Figure 2 Ninety-five percent tolerance intervals of plasma SMC by age for girls and boys based on one, two, three, and four determinations.

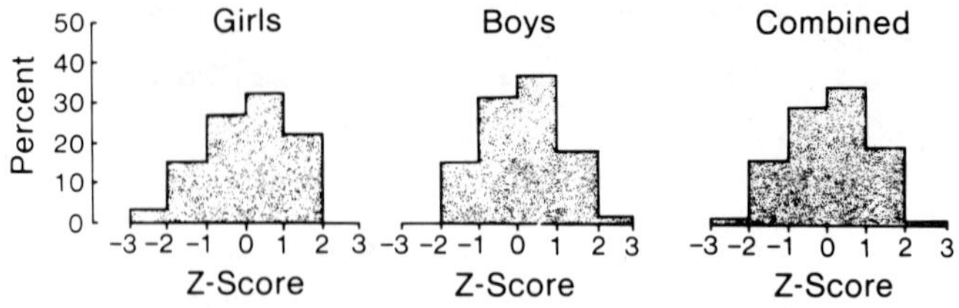

Figure 3 Empirical frequency distribution of the average of four SMC values for normal children standardized for age and sex.

The tolerance intervals developed here assume an underlying Gaussian distribution. Figure 3 displays the empirical distributions of the means, based on four determinations, standardized for age and sex. Means were converted to Z scores for each subject by subtracting the mean of means for the group and dividing by the standard deviation of the group mean (see Table 9 for specifics). Note the generally symmetrical, bell-shaped curves, which in probability content are consistent with a Gaussian distribution. Table 10 lists the observed and expected frequencies obtained by assuming a Gaussian distribution for the combined standardized data. Again note the generally close agreement between the observed and expected frequencies.

Figure 4 displays tolerance limits for girls and boys, aged 7–10, for commonly used percentiles based on the average of four determinations at 6-week intervals. These tolerance limits can be used to screen short children for GHD. If we define subnormal as the lower limit of the 95% tolerance intervals, then two or more SMC's per child are required before the lower limit of the 95% tolerance interval rises into a measurable range (>0.10 units/ml) (Fig. 2). Since the lower limit rises progressively using the average of two, three, and four determinations (Fig. 2), it is likely that more children with partial or total GHD would be identified according to the order (4) > (3) > (2) > (1). To empirically test this hypothesis, we present data taken at 6-week intervals on seven GHD children (Table 11). If we define subnormal SMC as below the lower limit of the 95% tolerance intervals, then based on the mean of one, two, three, or four determinations we would classify, *a posteriori*, zero, three, three, and seven of the GHD children as hyposomatomedinemic, respectively. Thus, no less than four determinations were necessary in order to identify all seven GHD children as having abnormally low SMC.

Table 10 Observed and Expected Frequencies of Z-Scores of Plasma SMC for Boys and Girls Combined (Phase 1)[a]

	<-2	-2 to -1	-1 to 0	0 to 1	1 to 2	>2	Total
Observed frequency	2	18	34	40	23	1	118
Expected frequency	2.7	16.0	40.3	40.3	16.0	2.7	118

[a] Chi square = 5.55 (not significant) ($p < 0.10$), for the comparison between observed and expected frequencies.

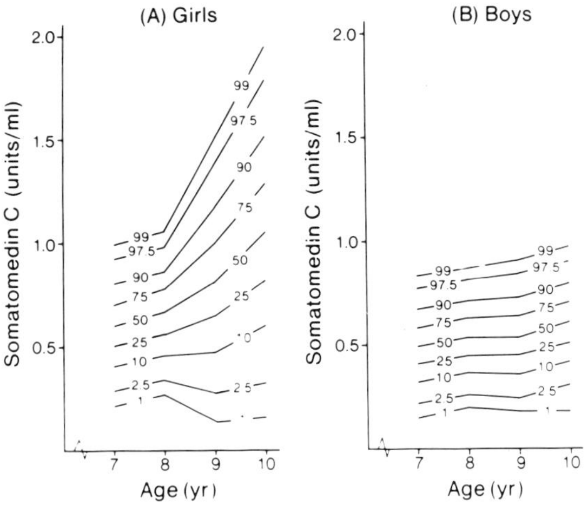

Figure 4 Percentiles for plasma SMC based on the average of four determinations in normal children by sex and age.

3.2. Phase 2

3.2.1. Distribution of SMC in 97 Short Children 7–10 Years Old

In each child, plasma SMC was measured four times at 6-week intervals, and the child's SMC status was represented as the average of the four values. Figure 5 compares the distribution of the mean of four SMC values in the 97 short children with those of the normal children as determined in Phase 1. A significant excess of low SMC was found. When SMC was less than the normal mean for chronological age (below the 2.5 percentile), the child was considered "hyposomatomedinemic." Forty-five percent of the short children (44 of 97) were in this category.

3.2.2. Classification of the 44 Hyposomatomedinemic Short Children

These children received the workup in Table 2. When the mean peak immunoreactive GH level was less than 4 ng/ml in four provocative tests, the

Table 11 SMC Values at 6-Week Intervals of Seven GHD Children (Phase 1)

Subject #	Individual SMC Values				Means (# of determinations) 2	3	4
1	0.14	0.26	0.29	0.19	0.20	0.23	0.22
2	0.13	0.14	0.26	0.15	0.14	0.18	0.17
3	0.25	0.19	0.28	0.16	0.22	0.24	0.22
4	0.25	0.24	0.21	0.19	0.25	0.23	0.22
5	0.20	0.22	<0.10[a]	0.22	0.21	0.17	0.19
6	0.11	0.25	0.36	0.20	0.18	0.24	0.23
7	0.33	0.14	<0.10[a]	0.24	0.24	0.19	0.20

[a] <0.10 was set equal to 0.10 for analysis.

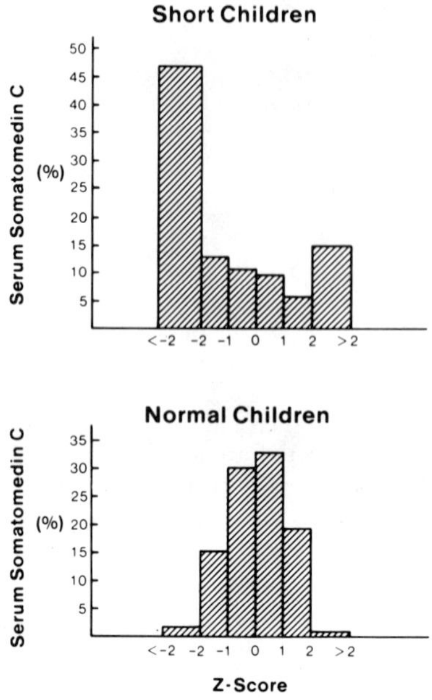

Figure 5 Distribution of plasma SMC values (average of four) among (1) 87 normal prepubertal children, aged 7–10, between 5% and 95% in height for chronological age, and (2) 97 prepubertal children of the same age, <3rd percentile in height for chronological age.

child was classified "GHD." If the mean value was 4–8 ng/ml, the child was classified "partial GHD." A mean value over 12 ng/ml was interpreted as "normal immunoreactive GH" status, and mean values between 18 and 12 were classified as "indeterminate." The numbers of children in the four groups were 12, 7, 5, and 20, respectively. The last group will be referred to as the "non-GHD" group. Children in the indeterminate group will not be considered further in this chapter.

3.2.3. Clinical Characteristics for the GHD, Partial GHD, and Non-GHD Groups

These characteristics are given in Table 3. Males predominated in all three groups. Anthropometric and auxological features were similar. TSH or ACTH deficiency was common in GHD, but was not found in partial GHD or non-GHD.

3.2.4. Responses to hGH by GHD and Non-GHD Children

To date there have been 65 6-month trials at one of the four standard doses as summarized in Fig. 6 and Table 12 (in these charts, GHD and partial GHD are considered as one group). GHD children responded to all four doses of hormone; the responses to 0.43 and 0.70 IU/kg per week were similar. In the non-GHD group, one trial at 0.16 IU/kg per week was ineffective; at 0.26, 0.43, and 0.70 IU/kg per week, average responses were 50–90% as great as in the GHD group. At the two intermediate doses, the responses by non-GHD children were significantly less than by GHD children.

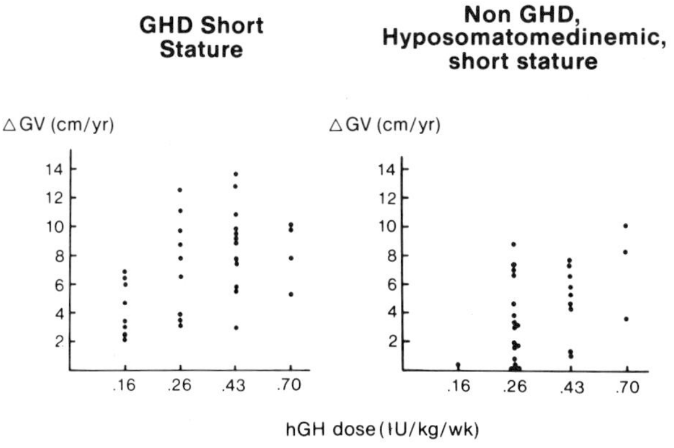

Figure 6 Linear growth responses (ΔGV) of growth hormone deficient (GHD) and non-growth-hormone-deficient (non-GHD) hyposomatomedinemic children to 6 months courses on four different doses of hGH.

3.3. Phase 3

3.3.1. Data on the Four RIA Systems

3.3.1a. Polyclonal (Kallestad) Assay. The control sera tested in 10 different kits gave less than 6% coefficient of variation (CV). The intraassay CV was 4%. No significant cross-reactivity was observed with hFSH, hLH, or hTSH. Human prolactin cross-reacted with 2.5% of the activity, and hPL cross-reacted with 0.004% of the activity of hGH. The sensitivity of the assay was 1.0 ng/ml, and the working range of the assay was 1–12 ng/ml.

3.3.1b. Biclonal (Hybritech) Assay. The control sera tested in 10 kits gave 6% CV. The intraassay CV was about 3%. Cross-reactivity of hLH, hFSH, or hPL was <0.01%, while cross-reactivity of human prolactin was 0.5%. The sensitivity of the assay was 0.2 ng/ml, but the working range was 1.0–25.0 ng/ml.

3.3.1c. Monoclonal RIA. Determination of Optimal Dilution of Antibody. Figure 7

Table 12 Linear Growth Responses of GHD and Non-GHD, Hyposomatomedinemic Short Children to 6-Month Courses of Four Different Doses of HGH (Phase 2)

Dose hGH (IU/kg per week)	Growth velocity (cm/year, average ± SEM)	
	GHD	non-GHD
0.16	4.35 ± 0.66 (n = 8)	0.2 (n = 1)
0.26	7.43 ± 1.15 (n = 9)	3.44 ± 0.69[a] (n = 16)
0.43	8.68 ± 0.89 (n = 12)	5.43 ± 0.66[a] (n = 12)
0.70	8.25 ± 1.09 (n = 4)	7.27 ± 1.96 (n = 3)

[a] Statistically different from GHD group ($p < 0.02$).

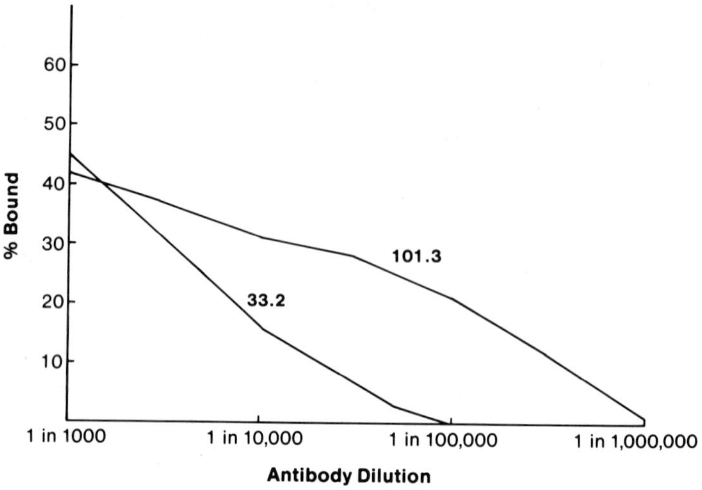

Figure 7 Dilution curve for monoclonal antibodies 33.2 and 101.3 showing % binding with $[^{125}I]hGH$.

shows a typical relationship between the dilution of monoclonal antibody (both 33.2 and 101.3) and absolute binding. A maximum binding of 50% for 33.2 and 42% for 101.3 was obtained at a dilution of 1 in 1000. The binding became undetectable (<4%) at a dilution of 1 in 100,000 for 33.2 and 1 in 1,000,000 for 101.3. At a dilution of 1 in 30,000, the absolute binding ranged from 30 to 35% for 101.3. At a dilution of 1 in 3000, the absolute binding ranged from 30 to 35% for 33.2. These dilutions of antibody were used for all subsequent assays.

The intraassay CV of binding was 3%; the interassay CV was about 6%. A stock solution of antibody #33.2 and 101.3 (diluted 1 in 3000) was kept frozen in 1-ml aliquots and diluted for use in each assay as required. No significant change in binding was noticed in frozen aliquots over a period of 12 weeks.

Displacement curves for each antibody. Figures 8 and 9 show typical displacement curves for each monoclonal antibody. The intraassay CV was 5%, and the interassay CV was 9% for both antibodies. The assay showed no significant cross-reactivity (<0.05%) with ACTH, hTSH, hFSH, hPL, or β-lipotropin. Human prolactin cross-reacted about 0.5%. For 33.2, 15 ng/ml hGH was required for 50% displacement. For 101.3, only 12 ng/ml was required for similar displacement.

The usable range for 33.2 was 16–32 ng/ml and for 101.3 it was 8–100 ng/ml. The sensitivities of the two monoclonal RIA's were 16 ng/ml and 8 ng/ml, respectively.

3.3.2. Sleep Tests

3.3.2a. Normal Children and Adolescents, 12–18 Years Old. A typical sleep test is shown in Fig. 10, and the group is summarized in Table 13. On retiring, the

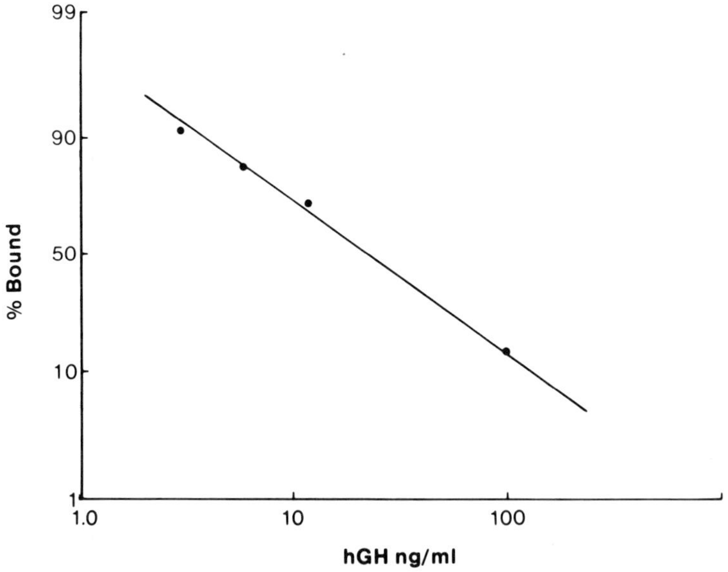

Figure 8 Displacement curve for monoclonal antibody 101.3 using immunochemical-grade hGH standards of 3, 6, 12, 25, and 50 ng/ml and [^{125}I]hGH.

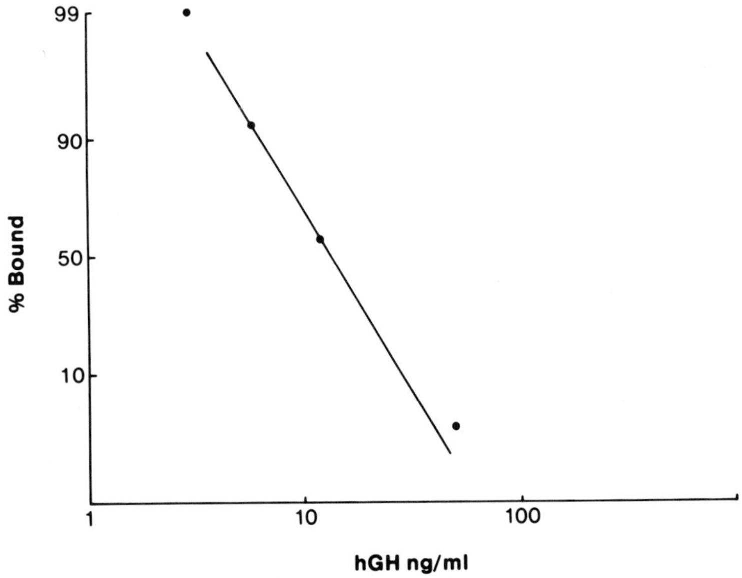

Figure 9 Displacement curve for monoclonal antibody 33.2 using immunochemical-grade hGH standards as defined in Fig. 8.

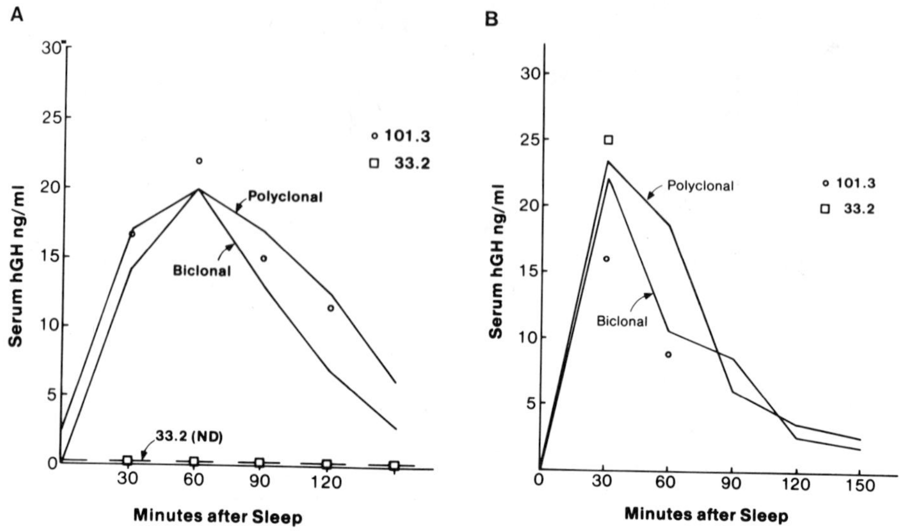

Figure 10 Typical sleep tests comparing nocturnal serum hGH levels of a normal (A) and a non-GHD, hyposomatomedinemic short child (B) as measured by four types of RIA.

polyclonal serum GH value was less than 1 ng/ml. At 30 and 60 min, it rose to 14 ng/ml, respectively, and then declined progressively during the next 90 min. Because the polyclonal and biclonal assays had similar thresholds, about 1 ng/ml, the biclonal to polyclonal ratio was calculated for all of the sera that contained 1 ng/ml or more of polyclonal immunoreactivity. At 30 and 60 min, this ratio was significantly less than 1, being 0.83 and 0.81, respectively.

At 30 and 60 min, the 33.2 monoclonal reactivity was below its threshold of 16 ng/ml in 75% and 67% of the samples, and the 101.3 assay was below its sensitivity of 8 ng/ml in 36% and 27% of the samples. For samples in which the 33.2 monoclonal assay was above its threshold sensitivity of 16 ng/ml, the 33.2 to polyclonal ratio averaged 1.14 (30 min) and 1.16 (60 min) (not significantly different than 1). For samples in which the 101.3 monoclonal assay was above its threshold of 8 ng/ml, the 101.3-to-polyclonal ratio averaged 1.00 (30 min) and 1.46 (60 min) (not significantly different from 1).

Biclonal-to-polyclonal ratios for individual normal children are shown in Table 14. The average ratio varied between 0.544 and 1.119. The 101.3- and 33.2-to-peak-polyclonal ratios for individual normal children are shown in Table 15. The average ratios varied between 0.521 and 1.637 for the 101.3 and between 1.021 and 1.27 for the 33.2.

3.3.2b. Hyposomatomedinemic, Short, Non-GHD Children, 7–10 Years Old. A representative sleep test is shown in Fig. 10B, and the results for the entire group are summarized in Table 13. The polyclonal GH immunoreactivity usually peaked at 30 min, occasionally at 60 min. The peak polyclonal value averaged 17.7 ng/ml. At 30 and 60 min, the biclonal-to-polyclonal ratio was

Table 13 Nocturnal Immunoractive Serum GH in Normal and Non-GHD Hyposomatomedinemic Short Children as Measured by Four Types of RIA (Phase 3)

	Normal (mean ± SD)	Non-GHD (mean ± SD)
Peak polyclonal (ng/ml)	20.0 ± 2.2 (n = 12)	17.7 ± 2.4 (n= 13)
Biclonal/peak polyclonal	0.84 ± 0.07[a]	0.81 ± 0.07[a]
33.2 monoclonal/peak polyclonal	1.28 ± 0.35 (n = 5)	1.18 ± 0.13 (n = 6)
101.3 monoclonal/peak polyclonal	0.90 ± 0.41 (n = 8)	0.98 ± 0.24 (n = 9)
33.2 monoclonal (% nd[b])	58 (7/12)	45 (5/11)
101.3 monoclonal (% nd[c])	27 (3/11)	10 (1/10)
30 min polyclonal (ng/ml)	13.9 ± 2.9 (n = 12)	15.1 ± 2.9 (n = 13)
30 min biclonal/polyclonal	0.83 ± 0.09	0.79 ± 0.07[a]
30 min 33.2 monoclonal/polyclonal	1.14 ± 0.11 (n = 3)	1.19 ± 0.57 (n = 5)
30 min 1091.3 monoclonal/polyclonal	1.00 ± 0.10 (n = 7)	1.06 ± 0.58 (n = 7)
33.2 monoclonal (% nd)	75 (9/12)	55 (6/11)
101.3 monoclonal (% nd)	36 (4/11)	30 (3/10)
60 min polyclonal (ng/ml)	14.4 ± 2.5 (n = 11)	12.6 ± 2.4 (n = 13)
60 min biclonal/polyclonal	0.81 ± 0.06[a]	0.69 ± 0.04[a]
60 min 33.2 monoclonal/polyclonal	1.16 ± 0.28 (n = 4)	0.98 ± 0.47 (n = 3)
60 min 101.3 monoclonal/polyclonal	1.46 ± 1.25 (n = 8)	1.72 ± 1.95 (n = 4)
33.2 monoclonal (% nd)	67 (8/12)	73 (8/11)
101.3 monoclonal (% nd)	27 (3/11)	60 (6/10)

[a] Significantly less than 1 ($p < 0.05$).
Significantly greater than 1 ($p < 0.05$).
[b] Not detectable: <16 ng/ml.
[c] Not detectable: < 8 ng/ml.

significantly less than 1, being 0.79 and 0.69, respectively. At 30 and 60 min, 55% and 73% of the samples, respectively, had undetectable monoclonal 33.2 levels (<16 ng/ml), while 30% and 60%, respectively, were undetectable in the monoclonal 101.3 assay (<8 ng/ml). For those samples above the thresholds in the 33.2 and 101.3 monoclonal assays, the 33.2/polyclonal and 101.3/polyclonal ratios averaged 1.19 and 1.06 at 30 min and 0.98 and 1.72 at 60 min, respectively (not significantly different from 1).

As shown in Table 14, the average biclonal-to-polyclonal ratio in the non-GHD, hyposomatomedinemic short children ranged from 0.520 to 1.175.

3.3.2c. Statistical Comparisons between Normal (Aged 12–18) and Non-GHD, Hyposomatomedinemic Short Children (Aged 7–10). No significant differences in any of the serum GH immunochemical levels or ratios shown in Table 13 were found between the groups using Student's t test or Fisher's exact test for comparing percentages.

3.3.3. Incubation Experiments

After the differing results in the four RIA's were recognized, the possibility was considered that the native hGH molecule was degraded in blood, with

Table 14 Peak Biclonal-to-Polyclonal Ratios of Nocturnal Serum HGH in Normal (N) and Non-GHD Hyposomatomedinemic Short Children (S) (Phase 3)

	N_1	N_2	N_3	N_4	N_5	N_6	N_7	N_8	N_9	N_{10}	N_{11}	N_{12}	
Sleep test 1	0.846	0.750	1.175	0.591	0.850	1.054	0.571	1.000	0.697	0.656	0.739	0.484	
2	—	0.583	0.962	0.979	0.611	1.296	1.130	0.636	0.591	0.762	—	0.489	
3	—	0.667	0.607	0.893	0.737	1.007	0.765	0.794	0.556	0.700	—	0.658	
4	—	—	—	—	0.711	—	—	—	—	—	—	—	
Average	0.846	0.667	0.915	0.821	0.727	1.119	0.822	0.810	0.614	0.706	0.739	0.544	
SD	—	0.083	0.287	0.204	0.098	0.155	0.284	0.182	0.073	0.053	—	0.099	
	S_1	S_2	S_3	S_4	S_5	S_6	S_7	S_8	S_9	S_{10}	S_{11}	S_{12}	S_{13}
Sleep test 1	0.547	0.520	0.691	0.650	1.014	1.175	0.652	0.925	0.981	0.628	1.147	0.636	1.409
2	—	—	0.950	—	—	—	0.538	0.779	0.610	0.440	—	0.650	0.949
3	—	—	0.781	—	—	—	—	—	—	—	—	0.603	0.808
Average	0.547	0.520	0.807	0.650	1.014	1.175	0.595	0.852	0.826	0.534	1.147	0.630	1.055
SD	—	—	0.131	—	—	—	0.081	0.103	0.220	0.133	—	0.024	0.314

Table 15 Peak Monoclonal-to-Polyclonal Ratios in Normal Children

	Peak 101.3-to-Polyclonal Ratios in Normal Children							
	N_1	N_2	N_3	N_4	N_5	N_6	N_7	N_8
Sleep test #1	0.950	0.571	0.939	0.900	1.538	0.600	1.150	1.455
#2	0.667	—	1.000	0.578	1.000	0.913	0.753	1.818
#3	<0.5	—	0.943	0.926	—	0.817	<1	—
#4	—	—	—	0.653	—	—	—	—
Average	0.809	0.521	0.961	0.764	1.269	0.795	0.927	1.637
SD	0.200	—	0.034	0.175	0.380	0.170	0.245	0.257

	Peak 33.2-to-Polyclonal Ratios in Normal Children				
	N_1	N_2	N_3	N_4	N_5
Sleep test #1	1.270	1.121	0.978	1.021	1.091
#2	0.857	1.183	1.368	1.021	—
#3	—	0.857	1.424	—	—
Average	1.064	1.056	1.273	1.021	1.091
SD	0.292	0.123	0.261	0.000	—

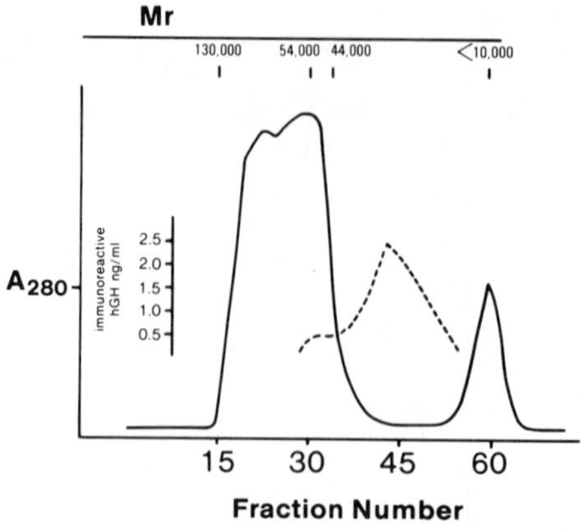

Figure 11 Gel-filtration profile of hGH-deficient serum spiked with clinical-grade hGH on Ultrogel ACA44. Solid line indicates absorption at 280 mm. Broken line represents immunoreactivity of fractions as determined by the polyclonal assay.

differential effects of the degradation process on immunoreactivity in the four assays. To investigate this question, we incubated clinical-grade hGH (20 ng/ml) in heparinized blood from a GHD child (<1.0 ng/ml serum GH by the polyclonal RIA), and in the corresponding plasma at 25°C or at 37°C, under either air or 95% O_2, 5% CO_2. In all four assays, the immunoreactive levels did not change significantly within 5 hr of incubation.

3.3.4. Gel-Filtration Analyses

Twenty-six nocturnal sera containing >10 ng/ml GH in the polyclonal RIA (12 normal subjects and 14 hyposomatomedinemic, non-GHD short children) were examined by gel-filtration chromatography. For comparison, zero GH serum from a GHD child, to which 20 ng/ml of pure hGH had been added, was also analyzed. Typical results are shown in Fig. 11. After chromatography of the serum containing clinical-grade hGH, the immunoreactivity eluted between fractions 36 and 42, with a peak at tube 40. The overall recovery was about 80%. No other immunoreactive peak was observed. The same type of chromatogram was obtained in all 26 nocturnal serum samples. The overall recovery of the serum polyclonal immunoreactivity ranged from 65 to 90% and was in fractions 36–42. The biclonal-to-polyclonal ratio of immunoreactive GH recovered did not differ significantly from that in the serum applied to the column.

4. Discussion

4.1. Defining the Normal Range for Plasma SMC

A study similar to the present one, but based on only one SMC determination per child, was published by Bala *et al.*[8] We have estimated from their graphic data the average ±1 SD for ages 7–9 to be 0.46 ± 0.30 units/ml for boys and 0.66 ± 0.31 for girls. Our data are in close agreement with those of Bala *et al.* For example, in the present study, single determinations at age 7.5–8.4 resulted in a mean value of 0.57 ± 0.30 units/ml for boys and 0.67 ± 0.30 for girls.

We identified four sources of variation in normal plasma SMC: intraassay, day-to-day, month-to-month, and intersubject. Since the month-to-month estimated variability was similar in magnitude to the intersubject variability, we investigated the idea of establishing tolerance limits based on multiple determinations per child.

Tolerance limits based on the average of four determinations at 6-week intervals were clearly preferable to fewer than four determinations. For example, the widths of the 95% tolerance intervals for the 7-year-old boys were 0.934, 0.886, 0.714, and 0.554 units/ml based on one, two, three, and four determinations, respectively. The width of the 95% tolerance interval based on five determinations at 6-week intervals was estimated to be 0.495 units/ml by assuming the mean of five determinations to be 0.495, and by modifying the estimate of the standard deviation to be based on five determinations instead of four determinations. Because the estimated width of the tolerance interval would be reduced by only about 0.06 units/ml by adding a fifth determination, we recommend tolerance intervals based on four determinations at 6-week intervals.

The tolerance limits are influenced not only by the number of determinations per child, but also by the number of children in each age group. Our tolerance limits are based on 14–15 subjects per group. The tolerance limits will in theory decrease with an increase in the number of subjects per group. For example, had we chosen 30 7-year-old boys instead of 15, then we estimate that the tolerance interval would be (0.235, 0.755), by assuming that the mean and standard deviation based on 15 subjects would remain the same if 30 subjects were observed, and by using 30 instead of 15 for n. Comparison of this interval with that based on 15 subjects (0.218, 0.772) shows that only a small reduction would be achieved by increasing the group size from 15 to 30.

Our data have two practical applications. First, identifying the child with a subnormal plasma SMC will help to screen short children for GHD. We demonstrated how this procedure could be applied with seven GHD children. In addition, the present analysis not only identifies the child with SMC <2.5 percentile, but also classifies other children according to their estimated percentile in the overall distribution. It now becomes possible, for example, to search for correlations between percentile SMC and percentile growth rate during a selected time period.

4.2. The Prevalence of Subnormal SMC among Short Children

Our data show that in children below the 3rd percentile in height, there is a 20-fold excess of low SMC levels for chronological age (45% versus 2%, Fig. 5). About 40% of the excess of low SMC in the present series is caused by either total or partial GHD. About 40% of the excess cannot be ascribed to GHD, and the following possibilities can be considered. (1) The low SMC could be related to the delayed maturation. As Bala *et al.*[8] showed and we have confirmed, plasma SMC increases in children with chronological age. The normal mean SMC level for 5-year-olds, for example, is substantially less than that for 10-year-olds. The excess of low SMC values in our short, non-GHD 7- to 10-year-olds, therefore, could be related to their retarded bone age (average bone age/chronological age, 0.66). If the SMC distribution were plotted versus bone age rather than chronological age, the excess of hyposomatomedinemia would probably be reduced. We could not examine this possibility because we have not determined average SMC ($n = 4$) for normal children with chronological and bone ages <7 years. (2) A possible failure of nocturnal secretion of hGH as described by Wise *et al.*[9] has been ruled out by the nocturnal serum GH data (Table 13). (3) Numerous other causes cannot be excluded: Impaired hepatic production of SMC, nutritional disorders,[10] abnormally rapid clearance of SMC, an altered SMC binding system,[11] or a bioinactive GH[3–5] are all possible, but are generally outside the scope of the present data. Average weight was appropriate for average height (Table 3), tending to rule out severe malnutrition. The profile of four RIA's did not differ significantly from normal, thereby failing to support a bioinactive GH hypothesis. However, the demonstration that at least 20% of the endogenous GH is altered both in the non-GHD group and in normal children and adolescents (see Section 4.3) means that the possibility of a structural difference in endogenous GH between the two groups is difficult to exclude at present.

4.3. Altered GH in Serum

The average biclonal/polyclonal immunoreactive ratio of nocturnal serum was significantly below 1 in both normal and hyposomatomedinemic, non-GHD, short children. What does this signify? If the endogenous GH in these sera was identical to the native hGH molecule which was used to standardize the four RIA's, then this ratio would have been 1.0. Because the average ratio was 0.7–0.8, at least 20–30% of the endogenous GH must differ in structure from the native hormone molecule. One possibility would be a mixture of about 75 parts of the native molecule, with biclonal/polyclonal ratio = 1.0, and 25 parts of an altered molecule with a ratio of 0. Alternatively, all the endogenous GH could be an altered form of hGH with a biclonal/polyclonal ratio of 0.75. Intermediate hypotheses are also possible. The extensive interindividual differences in the biclonal/polyclonal ratio (range, 0.5–1.1) (Table 14) suggest that endogenous serum GH is probably a complex mixture of altered forms.

Because the antigenic specificities of the five antisera in the four RIA's are not defined, we cannot speculate about where in the hGH molecule the alterations have occurred.

The structural alteration or heterogeneity responsible for the low biclonal/polyclonal ratio could arise in the pituitary gland, where two different GH genes are present, where the primary gene transcript is clipped and spliced, where the pre-GH product of the ribosomes is proteolytically shortened, and where structurally different forms of 22K GH have been recognized.[12] Alternatively, the low biclonal/polyclonal ratio of the serum GH could reflect alteration of the native GH molecules by peripheral tissues, with reentry of modified GH molecules into the blood. The incubation experiments in Phase 3 showed that the alteration does not occur in the circulation itself.

The absence of a statistically significant difference in any feature of the RIA profile, or in the gel-filtration chromatogram, between normal and non-GHD, hyposomatomedinemic short children, militates against the bioinactive GH hypothesis for the latter group as a whole. However, the considerable interindividual differences in the various RIA ratios within each group (Table 14) make it impossible to exclude this possibility in every case.

4.4. The Linear Growth Response

If we take a 5 cm/year or greater increment in growth velocity as a "satisfactory response," then at the 0.26 unit/kg per week dose about 25% of the non-GHD hyposomatomedinemic short children were "responders." At the next higher dose, the proportion of "satisfactory responses" increased to 60%. Thus, our data support the recent series of reports[2–6] that linear growth response to hGH is not a unique feature of GHD children, but can also be elicited in many non-GHD, hyposomatomedinemic short children. At conventional doses (up to 0.43 unit/kg per week), the magnitude of the response by such children was less than 60% as great as that of their GHD counterparts. At the highest dose of 0.70 unit/kg per week, the response by the non-GHD, hyposomatomedinemic short children resembled that of the GHD patients but only a few children were tested at this dose.

ACKNOWLEDGMENT. This work was supported in part by NIH Grants #AM 31149 and RR 00039.

References

1. R. W. Furlanetto, L. E. Underwood, J. J. Van Wyk, and A. J. D'Ercole, Estimation of somatomedin-C levels in normals and patients with pituitary disease by radioimmunoassay, *J. Clin. Invest.* **60**:648–657, 1977.
2. A. Hayek and G. T. Peake, Growth and somatomedin-C responses to growth hormone in dwarfed children, *J. Pediatr.* **99**:868–872, 1981.
3. A. A. Kowarski, J. Schneider, E. Ben-Galim, V. V. Weldon, and W. H. Daughaday, Growth failure with normal serum RIA-GH and low somatomedin activity: Somatomedin restoration and growth acceleration after exogenous GH, *J. Clin. Endocrinol. Metab.* **47**:461–464, 1978.
4. D. Rudman, M. H. Kutner, R. D. Blackston, R. A. Cushman, R. P. Bain, and J. H. Patterson, Children with normal-variant short stature: Treatment with human growth hormone for six months, *N. Engl. J. Med.* **305**:123–131, 1981.
5. T. Frazer, J. R. Gavin, W. H. Daughaday, R. E. Hillman, and V. V. Weldon, Growth hormone-dependent growth failure, *J. Pediatr.* **101**:12–15, 1982.
6. S. L. Blethen and F. I. Chasalow, Use of a two-site immunoradiometric assay for growth

hormone (GH) in identifying children with GH-dependent growth failure, *J. Clin. Endocrinol. Metab.* **57:**1031–1035, 1983.

7. S. D. Frasier, A review of growth-hormone stimulation tests in children, *Pediatrics* **53:**929–937, 1974.
8. R. M. Bala, J. Lopatka, A. Leung, E. McCoy, and R. G. McArthur, Serum immunoreactive somatomedin levels in normal adults, pregnant women at term, children at various ages, and children with constitutionally delayed growth, *J. Clin. Endocrinol. Metab.* **52:**508–512, 1981.
9. P. H. Wise, R. B. Burnet, T. D. Geary, and H. Berriman, Selective impairment of growth hormone response to physiological stimuli, *Arch. Dis. Child.* **50:**210–214, 1975.
10. R. L. Hintz, R. Suskind, K. Amatayakul, O. Thanangkul, and R. Olson, Plasma somatomedin and growth hormone values in children with protein–calorie malnutrition, *J. Pediatr.* **92:**153–156, 1978.
11. D. H. Morris and D. S. Schalch, Structure of somatomedin-binding protein: Alkaline pH-induced dissociation of an acid-stable, 60,000 molecular weight complex into smaller components, *Endocrinology* **111:**801–805, 1982.
12. R. K. Chawla, J. S. Parks, and D. Rudman, Structural variants of human growth hormone, *Ann. Rev. Med.* **34:**519–547, 1983.

13

Growth-Hormone-Resistant Syndromes

Zvi Laron, Aviva Silbergeld, Athalia Pertzelan, and Rina Eshet

1. Introduction

Resistance to a hormone is defined as a state of a cell or tissue in which a normal concentration of hormone evokes a lesser-than-expected (i.e., normal) response.

One distinguishes two types of resistance: (1) "decreased sensitivity"—when greater concentrations of hormone are required to elicit quantitatively normal responses, and (2) "decreased responsiveness"—when the maximal response is decreased without a necessary alteration in sensitivity.

The defects responsible for resistance to a given hormone may be at a "prereceptor," "receptor," or "postreceptor" level.

Syndromes of resistance to hormones have been described for insulin,[1] thyroid hormone,[2] androgens,[3] parathyroid hormone, vitamin D,[4] and others. A syndrome of resistance to human pituitary growth hormone is described in this chapter.

2. Laron-Type Dwarfism

In 1966[5] and subsequently in 1968[6] we described a syndrome of familial dwarfism which was indistinguishable from isolated growth hormone deficiency (IGHD) both clinically and with respect to many of the laboratory findings. Subsequently additional patients were described. The number of proven patients investigated by us so far is 29.[7] Following our reports, patients with this

Zvi Laron, Aviva Silbergeld, Athalia Pertzelan, and Rina Eshet • Institute of Pediatric and Adolescent Endocrinology, Beilinson Medical Center, Petah Tikva 49 100, Israel, and Sackler School of Medicine, Tel Aviv University, Tel Aviv, Israel.

syndrome, coined by Daughaday *et al.*[8] "Laron-type dwarfism" (LTD), have also been described in other countries.[7] Presented here is a detailed description of the genetic, clinical, metabolic, and endocrine features of this disease as well as the laboratory investigations that led us to discover its pathophysiology and etiology.

2.1. Ethnic Origin and Genetic Aspects

Of the 29 patients diagnosed to have LTD by us since 1964 all except one are of Oriental Jewish origin. The 25 living in Israel belong to 18 Oriental Jewish families—nine Iraqi, five Iranian, two Yemenite, one Afganistani, and one Syrian–Moroccan. Of the four patients investigated by us who live elsewhere, three are from Iran (one of them being non-Jewish) and one is from Turkey.

The group of 25 patients living in Israel is characterized by a high uncorrected segregation ratio of 0.36 (corrected: 0.20–0.26), a high rate of consanguinity (9 of the 17 families) (Fig. 1), and a normal sex ratio (10 males, 15 females).[9] These features support the conclusion that this disease is caused by an autosomal, fully penetrant recessive gene.[10]

The 18 families under our care have a total of 67 children of whom 24 are affected. In one family with two affected children, one of the parents is also affected.

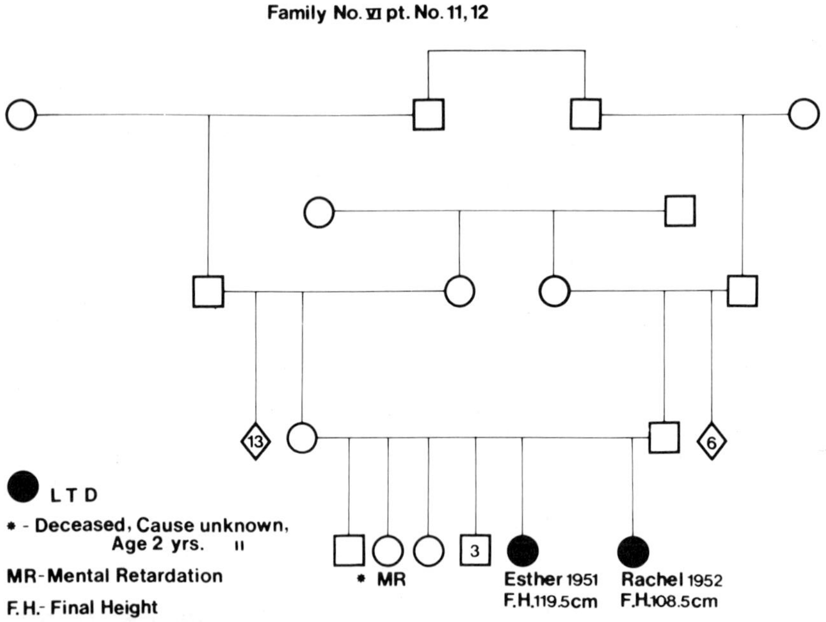

Figure 1 Pedigree of a family with two children with Laron-type dwarfism (LTD). Note marked degree of consanguinity.

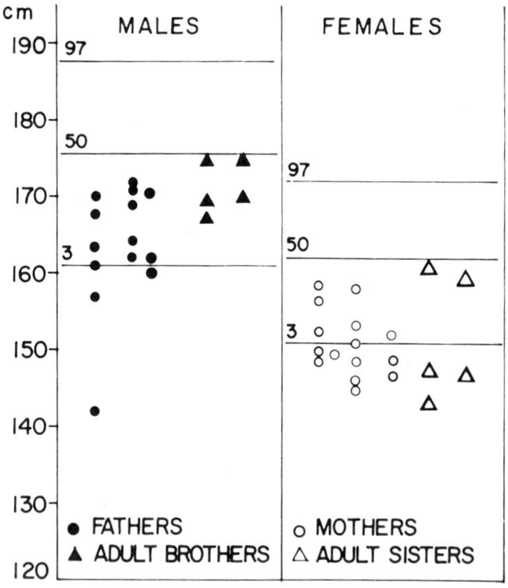

Figure 2 Height of parents and siblings of patients with Laron-type dwarfism.

The statures of the parents and siblings of affected children are below the 50th and even 3rd percentile for the respective sex (Fig. 2). They are short when compared to normal controls of the same ethnic groups.[11]

Around the world approximately 43 additional patients are known at present, some of them already reported in the literature.[7] Many of them are of Arab origin; others are of Spanish or Dutch origin (Table 1).

2.2. *Perinatal Features*

Birth weight was normal (>2500 g) in 18, low in three (1990–2120 g), and unknown in four children. Birth length was 2 SD or more below the mean normal length for sex and ethnic origin in 10 children, ranging from 42 to 46 cm.[11] In six children the birth length was within the normal range, and in the remaining it was unknown. Elders *et al.*[12] and Kastrup *et al.*[13] also mention a subnormal birth length of 47 cm.

2.3. *Congenital Malformations*

About half of the probands have some congenital malformations, including congenital dislocation of the hip joints, Perthes' disease, short 4th metatarsus, clinodactyly, and brachymesophalangy. Other anomalies reported are congenital strabismus, cataract, nystagmus, aortic stenosis, undescended testicle, and harelip.

Table 1 Laron-Type Dwarfism: Ethnic Origin of Known Patients[a]

Country	Origin	Number
Israel	Jewish Oriental	28
Israel	Iranian Moslem	1
Lebanon	Arab	7
Saudi Arabia	Arabs Moslem	4
Algeria	Moslem	1
Pakistan	Moslem	1
Netherlands	Dutch	1
Canada	Dutch	2
Canada	Eskimo	3
Spain	Spanish	3
Brazil	Spanish–Italian	2
Argentina	Spanish	1 (+ 1 ?)
United States	Spanish (partly)	2
Italy (N. Italy and Sicily)	Half Sicilian and Sicilian	2
Sardinia	Sardinian	1
France	French	2
	Italian	2
	Jewish–Iranian	1
Sweden	Scandinavian	1
Denmark	Scandinavian	1
Germany	German	1
Japan	Japanese	2
New Zealand	Anglo–Saxon	1

[a] The listed patients have been examined or brought to the attention of the senior author. Part of these data have already been published (for details see ref. 7).

2.4. Early Development

Motor development in general is slow. Skeletal maturation is retarded and the anterior fontanelle closes between 3 and 7 years. There is a characteristically high-pitched voice. As infants these children eat very little but are not thin. They have a craving for sugar and at night wake up crying and sweating until given a sweetened drink.(14)

2.5. Clinical Features

After infancy, the general appearance of these patients is identical to that of patients with IGHD who have received no treatment. No distinction can be made between these two types of dwarfs on the basis of the clinical findings (Fig. 3). Furthermore, there is a striking resemblance among patients from all over the world, no matter what their origin. The nutritional state of these children is good, and they appear obese. The small face and mandible give a false impression of a large head although the head circumference is below normal (Fig. 4). The disproportion between face and calvarium results in a saddle nose and the "sign of the setting sun." The small face is due to

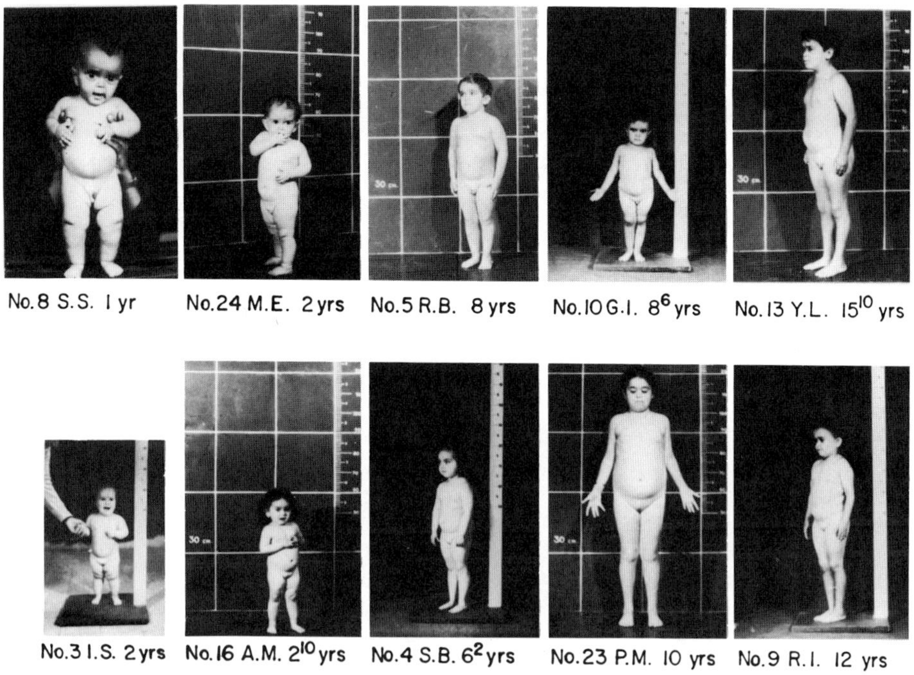

Figure 3 Appearance of Laron-type dwarfism patients of both sexes.

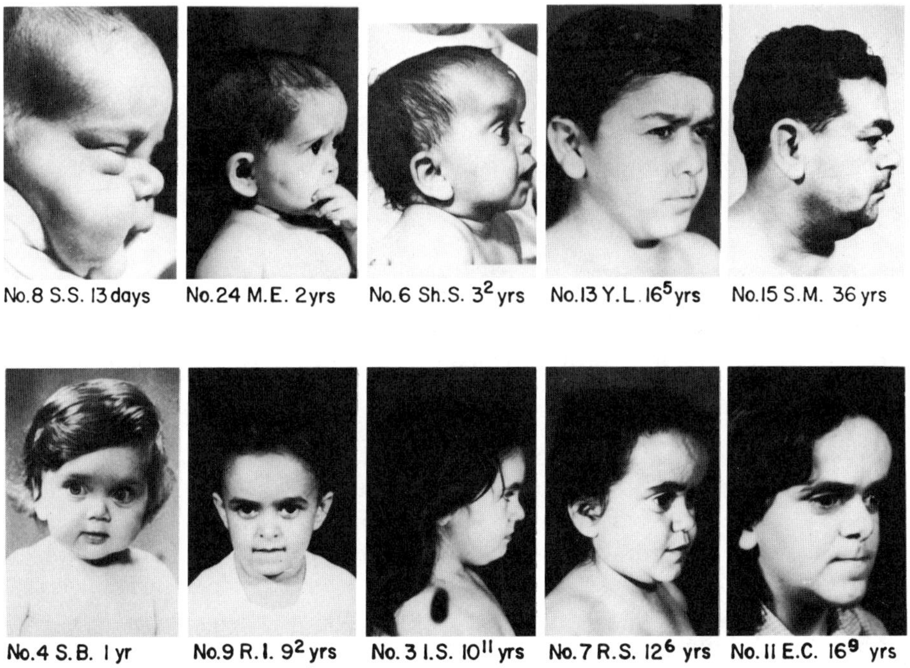

Figure 4 The facial appearance in Laron-type dwarfism in both sexes at different ages.

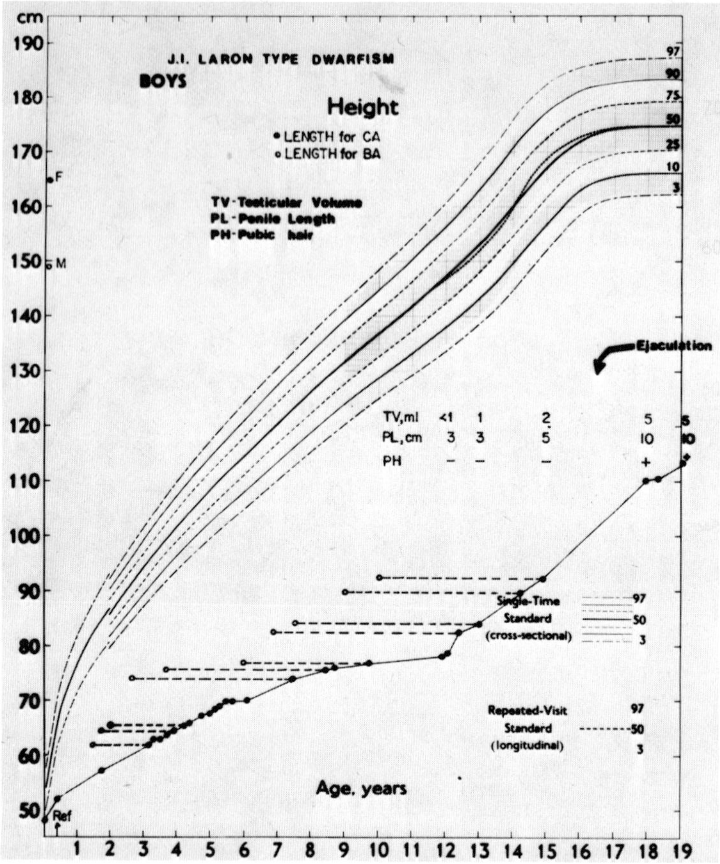

Figure 5 Linear growth pattern of a boy with Laron-type dwarfism. (Reproduced with permission from Laron.[7])

underdevelopment of the skull base.[15,16] The teeth are discolored and defective, and teething is slow. Owing to the smallness of the maxilla and mandible there is a tendency for crowding. Hair growth is also slow, and hair is sparse with deep temporal recessions.

From early infancy growth is slow, below the 3rd percentile (Figs. 5 and 6), and a pubertal growth spurt has been observed only in some of the boys.[17] Final heights registered up to the present in our patients are shown in Table 2. It is evident that they are very low. Body proportions are abnormal, with an abnormally high upper/lower body segment ratio[18] because of short limbs in comparison to the trunk (Fig. 7). The hands and feet are small (acromicria).

Measurement of skinfold thickness shows that subcutaneous fat tissue is well developed from early infancy, and these patients are obese at all ages (Fig. 3). The degree of obesity is not adequately expressed by the weight curve or calculation of the weight age since the bones are very thin and the muscular

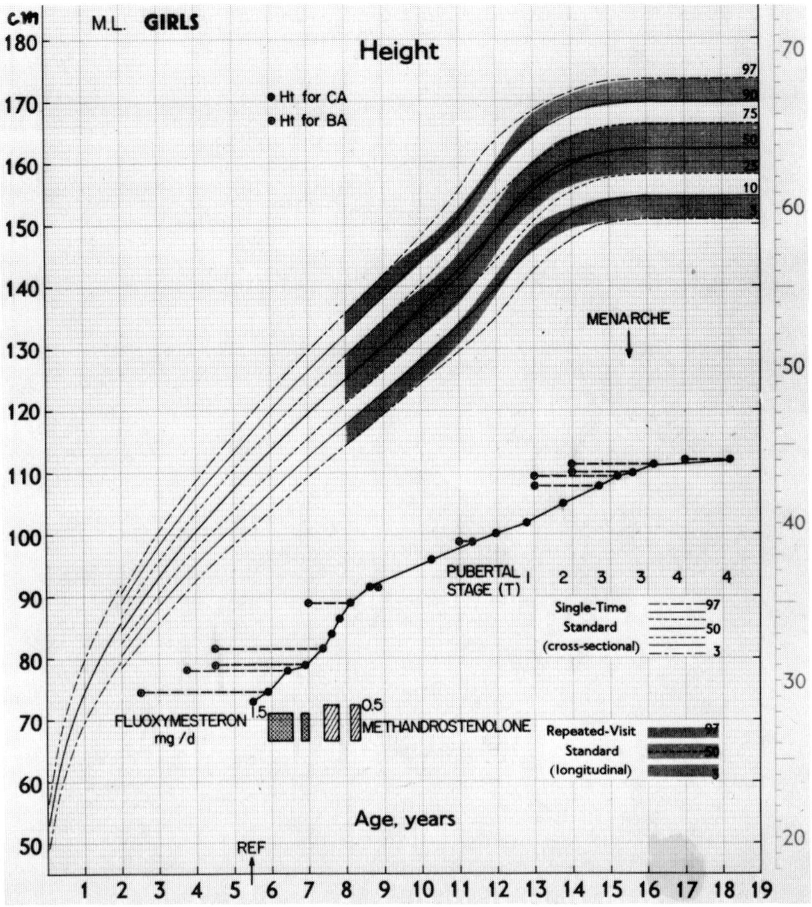

Figure 6 Linear growth pattern of a girl with Laron-type dwarfism. (Reproduced with permission from Laron.[7])

tissue underdeveloped. After puberty the skin has a fine texture and wrinkles, as in premature aging. The genitalia and gonads are small from early childhood[19]; this is easily observed in young boys (Fig. 8) but becomes evident in the girls at a later age. Sexual maturation is eventually achieved but with some delay.[17,20]

2.6. Laboratory Findings

In all of our patients routine blood counts and urinalysis, blood calcium, phosphorus, phosphatase, liver, and thyroid function were within the normal range. Chromosomal analyses revealed no abnormalities.

Skeletal maturation (bone age) was retarded until the late stages of puberty. X-rays of the skeleton revealed small, delicate long bones.

Table 2 Laron-Type Dwarfism—Final Heights

Males				Females			
		Height				Height	
Patient	Age at last examination	cm	SD	Patient	Age at last examination	cm	SD
B.R.	20:4	137.7	−6	R.S.	34:0	113.3	−8.7
S.S.	25:0	109.0	−10	B.S.	22:6	119.0	−7.7
S.S.	28:0	120.3	−8.5	R.S.	27:0	129.0	−4.7
G.J.	23:8	119.0	−6	R.J.	25:11	112.5	−9
Y.L.	30:0	128.3	−7.2	E.C.	32:0	119.0	−6.7
S.E.	20:2	125.1	−8	R.C.	31:0	100.8	−10
				M.L.	28:0	110.8	−8.3
				A.M.	17:2	123.0	−6.7
				S.C.	32:0	120.6	−7.2
				P.M.	22:6	136.3	−4.2

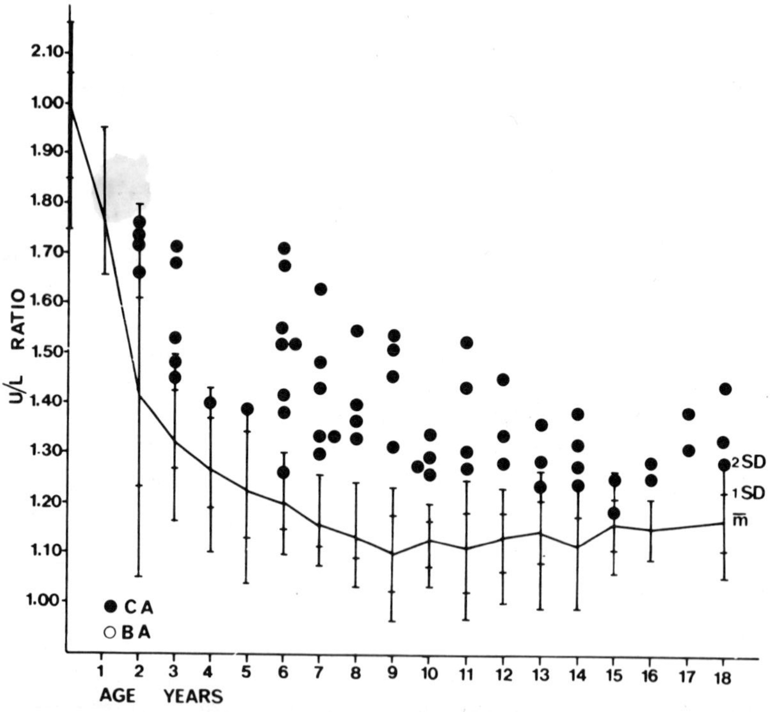

Figure 7 Upper/lower body segment ratios at repeated measurements in 11 girls with Laron-type dwarfism. (The boys have identical proportions.)

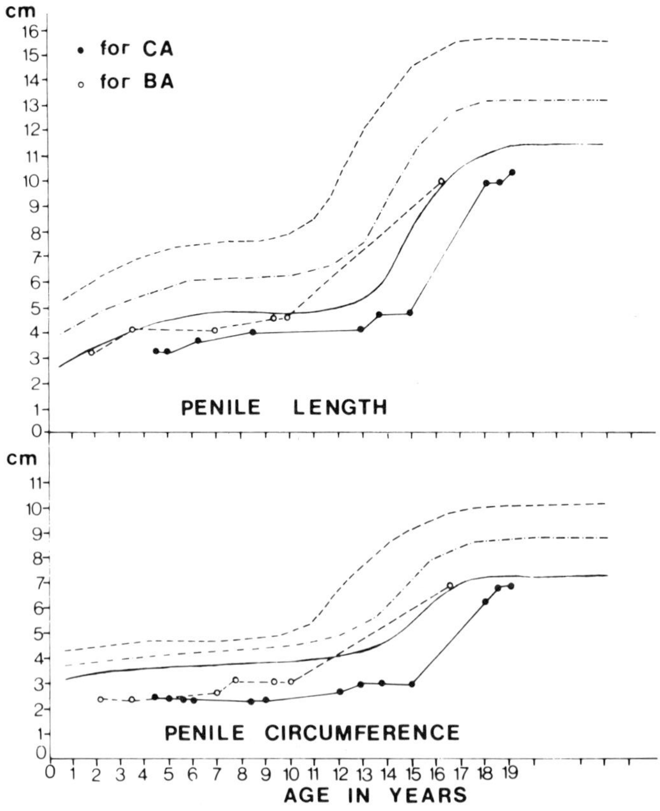

Figure 8 Penile and gonadal growth pattern in a male with Laron-type dwarfism.

Most patients have had low levels of blood glucose and a tendency to hypoglycemia at one time or another, occurring more often during the younger years.(14) The affected father developed type II diabetes.(21)

One of the main laboratory findings in this syndrome is the elevated level of plasma human growth hormone (hGH), particularly in early childhood, when levels in the acromegalic range are found.(6,7,14) Later in life, levels tend to fluctuate, even in individual patients (Table 3), and in some may approach the normal range. Figure 9 illustrates the increased secretion of hGH as registered by continuous monitoring. Stimulation tests also revealed a higher-than-normal rise in most of the patients studied.(7,14) The hGH secretion in the LTD patients is usually blocked by hyperglycemia,(14) corticosteroids,(6) and somatostatin.(22) The other main laboratory finding in this syndrome is the low level of serum somatomedin which fails to rise after one or two daily intramuscular injections of hGH (5–10 IU for 4 days)(8,23) (Figs. 10, 11). Prolonged treatment with exogenous hGH for several months did not significantly accelerate the linear growth rate.(23) Thus it became clear that LTD presents a condition of "resistance to human pituitary growth hormone."

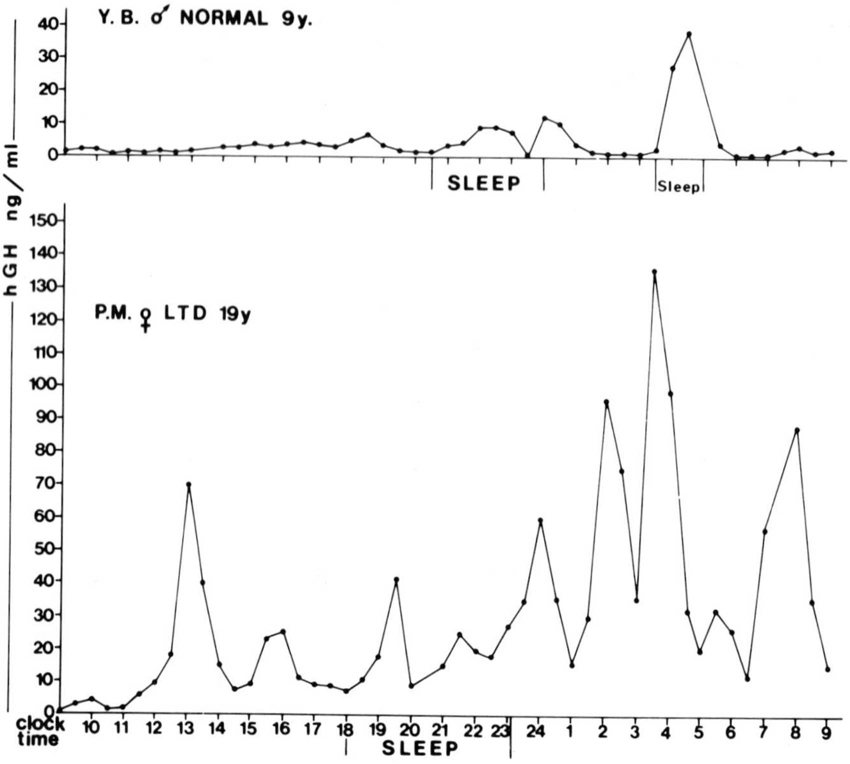

Figure 9 Diurnal variation of hGH secretion in a girl with Laron-type dwarfism compared to a normal control. Note increased number of human growth hormone spikes and high peaks. Normally young adults have lower peaks than children. (In collaboration with Z. Zadik and R. Keret.)

2.7. Pathogenesis

Resistance to a hormone may occur owing to defects at the prereceptor, receptor, or postreceptor level. Prereceptor defects may be caused by specific antibodies against hGH, abnormal hGH synthesis, or accelerated hGH degradation. Investigating the sera of 20 of the patients, we found no evidence for antibodies against hGH. The responsiveness of the pituitary of these patients to conventional stimuli and suppressors of hGH secretion as well as the elution behaviour on gel electrophoresis was normal.(24) There was a similar behavior to standard hGH of the plasma hGH of LTD patients in radioimmunoassays using homologous(25) or heterologous antisera.(26) The metabolic clearance rate of tracer [^{125}I]hGH in two LTD patients revealed a clearance of 73 and 75 ml/min per m^2, while in IGHD the clearance ranges from 74 to 160 and in normal controls from 90 to 125 ml/min per m^2. All these findings were considered as evidence for the integrity of the hGH molecule in the LTD patients. Unavailability of large quantities of hGH from LTD patients has made it impossible so

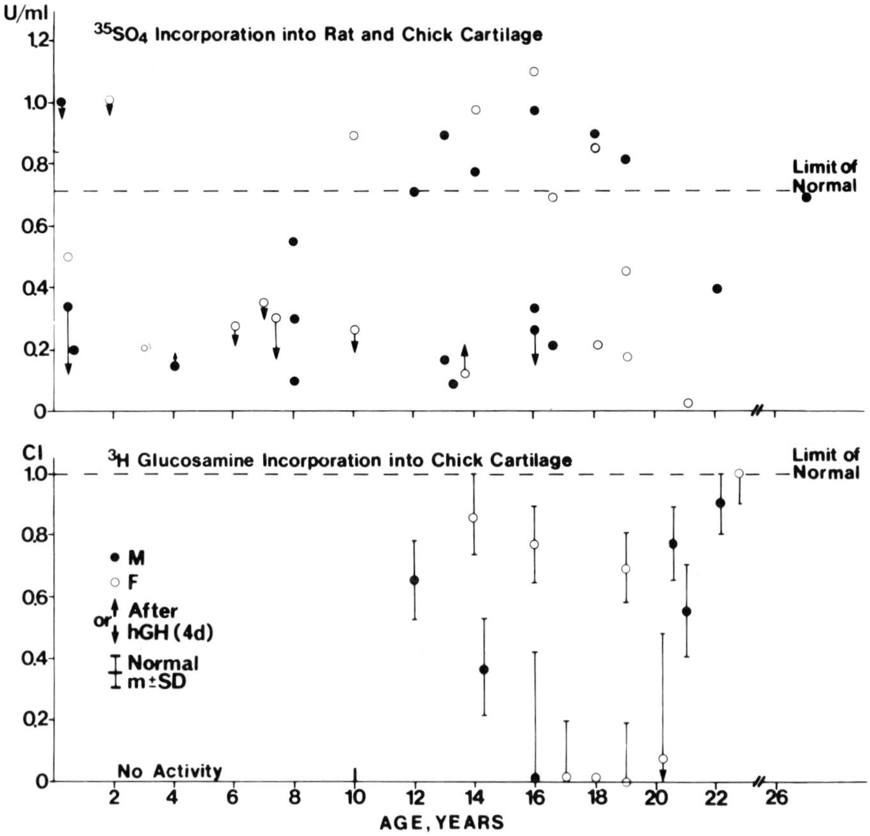

Figure 10 Serum bioassayable somatomedin levels in Laron-type dwarfism patients before and after human growth hormone (hGH) administration.

Table 3 Change in Fasting Human Growth Hormone Levels with Age in Two Patients with Laron-Type Dwarfism

Pt. R.J.—female		Pt. P.M.—female	
Age (years)	hGH (ng/ml)	Age (years)	hGH (ng/ml)
$12^{10}/12$	38.0	$4^3/12$	50.0
$14^8/12$	12.5	$6^{11}/12$	19.0
$18^7/12$	14.5	$13^4/12$	9.6

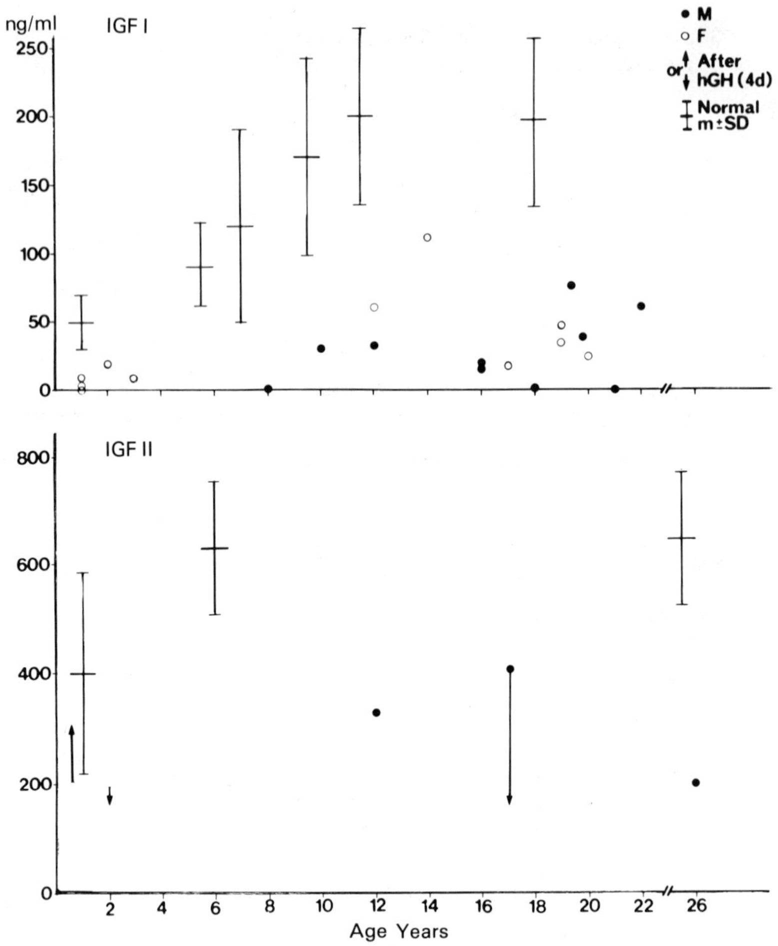

Figure 11 Radioimmunoassayable somatomedin levels in Laron-type dwarfism patients before and after human growth hormone (hGH) administration. (In collaboration with Dr. J. Zapf, Zurich).

far to test directly its biological activity, but measurement of binding capacity of plasma hGH from LTD patients to human liver microsomes in a radioreceptor assay revealed normal values.[27,28] Thus a prereceptor defect seemed excluded.

The next step was to study the possibility of a defect at the receptor level. Using radioreceptor methods we found no antibodies to the hGH receptor in the liver. Direct investigations of the hGH receptor were made possible when the families of two LTD patients and the older patient himself agreed to an open liver biopsy (approved also by the Hospital Ethical Committee). The patients were two males aged 4 and 24 years. One gram of liver tissue was obtained, and 100,000-g microsomal pellets were prepared and tested for

Table 4 *Percent Specific Binding of [^{125}I]hGH and [^{125}I]Insulin to Liver Microsomes from Two Patients with Laron-Type Dwarfism (LTD) as Compared to Normal Liver Microsomes*

				Percent binding of [^{125}I]hGH			Percent binding of [^{125}I]insulin		
Microsomes	Pt. no.	Sex	Age (years)	Total	Nonspecific with 5 μg/ml hGH	Specific	Total	Nonspecific with 10 μg/ml insulin	Specific
Liver from LTD[a]	1	M	4	7.0	6.5	0.5	47.3	13.9	33.4
	2	M	26	7.0	6.9	0.1	11.2	4.1	7.1
Normal liver[a] (concomitant human liver)	C	M	13	35.1	6.8	28.3	17.3	9.6	7.7
Normal human liver[b]	a	M	30		(n = 2)[c]	7.3 and 8.6[d]			
	b	M	16		(n = 7)	13.4 ± 3.0			
	c	M	13		(n = 10)	24.1 ± 4.3			
	d	F	20		(n = 5)	16.2 ± 4.5			
	e	M	19		(n = 2)	15.6 and 10.8			
	f	M	47		(n = 5)	12.2 ± 3.2			

[a] Liver tissue was tested per 12 mg/ml of microsomal fraction protein.
[b] Liver tissue was expressed per 12 mg/ml of microsomal fraction protein.
[c] n = number of hGH receptor binding assays.
[d] Mean ± SD.

binding with standard hGH and insulin (Table 4). The results were compared with those obtained with liver tissue of control subjects (the liver having been obtained from kidney transplantation donors pronounced clinically dead, with approval of their families. The results revealed no specific binding of [^{125}I]hGH by the liver microsomes of the LTD patients whereas those from normal subjects had a mean specific binding ranging from 8% to 24%. In contrast, it was found that the liver microsomes of the LTD patients bound [^{125}I]insulin, demonstrating that the hGH receptor defect was specific.[29] The plasma hGH levels at the time of the biopsy were 45 ng/ml in the 4-year-old and 20 ng/ml in the adult, indicating that there is no relationship between the level of hGH and binding by receptors.

Evidence indicating that this defect in hGH receptors may not be confined to the liver tissue is found in the investigations by Golde *et al.*,[30] who reported that "... erythroid colonies from 2 LTD patients grown in vitro were non-responsive to exogenous hGH in contrast to normal controls and patients with primary hGH deficiency." We found also that the addition of normal serum to cultures of fibroblasts from LTD patients enhanced the incorporation of [^{3}H]thymidine,[24] which could mean that fibroblasts and perhaps other target cells in LTD are responsive to growth factors. Thus it seems that the specific defect for the resistance to hGH in LTD is a lack of receptors for growth hormone in the liver (and possibly in other tissues), and this leads to an inability to generate somatomedin, which in turn causes the clinical signs and metabolic disturbances. This defect is hereditary, and its biochemical nature remains to be established. Recent investigations have revealed that Laron-type dwarfism patients possess the gene for IGF-I (somatomedin C).[33]

3. GH Resistance Other than LTD

A temporary and partial resistance to hGH has been noticed in newborns and acromegalics[23] in whom the high levels of endogenous hGH probably block the hGH receptrs and thus decrease the sensitivity to exogenous hormone.

Resistance to hGH due to postreceptor defects can be envisaged, but has not been definitely proven. The pygmies, who have been found to have a partial deficiency in IGF-I,[31] may belong to this group, but high values of plasma hGH, as would be expected in such a condition, have not been found.

Defects in target cells could occur at the level of the somatomedin receptors or at the postreceptor stage. We have observed an Iranian dwarfed girl with elevated serum somatomedin levels.[24] She did not resemble the LTD patients; she was thin and had low plasma hGH, as would be expected from a feedback regulation between somatomedins and hGH. The defect in IGF-I receptors in fibroblasts from a leprechaun child[32] may be evidence for another disease in this category.

4. Conclusions

Several types of resistance to hGH have been described. The most extensively studied disease has been LTD, the etiology of which has been found to be a defect in the GH receptor. Despite the elucidation of the basic defect in LTD, several problems remain to be solved. This will keep us and other investigators busy until the next symposium:

1. Delineation of the chemical nature of the defective hGH receptor.
2. Determination of whether the defect in hGH receptors is generalized or limited to specific tissues.
3. Explanation of the physiology of the existing linear growth in the absence of somatomedin.
4. Detection of so far undiagnosed patients around the world.
5. Discovery of therapeutic means.

References

1. L. C. Harrison, Insulin resistance: Clinical syndromes and molecular mechanisms, in: *Endocrinology 1980* (I. A. Cumming, J. W. Funder, and F. A. O. Mendelsohn, eds.), Australian Academy of Sciences, Canberra, 1980, pp. 571–574.
2. S. Refetoff, L. J. De Groot, B. Benard, and L. T. De Wied, Studies of a sibship with apparent hereditary resistance to the intracellular action of thyroid hormone, *Metabolism* **21:**723–756, 1972.
3. J. E. Griffin and J. D. Wilson, The syndrome of androgen resistance, *N. Engl. J. Med.* **302:**198–209, 1980.
4. S. E. Papapoulos, S. Adami, and J. L. H. O'Riordan, Hormonal resistance in disorders of calcium homeostasis, *Ann. Clin. Res.* **12:**254–258, 1980.
5. Z. Laron, A. Pertzelan, and S. Mannheimer, Genetic pituitary dwarfism with high serum concentration of growth hormone. A new inborn error of metabolism? *Isr. J. Med. Sci.* **2:**152–155, 1966.
6. Z. Laron, A. Pertzelan, and M. Karp, Pituitary dwarfism with high serum levels of growth hormone. *Isr. J. Med. Sci.* **4:**883–894, 1968.
7. Z. Laron, Laron type dwarfism (LTD)—Hereditary somatomedin deficiency—A review, in: *Advances in Internal Medicine and Pediatrics* (P. Frick, G. A. von Harnack, G. A. Martini, A. Prader, R. Schoen, and H. P. Wolff, eds.), Springer Verlag, Berlin-Heidelberg-New York, 1984, pp. 117–150.
8. W. H. Daughaday, Z. Laron, A. Pertzelan, and J. N. Heins, Defective sulfation factor generation: A possible link in dwarfism, *Trans. Assoc. Am. Physicians* **82:**129–138, 1969.
9. A. Adam, Z. Josefsberg, A. Pertzelan, Z. Zadik, J. M. Chemke, and Z. Laron, Occurrence of four types of growth hormone related dwarfism in Israeli communities, *Eur. J. Pediatr.* **137:**35–39, 1981.
10. A. Pertzelan, A. Adam, and Z. Laron, Genetic aspects of pituitary dwarfism due to absence or biological inactivity of growth hormone, *Isr. J. Med. Sci.* **4:**895–900, 1968.
11. R. Zaizov and Z. Laron, Body length and weight at birth and one year of age in different communities in Israel, *Acta Paediatr. Scand.* **55:**524–528, 1966.
12. M. J. Elders, J. T. Garland, A. Daughaday, D. A. Fisher, J. E. Whitney and E. R. Hughes, Laron's dwarfism: Studies on the nature of the defect, *J. Pediatr.* **83:**253–263, 1973.
13. K. W. Kastrup, H. Andersen, and K. F. Hanssen, Increased immunoreactive plasma and urinary growth hormone in growth retardation with defective generation of somatomedin A (Laron's syndrome), *Acta Paediatr. Scand.* **64:**613–618, 1975.
14. Z. Laron, M. Karp, A. Pertzelan, R. Kauli, R. Keret, and M. Doron, The syndrome of familial dwarfism and high plasma immunoreactive growth hormone (IR-hGH), in: *Growth and Growth Hormone* (A. Pecile and E. E. Muller, eds.), Excerpta Medica, Amsterdam, 1972, pp. 458–482.
15. R. Konfino, A. Pertzelan, and Z. Laron, Cephalometric measurements of familial dwarfism and high plasma immunoreactive growth hormone, *Am. J. Orthod.* **68:**196–201, 1975.

16. A. Scharf and Z. Laron, Skull changes in pituitary dwarfism and the syndrome of familial dwarfism with high plasma immunoreactive growth hormone. A roentgenologic study, *Hormone Metab. Res.* **4**:93–97, 1972.
17. Z. Laron, R. Sarel, and A. Pertzelan, Puberty in Laron type dwarfism, *Eur. J. Pediatr.* **134**:79–83, 1980.
18. I. Arad and Z. Laron, Standards for upper/lower body segment ratio sitting height—Sub-ischial leg length from birth to 18 years in girls and boys, in: *Proceedings of 1st International Congress of Auxology*, Centro Auxologico Italiano di Piancavallo, Milan, 1979, pp. 159–164.
19. Z. Laron and R. Sarel, Penis and testicular size in patients with growth hormone insufficiency, *Acta Endocrinol.* **63**:625–633, 1970.
20. Z. Laron, Arad J., R. Gurewitz, M. Grunebaum, and Z. Dickerman, Age at first conscious ejaculation: a milestone in male puberty, *Helv. Paediatr. Acta.* **35**:13–30, 1980.
21. Z. Laron and M. Karp, Carbohydrate metabolism in the syndrome of familial dwarfism and high plasma immunoreactive growth hormone (Laron type dwarfism), in: *Secondary Diabetes: The Spectrum of the Diabetic Syndrome* (S. Podolsky and M. Viswanathan, eds.), Raven Press, New York, 1980, pp. 363–371.
22. Z. Laron, A. Pertzelan, M. Doron, S. Assa, and R. Keret, The effect of dihydrosomatostatin in dwarfism with high plasma immunoreactive growth hormone, *Hormone Metab. Res.* **9**:338–339, 1977.
23. Z. Laron, A. Pertzelan, M. Karp, A. Kowadlo Silbergeld, and W. H. Daughaday, Administration of growth hormone to patients with familial dwarfism with high plasma immunoreactive growth hormone. Measurement of sulfation factor, metabolic and linear responses, *J. Clin. Endocrinol. Metab.* **33**:332–342, 1971.
24. Z. Laron, A. Kowadlo Silbergeld, R. Eshet, and A. Pertzelan, Growth hormone resistance, *Ann. Clin. Res.* **12**:269–277.
25. R. Eshet, Z. Laron, M. Brown, and R. Arnon, Immunoreactive properties of the plasma hGH from patients with the syndrome of familial dwarfism and high plasma IR-hGH, *J. Clin. Endocrinol. Metab.* **37**:819–821, 1973.
26. R. Eshet, Z. Laron, M. Brown, and R. Arnon, Immunological behaviour of hGH from plasma of patients with familial dwarfism and high IR-hGH in a radioimmunoassay system using the cross reaction between hGH and HCS. *Hormone Metab. Res.* **6**:80–81, 1974.
27. L. S. Jacobs, D. S. Sneid, J. T. Garland, Z. Laron, and W. H. Daughaday, Receptor active growth hormone in Laron dwarfism, *J. Clin. Endocrinol. Metab.* **42**:403–406, 1976.
28. R. Eshet, S. Peleg, Z. Josefsberg, C. Fuchs, R. Arnon, and Z. Laron, Some properties of the plasma hGH activity in patients with Laron type dwarfism determined by a radioreceptor assay using human liver tissue. *Horm. Res.* (in press).
29. R. Eshet, Z. Laron, A. Pertzelan, R. Arnon, and M. Dintzman, Defect of hGH receptors in the liver of two patients with Laron type dwarfism, *Isr. J. Med. Sci.* **20**:8–11, 1984.
30. D. W. Golde, N. Bersch, S. A. Kaplan, D. L. Rimoin, and C. H. Li, Peripheral unresponsiveness to human growth hormone in Laron dwarfism, *N. Engl. J. Med.* **303**:1156–1159, 1980.
31. T. J. Merimee, J. Zapf, and E. R. Froesch, Dwarfism in the pygmy, *N. Engl. J. Med.* **305**:965–968, 1981.
32. A. B. Knight, M. M. Rechler, J. A. Romanus, E. E. Van Obberghen Schilling, and S. P. Nissley, Stimulation of glucose incorporation and amino acid transport by insulin and an insulin like growth factor in fibroblasts with defective insulin receptors cultured from a patient with leprechaunism, *Proc. Natl. Acad. Sci. USA* **78**:2554–2558, 1981.
33. R. E. Russell, J. S. Parks, L. A. Sexton, R. Keret, M. Kelijman, A. Pertzelman, Z. Laron, and G. I. Bell, in: *The Endocrine Program and Abstracts* (67th Annual Meeting of the Endocrine Society, June 1985), Baltimore, 1985, p. 301.

14

Hormonal Regulation of Growth Hormone Gene Expression

Synthia H. Mellon

1. Introduction

The regulation of growth hormone (GH) production by thyroid and glucocorticoid hormones is controversial (for review see ref. 1). In a review of the literature, it appears that laboratory rats and human subjects may employ different regulatory mechanisms. Cell culture and gene transfer experiments seem to reflect the results seen *in vivo* in rats. However, since cell culture and gene transfer experiments are divorced from other regulatory influences, such as the hypothalamus, they should not necessarily be simplistically interpreted as reflecting *in vivo* physiology. Nevertheless, various lines of evidence support the hypothesis that thyroid and glucocorticoid hormones do regulate GH synthesis. Supraphysiological plasma concentrations of glucocorticoid hormones are associated with poor growth in human beings[2–6] and rats.[7–9] Whereas excess glucocorticoids appear to increase the GH content of rat pituitaries,[10] they seem to decrease GH content in human pituitaries,[11] as well as decreasing plasma GH concentrations.[12] Physiological concentrations of glucocorticoids increase GH production by cultured rat, monkey, or human pituitary cells *in vitro*[13–15]; supraphysiological concentrations of glucocorticoids increase GH production by cultured rat pituitary cells even further.[16–20] Thus, most laboratory data indicate a glucocorticoid-mediated enhancement of GH production in the rat, while extensive clinical data indicate the contrary in human beings.

Thyroid hormone may also play a major role in GH physiology. Hyperthyroidism appears to be directly correlated with increased pituitary GH content, while hypothyroidism is correlated with decreased pituitary GH content.[21–24]

Synthia H. Mellon • Metabolic Research Unit, Department of Medicine, University of California, San Francisco, California 94143.

Since pituitary GH content could be restored by treatment of hypothyroid rats with thyroxine,[10] a causal relationship of thyroid hormone concentration and GH production was inferred. However, human beings with hypo or hyperthyroidism seem to have variable responses of plasma GH in response to provocative stimuli. In addition, the degree of GH response does not necessarily correlate with the degree of hypo- or hyperthyroidism.[25]

Therefore, although studies presented in this paper using either cultured rat pituitary tumor cells or transfer of cloned rat or human GH genes into heterologous cells indicate that the GH gene is regulated by glucocorticoid and thyroid hormones, these studies should be viewed as a model of gene expression and should not necessarily be construed as reflecting human physiology.

2. *Cultured Cells as a Model for Rat GH Gene Expression*

Cultured rat pituitary tumor cells (GH cells) have been a model system for studying the hormonal regulation of GH production.[16,17] These cells secrete either GH, prolactin, or both hormones, into the culture medium. GH is not stored in or degraded by these cells, and thus the extracellular concentration of GH reflects the amount of GH synthesized by these cells. GH production increases in response to glucocorticoids and can also be modulated by other stimuli. Yu *et al.*[26] showed that when GH_3 cells (subline of GH cells) are stimulated with the synthetic glucocorticoid dexamethasone, a doubling of GH synthesis is observed after 6–8 hr. This stimulation increases linearly until a maximum 14-fold induction is attained by about 48 hr, while total cellular protein synthesis is not affected by glucocorticoids. In addition, an induction of GH synthesis is observed at dexamethasone concentrations as low as 10^{-9} M; this value is below the physiological range of plasma glucocorticoid concentrations ($3–5 \times 10^{-7}$ M)[27] even if dexamethasone is effective at a 10-fold lower concentration than the natural glucocorticoid hydrocortisone.[28]

Other laboratories have shown that glucocorticoid-mediated increases in GH production can only be observed in cells cultured in serum containing thyroid hormone (3,5,3′-triiodo-L-thyronine, T_3)[18,19,20,29,30] and not in cells grown in serum lacking T_3.[20,31] In contrast, a T_3 response can be observed in cells cultured in medium lacking glucocorticoids.[32] It has also been observed that when both thyroid and glucocorticoid hormones are given together, the production of GH is much greater than with either glucocorticoids or T_3 alone.[20,30–32] The synergistic effect of these two classes of hormones indicates that they act by independent mechanisms. These independent effects of thyroid and glucocorticoid hormones was clearly demonstrated by Ivarie *et al.*,[33] who examined the effects of thyroid and glucocorticoid hormones on protein synthesis in GH cells. They showed that GH production is increased when cells are treated with either glucocorticoid or thyroid hormone alone, or in combination. The effect of thyroid hormone is much greater than that of glucocorticoids, and the response to the two hormones together is much greater than with either hormone alone. They also demonstrated that these hormonal effects are not due to an overall increase in protein synthesis in GH cells, but are limited to less than 1% of the detectable proteins. Both classes of hormone act specifically, independently, and bidirectionally by increasing and decreasing the synthesis of a small fraction of the detected proteins. Thus, rat pituitary tumor cells are an excellent model for studying the mechanism of hormone action on the GH gene.

Protein synthesis is the final step in the expression of specific genes. At least five other steps must occur before the final protein is detected: (1) A gene must be transcribed into a precursor messenger RNA (pre-mRNA) in the nucleus of the cell; (2) this pre-mRNA is then processed into a mature mRNA; (3) the mature mRNA must be translocated from the nucleus to the ribosomes in the cytoplasm; (4) the mRNA is translated into a precursor protein; (5) the precursor protein is then processed into the mature protein; and finally, (6) the mature protein is either stored or secreted from the cell. Since GH cells do not store growth hormone, but rather secrete it constitutively into the medium, factors affecting cellular release of stored GH are not applicable to this system. The mechanism(s) by which thyroid and glucocorticoid hormones increase GH production have been determined by analyzing each step (from 6 back to 1) in the process of gene expression.

3. *Effects of Thyroid and Glucocorticoid Hormones on Rat GH Gene Expression*

3.1. *GH mRNA*

3.1.1. *Cell-Free Translation*

The first indirect evidence that thyroid hormone increases GH synthesis by increasing the cellular concentration of GH mRNA was provided by Samuels and Shapiro.[34] They showed that the thyroid hormone effect on increasing GH production is blocked when RNA synthesis is blocked; however, when protein synthesis is blocked, an intermediate necessary for stimulation of GH synthesis accumulates in response to thyroid hormone. Although these findings were indirect, they suggested that thyroid hormone may regulate the production of GH mRNA. To test this hypothesis, Seo *et al.*[29] isolated RNA from cultured rat pituitary tumor cells (GH_3 subline) and translated it in a cell-free rabbit reticulocyte lysate system. They showed that the T_3-stimulated GH synthesis was accompanied by a specific increase in GH mRNA activity. In addition, induction of GH synthesis and mRNA activity was observed as early as 6 hr after the addition of T_3 to cells cultured in thyroidectomized calf serum.

At the same time, Bancroft[35] studied the effect of glucocorticoids in inducing GH mRNA. This laboratory isolated RNA from GH_3 cells grown in the presence or absence of 10^{-5} M dexamethasone for 5 days and translated it in a wheat germ extract. They found that a 12-fold increase in GH synthesis was also accompanied by a 12-fold increase in GH mRNA.

Also at the same time, Martial *et al.*[19,20] measured the effects of both glucocorticoid and thyroid hormones on GH mRNA in another subline of cultured rat pituitary tumor cells (GC cells). They showed that T_3 and dexamethasone stimulated increases in GH production by 17- and 2.5-fold, respectively. Maximal responses were observed at 5–10 nM T_3 by 30 hr and 1 μM dexamethasone by 60 hr, with a lag of about 5 hr in both cases. In addition, there were comparable increases in the capacity of RNA from hormone-treated cells to direct synthesis of GH in a cell-free wheat germ translation system. When cells were stimulated with T_3 and dexamethasone simultaneously, the hormones are synergistic, resulting in a 45-fold induction. All these experiments suggested that thyroid and glucocorticoid hormones either increase the amount of GH mRNA or increase the translational efficiency of GH mRNA, or both.

Wegnez *et al.*[36] measured the effect of hormonal treatments on the translational efficiency of GH mRNA molecules in GH_3 cells. Their data demonstrated that T_3 had no influence on the translational efficiency of GH mRNA, while dexamethasone could either decrease translational efficiency (cells grown in serum-substitute medium) or increase translational efficiency (cells grown in serum-containing medium). However, these influences are not specific for GH mRNA, but appear to be a general effect on all mRNA's. Thus, the hormone-induced increases in translatable cytoplasmic GH mRNA's seem to arise from an increase in cytoplasmic GH mRNA sequences. It should be noted that the studies mentioned so far all measured hormonal effects on GH mRNA concentrations by indirect methods.

3.1.2. RNA/DNA Solution Hybridization and Northern Blot Analysis

A direct method for measuring the concentration of a specific mRNA is by RNA/DNA hybridizations in solution. This technique requires a very pure, radioactively labeled probe, i.e., DNA sequences complementary to GH mRNA (cDNA), to distinguish GH mRNA sequences from other mRNA's in low abundance. In one study, Martial *et al.*[19] used a cDNA probe prepared from partially purified GH mRNA and showed that the number of GH mRNA molecules per cell does parallel the mRNA activity assessed in cell-free translation assays. However, these experiments were not performed with a pure cDNA probe and could only estimate the amount of GH mRNA's. Pure probes became available only after cloning of the cDNA to rat GH mRNA.[37]

Wegnez *et al.*[36] used a pure radioactive probe in RNA/DNA solution hybridization experiments to estimate the glucocorticoid and thyroid hormone influences on GH mRNA molecules in cultured rat pituitary tumor cells (GH_3D_6 subline). They were able to overcome some problems arising from commonly used procedures for radioactive labeling of the DNA by labeling only the DNA strand of rat GH that hybridizes to rat GH mRNA (the cDNA strand).[38] In addition, they were able to overcome the problem of endogenous hormones in the medium in which the cells were grown. In cells cultured in the presence of 10% calf serum, GH production increases only minimally in response to added T_3, presumably because this hormone is already present at physiological concentrations in the serum.[19] In addition, in this medium, dexamethasone

Table 1 *Hormonal Effects on Growth Hormone (GH) mRNA Copies in Cultured Rat Pituitary Cells*[a]

	Molecules of GH mRNA/cell		
	GH_3D_6 cells		
Hormone treatment	Serum-substitute medium	Hypothyroid medium	GC cells (hypothyroid medium)
Control	<2	<5	45
+ Dexamethasone	8	20	125
+ T_3	12	130	1030
+ Dexamethasone + T_3	38	330	3900

[a] GH_3D_6 cell data are from Wegnez *et al.* (36); GC cell data are from Spindler *et al.* (42)

significantly increases GH production. Therefore, to bypass the problem of endogenous T_3 in calf serum, Wegnez *et al.* used two other media: one contained serum from a thyroidectomized calf (hypothyroid medium) which was depleted of thyroid hormones.[(39)] In this case, T_3 induces GH production greater than 10-fold, and dexamethasone has a much smaller influence.[(20,31,40)] The second medium contained a chemically defined serum substitute preparation[(41)] and therefore contained no endogenous T_3 or glucocorticoid. In addition, other hormones and factors present in serum may affect GH regulation, but are not present in serum substitute, permitting one to study the effects of glucocorticoid and thyroid hormones on GH regulation less ambiguously.

When rat pituitary tumor cells (GH_3 D_6 subline) were grown in serum substitute medium and stimulated for 72 hr with either dexamethasone or T_3, cytoplasmic GH mRNA content increased four- and six-fold, respectively (Table 1) (from <2 to 8 and from <2 to 12 molecules of GH mRNA/cell). When added together, the two classes of hormones act synergistically, causing a 19-fold increase in GH mRNA content (from <2 to 38 molecules of GH mRNA/cell). When similar experiments were performed with hypothyroid medium, dexamethasone increased cytoplasmic GH mRNA fourfold (Table 1) (from <5 to 20 molecules of GH mRNA/cell), T_3 increased GH mRNA 26-fold (to 130 molecules of GH mRNA/cell), and both hormones together increased GH mRNA 66-fold (to 330 molecules of GH mRNA/cell). Thus, while the results with the two different media were qualitatively similar, they were not quantitatively identical, suggesting that the hypothyroid medium contains other factors not present in serum substitute medium which modulate either GH mRNA synthesis or GH mRNA half-life.

Other experiments measuring cytoplasmic GH mRNA content were performed with another subline of rat pituitary tumor cells (GC cells).[(42)] GC cells grown in hypothyroid medium were induced for 72 hr with dexamethasone with or without T_3. The number of GH mRNA copies is increased approximately 2.8-fold by dexamethasone (Table 1) (from 450 to 125 molecules of GH mRNA/cell), 23-fold by T_3 (to 1030 molecules of GH mRNA/cell), and about 87-fold by both dexamethasone and T_3 (to 3900 molecules of GH mRNA/cell). Comparing these data to those of Wegnez *et al.*, dexamethasone and T_3 again exert a qualitatively similar effect of GH mRNA copy numbers, but quantitative differences may arise from the difference in the cell sublines used in each experiment. Nevertheless, the conclusions that can be drawn are that the effect of glucocorticoid and thyroid hormones on GH production may be explained by an increase in the number of cytoplasmic GH mRNA molecules.

The time required for this increase in GH mRNA by thyroid and glucocorticoid was determined by Northern blot analysis(43) of cytoplasmic RNA from GC cells. Cells grown in medium containing hypothyroid serum were stimulated with both T_3 (10^{-8} M) and dexamethasone (10^{-6} M) for various times; cytoplasmic RNA was prepared and separated according to size by agarose gel electrophoresis. The RNA was then blotted to nitrocellulose paper and probed for GH mRNA sequences with a radioactive rat GH cDNA probe. Figure 1 demonstrates that an increase in GH mRNA is detected as early as 4 hr after hormonal stimulation and increases up to 48 hr of hormonal stimulation. Thus the effects of thyroid and glucocorticoid hormones on increasing GH mRNA is very rapid and is consistent with the results of Yu *et al.*(26) and Seo *et al.*(29) of a 6-hr lag before GH production is noticed.

Although the RNA/DNA solution hybridizations and Northern blot analysis of cytoplasmic RNA's from GH cells indicate that both thyroid and glucocorticoid hormones increase GH mRNA, these results do not distinguish between an increase in GH mRNA synthesis and an increase in GH mRNA half-life.

3.2. GH Pre-mRNA

To determine whether this increase in cytoplasmic GH mRNA is due to an increase in the synthesis of the mRNA, Dobner *et al.*(44) and Lan *et al.*(45) examined the effects of glucocorticoids and T_3 on growth hormone pre-mRNA by Northern blot analysis.(43) This technique allows only a qualitative estimate of the synthesis of the primary GH transcript (pre-mRNA) and processing of this primary transcript into a mature of GH mRNA. Dobner *et al.* found that GH pre-mRNA was increased in GH_3 cells following glucocorticoid and/or T_3 treatment. Lan *et al.* found similar results in GC cells. Their data are shown in Fig. 2. Both groups detected a 2.2-kilobase (kb) nuclear GH-specific RNA sequence (Gh pre-mRNA) and a 1.0-kb cytoplasmic species of GH mRNA. Because the amount of these RNA's in control nuclei was too low to measure, they were not able to quantitate the amount of hormonal induction of these mRNA's Nevertheless, it could be concluded that part of the influence of glucocorticoid and thyroid hormones is to increase the synthesis of GH pre-mRNA. These results do not rule out effects on mRNA processing and/or stability.

3.3. Transcription of the Rat GH Gene

To determine if the effects of glucocorticoid and thyroid hormones on increasing GH pre-mRNA are due to an increase in the synthesis of this

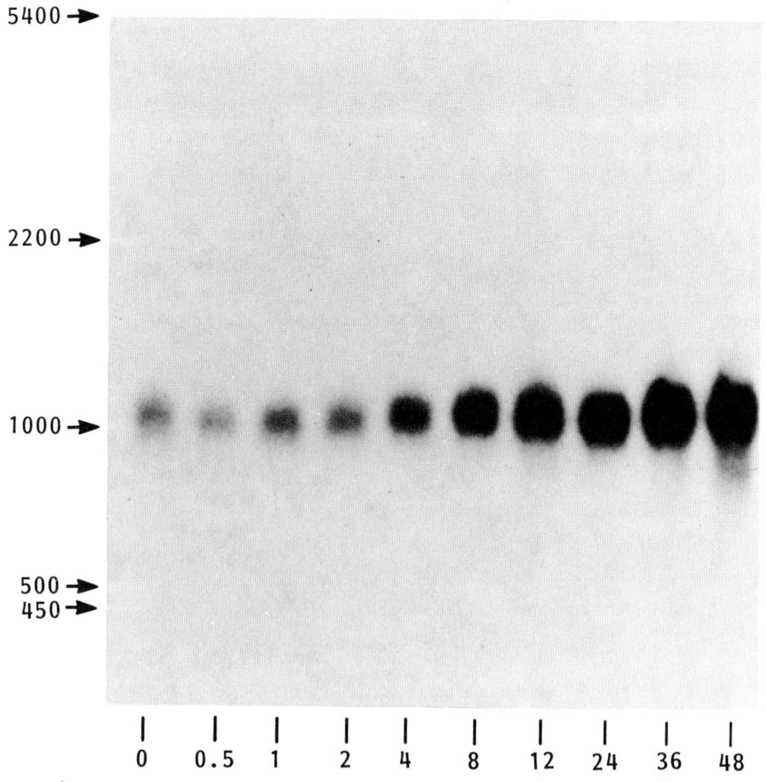

Figure 1 Northern blot[43] analysis of total RNA from control and hormone-treated GC cells. Total cellular RNA was prepared from GC cells grown in hypothyroid medium and treated with dexamethasone (10^{-6} M) and thyroid hormone (10^{-8} M) for various times. RNA (20 μg) was separated on a 1.5% agarose gel, transferred to nitrocellulose paper, hybridized with [^{32}P-labeled cDNA to rat GH mRNA,[37,38] and exposed to X-ray film. The size markers (in base pairs) on the left refer to the migration of PM2 DNA cleaved with *Hin*d III, and labeled with ^{32}P and treated identically to the RNA samples prior to electrophoresis.

molecule, it is necessary to measure its rate of synthesis, i.e., the rate of transcription of the GH gene, under different hormonal stimuli. We have quantitated the relative activity of the GH gene in GC cells. Nuclei from hormone-treated and control cells were isolated and incubated with radiolabeled ribonucleoside triphosphates under conditions in which the endogenously bound RNA polymerase molecules proceed with transcription and chain termination, but do not reinitiate RNA synthesis.[46,47] This "polymerase runoff" technique is illustrated in Fig. 3. Since the newly synthesized RNA's are labeled, their quantity can be assessed by hybridization to an unlabeled DNA probe, prepared as shown in Fig. 4.

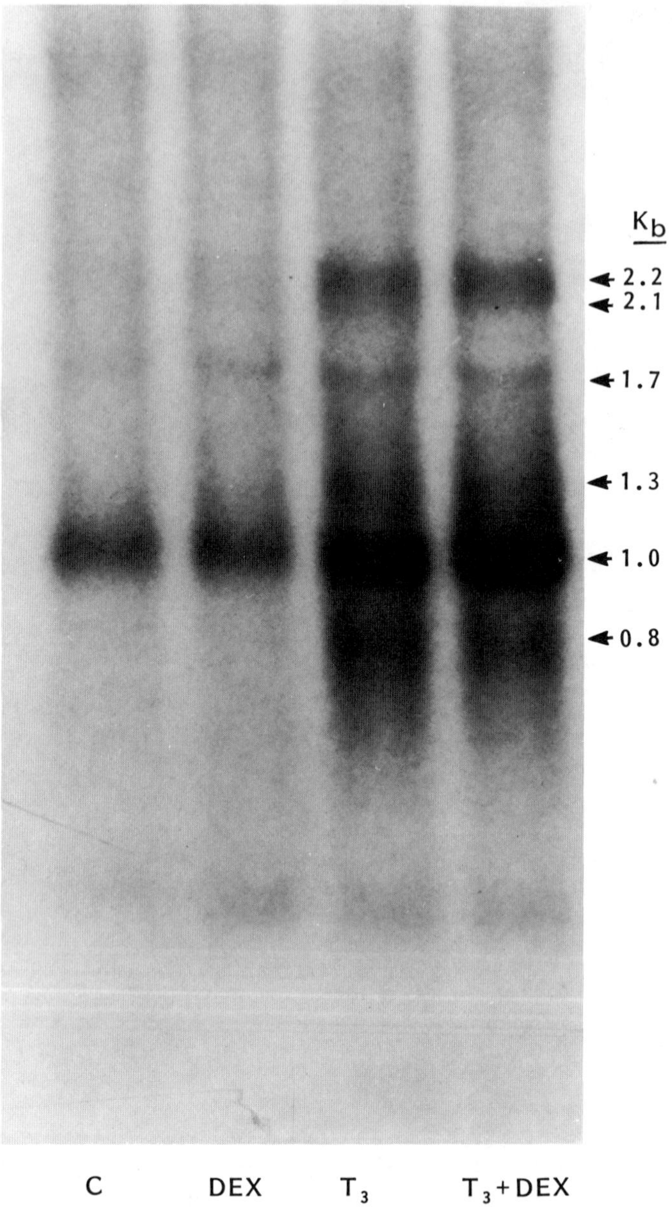

Figure 2 Effect of thyroid and glucocorticoid hormones on rat growth hormone (GH) pre-mRNA. Cellular RNA was prepared from GC cells grown in hypothyroid medium and stimulated with dexamethasone (DEX) (10^{-6} M), thyroid hormone (T_3) (10^{-8} M), or both hormones for 4 hr. Poly(A)-containing RNA's were prepared by binding total RNA to oligo(dT)-cellulose in 0.4 M KCl and eluting with water. Forty micrograms of Poly$(A)^+$ RNA was separated on a 1.5% agarose gel and analyzed by Northern blotting[43] and hybridization to a ^{32}P-labeled probe (intron D of the rat GH gene[54,55]).[37,38] The sizes of the GH-specific RNA species are noted on the right side of the autoradiogram. C is poly$(A)^+$ RNA from control cells (no hormone treatment). Data are from Lan *et al.*[45]

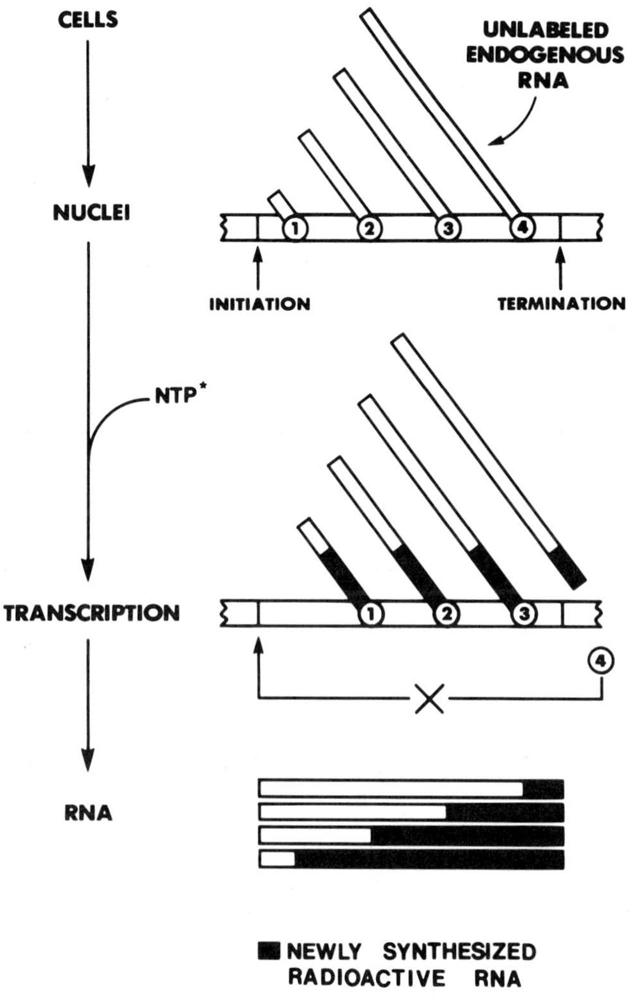

Figure 3 Diagrammatic representation of the cell-free transcription system used to estimate the RNA polymerases initiated *in vivo*. Nuclei were isolated and incubated in the presence of radioactive ribonucleoside triphosphates (NTP*). The nuclei contain the GH gene with molecules of RNA polymerase II which had been attached *in vivo*, thus initiating transcription. Transcription proceeds *in vitro*, resulting in the incorporation of radioactive nucleoside triphosphates into newly elongated RNA molecules (dark bars). The newly synthesized RNA is thus radioactive, and those molecules containing growth hormone (GH) mRNA sequences can be detected by hybridization with clonal cDNA to GH mRNA.

Cells grown to confluency were maintained for 7 days in medium containing hypothyroid serum. They were then induced for 4 or 72 hr with 10^{-6} M dexamethasone, 10^{-8} M T_3, or both dexamethasone and T_3. Cells were then harvested, and nuclei were separated from the remainder of the cellular contents. This nuclear pellet was then incubated in the presence of

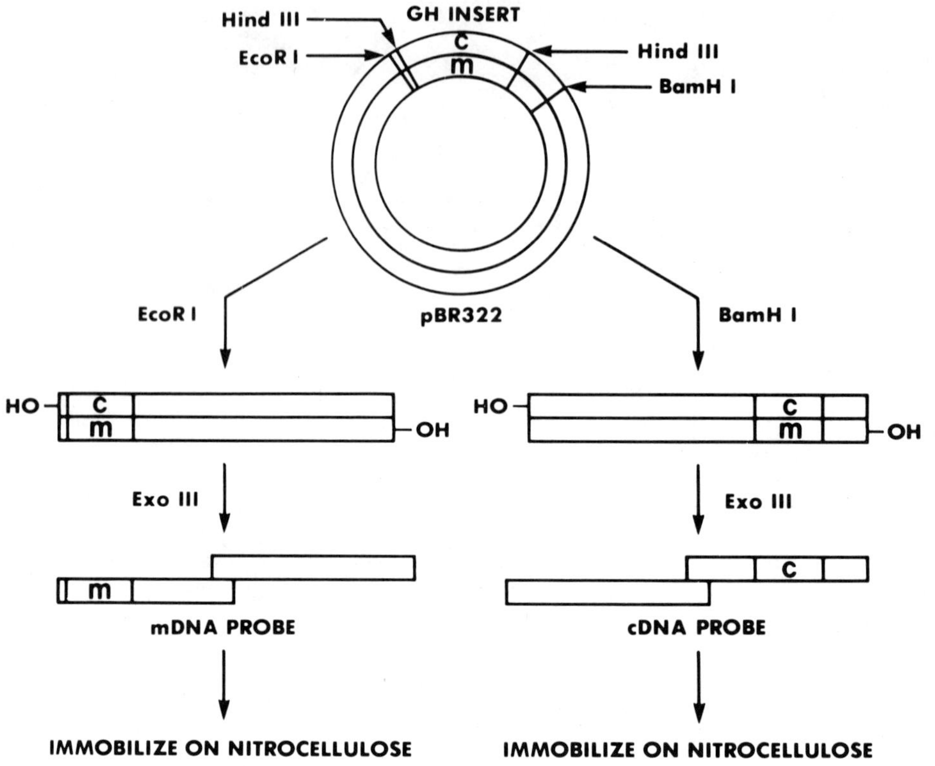

Figure 4 Preparation of unlabeled growth hormone (GH) gene sequences for use as hybridization probes. The plasmid-containing rat GH cDNA sequences cloned in the *Hin*d III site of pBR322 was cleaved with restriction endonucleases *Bam* HI or *Eco* RI, and the linear DNA was digested 3′ to 5′ with Exonuclease III. Clevage with *Bam* HI followed by Exonuclease III digestion exposes the DNA strand which is normally transcribed into RNA (labeled c, as it is the strand that is complementary to the mRNA). Cleavage with *Eco* RI followed by Exonuclease III exposes the nontranscribed strand (labeled m, as its sequence is equivalent to an mRNA molecule). These DNA's were then immobilized onto nitrocellulose filters for hybridization.

[^{32}P]ribonucleoside triphosphates, allowing endogenously initiated RNA polymerases to elongate RNA chains. Radiolabeled RNA was then purified and hybridized to nitrocellulose filters containing cloned cDNA to rat GH mRNA, or to the complement of the cDNA (GH mDNA probe, Fig. 4). The filters were then washed extensively, and the specifically bound RNA was quantitated by measuring the radioactivity on the filters.[48,49] The results from such an experiment are shown in Fig. 5.

After either 4 or 72 hr of hormonal stimulation, glucocorticoids only slightly increase GH gene transcription (i.e., increase the ability of nuclei to incorporate radioactivity into material that hybridizes specifically with the rat GH cDNA probe). After 72 hr of hormonal stimulation, T_3 increases GH gene transcription about 2.6-fold. There does not seem to be any synergism between T_3 and

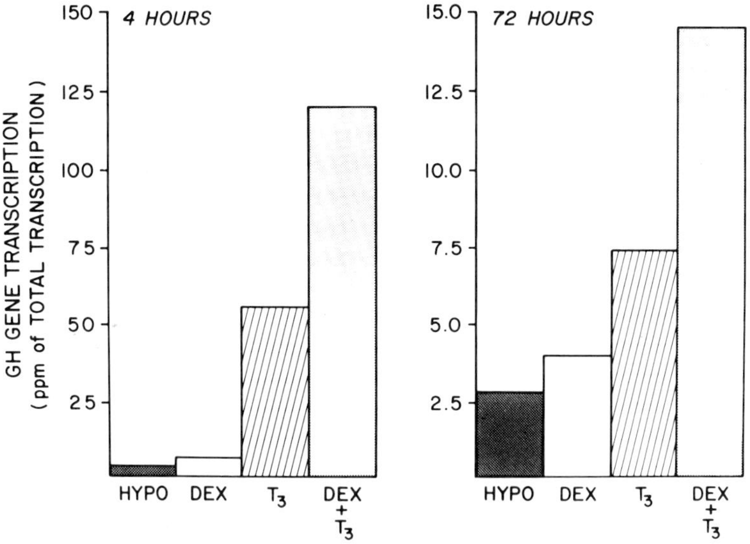

Figure 5 Transcription of the growth hormone (GH) gene in isolated nuclei. Cultured rat pituitary cells (GC subline) were grown to confluency in monolayer in media containing 10% calf serum and then maintained for 7 days in media containing 10% thyroidectomized calf serum (hypo serum). Cells were induced with dexamethasone (DEX) (1 μM), 3,5,3′ triiodo-L-thyronine (T_3) (10 nM), or both hormones for 4 hr or 72 hr. Nuclei were then isolated and incubated in the presence of [^{32}P]ribonucleoside triphosphates. The ^{32}P-labeled RNA was isolated and hybridized to filters containing the cloned rat GH cDNA strand. Values are represented as cpm of [^{32}P] RNA hybridized to the rat GH cDNA per million cpm in the hybridization reaction (ppm). Transcription data from 4-hr hormonal stimulations are shown in the left-hand graph, and from 72 hr of hormonal stimulation in the right-hand graph. Data are from Spindler *et al.*[42]

glucocorticoids after 72 hr of stimulation. After 4 hr of hormonal stimulation, T_3 increases GH gene transcription 17-fold, and together dexamethasone plus T_3 act synergistically, causing a 36-fold increase in GH gene transcription.

Evans *et al.*[50] performed experiments similar to those described here. Although they also used GC cells, they found that 4 hr of dexamethasone treatment alone, or in the presence of T_3 increased GH gene transcription 2.5- and five-fold, respectively. Thyroid hormone alone increased GH transcription only fourfold. Thus, they have observed a greater effect of glucocorticoids alone on GH gene transcription, while the effect of T_3 or T_3 plus dexamethasone was about one-fifth of the response we obtained. In addition, they did not observe any synergism between dexamethasone and T_3. Differences may be due to the medium in which the cells were cultured. We used medium containing serum from a thyroidectomized calf, and Evans *et al.* used charcoal-stripped fetal calf serum.

We have confirmed the hypothesis that RNA polymerase II is the enzyme involved in transcribing the GH gene. This enzyme is inactivated by 2 μg/ml α-amanitin (a fungal toxin).[51,52] When transcription experiments are performed in the presence of this concentration of toxin, the effects of thyroid and glucocorticoid hormones are abolished.[42]

Transcription of the GH gene is from one strand of DNA only. There were only background levels of hybridization with the mDNA probe, suggesting that the "antisense" strand of the rat GH gene is not transcribed.

The data therefore suggest that glucocorticoid and thyroid hormones increase the number of RNA polymerase II molecules engaged in transcribing the GH gene. Since the hormones also increase GH pre-mRNA, the results further imply that these two classes of hormones increase the rate of transcription of the gene, and that this increase is at the site of initiation of transcription. Since the increase in transcription can be observed by 4 hr, these data suggest that the effects of T_3 and dexamethasone are rapid. Therefore, the hormonal influences may be direct, rather than through the induction of some other protein(s).

To determine whether prior synthesis of some protein is required for initiating GH gene transcription, experiments were performed in the presence of 0.1 mM cycloheximide, which inhibits translation but not transcription. The results demonstrate that cycloheximide does not affect either the T_3- or dexamethasone-induced increase in GH gene transcription, implying that both T_3 and dexamethasone act directly on chromatin, presumably through their specific receptors, to increase GH gene transcription.[42]

These transcription experiments were also performed *in vivo* in rats. The hypothyroid cellular conditions used in experiments described previously were simulated by using thyroidectomized–adrenalectomized rats. These rats were then injected with dexamethasone, T_3, or both hormones. Under these experimental conditions, glucocorticoids and thyroid hormones alone or in combination greatly increased pituitary GH gene transcription.[53]

The regulation of GH gene transcription *in vivo* was also studied by Evans *et al.*[50] This group used adrenalectomized rats injected with dexamethasone and performed polymerase runoff experiments on isolated pituitary nuclei various times after hormone administration. They found that GH gene transcription was increased as early as 15 min after hormone treatment and increased from 27 to 107 parts per million in 2 hr. This fourfold increase in GH gene transcription was virtually identical with the effects they saw in GC cells treated with both dexamethasone and T_3, although the absolute rate of GH gene transcription *in vivo* was 10 times greater than that in GC cells (107 ppm *in vivo* versus 10 ppm in GC cells). Since the rats were only adrenalectomized and not thyroidectomized also, the response of the pituitaries to dexamethasone injection may have been, in fact, a response to both dexamethasone and endogenous thyroid hormone. Plasma concentrations of T_3 were not measured. Thus, from these experiments, one cannot conclude with certainty that glucocorticoids alone increase GH gene transcription *in vivo*.

Hormonal effects on GH gene transcription may not fully account for the increase in cellular GH mRNA content. When one compares the relative changes in GH gene transcription assayed by the experiments described previously with the changes in GH mRNA described in Section 3.1.2, a

Table 2 Growth Hormone (GH) Gene Transcription versus GH mRNA[a]

	Percent of control		
		GH gene transcription	
Hormonal treatment	GH mRNA copies (72 hours)	4 hr	72 hr
Control	100	100	100
+ Dexamethasone	300	200	140
+ T_3	2200	1700	260
+ Dexamethasone + T_3	8300	3600	520

[a] Data are from Spindler *et al.* (42)

discrepancy is found (Table 2). In the presence of T_3, dexamethasone induces a 36-fold increase in growth hormone gene transcription at 4 hr. However, the mRNA is increased 83-fold by 72 hr. This result may imply that the effects on transcription do not totally explain the glucocorticoid influence on GH mRNA. It is possible that a hormone–receptor complex (directly or indirectly) also affects GH mRNA stability.

4. *Hormone Regulation after Gene Transfer*

4.1. *Transfer of the Rat GH Gene into Mouse LTK⁻ Cells*

The regulation of the rat GH gene can also be studied by transferring this gene into a cell that does not normally express or contain the rat GH gene. By this method, one is able to determine what DNA sequencs within or flanking the rat GH gene are required for thyroid and/or glucocorticoid hormone regulation. The rat GH gene is approximately 2.2 kilobases between its transcription initiation and termination points and is contained within a 5.8-kb DNA fragment in the clone we used.[54] This clone contains approximately 1.7 kb of DNA flanking the start of transcription and 1.9 kb of DNA flanking the termination of transcription point. The structure and organization of all the GH genes have been reviewed in detail recently.[55]

Mouse L cells deficient in thymidine kinase (LTK^- cells) were cotransformed with the rat GH gene and the herpes simplex virus thymidine kinase gene (TK). Only those cells which have taken up and expressed the thymidine kinase gene will survive in selective medium.[56] Clonal cell lines stably grown in medium containing hypoxanthine amethopterin and thymidine (HAT medium) were then analyzed for GH gene expression by Northern blot analysis of the RNA from the cells. Cells that contained rat GH mRNA sequences were then tested for hormonal regulation of the transferred rat GH gene.[57]

The results from one cell line are shown in Fig. 6. Cells were treated with dexamethasone (10^{-6} M), or T_3 (10^{-8} M), or both hormones for 48 hr. Cytoplasmic RNA was prepared and analyzed by Northern blot techniques, using a radioactive rat GH cDNA as a probe. As shown in Fig. 6, dexamethasone

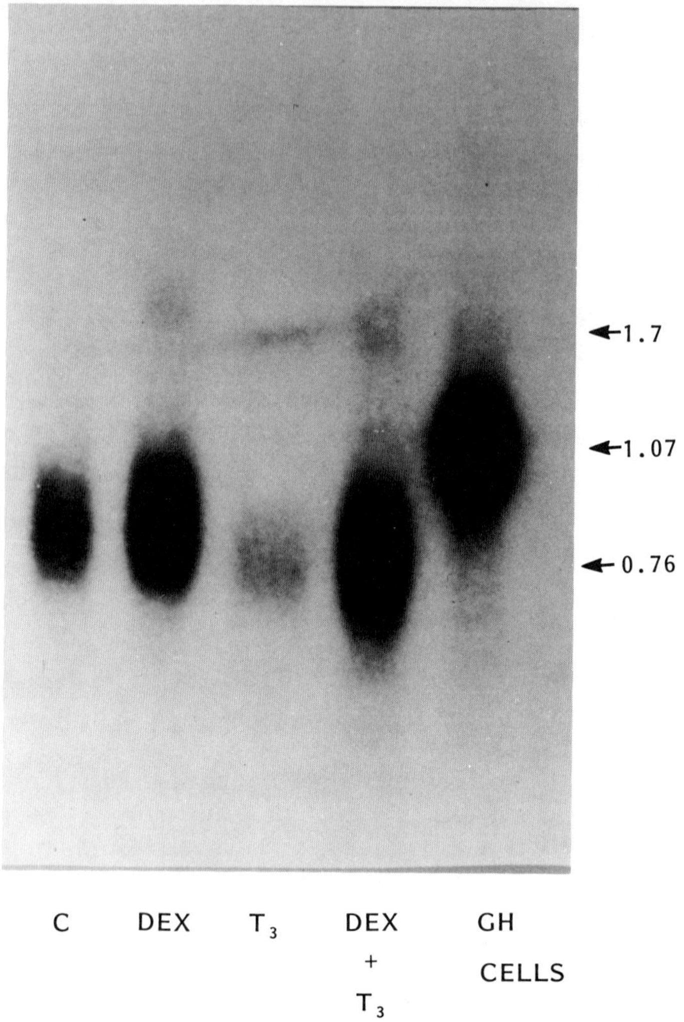

Figure 6 Hormonal induction of rat GH mRNA in mouse LTK^- cells cotransfected with the rat growth hormone (GH) gene and herpes simplex virus thymidine kinase gene. One LTK^+ cell line expressing the rat GH gene was treated with dexamethasone (DEX) (10^{-6} M) or thyroid hormone (T_3) (10^{-7} M) alone or in combination for 2 days. Cytoplasmic RNA (20 μg) was analyzed by Northern blotting techniques,(43) using [^{32}P]-rat GH cDNA as a probe, as described in Fig. 1. RNA from rat pituitary tumor cells (GH cells) was analyzed simultaneously. The sizes of the rat GH-specific RNA species are indicated (in kilobases) on the right. Data are from Karin *et al.*(57)

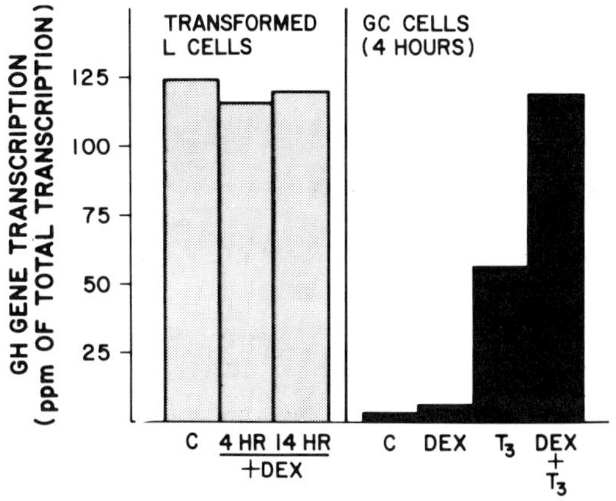

Figure 7 Transcription of the rat growth hormone (GH) gene in mouse LTK⁻ cells transfected with the rat GH gene. Results from the cell line shown in Fig. 6 are presented. These cells were treated with dexamethasone (DEX) (10^{-6} M) for 4 or 14 hr. Nuclei were isolated and incubated with [^{32}P]ribonucleoside triphosphates. Rat GH-specific transcripts in these transformed L cells (left-hand graph) were quantitated as described in Fig. 5. Rat GH transcription data from cultured rat pituitary tumor cells (GC cells) are shown for comparison in the graph on the right. Data are from Karin *et al.*[57]

increased GH mRNA about fourfold. Thyroid hormone did not increase the GH mRNA in the transformed cells. This may be due to the lack of appreciable numbers of T_3 receptors in L cells. (The specific T_3 binding capacity was found to be 5% of the value found in cultured rat pituitary cells.)

The major GH-specific transcript produced in the transformed L cells is about 800 nucleotides; this is approximately 200 nucleotides shorter than GH mRNA in GH_3 cells (1070 nucleotides). It is not known whether the shorter mRNA is due to aberrant processing of a pre-mRNA whose transcription was initiated at the normal site, or due to initiation of transcription at some other site within the GH gene.

Regardless of the size of the GH mRNA in transfected L cells, the data demonstrate that the information necessary for glucocorticoid hormone regulation of GH gene expression is contained within the 5.8-kb segment of DNA.

To determine whether the gulcocorticoid-mediated increase in GH mRNA sequences in these cells is due to an increase in the transcription of the GH gene, polymerase runoff experiments described in Section 3.3 were performed.[57] The data from one experiment are shown in Fig. 7 and are compared with the data from the GC cell experiment (data from Fig. 5). The basal level of GH gene transcription is about 125 ppm. This is much greater than that seen in cultured GC cells (about 1.5 ppm). Dexamethasone has no effect on GH gene transcription, after either 4 or 14 hr of hormonal stimulation. In GC cells, dexamethasone alone also has only a minor effect on GH gene transcription. However, T_3 or glucocorticoids in the presence of T_3 increase GH gene transcription substantially. Thus, it is possible that the glucocorticoid-mediated increase in GH mRNA in transfected L cells (as well as in GC cells) is due to an increase in this mRNA's stability.

4.2. Transformation of NIH 3T3 Cells by the Rat GH Gene via Retroviral Vector

Another laboratory has reported the expression and regulation of the rat GH gene after transfer into a different mouse cell line (NIH 3T3 cells).[58] The rat GH gene was cloned into a vector containing the entire genome of Moloney mouse sarcoma virus (MoMSV); MoMSV is able to transform 3T3 cells and thus provides a selection criterion. RNA from transformed cells containing rat GH DNA sequences was analyzed for the presence of rat GH-specific RNA transcripts; the size of this cytoplasmic mRNA species appeared to be larger than authentic GH mRNA from rat pituitaries, but the size of this mRNA was not reported. This laboratory also examined the regulation of the transfected rat GH gene by dexamethasone. Rather than analyze the rat GH-specific mRNA induction, they measured the amount of immunoreactive rat GH secreted from the cells after treatment with dexamethasone. This immunoreactive substance, presumably authentic rat GH, increased about two-fold after dexamethasone treatment. The glucocorticoid-mediated increase in "GH" secretion could be due to an increase in rat GH gene transcription, GH mRNA stability, GH mRNA translation, or GH secretion, and thus the mechanism of glucocorticoid hormone action in this system remains unknown.

4.3. Transfer of a Human GH Gene into Mouse LTK Cells

Robins *et al.* studied the regulation of a human growth hormone gene[59] after transfer into mouse LTK^- cells[60] analogous to the experiments described in Section 4.1. The clone of the human GH gene used contained the entire gene (1.6 kb) along with 500 base pairs of DNA flanking the start of transcription and 525 base pairs of DNA flanking the termination of transcription. They found that 10^{-6} M dexamethasone increased human GH-specific mRNA three- to fivefold, as determined by Northern blot analysis. However, the size of this mRNA species appears to be larger than human GH mRNA from a human pituitary (1000 nucleotides) as evidenced from Northern blots. In addition, another human GH-specific mRNA species (3000 nucleotides) was also synthesized in some cell lines. The origin of this extremely large transcript is unknown and may reflect aberrant initiation or termination of transcription or incorrect splicing of a correctly initiated primary transcript. However, this 3000-nucleotide species was not regulated by glucocorticoids.

This laboratory also performed human GH radioimmunoassays on the medium from these cells. They found that cell lines that demonstrated significant concentrations of human GH mRNA also secreted significant amounts of immunoreactive human GH into the tissue culture medium. In addition, the glucocorticoid-mediated increase in human GH mRNA was accompanied by a concomitant increase in the concentration of secreted human GH protein. Whether this protein was in fact authentic human GH was not determined.

Nevertheless, the data indicate that the human GH gene can be regulated by glucocorticoids, and that the information necessary for this regulation is contained within the 2.6-kb segment of the human GH gene.

To determine whether the sequence of DNA responsive to glucocorticoid induction is located within the 500 base pairs of DNA flanking the start of transcription of the human GH gene (5′ flanking DNA), Robins *et al.* fused this segment of DNA to another gene (TK) that is not normally regulated by glucocorticoids. This fusion gene was then introduced into LTK^- cells, and TK^+ cells (cells containing the fusion human GH–TK gene) were selected in HAT medium. In two of three TK^+ cell lines, TK mRNA was now induced by glucocorticoids. The data therefore suggest that DNA sequences in the human GH gene responsive to glucocorticoid induction are contained within 500 base pairs flanking the 5′ terminus of the human GH gene. Fusion of these sequences to another structural gene now renders the latter gene responsive to glucocorticoids.

5. Summary

Cultured rat pituitary tumor cells are an excellent model system for studying the regulation of the rat GH gene by thyroid and glucocorticoid hormones. These hormonal effects are specific, in that less than 1% of the expressed genes in these cells are affected. The hormone-mediated increase in GH production parallels an increase in GH mRNA. In addition, an increase in GH pre-mRNA precedes the increase in the mature mRNA. The increase in GH pre-mRNA is due to a direct effect on the transcription of the rat GH gene, both in cultured rat pituitary cells and *in vivo* in rats. Thyroid hormone alone rapidly increases rat GH gene transcription. Dexamethasone alone slightly increases transcription of this gene, and the action of both hormones together is synergistic. These increases in GH gene transcription are direct and do not involve *de novo* protein synthesis. From RNA/DNA solution hybridization experiments that measure rat GH mRNA copies per cell, it is apparent that the thyroid- and glucocorticoid-hormone-mediated increase in GH gene transcription does not account fully for the increased cytoplasmic GH mRNA. Thus, these hormones may also act posttranscriptionally to stabilize GH mRNA.

Gene transfer experiments demonstrate that the expression of both the rat and human GH genes is regulated by glucocorticoids. Whether this regulation is at the site of initiation of transcription or at some posttranscriptional site is not fully known. However, the data demonstrate that the information necessary for glucocorticoid hormone regulation of GH gene expression is contained within a 5.8-kb segment of the rat GH gene (1.7 kb of DNA flanking the start of transcription, 1.9 kb of DNA flanking the termination of transcription, and 2.2 kb of structural sequence) or a 2.6-kb segment of the human GH gene (1.6 kb of structural gene, with 500 base pairs of DNA flanking both ends). In addition, when 500 base pairs of DNA flanking the start of transcription of the human GH gene are fused to another gene not normally regulated by glucocorticoids, TK gene, the TK gene is now regulated by dexamethasone. Thus, the data suggest that DNA sequences in the human GH gene responsive to glucocorticoid induction are contained within these 500 base pairs flanking the 5′ terminus of the human GH gene.

The data presented here should not necessarily be interpreted as reflecting what occurs *in vivo* in human beings. Cell culture and gene transfer experiments merely provide models for studying the mechanisms of hormonal regulation of GH genes. As such, they are powerful tools for analyzing how hormones interact with chromatin at the molecular level.

ACKNOWLEDGMENTS. This work was supported by a National Research Service Award AM 06588. I wish to thank S. Spindler for collaboration on the transcription experiments, N. Lan for performing the experiment shown in Fig. 2, J. D. Baxter for guidance and support, D. Faber for typing the manuscript and W. L. Miller for perpetual guidance, support, and assistance in preparing the manuscript.

References

1. M. Zachman and A. Prader, Interactions of growth hormone with other hormones, in: *Human Growth Hormone* (A. S. Mason, ed.), William Heinemann Med. Books, Ltd., London, 1972, pp. 39–93.
2. J. D. Baxter, Mechanisms of glucocorticoid inhibition of growth, *Kidney Int.* **14:**330–333, 1978.
3. J. N. Loeb, Corticosteroids and growth, *N. Engl. J. Med.* **295:**547–552, 1976.
4. W. J. Daughaday, A. C. Herrington, and L. S. Phillips, The regulation of growth by endocrines, *Ann. Rev. Physiol.* **37:**211–244, 1975.
5. P. L. Ballard, Glucocorticoids and differentiation, in: *Glucocorticoid Hormone Action* (J. D. Baxter and G. G. Rousseau, eds.), Springer-Verlag, New York, 1979, pp. 493–515.
6. A. L. Strickland, Growth retardation in Cushing's Syndrome, *Am. J. Dis. Child.* **124:**207–213, 1972.
7. D. Bellamy, Effect of cortisol on growth and food intake in rats, *J. Endocrinol.* **31:**83–84, 1964.
8. M. Winick and A. Coscia, Cortisone induced growth failure in neonatal rats, *Pediatr. Res.* **2:**451–455, 1968.
9. S. Sawano, A. Arimura, A. V. Schally, T. W. Reding, and S. Schapiro, Neonatal corticoid administration: Effects upon adult pituitary growth hormone and hypothalamic growth hormone-releasing hormone, *Acta Endocrinol.* **61:**57–67, 1969.
10. U. J. Lewis, E. V. Cheever, and W. P. Van der Laan, Alterations of the proteins of the pituitary gland of the rat by estradiol and cortisol, *Endocrinology* **76:**362–368, 1965.
11. T. Suda, H. Demura, R. Demura, K. Jibiki, F. Tozawa, and K. Shizume, Anterior pituitary hormones in plasma and pituitaries from patients with Cushing's disease, *J. Clin. Endocrinol. Metab.* **51:**1048–1053, 1980.
12. J. B. Tyrrell, J. Wiener-Kronish, M. Lorenzi, R. M. Brooks, and P. H. Forsham, Cushing's disease: Growth hormone response to hypoglycemia after correction of hypercortisolism, *J. Clin. Endocrinol. Metab.* **44:**218–221, 1977.
13. P. O. Kohler, L. A. Frohman, W. E. Bridson, T. Vanha-Perttula, and J. M. Hammond, Cortisol induction of growth hormone synthesis in a clonal line of rat pituitary tumor cells in culture, *Science* **166:**633–634, 1969.
14. P. O. Kohler, W. E. Bridson, and P. L. Rayford, Cortisol stimulation of growth hormone production by monkey adenohypophysis in tissue culture, *Biochem. Biophys. Res. Commun.* **33:**834–840, 1968.
15. W. E. Bridson and P. O. Kohler, Cortisol stimulation of growth hormone production by human pituitary tissue in culture, *J. Clin. Endocrinol. Metab.* **30:**538–540, 1970.
16. A. H. Tashjian, Jr., Y. Yasumura, L. Levine, G. H. Sato, and M. L. Parker, Establishment of cloncal strains of rat pituitary tumor cells that secrete growth hormone, *Endocrinology* **82:**342–352, 1968.
17. P. S. Dannies and A. H. Tashjian, Jr., Growth hormone and prolactin from rat pituitary tumor cells, in: *Tissue Culture: Methods and Applications* (P. F. Kruse, Jr., and M. K. Patterson, Jr., eds.), Academic Press, New York, 1973, pp. 561–569.
18. F. C. Bancroft, L. Levine, and A. H. Tashjian, Jr., Control of growth hormone production by

a clonal strain of rat pituitary cells: Stimulation by hydrocortisone, *J. Cell Biol.* **43:**432–441, 1969.

19. J. A. Martial, J. D. Baxter, H. M. Goodman, and P. H. Seeburg, Regulation of growth hormone messenger RNA by thyroid and glucocorticoid hormones, *Proc. Natl. Acad. Sci. USA* **74:**1816–1820, 1977.
20. J. A. Martial, P. H. Seeburg, D. Guenzi, H. M. Goodman, and J. D. Baxter, Regulation of growth hormone gene expression: Synergistic effects of thyroid and glucocorticoid hormones, *Proc. Natl. Acad. Sci. USA* **74:**4293–4295, 1977.
21. G. T. Peake, C. A. Birge, and W. H. Daughaday, Alterations of radioimmunoassayable growth hormone and prolactin during hypothyroidism, *Endocrinology* **92:**487–493, 1973.
22. F. Hervas, G. Morreale de Escobar, and F. Escobar Del Rey, Rapid effects of single small doses of L-thyroxine and triiodo-L-thyronine on growth hormone, as studied in the rat by radioimmunoassay, *Endocrinology* **97:**91–101, 1975.
23. M. I. Surks and C. R. DeFesi, Determination of the cell number of each cell type in the anterior pituitary of euthyroid and hypothyroid rats, *Endocrinology* **101:**946–958, 1977.
24. C. R. DeFesi, H. S. Astier, and M. I. Surks, Kinetics of thyrotrophs and somamtotrophs during development of hypothyroidism and L-triiodothyronine treatment of hypothyroid rats, *Endocrinology* **104:**1172–1180, 1979.
25. H. P. Katz, R. Youlton, S. L. Kaplan, and M. M. Grumbach, Growth and growth hormone III. Growth hormone release in children with primary hypothyroidism and thyrotoxicosis, *J. Clin. Endocrinol. Metab.* **29:**346–351, 1969.
26. L-Y. Yu, R. J. Tushinski, and F. C. Bancroft, Glucocorticoid induction of growth hormone synthesis in a strain of rat pituitary cells, *J. Biol. Chem.* **252:**3870–3875, 1977.
27. R. J. B. King and W. I. P. Mainwaring, *Steroid–Cell interactions*, University Park Press, Baltimore, 1974.
28. H. H. Samuels and G. M. Tomkins, Relation of steroid structure to enzyme induction in hepatoma tissue culture cells, *J. Mol. Biol.* **52:**57–74, 1970.
29. H. Seo, G. Vassart, H. Brocas, and S. Refetoff, Triiodothyronine stimulates specifically growth hormone mRNA in rat pituitary tumor cells, *Proc. Natl. Sci. USA* **74:**2054–2058, 1977.
30. L. E. Shapiro, H. H. Samuels, and B. M. Yaffe, Thyroid and glucocorticoid hormones synergistically control growth hormone mRNA in cultured GH_1 cells, *Proc. Natl. Acad. Sci. USA* **75:**45–49, 1978.
31. H. H. Samuels, Z. D. Horwitz, F. Stanley, J. Casanova, and L. E. Shapiro, Thyroid hormone controls glucocorticoid action in cultured GH_1 cells, *Nature* **268:**254–256, 1977.
32. H. H. Samuels, D. Klein, F. Stanley, and J. Casanova, Evidence for thyroid hormone-dependent and independent glucocorticoid actions in cultured cells, *J. Biol. Chem.* **253:**5895–5898, 1978.
33. R. D. Ivarie, J. D. Baxter, and J. A. Morris, Interaction of thyroid and glucocorticoid hormones in rat pituitary tumor cells, *J. Biol. Chem.* **256:**4520–4528, 1981.
34. H. H. Samuels and L. E. Shapiro, Thyroid hormone stimulates *de novo* growth hormone synthesis in cultured GH_1 cells: Evidence for the accumulation of a rate limiting RNA species in the induction process, *Proc. Natl. Acad. Sci. USA* **73:**3364–3373, 1976.
35. R. J. Tushinski, P. M. Sussman, L-Y. Yu, and F. C. Bancroft, Pregrowth hormone messenger RNA: Glucocorticoid induction and identification in rat pituitary cells, *Proc. Natl. Acad. Sci. USA* **74:**2357–2361, 1977.
36. M. Wegnez, B. S. Schachter, J. D. Baxter, and J. A. Martial, Hormonal regulation of growth hormone mRNA, *DNA* **1:**145–153, 1982.
37. P. H. Seeburg, J. Shine, J. A. Martial, J. D. Baxter, and H. M. Goodman, Nucleotide sequence and amplification in bacteria of the structural gene for rat growth hormone, *Nature* **270:**486–494, 1977.
38. P. O'Farrell, Replacement synthesis method of labeling DNA fragments, *Focus* **3**(3)**:**1–3, 1981.
39. H. H. Samuels and J. S. Tsai, Thyroid hormone action in cell culture: Demonstration of nuclear receptors in intact cells and isolated nuclei, *Proc. Natl. Acad. Sci. USA* **70:**3488–3492, 1973.
40. H. H. Samuels, F. Stanley, and L. E. Shapiro, Control of growth hormone synthesis in cultured GH_1 cells by 3,5,3′-triiodo-L-thyronine and glucocorticoid agonists and antagonists: Studies on

the independent and synergistic regulation of the growth hormone response, *Biochemistry* **18**:715–721, 1979.

41. R. F. Bauer, L. O. Arthur, and D. L. Fine, Propagation of mouse mammary tumor cell lines and production of mouse mammary tumor virus in a serum-free medium, *In Vitro* **12**:558–563, 1976.
42. S. R. Spindler, S. H. Mellon, and J. D. Baxter, Growth hormone gene transcription is regulated by thyroid and glucocorticoid hormones in cultured rat pituitary tumor cells, *J. Biol. Chem.* **257**:11627–11632, 1982.
43. P. S. Thomas, Hybridization of denatured RNA and small DNA fragments transferred to nitrocellulose, *Proc. Natl. Acad. Sci. USA* **77**:5201–5205, 1980.
44. P. R. Dobner, E. S. Kawasaki, L-Y. Yu, and F. C. Bancroft, Thyroid or glucocorticoid hormone induces pre-growth hormone mRNA and its probably nuclear precursor in rat pituitary cells, *Proc. Natl. Acad. Sci. USA* **78**:2230–2234, 1981.
45. N. C. Lan, T. Nguyen, G. Cathala, S. K. Nordeen, M. E. Wolff, P. A. Kollman, S. Mellon, T. Nguyen, M. Karin, N. Eberhardt, and J. D. Baxter, Studies of the mechanism of glucocorticoid hormone action, in: *Gene Regulation by Steroid Hormones II* (A. K. Roy and J. H. Clark, eds.), Springer-Verlag, New York, 1983, pp. 119–133.
46. E. A. Zylber and S. Penman, Products of RNA polymerases in HeLa cell nuclei, *Proc. Natl. Acad. Sci. USA* **68**:2861–2865, 1971.
47. R. H. Reeder and R. G. Roeder, Ribosomal RNA synthesis in isolated nuclei, *J. Mol. Biol.* **67**:433–441, 1972.
48. M. I. Evans, L. J. Hager, and G. S. McKnight, A somatomedin-like peptide hormone is required during the estrogen-mediated induction of ovalabumin gene transcription, *Cell* **25**:187–193, 1981.
49. G. S. McKnight and R. D. Palmiter, Transcriptional regulation of the ovalbumin and conalbumin genes by steroid hormones in chick oviduct, *J. Biol. Chem.* **254**:9050–9058, 1979.
50. R. M. Evans, N. C. Birnberg, and M. G. Rosenfeld, Glucocorticoid and thyroid hormones transcriptionally regulate growth hormone gene expression, *Proc. Natl. Acad. Sci. USA* **79**:7659–7663, 1982.
51. R. Weinmann and R. G. Roeder, Role of DNA-dependent RNA polymerase III in the transcription of the tRNA and 5S RNA genes, *Proc. Natl. Acad. Sci. USA* **71**:1790–1794, 1974.
52. P. A. Weil and S. P. Blatti, Partial purification and properties of calf thymus deoxyribonucleic acid dependent RNA polymerase III, *Biochemistry* 14:1636–1642, 1974.
53. S. H. Mellon-Nussbaum, S. K. Nordeen, S. R. Spindler, and J. D. Baxter, Glucocorticoid receptors and effects on gene transcription, in: *Progress in Research and Clinical Applications of Corticosteroids* (H. L. Thomas and S. Fitzgerald, eds.), Heyden and Son, Inc. Philadelphia, 1982, pp. 1–3.
54. A. Barta, R. I. Richards, J. D. Baxter, and J. Shine, Primary structure and evolution of the rat growth hormone gene, *Proc. Natl. Acad. Sci. USA* **78**:4867–4871, 1981.
55. W. L. Miller and N. L. Eberhardt, Structure and evolution of the growth hormone gene family, *Endocrinol. Rev.* **4**:97–130, 1983.
56. M. Wigler, R. Sweet, G. F. Sim, B. Wold, A. Pellicer, E. Lacy, T. Maniatis, S. Silverstein, and R. Axel. Transformation of mammalian cells with genes from procaryote and eukaryotes, *Cell* **16**:777–785, 1979.
57. M. Karin, N. L. Eberhardt, S. H. Mellon, N. Malich, R. I. Richards, E. P. Slater, A. Barta, J. A. Martial, J. D. Baxter, and G. Cathala, Expression and hormonal regulation of the rat growth hormone gene in transfected mouse L cells *DNA* **3**:147–155, 1984.
58. J. Doehmer, M. Barinaga, W. Vale, M. G. Rosenfeld, I. M. Verma, and R. M. Evans, Introduction of rat growth hormone gene into mouse fibroblasts via a retroviral DNA vector: Expression and regulation, *Proc. Natl. Acad. Sci. USA* **79**:2268–2272, 1982.
59. F. M. DeNoto, D. D. Moore, and H. M. Goodman, Human growth hormone DNA sequence and mRNA structure: Possible alternative splicing, *Nucleic Acids Res.* **9**:3719–3723, 1981.
60. D. M. Robins, I. Paek, P. H. Seeburg, and R. Axel, Regulated expression of human growth hormone genes in mouse cells, *Cell* **29**:623–631, 1982.

15

Organization and Function of the Growth Hormone Gene Cluster

John S. Parks

1. Introduction

The human growth hormone (hGH) gene cluster contains genes coding for hGH and human chorionic somatomammotropin (hCS, also known as placental lactogen) together with related genes that do not code for familiar hormones. Intensive molecular genetic investigation over the past 6 years has established the mRNA sequences for both hormones, the chromosomal location and regional organization of the hGH gene cluster, and the DNA sequences of all genes within the cluster. Understanding of gene structure has explained the synthesis of 20K hGH and has suggested the possibility of a new peptide related to hGH. The molecular mechanisms responsible for tissue-specific expression of related genes have begun to be unraveled. Discovery of hGH and hCS gene deletions in several heritable disorders has clarified the roles of the hormones in human physiology.

2. Location and Regional Organization of the GH Gene Cluster

Analysis of cloned cDNA copies of hGH[1,2] and hCS[3] mRNA established the sequences of the signal peptides as well as the mature hormones. Several ambiguities of amino acid sequence were resolved, particularly those relating to glutamine versus glutamic acid and asparagine versus aspartic acid assignments. The hormones show 85% amino acid homology, and their mRNA sequences manifest 94% homology. Because of this high degree of homology,

John S. Parks • Department of Pediatrics, Emory University School of Medicine, Atlanta, Georgia 30322.

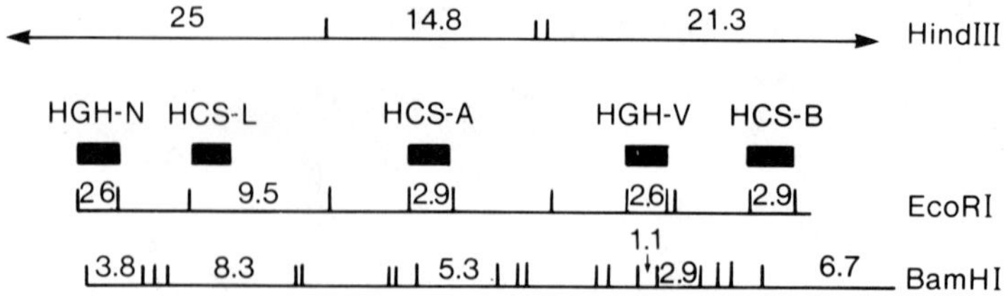

Figure 1 Regional organization of the growth hormone gene cluster. This regional map is based on the results of Barsh *et al.*(12) *Hind* III, *Eco* RI, and *Bam* HI restriction sites are indicated by vertical marks. The lengths in kilobases of hybridizing genomic fragments containing hGH or hCS genes are indicated between marks representing restriction sites.

radiolabeled cDNA's for either hGH or hCS can be used to detect related sequences in restriction endonuclease digests of genomic DNA, in cloned genomic DNA fragments, and in chromosome spreads. Analysis of genes retained in mouse–human somatic hybrid cells permitted localization of hGH and hCS genes to human chromosome 17(4) and, specifically, to a region between 17q21 and the end of the long arm.(5) *In situ* hybridization to chromosomes further located the genes in the region 17q22-q24.(6)

Early studies of hybridization of hGH or hCS cDNA probes to digests of whole cellular DNA showed that there were more hybridizing genes than were needed to code for the two known hormones.(7) The number of genes and their locations relative to each other have been a matter of interest and debate. A total of five nonallelic hGH or hCS genes have been cloned and sequenced.(8–11) Barsh *et al.*(12) have recently published a regional map of the hGH/hCS gene cluster which accounts for the five genes. Two overlapping genomic fragments of 43- and 39-kilobase (kb) pairs were generated by limited *Mbo* I digestion and isolated from a cosmid library. As shown in Fig. 1, the genes lie in the same transcriptional orientation in the order 5' hGN-N, hCS-L, hCS-A, hGH-V, hCS-B 3' over a distance of approximately 50 kb.

This chapter will employ the gene designations used by Barsh *et al.*(12) Different laboratory groups have used different designations for these genes; so a comment about terminology is in order. HGH-N codes for normal pituitary hGH. The *Eco* RI fragment containing this gene has been sequenced by DeNoto *et al.*(8) and by Seeburg.(9) The hCS-L or hCS-L-like gene has previously been referred to as a GH-like gene(4) and as the hPL-1 gene.(10) Details of its sequence have been alluded to,(12) but the complete sequence has not been published. The hGH-V or hGH-variant gene has been sequenced by Seeburg.(9) In other references, it has been termed the hGH-1(8) gene or the hGH-2(10) gene. HCS-A reflects the original hCS cDNA sequence.(3) Selby *et al.*(11) sequenced this gene and referred to it as the hCS-1 gene. The hCS-B gene has been sequenced by Seeburg,(9) who termed it the hCS gene. Barrera-Saldena *et al.*'s(10) hPL-4 cDNA sequence reflects the hCS-A gene, and their hPL-3 sequence is that of the hCS-B gene.

The map locations of *Hind* III, *Eco* RI, and *Bam* HI restriction sites shown in Fig. 1 predict hybridizing fragment lengths that are in complete agreement with the fragments generated by digestion of genomic DNA specimens from more than 50 normal individuals.[12–15] Cosmid libraries and genomic mapping do not support the existence of an hGH/hCS hybrid gene,[16] a pseudogene lacking the first exon,[17] a third hGH gene,[11,18] or genes with an opposite transcriptional polarity.[11,18] Some of the gene associations noted in genomic libraries generated in lambda bacteriophage appear to be artifacts of *in vitro* recombination between closely related sequences.

3. General Organization of hGH and hCS Genes

The various hGH and hCS genes are organized in a very similar manner. Each gene is transcribed to produce a large premessenger RNA containing 5' and 3' untranslated sequences, coding sequences or exons, and intervening sequences or introns. Intervening sequences are excised, and the codons are spliced or joined together to form a mature mRNA that codes for a prehormone of 217 amino acids. A signal peptide of 26 amino acids is removed to yield a mature hormone containing 191 amino acids. As illustrated in Fig. 2, the genes consist of five exons designated I to V. The exons are separated by four introns designated A through D. These introns occur at identical locations with respect to amino acid codons. Exon I contains 60 nucleotides of 5' untranslated sequences, codons -26 to -24 of the signal peptide, and the first nucleotide of

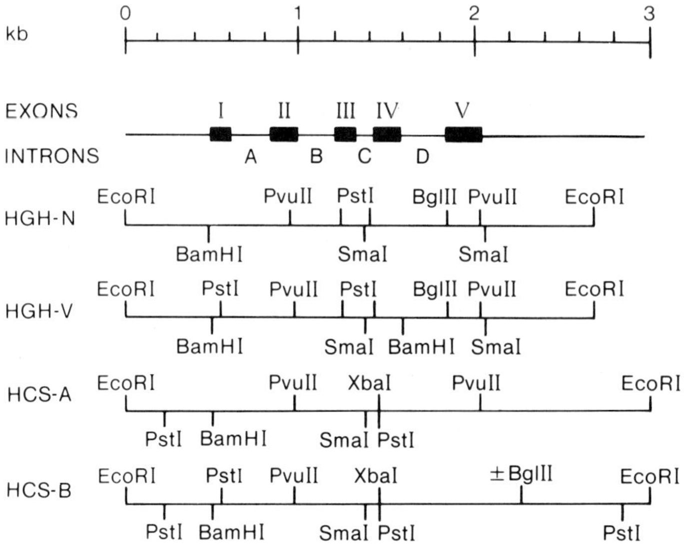

Figure 2 Organization of hGH and hCS genes. The regions of hGH and hCS genes that represent exons and introns are indicated in the upper panel. Restriction endonuclease recognition sites within the hGH-N, HGH-V, hCS-A, and hCS-B genes are indicated in the lower panel.

codon -23. The second exon codes for the remainder of the signal peptide and amino acid residues 1–31 of the mature hormones. Exons III and IV encode amino acids 32–71 and 72–126, respectively. The last exon codes for amino acids 127 to 191 and extends for approximately 100 nucleotides 3' to the translational termination signal, TAG. As in other eukaryotic genes, the 5' or "donor" portion of each intron begins with GT, and the 3' or "acceptor" portion ends with AG. In the hGH-N gene, intron A contains 256 base pairs, while introns B, C, and D are 208, 93, and 253 nucleotides in length, respectively. Intron A of the hGH-V gene and intron B of the hCS-A and hCS-B genes are longer than corresponding introns in the hGH-N gene. With these exceptions, the introns show greater than 90% homology among the four published gene sequences.

The 5' flanking and nontranslated sequences also show a great deal of homology. There are no systematic differences between hGH and hCS genes over a distance of at least 500 nucleotides 5' to the cap site. However, there are differences among the two classes of genes at the 3' end. The hGH-N and hGH-V genes contain an Alu middle repetitive sequence which begins approximately 100 nucleotides beyond the polyadenylation site and continues for about 270 nucleotides. In the hCS-A, hCS-B, and hCS-L genes, a similar sequence is truncated after 25 nucleotides.

4. *Structure and Function of Individual hGH and hCS Genes*

4.1. *The hGH-N Gene*

The hGH-N gene is solely responsible for encoding pituitary hGH. As discussed in greater detail by Dr. Phillips, individuals who are homozygous for a deletion of this gene are extremely short and have no immunoassayable hGH. The mature hormone encoded by this gene is 191 amino acids in length and has a molecular weight of approximately 22,000. This hormone accounts for approximately 90% of pituitary hGH content. The remaining 10% is accounted for by a smaller peptide with a molecular weight of 20,000. Lewis and associates[19] have reported that this 20K hGH lacks amino acid residues 32–46 of hGH. Wallis[20] observed that intron B of the hGH-N gene starts after residue 31 and suggested that 20K hGH could be a result of splicing the 5' end of exon I to a shorter Exon II which might begin after the AG of codon 46. DeNoto *et al.*[8] found an mRNA for 20K hGH in extracts of human pituitaries which could be explained by splicing codon 31 to a position 45 base pairs downstream from the normal 3' site of intron B at the start of codon 47. These authors concluded that 20K hGH could arise from alternative splicing of hGH-N premessenger RNA or from expression of a different hGH gene. With the realization that there is only one gene with the appropriate sequence, it is clear that the hGH-N gene codes for both 22K and 20K hGH.

4.2. The hGH-V Gene

As shown in Fig. 2, both hGH genes are contained in 2.6 kb *Eco RI* fragments of genomic DNA. The hGH-V gene is readily distinguished from hGH-N by the presence of a second *Bam* HI site 1.1 kb downstream from the first. Coding sequences of the two hGH genes and the two hCS genes are compared in Fig. 3. With hGH-V as with hGH-N, there is a possibility of alternative splicing between codons 31 and 47. There are two amino acid differences in the signal peptide and 13 in the mature hormone.(9) Many of the amino acid substitutions are nonconservative. In comparison to hGH-N, the hGH-V peptide has lost two acidic and gained three basic amino acids. The predicted isoelectric point is 8.9 rather than 5.5; so the two peptides should be easily resolvable by isoelectric focusing.

The hGH-V peptide is not known to be produced *in vivo*. It has been produced in monkey kidney cells infected with recombinant molecules consisting of the 2.6 kb *Eco* RI fragment of hGH-V linked to simian virus 40 DNA.(21,22) In this setting, hGH-V is the major secretory product of the tissue culture cells, and sufficient quantities are produced to examine the properties of the peptide by radioimmunoassay and radioreceptor assay. This hGH-V product cross-reacts weakly with antibodies to hGH-N. Displacement curves are nonparallel and show a potency 5–10% of standard hGH. In contrast, the hGH-V peptide displaces labeled hGH from IM-9 human lyphocytic cell and pregnant rabbit liver membrane receptors with a potency 50–100% of hGH.

The properties of hGH-V peptide produced in an artificial expression system suggest that it could have an important role as a biologically active but immunologically nonreactive hGH. Experiences with the hGH deletion syndromes imply that it does not make a major contribution to normal growth. Persons who lack the hGH-N gene but have an intact hGH-V gene grow poorly,(14) while persons who have an intact hGH-N gene but lack the hGH-V gene grow well.(13) However, it is possible that hGH-V contributes to the excess of GH that is produced in pituitary gigantism or acromegaly. Rarely, patients with acidophilic adenomas have low circulating concentrations of hGH by RIA but high somatomedin-C/IGF-I levels and clinically severe acromegaly. The GH extracted from some somatotropic tumors differs in its physical properties from standard hGH(23) and may represent an hGH-V peptide. Full understanding of the role of hGH-V will require development of a specific radioimmunoassay and application to acromegaly and other clinical situations.

4.3. The hCS-L Gene

This gene is contained in 9.5 kb *Eco* RI fragments of genomic DNA. It is the only member of the hGH gene cluster that has not been included in a recognized deletion. The fact that deletion of either the hGH-N gene(14) or all genes to the right of the hCS-L gene has little effect on fetal growth led to a suggestion that hCS-L might be responsible for a fetal GH.(13) However, most of the information available to date points away from a physiological role for

```
       -26 EXON I  '"  EXON II
hGH-N  ATG GCT ACA GGC TCC CGG ACG TCC CTG CTC CTG GCT TTT GGC CTG CTC TGC CTG CCC TGG CTT CAA GAG GGC AGT GCC
       met ala thr gly ser arg thr ser leu leu leu ala phe gly leu leu cys leu pro trp leu gln glu gly ser ala
hGH-V          GCA                                                             TCC
               ala                                                             ser
hCS-A          CCA                                         GCC     CTG                             GCT CGT
               pro                                         ala     leu                             ala gly
hCS-B          GCA                                         GCC     CTG                             GCT CGT

       1
N      TTC CCA ACC ATT CCC TTA TCC AGG CTT TTT GAC AAC GCT ATG CTC CGC GCC CAT CGT CTG CAC CAG CTG GCC TTT GAC ACC TAC
       phe pro thr ile pro leu ser arg leu phe asp asn ala met leu arg ala his arg leu his gln leu ala phe asp thr tyr
V                                                                          CGT CGC     TAC         GCA TAT         TAT
                                                                           arg         tyr             tyr
A      GTC CAA     GTT CCG                         CAC             CAA         CGC GCG                 ATT
       val gln     val                             his             gln             ala                 ile
B      GTC CAA     GTT CCG                         CAC             CAA         CGC GCG                 ATT

       29       ' "    EXON III
N      CAG GAG TTT GAA GAA GCC TAT ATC CCA AAG GAA CAG AAG TAT TCA TTC CTG CAG AAC CCC CAG ACC TCC CTC TGT TTC TCA GAG
       gln glu phe glu glu ala tyr ile pro lys glu gln lys tyr ser phe leu gln asn pro gln thr ser leu cys phe ser glu
V                                      CTG     GAG                                                     TGC
                                       leu
A                          ACC                 GAC         TCG         CAT GAC TCC             TTC TGC         GAC
                           thr                 asp         ser         his asp ser             phe cys         asp
B                          ACC                 GAC         TCA         CAT GAC TCC             TTC TGC         GAC

       57                                                     ' "  EXON IV
N      TCT ATT CCG ACA CCC TCC AAC AGG GAG GAA ACA CAA CAG AAA TCC AAC CTA GAG CTG CTC CGC ATC TCC CTG CTG CTC ATC CAG
       ser ile pro thr pro ser asn arg glu glu thr gln gln lys ser asn leu glu leu leu arg ile ser leu leu leu ile gln
V              CCA     CCT             GTG AAA ACG CAG         TCT
                                       val lys
A                                  ATG         ACG                 AAT                                             GAG
                                   met                                                                             glu
B                                  ATG         ACG                 AAT                                             GAG

       85
N      TCG TGG CTG GAG CCC GTG CAG TTC CTC AGG AGT GTC TTC GCC AAC AGC CTG GTG TAC GGC GCC TCT GAC AGC AAC GTC TAT GAC
       ser trp leu glu pro val gln phe leu arg ser val phe ala asn ser leu val tyr gly ala ser asp ser asn val tyr asp
V      TCA                         CTC         AGC                             TAT         TCG                     CGC
                                   leu                                                                             arg
A                              CGG                 ATG             AAC         TAT GAC ACC TCG             GAT GAC CAC
                               arg                 met             asn             asp thr                 asp asp his
B                              CGG                 ATG             AAC         TAT GAC ACC TCG             GAT GAC CAC

       113                                                ' "  EXON V
N      CTC CTA AAG GAC CTA GAG GAA GGC ATC CAA ACG CTG ATG GGG AGG CTG GAA GAT GGC AGC CCC CGG ACT GGG CAG ATC TTC AAG
       leu leu lys asp leu glu glu gly ile gln thr leu met gly arg leu glu asp gly ser pro arg thr gly gln ile phe lys
V      CAC                                                 TGG                                                     AAT
       his                                                 trp                                                     asn
A                                                                          GAC         CGC                     CTC
                                                                                       arg                     leu
B                                                                          GAC         CGC                     CTC

       141
N      CAG ACC TAC AGC AAG TTC GAC ACA AAC TCA CAC AAC GAT GAC GCA CTA CTC AAG AAC TAC GGG CTG CTC TAC TGC TTC AGG AAG
       gln thr tyr ser lys phe asp thr asn ser his asn asp asp ala leu leu lys asn tyr gly leu leu tyr cys phe arg lys
V          TCC             TTT         AAA TCG                     CTG
           ser                         lys
A                          TTT             TCG         CAT         CTG
                                                       his
B                          TTT             TCA         CAT         CTG

       169                                                                                     191
N      GAC ATG GAC AAG GTC GAG ACA TTC CTG CGC ATC GTG CAG TGC CGC TCT GTG GAG GGC AGC TGT GGC TTC
       asp met asp lys val glu thr phe leu arg ile val gln cys arg ser val glu gly ser cys gly phe
V

A                                              ATG
                                               met
B                                              ATG                     GTA     GGT
```

Figure 3 Codons and amino acid sequences for hGH and hCS genes and peptides. The hGH-N sequences are provided in full. Silent base differences and base differences resulting in amino acid substitutions with respect to hGH-N are provided for hGH-V, hCS-A, and hCS-B. End points of exons are indicated by ' and beginning points of exons are indicated by " above the hGH-N base sequence.

an hCS-L peptide. Barrera-Saldena, Seeburg, and Saunders[10] were unable to find an mRNA for hCS-L in extracts of term placenta. This left open the possibility that the gene is expressed by placenta at an earlier stage or by an organ other than the placenta. Sequence data mentioned by Barsh *et al.*[12] indicate that the hCS-L gene is potentially functional except for a G to A transition at the start of the intron B which abolishes a GT splice donor sequence. It is not known whether hCS-L can be expressed in a transgenomic expression system similar to that employed for the expression of hGH-V and hGH-N[21] or whether fetal tissues have the capacity to splice at sequences that deviate from consensus splice sequences. Until this information is known, it is reasonable to regard the hCS-L gene as a nonfunctional pseudogene.

4.4. The hCS-A and hCS-B Genes

Both hCS genes are contained in 2.9 kb *Eco* RI fragments of genomic DNA. The genes can be distinguished by the presence of a second *Pvu* II site and a third *Pst* I site in the hCS-A gene,[10] as shown in Fig. 2. There are four silent nucleotide differences in the coding regions of the genes, as shown in Fig. 3. Each codes for the same mature hCS peptide, but hCS-A specifies a proline and hCS-B specifies an alanine residue at position -24 of the signal peptide. Both genes are expressed in term placenta. Barrera-Saldena *et al.* originally studied expression of the genes in a single term placenta.[10] Three different experimental approaches led to the conclusion that hCS-A mRNA was more abundant than hCS-B mRNA by a ratio of 3:2. Subsequent studies of additional term placentas have shown that the ratio of hCS-A to hCS-B expression is highly variable.[24]

5. Tissue Specificity of hGH and hCS Gene Expression

Under normal circumstances, hGH is produced only by somatotrophic cells of the anterior pituitary, and hCS is produced only by syncytiotrophoblastic cells of the fetal placenta. Either hormone may be produced ectopically by certain malignant tumors. In contrast to human chorionic gonadotropin, hCS is generally not produced by hydatidiform moles.

The primary sequences of hGH and hCS genes are remarkably similar at the 5' end, but differ at the 3' end. It is not known whether these differences have anything to do with tissue specificity of expression. Other experimental approaches have been used to determine why the hGH-N gene is expressed in pituitary tissue and the hCS genes are expressed in placental tissue. They have addressed the question of whether hGH and hCS genes show characteristics of active genes in expressing tissues. Transcriptionally active genes generally show a lower extent of methylation of cytosine residues and exhibit regions of increased sensitivity to DNAse I digestion of extracted chromatin.

Hjelle *et al.*[25] have examined relative methylation of hGH and hCS genes in pituitary and placental tissues compared to human leukocytes and hydatidiform moles. They employed the enzymes *Hpa* II and *Hin* I, which recognize and cleave the sequences CCGG and GCGC only when the C residues are not methylated, and *Msp* I, which cleaves CCGG regardless of methylation. The hGH and hCS genes showed a greater degree of methylation in leukocyte DNA than in DNA extracted from the other three tissues. DNA from pituitary tissue showed somewhat more methylation than DNA from placentas or hydatidiform moles. It was not possible to demonstrate that hGH genes showed selective undermethylation in pituitary tissue or that hCS genes were less methylated in placental tissue. The authors concluded that undermethylation is a necessary but not sufficient requirement for hGH and hCS gene expression.

Barsh *et al.*[12] studied DNase I sensitivity of hGH and hCS genes in the chromatin of nuclei prepared from pituitary and placental tissue. The results suggested that sites in the 5' flanking region of hGH-N may be more sensitive in pituitary than in placental nuclei and that the 3' region of hCS-A and hCS-B genes may be more sensitive in placental than in pituitary nuclei. The possibility of a transcriptionally active chromatin configuration at the 5' end of the cluster in pituitary and an active configuration at the 3' end of the cluster in placental tissue is intriguing and requires further study.

6. Deletions of HCS-A, HGH-V, and HCS-B Genes

In 1979, Nielsen *et al.*[26] reported the total absence of immunoreactive hCS peptide in maternal serum and in extracts of placental tissue associated with an otherwise unremarkable pregnancy. Wurzel *et al.*[13] later studied hGH and hCS restriction patterns in genomic DNA from the child and his family. At the time of study, the child was 5 years old and in the 90th percentile for height. The hGH-N gene was present in fragments of normal size, and the hCS-L gene was represented in an abnormally large *Eco* RI fragment of 10.0 kb, but restriction fragments that normally contain the hCS-A, hGH-V, and hCS-B genes were absent. The child was homozygous for a large deletion of at least 30 kb, which includes all of the hGH gene cluster downstream from the hCS-L gene. This deletion is designated CSD1 in Fig. 4. Correlation of phenotype and genotype indicated that fetal production of hCS is not required for normal growth and development. Furthermore, the unanticipated finding of hGH-V gene deletion indicated that hGH-V gene expression is not necessary for postnatal growth.

Two additional children have been shown to have hCS gene deletions.[27,28] Each has inherited a growth hormone gene cluster similar to the original deletion cluster from one parent and a different deletion from the second parent. The new deletions are indicated as CSD2 and CSD3 in Fig. 4. Heterozygosity for hCS D1 and hCS D2 also resulted in complete antenatal deficiency of hCS production.[27] In hCS D2, the deletion apparently originated at a point closer to the hCS-L gene. Heterozygosity for hCS D1 and hCS D3 was present in a child whose mother had serum hCS concentrations on the order of 1 μg/ml, or somewhat less than one-quarter of normal hCS concen-

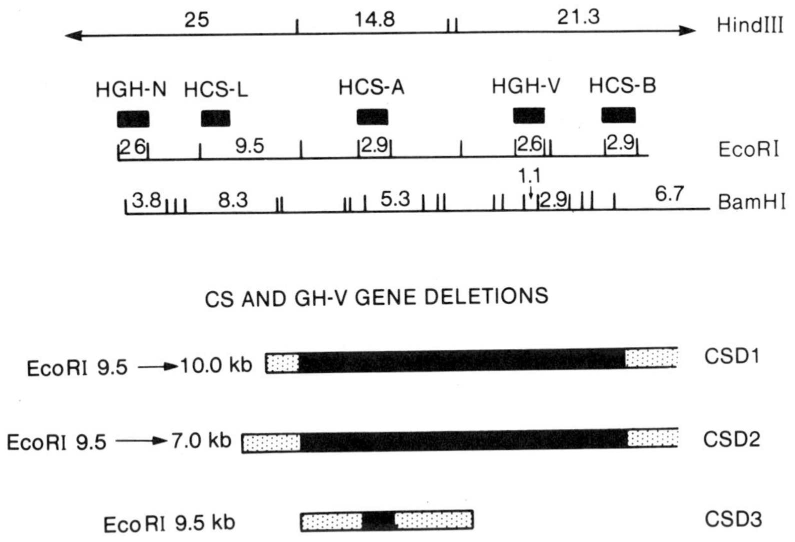

Figure 4 Deletions of hCS-A, hGH-V, and hCS-B genes. A regional map of the hGH gene cluster with locations of *Hin*d III, *Eco* RI and *Bam* HI sites is displayed in the upper panel. The lower panel shows three different deletions of hCS genes. The regions that are known to be deleted on the basis of genomic restriction mapping are indicating by solid bars. Regions which are not detected by an hCS cDNA probe, but which may be deleted, are indicated by stippled shading. Distinctive restriction fragment sizes associated with the hCS D1 and hCS D2 deletion clusters are also indicated.

trations during the 3rd trimester.[28] In hCS D3, the deletion is limited to hCS-A, so the child had only one of the normal complement of four functional hCS genes per diploid genome, and there was a proportional reduction in hCS production.

The hCS and hGH-V deletion syndromes are of significance in that they have provided conclusive evidence that hCS and hGH-V are not needed for development. They also indicate that gene deletion is a common mechanism for loss of hCS gene function and that different deletions can be responsible for the same biochemical phenotype.

ACKNOWLEDGMENTS. A portion of this work was supported by March of Dimes Birth Defects Foundation Clinical Research Grant 6-354. The author wishes to thank Mrs. Laura Sexton for critical review of the manuscript.

References

1. J. A. Martial, R. A. Hallewell, J. D. Baxter, and H. M. Goodman, Human growth hormone: complementary DNA cloning and expression in bacteria, *Science* **205:**602, 1979.
2. W. G. Roskam and F. Rougeon, Molecular cloning and nucleotide sequence of the human growth structural gene, *Nucleic Acids Res.* **7:**305, 1979.
3. J. Shine, P. H. Seeburg, J. A. Marital, J. D. Baxter, and H. M. Goodman, Construction and

analysis of recombinant DNA for human chorionic somatomammotropin, *Nature* **270**:494, 1977.

4. D. Owerbach, W. J. Rutter, J. A. Martial, J. D. Baxter, and T. P. Shows, Genes for growth hormone, chorionic somatomammotropin and growth hormone-like gene on chromosome 17 in humans, *Science* **209**:289, 1980.
5. D. L. George, J. A. Phillips, III, U. Francke, and P. H. Seeburg, The genes for growth hormone and chorionic somatomammotropin are on the long arm of human chromosome 17 in the region q21-qter, *Hum. Genet.* **57**:138, 1982.
6. M. E. Harper, H. A. Barrera-Saldena, and G. F. Saunders, Chromosomal localization of the human placental lactogen-growth hormone gene cluster to 17q22-24, *Am. J. Hum. Genet.* **34**:227, 1982.
7. J. C. Fiddes, P. H. Seeburg, F. M. DeNoto, R. A. Hallewell, J. D. Baxter, and H. M. Goodman, Structure of the genes for human growth hormone and chorionic somatomammotropin, *Proc. Natl. Acad. Sci. USA* **76**:4294, 1979.
8. F. M. DeNoto, D. D. Moore, and H. M. Goodman, Human growth hormone DNA sequence and mRNA structures: possible alternative splicing, *Nucleic Acids Res.* **9**:3719, 1981.
9. P. H. Seeburg, The human growth hormone gene family: nucleotide sequences show recent divergence and predict a new polypeptide hormone, *DNA* **1**:239, 1982.
10. H. A. Barrera-Saldena, P. H. Seeburg, and G. F. Saunders, Two structurally different genes produce the same human placental lactogen hormone, *J. Biol. Chem.* **258**:3787, 1983.
11. W. L. Miller and N. L. Gberhardt, Structure and evolution of the growth hormone gene family, *Endocrine Rev.* **4**:97, 1983.
12. G. S. Barsh, P. H. Seeburg, and R. E. Gelinas, The human growth hormone gene family: structure and evolution of the chromosomal locus, *Nucleic Acids Res.* **11**:3939, 1983.
13. J. Wurzel, J. S. Parks, J. E. Herd, and P. V. Nielsen, A gene deletion is responsible for absence of human chorionic somatomammotropin, *DNA* **1**:251, 1982.
14. J. A. Phillips, III, J. S. Parks, B. L. Hjelle, J. E. Herd, L. P. Plotnick, C. J. Migeon, and P. H. Seeburg, Genetic analysis of familial isolated growth hormone deficiency type I, *J. Clin. Invest.* **70**:489, 1982.
15. J. A. Phillips, III, B. L. Hjelle, P. H. Seeburg, and M. Zachmann, Molecular basis for familial isolated growth hormone deficiency, *Proc. Natl. Acad. Sci. USA* **78**:6372, 1981.
16. D. D. Moore, M. A. Conkling, and H. M. Goodman, Human growth hormone: A multigene family, *Cell* **29**:285, 1982.
17. D. D. Moore, M. D. Walker, D. J. Diamond, M. A. Conkling, and H. M. Goodman, Structure, expression and evolution of growth hormone genes, *Recent Prog. Hormone. Res.* **38**:197, 1982.
18. V. J. Kidd and G. F. Saunders, Linkage arrangement of human placental lactogen and growth hormone genes, *J. Biol. Chem.* **157**:10673, 1982.
19. U. J. Lewis, L. F. Bonewald, and L. J. Lewis, The 20,000 dalton variant of human growth hormone: location of the amino acid deletions, *Biochem. Biophys. Res. Commun.* **92**:511, 1980.
20. M. Wallis, Growth hormone: Deletions in the protein and introns in the gene, *Nature* **284**:512, 1980.
21. G. N. Pavlakis, N. Hizuka, P. Gorden, P. Seeburg, and D. H. Hamer, Expression of two human growth hormone genes in monkey cells infected by simian virus 40 recombinants, *Proc. Natl. Acad. Sci. USA* **78**:7398, 1981.
22. N. Hizuka, C. M. Hendricks, G. N. Pavlakis, D. H. Hamer, and P. Gorden, Properties of human growth hormone polypeptides: purified from pituitary extracts and synthesized in monkey kidney cells and bacteria, *J. Clin. Endocrinol. Metab.* **55**:545, 1982.
23. N. Hizuka, K. Takano, K. Asakawa, M. Miyakawa, and K. Shizume, Heterogeneity of hGH in pituitary adenoma of acromegaly (abstr.), *Endocrinology* **112**:533, 1983.
24. S. L. Fitzpatrick, R. Premont, H. A. Barrera-Saldena, and G. F. Saunders, Expression of two human placental lactogen genes in placentas from several individuals (abstr.), *Endocrinology* **112**:631, 1983.
25. B. J. Hjelle, J. A. Phillips, III, and P. H. Seeburg, Relative levels of methylation in human growth hormone and chorionic somatomammotropin genes in expressing and non-expressing tissues, *Nucleic Acids Res.* **10**:3459, 1982.

26. P. V. Nielsen, H. Pedersen and E. Kampmann, Absence of human placental lactogen in an otherwise uneventful pregnancy, *Am. J. Obstet. Gynecol.* **135:**322, 1979.
27. J. S. Parks, J. E. Herd, and P. V. Nielsen, A new deletion of the hGH-Variant and hCS genes (abstr.), **DNA** 2:78, 1983.
28. J. S. Parks, P. V. Nielsen, L. A. Sexton, and E. H. Jorgeusen, An effect of gene dosage on production on human chroionic somatomammotropin *J. Clin. Endocrinol. Metab.* **60:**994, 1984.

16

Defects of Growth Hormone Genes

Clinical Syndromes

John A. Phillips III, Angel Ferrandez, Herwig Frisch, Ruth Illig, and Klaus Zuppinger

1. *Introduction*

Human growth hormone (GH) is a 191-amino-acid polypeptide hormone whose production and release by somatotropes of the anterior pituitary is under complex hormonal control. Production of GH is increased by glucocorticoid, thyroid hormone, and growth-hormone-releasing factor (GRF) at least in part by stimulation of transcription of the GH gene.[1] Release of GH into the peripheral circulation is controlled by at least two hypothalamic factors: GRF, which stimulates, and somatostatin, which inhibits its release.[2–5] In turn, these hypothalamic factors are modulated by various influences originating in the central nervous system (CNS).[6] Because of the complexity of these interactions which control the synthesis and release of GH, GH deficiency could be caused by disorders at the level of the CNS, the hypothalamus, or the anterior pituitary. Thus, a wide variety of disorders including CNS developmental defects such as septooptic dysplasia, head trauma, histiocytosis X, tumors, or irradiation can cause GH deficiency.[6,7]

Although in most cases GH deficiency is assumed to be idiopathic or due to environmental factors, there are six known single gene disorders that are associated with deficiency or absence of GH (Table 1).[7–10] These disorders differ in their phenotypes (degree of GH as well as other hormonal deficiencies) as well as their modes of inheritance. This heterogeneity could be due to each

John A. Phillips III • Division of Genetics, Vanderbilt University School of Medicine, Nashville, Tennessee 37232. ***Angel Ferrandez*** • Hospital Infantil, E-S0009, Zaragosa, Spain. ***Herwig Frisch*** • Department of Endocrinology, University-Kinderklinik Stien, A-1090 Wien IX-Vienna, Austria. ***Ruth Illig*** • Universitatas Kinderklinik, CH-8032 Zurich, Switzerland. ***Klaus Zuppinger*** • Inselspital, CH-3010 Bern, Switzerland.

Table 1 Genetic Disorders with Growth Hormone Deficiency[a]

Disorder	Mode	Endogenous
IGHD		
1A	Autosomal recessive	GH absent
1B	Autosomal recessive	GH decreased
II	Autosomal dominant	GH decreased
III	X-linked	GH decreased
PIT dwarfism		
I	Autosomal recessive	GH, TSH, ACTH, LH, FSH decreased
II	X-linked	GH, TSH, ACTH, LH, FSH decreased

[a] GH = growth hormone; TSH = thyroid-stimulating hormone; LH = luteinizing hormone; FSH = follicle-stimulating hormone.

of these different disorders having their effect on different steps in GH synthesis or release. For example, one genetic disorder might alter differentiation of cells of the hypothalamus that are necessary for GRF production, a second change the GRF gene sequences, a third change the differentiation or function of the somatotropes, and finally the GH gene itself could be abnormal. The possible diversity in genetic defects causing GH deficiency suggested by Table 1 can be reduced to three simple mechanisms (Table 2). First, GH deficiency could be due to deletion of some or all of the GH gene sequences. Second, mutations (base substitutes, small insertions or deletions) within or near the GH genes could cause an altered GH gene product having decreased immunological and biological potency or, alternatively, a deficient amount of structurally normal GH peptide. Third, GH deficiency could be associated with normal GH genes that are rendered nonfunctional by mutations affecting distant loci such as GRF or somatostatin that encode factors important in GH production and/or its release.[10] Before presenting results of investigations of the GH genes in some of the disorders in Table 1, we shall briefly review some of what is known about the GH gene cluster.

Human chorionic somatomammotropin (CS) and GH are closely related polypeptide hormones that have 92% homology between their mRNA coding sequences.[11,12] Following digestion with *Eco* RI, the GH and CS genes are contained in 2.6- and 2.9-kilobase (kb) fragments of genomic DNA, respectively, while a third, uncharacterized fragment (9.5-kb) contains additional homologous or "CS-like" (CS-L) sequences (Fig. 1, top).[12–14] Studies of the nucleotide sequences of isolated genomic clones as well as large genomic fragments in

Table 2 Molecular Basis of Growth Hormone (GH) Deficiency

Deletion GH gene	— -- —
Mutation GH gene	—GH*—
Another gene	—GH— (crossed out, arrow from "Another gene")

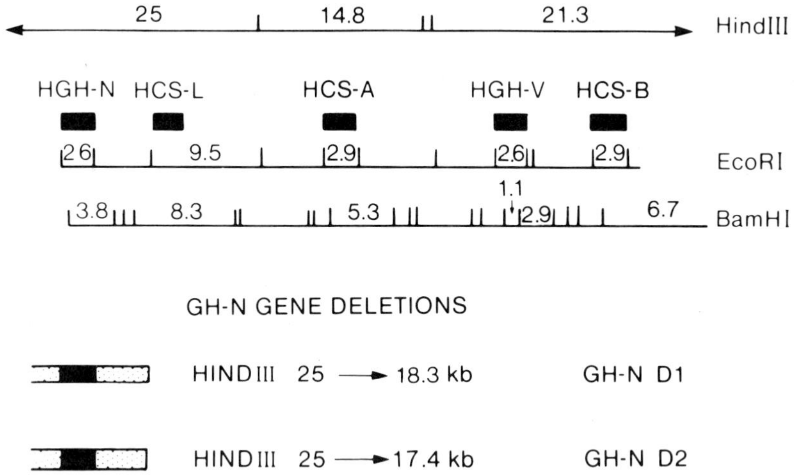

Figure 1 Regional map of the GH gene cluster. The size and location of the various genes are indicated by dark rectangles (introns not shown). The location of *Hind* III, *Eco* RI, and *Bam* HI recognition sites and selected genomic fragment sizes are indicated at the top.[12,14–16] The locations of the two sizes of GH-N gene deletions are shown at the bottom. The stippled boxes indicate the range of the endpoints, while the dark boxes indicate the obligatory deleted segments.

cosmid vectors indicate that five loci (two GH, two CS, and one CS-L) exist within the GH gene cluster.[12,14–16] In the case of the GH genes, one locus (GH-N) encodes the known protein sequence, while the other locus (GH-V) encodes a protein that differs by 14 amino acids.[17,18] The GH gene cluster spans about 50 kb (Fig. 1), and it has been regionally mapped to the long arm of chromosome 17, specifically bands q22–24.[13,19,20]

To determine whether selected clinical disorders listed in Table 1 were caused by genetic defects of the GH genes, we used restriction analysis of DNA from affected subjects and their families. The basis of isolated GH deficiency (IGHD)-1A was found to be complete deletion of the GH-N gene. In two other disorders studied, IGHD-1B and IGHD-II (Table 1), the GH-N genes were found to be intact. In IGHD-1B the cause of GH deficiency was found to be defects in distant genes not linked to the GH-N locus (Table 2, bottom), while in IGHD-II subtle mutations at the GH-N locus remain possible.

2. Results

2.1. IGHD-1A

Among the familial types of IGHD, Illig *et al.* described a subgroup, originally referred to as type A.[21–23] This subgroup was distinguished by the following clinical features: (1) familial incidence; (2) short body length at birth and early growth retardation resulting in extreme dwarfism in adulthood; (3)

typical facial appearance; (4) strong initial anabolic response to exogenous GH followed by (5) appearance of anti-GH antibodies in sufficient titer to inhibit the therapeutic effect and cause growth arrest.[23] These findings led to her hypothesis that these individuals had a prenatal deficiency of GH that caused a lack of immune tolerance to exogenous GH.

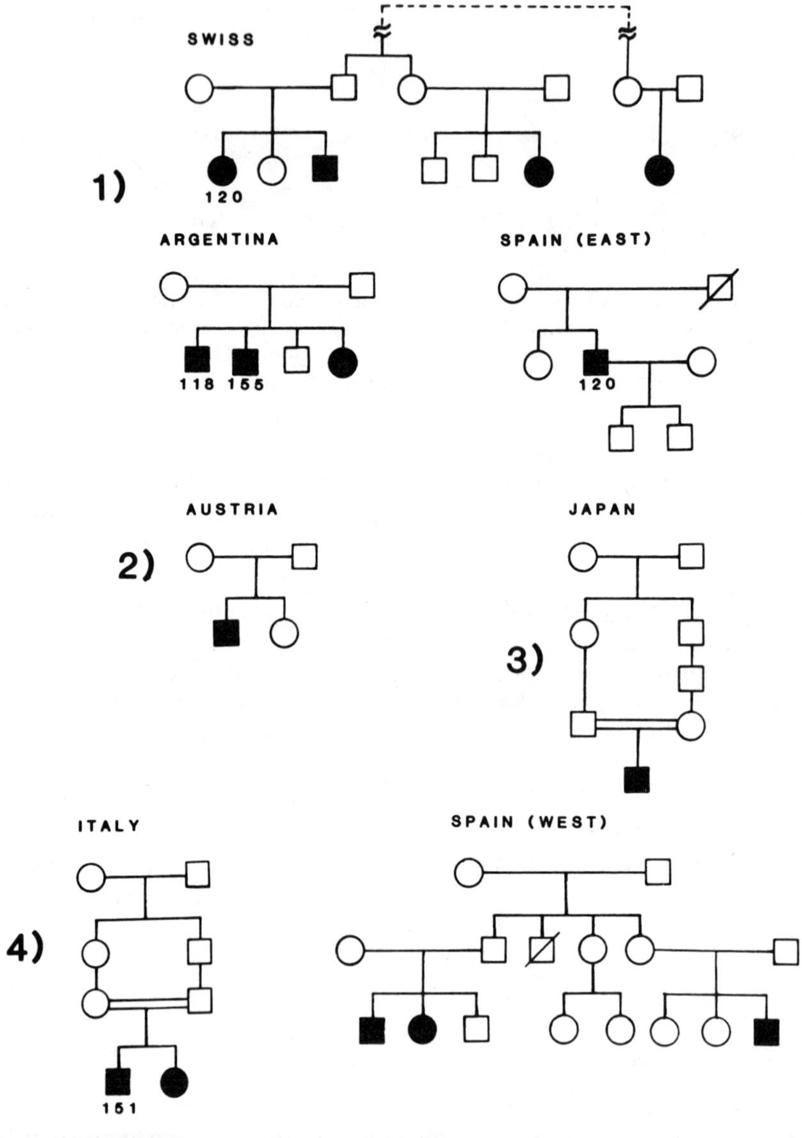

Figure 2 Pedigrees of families each of which has one or more individuals affected (solid symbols) with IGHD-1A. Country of origin and adult heights are given above and below the pedigrees, respectively.

The pedigrees of seven unrelated families each having one or more individuals with several of these clinical features are shown in Fig. 2. The affected persons are indicated by solid symbols, and the numbers below them indicate the adult heights. The pedigrees are divided into four groups based on the differences found in the genetic defect. To determine whether the GH-N genes were altered in these subjects, nuclear DNA was isolated from the leukocytes contained in 15 ml of peripheral blood.[24] The GH-N gene compliment of each was examined by cleaving the DNA with the restriction endonuclease *Bam* HI and separating the resulting fragments by size using agarose gel electrophoresis. The DNA fragments were then transferred to nitrocellulose filters and hybridized to a ^{32}P-labeled copy of complementary DNA to GH mRNA.[25] The filters were then washed and autoradiographed.[19,26,27] The resulting autoradiograph patterns, shown in Fig. 3, have multiple bands which correspond to DNA fragments containing the various genes that comprise the GH gene cluster. In order of size, the fragment and gene contained are 8.2 CS-L, 6.7 CS-B, 5.3 CS-A, 3.8 GH-N, and 3.0 + 1.2 GH-V (see Figs. 1 and

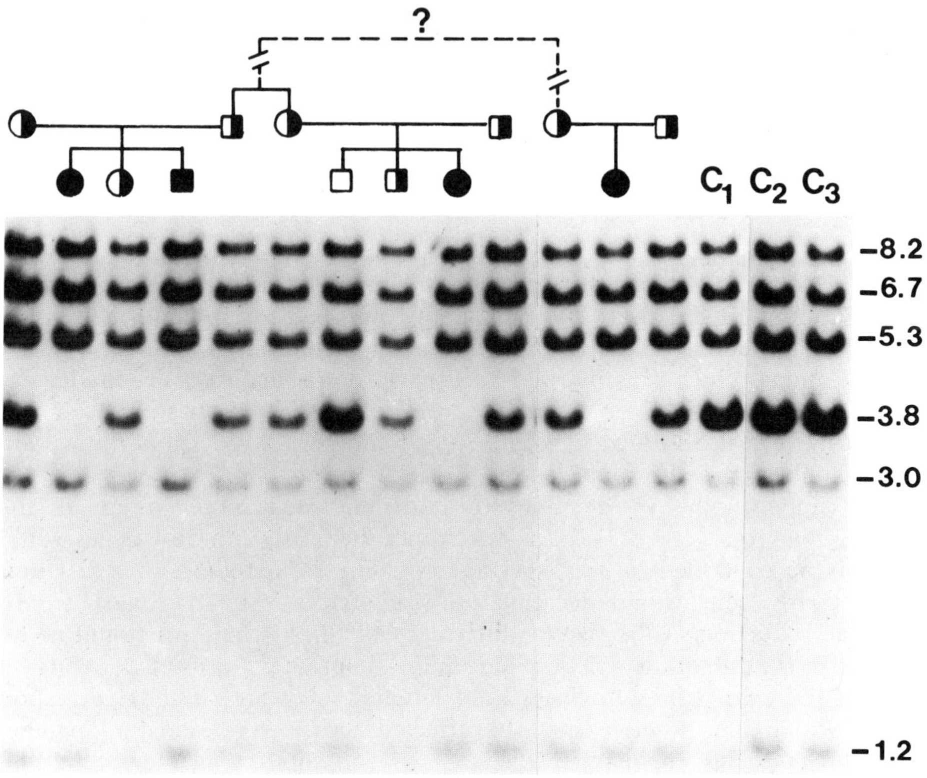

Figure 3 Autoradiogram patterns of DNA from three Swiss families (see Fig. 2) and three controls (C_{1-3}) after digestion with *Bam* HI restriction endonuclease and hybridization to the GH cDNA probe. Fragment sizes in kilobases are on right.[9]

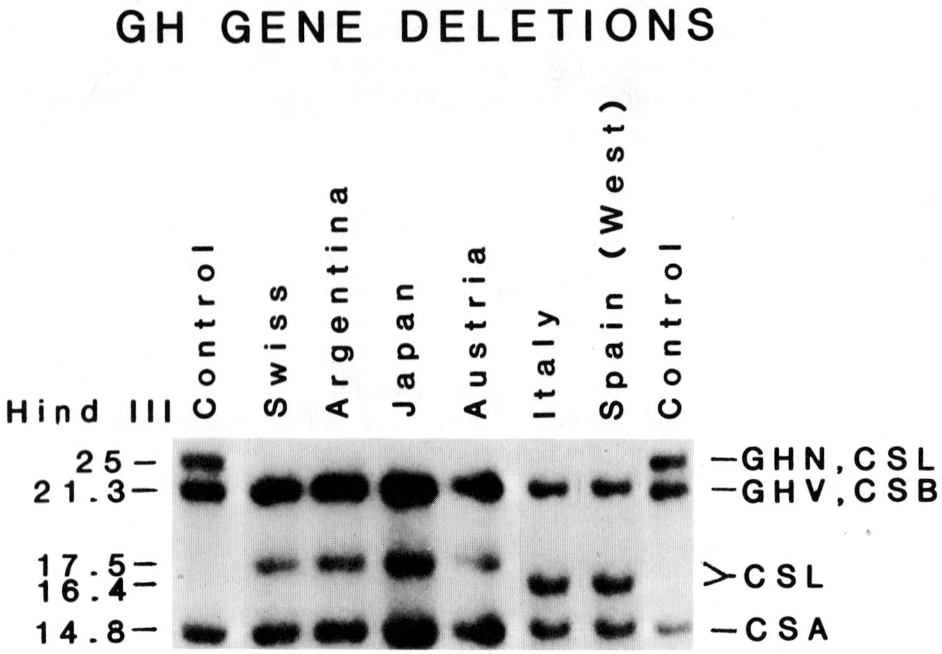

Figure 4 Autoradiogram patterns of DNA's from affected individuals in the various pedigrees shown in Fig. 2 and two controls following digestion with *Hin*d III endonuclease and hybridization to the GH cDNA probe. Fragment sizes are on the left and the gene(s) contained by the fragments are on the right.

3).[9,14] Note that the affected persons are missing, and their parents have decreased amounts of the 3.8-kb fragment compared to the controls (C_{1-3}.[9] This is in agreement with the affected persons being homozygous for GH-N gene deletions, which are inherited by a simple autosomal recessive mode.

Samples from additional families from Argentina, Austria, Italy, Japan, and Spain were also studied (Fig. 2). Affected individuals in each of these additional families were also homozygous for deletions of their GH-N genes.[28–29] To determine whether the deletions were the same size in all seven families, the DNA samples were cleaved with *Hin*d III. Since the GH-N and CS-L genes would normally be contained in a 25-kb fragment (Fig. 1), it follows that deletions of the GH-N gene would decrease the size of the 25-kb fragment.[9,14] In turn, the reduction in the size of this fragment (Fig. 1, bottom) would be an estimate of the size of the deletion. The autoradiogram patterns of DNA's from affected individuals in six of the seven families following *Hin*d III digestion are shown in Fig. 4. Note that the affected Swiss, Argentinean, Japanese, Austrian, and Eastern Spanish (data not shown) subjects all have a 18.3-kb fragment replacing the 25-kb fragment seen in controls. This implies that they all have deletions of approximately 6.7-kb (25 − 18.3) of DNA, which includes their GH-N gene. Also note that the Italian and Western Spanish subjects have

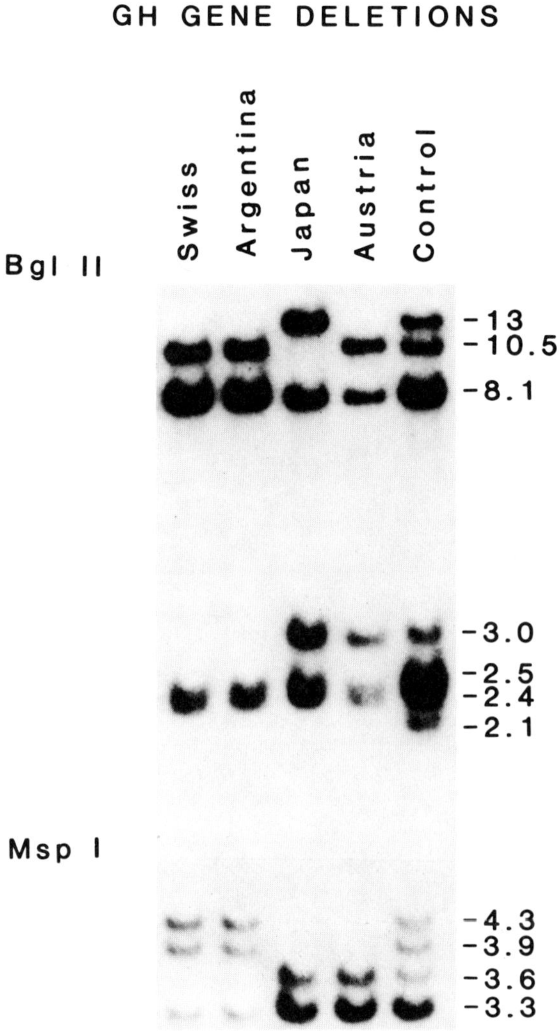

Figure 5 Autoradiogram patterns of DNA's from affected individuals in four pedigrees shown in Fig. 2 and one control following digestion with *Blg* II (top) or *Msp* I (bottom) and hybridization to the GII cDNA probe. Fragment sizes in kilobases are shown on the right (note that the *Msp* I fragments <3.3 kb are not shown).

a 17.4-kb fragment compatible with a larger 7.6-kb deletion. To determine whether all four of the 6.7-kb deletions were the same, the DNA's were digested with *Bgl* II or *Msp* I (Fig. 5). These two endonucleases have been useful in detecting four common polymorphic restriction sites that occur within or adjacent to components of the GH gene clusters in these families as well as in normal controls.[10,30] Interestingly, affected individuals from four different families yielded three different patterns of the polymorphic *Bgl* II or *Msp* I restriction sites linked to the remaining components of their GH gene clusters (Fig. 5). These patterns can be expressed as having the presence (+) or absence (−) of the site examined, and the resulting haplotypes are given in Table 3.[10,31]

Table 3 Growth Hormone Deletions[a]

Type	Size (BglII MspI)	Peak GH (ng/ml)	SMC (IU/ml)	GH Ab	Long-term response
1. Swiss	6.7 kb			+	Poor
Argentina	(+ − − −)	2.0	0.20	+	Poor–fair
Spain (East)			(0.46–1.5)		Poor
2. Austria	6.7	1.0	0.34	+	Poor
	(+ + + +)		(0.5–2)		
3. Japan	6.7	1.0	0.38	+	Poor
	(− + + +)		(0.5–2)		
4. Italy	7.6	0.81	0.20	+	Fair–good
Spain (West)	(− + + +)		(0.46–1.5)	−	Fair

[a] GH = growth hormone; SMC = somatomedin C.

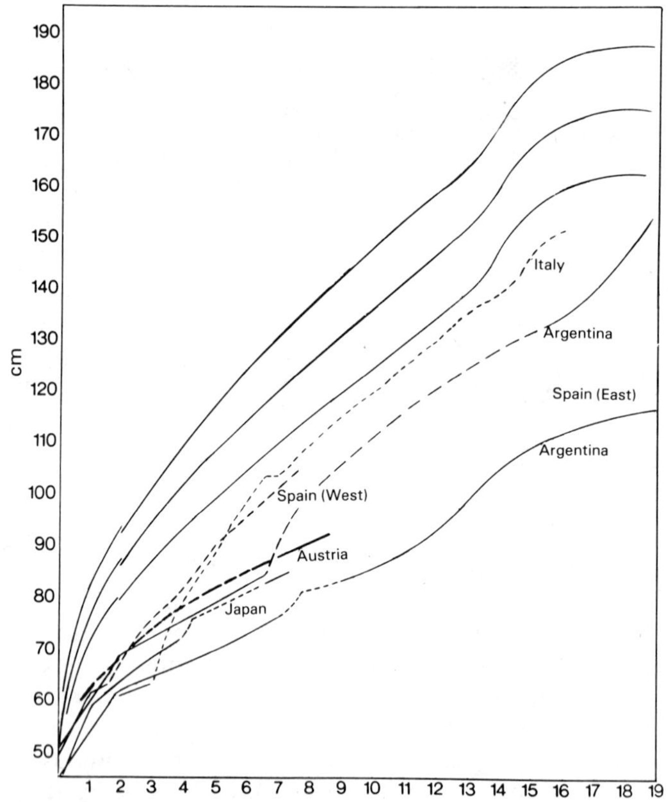

Figure 6 Composite growth charts of male individuals homozygous for GH-N gene deletions in various pedigrees shown in Fig. 2. Solid lines indicate growth without and dotted lines growth during GH treatment.

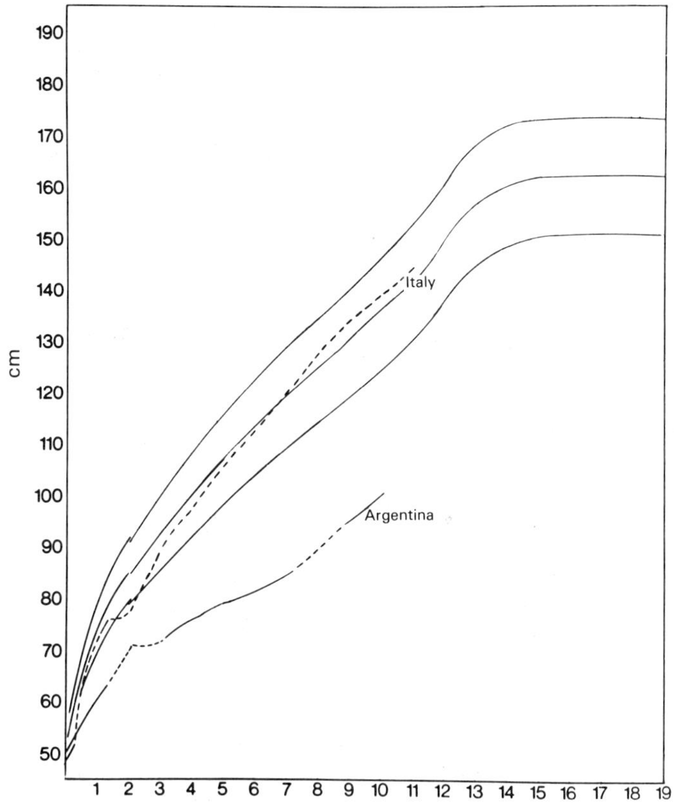

Figure 7 Subjects are females; otherwise see legend for Fig. 6.

Examples of the peak GH levels obtained after stimulation and the somatomedin C levels prior to stimulation are shown for one of the affected persons from each of the four different deletion types in Table 3. It should be noted these GH levels are extremely low and may, in fact, actually be zero. Also of interest is that all but one affected person (14 out of 15) developed high titers of anti-GH antibodies following treatment with exogenous GH. Finally, the clinical response to GH as determined by height increments ranged widely from poor to good (Table 3). These results are shown on composite growth charts (Figs. 6 and 7) in which the solid lines indicate growth without and dotted lines growth during GH treatment.[(28,29)] It is of interest that the best responses occurred in the subject whose treatment began in the first few months after birth (Fig. 7).

2.2. IGHD-1B

A second autosomal recessive type of IGHD (IGHD-1B) differs in that low but detectable levels of immunoassayable GH may be found in contrast to the

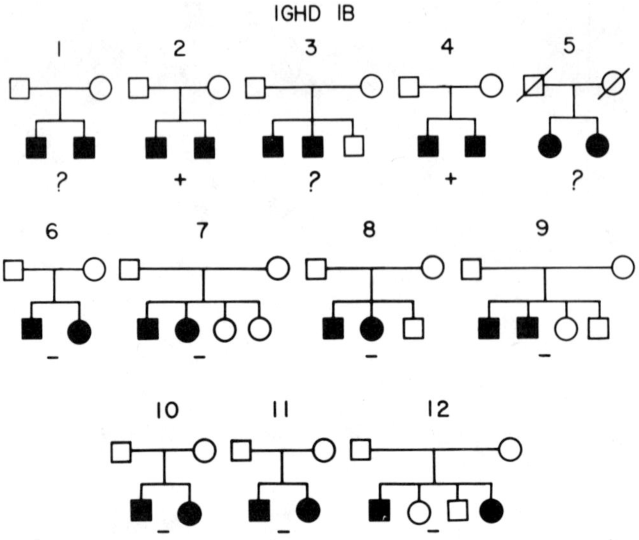

Figure 8 Pedigrees of 12 families each of which has two sibs (solid symbols) affected with IGHD-1B. Symbol below each pedigree indicates whether affected sib pairs have concordant (+), possibly concordant (?), or discordant (−) inheritance of the GH-N alleles.

absence of GH secretion that results in IGHD-1A from the GH-N gene deletions.[9,10] The group of IGHD-1B patients studied had parents of normal height and two sibs affected with IGHD by the following criteria: (1) no demonstrable anatomical cause for IGHD; (2) stature >2 SD below mean for age and sex; (3) significantly delayed bone age; (4) peak GH levels <7 ng/ml after standard pharmacological stimulation; (5) deficient growth velocity that responded to parenteral GH; (6) normal thyroid function; (7) spontaneous pubertal changes at an appropriate bone age; and (8) no known immunodeficiency or history of recurrent infections in cases from pedigrees having only affected males.[10]

In contrast to IGHD-1A we found the GH-N genes to be present and apparently normal in individuals affected with IGHD-1B.[10] To determine whether GH-N gene alterations were present that could not be detected by these methods, we used the *Msp*I and *Bgl* II and an additional *Hinc* II polymorphic restriction site as markers in linkage analysis of DNAs from family members. Among the 12 pairs of sibs affected with IGHD-1B whose pedigrees are shown in Fig. 8, those in families 6–12 were discordant or differed in the polymorphic restriction sites inherited from their parents (Fig. 9). Since these restriction sites are very close to the GH-N locus, they should be reliable indicators of which parental GH-N genes are inherited. From the discordance observed between the disease state (IGHD-1B) and the polymorphic restriction sites, hence GH-N alleles inherited by sib pairs in six families (6–12), we concluded that the mutation(s) causing the IGHD-1B phenotype and the GH-

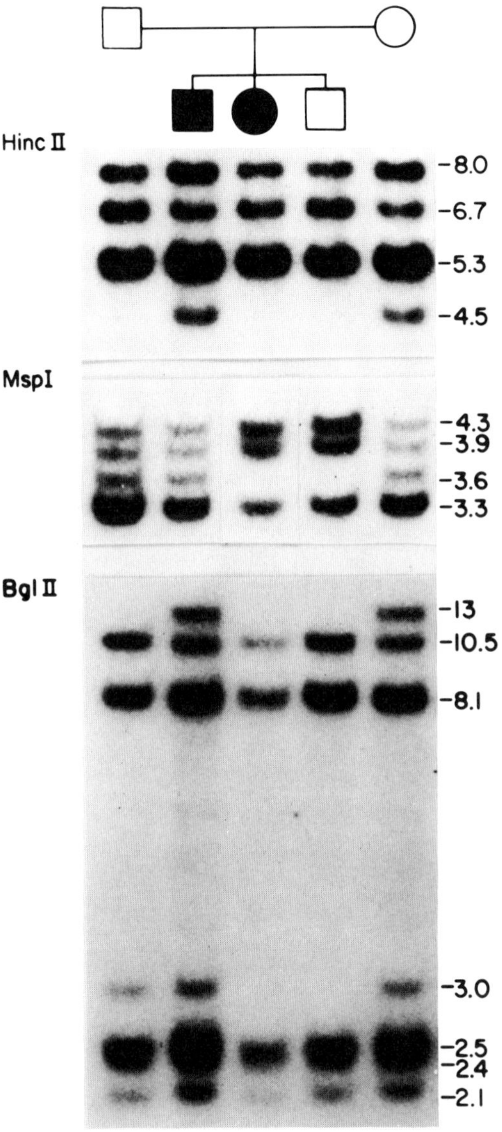

Figure 9 Autoradiogram patterns of DNA's from family 8, Fig. 8, after digestion with *Hinc* II (top), *Msp* I (middle), or *Bgl* II (bottom) and hybridization to the GH cDNA probe.[10] Fragment sizes in kilobases are shown on the right (note that the *Msp* I fragments <3.3 kb are not shown).

N gene assort independently. This suggests that the mutation causing IGHD-1B affects a nonlinked gene (possibly one important in GRF production or release, or important for development of somatotropic cell function), rather than the GH-N structural gene (Table 2, bottom).

In the remaining five families (families 1–5, Fig. 8) the affected sib pairs were shown to have the same two GH-N genes only in the case of families 2 and 4. In families 1, 3, and 5 the markers used were not sufficiently informative to prove whether the affected sibs definitely shared the same two GH-N genes or not. Thus, in a majority of these families (at least 7 out of 12) there was a

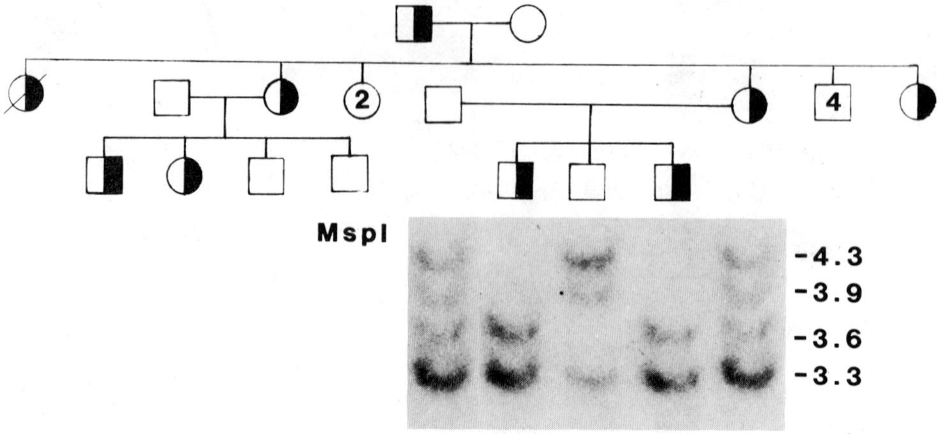

Figure 10 Pedigree of a family having multiple individuals affected with IGHD-II. Autoradiogram patterns of DNA's from five family members following digestion with *Msp* I and hybridization to the GH cDNA probe are shown below. Fragment sizes are shown on the right (note that the *Msp* I fragments <3.3 kb are not shown).

lack of association between the GH-N alleles inherited and IGHD-1B indicating that the genetic defect causing this disorder is distant from the GH-N locus.

2.3. *IGHD-II*

The final type of familial IGHD to be discussed (IGHD-II) has an autosomal dominant mode of inheritance. Affected persons in the various families reported differ in their propensity to develop hypoglycemia as well as their response to treatment with exogenous GH.[7] The clinical criteria used to diagnose IGHD-II differ from that used for IGHD-1B only in that the pattern of inheritance is different.

The pedigree of a large family having multiple individuals with IGHD-II in which the affected persons responded well to exogenous GH and in which an apparent autosomal dominant mode of inheritance occurs is shown in Fig. 10. Since the GH-N genes are retained in the three affected persons studied in this family (data not shown), it is possible that mutation(s) within one of their GH-N genes (GH*) prevents normal expression of this allele as well as the other presumably normal GH-N allele (Table 2, middle). Such a mechanism should be associated with consistent inheritance of the same GH allele (GH*) by all affected persons. To study this we used the polymorphic *Msp*I restriction sites as markers (Fig. 10). The mother's two GH-N bearing chromosomes give different patterns (one has 4.3- and 3.9-kb fragments containing components of the GH cluster while the other has a 3.6-kb fragment). Note that the mother's normal son inherited the GH-N gene associated with her 4.3- and 3.9-kb fragments. Interestingly, both her affected sons inherited her GH-N allele linked to the 3.6-kb fragment. Although these results suggest that the GH-N gene attached to the 3.6-kb fragment may be defective, samples from other family members will have to be studied to see whether all affected persons share this allele. If this allele were inherited by all affected persons but by none of the nonaffected persons, this allele would be a likely example of the GH* shown in Table 2. However, actual determination of the DNA sequence of such an allele would be required to localize the location of the mutation in the GH gene sequences as well as to clarify how such a change inactivates the other presumably normal GH-N allele.

3. Discussion

The molecular defects found in the various individuals affected with IGHD-1A were found to be examples of GH-N gene deletions as depicted at the top of Table 2. Since such deletions preclude the production of any GH-N peptide, the propensity of the affected homozygotes to form anti-GH antibodies is clarified. Their absence of GH-N peptide probably makes them immunologically intolerant of exogenous GH in analogy to other mutations that lack cross-reacting material. Interestingly, differences in the response of these individuals to exogenous GH ranged from poor to good in spite of the presence of high titers of anti-GH antibodies in virtually all. For some families this poor response to treatment might be sufficient indication for prenatal diagnosis, which could be done by DNA analysis of fetal amniotic fluid cells. Alternatively, DNA could be isolated from peripheral blood of such an infant at risk in the first few days of life to determine whether he were affected. If so, very early treatment with GH could be attempted in an effort to avoid immune intolerance of exogenous GH, which occurs later in most. The distributions of polymorphic restriction sites, which are linked to the nondeleted components of the GH gene clusters of homozygotes from each family (Figs. 2 and 4 and Table 3), suggest that three different haplotypes are associated with the more frequently detected 6.7-kb deletion. Although these deletions are of the same size, they may differ because different mutations causing β-thalassemia are often associated with specific polymorphic restriction site haplotypes.[32] If the deletions are in fact different, it implies that recurrent deletions of 6.7-kb which include GH-N have occurred within the GH cluster. An obvious mechanism of recurrent deletions of the same size would be unequal crossing over after mispairing of duplicated or analogous sequences such as has been documented for α-globin genes.[33] The larger, 7.6-kb deletion (Fig. 4, Table 3) unequivocally demonstrates that heterogeneity in the origin of such deletions exists. Further data on the organization of homologous sequences flanking the GH gene will be needed to establish the nature and location of such potential sites of recombination which generate these different deletions.

Our experience with IGHD-1B suggests a need for caution in the use of restriction site polymorphisms for diagnosis and prediction of inherited diseases. Such use should be predicated on study of a sufficient number of families to establish linkage and homogeneity of the disease. For example, conclusions based on definite or possible concordance between restriction patterns and disease in families 1–5, Fig. 8, would have led to erroneous prediction of the absence of disease in the second affected sibs in families 6–12.[10]

Analysis of polymorphic restriction patterns provides a powerful approach for establishing the absence of linkage between a particular disease and a specific DNA sequence within a single family. Using a relatively rapid and inexpensive technique, we determined that GH-N gene sequencing would not succeed in establishing the molecular basis of IGHD-1B in the families studied which were discordant (Figs. 8 and 9). In these families any sequence changes that might be found would probably be irrelevant to their disease. Other disorders involving growth for which this approach could be used to select individuals whose DNA would be appropriate for cloning and sequencing of the GH-N gene include IGHD-II (Fig. 10) and "biologically inactive" GH.[34]

4. Summary

There are six distinct Mendelian disorders that have associated GH deficiency. One of these (IGHD-1A) is caused by different deletions of the GH structural gene (GH-N). In the two remaining disorders examined (IGHD-1B and IGHD-II), the GH gene cluster is grossly intact, and for these polymorphic restriction sites within or adjacent to components of the GH cluster, they were used as "markers" in linkage analysis. In IGHD-1B most affected sib pairs were discordant for the "markers"; hence, the GH-N alleles inherited indicating the responsible mutation(s) were in distant genes rather than the GH-N locus. In a single sibship with IGHD-II an affected sib pair was concordant for the same allele which was not shared by their nonaffected sib. These results demonstrate the usefulness of linkage analysis in confirming that genetic disorders with GH deficiency may be due to alterations of the GH gene before DNA sequencing is done.

ACKNOWLEDGMENTS. The authors thank Drs. P. H. Seeburg and J. S. Parks for their collaboration in portions of these studies and Drs. J. Bailey, R. Blizzard, T. Kelly, L. Kirby, P. Malvaux, Y. Nishi, M. Rivarola, C. Sultan, J. Tze, and M. Zachmann for providing clinical samples. We thank Ms. Sandy Muscelli for her expert assistance in preparing the manuscript. We also thank Richard Mallonee for preparation of all the figures except Fig. 1, which was provided by Dr. J. S. Parks.

This work was supported in part by National Institute of Health grants AM00423 and AM35592 and RCDA AM01434.

References

1. M. Barinaga, G. Yamonoto, C. Rivier, W. Vale, R. Evans, and M. G. Rosenfeld, Transcriptional regulation of growth hormone gene expression by growth hormone-releasing factor, *Nature* **306:**84, 1983.

2. L. Krulich, A. P. S. Dhariwal, and S. M. McCann, Stimulatory and inhibitory effects of purified hypothalamic extracts on growth hormone release from rat pituitary in vitro, *Endocrinology* **83:**783, 1968.
3. N. W. Kasting, J. B. Martin, and M. A. Arnold, Pulsatile somatostatin release from the median eminence of the unanesthetized rat and its relationship to plasma growth hormone levels, *Endocrinology* **109:**1739, 1981.
4. M. Szabo, L. Chu, and L. A. Frohman, Biological effects of an ectopic growth hormone-releasing peptide in cultured adenohypophyseal cells: Comparison with growth hormone-releasing activity of porcine hypothalamus, *Endocrinology* **111:**1235, 1982.
5. J. Rivier, J. Spiess, M. Thorner, and W. Vale, Characterization of a growth hormone-releasing factor from a human pancreatic islet tumour, *Nature* **300:**276, 1982.
6. J. B. Martin, Neural regulation of growth hormone secretion, *N. Engl. J. Med.* **288:**1384, 1973.
7. D. L. Rimoin, Hereditary forms of growth hormone deficiency and resistance, *Birth Defects* **12:**15, 1976.
8. T. A. Fleisher, R. M. White, S. Broder, S. P. Nissley, R. M. Blaese, J. J. Mulvihill, G. Olive, and T. A. Waldmann, X-linked hypogammaglobulinemia and isolated growth hormone deficiency. *N. Engl. J. Med.* **302:**1429, 1980.
9. J. A. Phillips III, B. L. Hjelle, P. H. Seeburg, and M. Zachmann, Molecular basis for familial isolated growth hormone deficiency, *Proc. Natl. Acad. Sci. USA* **78:**6372, 1981.
10. J. A. Phillips III, J. S. Parks, B. L. Hjelle, J. E. Herd, L. P. Plotnick, C. J. Migeon, and P. H. Seeburg, Genetic analysis of familial isolated growth hormone deficiency type I, *J. Clin. Invest.* **70:**489, 1982.
11. H. D. Niall, M. L. Hogan, R. Sauer, I. Y. Rosenblum, and F. C. Greenwood, Sequence of pituitary and placental lactogenic and growth hormones: Evolution from a primordial peptide by gene reduplication, *Proc. Natl. Acad. Sci. USA* **68:**866, 1971.
12. J. C. Fiddes, P. H. Seeburg, F. M. DeNoto, R. A. Hallewell, J. D. Baxter, and H. M. Goodman, Structure of genes for human growth hormone and chorionic somatomammotropin. *Proc. Natl. Acad. Sci. USA* **76:**4294, 1979.
13. D. Owerbach, W. J. Rutter, J. A. Martial, J. D. Baxter, and T. B. Shows, Genes for growth hormone, chorionic somatomammotropin, and growth hormone-like gene on chromosome 17 in humans, *Science* **209:**289, 1980.
14. G. S. Barsh, P. H. Seeburg, and R. E. Gelinas, The human growth hormone gene family: Structure and evolution of the chromosomal locus, *Nucl. Acids Res.* **11:**3939, 1983.
15. H. M. Goodman, F. DeNoto, J. C. Fiddes, R. A. Hallewell, G. S. Page, S. Smith, and E. Tischer, Structure and evolution of growth hormone related genes, in: *Mobilization and Reassembly of Genetic Information. Miami Winter Symposium* (W. A. Scott, R. Werner, D. R. Joseph, and J. Schultz, eds.), Academic Press, New York, 1980, Vol. 17, pp. 155–179.
16. P. H. Seeburg, Structure and regulation of pituitary hormone genes, in: *Polypeptide Hormones* (R. F. Beers and E. G. Bassett, eds.), Raven Press, New York, 1980, pp. 19–31.
17. G. N. Pavlakis, N. Hizuka, P. Gorden, P. H. Seeburg, and D. H. Hamer, Expression of two human growth hormone genes in monkey cells infected by Simian virus 40 recombinants, *Proc. Natl. Acad. Sci. USA* **78:**7398, 1981.
18. P. H. Seeburg, The human growth hormone gene family: Nucleotide sequences show recent divergence and predict a new polypeptide hormone, *DNA* **1:**239, 1982.
19. D. L. George, J. A. Phillips III, U. Francke, and P. H. Seeburg, The genes for growth hormone and chorionic somatomammotropin are on the long arm of human chromosome 17 in region q21→qter, *Hum. Genet.* **57:**138, 1981.
20. M. E. Harper, H. A. Barrera-Saldana, and G. F. Saunders, Chromosomal localization of the human placental lactogen-growth hormone gene cluster to 17 q22–24, *Am. J. Hum. Genet.* **34:**227, 1982.
21. R. Illig, Growth hormone antibodies in patients treated with different preparations of human growth hormone (HGH), *J. Clin. Endocrinol. Metab.* **31:**679, 1970.
22. R. Illig, A. Prader, A. Ferrandez, and M. Zachmann, Hereditary prenatal growth hormone deficiency with increased tendency to growth hormone antibody formation in: *Endokrinologie der Entwicklung und Reifung* (J Kracht, ed.), Springer, Berlin, 1970, p. 246.

23. R. Illig, A. Prader, A. Ferrandez, and M. Zachmann, Hereditary prenatal growth hormone deficiency with increased tendency to growth hormone antibody formation ("A-type" isolated growth hormone deficiency), *Acta Pediatr. Scand. Suppl.* **60:**607, 1971.
24. L. M. Kunkel, K. D. Smith, S. H. Boyer, D. S. Borgaonkor, S. S. Wachtel, O. J. Miller, W. R. Breg, H. W. Jones, and J. M. Rary, Analysis of human Y-chromosome specific reiterated DNA in chromosome variants, *Proc. Natl. Acad. Sci. USA* **74:**1245, 1977.
25. J. A. Martial, R. A. Hallewell, J. D. Baxter, and H. M. Goodman, Human growth hormone: complementary DNA cloning and expression in bacteria, *Science* **205:**602, 1979.
26. E. M. Southern, Detection of specific sequences among DNA fragments separated by gel electrophoresis, *J. Mol. Biol.* **98:**503, 1975.
27. A. J. Jeffreys and R. A. Flavell, A physical map of the DNA regions flanking the rabbit β-globin gene, *Cell* **12:**429, 1977.
28. M. A. Rivarola, J. A. Phillips III, C. J. Migeon, J. J. Heinrich, and B. J. Hjelle, Phenotypic heterogeneity in familial isolated growth hormone deficiency (IGHD) type A, *J. Clin. Endocrinol. Metab.* **59:**34, 1984.
29. Y. Nishi, K. Aihara, T. Usui, J. A. Phillips III, R. L. Mallonee, and C. J. Migeon, Isolated growth hormone deficiency type 1A in a Japanese family, *J. Pediatr.* **104:**885, 1984.
30. J. M. Wurzel, J. A. Parks, J. E. Herd, and P. V. Nielsen, A gene deletion is responsible for absence of human chorionic somatomammotropin, *DNA* **1:**251, 1982.
31. J. A. Phillips III, The molecular basis of familial growth hormone deficiency, in: *Recombinant DNA and Medical Genetics* (A. Messer and I. H. Porter, eds.), Academic Press, New York, 1983, pp. 9–20.
32. S. H. Orkin, H. H. Kazazian, Jr., S. E. Antonarakis, S. C. Goff, C. D. Boehm, J. P. Sexton, P. G. Waber, and J. V. Giardinia, β-Thalassemia mutations, β-globin gene polymorphisms, and their linkage with DNA polymorphisms in the human β-globin gene cluster, *Nature* **296:**627, 1982.
33. J. A. Phillips III, T. A. Vik, A. F. Scott, K. E. Young, H. H. Kazazian, Jr., K. D. Smith, V. F. Fairbanks, and H. M. Koenig, Unequal crossing-over: A common basis of single alpha-globin genes in Asians and American blacks with hemoglobin-H disease, *Blood* **55:**1066, 1980.
34. A. A. Kowarski, J. Schneider, E. Ben-Galim, V. V. Weldon, and W. H. Daughaday, Growth failure with normal serum RIA-GH and low somatomedin activity: Somatomedin restoration and growth acceleration after exogenous GH, *J. Clin. Endocrinol. Metab* **47:**461, 1978.

III

BIOSYNTHESIS OF HUMAN GROWTH HORMONE

Clinical Studies

17

Standards for Growth Hormone

D. R. Bangham

1. Standards for Growth Hormone

The term "standard" is often confused in people's minds to mean either the reference material against which other preparations are assayed to estimate their biological activity ("potency") or a set of written specifications for the quality of products made for clinical use. I shall exploit this confusion and describe both. The international standard serves to identify and quantify what we mean by "growth hormone" (GH); the pharmacopoeia specification defines the product whose clinical effects are debated here. Both aspects were well discussed at the FDA–USP International Workshop in 1982.(1)

Normal natural GH is secreted from the pituitary as a group of polypeptides,(2,3) each of attested genetic origin.(4) Although the largest and predominant molecular form (191 residues, 22K-daltons) is customarily regarded as *the* hormone, the other peptides of this group may have some other important (but as yet unidentified) biological functions.(5) On the premise that they are there on purpose, they should also be regarded as components of GH. Furthermore, many preparations of human GH used for experimental and clinical studies also contain degraded forms—with their own pharmacological effects—many of which have been characterized and documented in detail.(6) Recombinant DNA procedures now make it possible to produce individual molecular forms in quantity so that their biological properties will be studied more closely.

2. The International Standard for GH, Human

It can be argued that until the establishment of the international standard (IS) for human growth hormone (hGH) in 1982, claims about the biological

D. R. Bangham • National Institute for Biological Standards and Control, London NW3 6RB, England.

potency of preparations were based on a "floating unit." The IS for GH, bovine, for bioassay (established in 1955) was available until about 5 years ago to estimate potency. The international reference preparation (IRP) of GH, Human, for Immunoassay, was established in 1968; it contains highly purified monomer and has been widely used to calibrate immunoassay systems. But hitherto many claims of precise potency figures may be invalid because (1) the supply of the IS ran out; (2) local standards were not calibrated against it with great precision by *in vivo* bioassay methods; and (3) the extent to which minor differences in the components of extracts even of so-called highly purified GH can influence the bias of estimates in various assay methods was not generally realized until the collaborative study in 1982.

The setting up of the IS[7] was delayed by various factors. The large amount of hormone required could be obtained only from large national pituitary collections—of the United States and the United Kingdom—at a time when supplies were badly needed for clinical use. Each of the two purified candidate preparations contained more than one molecular form of hGH. They were freeze-dried and ampuled in the same way. In the international collaborative study, coded ampules—and duplicates—were assayed against the IS of bovine GH in bioassays and against the IRP of hGH in immunoassays, by 22 expert laboratories in 10 countries, in the scheme used for IS.[8] A draft report on the standard was agreed to by the participants, and the Expert Committee on Biological Standardization of WHO accepted the unitage proposed, established the standard, and defined the international unit (IU).[9] The report has been published.[10] This study documents the first large-scale comparison of two hGH preparations by several assay methods. The results showed clear disparities of estimates made by *in vivo* and *in vitro* bioassays, and receptor and immunological binding assays (see Table 1). With the agreement of the participants, an arbitrary—but rationalized—figure of 4.4 IU/ampule was assigned by WHO. Furthermore, direct comparison of the two candidate materials revealed the substantial bias obtained with different assay methods (see Fig. 1) Yet one of these preparations contained only slightly more deamidated hGH and traces of other pituitary hormones and of pyrogens than the other. The extent to which pyrogens influence the biological response is not much studied.

These results emphasize that quite minor differences in preparations called "GH" influence the estimate of their relative potency. Hence, statements of estimated potency should always specify the precision of the estimate (as fiducial or confidence limits), the assay method, and the standard used and its unitage. The new IS provides a common, stable IU for such estimates from now on. But it remains to be determined whether individual molecular species of hGH give valid and reproducible estimates when assayed against it; it is quite possible that standards for certain single components may be needed.

An IS of a pure natural of 22K-daltons hGH will help to clarify certain issues. It will also be useful to check that products made by recombinant DNA procedures are identical in all respects to the natural molecule, by such methods as peptide finger print maps after enzyme hydrolysis.

Table 1 Geometrical Mean Potencies for Two Candidate Materials Given by the Various Assay Methods[a]

Results combined	Geometric mean 80/505	80/521
Body weight gain in hypo'xed rats		
All assays	3.67 IU IS (2.89–4.65)	4.59 IU IS (3.48–6.07)
Excluding Laboratory 24	4.30 IU IS (3.83–4.81)	5.49 IU IS (4.78–6.31)
Tibial epiphyseal width in hypo'xed rats		
All assays	2.57 IU IS (1.71–3.86)	2.73 IU IS (1.75–5.26)
Excluding Laboratory 5, assay 2	3.26 IU IS (2.57–4.14)	3.51 IU IS (2.62–4.72)
Receptor assays	4.80 IU IRP (4.60–5.01)	3.95 IU IRP (3.76–4.16)
Immunoassays		
Laboratory and method means	5.26 IU IRP (4.47–6.19)	4.50 IU IRP (3.96–5.12)
Final assigned potency 4.4 IU/ampule 80/505		

[a] 95% confidence limits about the mean are shown in brackets. Estimates are expressed as IU of the International Standard of GH, bovine, for Bioassay per ampule for bioassays (IU IS) and as IU of the International Reference Preparation of hGH for Immunoassay per ampule for receptor and immunoassays (IU IRP) (WHO Unpublished Working Document WHO/BS/82.1369).

3. Specifications for the Product Extracted from Pituitaries: hGH for Injection

3.1. The Value of Clinical Studies

The value of clinical studies with GH is much greater if the product used is precisely characterized. Preparations of GH extracted from human pituitaries have hitherto varied greatly; in the future, the use of products adhering to a common specification, such as a pharmacopeial monograph, could resolve differences in observations of clinical use. Since pituitary-derived GH is not an article of commerce as a bulk material, the monograph being formulated in Europe is for the product for injection, and the comments following reflect discussions for that purpose there, and at the International Workshop in 1982.[(1)]

3.2. Name and Definition

An official ("proper") name is required. The USAN term somatotropin is defined as the 191-amino-acid, 22K-dalton form; so "growth hormone, human" could be used for "preparations containing the active polypeptides of the anterior pituitary glands, that specifically stimulate the growth of bone, muscle, skin, and other tissues in growth hormone deficient children."

FIGURE

HGH 80/521 as % of 80/505

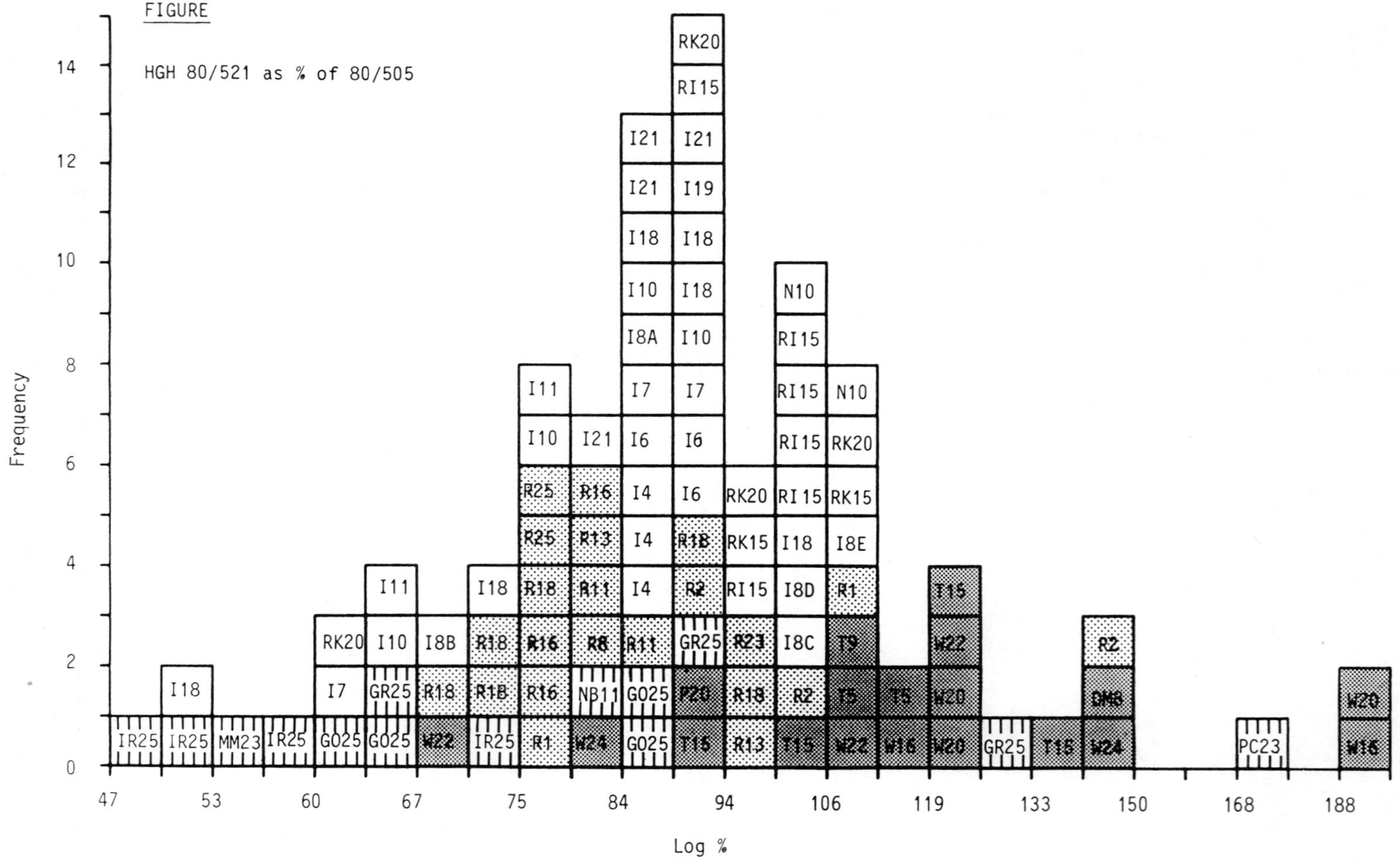

Figure 1 Estimates of potency by various assay methods of human growth hormone preparation 80/521 relative to 80/505 expressed as percent ampule/ampule. Number in square denotes code for laboratory in which estimate was obtained. ▨ W = weight gain in hypophysectomized rats; T = tibial width in hypophysectomized rats; DM = weight gain in dwarf mice; S = [^{35}S]sulfate uptake in hypophysectomized rats; P = [^{3}H]proline uptake in hypophysectomized rats. ▥ GO = glucose oxidation; GR = glycerol release; IR = refractoriness to insulin-like response; PC = pigeon crop sac assays; MM = mouse mammary gland; NB = NB2 cell growth. - ⚃ R = radioreceptor assays. □ I = immunoassays; RI = radial immunodiffusion assays; RK = rocket immunoassays; N = nephelometric assays.

3.3. Source Material

There is general agreement that the yield of unaltered monomer hGH proteins is better from frozen pituitaries, but preservation of glands in well-dried acetone is convenient and seems likely to continue to be used.

Selected pituitary glands should be obtained soon after death, from bodies of individuals not suffering from CNS infectious diseases transmissible by blood transfusion, or from pituitary tumors. In some countries the pathologist responsible for collecting the glands is required to sign a certificate to this effect. The possibility that the glands are contaminated with slow virus has received varied attention in different countries.* There is a theoretical but very small possibility that a gland carrying the slow virus associated with the Jakob–Creutzfeldt syndrome might be present among the tens of thousands of glands collected. In the United Kingdom, unpublished studies in collaboration with Dr. A. G. Dickenson of the MRC Neuropathogenesis Unit (Edinburgh) have shown that mouse scrapie virus used as a test model only just survived much of a current hGH extraction procedure and could not be detected after the final gel filtration. Bacterial filter membranes are believed to remove slow virus efficiently by adsorption by reason of its hydrophobic surface, but if detergent is used, the virus might lose protection of adsorbed material and alter its filter passing characteristics. The process of handling glands and the whole extraction procedure at the Centre for Applied Microbiology and Research in England has since 1980 been based on the assumption that the material may be potentially infective. But it is as yet unlikely that any test for slow virus applied to the end product can be discussed.

3.4. Specific Activity

Only an approximate figure (e.g., not less than 2.0 IU/mg) can be proposed until (1) the composition of the product is well defined; (2) several representative preparations have been bioassayed by weight gain and perhaps tibial plate assays against the IS, with assays that are valid and whose results combine to give confidence limits that are close enough to support a practicable lower limit; and (3) there is agreement on an 280-nm absorbance figure for the product as specified, an agreed factor for nitrogen determination, and an agreed method to determine moisture.

3.5. Identification

Complementary types of tests can be used: (1) physicochemical, electrophoretic, and high-performance liquid chromatography (HPLC) procedures, run in parallel with suitable reference preparation(s); (2) immunoreactivity with anti-hGH antibodies—provided they are sufficiently specific—but not too

* The contamination of pituitary-extracted hGH with slow virus has received intense attention since the deaths of treated patients from Jacob-Creutzfeld Disease in 1985. In many countries pituitary-derived hGH has stopped being used for treatment.

narrowly specific (e.g., for only one molecular form) as some monoclonal antibodies can be; and (3) biological activity, in one or more specified bioassay systems, using a suitable standard suitably calibrated, and to a given precision.

3.6. Characterization of Molecular Components

Tests should be aimed at excluding artifact forms such as aggregate and denatured proteins that could be potentially antigenic. Until more is known about the biological properties of the 20K and other minor native forms, there appears to be no rational basis to specify strict upper or lower limits for them.

Although hGH dimer is biologically active in animals, it is not known for certain whether it is undesirably antigenic in humans; a limit for dimer content should take into account the fact that the dimer content apparently increases with time, even in freeze-dried preparations. Deamidated forms occur in many products but may not be harmful. Hence, a strict low limit for components larger than 22K-daltons hGH could be made using gel filtration chromatography. A minimum content of not less than about 70% 22K-daltons monomer has been suggested.

Polyacrylamide gel electrophoresis could be used to detect deamidated forms, and SDS gels for other forms. Sephadex columns for gel separation tend to deteriorate with prolonged use, and aggregate tends to be retained at the entry to the column so that the total amount cannot be estimated from measurements on the elution pattern.

3.7. Total Protein

It may be desirable to stipulate the specific activity of hGH products to which a nitrogen-containing carrier is added, as X IU/mg of nitrogen before the addition of carrier. As mentioned previously, there is as yet no agreed factor for absorbance at 280 nm for a typical pituitary hGH product, or factor for hGH protein content measured by the Lowry method, or method of moisture determination.

3.8. Test for Proteins Other than Those of Human Origin

Certain manufacturers prudently test the first extract from each batch of glands with antisera to bovine, equine, ovine, and porcine albumin.

3.9. Moisture Content

A micromethod such as gas chromatography or microcoulometry is less wasteful of material than a test for loss on drying. It is believed that overdesiccation may denature hGH, but figures for upper and lower limits for moisture have not been agreed on.

3.10. Other Pituitary Hormones

Limits are required for both biologically active and denatured forms. Too much thyroid-stimulating hormone (TSH), luteinizing hormone (LH), vasopressin, and corticotropin could each result in serious cumulative clinically evident, effects. But the possible induction of autoimmune endocrine disease due to prolonged administration of denatured forms could be insidious, taking several years to become evident, and could be more difficult to treat.

The amount of such contaminant hormones, e.g., LH, allowed in a specification relates to the size of the dose of a given product and the frequency with which it is intended to be given. In the United Kingdom at present the usual dose is 4 IU two to three times a week. Much larger and more frequent doses may be given when the supply becomes easier and less costly, and also when preparations are used for treatment of conditions other than hGH deficiency in children.

The content of TSH and FSH is generally low in hGH preparations made by most extraction methods; despite the inaccuracy of radioimmunoassay (RIA) systems, the limits that can be set for TSH (about 2 mIU/IU hGH) and for FSH (0.1 IU/IU/hGH) are so low that the possible bias of estimates with RIA systems is not important. LH is more difficult to separate from hGH, and more than 20–30 IU injected three times weekly might have undesirable clinical effects.

The limit for LH requires accurate determination if it is more than about 10 IU/IU hGH and a dose of about 8 IU hGH/week is given. Evidence for the LH content of an hGH product should preferably be obtained initially with bioassays, and RIA used thereafter when estimates made with a particular RIA kit have been shown to correlate with those obtained by bioassay. All RIA kits used should, as a minimum, comply with the international requirements for immunoassay kits recommended by WHO.[11]

Vasopressin and corticotropin are normally excluded from the end product by the purification process; both can be bioassayed by pharmacopeial methods. Prolactin itself would probably not be harmful, but a few micrograms of denatured hormone injected repeatedly might lead to autoimmunity, and a limit is thus desirable. Similarly, native neurophysin by itself is presumably harmless, but if it still carried vasopressin—or was denatured—it would be desirable to specify low limits.

Decisions on the limits for several of these potential contaminants should thus be taken when evidence for the presence, or lack, of immunization or other effects has been documented in treated patients.

3.11. Bioassay

There appears to be general agreement[12] that the body weight gain assay is the defining bioassay for hGH; it is the only assay that reflects general body growth and the growth of bone, muscle, and skin. Its drawbacks are low precision, cost of hypophysectomized rats or dwarf or "little" mice, and susceptibility to traces of TSH and of vasopressin causing water retention (which may cause day-to-day variability in weights of individual animals). The tibial plate assay in hypophysectomized rats is more sensitive and more precise (the bias derived from oblique sections of the epiphysis can be avoided by cutting the tibia aligned in a jig); it is not influenced by vasopressin; but it does not reflect all the metabolic effects of hGH. Neither hormone receptor assays nor Nb2 cell assays reflect the survival time of hGH molecules in the circulation, or even the main metabolic effects *in vivo*. Indeed, in discussing the diverse biological activities of the various natural and artifact forms of hGH, Wilhelmi comments ". . . the limitations of the present bioassays and RRA based on binding to plasma membrane receptors of the blandest inscrutability are painfully obvious. The need for a team of sensitive cells that *do* something in response to hGH is acute."(12)

The precision required for the potency estimate for a batch of hGH is strongly influenced by the high cost of bioassays. Limits presently being considered are that the estimated potency is 80–125% of the stated potency and that the fiducial limits of error (p = 0.95) are 70–135% of the stated potency. To achieve precision of this order will probably require the combination of results of two to four 2 + 2 well-behaved bioassays, each of about 50 animals.

3.12. Hepatitis B Antigen

A test for the surface antigen, as used in the control of human blood products, should be applied to each batch. Manufacturers should test for themselves the stage in their extraction procedure at which hepatitis antigen is eliminated; in certain extraction processes this occurs only at the last gel filtration step.

3.13. General Safety Tests

The end product should be tested by the usual pharmacopeial tests for sterility, toxicity, and freedom from pyrogens. The last should be done in rabbits. *Limulus* pyrogen tests may not be standardized yet; so results of such tests should be interpreted with caution.(9)

3.14. Labeling

In addition to the usual national labeling requirements, it might be desirable to state the number of IU of LH present per IU of hGH activity, in case certain products intended to be given in small doses (e.g., 4 IU twice weekly) were given at a much higher dose rate when supplies are more obtainable.

4. Specifications for hGH Products Made by Recombinant DNA Methods

4.1. Name

A distinct proper name is needed for each molecular form of hGH made as a product for clinical use, together with a subscript to denote that it was made by an R-DNA procedure.

4.2. Control during Manufacture

Requirements for the control of manufacture of polypeptide hormones and their medicinal products are being formulated by various national control authorities. A useful discussion document based on a report of a WHO Consultation, held in Geneva in March 1983.[14] All such recommendations emphasize the importance of in-process control; the need for evidence for the correct identity, purity, and stability of the instructing plasmid DNA; and the extent of proof of the correct identity and purity of the end product. They also emphasize that R-DNA methodology is new and control procedures are being developed by current interchange between manufacturers and control authority laboratories.

4.3. Specifications for the End Product

The variety and labor of experimental procedures to prove the exact identity of a particular molecular species of hGH—without knowledge of its provenance—could be formidable (see ref. 12). Although such evidence should theoretically be obtained for each reconstructed plasmid seed, the full range of testing would be impracticable for routine production batches. For such purposes criteria could include a variety of physicochemical,[13] immunological, and biological tests, as described (in Sections 3.4, 3.5, 3.6. 3.7, 3.9, 3.11 and 3.13) for pituitary hGH. A test map (by HPLC or two-dimensional chromatography) of the peptides produced by treating the product with a highly purified proteolytic enzyme, and comparison of the pattern of peptides with that of a sample of a reference preparation of hGH similarly treated, is a very discriminating, qualitative analytical procedure. A test for the identity of end groups may be difficult to interpret if, as occurs with some R-DNA products, they are heterogeneous.

4.4. Tests for Other Proteins

For products like hGH which would be injected repeatedly over many months, freedom from traces (a few parts per million) of any extraneous protein (e.g., from the expression vectors used) is important. Various sensitive immunological techniques can be used to test for *Escherichia coli* proteins, for example, which may be isolated along with the hGH during the purification process. But such immunological systems detect only those antigens which the reagents are designed to, and proteins from other extraneous organisms that happened to contaminate a large production culture may escape detection. A complementary method which helps to detect such proteins is the use of silver staining of the gel after electrophoresis.

A test for the detection of nucleoprotein remnants is also essential, and sensitive hybridization procedures have been developed to do this.

The use of expression vectors other than bacteria (such as yeasts, mammalian cell tissue cultures) and the use of antibodies for purification require additional special control tests, for other proteins, nucleoproteins, or possible microbial contaminant organisms that derive from those materials.

4.5. Pyrogens

Experience with R-DNA hGH products has shown the importance of tests and strict limits for pyrogenic contaminants. The statutory test for pyrogens using rabbits should be suitable for the control test on the final product. The *Limulus* amebocyte lysate test is invaluable for in-process screening. But reliable and less costly alternative procedures are also needed.

In certain respects specifications for an R-DNA product, which should consist almost entirely of protein of a single molecular species, are simpler, insofar as there is no need for tests for other pituitary hormones or hepatitis B antigen.

References

1. *Hormone Drugs*, Proceedings of the FDA-USP Workshop on Drug and Reference Standards for Insulins, Somatropins and Thyroid-axis Drugs, USP Convention Inc, Rockville, Maryland, 1982.
2. U. J. Lewis, R. N. P. Singh, G. F. Tutwiler, M. B. Sigel, E. F. Vanderlaan, and W. P. Vanderlaan, Human growth hormone: A complex of proteins, *Rec. Adv. Hormone Res.* **36:**477, 1980.
3. R. K. Chawla, J. S. Parks, and D. Rudman, Structural variants of human growth hormone: biochemical, genetic and clinical aspects, *Ann. Rev. Med.* **34:**519, 1983.
4. M. J. Ross, W. J. Kohr, R. N. Harkins, and P. Seeburg, The human growth hormone gene family, in: *Hormone Drugs*, Proceedings of the FDA-USP Workshop on Drug and Reference Standards for Insulins, Somatropins and Thyroid-axis Drugs, USP Convention Inc, Rockville, Maryland, 1982, pp. 313–318.
5. D. Rudman and R. K. Chawla, Somatotropic assays in the rat and in man, in: *Hormone Drugs*, Proceedings of the FDA-USP Workshop on Drug and Reference Standards for Insulins, Somatropins and Thyroid-axis Drugs, USP Convention Inc., Rockville, Maryland, 1982, pp. 287–295.
6. A. E. Wilhelmi, Structure and function in growth hormone, in: *Hormone Drugs*, Proceedings of the FDA-USP Workshop on Drug and Reference Standards for Insulins, Somatropins and Thyroid-axis Drugs, USP Convention Inc., Rockville, Maryland, 1982, pp. 369–381.
7. D. R. Bangham, R. E. Gaines Das and D. Schulster, Selection of material, design of an international collaborative study and preliminary analysis of assays for a proposed international standard for human growth hormone for bioassay, in: *Hormone Drugs*, Proceedings of the FDA-USP Workshop on Drug and Reference Standards for Insulins, Somatropins and Thyroid-axis Drugs, USP Convention Inc, Rockville, Maryland, 1982, pp. 301–312.
8. Annex 4, 29th Report, WHO Expert Committee on Biological Standardization, *WHO Tech. Rep. Ser. No.* 626, 1978.
9. 33rd Report, WHO Expert Committee on Biological Standardization, *WHO Tech. Rep. Ser. No.* 687, 1983.
10. D. R. Bangham, R. E. Gaines Das, and D. Schulster, The international standard for human growth hormone for bioassay: calibration and characterization by international collaborative study, *Mol. Cell. Endocrinol.* **42:**269, 1985.
11. Annex 8, 31st Report, WHO Expert Committee on Biological Standardization, *WHO Tech. Rep. Ser. No.* 658, 1981.
12. Report of the Consensus-forming Session on Somatotropins, in: *Hormone Drugs*, Proceedings

of the FDA-USP Workshop on Drug and Reference Standards for Insulins, Somatropins and Thyroid-axis Drugs, USP Convention Inc., Rockville, Maryland, 1982, pp. 387–392.
13. A. J. S. Jones and J. V. O'Connor, Chemical characterization of methionyl human growth hormone, in: *Hormone Drugs*, Proceedings of the FDA-USP Workshop on Drug and Reference Standards for Insulins, Somatropins and Thyroid-axis Drugs, USP Convention Inc, Rockville, Maryland, 1982, pp. 335–351.
14. Quality control of biologicals produced by recombinant DNA techniques, *Bull. WHO* **61**:897, 1983.

18

Recombinant DNA Synthesis of Human Growth Hormone

Michael J. Ross, Kenneth C. Olson, Michael D. Geier, John V. O'Connor, and Andrew J. S. Jones

1. Introduction

1.1. Cloning and Expression

Human growth hormone (hGH) is a polypeptide of approximately 22,000 daltons in size that is synthesized in the pituitary. Mature hGH has been produced in *Escherichia coli* bacteria using recombinant DNA technology[1] using the scheme outlined in Fig. 1. The process involved cloning of cDNA to the pituitary hGH mRNA[2] and the subsequent adapting of the cloned gene for expression in *E. coli*. hGH, as it is synthesized in the pituitary, is made as a prehormone containing a hydrophobic leader peptide of some 20 amino acids in length. This leader peptide is proteolytically removed by the pituitary during the secretion of hGH. However, most *E. coli* do not have the biochemical machinery to do the same processing efficiently (Fig. 2), thus, 84 base pairs of double-stranded DNA were synthesized to tailor the hGH cDNA for direct expression. The process is illustrated in Fig. 3 and involved ligating the restriction fragment of the cDNA coding for amino acids 24 through 191 of growth hormone to a synthetic sequence which coded for a start codon (i.e., a methionine) and the first 23 amino acids of GH. Thus, the molecule synthesized in *E. coli* contained the full 191 amino acids of mature GH and one additional amino acid, an amino terminal methionine, from the start codon.

Michael J. Ross, Kenneth C. Olson, Michael D. Geier, John V. O'Connor, and Andrew J. S. Jones • Genentech, Inc., South San Francisco, California 94080.

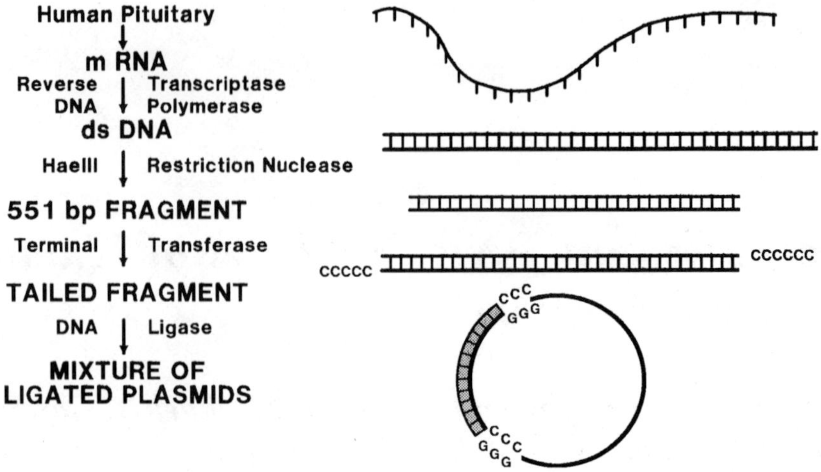

Figure 1 Cloning of the human growth hormone (hGH) message. Size fractionated messenger RNA isolated from human pituitaries was copied using reverse transcriptase to form a heteroduplex molecule. The RNA strand was then displaced with a second DNA strand with DNA polymerase I Klenow fragment. The resulting mixtures of double-stranded DNA molecules were cut with restriction endonuclease Hae III and fractionated on an agarose electrophoresis gel. The hGH-specific cDNA of 551 base pairs (bp) in length was "tailed" with deoxycytosine using deoxynucleotide terminal transferase and ligated into dG tailed cloning vectors using DNA ligase. The mixture of ligated plasmids was used to transform *E. coli* K-12. Clones containing the hGH cDNA fragment were isolated and the amplified 551-bp fragment purified.

1.2. Product Characterization

1.2.1. Chemical Structure

Material isolated from these *E. coli* has been extensively characterized, both at Genentech and at other institutions. The material synthesized in *E. coli* has been shown by tryptic analysis and Edman degradation to be identical to the sequence one would expect from the gene.[3,4] The methionyl hGH has been shown to have the correct two pairs of disulfide bridges between the four cysteines in the molecule.[3] The major isohormone found in pituitary hGH is identical in isoelectric point to methionyl hGH, and the minor acidic isohor-

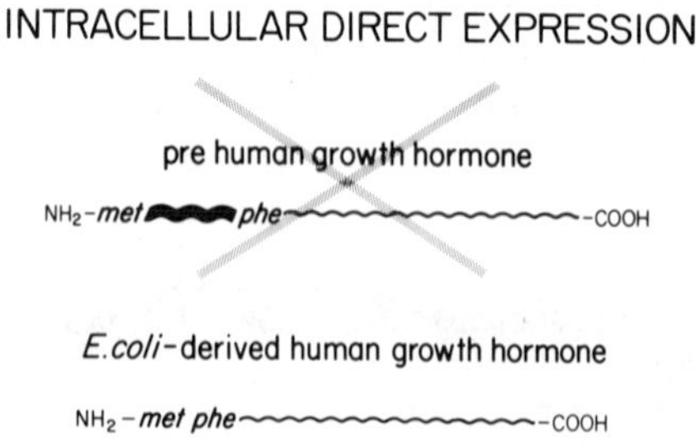

Figure 2 Intracellular direct expression of human growth hormone (hGH). GH is synthesized in secretory cells in the pituitary. There it is synthesized as a prehormone containing a hydrophobic leader sequence which is cleaved off by a protease as it is processed through the cell membrane into secretory granules. The hGH is expressed directly inside *E. coli* cells by deleting the hydrophobic amino acids between the methionine start codon and the first amino acid of the mature hormone, a phenylalanine.

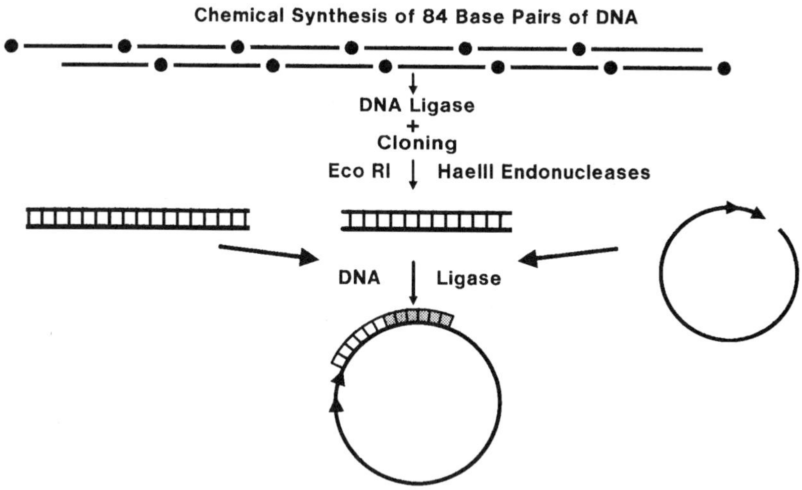

Figure 3 Expression of met-hGH. The Hae III–Eco R1 fragment of the hGH cDNA contains the coding sequence for amino acids 23–191 of hGH: an 84-base fragment of DNA which coded for a methionine start codon and amino acids 1–22 of hGH. These two fragments were then ligated together into an expression plasmid derived from pBR322. The expression plasmid contained promoters, selection markers, and ribosome binding site to allow efficient expression of hGH.

mones that are found in the methionyl hGH preparation are very similar in their isoelectric pattern to the isohormones found in pituitary-derived hGH.[5] The major difference observable using isoelectric focusing is that there are fewer isohormones present in met-hGH.

1.2.2. Folding

The three-dimensional structure of the molecule has been shown to be indistinguishable from pituitary hGH by circular dichroism studies.[6] Immunochemical studies comparing met-hGH and pit-hGH in sera from both hyperimmunized rabbits and patients have as yet found no new determinants in the methionyl compound, indicating that the amino terminus of the molecule is probably not accessible on the molecule's surface.

1.2.3. Bioassay

Finally, bioassay data using rat weight gain and rat tibia length models[4,7] as well as radioreceptor and receptor modulation assays show met-hGH to be equally potent with pit-hGH.[8]

2. The Production Process

The first step in the process of manufacturing met-hGH for clinical use is fermentation, a highly reproducible and easily scalable process which has been

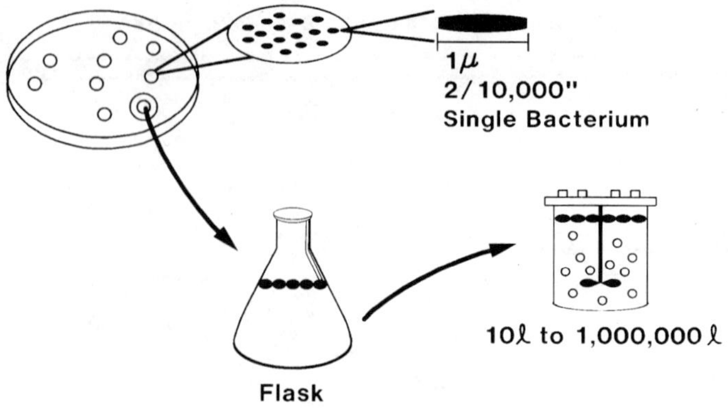

Figure 4 Controlled fermentation processes can be carried out at scales ranging from 10 liters to 10^6 liters. Each batch is derived from the single colony of bacteria which is grown up to yield a master culture. This master culture is exhaustively characterized and subdivided into identical vials and stored frozen. For each new batch, a single vial of this master culture is inoculated into a flask of nutrient broth where it is grown and used to inoculate a fermenter, which can then be used to inoculate larger fermenters.

used as a key tool in pharmaceutical manufacturing for many years. The process outlined in Fig. 4 shows production of shaking flask inoculum cultures and, from those flasks, stirred batch fermentations from single colonies of *E. coli* transformed with met-hGH production plasmid. After vegetative cell growth is complete, the bacteria are harvested by centrifugation and broken open to release the met-hGH. The purification steps are, however, the key to the ability to make acceptable clinical product. Modern chromatographic methods allow protein separation to be carried out at large scale with high resolution so that the complex mixture of *E. coli* proteins present with the met-hGH in the starting material can be removed. The successive removal of these proteins as the met-hGH progresses through the purification can be seen in Fig. 5.

In 1980, previous experiences with production of pharmaceutical peptides from bacteria such as streptokinase[(9)] and asparaginase[(10)] were limited to acute care indications, yet the first clinically tested products of rDNA technology were human insulin[(10)] and hGH. Our experience since 1980 with the development of the met-hGH, which Genentech, Inc., has chosen to call Protropin, has led us to the inescapable conclusion (not fully foreseen in 1980) that the details of the production process can have profound effects on the clinical performance of the drug. Increasing the expression levels of the product in the bacteria,[(11)] using process validation studies and final specifications to design and challenge the purification process,[(12)] and development of new analytical tools such as HPLC,[(3)] the endogenous pyrogen test,[(13)] and ECP assays[(14,15)] have all turned out to be key factors in producing clinically acceptable material. Genentech has developed three distinct manufacturing processes for manufacturing met-hGH over the past several years, and KabiVitrum has reported two more distinct processes. The clinical results from each of these do in fact differ

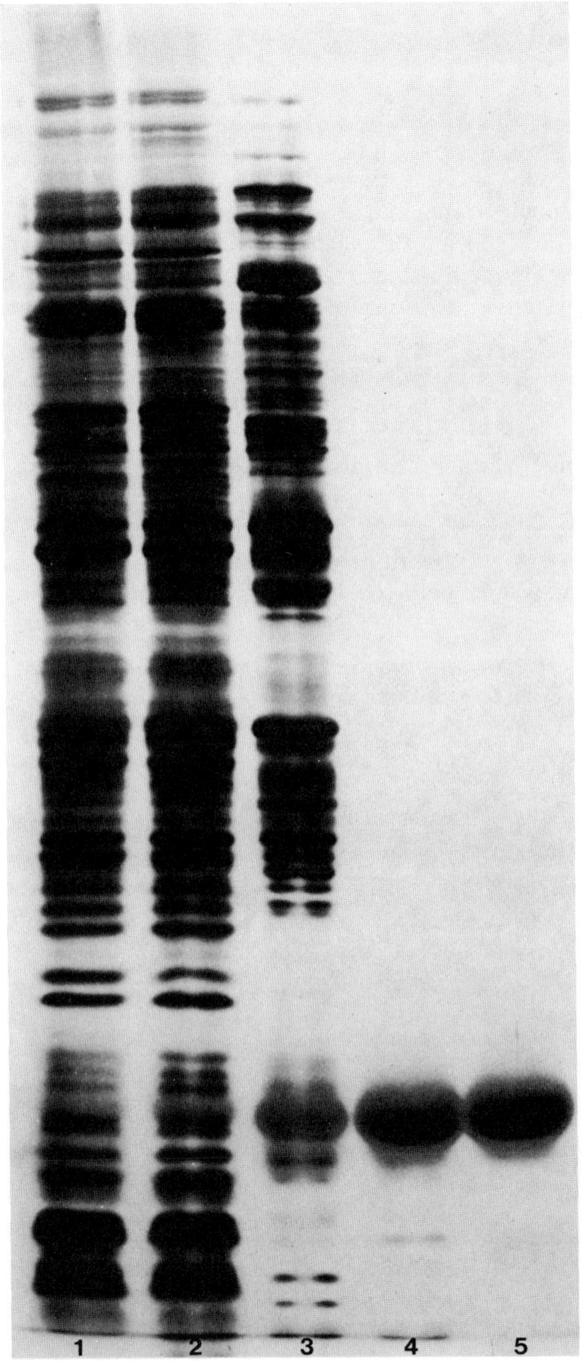

Figure 5 The purification of met-hGH. The met-hGH has *E. coli* proteins removed from it at each stage of the purification process. The figure shows a silver-stained SDS–PAGE analysis of pools made during the purification. Lane 1 is of crude lysate and lane 5 is of bulk product. Similar amounts of total protein were loaded in each lane. It should be noted that color differences in the silver stain cause the intensity of the hGH band to be far weaker than the *E. coli* bands.

markedly; so one must precisely define the manufacturing process used to produce hGH before evaluating the results of a clinical trial.

2.1. Process Design/Validation

2.1.1. G11002: The Use of SDS–PAGE

The first purification process that was developed at Genentech (Code G11002) was designed using Coomassie stained SDS–PAGE purity estimates as the primary tool. The process involved anion exchange chromatography and gel filtration chromatography.[5] Figure 6 shows an SDS–PAGE analysis across the fractions of a DEAE ion exchange column. The data from similar experiments were used to optimize the parameters for this chromatography step. Single protein contaminants of approximately 1% the amount of an overloaded met-hGH band could be detected using this kind of SDS–PAGE analysis, and thus final product of approximately 99% purity was achieved for Code G11002.

Met-hGH, code G11002 was formulated in glycine–phosphate and lyophilized. The material produced by this process appears homogeneous by SDS–PAGE, whereas in even the purest pituitary hGH preparation, one can detect minor bands of either other hGH forms or impurities (Fig. 7). When tested in adult human volunteers, the met-hGH behaved identically to pituitary hGH

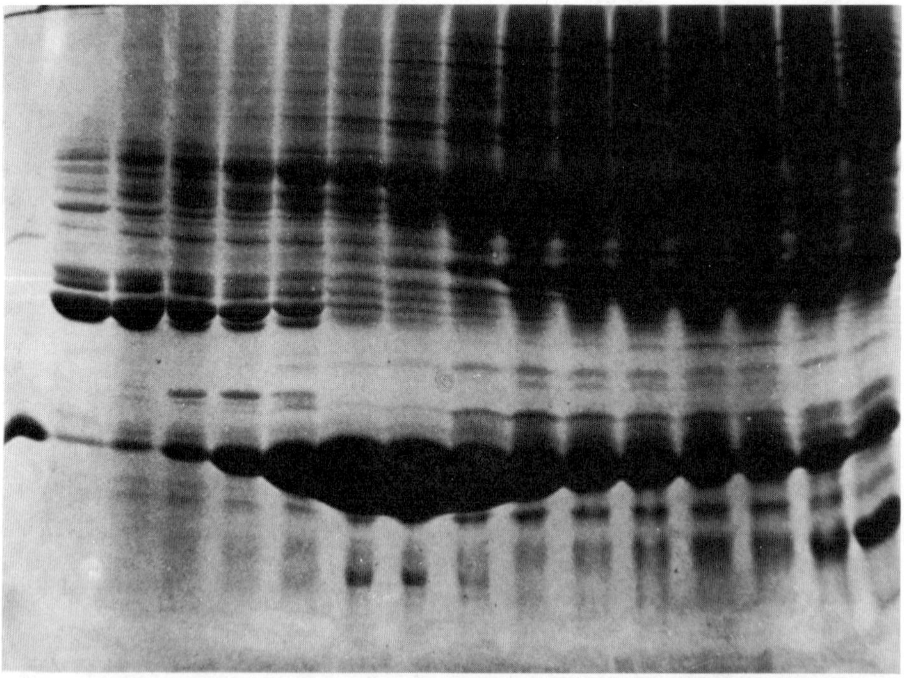

Figure 6 An SDS–PAGE analysis (Coomassie blue stain) of column fractions from the initial ion exchange column used to purify Protropin.

except that some classical signs of endotoxin contamination appeared in spite of the fact that the material passed both USP pyrogen and LAL tests (Hintz, *et al.*).[16] Repeat LAL and pyrogen assays yielded no clue as to the cause of the endotoxin reaction; so a material of considerably higher purity was prepared (see Code G11008 below) which eliminated the problem. Subsequent analysis using a human monocyte activation assay[12] revealed that LAL and rabbit (but not human) reactivity to *E. coli* endotoxin was irreversibly inhibited when lyophilized in the presence of glycine.

2.1.2. G11008: The Initial Preparation Used in Treating hGH Deficiency

A far more sensitive staining method (Silver versus Coomassie) for detecting proteins on SDS–PAGE[17,18] was applied to the process design problem. Using this technique, several new bands were detected in the G11002 preparation

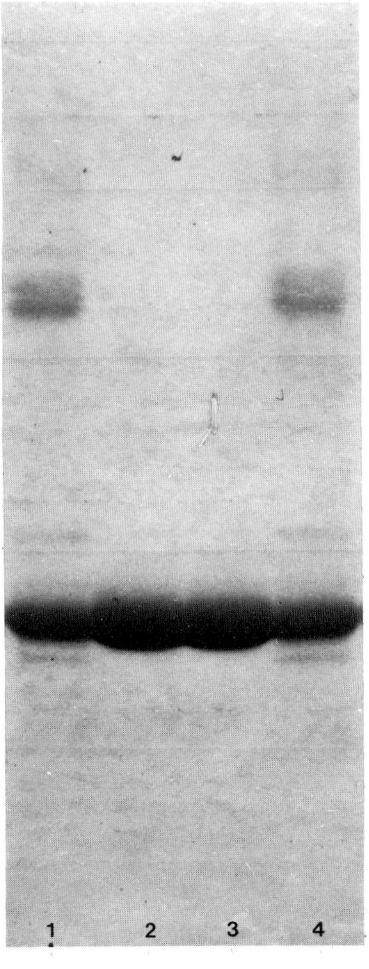

Figure 7 Analysis of Process Code G11002 met-hGH by SDS–PAGE. The original clinical preparation of met-hGH was denatured in SDS/BME and separated by SDS–PAGE and stained with Coomassie blue and compared with clinical grade pit-hGH. Lanes 1 and 4 are 20 μg of pit-hGH. Lanes 2 and 3 are 20 μg of met-hGH Proces Code G11002. Though several bands other than 22K Monomer hGH can be seen in the pituitary preparation, no bands except 22K hGH are visible in the met-hGH.

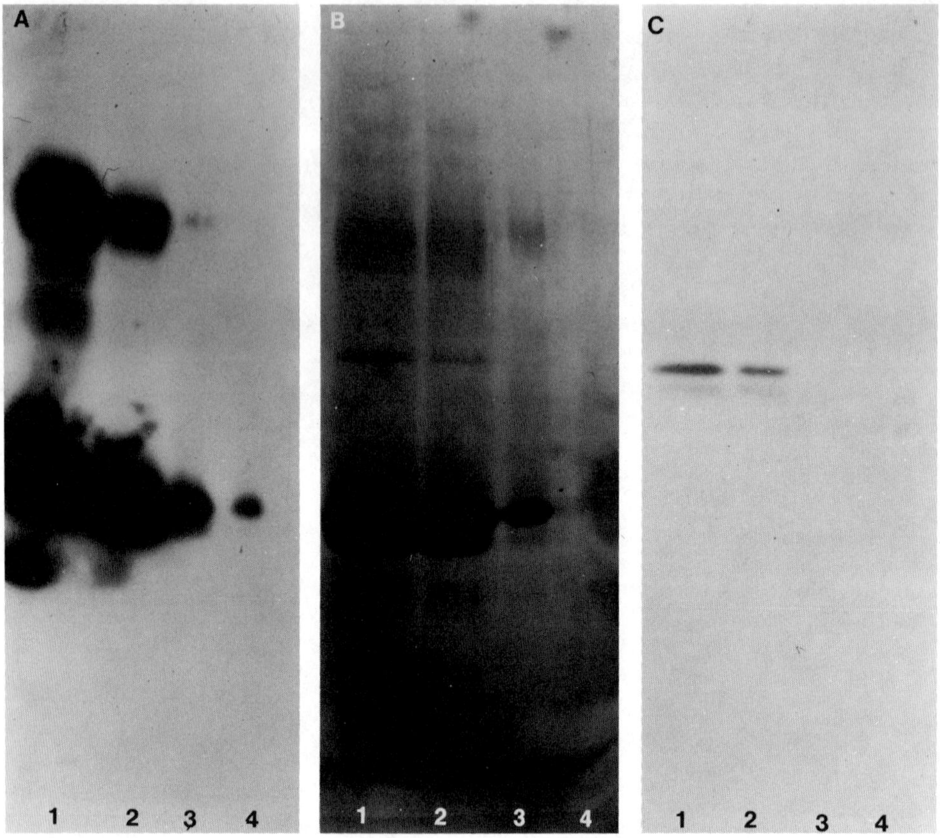

Figure 8 Met hGH (Process Code G11002) was analyzed by SDS–PAGE using silver staining and immunoblotting: 20 μg, 10 μg, 2 μg, and 1 μg of met-hGH (denatured in SDS/BME) were loaded onto three identical sections of the same gel in lanes 1, 2, 3, and 4, respectively. The proteins in the gel section in panel A were transferred to nitrocellulose and blotted with rabbit antipituitary hGH antibody. Antibodies bound to the blot were visualized by ^{125}I-labeled protein A followed by autoradiography. The major monomer band and the dimmer band can clearly be identified as hGH as well as some minor higher- and lower-molecular-weight species. Panel B is a section of gel stained with silver.[17] A contaminant between hGH monomer and dimer can be easily seen as some higher-molecular-weight species (larger than hGH dimer). Panel C is a section of gel which was transferred to nitrocellulose and blotted with anti-*E. coli* protein antibody. The contaminants can be clearly identified as being from *E. coli* origin.

(Fig. 8). Similar SDS gels were subjected to radioimmunological staining with both anti-*E. coli* and anti-hGH antibodies to unambiguously assign the origin of the newly detected bands. Two bands of *E. coli* origin were thus identified and the remaining higher-molecular-weight species (<1% of the total protein by Coomassie stain) were identified by anti-hGH antibody.

The purification process was then modified by monitoring SDS–PAGE silver gels and by using data from experiments in which radiolabeled *E. coli*

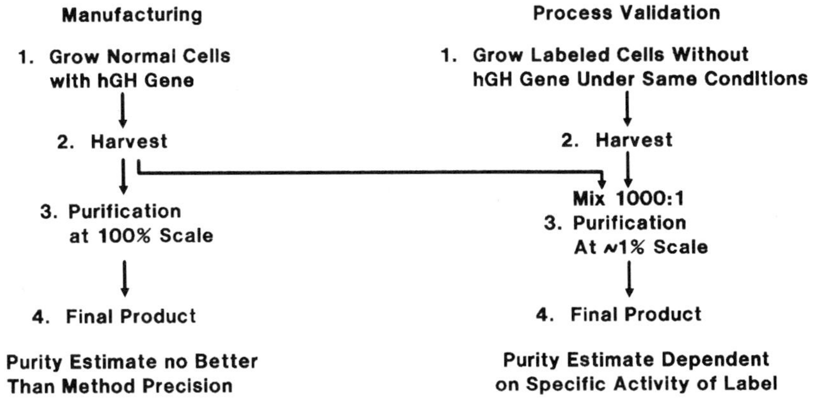

Figure 9 Radioactive spike process validation. Production organisms (*E. coli* missing the hGH gene) are grown in media containing ^{14}C- or ^{35}S-labeled amino acids. The labeled extract is added to hGH containing cell paste from a normal production run. The purification is carried out as it would be to produce Protropin except that the scale is reduced about 100X.

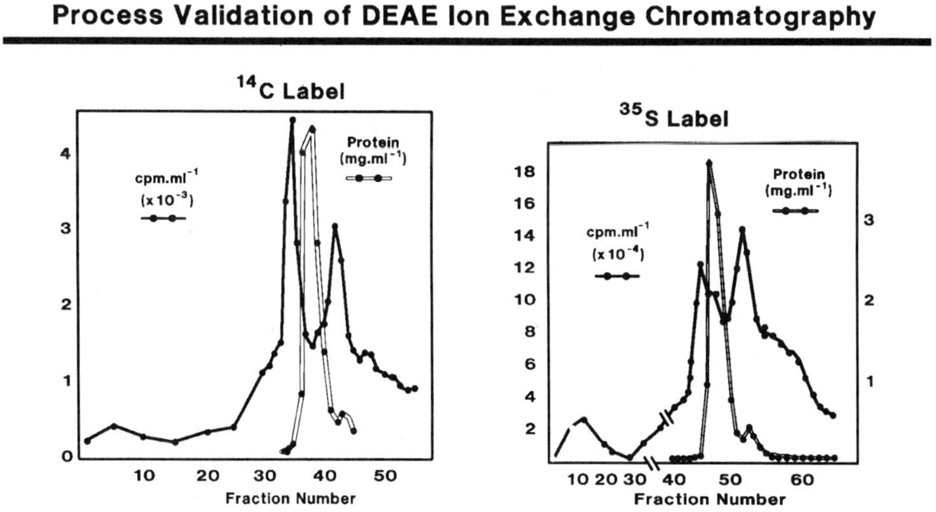

Figure 10 Monitoring of an ion exchange purification step using radioactive spikes. Radioactively labeled *E. coli* were added during the Protropin™ purification process as described in Fig. 9. The separation of contaminants (radioactivity) from met-hGH and the final step of the G11008 process when the met-hGH was 99.7% pure coming into this step are illustrated. Two separate experiments are shown; the experimental designs were similar, but in one experiment the *E. coli* were grown with $^{35}SO_4$ and in the other, grown with ^{14}C-labeled glucose.

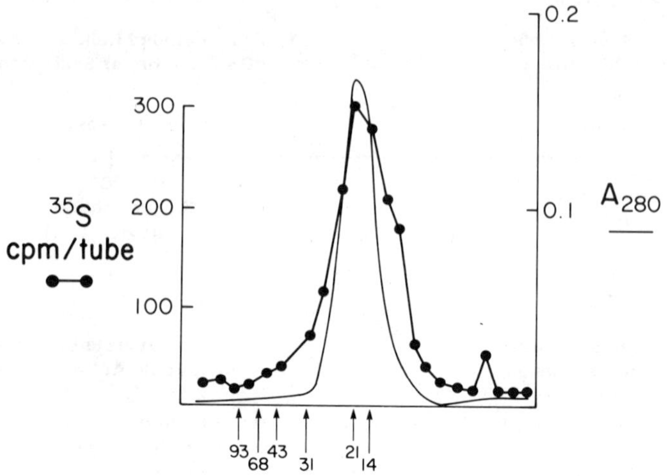

Figure 11 HPLC analysis of ^{35}S impurity in Protropin process validation. Size exclusion HPLC on a TSK-3000SW column run in 100 mM sodium phosphate 0.1% SDS, pH 7.0.

contaminants were added to lysate containing met-hGH. A schematic of the latter experimental protocol appears in Fig. 9. Both ^{14}C- and ^{35}S-labeled *E. coli* extracts were used in these experiments. The results of an optimization of a DEAE ion exchange column are shown in Fig. 10. The ^{14}C and ^{35}S spikes have almost superimposable profiles at each chromatography step indicating that molecules containing both sulfur and carbon (proteins) are the major contaminants being detected. Significant differences between the ^{35}S and ^{14}C profiles might indicate contamination by lipids, carbohydrates, or nucleic acids which contain little or no sulfur. This conclusion is supported by analysis of the product for the presence of *E. coli* DNA using hybridization techniques.(19) We were unable to detect any DNA in the product to the sensitivity of the assay (10 pg/mg of hGH) after the first ion exchange step in the process.

Though the spike experiments provided valuable guidance in designing the G11008 process, they do have their limitations: ^{14}C-labeled molecules have a very low specific activity owing to the long half-life of the isotope; so the number of counts left after a thorough purification is too small to characterize. ^{3}H-labeled molecules are so prone to label exchange with solvent that their use is precluded for this type of experiment. Even ^{35}S labels have been complicated by label migration, as can be seen with met-hGH.

We have characterized the ^{35}S counts remaining after the G11008 purification process was carried out to completion; these counts represented of the order of 0.2% contamination. However, we found to our surprise that the counts behaved like hGH when analyzed by HPLC (Fig. 11). Further characterization by a monoclonal antibody affinity column showed that about half of the label behaved identically to hGH while the other half of the labeled molecules were not recognized by this single antibody. Our postulate is that free radicals generated by the β particles and chlorine atoms which are the result of the sulfur decay cause low-molecular weight ^{35}S *E. coli* contaminants such as glutathione to become covalently attached to met-hGH. The damaged hGH could copurify with met-hGH and could have antigenic sites damaged. Since this process would be an artifact of the use of radioisotopes in the experiment, the similar met-hGH adducts are unlikely to exist in the clinical product. Therefore, absolute quantitation of impurities is probably not accurate if they represent less than ~0.1% of the major protein.

Phase II clinical trials were initiated in children with hGH deficiency. The detailed results of this 18-month trial are presented by S. Kaplan *et al.* (this volume, Chapter 20), but we will summarize the results very briefly: The met-hGH proved highly efficacious in promoting linear growth and no drug-related side effects were noted. However, a majority of the patients developed low-titer antibodies to hGH. In an effort to eliminate this immunological side effect of therapy further, process improvements were undertaken.

2.1.3. G11015: The Use of an anti-E. coli. Assay

A new tool was needed to improve the G11008 process since by the criteria of SDS–PAGE/Silver stain, no contaminants could be seen (Fig. 12). Immunological techniques have historically been used to quantitate small amounts of compounds and are particularly well suited to the relatively high-molecular-weight substances that are present. Baker *et al.*[15] and Ross *et al.*[14] have reported the development of an anti-*E. coli* protein (ECP) assay which was used in the assays for purity of human insulin produced in *E. coli*. We attempted to develop a similar tool for use with our hGH process but used different reagents since such assays have proven to be highly process specific (data not shown).

A solid-phase sandwich immunoradiometric assay for the ECP antigens likely to be present in the final steps of this met-hGH process was developed as described below. The principle of the assay was to coat a microtiter plate with anti-ECP antibody, adsorb the sample to the antibody, and quantitate the antigen using other anti-ECP antibody (^{125}I-labeled). In order for such an assay to be linear and accurate, there must be an excess of both coat and tracer antibody to every one of the antigens in the ECP mixture, or the assay would vary in sensitivity for different antigens.

E. coli cells that did not contain the hGH gene were grown, harvested, lysed, and carried through the first several steps of the met-hGH process to select a crude mixture of ECP's. This was done to reduce the complexity of the antigen mixture somewhat and therefore to improve the sensitivity of the assay. A crude *E. coli* lysate was used to hyperimmunize rabbits, and the antiserum was immunopurified on an ECP affinity column. A large excess of antiserum was passed over the column before wash and elution to adjust the population of antibodies used in the assay to better reflect the quantitites of the various antigens in the ECP preparation (Fig. 13). An assay was developed that measured ECP's accurately down to 10 ng ECP/mg hGH. The upper range of the assay was limited to 100 ng/mg to ensure conditions of antibody excess.

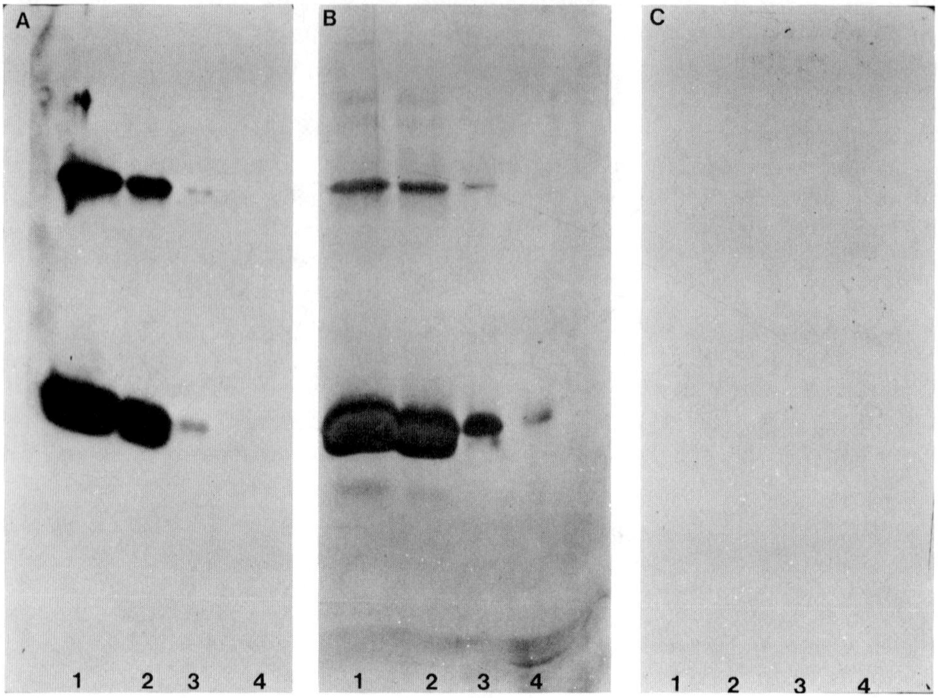

Figure 12 Analysis of Process Code G11008 met-hGH by SDS–PAGE. This figure is analogous to Fig. 8 in that 20 μg, 10 μg, 2 μg, and 1 μg of met-hGH (denatured in SDS/BME) were loaded on three identical sections of the same gel in lanes 1, 2, 3, and 4, respectively. The proteins in the gel section in panel A were transferred to nitrocellulose and blotted with anti-hGH antibody. Only met-hGH monomer, a small amount of dimer, and a hint of trimer can be seen in the resulting autoradiogram. Panel B is stained with silver. All of the visualized bands except a trace of a species with a molecular weight lower than hGH monomer in the overloaded 20 μg lane are identified in Panel A as being of hGH origin. In panel C the proteins on the gel were transferred to nitrocellulose and then blotted with anti-*E. coli* protein antibody. No bands can be seen on this autoradiogram.

The assay, however, was very process dependent. Only those antigens which could contaminate the hGH prepared by this manufacturing process (Code G11015) were selected during the preparation of those ECP's. The assay is therefore specific for these contaminants. A further complication is that the conditions (e.g., pH, buffer) for the extraction and handling of the ECP's have the potential for changing the tertiary structure of the antigen molecules and, hence, their antigenic determinants. This process with different first steps could give falsely low values in this ECP assay. In fact, Process Code G11002 met-hGH, which is prepared by a different process than G11015, gives an ECP assay value of about 100–200 ppm although in fact we know the preparation contains at least 10,000 ppm ECP contaminants.

This ECP assay was used to characterize the G11015 process for met-hGH which products contain less than 30 ppm ECP's. There are five major separation

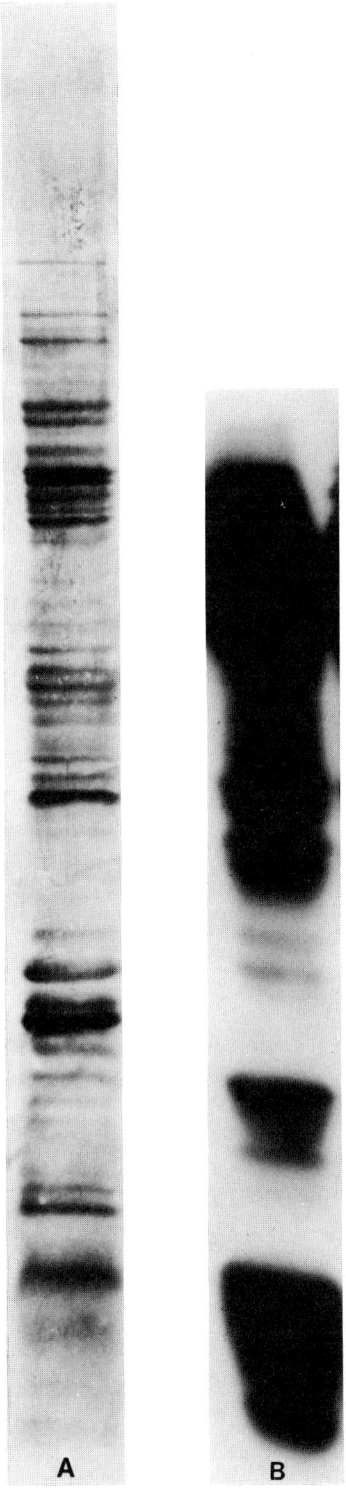

Figure 13 SDS–PAGE analysis of antigens and antibodies used in the ECP assay. ECP's were prepared as described in the text and run on a 12.5% acrylamide SDS gel. In one lane the proteins were strained with silver (panel A).[18] From a similar lane the proteins were transferred to a nitrocellulose by filter electroblotting.[21] The nitrocellulose was then incubated with the antibodies used in the ECP assay. Bound antibodies were visualized using ^{125}I-labeled protein A followed by autoradiography. This figure (panel B) represents a 60-hr exposure at −80°C using an image-enhancing screen.

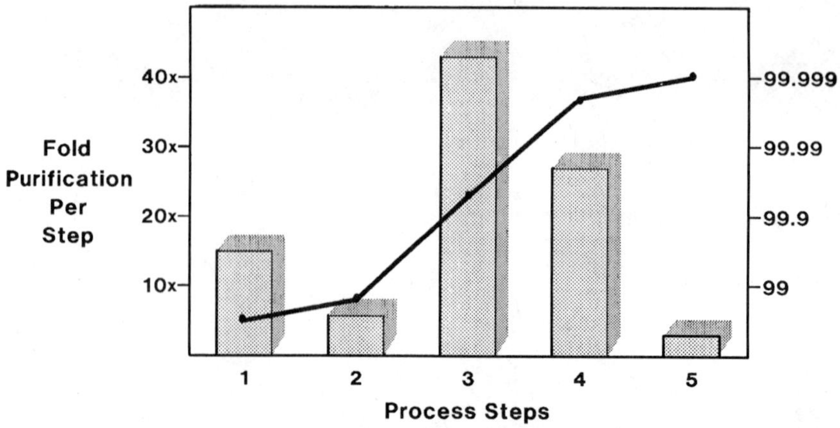

Figure 14 Prototropin purification. The five major purification steps used in the G11015 are analyzed to determine their efficiency in removing *E. coli* contaminants. The fold purifications (bars) are calculated by taking the ECP values/mg of hGH before process steps and dividing by the ECP values after those steps. The cumulative effect of the steps can be seen by plotting the overall purity at the completion of the steps. The material after step 5 is purified bulk ready for formulation.

steps in the process, and each contributes significantly to the overall purification, as can be seen in Fig. 14. The met-hGH produced by process G11015 has been tested in GS-deficient children for 6 months and has shown itself to be as efficacious as, but significantly less immunogenic than, material produced by process code G11008.

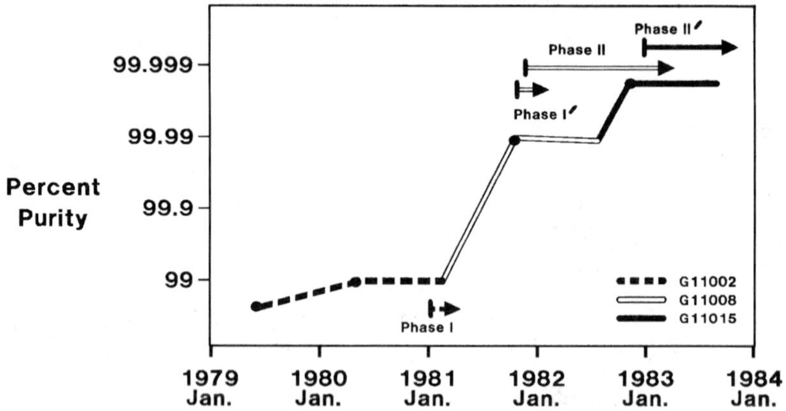

Figure 15 Purity of different hGH preparations. The approximate purity of various met-hGH preparations used clinically is plotted versus time. The clinical trials that were run with each preparation are indicated by the arrows. The purity of the G11002 was estimated by ^{35}S-labeled process validations and the G11008 and G11015 preparations using the immunoradiometric ECP assay.

2.2. Summary

Three different preparations of met-hGH have been tested by Genentech in human clinical trials. Each new process was significantly purer than its predecessor and required the development of new analytical tools before that new level of purity could be achieved (Fig. 15). Our experience has led to the conclusion that the different preparations of met-hGH behave very differently in the clinic. Consequently, it is very important to differentiate between manufacturers and between different processes from the same manufacturer when examining clinical results. We must rely on rigid control of a specific manufacturing process to guarantee the safety of the product.

ACKNOWLEDGMENTS. We thank Matt Field, Vince Anicetti, Kathy McKeown, Greg Bennett, Baron Reed, and Michele Sanda for their excellent technical assistance and Sara Wright for help in preparing this manuscript.

References

1. D. V. Goeddel, H. L. Heynecker, T. Hozumi, R. Arentzen, K. Itakura, D. G. Yansura, M. J. Ross, G. Miozzari, R. Crea, and P. H. Seeburg, Direct expression in *Escherichia coli* of a DNA sequence coding for human growth hormone, *Nature* **281**:544, 1979.
2. P. H. Seeburg, I. Shine, J. A. Martial, J. D. Baxter, and H. M. Goodman, Nucleotide sequence and amplification in bacteria of structural gene for rat growth hormone, *Nature* **270**:486, 1977.
3. W. J. Kohr, R. Keck, and R. N. Harkins, Characterization of intact and trypsin-digested biosynthetic human growth hormone by high-pressure liquid chromatography, *Anal. Biochem.* **122**:348, 1982.
4. K. Olson, J. Fenno, N. Lin, R. N. Harkins, C. Snider, W. H. Kohr, M. J. Ross, D. Fodge, G. Prender, and N. Stebbing, Purified human growth hormone from *E. coli* is biologically active, *Nature* **293**:408, 1981.
5. M. J. Ross, Production of medically important polypeptides using recombinant DNA technology, in: *Insulins, Growth Hormone, and Recombinant DNA Technology* (J. Gucriguian, ed.) Raven Press, New York, 1981, pp. 33–48.
6. A. J. S. Jones and J. V. O'Connor, Chemical characterization of methionyl human growth hormone, in: *Hormone Drugs, Proceedings of the FDA–USP Workshop on Drug and Reference Standard for Insulin, Somatotropins and Thyroid Axis Hormones*, U.S. Pharmacopeial Convention, Inc., Rockville, Maryland, 1981.
7. N. Stebbing, K. Olson, N. Lin, R. N. Harkins, C. Snider, M. J. Ross, F. Fields, L. May, J. Fenno, D. Fodge, and G. Prendor, Biological comparison of natural and recombinant DNA-derived polypeptides, in: *Insulins, Growth Hormone, and Recombinant DNA Technology* (J. Gueriguian, ed.), Raven Press, New York, 1981, pp. 117–132.
8. R. G. Rosenfeld, B. B. Aggarwal, R. L. Hintz, and L. A. Dollar, Recombinant DNA derived methionyl human growth hormone is similar in membrane binding properties to human pituitary growth, *Biochem. Res. Commun.* **106**:202, 1982.
9. J. L. Anderson, H. W. Marshall, B. E. Bray, J. R. Lutz, P. R. Frederick, F. G. Yanowitz, F. L. Datz, S. C. Klausner, and A. D. Hagan, A randomized trial of intracoronary streptokinase in the treatment of acute myocardial infarction, *N. Engl. J. Med.* **308**:1312, 1983.
10. E. L. Grinnan, L-Asparaginase: a case study of an *E. coli* fermentation project, in: *Insulins, Growth Hormone, and Recombinant DNA Technology*, (J. Gueriguian, ed.), Raven Press, New York, 1981, pp. 99–108.
11. H. DeBoer, H. Heyneker, L. Comstock, A. Wieland, M. Vasser, and T. Horn, Construction of three hybrid promoters and their properties in *Escherichia Coli*, in: *From Gene to Protein: Translation*

into Biotechnology (F. Ahmad, J. Schultz, E. E. Smith, and W. J. Whelan, eds.), Academic Press, New York, 1982, pp. 309–327.

12. A. J. S. Jones and J. V. O'Connor, Control of recombinant DNA produced pharmaceuticals by a combination of process validation and final specifications, in: Developments in Biological Standardization: *Symposium on Standardization and Control of Biologicals Produced by Recombinant DNA Technology*, Vol. S9, (S. Kerger, ed.), Internation Association of Biological Standardization, Basel, Switzerland, 1985, pp. 175–180.
13. C. A. Dinarello, J. V. O'Connor, G. LoPreste, and R. L. Swift, Human leukocytic pyrogen test for detection of pyrogenic material in growth hormone produced by recombinant *Escherichia coli, J. Clin. Micro.* **20:**323–329, 1984.
14. J. W. Ross, R. S. Baker, C. S. Hooker, I. S. Johnson, J. R. Scmidtke, and W. C. Smith, Procedure for detection of potential *E. coli* peptides (ECP's) in biosynthetic human insulin (BHI), antibodies to ECP's in patients treated with BHI, and measurement of bacterial endotoxins in BHI, in: *Hormone Drugs*, United States Pharmocopeial Convention, Inc., Rockville, Maryland, 1982, pp. 127–138.
15. R. S. Baker, J. R. Schmidtke, J. W. Ross, and W. C. Smith, Preliminary studies on the immunogenicity and amount of *Escherichia Coli* polypeptides in biosynthetic human insulin produced by recombinant DNA technology, *Lancet* **2:**1139, 1981.
16. R. L. Hintz, D. M. Wilson, J. Fenno, R. G. Rosenfeld, A. Bennett, B. McClellan, and R. Swift, Biosynthetic methionyl-human growth hormone is biologically active in adult men, *Lancet* **1:**1276, 1982.
17. J. H. Morrissey, Silver stained for proteins in polyacrylamide gels: A modified procedure with enhanced uniform sensitivity, *Anal. Biochem.* **117:**307, 1981.
18. B. R. Oakley, D. R. Kirsch, and N. R. Morris, A simplified ultrasensitive silver stain for detecting proteins in polyacrylamide gels, *Anal. Biochem.* **105:**361, 1980.
19. F. C. Kafatos, C. W. Jones, and A. Efstratiadis, Determination of nucleic acid sequence homologies and relative concentration by a dot hybridization procedure, *Nucelic Acid Res.* **7:**1541, 1979.
20. W. N. Burnette, Western blotting electrophoretic transfer of protein from sodium dodecyl sulfate polyacrylamide gels to unmodified nitrocellulose and radiographic detection of antibody and radioiodinated protein, *Anal. Biochem.* **112:**195, 1981.

19

Methionyl Human Growth Hormone (Somatonorm®)

Purity and Properties

Linda Fryklund, Johnny Brandt, Magnus Hagerman, Bohdan Pavlu, Bo Skoog, and Anders Wichman

1. Introduction

Recombinant DNA techniques have been employed to produce the pituitary human growth hormone (hGH) somatotropin in the bacterium *Escherichia coli* K12.[1] The product was found to be identical to the pituitary hormone in all aspects, with the exception of an additional methionine residue at the amino terminal.[2] Apparently the *E. coli* system is unable to recognize or else remove the methionine used to furnish the start instruction for translation.

An initial safety study of methionyl hGH (Somatonorm, met-hGH) produced no side effects in seven healthy male volunteers. This was followed by a 6-month clinical trial in London (this volume, Chapter 21) on 18 hypopituitary children in their last year of treatment. Somatonorm treatment resulted in a low antibody response in some children, no response in others, and a decrease in antibody levels in the two children who entered the study with antibodies to pituitary hGH. On the contrary, virtually all new patients in Germany produced hGH antibodies after some months of treatment, with, however, no effect on the therapeutic efficiency of Somatonorm (this volume, Chapter 22).

Identical results were obtained in radioimmunoassay regardless of whether hGH as Crescormon or Somatonorm was used as label or standard (unpublished data), when patient sera were analyzed for the presence of hGH antibodies.

Linda Fryklund, Johnny Brandt, Magnus Hagerman, Bohdan Pavlu, Bo Skoog, and Anders Wichman • KabiVitrum AB, Research and Development, Departments of Biochemistry and Analytical Chemistry, S-112 87 Stockholm, Sweden.

This indicates that the Somatonorm antibodies were being produced to the same antigenic sites as in pituitary hGH.

It might be thought that young children would be immunologically immature and therefore should respond less to antigenic stimuli. The converse could also be true; since they have not been exposed to GH earlier, they would therefore be hypersensitive and react by producing antibodies more readily than older children and adults.

The increased antigenicity of Somatonorm as compared to the pituitary hormone Crescormon can be ascribed to a variety of causes, where the additional amino terminal methionine is an obvious suspect. A further possibility is that unidentified *E. coli* components present in minute amounts act as adjuvants and provoke an antibody response or else raise antibodies that cross-react with Crescormon in radioimmunoassay. This chapter describes the work undertaken to reveal the causes of antigenicity and to provide means for its elimination from the product Somatonorm. Although the antibodies are of low titer and appear to lack clinical relevance, it was considered necessary to decrease the antigenicity.

2. *Materials and Methods*

Somatonorm, methionyl hGH and Crescormon, pituitary hGH, were standard production batches.

E. coli components (ECP) were purified from *E. coli* containing the plasmid without the hGH gene, essentially as for the hormone with the omission of the second ion-exchange step. The ECP preparation therefore contained substances of similar size and charge to Somatonorm.

Antisera against Crescormon, Somatonorm and ECP were produced in rabbits by standard techniques.

Electroimmunoassay (EIA) was performed according to the method of Weeke.[3] Polyacrylamide gel electrophoresis (PAGE) and polyacrylamide gel electrophoresis–sodium dodecyl sulfate (PAGE–SDS) were run according to Laemmli[4] and stained with either Coomassie or silver stain according to the method of Morrissey[5] or Merril *et al.*[6]

Western blotting was performed according to the method of Burnette[7] using anti-*E. coli*, anti-Somatonorm, anti-Crescormon, or sera from children treated with Somatonorm.

Patient sera, after 3, 6, or 9 months' treatment, were in general diluted 1:100 for Western blotting and 1:10 for immunosorption.

2.1. *Immunosorption*

Immunoadsorbed antisera were prepared by passing the anti-Crescormon serum through a column containing Somatonorm immobilized to Sepharose 4B gel and similarly anti-Somatonorm through a column containing Crescormon immobilized to Sepharose 4B gel, as described below. About 2 g immunoadsorbent containing about 1.4 mg Crescormon or Somatonorm was used for adsorption of 1 ml antiserum. The columns were equilibrated and eluted with 0.1 mole/liter phosphate buffer, pH 7.4. The various antisera, either adsorbed or nonadsorbed, were then fractionated by precipitation with 25% ammonium sulfate (wt./wt.) followed by dialysis against 0.1 mole/liter sodium bicarbonate buffer pH 8.0 to yield gamma-globulin fractions.

2.2. *Preparation of Adsorbents*

Preparation of solid-phase bound Crescormon and Somatonorm as well as the immunoadsorbents was performed by coupling, respectively, Crescormon, Somatonorm, or appropriate gamma-globulin fractions of adsorbed or non-adsorbed anti-Crescormon, anti-Somatonorm, or anti-human-gamma globulin to Sepharose 4B gel using the bromocyanate method.(8)

Coupling of preactivated Sepharose 4B gel (Pharmacia Fine Chemicals, Uppsala, Sweden) was performed according to the procedure recommended by the manufacturer including a final treatment of the gel derivatives with ethanolamine for effective blocking of remaining active groups. The amount of protein coupled to each gel derivative was about 0.7 mg Crescormon or Somatonorm and about 5 mg immunoglobulins per gram wet (excess liquid removed by suction) gel derivative.

2.3. *Preparation of ^{125}I-labeled Crescormon and Somatonorm*

The labeling of Crescormon and Somatonorm with ^{125}I was performed using lactoperoxidase essentially according to the procedure described by Thorell and Johansson.(9) The specific activity obtained was 120 μCi/μg protein for Crescormon and 102 μCi/μg protein for Somatonorm.

2.4. *Binding Assay*

For determination of cross-reactivity with *E. coli* or specific peptides, samples of immunosorbent (adsorbed or nonadsorbed anti-Crescormon or anti-Somatonorm coupled to Sepharose 4B gel) consisting of 0.;05 g wet gel derivative in 200 μl PBS–BSA buffer (0.1 mole/liter sodium phosphate, pH 7.4, containing 0.5% bovine serum albumin) were mixed with 50 μl PBS–BSA buffer only or PBS–BSA buffer containing 0.6 mg/ml of Crescormon or Somatonorm, ECP, or specific peptides in test tubes. Following incubation with gentle shaking of the tubes for 2 hr at room temperature, 25 μl of ^{125}I-labeled Crescormon or Somatonorm was added to each tube followed by further incubation and shaking for another hour. The amount of radioactivity bound by the gel derivatives was then determined in a gamma counter subsequent to careful washing by repeated centrifugation and resuspension of the gel derivatives in fresh PBS–BSA buffer.

For determination of specificity of patient antibodies to monomer, dimer, or polymer of Somatonorm or specific peptides, polymeric forms of Somatonorm were prepared by prolonged lyophilization with heating of the bottom plate. These were dissolved to 7 μg/ml (monomer) and 8 μg/ml (dimer or polymer). The method was otherwise as above, with the exception of a serial four-fold dilution of the respective solution. Patient antisera from 6-month samples were added (5 μg of a 1 : 10 dilution) together with tracer and incubated for 16 hr at room temperature without stirring or 2 hr with shaking for specific peptides; 0.1 ml of gel suspension of antihuman IgG was added, and the incubation was continued for 3 hr at room temperature with shaking.

3. Results

3.1. Possible Role of the Amino-Terminal Methionine in the Antigenic Response

The possibility that the amino-terminal methionine residue causes specific antibodies to be formed was investigated by specific immunosorption. Immunosorption of rabbit anti-Somatonorm on immobilized Crescormon causes removal of all antibodies that cross-react with hGH leaving those which are specific for met-hGH.

The specificity of the rabbit Somatonorm antibodies was examined and displacement could be achieved using the *N*-terminal tryptic nonapeptide from Somatonorm. The corresponding octapeptide from Crescormon did not react. When patient sera were used as the source of antibody, in contrast to rabbit sera, no cross-reaction whatsoever with *N*-terminal tryptic peptides 0–9, or synthetic peptides 0–15, 1–15 (Fig. 1), or other tryptic peptides from Somatonorm and Crescormon was observed, thus corroborating the observation from radioimmunoassay that patient antibodies cross-react equally well with Somatonorm and with Crescormon.

3.2. Possible Role of Dimer or Polymer in the Antigenic Response

Although dimer is virtually undetectable in Somatonorm and polymer present at less than 1%, it is conceivable that dimer or polymer could, however, be responsible. The reactivity of patient antisera to monomer, dimer, and polymer of met-hGH was tested in the binding assay. Monomer, dimer, and polymer were equivalent in displacing labeled Somatonorm or Crescormon from patient IgG, which indicates that higher-molecular weight forms do not have a greater affinity for the patient antibodies.

1 5 10 15

Met-Phe-Pro-Thr-Ile-Pro-Leu-Ser-Arg-Leu-Phe-Asp-Asn-Ala-Met-Leu-OH met-hGH

Phe-Pro-Thr-Ile-Pro-Leu-Ser-Arg-Leu-Phe-Asp-Asn-Ala-Met-Leu-OH hGH

Figure 1 Amino acid sequence of the *N*-terminal synthetic peptides.

Table 1 Binding of [^{125}I]hGH or [^{125}I]met-hGH to Various Immunadsorbents (Sepharose 4B Containing Coupled Anti-hGH or Anti-E. coli Antibodies) Subsequent to Incubation of the Immunadsorbents in Buffer Alone or Buffer Containing Excess Amounts of Non-labeled hGH or met-hGH

Sample	Immunadsorbent	Additives	Tracer	Binding of ^{125}I tracer (%)[a]
1	Anti-hGH	Buffer	[^{125}I]hGH	79.8
2	Anti-hGH	Buffer	[^{125}I]met-hGH	81.4
3	Anti-*E. coli*	Buffer	[^{125}I]hGH	0.70
4	Anti-*E. coli*	hGH	[^{125}I]hGH	0.52
5	Anti-*E. coli*	Met-hGH	[^{125}I]hGH	0.65
6	Anti-*E. coli*	Buffer	[^{125}I]met-hGH	1.19
7	Anti-*E. coli*	hGH	[^{125}I]met-hGH	0.97
8	Anti-*E. coli*	Met-hGH	[^{125}I]met-hGH	0.98

[a] All determinations were performed in duplicate. The deviation of each value from the mean value was less than 10%.

3.3. Possible Role of ECP in the Antigenic Process

An immunoradiometric assay has been developed using ECP (Johansson and Eriksson, unpublished observations), and sera from all patients in the Bierich study (this volume, Chapter 22) have been evaluated with this method. It was found that patients reacted to *E. coli* contaminants and produced antibodies toward these. Development of *E. coli* antibodies appeared to parallel the production of anti-hGH. The possibility that these ECP cross-reacted with hGH in radioimmunoassay was tested for in the binding assay. As shown in Table 1, no cross-reaction occurs.

The silver staining method for polyacrylamide gels has 50–100 times greater sensitivity than Coomassie blue. Silver staining reveals the presence of components other than hGH in Somatonorm which were not found in Crescormon (Fig. 2). Blotting of these gels against specific *E. coli* antisera showed that these extraneous bands originated from *E. coli*. The same result was observed for electroimmunoassay using these same antisera.

3.4. Elimination of ECP

Using all these specific methods, several production scale batches were evaluated; i.e., all fractions prior to, subsequent to, or within the GH fraction were tested from different chromatographic steps. It was found that two components, one with mol. wt. 30,000, another with mol. wt. less than 20,000, eluted in the latter half of the GH pool in ion exchange. These could be removed effectively by changing the chromatographic conditions slightly.

This change in production method permits removal of contaminants and results in an increase of more than 100-fold in the purity of the product, as shown in Fig. 2.

Western blotting of the same PAGE–SDS gel shown in Fig. 2 with patient

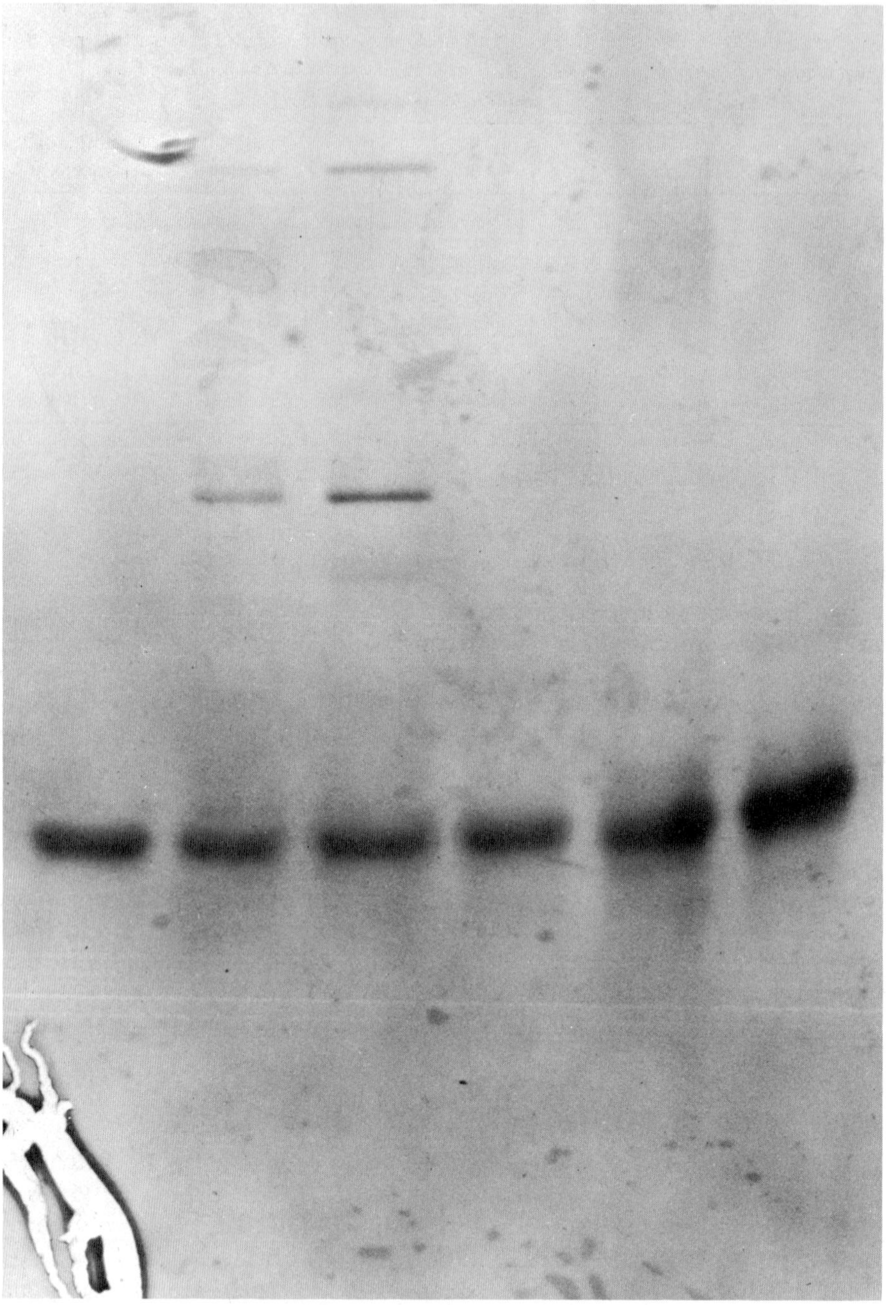

Figure 2 SDS–PAGE of hGH preparations followed by silver staining. Electrophoresis was run according to Laemmli with $T = 12.5\%$ and $C = 3.0\%$ in Tris-glycine (0.025 M Tris, 0.19 M glycine) buffer, pH 8.3. Silver staining was performed as described by Morrissey.[5] Applied from left to right: new met-hGH batch, two batches old met-hGH, new met-hGH batch, old batch of pituitary hGH, new batch of pituitary hGH.

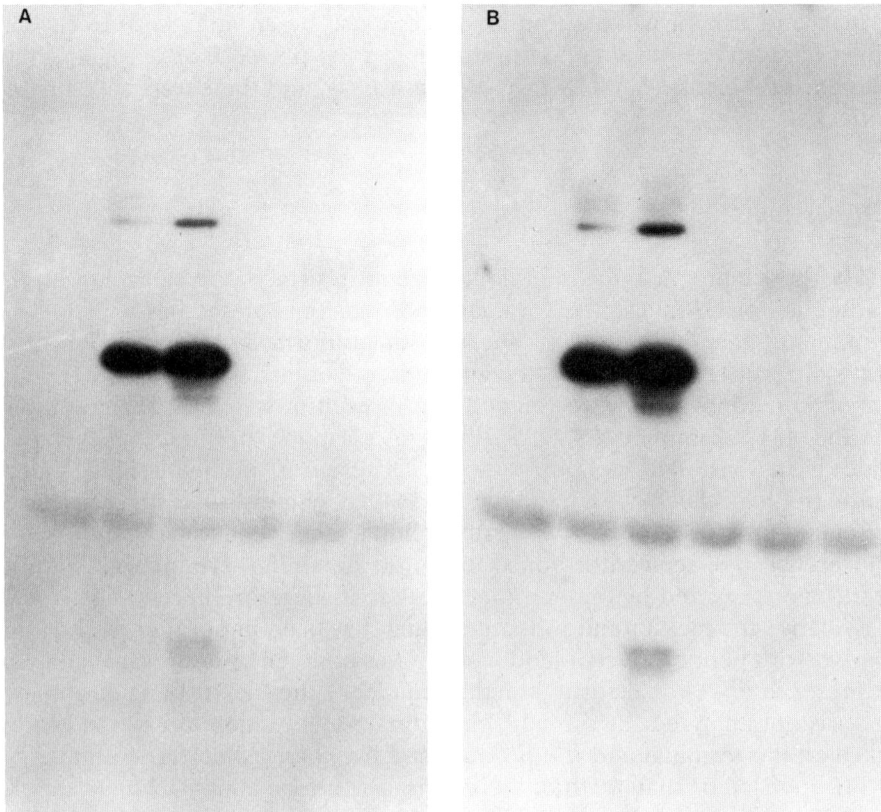

Figure 3A and B Autoradiographs of the gel in Fig. 2. A copy of the gel was transferred to nitrocellulose and blotted with anti-*E. coli* serum. The X-ray films were exposed for (1) 48 hr and (2) 4 days.

sera confirms these findings, since the sera only respond to the met-hGH bands for more purified Somatonorm and Crescormon, but pick up ECP in the earlier preparations (Fig. 3A,B).

3.5. Other Possible Components

We have also been able to examine final product in a PAGE–SDS system that distinguishes between outer and inner membrane and cytosol components (K.-E. Johansson, unpublished data), which has shown that the extraneous proteins seen in Somatonorm are derived only from the cytosol and are absent from more purified Somatonorm. This information is also of importance since outer-membrane components are known to be especially antigenic and can stimulate the immune system.

Outer-membrane lipopolysaccharides are known to be pyrogenic in animals and man and can provoke an antibody response by an adjuvant effect. The product has also been examined thoroughly for the presence of pyrogens, both by standard rabbit assay and by *Limulus* lysate assay, and these tests have proved negative.

4. Discussion

The possibility that the methionine residue affects the behavior of the biosynthetic hormone has been evaluated. Monoclonal antibodies with specific determinants have been used to analyze Somatonorm and Crescormon,(10,11) and no differences are found. Retegui and co-workers have also studied the effect of Somatonorm and Crescormon in two receptor systems, pregnant rabbit liver and IM-9 lymphocytes, and the two hormone preparations behaved identically in these two receptor systems. Differences in chemical, biological, and immunological behavior in radioimmunoassay systems between Crescormon and Somatonorm have been investigated, and as other reports have shown,(2,12) no differences are seen with the exception of the amino-terminal methionine. Any influence exerted by the extra methionine is therefore necessarily subtle.

Evidence for methionine sulfoxide, which has been implicated in antigenic responses with peptides such as endorphins, was not seen by amino acid analysis. The presence of residual formyl methionine from the biosynthesis of Somatonorm was not analyzed for directly, but Edman degradation did not indicate a blocked amino terminal and the product had the correct isoelectric point.

The possibility that methionine exerts a comformational change on the GH molecule is not borne out by studies with circular dichroism (CD) or fluorescence where the b and c components of Crescormon and Somatonorm were analyzed. These studies revealed slight differences between b and c components irrespective of whether the source was Crescormon or Somatonorm.(13) Components b and c differ only by one negative charge, presumably as the result of deamidation, which is, however, sufficient to give rise to a slight conformational change detected by CD. This indicates that the effects of *N*-terminal methionine on secondary or tertiary structure are not profound.

The experimental results described here indicate that the *N*-terminal methionine residue is not directly responsible for antigenicity of Somatonorm. No specific antibodies are produced toward the *N*-terminal.

Polymer and dimer, prime suspects for causing antibodies to GH, do not appear to bind better to patient antibodies when checked in the binding assay, which suggests that the antibodies are not directed toward them. ECP do not cross-react nonspecifically in radioimmunoassay for hGH or met-hGH.

The most satisfactory explanation for the enhanced antigenicity would appear to be that the hydrophobic protein met-hGH has one or several ECP's bound by noncovalent bonds to its surface. These components can act as adjuvants to promote formation of antibodies to hGH in patients not previously exposed to the hormone. These antibodies do not affect the biological activity of hGH in any way since the children grow normally without side effects.

Analysis of the final product by sensitive and specific methods such as silver stain of electrophoresis gels, EIA, and Western blotting with several antisera including patient sera revealed the presence of minute amounts of ECP similar in size and charge to met-hGH. These contaminants can be effectively removed by changing the chromatographic conditions in the purification process.

Evaluation of the new more purified product is underway.

5. Summary

A clinical trial with the recombinant DNA produced human growth hormone, Somatonorm, indicated that the product was more antigenic than its pituitary counterpart Crescormon. However, no negative effects on growth were observed.

Somatonorm contains an extra residue of methionine at the aminoterminus but this has not been found to be antigenic in itself or to result in a conformational change. Sensitive analytical techniques revealed the presence of small amounts of *E. coli* proteins in the final product which are presumed to exert an adjuvant effect on Somatonorm, resulting not only in producing antibodies to themselves but also towards the growth hormone molecule. These proteins originate from the cytosol and can be excluded from the final product by chromatographic means.

References

1. D. V. Goeddel, H. C. Heyneker, T. Hozumi, R. Arentzen, R. Arentzen, K. Itakura, D. G. Yansura, M. J. Ross, G. Moizzari, R. Crea, and P. H. Seeburg, Direct expression in Escherichia coli of a DNA sequence coding for human growth hormone, *Nature (London)* **281**:544–548, 1979.
2. L. Fryklund, J. Brandt, G. Eketorp, K. Fhölenhag, B. Skoog, A. Skottner-Lundin, and A. Wichman, *A Comparison of Human Biosynthetic and Pituitary Somatotropins: Purity and Potency, Insulins, Growth Hormone, and Recombinant DNA Technology* (J. L. Gueriguian, ed.), Raven Press, New York, 1981, pp. 329–326.
3. B. Weeke, *Quantitative Immunoelectrophoresis* (N. H. Axelson, J. Kroll, and B. Weeke, eds.), Universitetsforlaget, Oslo, Norway, 1973.
4. U. K. Laemmli, Cleavage of structural proteins during the assembly of the head of Bacteriophage T4, *Nature* **227**:680–685, 1970.
5. J. H. Morrissey, Silver stain for proteins in polyacrylamide gels: A modified procedure with enhanced uniform sensitivity, *Anal. Biochem.* **117**:307–310, 1981.
6. C. R. Merril, D. Goldman, S. A. Sedman, and M. H. Ebert, Ultrasensitive stain for proteins in polyacrylamide gels shows regional variation in cerbrospinal fluid proteins, *Science* **222**:1437–1438, 1981.
7. W. N. Burnette, "Western blotting": Electrophoretic transfer of proteins from sodium dodecyl sulfate–polyacrylamide gels to unmodified nitrocellulose and radiographic detection with antibody and radioiodinated protein A, *Anal. Biochem.* **112**:195–203, 1981.
8. R. Axén, J. Porath, and S. Ernback, Chemical coupling of peptides and proteins to polysaccharides by means of cyanogen halides, *Nature (London)* **214**:1302–1304, 1967.

9. J. I. Thorell and B. G. Johansson, Enzymatic iodination of polypeptides with ^{125}I to high specific activity, *Biochim. Biophys. Acta* **251:**363–369, 1971.
10. L. A. Retegui, L. M. Keefer, P. De Meyts, and L. Fryklund, Monoclonal antibodies and specific cell surface receptors do not discriminate between human growth hormone prepared by DNA recombinant techniques and the native hormone, *Acta Physiol. Pharmacol. Latinoam.* **34:**193–197, 1984.
11. I. Jónsdóttir, H. P. Ekre, and P. Perlmann, Comparative study of pituitary and bacteria-derived human growth hormone by monoclonal antibodies, *Mol. Immunol.* **20:**871–876, 1983.
12. A. Skottner, A. Forsman, K. Fhölenhag, A. Helleberg, E. Löfberg, L. Fryklund, B. Vangbo, and B. Skoog, Human *Growth Hormone Produced by* E. coli: *A Preliminary Study of Effects on Hypophysectomized Rats, Insulins, Growth Hormone, and Recombinant DNA Technology* (John L. Gueriguian, ed.), Raven Press, New York, pp. 109–115, 1981.
13. H. Larhammar, I. Sjöholm, B. Skoog, and A. Wichman, Comparative conformational analysis of human somatotropin and biosynthetic methionyl human somatotropin by spectroscopic titrations, *International Journal of Pharmaceutics* **23:**13–23, 1985.

20

Clinical Studies with Recombinant-DNA-Derived Methionyl-Human Growth Hormone in Growth Hormone-Deficient Children

S. L. Kaplan, L. E. Underwood, G. P. August, J. J. Bell, S. L. Blethen, D. R. Brown, T. P. Foley, R. L. Hintz, N. J. Hopwood, A. J. Johanson, R. T. Kirkland, and L. P. Plotnick

1. Introduction

The advent of recombinant-DNA technology provides the prospect of synthesis of human growth hormone (hGH) in unlimited quantitites. Since 1% of protein synthesized by *Escherichia coli* by this procedure is hGH, the estimated yields are predicted to be substantial. The details of isolation of the mRNA and

S. L. Kaplan • Department of Pediatrics, University of California–San Francisco, San Francisco, California 94143. ***L. E. Underwood*** • Department of Pediatrics, University of North Carolina School of Medicine, Chapel Hill, North Carolina 27514. ***G. P. August*** • Department of Pediatrics, Children's Hospital, National Medical Center, Washington, D.C., 20010. ***J. J. Bell*** • Department of Pediatrics, College of Physicians and Surgeons, Columbia University, New York, New York 10032. ***S. L. Blethen*** • Department of Pediatrics, Washington University School of Medicine, Saint Louis, Missouri 63110. ***D. R. Brown*** • Department of Pediatrics, University of Minnesota, Minneapolis, Minnesota 55455. ***T. P. Foley*** • Department of Pediatrics, University of Pittsburgh, Pittsburgh, Pennsylvania 15213. ***R. L. Hintz*** • Department of Pediatrics, Stanford University, Stanford, California 94305. ***N. J. Hopwood*** • Department of Pediatrics, University of Michigan, Ann Arbor, Michigan 48109. ***A. J. Johanson*** • Department of Pediatrics, University of Virginia School of Medicine, Charlottesville, Virginia 22901. ***R. T. Kirkland*** • Department of Pediatrics, University of Texas, Houston, Texas 77030. ***L. P. Plotnick*** • Department of Pediatrics, Johns Hopkins University School of Medicine, Baltimore, Maryland 21205.

transcription of the DNA for hGH have been described earlier.[1–3] hGH has been cloned as a methionyl analogue.

Biosynthetic methionyl-hGH derived from *E. coli* has essentially the same amino acid analysis, chromatographic pattern, and primary structure as pituitary hGH.[4] The presence of an extra *N*-terminal methionine arises from the method of cloning, but the tertiary structure by circular dichroism is comparable to that of pituitary hGH.

The biological activity of methionyl-hGH in stimulation of body weight gain and increase in width of tibial epiphysis in hypophysectomized rats is comparable on a weight basis to that of pituitary hGH; biopotency is 2 U/mg.[4] Neither methionyl-hGH nor purified pituitary hGH stimulate glycerol release in adipose tissue from fasted rats. A significant increase in serum free fatty acids is induced by intraperitoneal administration of methionyl-hGH or pituitary hGH to hypophysectomized rats.[5] Equivalent displacement of [^{125}I]met-hGH or [^{125}I]pituitary hGH by the heterologous hGH has been demonstrated in a radioimmunoassay system and in a liver membrane assay.[6,7]

A double-blind cross-over study in human adult volunteers using methionyl hGH (code G11008) and pituitary hGH (Serono) was carried out in 1981 by Hintz *et al.* The effects of methionyl hGH (Genentech) and pituitary hGH (Serono) in human adults were comparable on a weight basis in the stimulation of somatomedin release, insulin secretion, and nitrogen retention,[7,8] Similar results were obtained with two subsequent preparations of methionyl-hGH (codes G11014 and G11015) tested in 1982.

Administration of preparations G11008, G11014, and G11015 was not associated with alterations in hemogram or serum chemistries or with febrile reactions. Takano *et al.*[6] have reported comparable results in normal adult males using met-hGH obtained from Kabi, Sweden.

This chapter describes clinical trials of methionyl-hGH (Genentech) to assess its efficacy in stimulation of growth in GH-deficient children.

2. *Patients and Methods*

Thirty-six children with documented GH deficiency have been treated for periods of 12–24 months with methionyl-hGH. A control group of 10 children with GH deficiency received pituitary hGH (Kabi) for 12 months. The study was approved by the ethical committee of the participating medical schools, and written consent was obtained from the parents and child (when of appropriate age).

Clinical details of the patients are indicated in Table 1.

Twenty-two children were included in Group A (September 1981) and received methionyl-hGH (Code G11008) initially; glucose and insulin levels following intravenous glucose (0.5 gm/kg glucose) and measurements of somatomedin C were determined before and following three injections (0.1 mg/kg) of methionyl-hGH (Code G11008). Fourteen children (Group B) in whom treatment with methionyl-hGH (Code G11015) was started in January 1983

Table 1 Clinical Data of GH-Deficient Children Pretreatment

hGH preparation	Sex (F/M)	Chronological age[a] (years)	Bone age[a] (years)	Height[a] (SD for age)	Weight[a] (SD for age)	Growth rate (cm/yr) (mean ± SD)
Group A (G11008)	10/12	9.1 (3.3–14.5)	5.4 (1.6–10.7)	−3.7 (−1.5 to −5.0)	−1.6 (+0.1 to −3.5)	3.2 ± 1.1
Group B (G11015)	6/8	8.8 (3.1–13.9)	6.6 (2.1–11.0)	−3.6 (−1.8 to −4.7)	−1.3 (+0.6 to −3.5)	3.0 ± 1.0
Group C Pituitary hGH	4/6	6.1 (4.1–12.2)	5.0 (1.5–11.4)	−3.7 (−2.1 to −5.5)	−1.4 (+0.5 to −3.1)	3.8 ± 1.0

[a] Median (range).

and 10 children (Group C) who initiated treatment with pituitary hGH in April 1983 had laboratory studies performed at bimonthly intervals during treatment. None of the patients had received hGH prior to entrance into the study. All were prepubertal by physical examination.

Three preparations of methionyl-hGH were used for treatment of GH-deficient children: G11008 (September 1981 to June 1983) was used for 11 of the original 22 subjects; the remaining 11 subjects were crossed over to Preparation G11014 in December 1982 and treated through August 1983. All Group A patients were changed to Preparation G11015 in October 1983 after a 2-month period off hGH. Group B children received G11015 (January 1983 to December 1983). No significant differences in the data between the continuous 11008-treated versus 11008/11014 were noted.

The hormone was dispensed in sterile vials containing 5 mg of methionyl-hGH. Reconstitution was carried out by the addition of sterile distilled water containing 0.5% benzyl alcohol. Vials of pituitary hGH contained 2 mg of hormone.

Intramuscular injections of 0.1 mg/kg methionyl-hGH and pituitary hGH were given three times weekly, usually by a parent and on occasion by medical personnel.

Antibodies to hGH were determined by a modification of the method described by Underwood *et al.* using [^{125}I]hGH and propylene ethyl glycol as precipitant.[9] Binding was considered positive if greater than 2X nonspecific binding; the calculated numerical value is based on maximum serum dilution at which significant binding occurs, i.e., binding at 1:100 designated as a value of 2.0.

3. Results

3.1. Effect on Height and Weight

The growth rate of children in Groups A and B during the first year of treatment with methionyl-hGH increased threefold compared to baseline growth rate (Table 2, Fig. 1). This is comparable to the effect of pituitary hGH on the growth rate of children in Group C (Table 2, Fig. 1).

During the second year of treatment with methionyl-hGH in Group A, the growth rate decreased slightly but was 2.5-fold greater than the baseline rate of growth.

The increment in weight during the first year of therapy was comparable in Groups A, B, and C (Table 2, Fig. 1). The advancement in bone age was equivalent to that of chronological age in Group A.

3.2. Biochemical Studies

In group A children, an intravenous glucose tolerance test was carried out before and after three intramuscular injections of methionyl-hGH (0.1 mg/kg).

Table 2 Effect of Methionyl-hGH and Pituitary hGH on Height and Weight of GH-Deficient Children[a]

Mean ± SD	Group A methionyl hGH (G11008)	Group B methionyl hGH (G11015)	Group C pituitary hGH Kabi
Basal height (cm)	107.5 ± 16.1	111.9 ± 17.8	102.9 ± 14.9
6 months	112.9 ± 15.5	117.8 ± 17.2	108.8 ± 14.2
12 months	117.8 ± 15.4	122.2 ± 16.5	113.1 ± 14.1
18 months	122.0 ± 15.6	NA	118.2 ± 14.3
24 months	124.5 ± 15.7	13.0 ± 24.3	120.2 ± 14.5
Basal weight (kg)	20.1 ± 7.7	22.9 ± 10.4	18.3 ± 5.3
6 months	21.9 ± 8.3	24.2 ± 10.6	19.9 ± 5.5
12 months	24.3 ± 8.7	26.7 ± 11.3	21.3 ± 6.2
18 months	26.7 ± 9.2	NA	24.8 ± 6.0
24 months	27.5 ± 7.8	30.6 ± 10.1	25.4 ± 7.1

[a] NA = not available.

The peak values of glucose and of insulin were not significantly different before and after administration of met-hGH (Fig. 2).

After administration of three doses of methionyl-hGH to Group A GH-deficient children, somatomedin C increased from 0.26 ± 0.23 U/ml to 0.96 ± 0.72 U/ml. A significant increase in serum somatomedin C (SMC) occurred in all groups during chronic treatment with either methionyl-hGH or pituitary hGH (Table 3).

In this study, four children from Group A had significantly decreased serum thyroxine and free thyroxine levels after 4–12 months of treatment with methionyl-hGH. Growth rate improved after institution of L-thyroxine therapy.

A significant decrease in serum blood urea nitrogen (BUN) values and an increase in serum phosphorus and alkaline phosphatase was observed in all

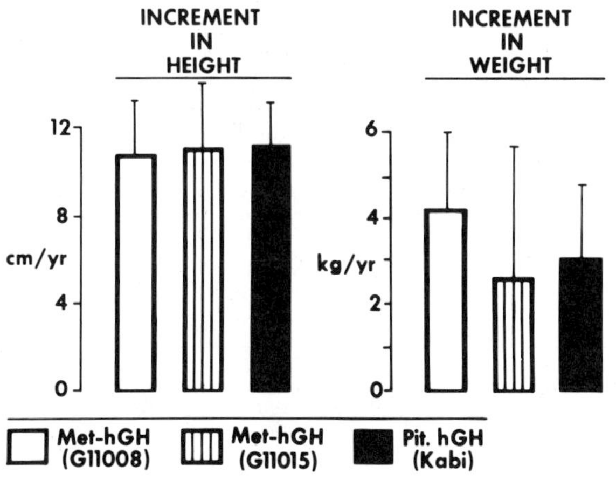

Figure 1 The mean (±SD) increment in height (cm/year) and weight (kg/year) in GH-deficient children following administration of two preparations of methionyl-hGH is compared with that of pituitary hGH. No significant differences were determined between these values in any group.

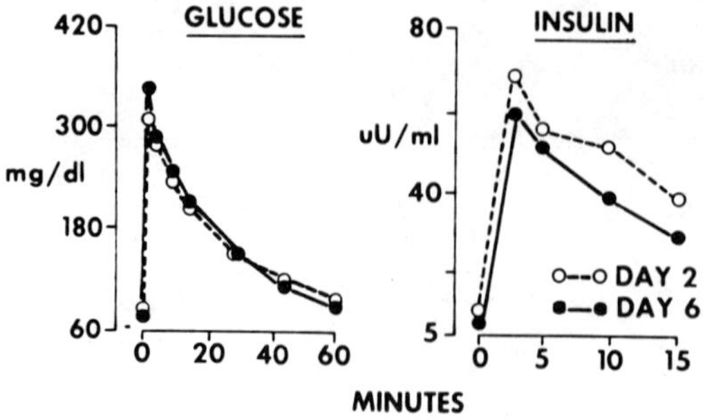

Figure 2 The serum glucose (mg/dl) (left panel) and plasma insulin (right panel) response to i.v. glucose (0.5 g/kg) are shown before and after administration of methionyl-hGH. Three daily intramuscular doses of 0.1 mg/kg were given. No significant differences in levels or disappearance rate of glucose and insulin were demonstrated after administration of methionyl-hGH.

GH-deficient children after administration of either methionyl-hGH or pituitary hGH. No significant changes of serum electrolytes, liver chemistries, white blood count, red blood count, platelets, or urinalysis were demonstrated after treatment with methionyl-hGH. There were no febrile reactions or change in blood pressure or heart rate.

3.3. Development of Antibodies to hGH

In the Group A GH-deficient children (treated with methionyl-hGH; Preparation G11008), 21 of 22 children developed IgG antibodies to hGH within 4–6 months of therapy. With binding, 16 subjects (73%) developed levels of IgG antibodies to hGH (binding greater than 2X nonspecific binding at a dilution of 1:1000); five subjects (23%) had sera that showed low-level binding of hGH (titers less than 1:50) and one subject remained anti-hGH negative

Table 3 Somatomedin C Levels (Units/ml-RIA) (Mean ± SD) before and after Administration of met-hGH and Pituitary hGH

	Group A met-hGH (G11008)	Group B pituitary hGH (NHPP)[b]
Basal SMC	0.26 ± 0.23	0.27 ± 0.16
SMC generation test[c]	0.96 ± 0.72	0.84 ± 0.57
	Group A met-hGH (G11008)	**Group B Pituitary hGH (Kabi)**
Basal SMC	0.26 ± 0.23	0.36 ± 0.3
At 6 months[d]	0.79 ± 0.53	0.98 ± 0.62
At 12 months[d]	1.16 ± 0.64	1.1 ± 1.0
At 18 months[d]	1.4 ± 0.80	NA

[a] NA = not available.
[b] Ref. Van Vliet.[33]
[c] Sample obtained 20–24 hr after three injections of 0.1 mg/kg hGH.
[d] hGH 0.1 mg/kg t.i.w.

Table 4 Relationship of Binding Affinity and Binding Capacity to Growth Rate and Increment in Height in Representative GH-Deficient Children during Treatment with Methionyl-hGH[a]

		At 12 Months		Antibodies to met-hGH	
Subject	Pre-Rx (cm/yr)	Growth rate (cm/yr)	Increment in height (cm/yr)	Binding capacity (mg/liter)	Binding affinity (10 liter/m)
4001	3.5	7.8	4.3	1.2	4.0
6001	4.9	8.4	3.5	1.8	2.6
8002	4.7	15.6	10.9	1.8	3.6
9002	3.0	9.4	6.4	0.4	5.7
10002	3.0	10.8	7.8	0.3	8.8
12001	2.6	11.4	8.8	2.4	3.2
Mean	3.6 ± 1.0	10.6 ± 2.8	7.0 ± 2.8	0.98 ± 0.7	4.7 ± 2.3
1002	2.3	9.7	7.4	Neg	Neg
3001	1.9	9.1	6.8	Neg	Neg
4002	2.2	10.5	8.3	0.05	Neg
7002	4.1	9.2	5.1	0.01	Neg
8001	1.8	11.0	9.2	0.01	Neg
10003	1.6	10.8	9.2	Neg	Neg
Mean	2.3 ± 0.9	10.1 ± 0.8	7.7 ± 1.6		

[a] Neg = negative.

(1:10) through 24 months of therapy. A subsequent group of 14 GH-deficient children (Group B) received a more purified preparation of methionyl-hGH (Code G11015), which was associated with a lower incidence of antibody formation (43%). Two subjects (14%) on methionyl-hGH (Code G11015) had titers greater than 1:10,000 and high binding capacity (>1 mg/liter; an additional five subjects had measurable hGH binding by their sera, but titers were less than 1:100 (most 1:50) and binding capacities less than 0.1 mg/liter. One of the subjects with a low-level titer (on Code G11015) became anti-hGH negative at month 8 and remained so through month 10. The Group A children received the new preparation (G11015) after a 2-month washout period. The antibody titers were sustained in 62% (13 patients) and decreased in three patients to low titers (less than 1:30).

The acceleration in growth rate of the 35/36 children receiving methionyl hGH was not altered by the presence of antibodies to hGH. There was no relationship of binding capacity of hGH antibodies to growth rate in 35 of 36 patients (Table 4). Their growth rate and increment in height was not significantly different from those of 10 GH-deficient children who received purified pituitary hGH (Kabi AB) (Table 2); none of the children developed antibodies to this preparation of pituitary-derived hGH. One child (G11015 group) developed high binding capacity antibodies which led to growth attenuation.

No clinical evidence of an allergic diathesis was demonstrated in any of the children with detectable antibodies to hGH. None of the children developed antibodies to *E. coli* proteins.

4. Discussion

The clinical efficacy of methionyl hGH has been substantiated in GH-deficient children. Its equivalent biopotency to purified pituitary GH in stimulation of growth in GH-deficient children is demonstrated by these studies.[10–13].

Methionyl-hGH induced effects on BUN, phosphorus, and alkaline phosphatase levels in children and in adults similar to those reported earlier for pituitary growth hormone.[13,14]

No adverse drug-related systemic effects or hematological, hepatic, or renal abnormalities were observed during the 24-month administration period.

The induction of SMC release is an expression of the biological activity of growth hormone. Short-term administration of high doses of hGH is associated with significant increases in serum SMC.[15–18] Similar effects on serum SMC levels were observed after administration of methionyl-hGH to GH-deficient children in this study and to adults in earlier studies.[7]

The diabetogenic action of GH has been documented in animals, but the effect on carbohydrate metabolism in GH-deficient subjects is variable.[19–22] Acute administration of GH within a 4-hr period before a glucose tolerance test invariably results in glucose intolerance associated with increased insulin secretion. Prolongation of the interval between GH administration and the glucose tolerance test in GH-deficient patients to more than 12 hr is associated with minimal alterations in either glucose utilization or insulin secretion. Similar results have been obtained in studies with methionyl hGH. In the GH-deficient children, no significant differences in insulin secretion or glucose disappearance rate were demonstrated when methionyl hGH was administered 12–18 hr before intravenous glucose. In contrast, in normal adults, higher doses of pituitary or methionyl-hGH administered within 18 hr of an oral glucose tolerance test were associated with glucose intolerance and hyperinsulinism.[8] Based on these findings, methionyl-hGH and pituitary hGH have comparable dose-related effects on carbohydrate metabolism.

Alterations in thyroid hormone metabolism occur after administration of GH to GH-deficient patients,[9] as in this study. Decreased serum thyroxine may be associated with diminished thyroidal uptake of radioactive iodine and suppression of thyroid-stimulating hormone response to thyroid-releasing factor.[23–25] Peripheral conversion of thyroxine to triiodothyronine is stimulated by hGH and may be a factor in the observed changes in serum thyroxine. Whether low serum thyroxine reflects a change in intracellular metabolism of thyroxine induced by hGH or indicates development of tertiary hypothyroidism may complicate clinical assessment and appropriateness of replacement therapy with exogenous thyroxine.

A high proportion of GH-deficient patients (30%) developed antibodies to earlier pituitary hGH preparations which were less purified and contained aggregated components of hGH.[26–30] Utilization of predominantly monomeric preparations reduced the incidence of antibody formation to about 5%.[30–32]

The incidence of antibodies to hGH induced in GH-deficient children (90%) by the first preparation of methionyl hGH (Code G11008) was unexpected since this preparation was highly purified and monomeric. Further changes in the purification of methionyl-hGH led to a significant reduction (43%) in the incidence of antibody formation in the second group of GH-deficient patients. In contrast, none of the GH patients treated with pituitary hGH (Kabi) developed antibodies to hGH.

Does the additional methionine increase the antigenicity of hGH? Immunogenic studies in animals and *in vitro* immunological assays have failed to demonstrate intrinsic differences between pituitary and methionyl-hGH at present. Furthermore, the stability of the pituitary-derived and methionyl-hGH are similar. The growth rate of 35/36 children on methionyl-hGH treatment was comparable to those on therapy with pituitary hGH. Growth attenuation correlates best with the antibody-binding capacity rather than with the antibody titer or antibody affinity.[27,28] No other manifestations of an allergic reaction such as hives, erythema, or rash were demonstrated in these patients. No increase over basal levels of antibodies to *E. coli* proteins were detected in any of the patients treated with methionyl-hGH.

5. Summary

Thirty-six children with GH deficiency have been treated with methionyl-hGH (Genentech) synthesized by DNA recombinant methods for periods up to 24 months.

The growth rate in cm/year for these 36 children increased threefold from 3.5 ± 1.4 to 10.4 ± 2.3 (mean ± SD), similar to the effect of pituitary hGH in 10 GH-deficient children (4.7 ± 0.7 to 10.9 ± 1.9). Serum SMC rose from 0.26 ± 0.23 IU/ml to 0.79 ± 0.53 IU/ml after 6 months of methionyl-hGH therapy and from 0.36 ± 0.3 to 0.98 ± 0.62 IU/ml after 6 months of pituitary hGH therapy.

The decrease in serum BUN and rise in serum phosphorus and alkaline phosphatase were comparable following administration of methionyl hGH or pituitary hGH to GH-deficient patients.

Methionyl-hGH had no significant effect on glucose tolerance or insulin secretion in GH-deficient children.

A higher incidence of antibody formation to methionyl-hGH was observed when compared to pituitary hGH (Kabi). The presence of antibodies to hGH did not affect the growth rate of the GH-deficient children. No consistent changes in levels of antibodies to *E. coli* proteins were detected. No other allergic manifestations or systemic side effects were demonstrable.

References

1. D. V. Goeddel, H. L. Hyeneker, T. Hozum, R. Arentzen, K. Itakura, D. G. Yansura, M. J. Ross, G. Miozzari, R. Crea, and P. H. Seeburg, Direct expression in *Eschericia coli* of a DNA sequence coding for human growth hormone, *Nature* **281**:544–548, 1979.

2. J. H. Martial, R. A. Hallewell, J. D. Baxter, and H. M. Goodman, Human growth hormone: Complementary DNA cloning and expression in bacteria, *Science* **205:**602–607, 1979.
3. K. C. Olson, J. Fenno, N. Lin, R. N. Harkins, C. Snider, W. H. Kohr, M. J. Ross, D. Fodge, G. Prender, and N. Stebbing, Purified growth hormone from E. coli is biologically active, *Nature* **293:**408–411, 1981.
4. R. G. Rosenfeld, B. B. Aggarwal, R. L. Hintz, and L. A. Dollar, Recombinant DNA-derived methionyl human growth hormone is similar in membrane binding properties to human pituitary growth hormone, *Biochem. Biophys. Res. Commun.* **106:**202–209, 1982.
5. L. G. Frigeri, G. Robel, and N. Stebbing, Bacteria-derived human growth hormone lacks lipolytic activity in rat adipose tissue, *Biochem. Biophys. Res. Commun.* **104:**1041–1046, 1982.
6. K. Takano, N. Hizuka, K. Shizume, K. Asakawa, and M. Kogawa, Short-term study of biosynthesized hGH in man, *Endocrinol. Japan* **30:**(1):79–84, 1983.
7. R. L. Hintz, R. G. Rosenfeld, D. M. Wilson, A. Bennett, J. Finno, B. McClellan, and R. Swift, Biosynthetic methionyl human growth hormone is biologically active in adult man, *Lancet* **1:**1276––1279, 1982.
8. R. G. Rosenfeld, D. M. Wilson, L. A. Dollar, A. Bennett, and R. Hintz, Both human pituitary growth hormone and recombinant DNA-derived human growth hormone cause insulin resistance at a postreceptor site, *J. Clin. Endocrinol. Metab.* **54:**1033–1038, 1982.
9. L. E. Underwood, S. J. Voina, and J. J. Van Wyk, Restoration of growth by human growth hormone (Roos) in hypopituitary dwarfs immunized by other human growth hormone preparations: Clinical and ummunological studies, *J. Clin. Endocrinol. Metab.* **38:**288–297, 1974.
10. H. G. Goodman, M. M. Grumbach, and S. L. Kaplan, Growth and growth hormone. II. A comparison of isolated growth-hormone deficiency and multiple pituitary-hormone deficiencies in 35 patients with idiopathic hypopituitary dwarfism, *N. Engl. J. Med.* **278:**57–68, 1968.
11. M. A. Preece, J. M. Tanner, R. H. Whitehouse, and N. Cameron, Dose dependence of growth response to human growth hormone in growth hormone deficiency, *J. Clin. Endocrinol. Metab.* **42:**477–483, 1976.
12. S. D. Frasier, Human pituitary growth hormone (HGH) therapy in growth hormone deficiency, *Endocrine Rev.* **4:**155–170, 1983.
13. T. Aceto, Jr., S. D. Frasier, A. B. Hayles, H. I. L. Meyer-Bahlburg, M. L. Parker, R. Munschauer, and G. Di Chiro, Collaborative study of the effects of human growth hormone in growth hormone deficiency. I. First year of therapy, *J. Clin. Endocrinol. Metab.* **35:**483–496, 1972.
14. D. M. Bergenstal, and M. B. Lipsett, Metabolic effects of human growth hormone and growth hormone of other species in man, *J. Clin. Endocrinol. Metab.* **20:**1427–1436, 1960.
15. A. J. D'Ercole, M. D. Underwood, and J. J. Van Wyk, Serum somatomedin-C in hypopituitarism and in other disorders of growth, *J. Pediatr.* **90:**375–381, 1977.
16. K. C. Copeland, L. E. Underwood, and J. J. Van Wyk, Induction of immunoreactive somatomedin C in human serum by growth hormone: dose-response relationships and effect of chromatographic profiles, *J. Clin. Endocrinol. Metab.* **50:**690–697, 1980.
17, R. G. Rosenfeld, S. F. Kemp, and R. L. Hintz, Constancy of somatomedin response to growth hormone treatment of hypopituitary dwarfism, and lack of correlation with growth rate, *J. Clin. Endocrinol. Metab.* **53:**611–617, 1981.
18. S. L. Blethen, W. H. Daughaday, and V. V. Weldon, Kinetics of the somatomedin C/insulin-like growth factor I: Response to exogenous growth hormone (GH) in GH deficient children, *J. Clin. Endocrinol. Metab.* **54:**986–990, 1982.
19. L. A. Frohman, M. H. MacGillivray, and T. Aceto, Jr., Acute effects of human growth hormone on insulin secretion and glucose utilization in normal and GH deficient states, *J. Clin. Endocrinol. Metab.* **27:**561–567, 1967.
20. H. Gold, S. Spector, N. A. Samaan, and O. Pearson, Effect of growth hormone on carbohydrate metabolism in hypopituitary dwarfs, *Metabolism* **17:**74–83, 1983.
21. G. Costin, M. D. Kogut, and S. D. Frasier, Effect of low dose human growth hormone on carbohydrate metabolism in children with hypopituitarism, *J. Pediatr.* **80:**796–803, 1972.
22. L. E. Underwood, J. L. Van der Brande, G. J. Anthony, S. J. Voina, and J. J. Van Wyk, Islet cell function and glucose homeostasis in hypopituitary dwarfism: Synergism between growth hormone and cortisone, *J. Pediatr.* **82:**28–37, 1973.

23. B. M. Lippe, S. A. Kaplan, M. P. Golden, S. A. Hendricks, and M. L. Scott, Carbohydrate tolerance and insulin binding in children with hypopituitarism: Response after acute and chronic human growth hormone administration, *J. Clin. Endocrinol. Metab.* **53**:507–513, 1981.
24. A. W. Root, A. M. Bongiovanni, and W. R. Eberlein, Inhibition of thyroidal radioiodine uptake by human growth hormone, *J. Pediatr.* **76**:422–429, 1970.
25. B. A. Porter, S. Refetoff, R. L. Rosenfeld, L. J. De Groot, V. S. Fang, and V. Stark, Abnormal thyroxine metabolism in hyposomatotrophic dwarfism and inhibition of responsiveness to TRH during GH therapy, *Pediatrics* **51**:668–674.
26. B. M. Lippe, A. J. Van Herle, S. H. La Franchi, R. P. Uller, N. Lavin, and S. A. Kaplan, Reversible hypothyroidism in growth hormone deficient children treated with human growth hormone, *J. Clin. Endocrinol. Metab.* **40**:612–618, 1975.
27. A. Prader, J. Szeky, H. Wagner, R. Illig, J. L. Touber, and D. Maingoy, Acquired resistance to human growth hormone caused by specific antibodies, *Lancet* **2**:378–382, 1964.
28. R. Illig, Growth hormone antibodies in patients treated with different preparations of human growth hormone (HGH), *J. Clin. Endocrinol.* **31**:679–688, 1970.
29. S. L. Kaplan, D. C. L. Savage, S. Suter, R. Wolter, and M. M. Grumbach, Antibodies to human growth hormone arising in patients treated with human growth hormone: incidence, characteristics, and effects on growth, in: *Advances in Human Growth Hormone Research; A Symposium,* U. S. Department of Health, Education and Welfare, Publication No. NIH 74–612, 725747, 1974.
30. S. D. Frasier, T. Aceto, Jr., A. B. Hayles, M. L. Parker, and H. F. L. Meyer-Bahlburg, Collaborative study of the effects of human growth hormone in growth hormone deficiency. II. Development and significance of antibodies to human growth hormone during the first year of therapy, *J. Clin. Endocrinol. Metab.* **38**:14–18, 1974.
31. W. V. Moore and P. Leppert, Role of aggregated human growth hormone (hGH) in development of antibodies to hGH, *J. Clin. Endocrinol. Metab.* **51**:691–697, 1980.
32. O. Westphal and P. Roos, Treatment of pituitary dwarfism with human growth hormone, *Acta. Paediatr. Scand* **182(Suppl)**:52–81, 1968.
33. G. Van Vliet, D. M. Styne, S. L. Kaplan, and M. M. Grumbach, Growth hormone treatment for short stature, *New Engl. J. Med.* **309**:1016–1022, 1983.

21

Experience with Methionyl Growth Hormone (Somatonorm I) in the United Kingdom

M. A. Preece and J. M. Tanner

1. Introduction

Our experience with recombinant DNA methionyl human growth hormone (hGH) in the United Kingdom is confined at present to a single study, extending over 6 months, and aimed entirely at determining whether or not the substance was safe to administer to GH-deficient (GHD) children. In particular, we were worried that the extra amino-terminal methionine group in the biosynthetic molecule might cause antibodies (possibly growth-inhibiting) to develop. In a situation where there was (just) enough cadaver GH (p-hGH) to meet the needs of all children demonstrated to have GHD, we did not feel justified in proceeding at once to what we felt was the second stage of a trial, that is, a test of the ability of the substance to cause growth. Thus our subjects were children with GHD, treated with cadaveric hormone for so long that they had practically stopped growing and were soon to be taken off treatment. We asked them during their last 6–9 months of treatment to switch to the biosynthetic material. Thus if growth-inhibiting antibodies did appear, the practical effect would be negligible. The met-hGH used was KabiVitrum Somatonorm I.

The amount that these 13 male and six female subjects grew during the first 6 months of the met-hGH is interesting and is shown in Figs. 1 and 2. Plotted against TW2 RUS bone age (rather than chronological age) the velocity of height during the 6 months prior to the switch in treatment was at about the 30–35th percentile and it remained at exactly the same percentile during the 6 months of met-hGH treatment.

M. A. Preece and J. M. Tanner • Department of Growth and Development, Institute of Child Health, University of London, London WC1N 1EH, England.

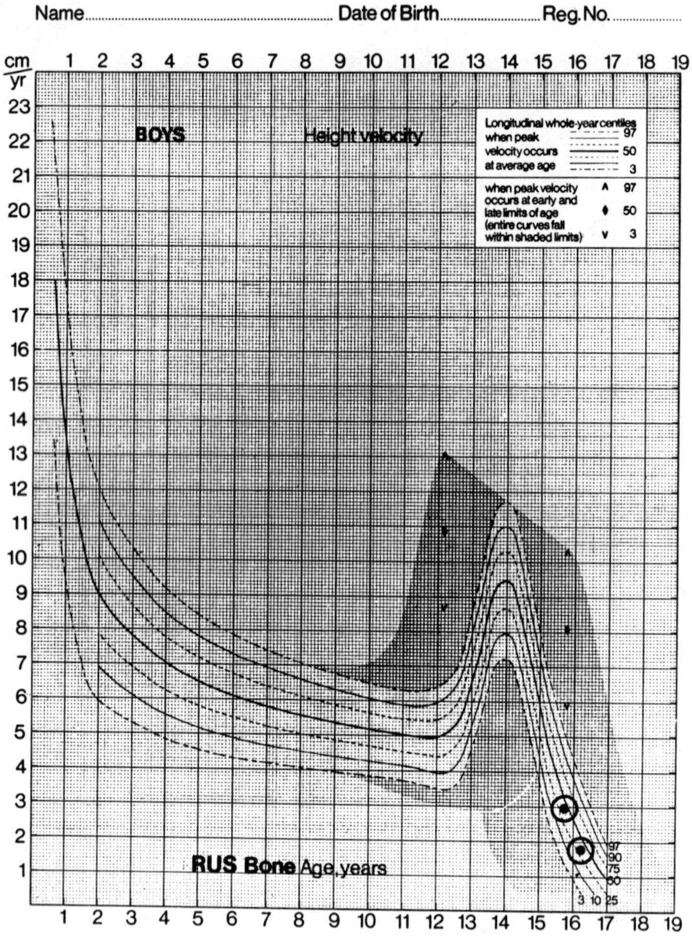

Figure 1 Mean velocity of height in 13 boys with growth hormone deficiency (hGH) plotted against RUS bone age. First point represents last 6 months of cadaveric hormone treatment; second point, first 6 months of biosynthetic met-hGH treatment.

2. *Subjects and Methods*

Nineteen patients, 13 male and six female, fulfilled our criteria for inclusion in the trial. Details of diagnosis, age, bone age, and height are shown in Tables 1 and 2. In all, 12 patients were deemed as having isolated GHD as all had passed through puberty spontaneously and had no evidence of thyroid-stimulating hormone (TSH), ACTH, or posterior pituitary hormone deficiencies. Two boys had combined GH and gonadotropin deficiency; one boy and one girl had GH, gonadotropin, and TSH deficiency; and a single girl had the combination of GH and TSH deficiency only. There were, in addition, two patients with craniopharyngiomas, both of whom had panhypopituitarism.

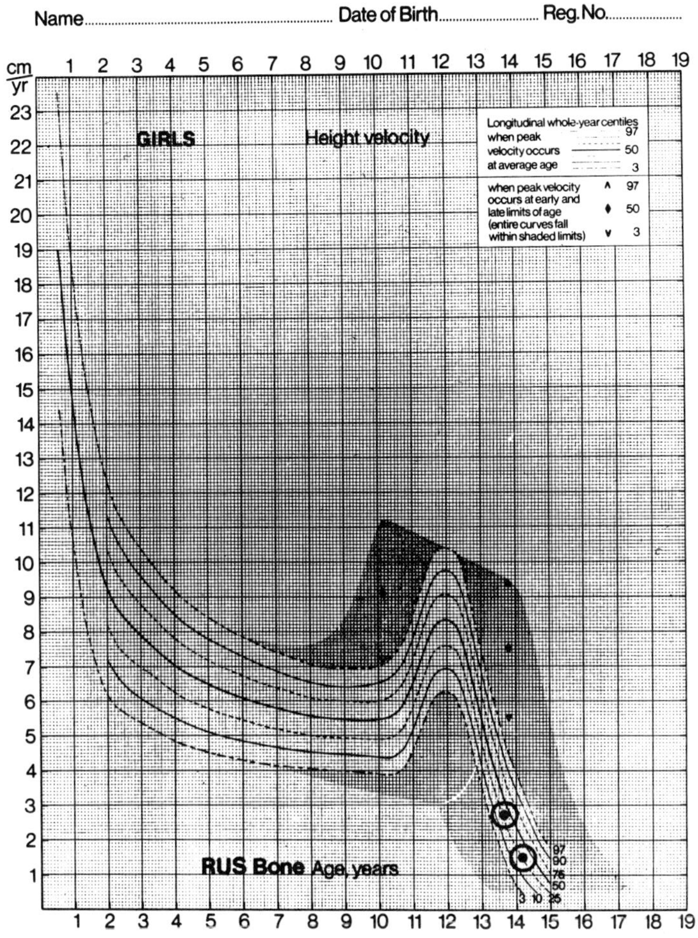

Figure 2 Mean velocity of height in six girls with growth hormone deficiency (hGH) plotted against RUS bone age. First point represents last 6 months of cadaveric hormone treatment; second point, first 6 months of biosynthetic met-hGH treatment.

Table 1 Diagnoses of Patients[a]

	Male	Female
Isolated GH deficiency	8	4
GH and Gn deficiency	2	0
GH, TSH, and Gn deficiency	1	1
GH and TSH deficiency	0	1
Craniopharyngioma	2	0

[a] GH = growth hormone; Gn = gonadotropin; TSH = thyroid-stimulating hormone.

Table 2 Ages and Heights of Patients, Mean (SD)

	Males (13)	Females (6)
Chronological age	19.2 (1.9)	16.5 (1.4)
Bone age (RUS, TW2)	16.0 (0.7)	13.9 (0.5)
Height SD_{CA}	−2.3 (0.6)	−2.2 (1.1)
SD_{BA}	−1.7 (0.9)	−1.6 (1.1)

Those patients who were gonadotropin deficient had already been through an artificially induced puberty using the appropriate sex steroids, and the TSH-deficient patients were receiving thyroxine. In the two craniopharyngioma patients supplementation was given also with hydrocortisone and DDAVP.

The protocol of the study was that each of these patients should change from their conventional p-hGH therapy (4 IU three times weekly) to met-hGH therapy without a preliminary washout period. They were seen at 1-, 3-, and 6-month intervals when routine clinical examination and measurements were carried out, and blood and an early morning urine were collected. The following tests were performed:

1. *Hematology*: hemoglobin, white cell count, red blood cell count, and packed cell volume; the usual red cell indices, platelet count, and differential white cell count.
2. *Blood chemistry*: serum electrolytes, blood urea and creatinine; measurements of liver function including alkaline phosphatase, aspartate transaminase, and alanine transaminase, serum proteins, and bilirubin.
3. *Endocrinology*: somatomedin and anti p-hGH and anti-met-hGH antibodies.
4. *Urine analyses*: routine microscopy plus osmolarity and determination of the lysozyme/albumin ratio.

All methods were by standard published techniques. Somatomedin was measured by bioassay using the porcine rib cartilage method.[1] The hGH antibodies were measured by a polyethylene glycol precipitation method using a lactoperoxidase method for labeling the hGH.

Full informed consent was obtained from the patients or their parents depending on their age.

3. Results

The blood chemistry results and the tests of renal and liver function showed no trend over the 6-month study; so we do not report them in detail. In three systems only were there any results indicative of an effect of the met-hGH.

3.1. Hematology

Table 3 gives the results for hemoglobin, red blood cell count, and platelet counts. There was no significant change in hemoglobin level or red cell count

Table 3 Hemoglobin, Erythrocyte, and Platelet Results: Mean and SEM

	Months			
	0	1	3	6
Hemoglobin (g/dl)	14.08	14.07	13.90	13.83
	1.08	0.93	0.94	1.39
Erythrocytes ($\times 10^{12}$/liter)	4.86	4.87	4.77	4.80
	0.40	0.36	0.36	0.41
Platelets ($\times 10^{9}$/liter)	266	229[a]	228[a]	238
	50	54	44	38

[a] Indicate mean values significantly different to baseline at $p < 0.001$.

over the period of the study, but there was a highly significant fall in platelet count between 0 and 1 months, which was maintained to 3 months. The initial mean platelet count of 266 $\times$ 10^9/liter fell to a nadir of 228 $\times$ 10^9/liter at 3 months. After this there was a small rise toward the initial values. No individual child became overtly thrombocytopenic, and there were no symptoms attributable to the change in platelets.

Table 4 shows the hematological results referring to the white cells. There was a significant ($p < 0.05$) fall in total white cell count from 6.78 $\times$ 10^9/liter at the start of the study to 5.97 $\times$ 10^9/liter at 1 month and further to 5.84 $\times$ 10^9/liter at 3 months. By 6 months these values were returning toward normal such that the 6-month value was not significantly different from the starting value. It would seem that most of this effect was due to a fall in neutrophil leukocytes, which decreased from a mean count of 3.73 $\times$ 10^9/liter at the start to a nadir of 2.88 $\times$ 10^9/liter ($p < 0.01$) at 3 months, returning to normal values by the end of the 6 months. There was a similar small fall in lymphocyte count, but this was not statistically significant and not maintained beyond 1 month.

Table 4 Differential White Cell Counts: Mean and SEM

WBC[a]	Months			
($\times 10^9$/liter)	0	1	3	6
Total	6.78	5.97[b]	5.84[c]	6.33
	1.22	1.64	1.15	1.91
Neutrophils	3.73	3.15	2.88[c]	3.51
	1.07	1.50	0.95	1.67
Lymphocytes	2.33	2.21	2.35	2.12
	0.77	0.64	0.75	0.58
Monocytes	0.48	0.38	0.42	0.45
	0.18	0.16	0.23	0.26
Eosinophils	0.22	0.22	0.18	0.19
	0.21	0.14	0.14	0.19

[a] WBC = white blood cell count.
[b] Mean values significantly different from baseline at $p < 0.05$.
[c] Mean values significantly different from baseline at $p < 0.01$.

Table 5 Ratio of Binding of $[I^{125}]$p-hGH by the Test Sera to that of a Blank Serum

	Months			
	0	1	3	6
Antibody binding	1.93 ± 0.27	2.0 ± 0.23	1.96 ± 0.23	3.68 ± 0.44 $p < 0.001$
Proportion greater than 2	6/19	8/19	7/19	13/18

3.2. GH Antibodies

The reports given here are only for antibodies to p-hGH, but in the few cases studied the antibodies recognized met-hGH equally. Table 5 gives the results of the antibody-screening test expressed as the ratio of GH bound by the test sample to that bound by a serum blank. The mean antibody-binding ratio was 1.93 ± 0.27 at time 0. It remained the same at 3 months but by 6 months had increased to 3.68 ± 0.44. This change is highly significant, with $p<0.001$. The slightly high antibody binding at time 0 was due to the presence of two individuals who turned out to have low but detectable levels of antibody, presumably resulting from their earlier treatment with p-hGH.

A more detailed analysis of these antibodies using Scatchard analysis is given in Table 6. Here, the seven sera with the highest antibodies in the screening test were further analyzed. The affinity constants ranged from 1.3 to 8.5×10^9 M^{-1} with a mean of 5.6×10^9 M^{-1}. Binding capacity, in mU/liter, ranged from 30 to 280 with a mean of 110 mU/liter. In Table 6 these are compared with a single result, using the same antibody analysis on one child, R.K., who developed the antibodies many years previously to an early preparation of p-hGH.[2] The affinity constant in his case was 16.0×10^9 M^{-1}; the binding capacity was 13,400 mU/liter. R.K.'s affinity constant is of the same order as those of the met-hGH-treated patients (even if slightly higher), but his binding capacity is substantially greater. In his case the antibody completely blocked the effect of GH such that he has remained resistant to further GH treatment from any source. He was recently discharged from our clinic at the age of 20 years with a mature height of 139 cm.

3.3. Somatomedin

Table 7 gives the results for the mean somatomedin activity at each of the visits. At the time of the switch from p-hGH to met-hGH the somatomedin

Table 6 GH Antibodies: Scatchard Analysis of Sera of Seven Patients on Met-hGH Compared with Subject R.K.

	Affinity K_a (M^{-1})	Binding capacity (mU/liter)
Met-hGH	5.6 (range 1.3–8.5) × 10^9	110 (range 30–280)
p-hGH (R.K.)	16.0 × 10^9	13,400

Table 7 Somatomedin Activity

	Months			
	0	1	3	6
Somatomedin[a] (IU/ml)	0.9 ± 0.1	1.1 ± 0.2	1.1 ± 0.1	1.0 ± 0.1

[a] Porcine rib cartilage bioassay.

level was 0.9 ± 0.1 mU/liter, and this did not significantly change over the course of the 6 months of the study.

4. *Discussion*

It appears from these results that met-hGH produced a small effect on platelet and white blood cell count but that these changes were reversing well before 6 months despite continued use of the preparation. Whether these effects reflect some slight toxicity, perhaps due to *Escherichia coli* protein contamination, cannot be assessed at the present time. None of the changes were associated with any symptoms, and it seems likely that they are of little importance.

The fact that the somatomedin levels remained at the normal level for treated GHD children supports that supposition that the met-hGH was biologically active. If it had had reduced biological activity, for whatever reason, we would have expected that the somatomedin levels would have declined by 3 months and certainly by 6 months. This did not happen; so we must interpret this as indicating that at least up to 6 months of treatment met-hGH was fully biologically active in the human.

Unfortunately, there does seem to be clear evidence of the appearance of anti-p-hGH antibodies (which seem to cross-react with met-hGH). Although the mean levels for antibody binding were the same at 0, 1, and 3 months, so that on mean values the rise did not occur until after the third month, there were seven individuals who at 3 months did have suggestively elevated antibody binding titres. The Scatchard analysis results indicate that in these seven individuals, the affinity constants of the antibodies were comparable to those seen in a patient whose antibody was developed against some early preparations of p-hGH and whose growth response became totally inhibited. However, the binding capacities of the antibodies in the present study were much lower than those seen in this individual. The interpretation of these antibody results remains difficult; in the present study none were of a binding capacity at which we would expect to see an effect on growth. Equally, it is not clear exactly why the antibodies have appeared. There seem to be three possible reasons: first, the *N*-terminal methionine may itself be antigenic; second, the presence of the *N*-terminal methionine may disturb the tertiary structure of the met-hGH making some other part of this molecule antigenic; and third, *E. coli* protein contaminants may be acting as adjuvants increasing the antigenicity of the met-hGh molecule. Investigation of each of these possibilities and, one hopes, the prevention of the antibody formation is now an important task.

ACKNOWLEDGMENTS. The met-hGH used was the gift of Kabi Vitrum.

References

1. G. S. G. Spencer and A. M. Taylor, A rapid simplified bioassay for somatomedin, *J. Endocrinol.* **78:**83–88, 1978.
2. S. R. Chalkley and J. M. Tanner, Incidence and effects on growth of antibodies to human growth hormone, *Arch. Dis Childh.* **46:**160–166, 1971.

22

Clinical Experience with Biosynthetic Growth Hormone (III)

Juergen R. Bierich

1. Introduction

Between 1979 and 1981 the research laboratories of Kabi Vitrum AB in Stockholm developed a biosynthetic polypeptide that chemically corresponded to human growth hormone (hGH), apart from an additional methionyl group at the *N*-terminal end of the molecule. The toxicological division of Kabi Vitrum performed numerous tolerance tests in rats and dogs which showed no toxic side effects. Furthermore, a series of examinations in man was carried out which again showed a good compatibility. Accordingly, the Swedish as well as the German authorities released the preparation for clinical trial.

Two years ago we had published a report on a large series of pituitary dwarfs in Tubingen who for several years had been successfully treated by the extractive hGH preparation of Kabi Vitrum, Crescormon.[1] In 1981 we started to build up a prospective study with Somatonorm, the new biosynthetic preparation, and to compare the results with those obtained with Crescormon®. Our first step was to gather a large multicenter working group in order to warrant a broad basis for the trial and to collect as many fresh cases as possible. At present the study group consists of 14 investigators from 12 university clinics, 11 from West Germany, and one from Austria (Table 1).

2. Diagnostic Methods

Auxologically, the criteria for accepting a patient for the trial were body height 2.5 SD or more below average for the age group, according to Tanner,

Juergen R. Bierich • Department of Pediatrics, University of Tubingen, 74 Tubingen, Federal Republic of Germany.

Table 1 Participants in the Study Group

J. R. Bierich and M. B. Ranke, Tübingen
P. Beyer, Berlin
P. Bittner, Giessen
M. Borkenstein, Graz
J. Brämswig, Münster
O. Butenandt, München
D. Schönberg and U. Heinrich, Heidelberg
W. Sippell, Kiel
N. Stahnke, Hamburg
H. Stolecke, Essen
P. Stubbe, Göttingen
R. Ziegler, Heidelberg

and growth velocity 2 SD or more below average, or a velocity below 4 cm/year at age 6–12. Only prepubertal children with bone ages of 11 years in girls and 12 years in boys were accepted. All bone ages were estimated by Dr. Stöver from Frankfurt according to Greulich and Pyle and to Tanner-Whitehouse 2.

Endocrinologically, all functions of the adenohypophysis were examined in a standardized manner. In every case two provocation tests for hGH were performed, the arginine test and the insulin tolerance test, using 0.1 IU/kg. The hGH peaks after arginine and insulin did not exceed 10 ng/ml. Depending on peak height, two groups of responders were defined, the first with peaks from 0 to 5, the second from 5 to 10 ng/ml. Moreover, somatomedin A and somatomedin activity in the plasma were determined, basically as well as after 2 days' administration of 4 IU hGH.

The *compatibility* of Somatonorm was evaluated by measuring the antibody formation against the preparation as a whole as against a special *Escherichia coli* polypeptide (ECP), which was extracted from *E. coli* strain K12 used for the biosynthesis. Furthermore, intracutaneous skin tests with Somatonorm® in three dilutions were carried out in all cases.

3. Therapy

Therapeutically we administered a standard dose of 12 IU GH per m^2 per week, divided into three intramuscular injections. This was the same dose as we had employed in our former study with extractive hGH. If secondary hypothyroidism was present, replacement therapy with thyroxine was given. Control investigations were performed 6 weeks after the start of treatment, and then every 3 months. The whole protocol was submitted to and accepted by the ethical committees of the participating clinics. Informed consent from the parents was obtained in every case.

Table 2 Multicenter Study of Somatonorm

Patients	
Registered	31
Male	19
Female	12
Cause	
Midline defect	2
Congenital rubella	1
Breech presentation	10
Deflexion presentation	1
Additional birth complications	3
Unknown	14
Age	8.2 ± 5.0 years
Bone age	4.5 ± 3.6 years
Height	−3.3 ± 0.9 SDCA
Mid parents' height	0.2 ± 1.0 SD
Height for MPH	−3.4 ± 0.8 SDCA

4. Basal Data

At present our study consists of 41 patients. Two groups are differentiated. The first 31 children have been treated with the first hormone preparation, Somatonorm I, which was used until April 1983. After that, treatment was continued with Somatonorm II, a much more purified preparation. The second group of patients consisting of 10 children received Somatonorm II from the beginning. The basic data of the first group of children are listed in Table 2. There were 19 boys and 12 girls, all of whom had never been treated before. In order to make the material as homogenous as possible, patients with genetic forms of hypopituitarism and with craniopharyngiomas had been excluded from the study.

The following causative factors of the disorder in the 31 children were ascertained:

- In two patients midline defects were found.
- One child had had intrauterine rubella with typical clinical symptoms. There have been several previous observations of this type in the literature.
- In 14 children an abnormal birth history was reported. Ten infants were born in breech or foot presentation, and one baby in a deflexion presentation. Twenty years ago I drew attention to the important role of pelvic presentations for the pathogenesis of pituitary dwarfism. Three additional infants had been asphyxiated and cyanotic at birth without being born in pelvic presentation.
- In the remaining 14 cases no special pathogenic factors could be found.

The chronological age of the patients at admission ranged from 2 to 18 years, with a mean of 8.2 ± 5 years. Mean height deficit, expressed with the

SDS, amounted to −3.3 ± 0.9 SD. Bone age was on the average 4.5 ± 3.6 years.

Endocrinologically, the provocation tests resulted in GH peaks from 0 to 5 ng/ml in 27 patients, from 5 to 10 ng in four. Eleven patients showed either initially, or in the course of the treatment, secondary hypothyroidism which had to be treated with thyroxine. Hints of a gonadotropic deficiency were noted in 10 cases. No patient had corticotropic deficiency.

The intracutaneous scratch tests with Somatonorm® were all negative, as well as the serum antibodies against the preparation.

5. *Results of Treatment*

In this section I shall report on the results in the 26 children who entered long-term treatment with Somatonorm I. The data have already been presented

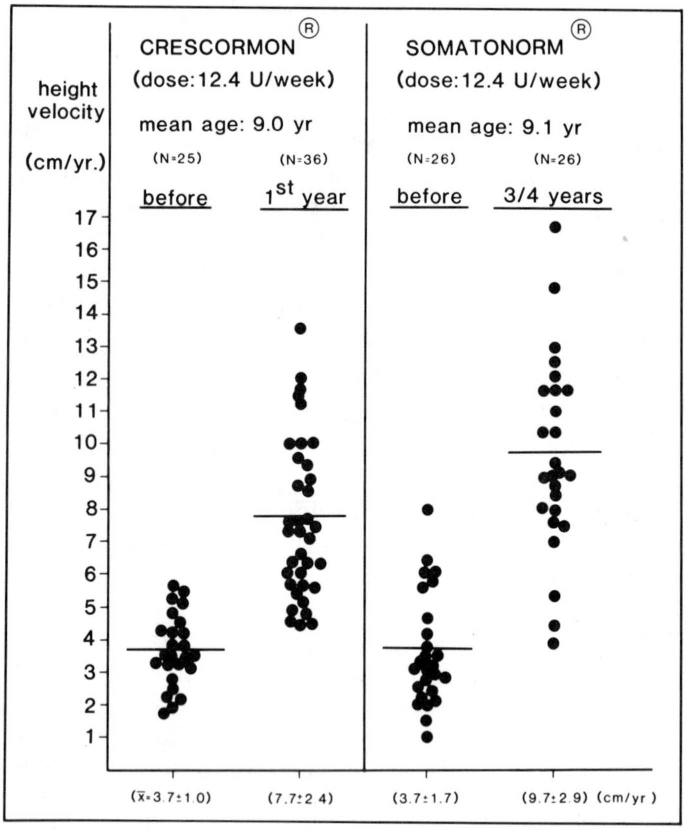

Figure 1 Height velocity of pituitary dwarfs prior to and during 9 months of treatment; (left panel) with extractive human growth hormone (Crescormon), (right panel) with a biosynthetic preparation (Somatonorm).

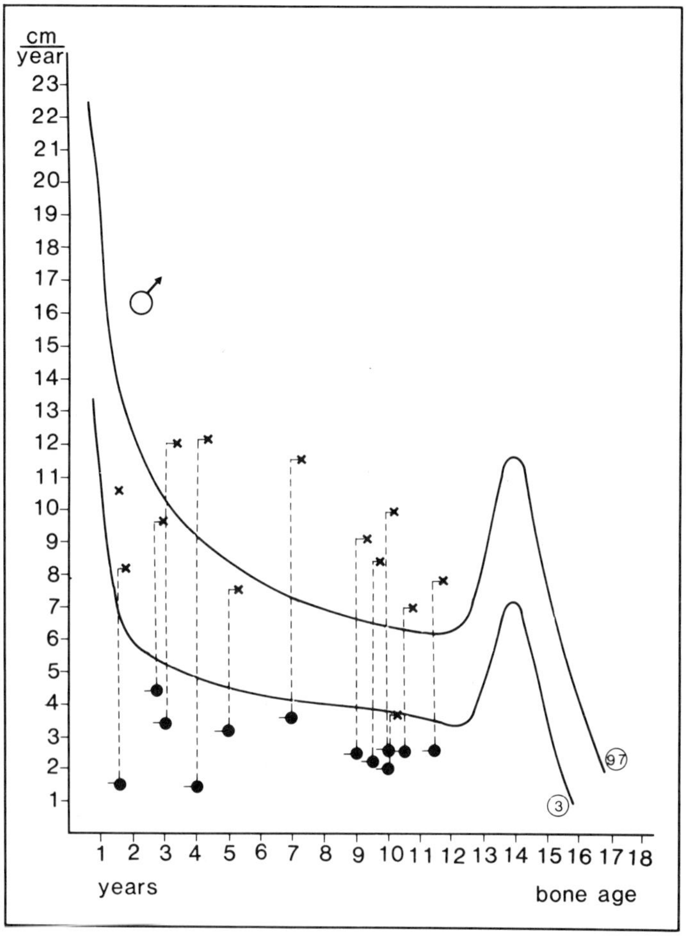

Figure 2 Height velocity of 13 male pituitary dwarfs prior to (●) and under (x) Somatonorm therapy, related to bone age.

at the Acta Endocrinologica Congress in Stockholm and comprise on the average a treatment of 8 months. The right panel of Fig. 1 shows the results of Somatonorm® therapy; for comparison, the left panel shows results with Crescormon therapy, which we published 4 years ago. The exit values for growth velocity prior to treatment, were 3.8 cm/year. With Somatonorm treatment growth velocity rose to 9.7 cm; with Crescormon treatment, to 7.7 cm/year.

Figures 2 and 3 show the growth rates of the patients in relation to their development, expressed as bone age on the abscissa—Fig. 2. for the boys, Fig. 3 for the girls. The enormous height gain is evident. If the values of all cases are compiled, we arrive at a mean SDS of + 2.7 SD in contrast to a mean of + 0.7 SD in the Crescormon series. An additional comparison is possible with

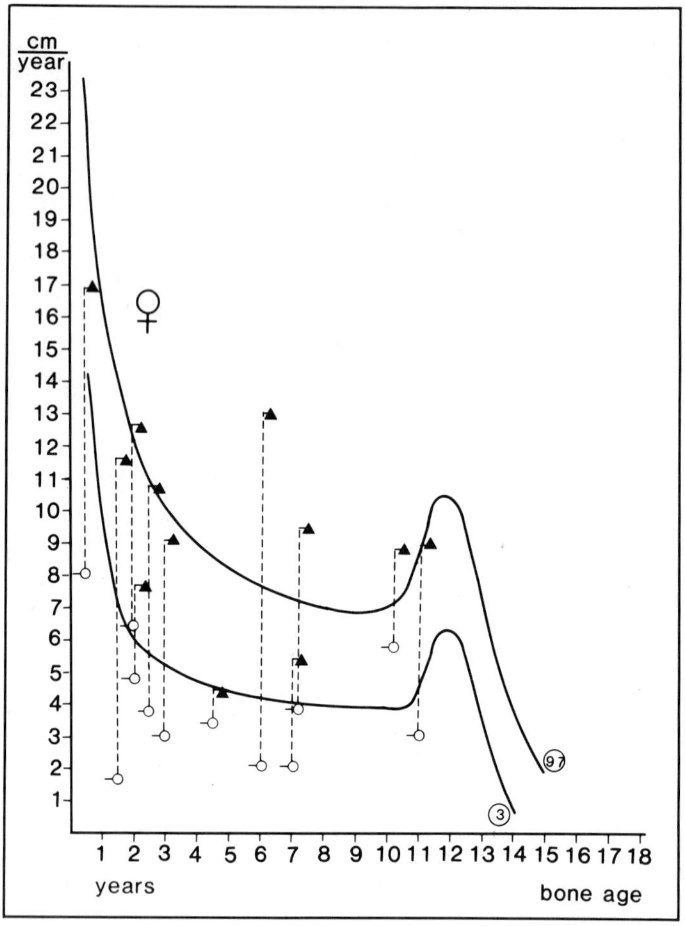

Figure 3 Height velocity of female pituitary dwarfs prior to (○) and under (▲) Somatonorm®.

the values of the Children's Hospital of Zurich, which were published 1 year ago, employing the same preparation and an identical dosage. The authors arrived at a mean velocity of 2.1 SD for the first half-year. In the meantime 15 patients of the group have gone beyond their first year of treatment. So we are able to report precise statements for the growth within a full year. In Fig. 4 the results of Somatonorm treatment are again compared with those obtained with Crescormon.

Although one could expect that the first intensive growth, the catchup growth was now over, the average growth velocity was not much slower than that shown in Fig. 1; the mean velocity was now 9.5 cm/year, instead of 9.7 cm. Figure 5 demonstrates the course of the growth velocity curves of patients with long-term treatment. Catchup growth starts early. From the third or fourth 3-months period on, growth velocity gradually decelerates.

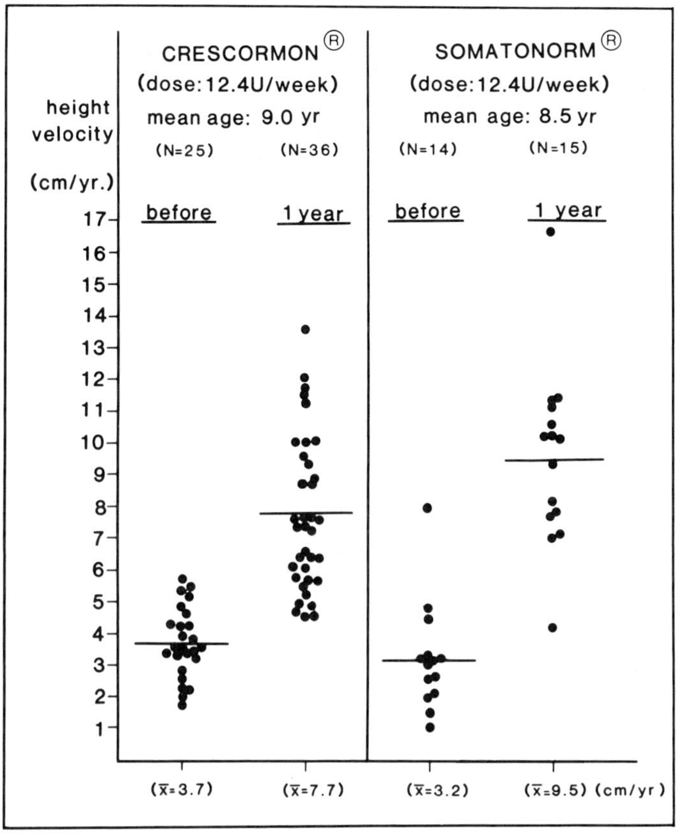

Figure 4 Height velocity of pituitary dwarfs prior to and during 1 year of therapy; see Fig. 1.

In conclusion, we can say that the effects of Somatonorm on growth are no worse but rather better than the previous results with extractive hGH.

Another useful parameter for the effectiveness of the preparation is somatomedin. Figure 6 presents the dynamics of somatomedin activity. Prior to therapy we obtained a mean of 0.27 U/ml, which is very low and rather characteristic for pituitary dwarfs. After 2 days' administration of 4 IU Somatonorm I, the levels reached double the basal values. Later, a supranormal level, namely 1.8 U/ml, was attained.

Figure 7 demonstrates a third parameter for the evaluation of the efficacy of the hormone preparation, the changes in alkaline serum phosphatase. Six weeks after start of treatment the initially very low enzyme levels had doubled and remained so during the observation period.

6. *Antibody Formation*

For evaluation of the preparation used, compatibility is of equal importance to efficacy, in particular in relation to antibody formation. As previous inves-

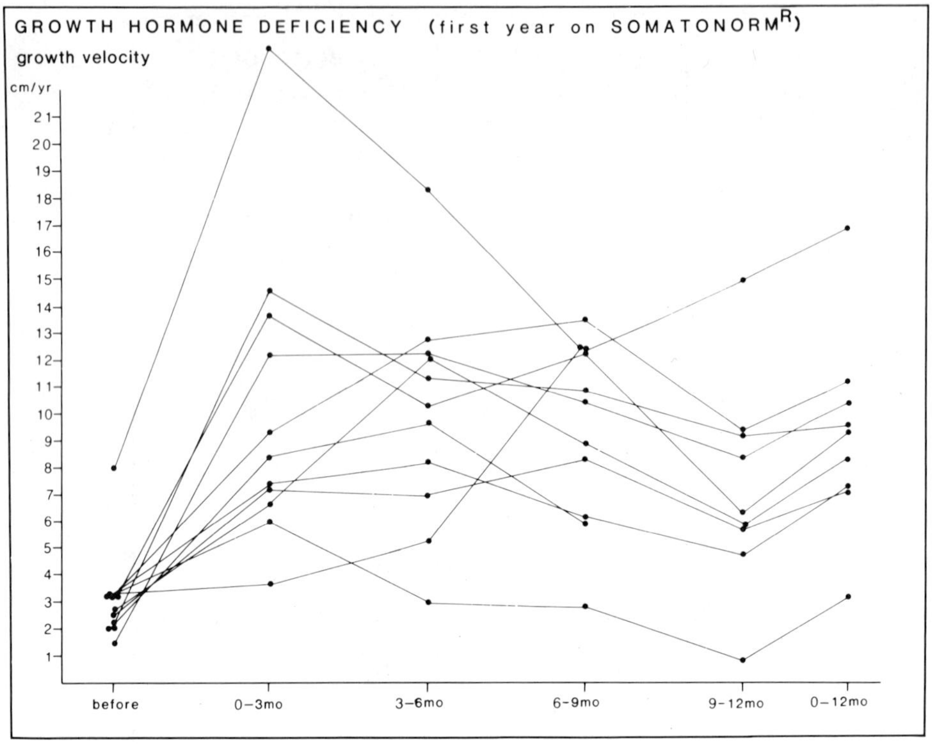

Figure 5 Eleven growth velocity curves under treatment with Somatonorm.

tigations have shown, antibodies against extractive hGH did not occur too often, were usually harmless, but were sometimes harmful, especially when they possessed a high binding capacity.[2–8]

In the present study all antibody determinations were performed in the laboratory of Dr. Karlén at the Kabi Vitrum Research Department. For screening, precipitating antibodies were measured by radioimmunoassay. If bound tracer over blank exceeded 2, then antibody proof was termed positive; titration was carried out, and binding capacity and affinity constants were determined. In addition, specific antibodies against ECP were measured by means of an ELISA immunoassay. Furthermore, intradermal scratch tests were carried out. The first hint of the presence of antibodies was obtained in two children who exhibited allergic skin reactions. After the fourth intramuscular injection of Somatonorm I given on the 12th day of treatment, painful skin erythemas appeared at the site of the injections, resembling Arthus reactions in type III allergy. This was confirmed by positive intradermal tests with Somatonorm whereas similar tests with extractive hGH remained negative. It is noteworthy that the two children were atopic. Both were later successfully treated with extractive hGH, which was well tolerated.

The course of the antibody titers against Somatonorm is depicted in Fig.

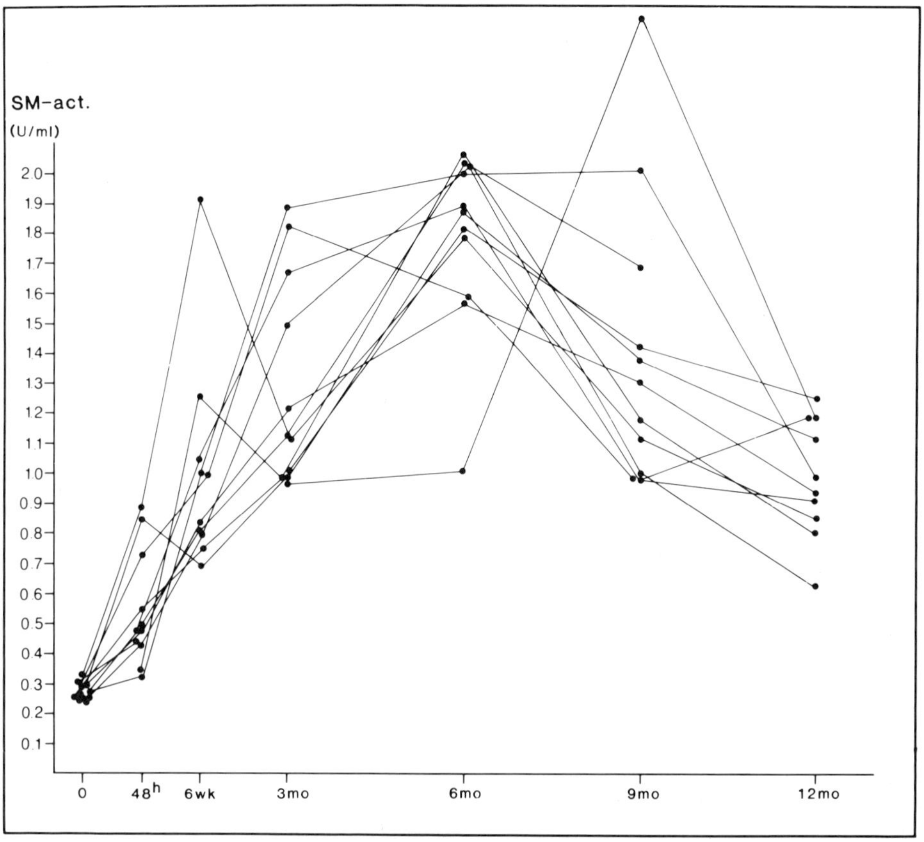

Figure 6 Somatomedin activity curves of the 11 patients of Fig. 5.

8. The abscissa represents the time; determinations of antibodies were carried out every 3 months. The ordinate signifies the antibody titer in logarithmic form; the ciphers denote the logarithm (base 10) of the final serum dilution. After 3 months a mean antibody titer of 2.4 was reached; after 6 months, 3.0; after 9 months, 3.4. The increment from 3 to 9 months is statistically significant. Five children had titers between 4 and 5, corresponding to serum dilutions of 1:10,000 to 1:100,000.

As already mentioned, high antibody titers have been found particularly in patients treated by hGH preparations with ample amounts of polymers as present in the older Raben preparations. Crescormon Kabi, which we have used for many years, is extracted according to Roos. During the last year 150 serum samples from Crescormon-treated children have been collected and investigated for antibodies by Dr. Karlén in Stockholm. Antibodies against hGH were not detectable in these samples, a fact that speaks, on one hand, in favor of the purity of Crescormon and, on the other, for the principal difference of the synthetic preparation.

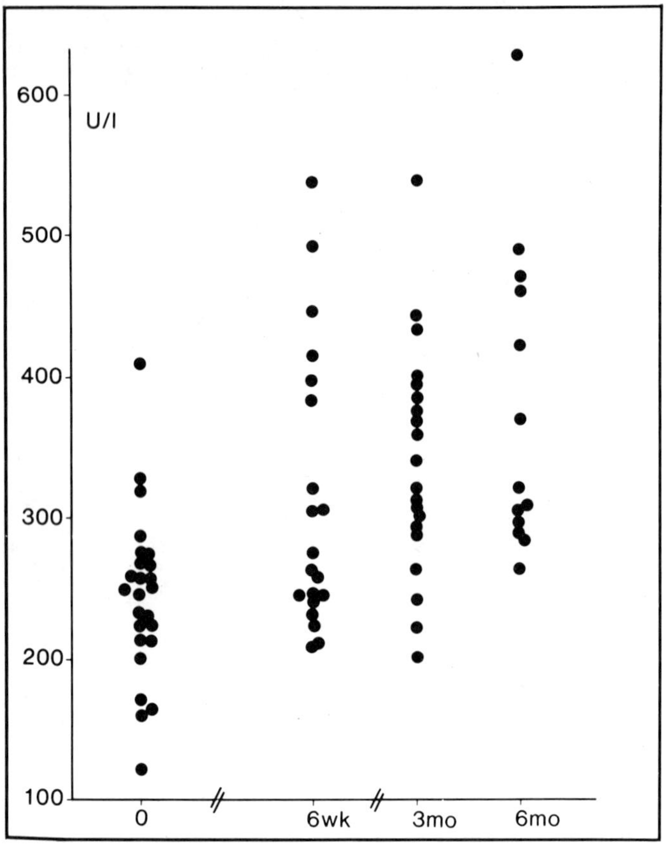

Figure 7 Alkaline phosphatase under treatment with Somatonorm.

Whether antibodies induce inactivation of the hormone against which they have been formed depends not only on their titer but also on their binding capacity. High binding capacity causes loss of efficiency, as a rule. Five of our twenty-nine patients had antibodies with rather high binding capacities, binding more than 4000 μ g = 4 mg/liter. All these children also had high antibody titers, the log titers ranging between 3.8 and 4.7.

Considering the clinical significance of these findings, we arrive at the surprising conclusion that it is rather slight and almost negligible. As I have shown, the growth rates of our children were absolutely satisfactory. There is, however, one exception. One of our patients from Tubingen, a 14-year-old boy with known atopia, exhibited a rather slow growth velocity—as a matter of fact, the slowest growth rate of all patients, i.e., 3.1 cm/year. This boy, the patient with the highest antibody production, attained a log titer of 4.7 and a binding capacity of 6 mg/liter. When we transferred the boy to extractive hGH, he immediately started to grow.

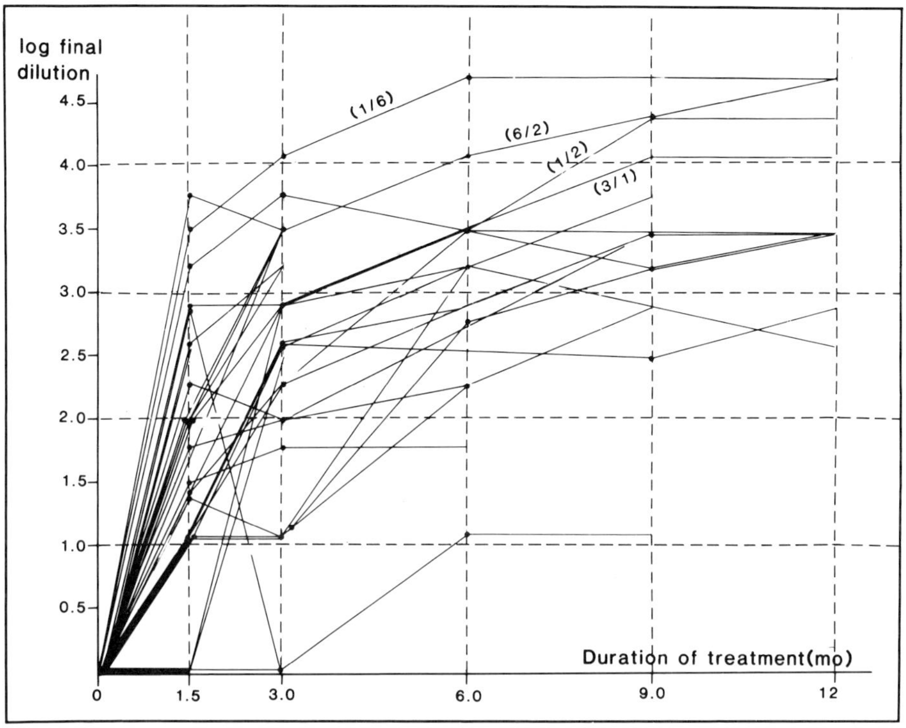

Figure 8 Anti-hGH antibodies of children under treatment with Somatonorm I.

In Fig. 9 the growth rate of all patients (on the ordinate) is plotted against the log titers of the hGH antibodies on the abscissa. No correlation between the two parameters was found, except for the boy just mentioned who is represented in the lower right corner.

A second important investigation concerned the antibodies formed against the ECP in the medium of the hormone preparation, which was determined by an ELISA immunoassay. In the course of treatment increasing titers of anti-ECP antibodies were measured in all patients. Figure 10 demonstrates the simultaneous changes of the log titers of the hGH antibodies and the titers of the ECP antibodies. The correlation is evident. The formation of both types of antibodies is progressing in the same way, and one gains the impression that both processes are connected not only as to time but also causally.

7. *Highly Purified Methionyl-hGH*

In view of the allergic reactions and the antibody formation against the preparation as a whole and against ECP, Kabi Vitrum decided to radically change the extraction procedure. Dr. Linda Fryklund, who did the main work,

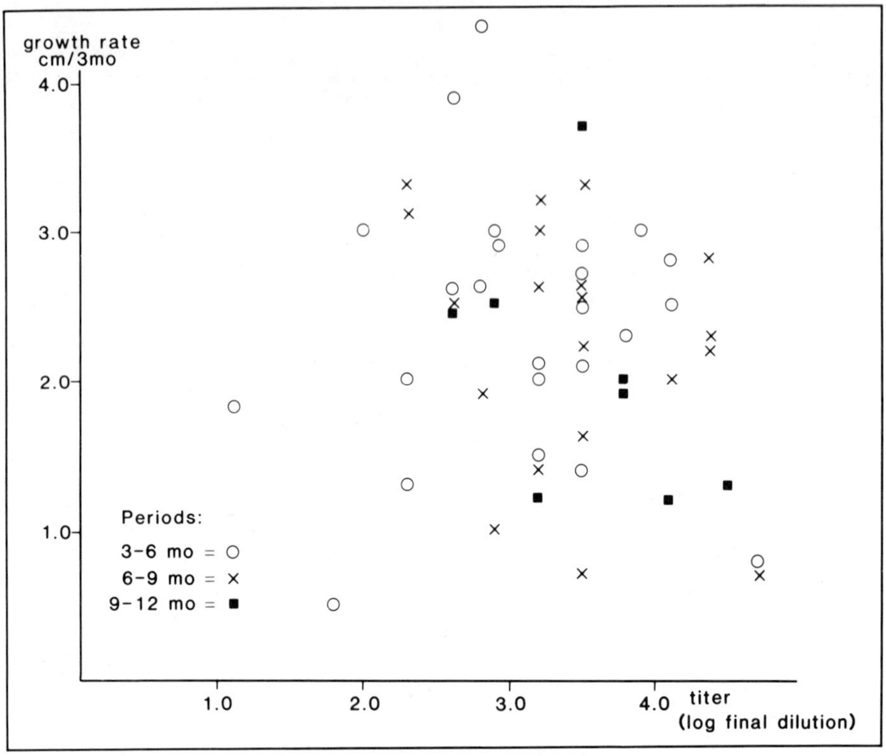

Figure 9 Growth velocity in cm/3 month, plotted against log titer of anti-human growth hormone antibodies.

reports on the methods used for this purpose. The new preparation, Somatonorm II, exhibits a degree of purity that is approximately 100-fold, compared with Somatonorm I. As already mentioned, from May 1983 on, the old preparation was withdrawn, and all patients were transferred to the new preparation. In addition, 10 new patients were admitted to the study and are being treated exclusively with Somatonorm.

In seven of these children the duration of treatment now exceeds 3 months. The time is too short for any auxological evaluation. The immunological investigations are, however, very promising and show satisfactory results. Allergic Arthus reaction did not occur. HGH antibodies are either not detectable or rather low. ECP antibodies have not been found at all, thus far.

Up to now, it has been assumed that radical purification of the hormone has led to the removal of the bulk of bacterial antigens. Certainly this step has been necessary, although the old preparation also exhibited good auxological effects, even though the purification led to a considerable diminution of extractable GH.

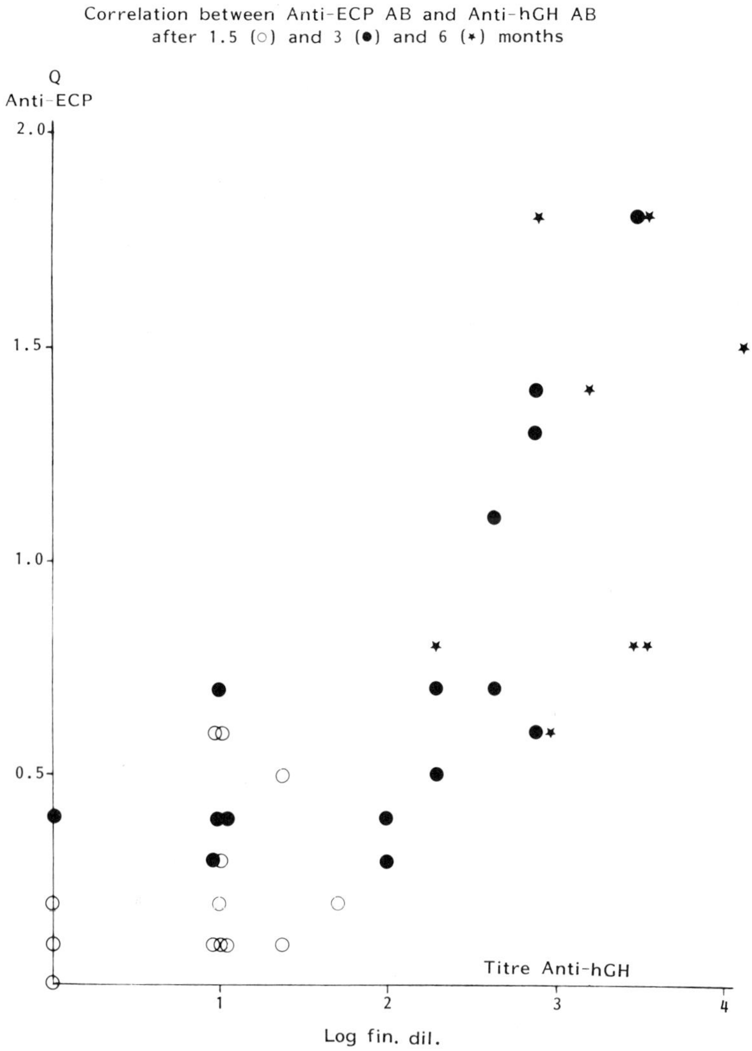

Figure 10 Log titer anti-hGH antibodies, plotted against titer of anti-*E. coli* polypeptide antibodies.

References

1. M. B. Ranke, J. Weber, and J. R. Bierich, Long-term response to human growth hormone in 36 children with idiopathic growth hormone deficiency, *Eur. J. Pediatr.* **132:**221–238, 1979.
2. R. Illig, A. Prader, H. Ferrandez, and M. Zachmann, Familiärer pränataler Wachstumshormonmangel mit erhöhter Bereitschaft zur Bildung von Wachstumshormon-Antikörpern, in: *Endokrinologie der Entwicklung und Reifung. 16. Symposium Dtsch. Ges. Endokrinologie,* (J. Kracht, ed.), Springer-Verlag, Berlin-Heidelberg-New York, 1970, pp. 246–247.
3. S. R. Chalkley and J. M. Tanner, The incidence and effects on growth of antibodies to human growth hormone, *Arch. Dis. Childh.* **46:**160, 1971.

4. K. Hall and P. Olin, Sulphation factor activity and growth rate during longterm treatment of patients with pituitary dwarfism with human growth hormone, *Acta Endocrinol.* (*Kbh*) **69:**417, 1972.
5. S. L. Kaplan, D. C. L. Savage, S. Suter, R. Wolter, and M. M. Grumbach, in: S. Raiti (Ed.), *Advances in Human Growth Hormone Research* (S. Raiti, ed.), DHEW Publ. No (NIH) 74–162, 1973, pp. 725–756.
6. L. E. Underwood, S. J. Voina, and J. J. Van Wyk, Restoration of growth by human growth hormone (Roos) in hypopituitary dwarfs immunized by other growth hormone preparations, *J. Clin. Endocrinol. Metab.* **38:**288, 1974.
7. M. V. Moore, The role of aggregated hGH in the therapy of hGH-deficient children, *J. Clin. Endocrinol. Metab.* **46:**20, 1978.
8. F. Purrello, R. Vigneri, A. Belfiore, V. Pezzino, S. Squatrito, and P. Polosa, High incidence of anti-GH antibodies in subjects treated with the GH clinical preparation available in Italy, *J. Endocrinol. Invest.* (*Ital.*) **3:**313, 1980.

IV

CONTROL OF GROWTH HORMONE SECRETION, GROWTH HORMONE-RELEASING FACTOR, SOMATOSTATIN

23

Regulation of Growth Hormone Secretion

Joseph B. Martin

1. Introduction

Human growth hormone (hGH) is a 191-amino-acid, single-chain polypeptide containing two intramolecular disulfide bonds which has a molecular weight of 21.5K daltons.[1,2] Growth hormone (GH) is the most abundant hormone in the anterior pituitary. Approximately 15% of total pituitary hGH is represented by a smaller 20K-dalton molecule with a 15-amino-acid deletion between residues 32 and 46. Although this variant has been shown to possess growth-promoting activity and lacks insulinlike activity, it is uncertain whether it is a physiologically important secretory product of the somatotrope.

GH circulates unbound in the plasma. Its half-life of disappearance is between 20 and 45 min, and its estimated rate of secretion is approximately 17 μg/hr or 400 μg/day.[2] GH does not act on a specific target organ, but exerts effects at many different sites. It has an insulinlike effect on glucose metabolism, a powerful anabolic action promoting enhanced incorporation of amino acids into protein, and lipolytic effects.[3] The biological actions of GH are thought to be mediated primarily by the generation in the liver and other tissues of a class of lower-molecular-weight substances, the *somatomedins*. Of these, somatomedin C (also called insulin growth factor, IGF/I) and IGF/II are the most important. Somatomedins stimulate sulfate incorporation into cartilage ("sulfation factor") and thymidine uptake in liver and also have insulinlike effects.[4] The latter, because of their failure to suppress after glucose administration, are sometimes referred to as nonsuppressible insulinlike activity.

Joseph B. Martin • Neuroendocrinology Laboratory, Department of Neurology, Massachusetts General Hospital, Harvard Medical School, Boston, Massachusetts 02114.

2. Patterns of GH Secretion

Basal levels of plasma hGH in unstressed adult humans are undetectable most of the day. Sequential determinations at frequent intervals throughout the day and night show spontaneous fluctuations in hGH levels at least until age 50–60 years. Surges of secretion are particularly prominent during the pubertal growth period and may reach plasma levels of 20–60 ng/ml; the largest peaks occur during the first 2 hr of nocturnal sleep.[5] These surges in adolescent human males, in the baboon[6] and the rhesus monkey,[7] and in lower species including dogs, rabbits, and rats[8] all share a common frequency of 3–5 hr between individual pulses.

A number of studies have been undertaken to determine the basis of these spontaneous variations in GH concentration. Factors such as sleep, exercise, and stress may stimulate GH secretion, but do not account for most of the surges. Changes in plasma GH levels are not caused by or related to variations in plasma concentrations of the common nutrients, glucose, amino acids, or free fatty acids. The amplitude, frequency, and entrainment of the GH secretory episodes and their independence of metabolic or homeostatic mechanisms indicate that they are generated by the hypothalamus; pituitaries transplanted to the renal capsule and anterior pituitaries incubated *in vitro* do not secrete GH in an episodic manner.[9]

Experimental observations support the concept that surges of GH secretion result from neural influences, mediated by excitatory and inhibitory regions of the hypothalamus. These hypothalamic regions are in turn regulated by neural connections from the limbic system and the brain stem; thus, many neurotransmitters and neuropeptides are involved in the mediation of GH secretion.

3. Hypothalamic Regulation of GH Secretion

The signals that reach the anterior pituitary to determine minute-to-minute GH secretion are derived from hypothalamic peptidergic neurons that secrete somatostatin (SS) or GH-releasing factor (GRF or somatocrinin). These neurons are localized in particular anatomical regions of the hypothalamus, and certain intrahypothalamic connections are important for maintaining normal secretion.

Experimental studies in rats, using electrical stimulation techniques and destructive focal lesions, demonstrated two essential functional regions of the hypothalamus.[8] The first region, the medial preoptic area (MPOA), contains SS neurons. The second region, comprised of the ventromedial nucleus (VMN) and the arcuate nucleus (ARC), has an excitatory role in GH secretion and has recently been shown to be the location of GRF (see below). The lateral hypothalamus, which contains the ascending and descending fibers of the medial forebrain bundle and its intercalated cells, represents a link between the medial hypothalamus (both the MPOA and the VMN–ARC) and other regions of the central nervous system, particularly the limbic system and the brain stem.

3.1. Regions of the Hypothalamus Important for GH Regulation

3.1.1. Preoptic Area

The preoptic region is divided into medial and lateral regions. Anatomical studies show that the medial preoptic nucleus has neural connections with many other hypothalamic nuclei and with the median eminence (ME).[10,11] Direct connections have been shown between the medial preoptic nucleus and the ARC and VMN nuclei.[12,13] The periventricular region of the medial preoptic area has similar hypothalamic connections.[10,11] Some neurons positioned in this region contain SS and send axonal projections to the ME. A few dopaminergic cells can be found in the periventricular preoptic nucleus. These cells are believed to innervate the preoptic region.[14]

3.1.2. Periventricular Nucleus

This thin nucleus is located on either side of the full extent of the 3rd ventricle. It consists of several layers of nerve cells enmeshed in a delicate fiber lattice of extreme density. Some of the cells contain dopamine.[14] More important, in terms of GH regulation, is the finding that they contain a heavy concentration of SS-positive cell bodies. Lesions of this nucleus result in terminal axonal degeneration in the ARC and in both the external and internal layers of the ME[15] and lead to considerable depletion of SS concentration in the ME.[16]

3.1.3. Suprachiasmatic Nucleus (SCN)

This nucleus is important in the generation of circadian rhythms. The regulatory role of SCN in the light–dark entrainment of the GH rhythm in the rat has been demonstrated.[17] The SCN is innervated by serotonergic afferents. Its cells contain several neuropeptides, including vasoactive intestinal polypeptide and vasopressin. The efferent connections of the SCN remain to be defined.

3.1.4. Ventromedial Nucleus (VMN)

The VMN consists of six subdivisions and is surrounded by a uniform capsule formed by the dendrites of the VMN neurons as well as the nerve terminals of numerous afferent pathways.[15,18] A few fibers of the VMN appear to go to the ME. In most species, a few cells of this nucleus stain with antibodies to GHRF (see below).

3.1.5. Arcuate (Infundibular) Nucleus

The main projection field of the ARC is the ME.[19] The ARC also contains dopaminergic,[20 cholinergic,[21] and endorphinergic cells that are important for pituitary regulation. This nucleus is the primary site of these transmitters. Neurons in the ARC that project to the ME contain GRF but not SS.

3.2. The Structure of the ME

The ME contains the anatomical site of the neurovascular anastamosis which is essential for hypothalamic regulation of hormone secretion by the anterior pituitary. The ME contains (1) nerve terminals that end on the capillaries of the pituitary portal systems, (2) axons of the hypothalamoneurohypophyseal system in transit to the infundibulum and neurohypophysis, and (3) specialized secretory ependymal cells (tanycytes) of unknown function. There are no nerve cell bodies in the ME. The capillaries of the ME are fenestrated so that blood-borne hormones and other substances penetrate freely into the extracellular space of the ME and might thereby exert feedback effects. Axons that enter the ME arise principally from neurons of the medial basal hypothalamus but also from several extrahypothalamic regions. Neurochemical micromethods have revealed that dozens of neurotransmitters, peptides, enzymes, and hormones are localized in axons of the ME. Many of these substances influence GH secretion (see below).

3.3. Hypothalamic Structure–Function Relationships in GH Regulation

Lesions of the ME and mediobasal hypothalamus suppress hypoglycemia- and stress-induced GH release in primates, including man.[2] Pulsatile GH secretion in the rat is abolished by bilateral lesions of the hypothalamic VMN and ARC nuclei and persists, in somewhat altered form, in rats with hypothalamic islands.[8] In contrast, small, discrete lesions of the hypothalamic periventricular SS-containing region result in a sustained increase in plasma GH levels in rats and under certain conditions cause an increase in growth rates.[8] These effects appear to result from ablation of somatostatinergic pathways that originate in the periventricular nuclei and project to the ME, since SS levels in the ME are reduced by more than 80%[16,22] by such lesions. The importance of this area in inhibiting GH secretion is also shown by electrical stimulation studies in anesthetized rats which result in decreased pituitary GH secretion and increased SS concentration in the portal blood.[23]

Hypothalamic centers for stimulation of GH release appear to be confined to the VMN–ARC region. Electrical stimulation of these nuclei elicits GH secretion in rabbits, rats, cats, and sheep.[23] Stimulation of the lateral or anterior hypothalamus, the supraoptic or paraventricular nuclei, or the mammillary bodies has no effect on plasma GH levels.[23]

Although the frequency, pattern, and light–dark entrainment of the GH secretory rhythm are altered in rats with complete hypothalamic deafferentation, periodic rises of plasma GH persist.[24] These data suggest that the neural mechanism for episodic secretion of GRF resides within the mediobasal hypothalamus, and that this region of the hypothalamus does not require extra hypothalamic inputs to activate secretion.

Clinical studies support the importance of particular hypothalamic areas in GH regulation. Basal hypothalamic destruction and pituitary stalk section result in reduced plasma GH levels in man and in suppression of responses to insulin-induced hypoglycemia.[2] Hypothalamic lesions inhibit sleep-associated GH release.[25] Occasionally patients with anterior hypothalamic–preoptic area tumors show increased GH secretion. Taken as a whole, these data indicate that specific neural systems within the hypothalamus are responsible for the regulation of rhythmical GH secretion. Thus there appears to be a dual regulatory system for generation of GH bursts (the VMN–ARC nuclei) and ebbs (the periventricular SS-containing nuclei) with entrainment to the light–dark cycle mediated through retinosuprachiasmaticohypothalamic connections.[8]

4. Hypothalamic Peptides and GH Secretion

4.1. GRF

Although the precise structure of *hypothalamic* human GRF has still not been elucidated, recent studies of GRF peptides isolated from human pancreatic tumors in patients with acromegaly have revealed the structure of a specific, potent 44-amino-acid peptide.[26,27] This is the first example of the structure of a hypophysiotropic hormone being determined by isolation of a peptide responsible for a paraneoplastic syndrome.

The active peptide was given the name human pancreas growth hormone-releasing factor (hpGRF). Guillemin proposes to replace the acronym GRF by the name *somatocrinin* from the Greek *somato*, abbreviated from somatotropin, a trivial name for GH, and *crinin*, meaning to secrete.[26] Somatocrinin would thus be the terminological counterpart of somatostatin.

A comparison of the properties of tumor-derived GRF with purified hypothalamic GRF has shown identical biological activity including parallel GH-release dose–response curves and lack of effect on other anterior pituitary hormones. Recent studies have confirmed that the amino acid constituents of human hypothalamic GRF are identical to those of tumor GRF. Hypothalamic GRF and hpGRF have similar molecular size in several chromatographic separation systems and identical interaction with SS. Additional evidence that hpGRF is similar to or identical with hypothalamic GRF is the finding that antibodies to hpGRF are cross-reactive with a specific population of hypothalamic nerve cells. In rats, monkeys, and human brains these cells are located predominantly in the ARC nuclei with scattered cells elsewhere in the hypothalamus including the VMN (ref. 28 and J. B. Martin and P. J. Marshall, unpublished findings). Fibers in the ME stain for GRF also (Fig. 1).

It is known that brain–gut peptides are commonly grouped in "families" with homologous amino acid sequences that reveal evolutionary relationships. Comparison of the structure of hpGRF with other gut peptides reveals a close homology with the glucagon–secretin family, most strikingly with the recently described peptide designated PHI-27 isolated by Tatemoto and Mutt[29] (Fig. 2). These two molecules have 12 identical amino acids in an equivalent position

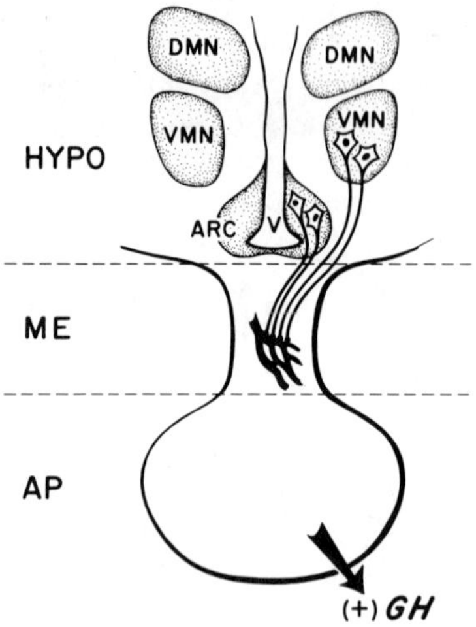

Figure 1 Hypothalamic regions important for facilitation of GH release. Neurons containing GHRF immunoreactivity are found in large numbers in the arcuate (ARC) nucleus in human and animal brains. A few cells are also found in the ventromedial (VMN) nucleus. Terminations in the median eminence (ME) are prominent. AP = anterior pituitary; DMN = dorsomedial nucleus; HYPO = hypothalamus. Reproduced with permission from Martin.[120]

and also possess 12 other residues that could represent mutations in a single base. Strong homologies with VIP, glucagon, secretin, and gastric inhibitory peptide can also be demonstrated. Each of the peptides that appear to be related to hpGRF possess intrinsic GRF activity in one or another physiological test system, but none has a potency remotely similar to that of hpGRF, and none is effective in pituitary cell cultures. Rat GRF contains 43 amino acids and has 13 amino acid substitutions when compared to hpGRF.[30]

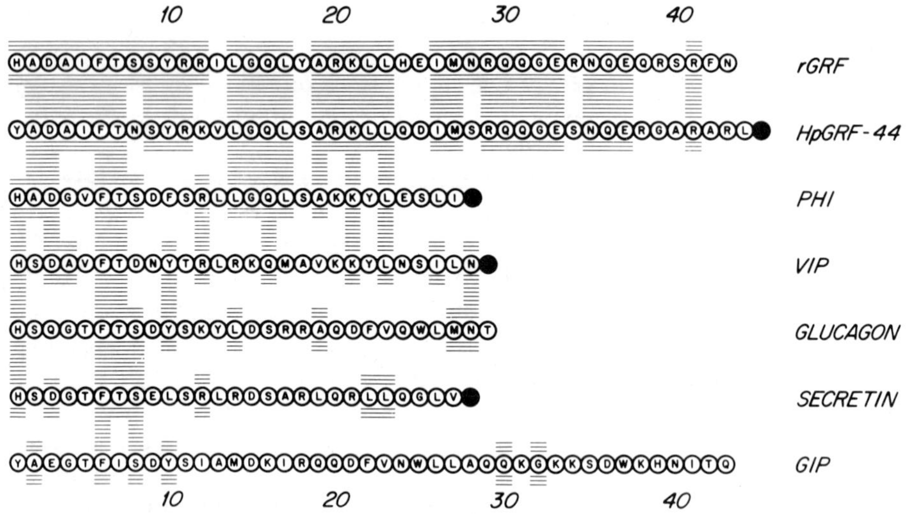

Figure 2 Comparison of the structures of peptides comprising the VIP–glucagon–GHRF family. PHI = peptide with histidine as *N*-terminus and isoleucine amide as *C*-terminus; VIP = vasoactive intestinal polypeptide; GIP = gastric inhibitory peptide. Modified from Fig. 4 in Spiess *et al.*[30]

Synthetic GRF has been used to elucidate a number of the aspects of the hypothalamic control of GH secretion, and its discovery has already explained several phenomena recognized for many years. One of the most interesting insights has been the explanation of why it was impossible to obtain enough material from hypothalamic extracts to identify the nature of GRF. The concentrations in hypothalamus are several orders of magnitude less than that present in the tumors. hpGRF is exceedingly potent and is probably the most potent on a molar basis of any of the hypophysiotropic factors.

Actions of GRF

Synthetic GHRF binds to membrane receptors and stimulates the accumulation of 3′,5′-cAMP in rat pituitary cells.[31] GRF effects are inhibited by SS, but GRF does not interfere with binding of SS. Hence, it can be concluded that they interact on different receptors. Thus far, based on studies in man, GRF appears to have extreme specificity of action. It does not release other anterior pituitary hormones. About 20–30% of human dwarfs with GH deficiency respond to a single injection of GRF with release of GH.[32]

Antibodies against rat hypothalamic GRF suppress spontaneous bursts of GH secretion in the rat, providing strong evidence that the GH surges are brought about by bursts of GRF secretion.[33]

4.2. SS

SS, a tetradecapeptide, was first isolated from sheep hypothalamic extracts by its demonstrated ability to inhibit GH secretion.[34] SS inhibits GH secretion in all species tested and also inhibits basal thyrotropin (TSH) secretion. It blocks TSH-releasing hormone (TRH) stimulation of TSH but not prolactin.[35] It suppresses plasma GH levels in acromegaly and in diabetes. SS has a rapid onset of effect and a short duration of action. GH levels rapidly rebound after cessation of SS administration, making it difficult to use in the treatment of elevated GH levels.

The postinhibitory GH rebound caused by SS is not an adequate explantion for episodic GH secretion however. First, if this was the case, ablation or interruption of SS systems should abolish rhythmical release and cause persistently high GH levels. This does not occur: although altered in pattern, episodic GH secretion continues after lesions that result in depletion of hypothalamic SS.[8,23] Second, electrical stimulation of the preoptic and corticomedial amygdalar nuclei, areas rich in somatostatinergic neurons and fibers, causes GH inhibition without rebound. Third, lesions of the VMN which obliterate episodic GH release cause persistently low levels of GH.[8] Fourth, activation of specific afferent hypothalamic connections such as the hippocampus and basolateral amygdala cause a rapid rise in plasma GH levels during stimulation, a response that cannot be attributed to a postinhibitory rebound effect.[8] Finally, SS infusions do not cause postinhibitory GH release when administered (1) during a GH ebb in normal rats, or (2) to animals with VMN lesions.[8,23]

The availability of specific antisera against SS provides another method by which to study the function of the peptide in the regulation of GH secretion. Effective neutralization of circulating SS can be accomplished by passive immunization with such antiserum. Results of such studies in rats show that administration of SS antiserum causes elevation in GH trough levels without blunting of peak levels.[23] Active immunization of baboons with SS also results in elevation of basal GH levels.[36] The rich innervation of the ME with SS-containing fibers provides an anatomical basis for its role in GH regulation. Both SS and its larger form, SS-28, are released from ME fragments *in vitro*.[37] Administration of cysteamine to rats, which causes depletion of SS in the hypothalamus, also results in an elevation of baseline GH levels, providing further evidence for a tonic inhibitory effect of SS on GH secretion.[38]

4.3. *Other Peptides with GH-Releasing Activity*

4.3.1. *TRH and Luteinizing Hormone-Releasing Hormone (LHRH)*

TRH has no effects on GH in normal human subjects but is remarkably potent in acromegaly[39] and in some patients with renal failure, ectopic GRF-producing neuroectodermal tumors (with increased GH secretion),[40] depression,[41] polyostotic fibrous dysplasia,[42] anorexia nervosa,[43] and liver disease.[44] TRH stimulates GH secretion *in vitro* from bovine, ovine, and rat pituitaries.[23] LHRH also stimulates hGH in some cases of acromegaly but is generally less potent than TRH.[23]

4.3.2. *Vasopressin*

Vasopressin releases GH in man, monkey, and rat.[23] This effect was considered to be of little physiological significance prior to the observations of Zimmerman *et al.*[45] that vasopressin (and neurophysin) is present in high concentrations in the pituitary portal blood of surgically stressed rhesus monkeys. This raised the possibility that vasopressin might act as a "GRF" to stimulate GH release during stress. That vasopressin was not the only GRF was demonstrated by the finding that rats with hereditary vasopressin deficiency (Brattleboro strain) have normal plasma GH levels.[23] Furthermore, Brattleboro rats show normal GH release after hypothalamic electrical stimulation (J. B. Martin, unpublished observations), and stimulation of the supraoptic nuclei (in normal rats) has no effect on GH release. Thus, although vasopressin in pharmacological doses is able to elicit GH secretion, its function, if any, in physiological control remains to be demonstrated. It is possible that as in the case of CRF, vasopressin may facilitate the effects of GRF.

4.3.3. *Melanocyte Stimulating Hormone (α-MSH)*

Several papers have appeared that suggest that α-MSH is effective in releasing GH in man.[23] The specificity and significance of this response are unknown. Neurons of the ARC nucleus stain with antibodies to α-MSH, and terminals containing immunoreactivity are found in the ME. Whether the peptide has any physiological function in hGH regulation is unknown but seems unlikely.

4.3.4. *Glucagon*

The effect of glucagon to stimulate release of GH in man is well established.[46] The site of action of glucagon is unknown; its effects are believed by some to be secondary to changes in plasma glucose, although the rapidity of the response, particularly in children, argues against this explanation. The effect of glucagon is potentiated by β-receptor blockade, suggesting that it acts on hypothalamic noradrenergic pathways.[43]

4.3.5. *Substance P and Neurotensin*

Substance P and neurotensin cause GH release in rats[23,47] but have no effects in man.

4.3.6. *Opioid Peptides*

The identification of methionine- and leucine-enkephalin and the endorphins has led to extensive investigations of their neuroendocrine effects. Morphine sulfate is a potent stimulant of GH release in the rat,[48] but under most circumstances it has no effect on GH in the human. Longer-acting enkephalin analogues release GH in man, and their effects are blocked by naloxone.[49] Naloxone alone does not alter sleep-associated hGH secretion,[50] raising doubts about the importance of the opioids in physiological GH secretion. Recent reports give evidence that naloxone decreases exercise-stimulated GH secretion but does not affect insulin–hypoglycemia-induced release.[49]

4.3.7. *Other Peptides*

In animals, bombesin[51] and vasoactive intestinal polypeptide (VIP)[52] are reported to cause GH secretion after either systemic or intracerebroventricular injection. VIP is reported to cause *in vitro* GH secretion from pituitary tumors removed from a patient with acromegaly.[53] *In vitro* studies using hypothalamic fragments show that VIP causes inhibition of SS release.[54]

In summary, many peptides found in the ME stimulate GH secretion. Their role in the physiological regulation of GH (except for GRF and SS) remains unclear.

5. *Neuropharmacological Regulation of GH Secretion (Table 1)*

The full integration of pituitary secretory functions is accomplished by the central nervous system through neuronal afferents that synapse on hypothal-

Table 1 Pharmacological Regulation of Growth Hormone (GH) Secretion[a]

I. Agents that stimulate GH secretion
   A. Biogenic amines
      1. Dopaminergic
         Apomorphine
         Bromocriptine
      2. α-Adrenergic
         Clonidine
      3. Serotonergic
         Tryptophan
         5-Hydroxytryptophan
   B. Other neurotransmitters
      1. Histamine H_2 receptor antagonists
      2. GABA and baclofen
      3. γ-Aminohydroxybutyric acid
      4. Cholinergic agents—nicotinic receptor agonists
   C. Amino acids
      1. Arginine
      2. Leucine
      3. Glycine
      4. Isoleucine
      5. Ornithine
   D. Prostaglandin E_2
   E. Peptides
      1. GRF
      2. Vasopressin
      3. Glucagon
      4 Insulin—by inducing hypoglycemia
      5. α-Melanotropin
      6. Enkephalins—including synthetic opioid agonists (SK-33-824)
      7. TRH[b]
      8. LHRH (in acromegaly)
      9. VIP[c]
      10. Substance P[d], neurotensin[d], CCK[d], bombesin[d]
II. Agents that inhibit GH secretion
   A. Biogenic amines
      1. β-Adrenergic agonists
      2. Melatonin (in some studies)
      3. Dopamine (by direct effects on the pituitary in normal subjects)
      4. Bromocriptine—in acromegaly
   B. Peptides
      1. Somatostatin
   C. Alcohol

[a] Modified from Martin.[120]
[b] In pathological states including acromegaly, renal failure, liver failure, depression, anorexia nervosa, and chronic alcoholism.
[c] In acromegaly (*in vitro* studies).
[d] In rodents, but not man.

amic neurosecretory cells. Converging neuronal afferents may act to either stimulate or inhibit peptidergic neurosecretory cells through interactions in the ME or on their cell bodies located in other brain areas. A host of putative neurotransmitter substances have been investigated to determine their role in the hypothalamic regulation of GH secretion. This section will focus on the most widely studied neurotransmitter candidates, the catecholamines norepinephrine (NE) and dopamine and the indoleamine 5-hydroxytryptamine (5-HT or serotonin).[55] Attention will be drawn to identifying possible interactions of these neurotransmitters with particular hypothalamic peptidergic cells.

5.1. NE

Noradrenergic systems act to enhance pituitary GH secretion by exerting an excitatory effect at the level of the hypothalamus. The noradrenergic neurons that give rise to hypothalamic inputs are located in several brain stem nuclei,[56] including the locus ceruleus, the nucleus tractus solitarii, and the lateral reticular nucleus of the medulla.

Axons arising from these cells groups form two ascending bundles, called the dorsal and ventral noradrenergic bundles.[14,57] Locus ceruleus fibers contribute to both bundles, although mainly to the dorsal one, whereas the axons of the other cell groups run primarily within the ventral bundle. Both bundles course to the lateral hypothalamus and enter the medial forebrain bundle where they can no longer be traced as individual tracts.

The NE-containing fibers innervating the medial hypothalamic nuclei reach them from lateral zones of the hypothalamus. Based on fluorescent microscopic observations, the periventricular, medial preoptic, anterior hypothalamic, paraventricular, supraoptic, and retrochiasmatic regions and the ME have the most dense noradrenergic innervations. The ARC nucleus receives only a moderate innervation.[57]

In most species studied, α-adrenergic mechanisms stimulate, whereas α-adrenergic mechanisms either inhibit or have no effect on, physiological GH secretion.[43] NE has no effect on *in vitro* GH secretion from the pituitary. Indeed, most of the stimuli that cause GH release in man appear to act via central α-adrenergic receptors.[43] Thus, GH release induced by insulin hypoglycemia, arginine, exercise, vasopressin, L-dopa, and certain stresses is prevented by phentolamine. On the other hand, propranolol, a β-adrenergic blocker, enhances GH release induced by glucagon, vasopressin, and levodopa.

The administration of levodopa, the precursor of both dopamine and NE, stimulates GH release in man[58] and in rhesus monkeys.[59] The stimulatory effect of levodopa in man is blocked by phentolamine,[60] an α-adrenoceptor antagonist, suggesting mediation by α-NE receptors. Similarly, the administration of the tyrosine hydroxylase inhibitor α-methyl-para-tyrosine to rats markedly inhibits pulsatile GH release, an effect reversed by clonidine but not by apomorphine.[61] The administration of the α-adrenoceptor agonist clonidine also stimulates GH release in man,[62] dogs,[63] and rhesus monkeys.[64] On the other hand, the administration of the α-adrenoceptor antagonist propranolol potentiates exercise-induced GH release in man[65] but does not effect pulsatile GH secretion in the rat.[66] NE has no effect on GH secretion from the pituitary *in vitro*.

We have shown in the rat that GH release after ARC–VMN electrical stimulation is not prevented by agents that block catecholamine synthesis.[66] On the other hand, the secretion of GH following electrical stimulation of several extrahypothalamic areas is dependent on the release of catecholamines by brain neurons. Thus, GH release after stimulation of the hippocampus or the basolateral amygdala is blocked by pretreatment with α-methyl-paratyrosine. Further, amygdaloid-induced GH release is blocked by phenoxybenzamine but not by propranolol. These studies suggest that α-adrenoceptors act to mediate extrahypothalamic influences on GH regulation.

Our current hypothesis is that α-adrenergic effects are exerted directly on GRF neurons to stimulate GRF release. There is evidence that NE modifies pituitary GH release by effects within the hypothalamus. Toivola and Gale[67,68] showed that the microinjection of NE into the VMN of conscious baboons stimulates GH release. Furthermore, they showed that other hypothalamic areas were unresponsive to NE infusion—including the preoptic area, the anterior hypothalamus, the mammillary bodies, and the area dorsal to the VMN.[68] In contrast, phentolamine infusion into the third ventricle or anterior hypothalamus reduced GH secretion in this species.[68] That NE exerts effects on GRF and not SS is suggested by the experiments of Arnold and Fernstrom,[69] who showed that inhibition of episodic GH release in the rat by an α-adrenoceptor antagonist (yohimbine) was not affected by administration of anti-SS serum.

Despite these interesting studies, which point to the important role of NE in physiological GH regulation, α-adrenergic blockage is not effective in the treatment of GH hypersecretory states in man.

5.2. *Dopamine*

Dopamine stimulates GH secretion in normal subjects and inhibits GH secretion in many acromegalics. The role of dopamine in GH regulation is complicated but important to consider since dopamine agonists are used in the treatment of acromegaly.

High concentrations of dopamine are found in the hypothalamus and in the ME.[70] The most important group of dopaminergic cells is located in the ARC and in the periventricular nuclei.[56] A smaller cell group is located in the periventricular preoptic nucleus.[71,72] These two dopaminergic cell groups are believed to innervate other hypothalamic nuclei and to send fibers to the external zone of the ME. Dopamine concentrations of the hypothalamus do not change after complete hypothalamic deafferentation, indicating their intrinsic origin.[22]

Dopamine-containing cells are also found in the subthalamus and in the medial part of the zona incerta. Their axons form the incertohypothalamic dopaminergic system, which innervates certain regions of the hypothalamus and may send a few fibers to the ME. The role of this system in GH regulation is unknown. A large group of dopaminergic cells is found in the mesencephalon, mostly within the substantia nigra and the ventral tegmental area.(56) Fibers arising from these cells form the nigrostriatal dopaminergic pathway, which innervates the striatum, the limbic system, and the cerebral cortex. Few, if any, of these dopaminergic fibers reach the hypothalamus or the ME.

Dopamine neurons in the brain have been implicated in GH secretion in man. Apomorphine, a centrally active dopamine receptor agonist, is a potent stimulant for hGH release.(73) Glucose administration reduces the hGH rise in plasma following levodopa or apomorphine administration,(74) suggesting that glucoreceptor stimulation can override the dopaminergic stimulus for GH secretion. Other dopamine agonists such as bromocriptine and piribedil also elicit hGH stimulation in normal subjects.

Recent evidence obtained by examining the effects of intravenous dopamine on hGH secretion in normal subjects suggests an additional *direct* dopaminergic *inhibition* of GH secretion on the pituitary.(75) Dopamine administered intravenously does not cross the blood–brain barrier (whereas apomorphine does), although it does reach nerve terminals in the ME. *In vitro* studies of normal human pituitary tissue have confirmed the inhibitory effects of dopamine.(76,77) These observations are important because they may help explain the known efficacy of dopaminergic drugs in the suppression of GH in acromegaly.

About 70% of cases of acromegaly show a "paradoxical" inhibition of GH secretion after administration of dopamine receptor agonists such as L-dopa, apomorphine, or bromocriptine.(78) This inhibitory effect is the basis for its use in the treatment of acromegaly. Current evidence is most consistent with the view that the suppression of GH secretion that occurs in acromegaly results from direct effects on pituitary adenoma cells. Clinical evidence in support of this has been documented in several studies in which GH responses were evaluated before and after pituitary surgery.(79) GH secretory dynamics return to normal after successful removal of the adenoma. Moreover, *in vitro* studies of pituitary tumor tissue removed at transsphenoidal surgery indicate that both apomorphine and bromocriptine inhibit GH secretion.(80,81) These studies provide convincing evidence that hGH-producing adenoma cells respond directly to these agents and suggest that the effects are mediated via cell surface receptors.

These hGH secretory responses are similar to those that occur in the normal prolactin secreting cell (lactotrope).(80,81) Indeed, up to 30% of patients with acromegaly have elevated prolactin levels, which in many cases are due to concomitant hypersecretion of prolactin by the tumor. The tendency for these two cell types to occur together in the same adenoma suggests a possible common pathophysiological process. One possibility is that both prolactin and hGH-secreting cells arise from a common precursor stem cell and that the adenomas represent a dedifferentation to a common cell type. In the case of the hGH-secreting tumor, changes in membrane receptor characteristics may occur as part of the adenomatous dedifferentation to give rise to cellular responses similar to those that occur in the normal prolactin-secreting cell.

5.3. 5-HT

Considerble evidence has accumulated to support a facilitatory function of 5-HT on GH secretion. The administration of the 5-HT precursor tryptophan is reported to increase plasma hGH levels.[82] In one study, pretreatment with the serotonin antagonist cyproheptadine blocked the rise in plasma hGH following the oral administration of tryptophan to women.[83] The administration of the immediate precursor of 5-HT, 5-hydroxytryptophan, also stimulates GH secretion in man.[84] As 5-hydroxytryptophan (but not tryptophan) increases 5-HT synthesis in both serotonergic and adrenergic neurons, caution is advised in interpreting these data.

Serotonin fibers that reach the hypothalamus are believed to originate exclusively from the brain stem raphe system. Interference with this pathway can alter GH responses to phamacological stimuli. Thus, the putative 5-HT receptor antagonists cyproheptadine and methysergide are effective in blocking hGH release from oral 5-hydroxytryptophan, [85] insulin-induced hypoglycemia,[86,87] arginine infusion,[85] physical exercise,[87] and during sleep.[88] Conversely, in another report, sleep-related hGH secretion was shown to be *potentiated* by methysergide,[89] and the 5-HT receptor antagonist metergoline had no effect on arginine-induced hGH release.[87] Imura (personal communication) recently found cyproheptadine effective in suppressing GH in four patients with acromegaly who were refractory to bromocriptine. In two patients the response was particularly dramatic with restoration of plasma GH to levels less than 5 ng/ml.

In rats, the intraventricular injection of 5-HT[90] or the intraperitoneal injection of 5-hydroxytryptophan[91] stimulates GH secretion. The pulsatile release of GH was shown to be diminished by treatment with 5-HT receptor antagonists[92] while it was enhanced by treatment with tryptophan.[92]

Systemic administration of neuropharmacological agents does not provide information regarding the exact site of action of these drugs. A direct effect on the pituitary is unlikely, however, as application of 5-HT to the pituitary *in vitro* does not modify GH release.[93] Serotonin may act on GH secretion by altering SS secretion. In the rat, GH inhibition induced by a 5-HT receptor antagonist, metergoline, is completely reversed by treatment with anti-SS serum.[91] A possible interpretation of these findings is that metergoline blocks GH release by stimulating SS secretion without impairing GRF release. These data support the view that episodic GH secretion in the rat is the result of an interaction between α-adrenoceptor-modulated GRF release and 5-HT receptor-modulated SS secretion. Further studies will be required to confirm this hypothesis.

The effect of putative neurotransmitter substances on hypothalamic SS release has been studied directly. Intraventricular administration of dopamine, NE, and acetylcholine was reported to increase the concentration of SS in the hypophyseal portal blood of urethane-anesthetized rats.[94] In this study plasma GH levels were not reported, so the effect of these agents on GRF release cannot be estimated.

5.4. Other Neurotransmitters

A number of other agents, presumed to act via hypothalamic mechanisms, have been shown to stimulate GH secretion in man. These include histamine receptor antagonists,[95,96] γ-amino butyric acid (GABA)[97] and baclofen (a GABA agonist),[98] nicotinic cholinergic agents,[99] and various amino acids.[100–102] Melatonin, an indoleamine derived from serotonin, is reported to both stimulate and inhibit[87,103] or have no effect[104] on hGH secretion. Prostaglandin E_2 stimulates hGH in children.[105] The importance of these substances in physiological regulation, in stimulatory testing, or potentially in treatment is not presently clear. Many drugs that cause GH secretion may exert their effects by nonspecific "stress" mechanisms.

5.5. Insulin-Induced GH Secretion

This form of GH stimulation remains the most widely used for testing pituitary reserve and is by and large safe. It is likely that falling blood glucose levels act on glucoreceptors in the hypothalamus to trigger GRF secretion. Recent papers have made comparisons between insulin hypoglycemia and levodopa, apomorphine and clonidine, but no compelling evidence exists to displace insulin as the principal agent for GH testing.

6. Feedback Regulation of GH Secretion

6.1. Effects of GH and Somatomedins on GH Secretion

There is considerable evidence, reviewed elsewhere,[1,4] that GH administration causes GH suppression by a short-loop feedback effect. Episodic GH secretion in the rat is suppressed by hGH administration.[106,107] Recent evidencc also supports an inhibitory effect of somatomedins on GH secretion in the rat.[106,108,109]

6.2. Effects of GH and Somatomedins on Hypothalamic SS Release

SS secretion appears to be at least one of the factors mediating short-loop feedback control of GH secretion. GH deficiency (hypophysectomy) decreases SS content in the ME, which is reversed by GH administration.[106,110,114] SS release from hypothalamic tissue, studied *in vitro*, is stimulated by application of GH[115] and of somatomedin C.[116] Intraventricular administration of GH stimulates SS release into portal blood in the rat.[117] Paradoxically, intraventricular administration of SS causes stimulation of GH secretion in the rat.[118] These effects are difficult to interpret from a physiological perspective but point to the possibility that both GH and somatomedins suppress GH secretion by effects exerted on the hypothalamus (via SS) or on the pituitary, and that SS may inhibit its own or GRF secretion by an effect on the hypothalamus, thereby causing a stimulation of GH secretion. The effects of somatomedin C on GRF-stimulated GH secretion have been reported recently. Inhibition of GRF stimulation at the level of the pituitary was found.[119]

ACKNOWLEDGMENTS. I thank Kathy Kreatz for typing the manuscript. This work was supported by NIH grants AM 26252 and NS 16367.

References

1. J. B. Martin, Brain regulation of growth hormone secretion, in: *Frontiers in Neuroendocrinology* (L. Martini and F. Ganong, eds.), Raven Press, New York, 1976, pp. 129–168.
2. J. B. Martin, S. Reichlin, and G. Brown, *Clinical Neuroendocrinology*, Davis, Philadelphia, 1977.
3. T. J. Merimee, Growth hormone: secretion and action, in: *Endocrinology*, vol. 1 (L. Degroot, ed.), Grune & Stratton, New York, 1979, pp. 123–132.
4. L. S. Phillips and R. Vasilopoulou-Sellin, Somatomedins I and II, *N. Engl. J. Med.* **302:**371–376, 438–444, 1980.
5. J. W. Finkelstein, H. P. Roffwarg, R. M. Boyar, J. Kream, and L. Hellman, Age-related changes in the twenty-four hour spontaneous secretion of growth hormone, *J. Clin. Endocrinol. Metab.* **35:**665–670, 1972.
6. R. A. Steiner, J. K. Stewart, J. Barber, D. Koerker, C. J. Goodner, A. Brown, P. Illner, and C. C. Gale, Somatostatin: a physiological role in the regulation of growth hormone secretion in the adolescent male baboon, *Endocrinology* **102:**1587–1592, 1982.
7. H.-J. Quabbe, M. Gregor, C. Bunke-Vogt, A. Eckhof, and J. Witt, Twenty-four hour pattern of growth hormone secretion in the rhesus monkey: Studies including alterations of the sleep/wake cycle and sleep stage cycle, *Endocrinology* **109:**513–522, 1981.
8. J. B. Martin, Bowditch Lecture: Brain mechanisms for integration of growth hormone secretion, *Physiologist* **22:**23–29, 1978.
9. H. E. Carlson, I. K. Mariz, and W. H. Daughaday, Thyrotropin-releasing hormone stimulation and somatostatin inhibition of growth hormone secretion, *Endocrinology* **94:**1709–1713, 1974.
10. L. C. A. Conrad and D. W. Pfaff, Efferents from medial basal forebrain and hypothalamus in the rat. I. An autoradiographic study of the medial preoptic area, *J. Comp. Neurol.* **169:**185–226, 1976.
11. L. W. Swanson, An autoradiographic study of the efferent connections of the preoptic region of the rat, *J. Comp. Neurol.* **167:**227–256, 1976.
12. B. Halasz, K. Koves, M. Rethelyi, M. Bodiky, and S. Koritsanszky, Recent data on neuronal connections between nervous structures involved in the control of the adenohypophysis, in: *Anatomical Neuroendocrinology* (W. E. Stumpf and L. D. Grant, eds.), Karger, Basel, 1975, pp. 9–14.
13. K. Koves and M. Rethelyi, Direct neural connection from the medial preoptic area to the hypothalamic arcuate nucleus of the rat, *Exp. Brain Res.* **25:**529–539, 1976.
14. M. Palkovits and L. Zaborszky, Neural connections of the hypothalamus, in: *Anatomy of the Hypothalamus* (P. J. Morgane and J. Panskepp, eds.), Marcel Dekker, Inc., New York, 1979, pp. 379–501.
15. L. Zaborszky and G. B. Makara, Intrahypothalamic connections: An electron microscopic study in the rat, *Exp. Brain Res.* **34:**201–215, 1979.
16. J. Epelbaum, J. O. Willoughby, P. Brazeau, and J. B. Martin, Effects of brain lesions and hypothalamic deafferentation on somatostatin distribution in the rat brain, *Endocrinology* **101:**1495–1502, 1977.
17. J. O. Willoughby and J. B. Martin, The suprachiasmatic nucleus synchronizes growth hormone secretory rhythms with the light:dark cycle, *Brain Res.* **151:**413–417, 1978.
18. O. E. Millhouse, Certain ventromedial hypothalamic afferents, *Brain Res.* **55:**89–105, 1973.
19. J. Szentagothai, B. Flerko, B. Mess, and B. Halasz, *Hypothalamic Control of the Anterior Pituitary*, Academiai Kiado, Budapest, 1968.
20. T. Hokfelt, R. Elde, K. Fuxe, O. Johanssen, A. Ljungdahl, M. Goldstein, R. Luft, S. Efendic,

G. Nilsson, L. Terenius, D. Ganten, S. L. Jeffcoate, J. Rehfeld, S. Said, M. Perez de la Mora, L. Posaani, R. Rapia, L. Teran, and R. Pelacious, Aminergic and peptidergic pathways in the nervous system with special reference to the hypothalamus, in: *The Hypothalamus* (S. Reichlin, R. Baldessarini, and J. B. Martin, eds.), Raven Press, New York, 1978, pp. 69–135.

21. T. Meszaros, I. J. Csuri, J. Hazas, and M. Palkovits, Esterase activity in the hypothalamus, *Acta Morph. Hung.* **17:**201–215, 1969.
22. M. J. Brownstein, A. Arimura, R. Fernandez-Durango, A. V. Schally, M. Palkovits, and J. S. Kizer, The effect of hypothalamic deafferentation on somatostatin-like activity in the rat brain, *Endocrinology* **100:**246–248, 1977.
23. J. B. Martin, P. Brazeau, G. S. Tannenbaum, J. O. Willoughby, J. Epelbaum, L. C. Terry, and D. Durand, Neuroendocrine organization of growth hormone regulation, in: *The Hypothalamus* (S. Reichlin, R. Baldessarini, and J. B. Martin, eds.), Raven Press, New York, 1978, pp. 329–357.
24. J. O. Willoughby, L. C. Terry, P. Brazeau, and J. B. Martin, Pulsatile growth hormone, prolactin and thyrotropin secretion in rats with hypothalamic deafferentation, *Brain Res.* **127:**137–152, 1977.
25. D. T. Krieger and S. M. Glick, Sleep EEG stages and plasma growth hormone concentration in states of endogenous and exogenous hypercortisolemia of ACTH elevation, *J. Clin. Endocrinol. Metab.* **39:**986–1000, 1974.
26. R. Guillemin, P. Brazeau, P. Bohlen, F. Esch, N. Ling, and W. B. Wehrenberg, Growth hormone-releasing factor from a human pancreatic tumor that caused acromegaly, *Science* **218:**585–587, 1983.
27. J. Rivier, J. Spiess, M. Thorner, and W. Vale, Characterization of a growth hormone-releasing factor from a human pancreatic tumor, *Nature* **300:**276–278, 1982.
28. B. Bloch, P. Brazeau, N. Ling, P. Bohlen, F. Esch, W. Wehrenberg, R. Benoit, F. Bloom, and R. Guillemin, Immunohistochemical detection of growth-hormone releasing factor in brain, *Nature* **301:**607–608, 1983.
29. K. Tatemoto and V. Mutt, Isolation and characterization of the intestinal peptide porcine PHI (PHI-27), a new member of the glucagon–secretin family, *Proc. Natl. Acad. Sci. USA* **78:**6603–6607, 1981.
30. J. Spiess, J. Rivier, and W. Vale, Characterization of rat hypothalamic growth hormone-releasing factor, *Nature* **303:**532–535, 1983.
31. J. B. Martin, Discovery of a new peptide with growth hormone releasing activity, *Trends Neurosci.* **6:**1–3, 1983.
32. J. L. C. Borges, R. M. Blizzard, M. Gelato, H. D. Furlane, A. D. Rogol, W. S. Evans, M. L. Vance, D. L. Kaiser, R. M. MacLeod, G. R. Merriam, L. Loriaux, J. Spiess, J. Rivier, W. Vale, and M. O. Thorner, Effects of human pancreatic tumor growth hormone releasing factor on growth hormone and somatomedin C levels in patients with idiopathic growth hormone deficiency, *Lancet* **2:**119–123, 1983.
33. W. B. Wehrenberg, P. Brazeau, R. Luben, P. Bohlen, and R. Guillemin, Inhibition of the pulsatile secretion of growth hormone by monoclonal antibodies to the hypothalamic growth hormone releasing factor (GRF), *Endocrinology* **111:**2147–2148, 1982.
34. P. Brazeau, W. Vale, R. Burgus, N. Ling, M. Butcher, J. Rivier, and R. Guillemin, Hypothalamic polypeptide that inhibits the secretion of immunoreactive pituitary growth hormone, *Science* **179:**77–79, 1973.
35. T. M. Siler, S. S. C. Yen, W. Vale, and R. Guillemin, Inhibition by somatostatin of the release of TSH induced in man by thyrotropin-releasing factor, *J. Clin. Endocrinol. Metab.* **38:**742–745, 1974.
36. R. A. Steiner, J. K. Stewart, J. Barber, D. Koerker, C. J. Goodner, A. Brown, P. Illner, and C. C. Gale, Somatostatin: a physiological role in the regulation of growth hormone secretion in the adolescent male baboon, *Endocrinology* **102:**1587, 1978.
37. C. Bakhit, R. Benoit, and F. E. Bloom, Release of somatostatin-28(1–12) from rat hypothalamus *in vitro*, *Nature* **301:**524–526, 1983.
38. W. J. Millard, S. M. Sagar, T. M. Badger, and J. B. Martin, Cysteamine effects on growth hormone secretion in the male rat, *Endocrinology* **112:**509–514, 1983.

39. M. Irie and T. Tsushima, Increase of serum growth hormone concentration following thyrotropin-releasing hormone injection in patients with acromegaly or gigantism, *J. Clin. Endocrinol. Metab.* **35:**7–100, 1972.
40. M. O. Thorner, R. L. Perryman, M. J. Cronin, A. D. Rogol, M. Draznin, A. Johanson, R. Grindeland, W. Vale, E. Horvath, and K. Kovacs, Acromegaly with somatotroph hyperplasia: Successful treatment by resection of a pancreatic tumor secreting a growth hormone (GH) releasing factor (GRF), *J. Clin. Invest.* **70:**965–977, 1982.
41. A. Winokur, J. Amsterdam, S. Caroff, P. J. Snyder, and D. Brunswick, Variability of hormonal responses to a series of neuroendocrine challenges in depressed patients, *Am. J. Psychiatry* **139:**39–44, 1982.
42. J. Albin and R. Wu, Abnormal hypothalamic–pituitary function in polyostatic fibrous dysplasia, *Clin. Endocrinol.* **14:**435–443, 1981.
43. J. B. Martin, Neural regulation of growth hormone secretion, medical progress report, *N. Engl. J. Med.* **288:**1384–1393, 1973.
44. M. Musseo, A. Tiengso, D. Fedele, and G. Crepaldi, Altered control of growth hormone secretion in patients with cirrhosis of the liver, *Arch. Intern. Med.* **139:**1157–1160, 1979.
45. E. A. Zimmerman, P. W. Carmel, M. K. Husain, M. Ferin, M. Tannenbaum, A. G. Frantz, and A. G. Robinson, Vasopressin and neurophysin: high concentrations in monkey hypophyseal portal blood, *Science* **182:**925–927, 1973.
46. R. L. Eddy, P. F. Gilliland, J. D. Ibarra, J. F. McMurray Jr., and J. Q. Thompson, Human growth hormone release—Comparison of provocative test procedures, *Am. J. Med.* **56:**179–185, 1974.
47. Y. Kato, K. Chihara, S. Ohgo, Y. Iwasaki, Y. Abe, and H. Imura, Growth hormone and prolactin release by substance P in rats, *Life Sci.* **19:**441–446, 1976.
48. J. B. Martin, J. Audet, and A. Saunders, Effect of somatostatin and hypothalamic ventromedial lesions on GH release induced by morphine, *Endocrinology* **96:**839–847, 1975.
49. W. J. Millard and J. B. Martin, Opioid modulation of human growth hormone secretion, in: *Opioid Modulation of Endocrine Function* (G. Guisti, L. Martin, C. G. Pepeu, M. Serio, and G. Delitala, eds.), Raven Press, New York, 1984, pp. 111–123.
50. J. B. Martin, G. Tolis, I. Woods, and H. Guyda, Failure of naloxone to influence physiological growth hormone and prolactin secretion, *Brain Res.* **168:**210–215, 1979.
51. C. Rivier, J. Rivier, and W. Vale, The effects of bombesin and related peptides on prolactin and growth hormone secretion in the rat, *Endocrinology* **102:**519–523.
52. E. Vijayan, W. K. Samson, S. I. Said, and S. M. McCann, Vasoactive intestinal peptide: Evidence for a hypothalamic site of action to release growth hormone, luteinizing hormone, and prolactin in conscious ovariectomized rats, *Endocrinology* **104:**53–57, 1979.
53. N. Matsushita, Y. Kato, H. Katakami, A. Shimatsu, N. Yanihara, and H. Imura, Stimulation of growth hormone release by vasoactive intestinal polypeptide from human pituitary adenomas *in vitro*, *J. Clin. Endocrinol.* **53:**1297–1300, 1981.
54. J. Eppelbaum, L. Tapia-Arancibia, J. Besson, W. H. Rotsztejn, and C. Kordon, Vasoactive intestinal polypeptide inhibits release of somatostatin from hypothalamus *in vitro*, *Eur. J. Pharmacol.* **58:**493–497, 1979.
55. J. B. Martin, Functions of central nervous system neurotransmitters in regulation of growth hormone secretion, *Fed. Proc.* **39:**2902–2905, 1980.
56. A. Dahlstrom and K. Fuxe, Evidence for the existence of monoamine neurons in the central nervous system. I. Demonstration of monoamines in the cell bodies of brainstem neurons. *Acta. Physiol. Scand.* **64**(Suppl. 232)**:**1–55, 1964.
57. R. Y. Moore and F. E. Bloom, Central catecholamine neuron systems: Anatomy and physiology of the norepinephrine and epinephrine systems, *Ann. Rev. Neurosci.* **2:**113–168, 1979.
58. A. Boyd, H. Lebovittz, and J. B. Pfeiffer, Stimulation of human-growth hormone secretion by L-dopa, *N. Engl. J. Med.* **283:**1425–1429, 1970.
59. J. H. Jacoby, M. Greenstein, J. F. Sassin, and E. D. Weitzman, The effect of monoamine precursors on the release of growth hormone in the rhesus monkey, *Neuroendocrinology* **14:**95–102, 1974.

60. P. C. Kansel, J. Buse, O. R. Talbert, and M. G. Buse, Effect of L-dopa on plasma growth hormone, insulin, and thyroxine, *J. Clin. Endocrinol. Metab.* **34:**99–105, 1973.
61. J. B. Martin, D. Durand, W. Gurd, G. Faille, J. Audet, and P. Brazeau, Neuropharmacologic regulation of episodic growth hormone and prolactin secretion in the rat, *Endocrinology* **102:**106–113, 1978.
62. S. Lal, G. Tolis, J. B. Martin, G. M. Brown, and H. Guyda, Effects of clonidine on growth hormone, prolactin, luteinizing hormone, follicle-stimulating hormone and thyroid-stimulating hormone in the serum of normal men, *J. Clin. Endocrinol. Metab.* **41:**1703–1708, 1975.
63. R. D. Lovinger, J. Holland, S. Kaplan, M. Grumbach, A. T. Boryczka, R. Shackleford, J. Salmon, I. A. Reid, and W. F. Ganong, Pharmacological evidence for stimulation of growth hormone secretion by a central noradrenergic system in dogs, *Neuroscience* **1:**443–450, 1976.
64. J. W. Chambers and G. M. Brown, Neurotransmitter regulation of growth hormone and ACTH in the rhesus monkey: Effects of biogenic amines, *Endocrinology* **98:**420–428, 1976.
65. J. Sutton and L. Lazarus, Effect of adrenergic blocking agents on growth hormone responses to physical exercise, *Horm. Metab. Res.* **6:**428–429, 1974.
66. J. B. Martin, J. Kontor, and P. Mead, Plasma growth hormone responses to hypothalamic, hippocampal and amygdaloid electrical stimulation: Effects in variations in stimulation of parameters and treatment with alpha-methyl-p-tyrosine (α-MT), *Endocrinology* **92:**1354–1361, 1973.
67. P. T. K. Toivola and C. C. Gale, Stimulation of growth hormone release by microinjection of norepinephrine into hypothalamus of baboons, *Endocrinology* **90:**895–902, 1972.
68. P. T. K. Toivola, C. C. Gale, C. J. Goodner, and J. H. Werbach, Central alpha-adrenergic regulation of growth hormone and insulin, *Hormones* **3:**193–213, 1972.
69. M. A. Arnold and J. D. Fernstrom, Administration of antisomatostatin serum to rats reverses the inhibition of pulsatile growth hormone secretion produced by injection of metergoline but not yohimbine, *Neuroendocrinology* **31:**194–199, 1980.
70. M. Palkovits, M. Brownstein, J. M. Saavedra, and J. Axelrod, Norepinephrine and dopamine content of hypothalamic nuclei of rats, *Brain Res.* **77:**137–149, 1974.
71. A. Bjorklund, R. Y. Moore, A. Nobin, and U. Steineri, The organization of tuberohypophyseal and reticulo-infundibular catecholamine neuron systems in the rat brain, *Brain Res.* **51:**171–191, 1973.
72. A. Bjorklund and A. Nobin, Fluorescence histochemical and microspectrofluorometric mapping of dopamine and noradrenaline cell groups in the rat diencephalon, *Brain Res.* **51:**193–205, 1973.
73. S. Lal, C. E. de la Vega, T. L. Sourkes, and H. G. Friesen, Effect of apomorphine on growth hormone, prolactin, luteinizing hormone and follicle stimulating hormone levels in human serum, *J. Clin. Endocrinol. Metab.* **37:**719–724, 1973.
74. P. Ettigi, S. Lal, J. B. Martin, and H. Friesen, Effects of sex, oral contraceptives and glucose loading on apomorphine induced growth hormone secretion, *J. Clin. Endocrinol. Metab.* **40:**1094–1098, 1975.
75. S. A. Bansal, L. A. Lee, and P. D. Woolf, Dopaminergic stimulation and inhibition of growth hormone secretion in normal man: studies of the pharmacologic specificity, *J. Clin. Endocrinol. Metab.* **53:**1273–1277, 1981.
76. D. Tallo and W. B. Malarkey, Adrenergic and dopaminergic modulation of growth hormone and prolactin secretion in normal and tumor bearing human pituitaries in monolayer culture, *J. Clin. Endocrinol. Metab.* **53:**1279–1284, 1981.
77. S. Marcovitz, C. G. Goodyer, H. Guyda, R. J. Gardiner, and J. Hardy, Comparative study of human fetal, normal adult, and somatotropic adenoma pituitary function in tissue culture, *J. Clin. Endocrinol. Metab.* **54:**6–16, 1972.
78. P. Riskind and J. B. Martin, Management of pituitary secretory adenomas, in: *Harrison's Principles of Internal Medicine, Update III* (K. J. Isselbacher, R. D. Adams, E. Braunwald, J. B. Martin, R. G. Petersdorf, and J. D. Wilson, eds.), McGraw-Hill, New York, 1982.
79. K. M. Hoyte and J. B. Martin, Recovery from paradoxical growth hormone responses in acromegaly after transsphenoidal selective adenomectomy, *J. Clin. Endocrinol. Metab.* **41:**656–659, 1975.

80. E. F. Adams, J. F. Brajkovich, and K. Mashiter, Hormone secretion by dispersed cell cultures of human pituitary adenomas: Effects of theophylline, thyrotropin-releasing hormone, somatostatin, and 2-bromo-α- ergocryptine, *J. Clin. Endocrinol. Metab.* **49:**120–126, 1979.
81. F. Peillon, F. Cesselin, D. Bression, N. Zygelman, A. M. Brandi, A. Nousbaum, and A. Maugorgne, *In vitro* effect of dopamine and L-dopa on prolactin and growth hormone release from human pituitary adenomas, *J. Clin. Endocrinol. Metab.* **49:**737–741, 1979.
82. P. Woolf and L. Lee, Effect of the serotonin precursor tryptophan on pituitary hormone secretion, *J. Clin. Endocrinol. Metab.* **45:**123–133, 1977.
83. W. M. Fraser, H. S. Tucker, S. R. Grubb, J. P. Wigand, and W. G. Blackard, Effect of L-tryptophan on growth hormone and prolactin release in normal volunteers and patients with secretory pituitary tumors, *Horm. Metab. Res.* **11:**149–155, 1979.
84. I. Lancranjan, A. Wirz-Justice, W. Puhringer, and E. Del Pozo, Effect of L-5-hydroxytryptophan infusion on growth hormone and prolactin secretion in man, *J. Clin. Endocrinol. Metab.* **45:**583–593, 1977.
85. Y. Nakai, H. Imura, H. Sakurai, H. Kurahachi, and T. Yoshimi, Effect of cyproheptadine on human growth hormone secretion, *J. Clin. Endocrinol. Metab.* **38:**446–449, 1974.
86. G. A. Smythe and L. Lazarus, Suppression of human growth hormone by melatonin and cyproheptadine, *J. Clin. Invest.* **54:**116–121, 1974.
87. G. A. Smythe and L. Lazarus, Growth hormone regulation by melatonin and serotonin, *Nature* **244:**230–231, 1973.
88. K. Chihara, Y. Kato, K. Maeda, S. Matsukura, and H. Imura, Suppression by cyproheptadine of human growth hormone and cortisol secretion during sleep, *J. Clin. Invest.* **57:**1393–1402, 1976.
89. W. B. Mendelson, L. S. Jacobs, J. D. Reichman, E. Othmer, P. E. Cryer, B. Trivedi, and W. H. Daughaday, Suppression of sleep-related prolactin secretion and enhancement of sleep-related growth hormone secretion, *J. Clin. Invest.* **56:**690–697, 1975.
90. R. Collu, F. Franschini, P. Visconti, and L. Martini, Adrenergic and serotonergic control of growth hormone secretion in adult rats, *Endocrinology* **90:**1231–1237, 1972.
91. M. A. Arnold and J. D. Fernstrom, Serotonin receptor antagonists block a natural, short-term surge in serum growth hormone levels, *Endocrinology* **103:**1159–1163, 1978.
92. M. A. Arnold and J. D. Fernstrom, L-Tryptophan injection enhances pulsatile growth hormone secretion in the rat, *Endocrinology* **108:**331–335, 1981.
93. R. M. MacLeod, Influence of norepinephrine and catecholamine-depleting agents on the synthesis and release of prolactin and growth hormone, *Endocrinology* **85:**916–923, 1969.
94. K. Chihara, A. Arimura, and A. V. Schally, Effect of intraventricular injection of dopamine, norepinephrine and 5-hydroxytryptamine immunoreactive somatostatin release into rat hypophyseal portal blood, *Endocrinology* **104:**1656–1662, 1979.
95. U. Knigge, F. Wollesen, A. Dejgaard, and B. Thuesen, Comparison between dose-responses of prolactin, thyroid stimulatory hormone and growth hormone to two different histamine H_2 receptor antagonists in normal man, *Clin. Endocrinol.* **15:**585–592, 1981.
96. H. E. Carlson and R. J. Chang, Studies on the role of histamine in human pituitary function, *Clin. Endocrinol.* **12:**461–466, 1980.
97. F. Cavagnini, C. Invitti, M. Pinto, G. Maraschini, A. Di-landro, A. Dubini, and A. Marelli, Effect of acute and repeated administration of gamma-amino butyric acid (GABA) on growth hormone and prolactin secretion in man, *Acta Endocrinol.* **93:**149–154, 1980.
98. M. Koule, R. Lammintausta, and S. Dahlstrom, Stimulatory effect of acute baclofen administration on human growth hormone secretion, *J. Clin. Endocrinol. Metab.* **48:**1038–1040, 1979.
99. B. W. Mendelson, R. A. Lantigue, R. J. Wyatt, J. C. Gillin, and L. S. Jacobs, Piperidine enhances sleep-related and insulin-induced growth hormone secretion: further evidence for a cholinergic secreting mechanism, *J. Clin. Endocrinol. Metab.* **52:**409–415, 1981.
100. S. Reichlin, Regulation of somatotropic hormone secretion, in: *Handbook of Physiology*, vol. 7 (R. Greep and W. Sawyer, eds.), American Physiological Society, Washington, D.C., 1974, pp. 405–447.
101. K. Kasai, H. Susuki, T. Nakamura, H. Shiing, and S. I. Shimoda, Glycine stimulated growth hormone release in man, *Acta Endocrinol.* **93:**282–286, 1980.

102. A. Besset, A. Bonardet, G. Rondovin, B. Descomps, and P. Passouant, Increase in sleep related GH and Prl secretion after chronic arginine aspartate administration in man, *Acta Endocrinol.* **99:**18–23, 1982.
103. M. Koule and R. Lammintausta, Effect of melatonin on L-tryptophan- and apomorphine-stimulated growth hormone secretion in man, *J. Clin. Endocrinol. Metab.* **49:**70–72, 1979.
104. U. Weinberg, E. D. Weitzman, A. D. Horowitz, and A. C. Burg, Lack of an effect of melatonin on the basal and L-dopa stimulated growth hormone secretion in man, *J. Neural Transm.* **52:**117–121, 1981.
105. W. Hamilton and D. M. Hussein, Growth hormone response to prostaglandin E_2, *Acta Paediatr. Scand.* **68:**25–156, 1979.
106. G. S. Tannenbaum, Evidence for autoregulation of growth hormone secretion via the central nervous system, *Endocrinology* **107:**2117, 1980.
107. J. O. Willoughby, M. Menadue, P. Zeegers, P. H. Wise, and J. R. Oliver, Effects of human growth hormone on the secretion of rat growth hormone, *J. Endocrinol.* **86:**165–169, 1980.
108. M. Molitch, H. Abe, J. Vanwyk, and L. Underwood, Human growth hormone (hGH) and somatomedin C (SMC) suppress the spontaneous release of GH in unanesthetized rats, Endocrine Society Abstracts, 64th Annual Meeting, 1982, p. 126.
109. G. S. Tannenbaum, H. J. Guyda, and B. I. Posner, Insulin-like growth factors: in vivo evidence for a role in growth hormone negative feedback and body weight regulation via the central nervous system, Endocrine Society Abstracts, 64th Annual Meeting, 1982, p. 126.
110. R. Fernandez-Durango, A. Arimura, J. Fishback, and A. V. Schally, Hypothalamic somatostatin and LHRH after hypophysectomy, in hyper- or hypothyroidism, and during anesthesia in rats, *Proc. Soc. Exp. Biol. Med.* **157:**235–240, 1978.
111. B. L. Baker and Y. Y. Yen, The influence of hypophysectomy on the stores of somatostatin in the hypothalamus and pituitary stem, *Proc. Soc. Exp. Biol. Med.* **151:**599–602, 1976.
112. D. L. Hoffman and B. L. Baker, Effect of treatment with growth hormone on somatostatin in the median eminence of hypophysectomized rats, *Proc. Soc. Exp. Biol. Med.* **156:**265–271, 1977.
113. Y. C. Patel, Growth hormone stimulates hypothalamic somatostatin, *Life Sci.* **24:**1589–1594, 1979.
114. M. C. Sheppard, S. Kronheim, and B. L. Pimstone, Stimulation by growth hormone of somatostatin release from the rat hypothalamus *in vitro, Clin. Endocrinol.* **9:**583–586, 1978.
115. M. Berelowitz, S. L. Firestone, and L. A. Frohman, Effects of growth hormone excess and deficiency on hypothalamic somatostatin content and release on tissue somatostatin distribution, *Endocrinology* **109:**714–719, 1981.
116. M. Berelowitz, M. Szabo, L. A. Frohman, S. L. Firestone, L. Chu, and R. L. Hintz, Somatomedin-C mediates growth hormone negative feedback by effects on both the hypothalamus and the pituitary, *Science* **212:**1279–1281, 1981.
117. K. Chihara, N. Minamilani, H. Kaji, A. Arimura, and T. Fujata, Intraventricularly injected growth hormone stimulates somatostatin release into rat hypophysial portal blood, *Endocrinology* **109:**2279–2281, 1981.
118. M. D. Lumpkin, A. Negro-Vilar, and S. M. McCann, Paradoxical elevation of growth hormone by intraventricular somatostatin: possible ultrashort-loop feedback, *Science* **211:**1072–1074, 1981.
119. R. Guillemin, A summary of current studies with somatocrinin, growth hormone releasing factor, *Clin. Res.* **31:**338–341, 1983.
120. J. B. Martin, Neuroendocrine regulation of growth hormone secretion, *Pediatric and Adolescent Endocrinology*, vol. 12 (Z. Laron, ed.), Karger, Basel, 1983, pp. 1–26.

24

Chemical, Anatomical, and Physiological Studies on Human Pancreatic and Rat Hypothalamic Growth Hormone-Releasing Factor

Wylie Vale, Joan Vaughan, Paul Sawchenko, Hans Seifert, Louise Bilezikjian, Marilyn Perrin, Larry Swanson, Michael O. Thorner, Joachim Spiess, and Jean Rivier

The hypothalamic regulation of growth hormone (GH) secretion is mediated by stimulatory and inhibitory factors, both of which reach the adenohypophysis via the hypothalamic–hypophysial portal system.[1,2] Somatostatin, which exerts inhibitory control over the secretion of GH, was purified from ovine hypothalamic extracts and characterized over 10 years ago[3] as a tetradecapeptide, and subsequently a biologically active, *N*-terminally extended, 28-amino-acid long form was identified in duodenal[4] and hypothalamic extracts.[5–7] Prior to 1982, several agents with GH-releasing activity had been described; none, however, were of high potency or otherwise fulfill the criteria expected of a physiological GH-releasing factor (GRF).

Several groups including our own were working to purify GRF from animal hypothalamus, and we had obtained a partial sequence of rat hypothalamic GRF in early 1982. The first fully characterized peptides with high GH-releasing potency and intrinsic activity were, however, isolated from two pancreatic

Wylie Vale, Joan Vaughan, Hans Seifert, Louise Bilezikjian, Marilyn Perrin, Joachim Spiess, and Jean Rivier • The Clayton Foundation Laboratories for Peptide Biology, The Salk Institute, La Jolla, California 92037. ***Paul Sawchenko and Larry Swanson*** • Developmental Neurobiology Laboratory, The Salk Institute, La Jolla, California 92037. ***Michael O. Thorner*** • Department of Pediatrics, University of Virginia School of Medicine, Charlottesville, Virginia 22901.

tumors removed from patients with acromegaly.[8,9] The presence of ectopic GRF-like activity in certain tumors removed from acromegalic subjects had been reported previously by Frohman, Daughaday, and others.[10,11]

The tumor we processed was obtained at the University of Virginia School of Medicine from a patient who presented symptoms of acromegaly and high GH levels that rapidly abated after surgical removal of the pancreatic tumor.[12] Acid extracts of this tumor were gel-filtered on G-50 Sephadex, and a zone of GH-releasing activity was found to elute in the 4K-5K-molecular-weight region as we had previously observed for rat hypothalamic GRF.[13] The active fraction was then subjected to multiple reverse-phase high-pressure-liquid-chromatographic steps,[8] and GH-releasing activity was identified by ability to stimulate GH secretion from dispersed rat anterior pituitary cells in monolayer cultures.[13–15] The isolated human pancreatic (tumor) GHF (hpGRF) was characterized by sequence analysis and peptide mapping techniques[16] to be a 40-amino acid, straight-chain peptide with free *N*- and *C*-termini (Table 1).

From a tumor removed from an acromegalic patient in Lyon, Guillemin and colleagues[9] purified this same 40-residue peptide (hpGRF-40), which accounted for two-thirds of the GH-releasing peptides of that tumor. The remaining one-third consisted of a *C*-terminally extended, amidated 44-amino-acid peptide [hpGRF(1-44)-NH_2 or hpGRF-44] and a *C*-terminally shortened 37-amino-acid peptide [hpG RF(1-37)-OH].

Recently, Mayo *et al.*[17] and Gubler *et al.*[18] have deduced the structure of the hpGRF precursors from the cDNA sequence of clones produced from mRNA purified from the Charlottesville or Lyon tumors, respectively. The predicted precursors in the tumors contain the 44-amino-acid peptide flanked by Arg–Arg on the *N*-terminus and Gly–Arg on the *C*-terminus with this Gly presumably providing the *C*-terminal amide. Thus, both tumors contain mRNA encoding the precursor for hpG RF(44)-NH_2 although most in one tumor and all in another tumor of hpGRF(44)-$NH_2$was further processed. It was pointed out[17] that all of the reported naturally occurring forms of hpGRF could be generated by cleavages involving a single arginine residue.

Both hpGRF(40)-OH and hpGRF(44)-NH_2 are potent GH secretagogues *in vivo* and *in vitro*. Although Guillemin *et al.*[9] reported hpGRF(1-44)-NH_2 to have more than three times the potency of hpGRF-40, we find the potency difference *in vitro* to be only 20–40% and that we ascribe to the *C*-terminal amidation rather than the extra four amino acids. Our structure function studies have shown that the entire *C*-terminal region is not critical for activity, as hpGRF(1-29)-NH_2 and other shortened analogues are fully potent (Table 2). Thus it would not be surprising to find *C*-terminal heterogeneity. It is possible that more than one form of GRF might be secreted as has been shown to be the case for somatostatin.[19]

In order to establish the relationship between hpGRF and human hypothalamic GRF, we have recently processed 10,000 human hypothalamic fragments provided to us by Dr. Seymour Reichlin, Tufts University. From several thousand of these fragments we were able to isolate two GRF's.[20,21] Final purification of these two peptides was accomplished after oxidation of both

Table 1 *GRF Family*[a]

	1 5 10 15 20 25 30 35 40
hpGRF-40	Y A D A I F T N S Y R K V L G Q L S A R K L L Q D I M S R Q Q G E S N Q E R G A
hpGRF-44	Y A D A I F T N S Y R K V L G Q L S A R K L L Q D I M S R Q Q G E S N Q E R G A R A R L ■
rhGRF	H A D A I F T S S Y R R I L G Q L Y A R K L L H E I M N R Q Q G E R N Q E Q R S R F N
pPHI	H A D G V F T S D F S R L L G Q L S A K K Y L E S L I ■
hPHM	H A D G V F T S D F S K L L G Q L S A K K Y L E S L M ■
VIP	H S D A V F T D N Y T R L R K Q M A V K K Y L N S I L N ■
Glucagon	H S Q G T F T S D Y S K Y L D S R R A Q D F V Q W L M N T
Secretin	H S D G T F T S E L S R L R D S A R L Q R L L Q G L V ■
GIP	Y A E G T F I S D Y S I A M D K I R Q Q D F V N W L L A Q Q K G K K S D W K H N I T Q

[a] hpGRF, human pancreatic growth hormone releasing factor; rhGRF, rat hypothalamic growth hormone releasing factor; PHI, peptide histidyl isoleucine; PHM, peptide histidyl methionine; VIP, vasoactive intestinal polypeptide; GIP, gastric inhibitory polypeptide.

Table 2 Relative Potencies of Analogues of hpGRF and rhGRF to Stimulate GH Secretion by Cultured Rat Anterior Pituitary Cells[a]

	Relative potency to stimulate GH secretion by cultured rat pituitary cells
hpGRF(1-40)-OH	100%
hpGRF(1-40)-NH_2	120
hpGRF(1-44)-OH	100
hpGRF(1-44)-NH_2	130
hpGRF(1-32)-OH	80
hpGRF(1-32)-NH_2	140
hpGRF(1-29)-OH	30
hpGRF(1-29)-NH_2	120
[Nle^{27}]-hpGRF(1-29)-NH_2	140
hpGRF(1-27)-OH	<1
hpGRF(1-27)-NH_2	15
rhGRF(1-43)-OH	260
rhGRF(1-32)-NH_2	700
rhGRF(1-29)-NH_2	510
[Nle^{27}]-rhGRF(1-29)-NH_2	830

[a] Relative potencies of hpGRF and rhGRF and analogues based on abilities to stimulate GH secretion by cultured anterior pituitary cells.

peptides to the respective methionine sulfoxide forms. These chemically modified peptides coeluted in reverse-phase high-performance liquid chromatography (HPLC) with the oxidized forms of synthetic hpGRF-44 and hpGRF-40, respectively.[21] On the basis of peptide mapping on HPLC after digestion of the methionine sulfoxide forms of the two native peptides with clostripain, it was indicated by retention times and amino acid analysis that both native peptides contained the *N*-terminal sequence of hpGRF(1-29) observed to be responsible for the full intrinsic activity of hpGRF. The analytical data obtained with the two native human hypothalamic GRF's (hhGRF's) and their digests are strongly suggestive for structural identity of hhGRF and hpGRF. It is probable, however, that the pancreatic tumors were making peptides identical to the hypothalamic ones, and it is likely that hpGRF indeed is the same as human hypothalamic GRF. Similar conclusions have been reached by Bohlen *et al.*[22] and by Reichlin and colleagues (personal communication).

Antibodies raised against hpGRF-40 have been used to detect GRF-like activity by immunocytochemistry[23,24] and radioimmunoassay (RIA) (Fig. 1) in the primate hypothalamus. Antibodies were produced in rabbits[15] in response to a variety of immunogens (Table 3). Routinely we use anti-hpG-56 serum[25] for RIA's of hpGRF which are performed in a manner similar to that described for corticotropin-releasing factor.[15] hpGRF-40 is iodinated by a mild chloramine T method, and [^{125}I]hpGRF-40 is then purified by adsorption and elution from a Bond-Elute C_{18} cartridge followed by HPLC using a Vydac C_{18} column with a 0.1% trifluoroacetic acid/acetonitrile system. Several peaks of radioactivity are reproducibly eluted, and the first major zone routinely displays the highest

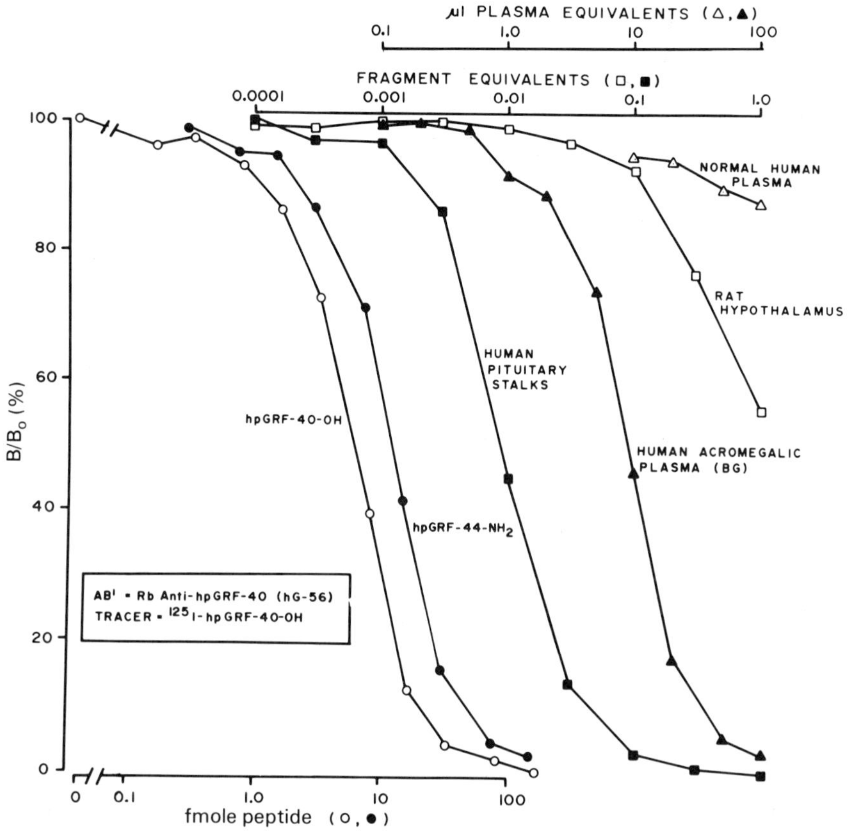

Figure 1 Radioimmunoassay of hpGRF. Displacement of [^{125}I]hpGRF(1-40)-OH from binding to rabbit anti-GRF-40 serum (hG-56) by hpGRF-40, hpGRF-44-NH_2, zone from gel filtrate of human infundibular extracts, zone from gel filtrate rat hypothalamic extracts, and CH Bond-Elute cartridge extracted human acromegalic plasma.

specific binding to the antibody used in the RIA. The RIA is carried out at 0°–4° C over a 2- to 3-day period using a pH 7.4 phosphate-buffered saline with 0.1% BSA, 0.1% Triton × 100, and 25 mM EDTA. Trace addition is delayed for 18 hr and bound from free tracer is separated by fixed *Staphococcus aureus* or goat anti-rabbit α globulin. The RIA's with hpG-56 detect both hpGRF-40 and hpGRF(1-32)-OH equivalently while hpGRF-44-NH_2 has slightly lower activity. Acid extracts of human pituitary stalks provided by S. Reichlin contain approximately 1 pmole hpGRF-like immunoactivity (hpGRF-LI) per fragment equivalent.

We extract plasma and other biological fluids such as cerebral spinal fluid by adsorption to cyclohexyl-derivatized silica cartridges (CH Bond-Elute columns from Analytichem International). hpGRF-LI is then eluted from the cartridges with a triethyl ammonium formate/isopropranol solvent system.[25]

Table 3 List of Immunogens Used for Production of Antisera to hpGRF and rhGRF

Rabbit/ antiserum no.	Immunogen
hpG-50	E^{18}-hpGRF(1-20)-OH coupled to h α-globulins with glutaraldehyde
hpG-51	Same as hpG-50
hpG-52	Same as hpG-50
hpG-56	hpGRF(1-40)-OH coupled to h α-globulins with glutaraldehyde
hpG-57	Same as hpG-56
hpG-58	Same as hpG-56
hpG-59	hpGRF(1-40)-OH coupled to h α-globulins with *bis*-diazotized benzidine
hpG-60	Same as hpG-59
hpG-61	Same as hpG-59
hpG-62	hpGRF(1-40)-OH coupled to keyhole limpet hemocyanin by glutaraldehyde
hpG-63	Same as hpG-62
hpG-64	Same as hpG-62
rG-71	rGRF(1-43)-OH coupled to h α-globulins with glutaraldehyde
rG-72	Same as rG-71
rG-73	Same as rG-71
rG-74	rGRF(1-43)-OH coupled to h α-globulins with *bis*-diazotized benzidine
rG-75	Same as rG-74
rG-76	Same as rG-74
rG-77	rGRF(38-43)-OH coupled to h α-globulins with glutaraldehyde
rG-78	Same as rG-77
rG-79	Same as rG-77

Plasma from the Charlottesville patient with the GRFoma contains approximately 0.7 nM hpGRF-LI, a concentration that would be maximally active in our *in vitro* GRF assay.

Rat hypothalamic extracts are very poorly detected by the hpG-56 RIA. These results along with our earlier data on the *N*-terminal sequence of rat GRF and its HPLC behavior suggested that rat GRF may be quite different from hpGRF.

The purification of GRF from rat hypothalamic tissue provided by Dr. A. Parlow and the NIADDK was monitored for GH-releasing activity with the rat anterior pituitary cell culture assay. Several batches totaling approximately 80,000 hypothalami were defatted with acetone, extracted under acidic reducing conditions, and gel filtered through a Sephadex G-50 column as previously described.[13] GH-releasing bioactivity as followed by the pituitary cell culture assay was found in the 4K- to 5K-dalton zone with CRF-LI and was separated from SS-28 or SS-14, which would have obscured the biological activity. Purification of rGRF was accomplished by successive HPLC steps, and the sequence of rat GRF was determined to be a 43-amino-acid peptide with a free *C*-terminus, differing from the corresponding residues of hpGRF-44 at 14 positions[26] (Table 1).

That the characterized 43-residue polypeptide was responsible for the assayed biological activity was supported by the finding that native rhGRF and its synthetic replicate did not differ significantly in their biological activities *in vitro*. Native and synthetic rhGRF were highly potent stimulators of the GH secretion *in vitro*; rhGRF appears to be slightly more potent than hpGRF in the rat anterior pituitary cell culture assay.(26) The 14 amino acid differences between rhGRF and hpGRF-44 are distributed over the whole molecule, but especially accumulated in the *C*-terminal end. All differences except two (phenylalanine-42 versus alanine-12 and asparagine-43 versus arginine-43) could be explained as the result of single-base mutations.

Because this was the only GRF-like peptide in our rat hypothalamic extracts and in view of our recent results with *C*-terminally directed antirat GRF antisera, we consider it highly likely that the mature form of rat GRF is not *C*-terminally amidated and that marked heterogeneity in this region of GRF may occur both between and within species (Table 1).

rhGRF and hpGRF belong structurally to the glucagon–secretin family. Although peptides of this group are of different sizes, a comparison of the sequences of the smaller peptides with corresponding *N*-terminal sequences of the larger peptides shows significant homologies. Thus, the sequence homologies of rhGRF with hpGRF-40, hpGRF-44, PHI-27, VIP, glucagon, secretin, and GIP are 70, 67, 52, 43, 31, 30, and 16%, respectively. None of the members of this family, except rhGRF and hpGRF, exhibit significant GH-releasing activity of their own at the pituitary level. It is noteworthy that rhGRF is slightly more closely related than is hpGRF to the other members of the family. The observation that rhGRF has an *N*-terminal histidine like most of the other peptides has led us to explore the possibility that rhGRF but not hpGRF might have VIP-like activity in other systems.(27)

Analogues of both GRF's have been synthesized and tested biologically in several systems. While *C*-terminally shortened peptides in the human series are very active, the peptides based on the rat structure are even more potent to stimulate GH secretion *in vitro* (Table 2). Substitution of the methionine^{27} by norleucine is tolerated by all GRF peptides and improves the chemical stability of the molecule. Clinical studies are underway with the [Nle^{27}]-hp[GRF(1-29)]-NH_2.

Antibodies have also been developed in rabbits to rhGRF(25) and have been used for RIA's and for immunocytochemical studies.(24) The rhGRF RIA using anti-GRF serum rG-71 and HPLC-purified [^{125}]I[Nle^{27}]-rhGRF can be used to detect GRF-LI in rat hypothalamic extracts (Fig. 2). Following acid extraction and gel filtration we measure approximately 110 fmoles rat GRF-LI per rat hypothalamic equivalent. This assay detects hpGRF very poorly.

Conventional indirect immunohistochemical methods with antiserum rG-71 were used to chart the distribution of rhGRF immunoreactivity in the rat brain(24) (Fig. 3). The results suggest the existence of at least two distinct GRF containing systems, one responsible for stimulating GH secretion from the anterior pituitary, and one whose relationship, if any, to hypophysiotropic function is less direct. First, a massive plexus of rhGRF-stained fibers was found that spanned the entire width of the portal capillary (external) zone of the median eminence. This projection almost certainly arises from a prominent group of immunoreactive neurons centered in the ventral part of the arcuate

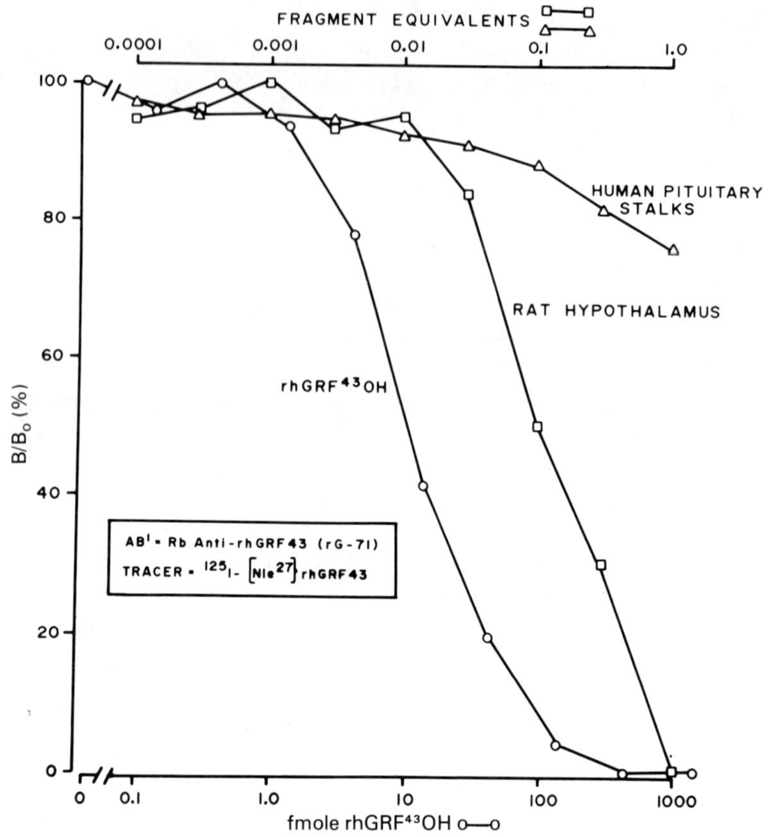

Figure 2 Radioimmunoassay of rhGRF. Displacement competition for binding of [^{125}I]rhGRF to rabbit anti-rhGRF (rG-71) by rhGRF, zone from gel filtrate of human infundibular extracts, and zone from gel filtrate of rat hypothalamic extracts.

nucleus of the hypothalamus. In colchichine-treated rats some 1000–15,000 rhGRF-stained cells were counted in this region on each side of the brain. Staining for rhGRF in this system was abolished by addition of synthetic rhGRF (1 μg/ml) to the primary antiserum, but was not diminished by preabsorbtion with higher concentrations (1 mg/ml) of any of the other known releasing factors of the hypothalamus, or of various proopiomelanocortin-derived peptides, which are known to be localized in neurons in the arcuate nucleus.

Apart from those in the arcuate region, cell bodies stained for the presence of rhGRF have been detected only in the region that immediately encapsulates the ventromedial nucleus of the hypothalamus. Cells in this area are not known to project to the median eminence and thus may be assumed to contribute to the extrahypophysiotropic projections described later.

From the level of the arcuate and ventromedial nuclei, rhGRF-immunoreactive fibers could be traced along the base of the brain and through the

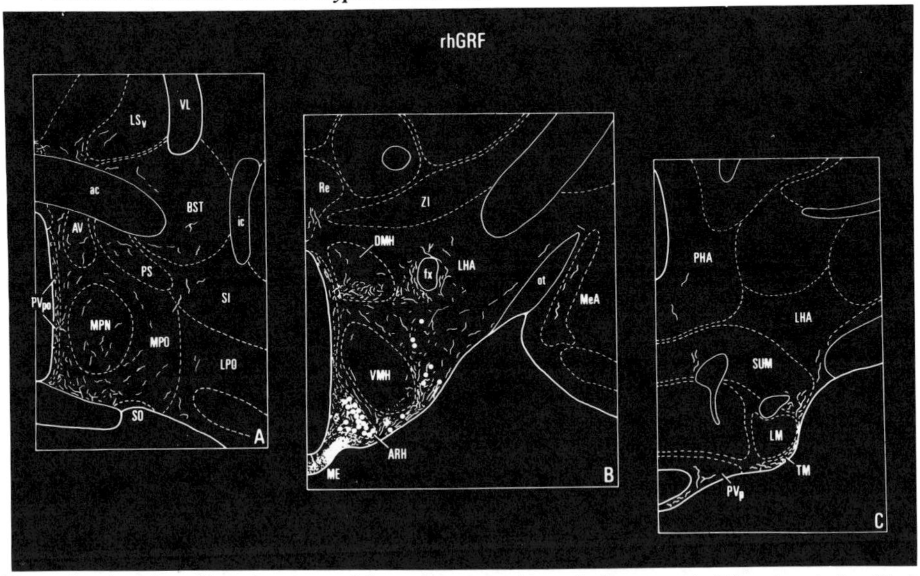

Figure 3 A series of line drawings of frontal sections through the preoptic (A), tuberoinfundibular (B), and mammillary (C) regions of the hypothalamus that summarize the distribution of rhGRF-immunoreactive cells and fibers in the rat brain. Cell bodies stained for rhGRF have been localized in just two regions. First, a prominent group of cells is found within and immediately adjoining the ventral part of the arcuate nucleus of the hypothalamus (ARH, see panel B) that most certainly correspond to the cells of origin of hypophysiotropic projections to the median eminence (ME). The second group of rhGRF-immunoreactive neurons is localized in the area that surrounds the ventromedial nucleus of the hypothalamus (VMH). Extrahypophysiotropic projections of one or both of these cell groups are distributed almost exclusively to structures within the hypothalamus and adjacent regions. Among the most prominent of these are the anterior part of the periventricular region, where rhGRF-immunoreactive fibers may interact with somatostatin-containing neurons, the medial preoptic area (panel A), the dorsomedial nucleus (DMH), the perifornical part of the lateral hypothalamic area (LHA; panel B), No rhGRF-stained fibers were detected caudal to the mammillary complex of the hypothalamus (plate C). ac = anterior commissure; BST = bed nucleus of the stria terminalis; fx = fornix; ic = internal capsule; LHA = lateral hypothalamic area; LM = lateral mammillary nucleus; LPO = lateral preoptic area; LS_v = lateral septal nucleus, ventral part; MeA = medial nucleus of the amygdala; MPN = medial preoptic nucleus; MPO = medial preoptic area; PHA = posterior hypothalamic area; PS = parastrial part of the medial preoptic area; PV_p, PV_{po} = periventricular nucleus (posterior, preoptic parts); Re = nucleus reuniens (thalamus); SI = substantia innominata; SO = supraoptic nucleus; SUM = supramammillary nucleus; TM = tuberomammillary nucleus; VL = lateral ventricle. From Sawsenko *et al.*[24]

periventricular region to discrete terminal fields in cell groups found almost exclusively in the hypothalamus and adjoining parts of the basal telencephalon. Projections were traced to all parts of the periventricular region of the hypothalamus, including the preoptic and anterior parts in which somatostatin-containing neurons that project to the median eminence are clustered. This raises the possibility of a synaptic interaction between rhGRF-containing fibers and somatostatin-containing neurons at the level of the hypothalamus. Other prominent terminal fields were found in discrete parts of the paraventricular, dorsomedial, suprachiasmatic, and premammillary nuclei, the medial preoptic area, and the perifornical part of the lateral hypothalamic area. Beyond the hypothalamus, relatively sparse projections have been traced to the lateral septal nuclei, the bed nucleus of the stria terminalis, and the medial nucleus of the amygdala, cell groups that comprise a part of the limbic region of the telencephalon. No rhGRF-stained cells or fibers have been detected in the neocortex, or in the brain stem caudal to the level of mammillary complex of the hypothalamus.

Synthetic rGRF and hpGRF are very powerful GH secretagogues *in vitro*, stimulating GH secretion minimally at 3–8 pmole, half-maximally at 20–100 pmole and maximally at ≤1 nmole. Somatostatin noncompetitively inhibits the GRF-induced stimulation of GH secretion.[8,13,28] In keeping with results with GH-secreting tumor cells,[29,30] GH-releasing activity of hpGRF or rGRF on normal pituitary cells is potentiated by pretreatment with thyroid hormones[13] and glucocorticoids.[13,31,32] Other groups have reported that somatomedins inhibit GH secretion.[28,33]

High-affinity binding sites for hpGRF and rGRF have been characterized on rat anterior pituitary tissue.[34] The radioligand for the receptor studies is the iodinated analogue of [His^1,Tyr^{10},Nle^{27}]-hpGRF(1-32)-NH_2, a modified analogue of human pancreatic GRF, which showed high *in vitro* potency in releasing GH and which is not susceptible to methionine oxidation during chloramine T iodination. The dissociation constant for [His^1,Nle^{27}]-hpGRF(1-32)-NH_2 is 0.7 nmole. The *C*-terminally amidated shortened analogues of 29 or 32 amino acids have the highest binding affinity. Homologous peptides from the glucagon–secretin series such as PHI, glucagon, and VIP do not compete for binding, and there is good correlation between the radioreceptor potencies and the *in vitro* biological potencies. Glucocorticoids increase GRF receptor numbers.[34] Physiologically the secretion of GH may be under positive control of a hypothalamic GRF closely related to hpGRF, thyroid hormones, and glucocorticoids and under negative regulation by the neuropeptide somatostatin and to a lesser extent somatomedins.

Both hpGRF and rhGRF increase production of cyclic AMP by pituitary cells.[35,36] This effect is rapid, but the half-maximally effective concentration for cyclic AMP production are circa 10-fold greater than those for GH release.[35] Somatostatin blocks GH release due to GRF during a 1-hr incubation and partially attenuates cyclic AMP. The calcium-dependency of GRF mediated GH release is indicated by the ability of $CoCl_2$ and verapamil to block GH release.[8,35,37] Eliminating calcium from the incubation medium also blocks GH release. Under these conditions, however, GRF still stimulates cyclic AMP production to the same extent.

Consistent with the possibility that GRF can stimulate GH synthesis, the incubation of cultured pituitary cells with GRF for 3 days increases the total amounts of GH (medium + cells) in the cultures.[13] Barinaga *et al.* have found that exposure of cells to GRF for 1 hr results in a two- to three-fold increase of GH gene transcriptional rates.[38] Similar effects on GH gene transcription were observed following injection of GRF *in vivo*.[38]

Although GRF can stimulate the rate of GH synthesis, cells continuously exposed to GRF do not maintain GH secretory rates at the high level initially induced by GRF. Following a 24-hr pretreatment of cells with various doses of GRF, GH secretory response is greatly attenuated to a subsequent challenge with the peptide.[39] The interpretation of these results, however, is complicated because under these conditions, cells are greatly depleted of their GH stores. To provide additional information, cyclic AMP levels have also been monitored. Results suggest that pretreatment with GRF decreases the responsiveness of adenylate cyclase to subsequent GRF treatment and that both desensitization and depletion of GH stores may be involved.

The determinants (stimulus concentration, duration, and frequency) and extent of the development of pituitary refractoriness to GRF will ultimately define the strategy for the exploitation of this new group of peptides. hpGRF, rhGRF, and their analogues are powerful GH secretagogues *in vivo* in numerous species including human beings, monkeys, rats, dogs, sheep, cattle, pigs, chickens, ducks, turkeys, and carp.[8,40–44] The use of GRF's to stimulate GH secretion for the purpose of promoting growth or anabolic processes is under consideration in many laboratories as to the possibility that GRF can be applied to the investigation and ultimately the management of certain human growth disorders.

ACKNOWLEDGMENTS. This work supported in part by NIH grants AM26741, HD13527, AA03504, and The Rockefeller Foundation. This research was conducted in part by the Clayton Foundation, California Division. Drs. Vale, Swanson, and Spiess are Clayton Foundation Investigators. The authors thank Susan O. McCall for manuscript preparation.

References

1. L. Krulich, P. Illner, C. P. Fawcett, M. Quijada, and S. M. McCann, in: *Growth and Growth Hormone* (A. Pecile and E. Muller, eds.), Excerpta Medica, Amsterdam, 1972, p. 306.
2. J. B. Martin, in: *Frontiers in Neuroendocrinology*, Vol. 4 (L. Martini and W. F. Ganong, eds.), Raven Press, New York, 1976, p. 129.
3. P. Brazeau, W. Vale, R. Burgus, N. Ling, M. Butcher, J. Rivier, and R. Guillemin, *Science* **179:**177, 1973.
4. L. Pradayrol, H. Jornvall, V. Mutt, and A. Ribet, *FEBS Lett* **109:**55–58, 1980.
5. J. Spiess, J. Villarreal, and W. Vale, *Biochemistry* **20:**1982–1988, 1981.
6. F. Esch, P. Bohlen, N. Ling, R. Benoit, P. Brazeau, and R. Guillemin, *Proc. Natl. Acad. Sci. USA* **77:**6827–6831, 1980.
7. A. V. Schally, W. Y. Huang, L. C. Chang, A. Arimura, T. W. Redding, R. P. Millar, M. W. Hunkapiller, and L. E. Hood, *Proc. Natl. Acad. Sci. USA* **77:**4489–4493, 1980.
8. J. Rivier, J. Spiess, M. Thorner, and W. Vale, *Nature* **300:**276–278, 1982.
9. R. Giullemin, P. Brazeau, P. Bohlen, F. Esch, N. Ling, and W. B. Wehrenberg, *Science* **218:**585–587, 1982.
10. L. Frohman, M. Szabo, M. Berelowitz, and M. Stachura, *J. Clin. Invest.* **65:**43, 1980.
11. S. Leveston, M. McKeel, D. Buckley, K. Deschryver, M. Greider, B. Jaffe, and W. Daughaday, *J. Clin. Endocrinol. Metab.* **53:**682, 1981.
12. M. O. Thorner, P. L. Perryman, M. J. Cronin, A. D. Rogol, M. Draznin, A. Johanson, R. Grindeland, W. Vale, E. Horvath, and K. Kovacs, *J. Clin. Invest.* **70:**965, 1982.

13. W. Vale, J. Vaughan, G. Yamamoto, J. Spiess, and J. Rivier, *Endocrinology* **112:**1553–1555, 1983.
14. W. Vale, G. Grant, M. Amoss, R. Blackwell, and R. Guillemin, *Endocrinology* **91:**562–572, 1972.
15. W. Vale, J. Vaughan, G. Yamamoto, T. Bruhn, C. Douglas, D. Dalton, C. Rivier, and J. Rivier, in: *Methods in Enzymology: Neuroendocrine Peptides,* Vol. 103 (P. M. Conn, ed.), Academic Press, New York, 1983, pp. 565–577.
16. J. Spiess, J. Rivier, M. Thorner, and W. Vale, *Biochemistry* **24:**6037–6040, 1982.
17. K. E. Mayo, W. Vale, J. Rivier, M. G. Rosenfeld, and R. M. Evans, *Nature* **306:**86–88, 1983.
18. U. Gubler, J. J. Monahan, P. T. Lomedico, R. S. Bhatt, K. J. Collier, B. J. Hoffman, P. Bohlen, F. Esch, N. Ling, F. Zeytin, P. Brazeau, M. S. Poonian, and L. P. Gage, *Proc. Natl. Acad. Sci. USA* **43:**4311–4314, 1983.
19. R. A. Peterfreund, and W. Vale, *Brain Res.* **239:**463–477, 1982.
20. J. Rivier, J. Spiess, and W. Vale, in: *Peptides: Structure and Function* (V. J. Hruby and D. H. Rich, eds.), Pierce Chemical Co., Rockford, IL, 1983, pp. 853–856.
21. J. Spiess, J. Rivier, and W. Vale, in: *Biochemical and Clinical Aspects of Neuropeptides–Synthesis, Processing, and Gene Structure* (G. Koch and D. Richter, eds.), Academic Press, New York, 1983, pp. 225–233.
22. P. Bohlen, P. Brazeau, B. Bloch, N. Ling, R. Gaillard, and R. Guillemin, *Biochem. Biophys. Res. Commun.* **114:**930–936, 1983.
23. B. Bloch, P. Brazeau, N. Ling, *et al., Nature* **301:**607–608, 1983.
24. Sawchenko, P. E., L. W. Swanson, J. Rivier and W. W. Vale, *J. Comp. Neurol.,* **237:**100–115.
25. W. Vale, J. Vaughan, D. Jolley, G. Yamamoto, T. Bruhn, H. Siefert, M. Perrin, and J. Rivier, in: *Methods in Enzymology: Neuroendocrine Peptides.* (P. M. Conn, ed.), Academic Press, Orlando, Florida, 1985, in press.
26. J. Spiess, J. Rivier, and W. Vale, *Nature* **303:**532–535, 1983.
27. S. J. Pandol, H. Seifert, M. W. Thomas, J. Rivier, and W. Vale. *Science* **225:**326–328, 1984.
28. P. Brazeau, N. Ling, F. Esch, P. Bohlen, C. Mougin, and R. Guillemin, *Biochem. Biophys. Res. Commun.* **109:**588, 1982.
29. Shapiro, L., H. Samuels and B. Yaffee, J. *Biol. Chem.* **252:**1977, 1979.
30. C. Bancroft, L. Levine, and A. Tashjian, *J. Cell. Biol.* **43:**432, 1969.
31. C. Webb, M. Szabo, and L. Frohman, *Endocrinology* **113:**1191, 1983.
32. W. B. Wehrenberg, N. Ling, P. Bohlen, F. Esch, P. Brazeau, and R. Guillemin, *Biochem. Biophys. Res. Commun.* **109:**562–567, 1982.
33. M. Berelowitz, M. Szabo, L. A. Frohman, S. Firestone, L. Chu, and R. L. Hintz, *Science* **212:**1279, 1981.
34. H. Siefert, M. Perrin, J. Rivier, and W. Vale, *Nature* **313:**487–489, 1985.
35. L. M. Bilezikjian, and W. W. Vale, *Endocrinology* **113:**1726, 1983.
36. M. Cronin, A. Rogol, L. Dabney, and M. O. Thorner, *J. Clin. Endocrinol. Metab.* **55:**381, 1982.
37. M. Szabo, L. Chu, and L. Frohman, *Endocrinology* **111:**1235, 1982.
38. M. Barinaga, G. Yamamoto, C. Rivier, W. Vale, R. Evans, and M. G. Rosenfeld, *Nature* **306:**84–85, 1983.
39. L. M. Bilezikjian and W. W. Vale, *Endocrinology* **115:**2032–2034, 1984 (Rapid Communication).
40. M. O. Thorner, J. Rivier, J. Spiess, J. L. Borges, M. L. Vance, S. S. Bloom, A. D. Rogol, M. J. Cronin, D. L. Kaiser, W. S. Evans, J. D. Webster, R. M. MacLeod, and W. Vale, *Lancet* **1:**24–28, 1983.
41. L. L. Anderson, J. J. Ford, J. Klindt, W. W. Vale, and J. Rivier, *Proceedings 7th International Congress of Endocrinology* Quebec, Canada, July 1–7, 1984, abstract #129, p. 325.
42. C. G. Scanes, R. V. Carsia, T. J. Lauterio, L. Huybrechts, J. Rivier, and W. Vale, *Life Sci.* **34:**1127–1134, 1984.
43. R. E. Peter, C. S. Nahorniak, W. W. Vale, and J. E. Rivier, *J. Exp. Zool.* **231:**161–163, 1984.
44. W. B. Wehrenberg, A. Baird, and N. Ling, *Science* **221:**556–558, 1983.

25

Growth Hormone-Releasing Factor

Isolation, Characterization, and Physiology

P. Brazeau, P. Bohlen, F. Esch, N. Ling, W. B. Wehrenberg, A. Briskin, and R. Guillemin

1. Isolation and Characterization

In June 1982, we isolated and characterized from aqueous extracts of a human pancreas tumor (obtained from Dr. Sassolas of Lyon) three molecules of growth hormone-releasing factor (GRF): hpGRF37(OH), hpGRF40(OH), and hpGRF44NH_2.[1] A few months later, with the use of our own two polyclonal antibodies reading specific portions of the molecule, we were able to isolate and partially characterize two of the molecules from human hypothalamiuses, hGRF40(OH) and hGRF44NH_2.[2] This work was accomplished by monitoring both immuno- and bioactivities of fractions eluted from gel filtration (Sephadex G-75) columns and other fractions obtained from reverse-phase high-pressure liquid chromatographies. More recently, by adding a prepurification affinity–chromatography step to our isolation scheme, we successfully isolated and characterized two other hypothalamic GRF's: one of porcine origin, the other of bovine origin.[3,4] The characterization was accomplished by the Edman degradation using gaz phase sequencing on 6.8 and 2.2 nmoles of pure material, respectively. Although both primary structures of these two new GRF molecules have a high degree of homogeneity with the human GRF44NH_2, they still show some heterogeneity.

The message coding for human GRF44NH_2 was recently characterized in collaboration with the research team of Dr. Gubler at Roche Institute.[5] Since the message code for residue Gly–Arg following the Leu_{44} is a typical clipping

P. Brazeau • Laboratoire de neuroendocrinologie expérimentale, Hôpital Notre-Dame, Montréal, Quebec, H21 4K8 Canada. ***P. Bohlen, F. Esch, N. Ling, W. B. Wehrenberg, A. Briskin, and R. Guillemin*** • Neuroendocrinology Laboratories, Salk Institute, San Diego, California 92037.

site of a precursor to generate an amidated leucine residue, in addition to the fact of the higher specific activity of $hGRF44NH_2$ over hGRF44(OH) and any shorter form of the molecule,[6] we believe that a $GRF44NH_2$ molecule in these three different species is the mature form of GRF.

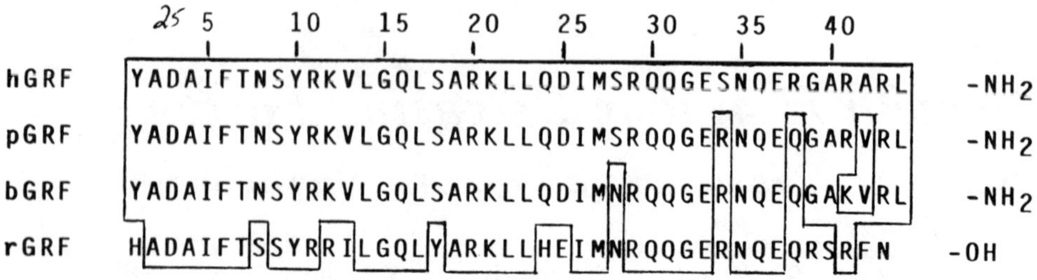

2. Physiological Studies In Vitro

Rat GRF is equipotent to human GRF *in vitro* when assayed on rat pituitary cells using the same defined media conditions. Under the same parameters synthetic porcine and bovine GRF's show 75% and 62% of the specific activity of the human GRF (Table 1). This significant difference can be best explained by variations in the molecular structure within the *C*-terminal portion from residue 24–44. We know by experience that deletion of residues in that portion of the molecule will affect the specific activity proportionaly from 100% to less than 5% (Tables 2 and 3). Although this is not a complete explanation in itself, it led us to believe that variations of the molecule in this segment of the molecule, as is the case for porcine and bovine GRF's, may produce some variations of the specific activity but not of the intrinsic activity. However, a note of caution must be introduced here in interpreting these results obtained

Table 1 Percent of Secretion (as Compared to Control)

GRF (moles/ ml)	$hGRF44NH_2$ (synthetic)	$pGRF44NH_2$ (synthetic)	$bGRF44NH_2$ (native)	rGRF43(OH) (synthetic)
0.0	100[a]	100	100	100
3.1	323	271	254	308
6.3	421	340	327	415
12.5	527	495	452	524
25.0	625	608	554	638
50.0	702	689	684	709
100.0	727	742	732	725
Potency[b]	1.00	0.76	0.62	0.99
(95% confidence limit)		(0.64–0.83)	(0.50–0.70)	(0.92–1.07)

[a] Mean results of three assayed wells immunoassayed for rat GH in duplicate.
[b] Calculated as a six-point bioassay from 3.1 to 50.0 fmoles/ml using our "Exbiol" program.

Table 2 Relative Potencies of CO-OH-Terminal Deletion Analogues of hpGRF-44-NH$_2$

Analogues	Potency	95% confidence limits
hpGRF-44-NH$_2$	100	
hpGRF(1-44)OH	61	50–75
hpGRF(1-40)NH$_2$	49	38–62
hpGRF(1-40)OH	30	25–37
hpGRF(1-37)NH$_2$	28	23–33
hpGRF(1-37)OH	12	9–16
hpGRF(1-34)OH	17	12–25
hpGRF(1-31)OH	9	6–12
hpGRF(1-28)OH	6	4–9
hpGRF(1-24)OH	0.014	—
hpGRF(1-21)OH	<0.002[a]	—
hpGRF(1-19)OH	<0.001[a]	—

[a] Dose–response curve not parallel to standard.

from cultured rat pituitary cells; they might be different if they were obtained from cultured human, porcine, or bovine pituitary cells; different GRF structure can also mean different receptors and different specific activities.

We know that the intrinsic activity of human GRF is contained in the molecule from residue 1–24; any deletion within this portion of the molecule will not only alter the specific activity by more than 99% but will also irreversibly affect the intrinsic activity. We have data confirming that an alteration of the *N*-terminal part of the molecule, by deletion of Tyr$_1$ and/or Ala$_2$ and deletion of residues 1–24 will abolish the intrinsic activity of the molecule. Moreover, the deletions at the *N*-terminus will convert these analogues into partial antagonists when assayed against a sub E_{max} dose of hGRF44NH$_2$. These findings lead us to rationalize that the *N*-terminus-deleted analogues are still recognizable by the receptors of rat somatotrophs. But here again a note of

Table 3 Relative Potencies of Position 27 or 1 Substituted Analogues of hpGRF-40-OH

Analogue	Potency	(95% confidence limits)
hpGRF-44-NH$_2$	100	
[Normal^{27}]hpGRF-40-OH	99	92–118
[Ala–Tyr^1]hpGRF-40-OH	4.2	2.4–6.5
[Arg–Tyr^1]hpGRF-40-OH	0.4	0.1–1.1
[Ac–Tyr^1]hpGRF-40-OH	10	7–14
[D–Tyr^1]hpGRF-40-OH	0.006[a]	—
[Ala^1]hpGRF-40-OH	0.84	0.7–1.0
[Phe^1]hpGRF-40-OH	3.1	2.5–3.9
[His^1]hpGRF-40-OH	54	46.5–63.2
[Trp^1]hpGRF-40-OH	0.006[a]	—
Rat GRF	93.7	82.6–106.3

[a] Dose–response curve not parallel to standard.

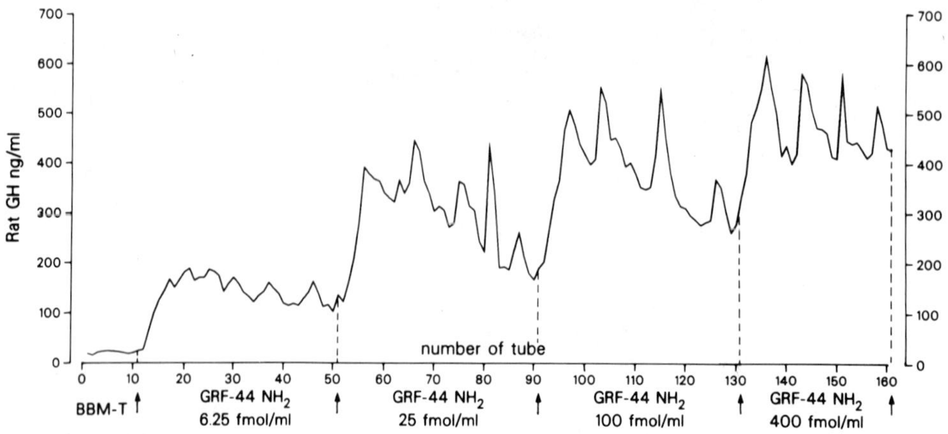

Figure 1 Dose–response effect of human $GRF44NH_2$ on rGH-secreting pattern of two perfused rat pituitaries. (Each tube corresponds to a 2-min collection.)

caution should be given before extrapolating the significance of these findings to possible human applications.

In a perifusion system, using redispersed cultured rat pituitary cells, we obtained some interesting results. First, we found that a 2-min pulse stimulation with $hGRF44NH_2$ resulted in a prompt monospecific secretion of GH within 30 sec, which is a characteristic response for a hypothalamic factor.[6] Second, a 1-hr continuous stimulation with a sub E_{max} dose (80%) of hGRF resulted in a secretion of GH in an oscillating harmonic fashion[7]; the harmonic secretion varied in its periodicity from 5–8 to 13–20 min between intervals depending on the intensity of dose perifused (Fig. 1); as we increased the dose, the periodicity of the oscillations became slower but the intensity and amount of hormone secreted increase (Fig. 1). These phenomena will hold true up to a point of hysteresis mimicking a pseudotetanic pattern leading to slowly diminished secretion. We were at first surprised by these results, but they were reconfirmed many times. We expected a smooth, plateaulike response of the somatotrophs, as suggested in the literature, but none of the results described in the literature were based on a minute-to-minute collection of the perfused medium, thus missing the variations that we observed. Trying to explain this oscillating secreting response, which is also true for luteinizing-releasing factor on luteotrophs and naturally occurring prolactin secretion of lactotrophs, we hypothesized the existence of a common biochemical denominator inducing such a harmonic secretion.[7] cAMP was the first substance that we suspected could be responsible. With 100 femtomoles GRF/ml (an 80% E_{max} dose), perfused pituitaries show no oscillations but a smooth plateau when 8Br-cAMP or phosphodiesterase is added in a sufficient amount to the perfused medium. These findings demonstrate that cAMP secreted by specific cells responding to specific stimuli is the mediator between cells harmonizing their oscillating secretion.

3. Physiological Studies In Vivo

It is commonly known that GH secretion is under the control of two hypothalamic hormones, one with a negative effect (somatostatin) and the other with a positive effect (somatocrinin, GRF). This duality of control makes any *in vivo* studies more complex than the *in vitro* studies where cultured pituitary cells are separated from this dual control.

It is also common knowledge that mammals, including primates, will secrete GH in an episodic fashion.[8] This is even more so in the rat where there are seven 3-hr, 15-min pulses of GH (300–600 ng/ml) secretion a day, alternating with trough periods of 60–90 min. Many experimental data lead us to believe that these troughs of GH secretion are due to periodical secretion of somatostatin. Therefore, if one wants to study *in vivo* the releasing activity of a GRF preparation with homogeneity of the response, one must use an experimental model where somatostatin secretion is eliminated, since pulses of GH are not synchronized from one rat to another. One method of doing this is the use of a passive immunization approach, whereby, for a few days, corresponding to the half-life of γ-globulins, somatostatin secretion is neutralized by an antisomatostatin serum. Another method by which endogenous somatostatin secretion can be reduced is the use of an anesthetic dose of Na–pentobarbital (50 mg/kg); however, this approach has its limitation in time (few hours).

Using such preparations, Dr. Wehremberg and our group at the Salk Institute were able to accomplish a variety of physiological studies in rats.

With the Na–pentobarbital model, it was demonstrated first that intravenous administration of hGRF will have its peak effect on rGH secretion at 3–4 min after injection and that the effect will end 20–30 min later depending on the dose administered.[9] Empirically the E_{max} dose of GRF's to be administered (i.v.) in the rat is approximately 0.5 μg/kg body weight; beyond that dose (1, 3, 5, and 10 μ/kg) no substantial increase of GH secretion can be obtained.

When GRF is administered subcutaneously in the rat, the peak effect on GH secretion is delayed up to 20 min, terminating 90 min later, which is approximately 5 times later compared to i.v. injection. However, the amount of rGH secreted in response to a high dose of GRF, 500 ng/kg, is much less marked, ≈ 20 times less compared to i.v. injection. We have the impression that there is either adsorption and/or degradation of the peptide before it reaches the pituitary gland. More studies are needed in order to find a suitable galenic preparation to eliminate those losses of activity. This concept of degradation of the peptide into a less active form is reinforced by our structure activity studies done *in vivo*. Deletion of amino acid residues at the carboxy-terminal is seen by *in vitro* analysis to cause a substantial reduction in specific activity; such a reduction of activity cannot be as clearly defined *in vivo* at least for a deletion down to GRF34(OH) even if the GRF31(OH) and the GRF28(OH) analogues show definite reduction of specific activity (unpublished data). We hypothesized that this discrepancy could be due to tissue adsorption of the

Table 4 HGRFs Degradation Analysis

Time exposure to serum	Bioactivity (% recovery)	Immunoreactivity (% bound to 456-10)
0 min[a]	96	—
1 min[a]	101	—
2 min[a]	98	—
4 min[a]	92	—
6 min[a]	93	—
8 min[a]	95	—
10 min[a]	101	99
12 min[a]	89	—
15 min[a]	93	—
30 min[a]	97	96
18 hr[b]	98[c]	97[c]
24 hr[b]	99[c]	96[c]
30 hr[b]	101[c]	98[c]
72 hr[b]	101[c]	96[c]

[a] To human serum.
[b] To rat serum.
[c] Frozen −20°C.

longer GRF molecules and/or enzymatic degradation in blood; this latter possibility, however, does not hold firm because we have demonstrated that GRF40(OH) and $GRF44NH_2$ are stable in human and rat serum (Table 4). Nevertheless, we believe that this point of discrepancy needs more in-depth investigation.

Both *in vivo* and *in vitro*, rat GRF is equipotent to human $GRF44NH_2$ (Table 1).

When perfused i.v. over a 30 min period, a high dose (500 ng/min) of GRF will drain the pituitary of its readily available GH while a small dose (20 ng/min) will not create this exhaustion of the GH supply even though a substantial secretion will be maintained (10- to 15-fold over control). This difference in response raises a very important question. What is the physiological dose to be used to initiate physiological response from the pituitary?

Since it is possible that overstimulation by GRF can cause a substantial depletion of the GH content at the pituitary level, a slow reversible process by *de novo* synthesis, and because of the fact that such an induced secretion of GH can trigger a negative feedback through the secretion of either somatostatin or somatomedins, one should be very cautious about the dosage to be used for further clinical and fundamental investigations.

If one infuses 15 μg/hr or 1.5 μg/hr of $GRF44NH_2$ to rats for 9 hr, in both cases a substantial elevation of rGH can be observed. However, one can also observe that at the high dose, a reduction in the secretion response is seen even before the end of the infusion period, while at the low dose, the elevation of the secretion of GH, though more modest, lasts throughout the period of infusion. If, on the other hand, one infuses GRF at a high dose for 24 hr, the following day a 2-μg bolus injection of GRF in the same animal will have a very diminished secretory response. This may be explained by the depletion of the pituitary GH content by more than 85%–95% the day before (unpublished data). We assume that the pituitary can replenish itself over a few days, but during this lengthy process the somatotrophs may be less responsive to further stimulation with GRF and probably more so owing to stimulation of somatostatin and somatomedin secretion.

Using a passive immunization (antisomatostatin) approach in our *in vivo* studies, we first demonstrated that this preparation enables us to stimulate in a dose–response manner the secretion of rGH with GRF even at a time when a trough period is expected. These findings seem to confirm that these trough periods are due to endogenous somatostatin secretion.

Finally, we have shown that, with the use of two antibodies, one monoclonal raised *in vitro* against native rat GRF[(10)] and the other raised against the sequenced rat GRF (by Vale *et al.*), in both cases pulses of rGH could be abolished (unpublished data). However, administration of the monoclonal antibody with exogeneous rat GRF did not abolish or diminish the rGH secretion response, while it did so *in vitro*.[(10)]

4. Human Studies

In human studies, the dose–response results obtained are in keeping with those obtained in laboratory animals; the physiological intravenous dose to be used should not exceed 0.5 μg/kg body weight. A higher dose will not increase the response and can only induce a negative feedback effect through secretion of somatostatin and somatomedins.[(6)]

Hypopituitary dwarfs, as expected, respond poorly or not at all to hGRF. Whether or not the GH secretion in this case can be improved by chronic injection is still speculative.

In normal volunteers, the response to a bolus injection (100 μg, i.v.) of $hGRF44NH_2$ will vary with the age of the subject. If the average response is strong for persons between 20 and 30 years of age, the response is strongly diminished in persons 40 years of age or more.

In order to elucidate the physiological reasons for a lower responsiveness of older persons, we went back to our *in vitro* model using perifused rat pituitary cells. By comparing the responsiveness of cultured rat pituitary cells of an 8-week-old rat to the response of a 6-month-old rat, we found the following: young rat pituitary cells will respond strongly (20- to 30-fold increase) to an 80% E_{max} dose of hGRF (40 femtomoles/pituitary) (Fig. 2), while the response of old rat pituitary cells to the same dose will be poor (three- to fivefold at the most) (Fig. 3). Based on findings by others (Dr. P. M. Dubois, in France), claiming the presence of immunoreactive somatostatin in somatotrophs, we hypothetized that the lower responsiveness of older persons could be due to an age-related accumulation of somatostatin in somatotrophs. To test this hypothesis, we perifused with the same constant dose of hGRF a high-affinity specific antisomatostatin serum; the results obtained were clear. Addition of

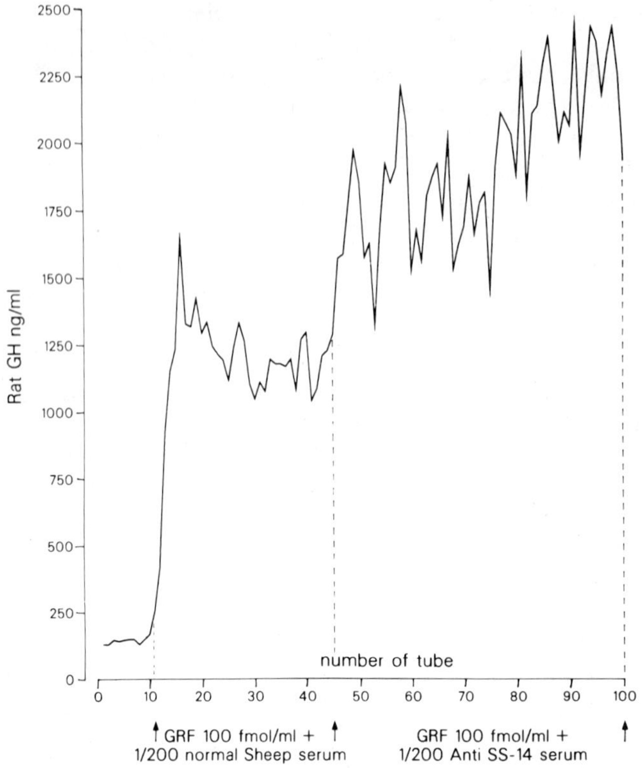

Figure 2 Effect of $hGRF44NH_2$ on the rGH-secreting pattern of two 8-week-old perfused rat pituitaries, with and without anti SS_{14} serum. (Each tube correspond to a 2-min collection.)

the antiserum initiated some sort of a "rejuvenation" of the response to GRF in a timewise fashion (Figs. 2 and 3); as if progressively, the action of somatostatin present in somatotrophs was negated by the presence of the antiserum. With proper control, using normal serum, we have eliminated the possibility of a nonspecific protein effect. To further elucidate this physiological phenomenon we are now investigating by use of RIA and bioassay the content and/or secretion of somatostatin of somatotrophs of different species at different ages.

5. Conclusion

In conclusion, I would like to stress a few physiological facts that I now believe are of great importance for the further use of GRF's both clinically and fundamentally.

1. Growth hormone, contrary to other pituitary hormone, is under a dual hypothalamic control[11]: somatostatin and somatocrinin (GRF). Because of this physiological aspect, one should be extremely cautious about the methodology

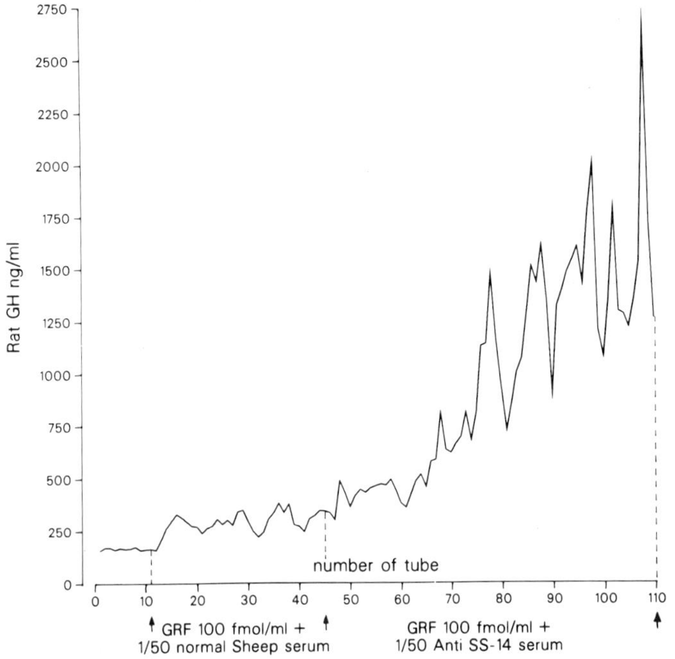

Figure 3 Effect of $hGRF44NH_2$ on the rGH-secreting pattern of two 6-month-old perfused rat pituitaries, with and without anti SS_{14} serum. (Each tube corresponds to a 2-min collection.)

and dosage to be used in these studies. We have to think in terms of a "limited physiological window" for GRF.

2. Overstimulation of somatotrophs with hGRF may result in a depletion of the readily secretable pool of GH; this will be followed by a slow process of replenishing, a period when the somatotrophs are to be considerably less responsive.

3. Overstimulation of the somatotrophs with GRF will also lead to a negative feedback, triggering an overproduction of both somatostatin and somatomedins; this will also have a time-delaying effect for the chronic use of GRF.

4. If I were to suggest a preferred route of administration, I would strongly suggest the intravenous route; the subcutaneous and intramuscular routes lead to losses of activity that are not yet explained. Until properly explained and a proper galenic preparation is made available to ensure reproduction of the pulsatile release of GH, these modes of administration should be avoided clinically; they may engender long-term adverse physiological effects.

References

1. R. Guillemin, P. Brazeau, P. Böhlen, F. Esch, N. Ling, and W. Wehrenberg, Growth hormone releasing factor from a human pancreatic tumor that caused acromegaly, *Science* **218:**585, 1982.

2. P. Böhlen, P. Brazeau, B. Bloch, N. Ling, R. Gaillard, and R. Guillemin, Human hypothalamic growth hormone releasing factor (GRF): Evidence for two forms identical to tumor derived GRF44NH_2 and GRF40(OH), *Biochem. Biophys. Res. Commun.* **114:**930, 1983.
3. P. Bohlen, F. Esch, P. Brazeau, N. Ling, and R. Guillemin, Isolation and characterization of the porcine hypothalamic growth hormone releasing factor, *Biochem. Biophys. Res. Commun.* **116:**726, 1983.
4. F. Esch, P. Bohlen, N. Ling, P. Brazeau, and R. Guillemin, Isolation and characterization of the bovine hypothalamic growth hormone releasing factor, *Biochem. Biophys. Res. Commun.*, **16:**726–734, 1983
5. U. Gubler, J. J. Monahan, P. T. Lomedico, R. S. Bhatt, K. J. Collier, B. J. Hoffman, P. Bohlen, F. Esch, N. Ling, F. Zeytin, P. Brazeau, M. S. Poonian, and L. P. Gage, Cloning and sequence analysis of cDNA for the precursor of human growth hormone releasing factor, somatocrinin, *Proc. Natl. Acad. Sci. USA* **80:**4311, 1983.
6. P. Brazeau, N. Ling, P. Böhlen, F. Esch, S.-Y. Ying, and R. Guillemin, Somatocrinin, growth hormone releasing factor, releasing pituitary growth hormone *in vitro, Proc. Natl. Acad. Sci. USA* **79:**7909, 1982.
7. R. Guillemin, P. Brazeau, A. Briskin, and A. J. Mandell, Evidence for synergetic dynamics in a mammalian pituitary cell perifusion system, in: *Synergetics of Brain* (H. Haken, ed.), Springer-Verlag, New York, (in press), 1983, pp. 155–162.
8. G. S. Tannenbaum and J. B. Martin, Evidence for an endogenous ultradian rhythm governing growth hormone secretion in the rat, *Endocrinology* **98:**562, 1976.
9. W. B. Wehrenberg, N. Ling, P. Brazeau, F. Esch, P. Böhlen, A. Baird, S-Y. Ying, and R. Guillemin, Somatocrinin, growth hormone releasing factor, stimulates secretion of growth hormone in anesthetized rats, *Biochem. Biophys. Res. Commun.* **109:**382, 1982.
10. R. A. Luben, P. Brazeau, P. Böhlen, W. B. Wehrenberg, and R. Guillemin, Monoclonal antibodies to growth hormone releasing factor by *in vitro* immunization with picomoles of antigen. Generality of the method, *Science* **218:**887, 1982.
11. J. B. Martin, P. Brazeau, G. S. Tannenbaum, J. O. Willoughby, J. Epelbaum, L. C. Terry, and D. Durand, Neuroendocrine organization of growth hormone secretion, in: *The hypothalamus*, Vol. 56 (S. Reichlin, R. Baldasserini, and J. B. Martin, eds.), Raven Press, New York, 1977, pp. 329–357.

26

Ectopic Growth Hormone-Releasing Factor Syndromes

Lawrence A. Frohman, Jennifer L. Thominet, and Marta Szabo

1. Historical Aspects

The association of carcinoid tumors of the bronchial tree and the intestine with acromegaly was first reported in 1960,[1] and selective case reports appeared throughout the next two decades.[2–5] The rationale for proposing that growth hormone-releasing factor (GRF) might be produced by tissues other than the hypothalamus was based on these observations which suggested that the association was more than fortuitous. The possibility that the extrapituitary tumors secreted growth hormone (GH) was suggested by the report of Beck *et al.*[6] in which extracts of human lung tumors stimulated GH release from superfused rat pituitaries. The patients from whom the tumors were removed, however, did not exhibit clinical evidence of GH hypersecretion. In the mid 1970s, two reports[4,5] appeared describing four patients with acromegaly associated with bronchial carcinoid tumors who had remission of the clinical features of their disorders and/or a return of GH levels to normal after removal of the bronchial tumor and without therapy directed to the hypothalamic–pituitary region. Immunoreactive GH was demonstrated in one of these tumors but not in the others. The possibility that GH hypersecretion and pituitary tumor formation could occur on the basis of excessive stimulation of somatotrophs by GRF was further supported by preliminary reports demonstrating the presence of GH-releasing activity in plasma of acromegalic subjects unassociated with carcinoid tumors.[7,8]

Lawrence A. Frohman and Jennifer L. Thominet • Division of Endocrinology and Metabolism, Department of Internal Medicine, University of Cincinnati College of Medicine, Cincinnati, Ohio 45267. *Marta Szabo* • Division of Endocrinology and Metabolism, Michael Reese Hospital, Chicago, Illinois 60616.

In 1979 we reported the presence of GRF-like activity in an extract of a bronchial carcinoid removed from a patient previously hypophysectomized because of acromegaly.[9] The patient continued to exhibit GH hypersecretion following the neurosurgical procedure despite development of panhypopituitarism, and GH levels became undetectable only after removal of the bronchial tumor 10 years later. The tumor exhibited characteristic histological features of a bronchial carcinoid, was devoid of GH, and a 2N acetic acid extract of the tumor stimulated GH release from dispersed rat pituitary cells in primary culture. The response was hormone-specific, could not be attributed to the presence of biogenic amines, and the activity was shown to be peptidergic in nature. Within the next few years, additional tumors with similar GRF-like properties were identified by us and by others.[10–12]

2. *Biological Characterization*

Biological activity was demonstrable in a variety of *in vivo* and *in vitro* systems including dispersed pituitary monolayer cultures,[9,11] perifused intact pituitaries,[10,11,13] anesthetized rats *in vivo*,[14,15] and human GH-secreting adenomas in culture.[16]

Using crude or partially purified tumor tissue extracts, considerable information was obtained concerning the mechanism of action of GRF, all of which has been subsequently confirmed using synthetic peptide. GH release is stimulated within 5 min of exposure to GRF both *in vitro*[11,13] and *in vivo*,[14,15] and the effects are dose-responsive. The *in vitro* effects persist for at least 24 hr and consist of a stimulation of hormone synthesis as well as release.[17] The effects of GRF are calcium-dependent and are suppressed or abolished by the absence of extracellular Ca^{2+} or Ca^{2+} channel blockers.[17] GRF stimulates cAMP formation though its role as an obligatory intermediate in GH-releasing effects is uncertain.[18] The effects of GRF require the presence of glucocorticoids and are enhanced by estrogen.[19] GRF stimulation of GH release is inhibited by somatostatin[9] in a noncompetitive manner. In addition, the GH-releasing effects are also inhibited by somatomedin-C.[20] Partially purified GRF also stimulates the release of GH from human somatotroph tumor tissue *in vitro*.[16] This observation was the first demonstration of the responsiveness of adenomatous pituitary tissue to GRF and provides support for the hypothesis that GRF exerts a role in the pathogenesis of GH-secreting tumors.

3. *Spectra of GRF-Secreting Tumors*

Initial studies were conducted using rat pituitary monolayer cultures as a GRF bioassay system. Acetic acid extracts of 46 tumors were examined for GH-releasing activity (Table 1); the results are shown in Fig. 1. The largest number of bioactive tumors were carcinoid tumors (8 out of 15), pancreatic tumors (2 out of 4), and small cell carcinomas (3 out of 6). One of eight neural tumors,

Table 1 Ectopic GRF Production—Tumors Examined

Tumors	Total		GH hypersecretion	
Carcinoid	15		6	
Bronchial		9		3
Intestinal		5		3
Other		1		
Pancreatic	4		2	
Small cell carcinoma of lung	7		1	
Neural tumors	8			
Neuroblastoma		3		
Pheochromocytoma		3		
Glioblastoma		1		
Neurofibroma		1		
Other tumors	12		1	
Sarcoma		7		
Adrenal adenoma		1		1
Miscellaneous		4		
Total	46		10	

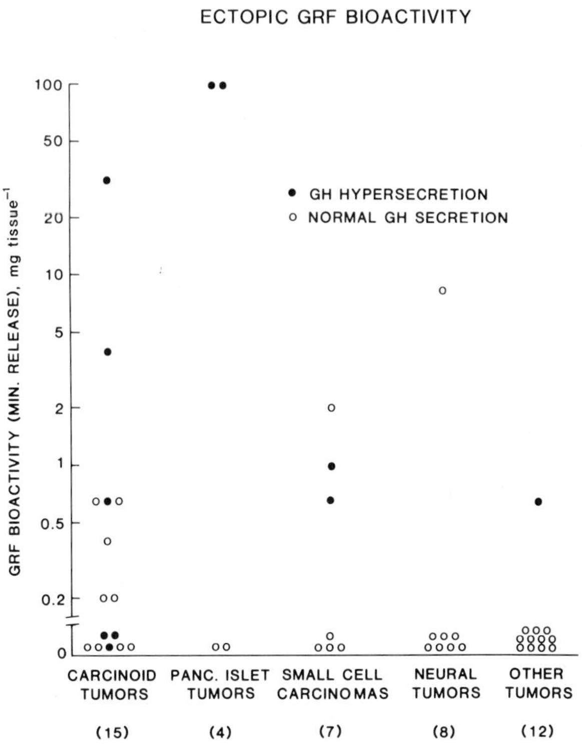

Figure 1 GH-releasing activity in tumors. Shown is the reciprocal of the least quantity of tissue required for demonstration of biological activity. Tumors from patients with clinical or biochemical evidence of GH hypersecretion are shown in solid circles and those from patients with normal GH secretion in open circles. Reprinted with permission from Frohman.[21]

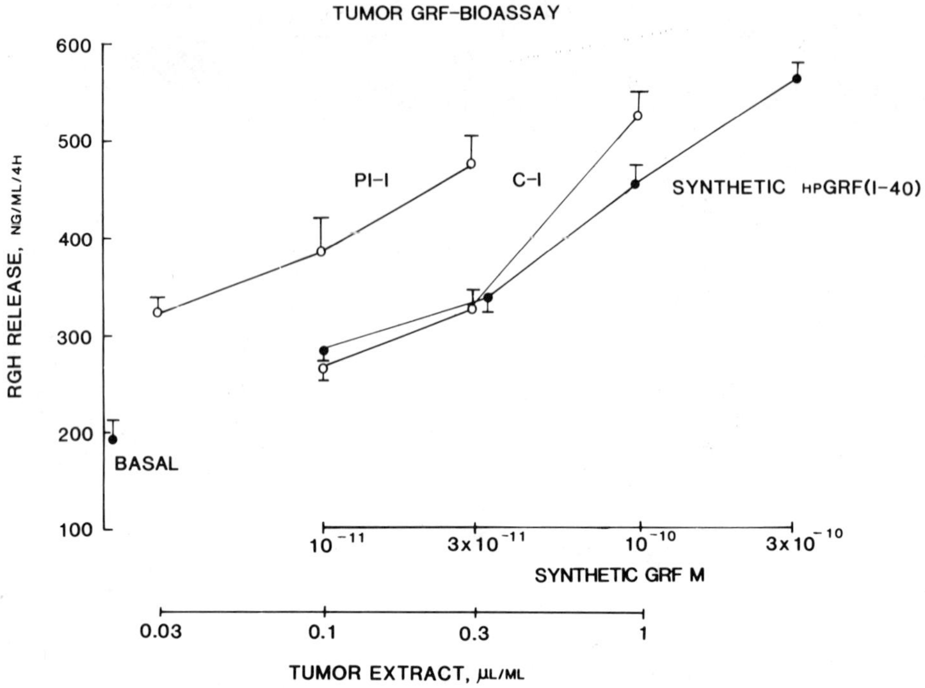

Figure 2 Comparison of GH-releasing effects of synthetic hpGRF(1–40) and extracts of a pancreatic islet (PI-I) and carcinoid (C-I) tumor in the dispersed rat pituitary cell culture system. Shown are the mean ± SE of four replicate cultures.

a neurofibroma, and 1 of 12 other tumors, an adrenal adenoma, were also bioactive. In addition, media from 10 small cell carcinoma cell lines grown in continuous culture were evaluated, and three exhibited significant GRF bioactivity.[14] Where examined, dose–response curves of tumor extracts were found parallel to that of synthetic GRF (1–40) (Fig. 2). The levels of bioactivity tended to be greatest in those patients with clinical evidence of GH hypersecretion, and the highest levels of bioactivity were present in pancreatic islet tumors. An increased awareness of this fact led to the preoperative identification of two pancreatic islet tumors which subsequently were used for the isolation and structural characterization of GRF[22] (see also Chapter 28, this volume).

Following the synthesis of GRF, a radioimmunoassay for the peptide was developed using a rabbit anti-GRF(1–20)-α-globulin conjugant, prepared by Drs. J. Rivier and W. Vale. The details of the radioimmunoassay have been described elsewhere.[23] The assay exhibits a sensitivity of 5–10 pg and serial dilutions of a human hypothalamic extract are parallel to those of synthetic GRF(1–40) standard (Fig. 3).

Radioimmunoassay of the tumor extracts described in Fig. 1 revealed detectable levels of GRF in 42 of the 46 tissues (Fig. 4). The highest levels were once again seen in tumors from patients with clinical evidence of GH hyperse-

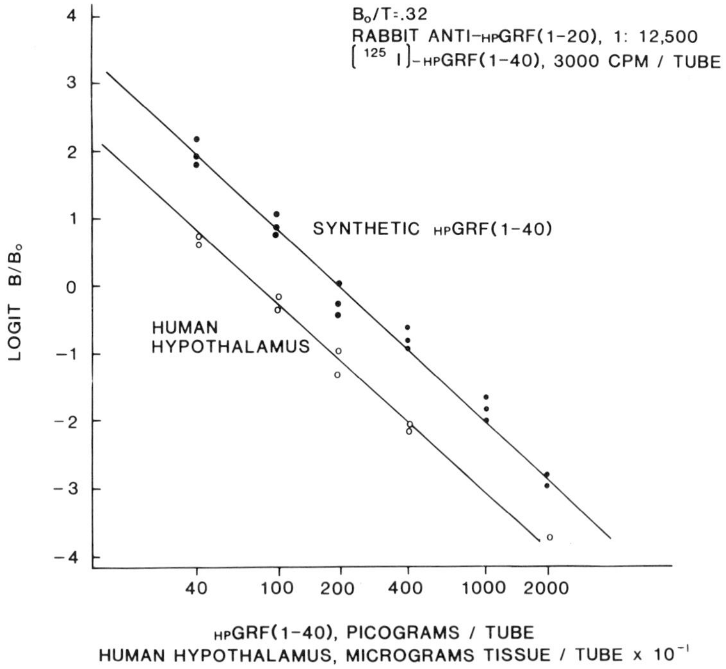

Figure 3 Radioimmunoassay standard curve for hpGRF(1–40) and an acetic acid extract of human hypothalamus. Shown is the log–logit plot. Reprinted with permission from Frohman.[21]

cretion, and as with bioassay, the highest levels were present in the two pancreatic islet adenomas. Serial dilution curves of the six tumors with the highest levels of immunoreactive GRF were parallel to those of GRF(1–40) (Fig. 5). Correlation of bioassay and radioimmunoassay values for GRF revealed a highly significant correlation ($r = 0.879$, $p < 0.01$) (Fig. 6). The ratios of bioactivity to immunoreactivity in five tumors in which bioassay potency estimates could be obtained all ranged from 0.31 to 0.58. The relative decrease in biological activity in the tumors as compared with synthetic GRF can be attributed to the presence of inhibiting factors in the crude tumor extracts (somatostatin was present in each of the tumors studied), the loss of biological activity through oxidation of GRF (which destroys greater than 95% of the biological activity), the presence of immunoreactive but bioinactive precursors or fragments, or other as-yet-undetermined factors.

Immunohistochemical studies performed with anti-GRF(1–40) serum prepared in our laboratory revealed intense staining of both bronchial carcinoid and pancreatic islet carcinomas.[24] No immunoreactivity was detected in the media in which the small cell carcinoma cell lines were grown or in solid tumors that developed in nude mice injected with the tumor cells.

A series of 46 tumors of neural crest origin frozen for periods of from 8 to 20 years provided by Dr. E. LaBrosse were examined by both bioassay and

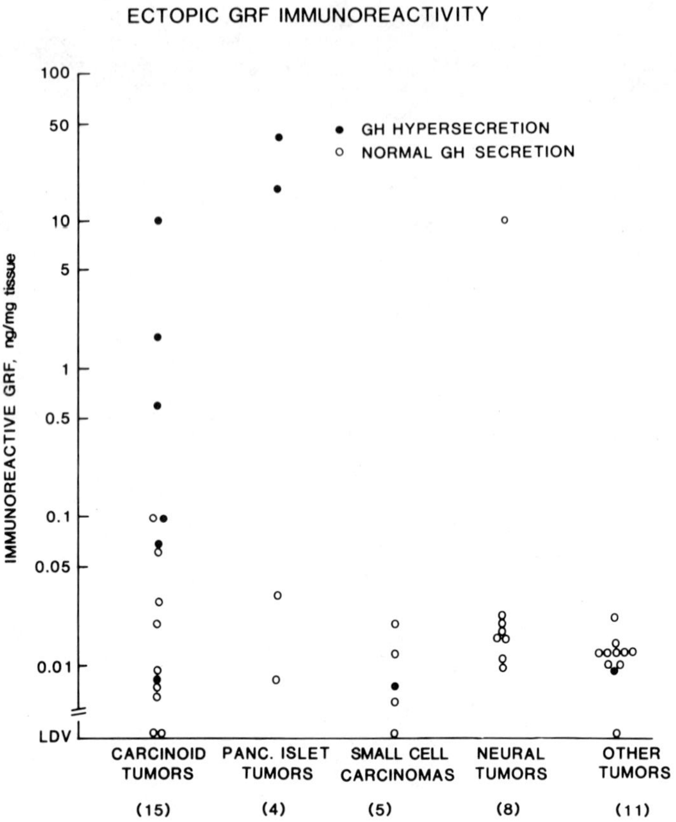

Figure 4 GRF immunoreactivity in tumors. All tumors were extracted in 2 N acetic acid and the aqueous soluable portion of the lyophilized residue assayed. Symbols as in Fig. 1. Reprinted with permission of Frohman.[21]

Table 2 GRF in Neural Crest Tumors[a]

Tumor	No.	Bioactive[b]	Immunoreactive GRF (pg/mg tissue)	
			>50	10–40
Neuroblastoma	33	22	1	11
Pheochroocytoma	8	3	—	7
Ganglioneuroblastoma	3	1	—	1
Total	44	26	1	19

[a] Provided by Dr. E. LaBrosse, University of Maryland.
[b] Stimulation of GH release by dispersed rat anterior pituitary cells in monolayer culture.

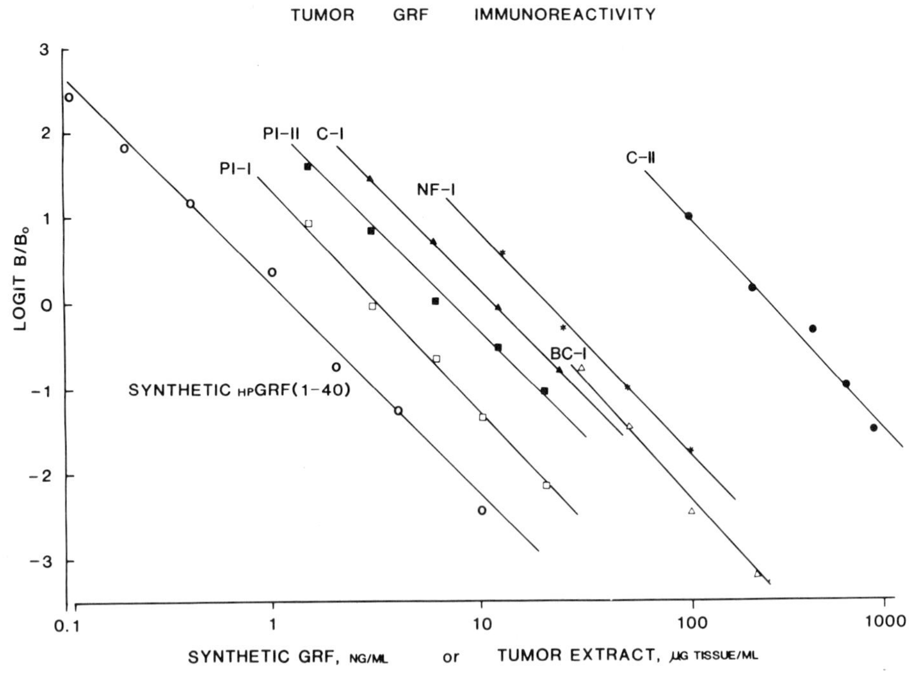

Figure 5 Comparison of immunoreactivity of extracts of GRF-containing tumors with that of synthetic hpGRF(1–40) in the hpGRF radioimmunoassay shown in Fig. 3. PI = pancreatic islet tumor; C = (intestinal) carcinoid tumor; BC = bronchial carcinoid tumor; NF = neurofibroma.

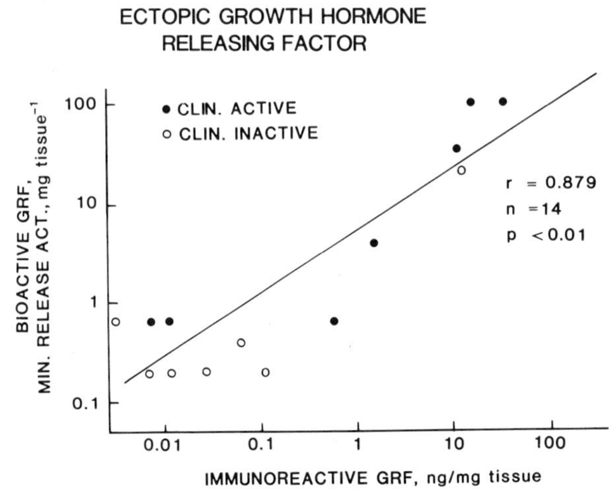

Figure 6 Correlation of bioactivity and immunoactivity in GRF-containing tumors. Bioactivity was measured in the dispersed rat pituitary cell culture system and is expressed as the reciprocal of the quantity of tissue required for minimal releasing activity. Reprinted with permission from Frohman.[21]

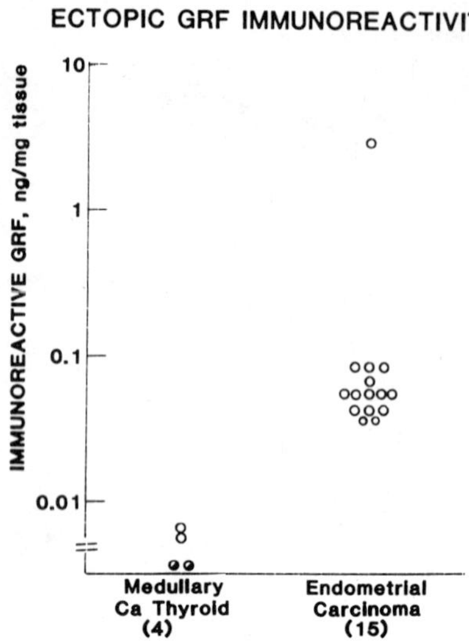

Figure 7 GRF immunoreactivity in medullary carcinoma of thyroid, endometrial carcinoma, and C-cell hyperplasia (half-closed circles). The lower limit of detectability is indicated by the break in the ordinate.

radioimmunoassay. Slightly more than half of the tumors exhibited moderate degrees of bioactivity though comparable levels of immunoreactive GRF were found in only one tumor and barely detectable levels were found in an additional 19 (Table 2). Thus, it appears that not all tumors with GH-releasing activity exhibit GRF-like immunoreactivity and the latter may represent other chemically dissimilar GRF's.

We have more recently had an opportunity to examine a series of medullary carcinomas of the thyroid and endometrial carcinomas (Fig. 7). Low levels of GRF immunoreactivity were found in two medullary carcinomas of the thyroid, but two thyroid glands containing C-cell hyperplasia were devoid of immunoreactive GRF. Among a group of 15 endometrial carcinomas, 14 exhibited low, but readily detectable levels of GRF immunoreactivity, whereas one tumor contained a level of 3 ng/mg tissue, comparable to that in the carcinoid tumors associated with acromegaly. This observation is of significance in that it may explain the previously reported elevations of basal GH levels and paradoxical responses to glucose suppression in patients with endometrial carcinomas.[25]

4. *Diagnosis of Ectopic GRF Syndrome*

In the patient with clinical manifestations of GH hypersecretion, a pituitary adenoma, and an extrapituitary tumor, particularly of the carcinoid or pancreatic islet type, the diagnosis of the ectopic GRF syndrome is readily established. The question to be addressed, however, is the diagnosis of this syndrome in the absence of obvious clinical manifestations. Three approaches will be addressed: (1) pituitary histology, (2) GH secretory responses to provocative stimuli, and (3) immunoreactive GRF levels in peripheral plasma.

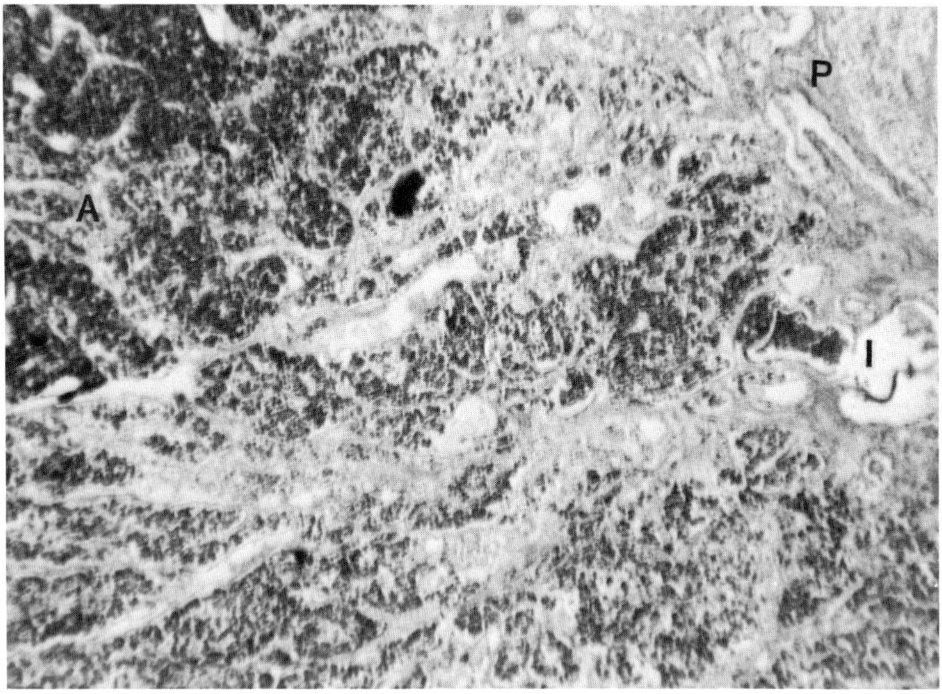

Figure 8 Multifocal hyperplastic changes in the pituitary from a patient with a GRF-secreting small cell carcinoma of the lung. Nests of acidophilic cells in the anterior lobe (A) are located adjacent to the posterior lobe (P) and in cystic cavities of the pars intermedia (I). Immunohistochemical stains revealed all of the cells to contain GH, and the reticulin pattern is deficient in the areas of hyperplasia. Reprinted with permission from Frohman.[24]

It has recently been suggested that somatotroph hyperplasia is a characteristic finding of the ectopic GRF syndrome (22) and that its presence should suggest the existence of an ectopic GRF-producing tumor. Figure 8 shows the pattern of somatotroph hyperplasia in a patient with a GRF-containing small cell carcinoma of the lung. Sheets of somatotropic cells are present which show no clear separation from normal pituitary architecture. However, this change is not always observed in patients with the ectopic GRF syndrome. There are several well-documented cases in the literature[4,9,26,27] in which pituitary tumors were observed as a result of ectopic GRF overproduction. Thus, pituitary hyperplasia should best be viewed as an intermediate stage which eventually leads to tumor formation. Its presence certainly suggests the possibility of GRF overproduction, but the absence of hyperplasia cannot be used to exclude a GRF-mediated mechanism. Furthermore, this information is available only after pituitary tissue has been removed surgically while the preferable time for diagnosis is prior to such a procedure.

The classical alterations in plasma GH responses to provocative stimuli in patients with acromegaly include an absence of suppression or stimulation

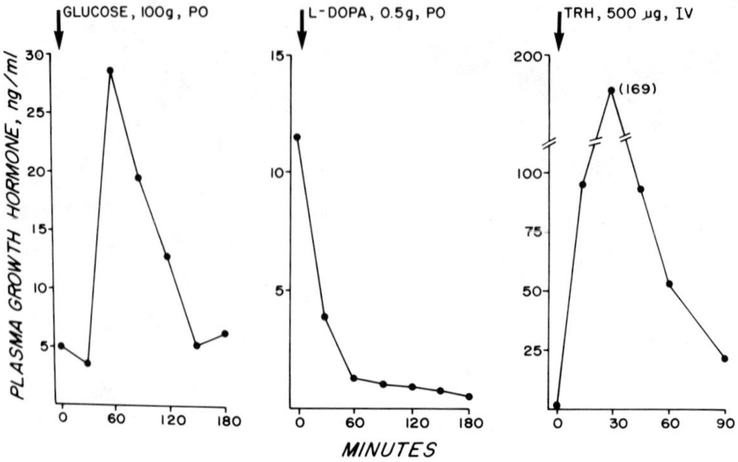

Figure 9 Plasma GH responses to TRH, L-dopa, and glucose in a patient with a pancreatic islet carcinoma and elevated plasma GH levels. There was no evidence of a pituitary tumor on sellar tomography. Reprinted with permission from Frohman.(24)

following oral glucose, suppression following L-dopa or dopamine, and stimulation of GH secretion in response to TRH. Responses to these three agents in a patient with the ectopic GRF syndrome are shown in Fig. 9, and are typical of those seen in acromegaly unassociated with ectopic GRF production. A compilation of our own results as well as those reported in the literature is shown in Table 3 and indicate that there is no distinction between patients with and without ectopic GRF production.

In collaboration with Dr. M. Thorner, sera was collected from patients with known, presumed, and suspected ectopic GRF syndromes and from patients with acromegaly not suspected of harboring GRF-producing tumors. Plasma was extracted on C-18 reverse-phase cartridges which allowed removal of nonspecific interfering substances as well as concentration of peptide.

In 18 normal subjects plasma IR-GRF levels were undetectable with an assay sensitivity of 110 pg/ml. In a second assay, greater concentration of plasma was achieved, and in 12 normal subjects no detectable immunoreactivity was found with a sensitivity of 25 pg/ml. In one patient with a proven ectopic GRF

Table 3 Plasma Growth Hormone Responses to Stimuli in Patients with Ectopic GRF Production

Stimulus	No. tested	No. responses	
TRF	6	5	(stimulation)
L-Dopa/dopamine	5	4	(suppression)
Glucose	5	4	(stimulation or lack of suppression)

syndrome, initial plasma IR-GRF was 4.8 ng/ml and fell to 0.5 ng/ml following removal of the tumor, while in a second patient with metastatic disease, levels from 2 to 6 ng/ml were observed. In two patients with presumed ectopic GRF production (patients with GH hypersecretion and carcinoid tumors, though unconfirmed by immunohistochemical or radioimmunoassay of tumor tissue) similar levels of plasma IR-GRF (2–5 ng/ml) were present. In six patients with suspected ectopic GRF secretion (acromegaly unresponsive to standard therapy or unassociated with anatomical evidence of pituitary abnormalities but without clinical evidence of an extrapituitary tumor) plasma IR-GRF levels were undetectable in five and at the lower limits of sensitivity in one. In patients with acromegaly unsuspected of harboring GRF-secreting tumors, plasma IR-GRF levels were either undetectable or within 2.5 times lower limit of sensitivity in 137 of 138 patients, with an assay exhibiting a least detectable value of 110 pg/ml. One patient with a value of 580 pg/ml is currently being further evaluated. In a second assay with a sensitivity of 25 pg/ml, 24 of 25 acromegalic subjects exhibited undetectable plasma IR-GRF levels, and in one, a barely detectable level was present.

Thus, it appears that in patients with the clinically recognized ectopic GRF syndrome, plasma values are in the nanogram/ml range. In contrast, acromegalic subjects not suspected of ectopic GRF secretion exhibit undetectable or barely detectable levels. It is of interest that the smallest dose of GRF(1–40) given as a constant intravenous infusion which reliably stimulates GRF release results in a plasma IR-GRF level of approximately 300 pg/ml,[28] suggesting this may be near the threshold level required for stimulation of GH release. It further suggests that the diagnosis of ectopic GRF secretion will be made infrequently by measurement of plasma IR-GRF levels in patients with acromegaly not clinically suspected of having the disease.

5. *Hypothalamic–Pituitary GRF Hypersecretion*

In addition to the tumors already described, ectopic production of GRF has also been observed within the hypothalamic–pituitary region. A series of patients have recently been reported with intrapituitary neuronal gangliocytomas in juxtaposition to GH-secreting somatotroph adenomas.[27] The tumors consist of randomly oriented large ganglion cells similar to those in the hypothalamic magnocellular nuclei with axonal networks which can be shown by both light and electron microscopy to terminate on somatotroph cells. Immunohistochemical stains reveal the neuronal cell bodies to contain GRF. In addition, a hypothalamic hamartoma consisting of masses of redundant, partially disoriented glial and neuronal cells has been identified in a patient with a GH-secreting pituitary tumor.[26] The hamartoma similarly has subsequently been shown to contain GRF. Vessels in the hamartoma have fenestrations characteristic of those in the median eminence suggesting a secretory process comparable to that from hypophysiotropic hormone-containing nerve terminals in the median eminence into the pituitary portal system.

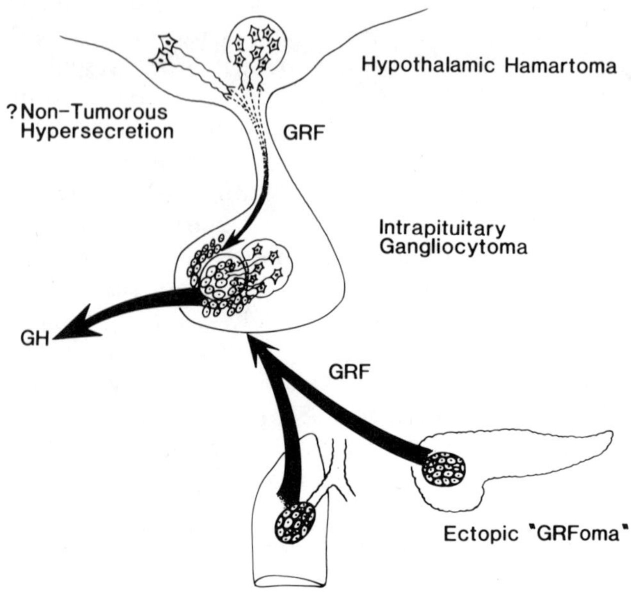

Figure 10 Spectrum of secondary hypersomatotropism involving ectopic production of GRF from peripheral, intrapituitary, and central nervous system sources and ectopic hypersecretion of GRF from hypothalamic neurons. Reprinted with permission from Frohman.(21)

Thus, the full spectrum of secondary hypersomatotropism has now been identified (Fig. 10). In patients with ectopic GRF-secreting tumors, GRF acts as a classical hormone, being released into the systemic circulation and transported to the pituitary; in patients with intrapituitary GRF-secreting tumors, GRF exerts a paracrine or neurotransmitter effect, being released from nerve terminals directly at the surface of somatotroph; and in patients with hypothalamic GRF-secreting tumors, true neurosecretion occurs with release of GRF from tumor tissue into portal vessels. Since hypersecretion of GRF by any of these mechanisms results not only in GH hypersecretion but also in somatotroph hyperplasia and tumor formation, it is tempting to postulate that nontumorous hypersecretion of GRF from neuronal elements in the hypothalamus may exert a similar role in the pathogenesis of acromegaly in at least some patients with this disorder. The challenge now remains to provide convincing evidence for this hypothesis.

Acknowledgments. These studies were supported in part by USPHS Grants AM-30667 and RR-0068. The excellent technical assistance of Jeanne Hirth is gratefully acknowledged.

References

1. A. L. Southern, Functioning metastatic bronchial carcinoid with elevated levels of serum and cerebrospinal fluid serotonin and pituitary adenoma, *J. Clin. Endocrinol. Metab.* **20:**298, 1960.
2. L. Weiss and M. Ingram, Adenomatoid bronchial tumors: A consideration of the carcinoid tumors and the salivary tumors of the bronchial tree, *Cancer* **14:**161, 1961.
3. J. Buse, M. G. Buse, and W. J. Roberts, Eosinophilic adenoma of the pituitary and carcinoid tumors of the rectosigmoid area, *J. Clin. Endocrinol. Metab.* **21:**735, 1961.
4. J. T. Dabek, Bronchial carcinoid tumor with acromegaly in two patients, *J. Clin. Endocrinol. Metab.* **38:**329, 1974.
5. P. H. Sonksen, A. B. Ayres, M. Briambridge, B. Corrin, D. R. Davies, G. M. Jeremiah, S. W. Oaten, C. Lowy, and T. E. T. West, Acromegaly caused by pulmonary carcinoid tumors, *Clin. Endocrinol.* **5:**503, 1976.
6. C. Beck, R. G. Larkins, T. J. Martin, and H. G. Burger, Stimulation of growth hormone release from superfused rat pituitary by extracts of hypothalamus and of human lung tumors, *J. Endocrinol.* **59:**325, 1973.
7. M. E. Stachura, L. A. Frohman, and A. P. S. Dhariwal, Effect of purified hypothalamic extract and acromegalic plasma on growth hormone synthesis and release *in vitro*, *Proceedings of the Endocrine Society Annual Meeting*, San Francisco, 1971, p. A-81.
8. T. C. Hagen, A. M. Lawrence, and L. Kirstens, *In vitro* release of monkey pituitary growth hormone by acromegalic plasma, *J. Clin. Endocrinol. Metab.* **33:**448, 1971.
9. M. S. Zafar, R. C. Mellinger, G. Fine, M. Szabo, and L. A. Frohman, Acromegaly associated with a bronchial carcinoid tumor: Evidence for ectopic production of growth hormone-releasing activity, *J. Clin. Endocrinol. Metab.* **48:**66, 1979.
10. S. M. Shalet, C. G. Beardwell, I. A. Macfarlane, M. L. Ellison, C. M. Norman, L. H. Rees, and M. Hughes, Acromegaly due to production of a growth hormone releasing factor by a bronchial carcinoid tumour, *Clin. Endocrinol.* **10:**61, 1979.
11. L. A. Frohman, M. Szabo, M. Berelowitz, and M. E. Stachura, Partial purification and characterization of a peptide with growth hormone-releasing activity from extrapituitary tumors in patients with acromegaly, *J. Clin. Invest.* **65:**43, 1980.
12. S. A. Leveston, D. W. McKeel, P. J. Buckley, K. Deschryver, M. H. Greider, B. M. Jaffe, and W. H. Daughaday, Acromegaly and Cushing's syndrome associated with a foregut carcinoid tumor, *J. Clin. Endocrinol. Metab.* **53:**682, 1981.
13. L. A. Frohman, M. Szabo, M. E. Stachura, and M. Berelowitz, Growth hormone-releasing activity in extracts of bronchial carcinoid and pancreatic islet tumors associated with acromegaly, *Trans. Am. Assoc. Physicians* **92:**77, 1979.
14. M. Szabo and L. A. Frohman, Ectopic production of a growth hormone-releasing factor by tumors: partial isolation and characterization, in: *Hormones in Normal and Abnormal Human Tissues* (K. Fotherby and S. B. Pal, eds.), Walter de Gruyter, New York, 1981, p. 415.
15. M. Szabo, D. Dudlak, J. Thominet, and L. A. Frohman, Ectopic growth hormone-releasing factor (GHRF) stimulates growth hormone secretion in the urethane-anesthetized rat *in vivo*, *Neuroendocrinology* **37:**328, 1983.
16. C. Webb, J. L. Thominet, and L. A. Frohman, Ectopic growth hormone releasing factor stimulates growth hormone release from human somatotroph adenomas *in vitro*, *J. Clin. Endocrinol. Metab.* **56:**417, 1983.
17. M. Szabo, L. Chu, and L. A. Frohman, Biological effects of an ectopic growth hormone-releasing peptide in cultured adenohypophyseal cells: Comparison with growth hormone-releasing activity of porcine hypothalamus, *Endocrinology* **111:**1235, 1982.
18. M. J. Cronin, A. D. Rogol, L. G. Dabney, and M. O. Thorner, Selective growth hormone and cyclic AMP stimulating activity is present in a human pancreatic islet cell tumor, *J. Clin. Endocrinol. Metab.* **55:**381, 1982.
19. C. B. Webb, M. Szabo, and L. A. Frohman, Ectopic growth hormone releasing factor and dibutyryl cyclic adenosine monophosphate-stimulated growth hormone release in vitro: Effects of corticosterone and estradiol, *Endocrinology* **113:**1191, 1983.

20. M. Berelowitz, M. Szabo, L. A. Frohman, S. Firestone, L. Chu, and R. L. Hintz, Somatomedin-C mediates growth hormone negative feedback by effects on both the hypothalamus and the pituitary, *Science* **212:**1279, 1981.
21. L. A. Frohman, Growth hormone releasing factor. A neuroendocrine perspective, *J. Lab. Clin. Med.* **103:**819, 1984.
22. M. O. Thorner, R. L. Perryman, M. J. Cronin, A. D. Rogol, M. Draznin, A. Johanson, M. Vale, E. Horvath, and K. Kovacs, Somatotroph hyperplasia: Successful treatment of acromegaly by removal of a pancreatic islet tumor secreting a growth hormone releasing factor, *J. Clin. Invest.* **70:**965, 1982.
23. L. A. Frohman, J. L. Thominet, C. B. Webb, M. L. Vance, H. Uderman, J. Rivier, M. Vale, and M. O. Thorner, Metabolic clearance and plasma disappearance rates of human pancreatic tumor growth hormone releasing factor in man, *J. Clin. Invest.*, **73:**1304, 1984.
24. L. A. Frohman, Ectopic production of peptide hormones by human tumors, in: *Neuroendocrine Perspectives*, Vol. 3 (E. E. Muller, and R. M. MacLeod, eds.), Elsevier Biomedical Press, The Netherlands, 1984, p. 201.
25. F. Benjamin, D. J. Casper, L. Sherman, and H. D. Kolodny, Growth-hormone secretion in patients with endometrial carcinoma, *N. Engl. J. Med.* **281:**1448, 1969.
26. S. L. Asa, J. M. Bilbao, K. Kovacs, and J. A. Linfoot, Hypothalmic neuronal hamartoma associated with pituitary growth hormone cell adenoma and acromegaly, *Acta Neuropathol.* **52:**231, 1980.
27. S. L. Asa, B. W. Scheithauer, J. M. Bilbao, E. Horvath, N. Ryan, K. Kovacs, R. V. Randall, E. R. Laws Jr., W. Singer, J. A. Linfoot, M. O. Thorner, and W. Vale, A case for hypothalamic acromegaly: Hypothalamic ganglicytomas producing somatoliberin. A clinical study of six cases, *J. Clin. Endocrinol. Metab.*, 1984, **58:**796.
28. C. B. Webb, M. L. Vance, M. O. Thorner, G. Perisutti, J. Thominet, J. Rivier, W. Vale, and L. A. Frohman, Plasma growth hormone responses to constant infusions of human pancreatic growth hormone-releasing factor: Intermittent secretion or response attenuation, *J. Clin. Invest.* **74:**96, 1984.

27

Growth Hormone-Releasing Factor

Michael O. Thorner, William S. Evans, Mary Lee Vance, Robert M. Blizzard, Alan D. Rogol, Donald L. Kaiser, Ann J. Johanson, Lawrence A. Frohman, Jennifer L. Thominet, Richard Furlanetto, Marie C. Gelato, George R. Merriam, James H. Goldman, Seymour Reichlin, Jean Rivier, and Wylie Vale,

1. Introduction

We have previously described in detail the clinical and laboratory findings of a patient who harbored a pancreatic tumor secreting a growth hormone releasing factor (GRF).[1] This 40-amino-acid-containing peptide (hpGRF-40) was extracted, isolated, sequenced, and a synthetic replicate produced by the Peptide Biology Laboratory at the Salk Institute.[2,3] A 44-amino-acid peptide (hpGRF-44) was identified from another pancreatic tumor and subsequently synthesized by the Laboratories for Neuroendocrinology at the Salk Institute.[4] One year later convincing immunological, biochemical, and genetic information appears to have confirmed that human hypothalamic growth hormone-releasing hormone (GHRH) is identical to human pancreatic tumor GRF(s) and has approx-

---

Michael O. Thorner, William S. Evans, Mary Lee Vance, Robert M. Blizzard, Alan D. Rogol, Donald L. Kaiser, and Ann J. Johanson • Department of Pediatrics, University of Virginia School of Medicine, Charlottesville, Virginia 22901. ***Lawrence A. Frohman and Jennifer L. Thominet*** • Division of Endocrinology and Metabolism, Department of Internal Medicine, University of Cincinnati College of Medicine, Cincinnati, Ohio 45267. ***Richard Furlanetto*** • Children's Hospital of Philadelphia, Philadelphia, Pennsylvania 19104. ***Marie C. Gelato and George R. Merriam*** • Developmental Endocrinology Branch, National Institutes of Health, Bethesda, Maryland 20205. ***James H. Goldman and Seymour Reichlin*** • Endocrine Division, Department of Medicine, Tufts University School of Medicine, New England Medical Center, Inc., Boston, Massachusetts 02111. ***Jean Rivier and Wylie Vale*** • The Clayton Foundation Laboratories for Peptide Biology, The Salk Institute, La Jolla, California 92037.

imately 70% homology with rat hypothalamic GHRH.[5-8] In this chapter we will review our first 11 months of clinical experience with GRF and compare our results to those reported by other groups.

The studies were performed after obtaining approval from the Human Investigation Committees of our respective institutions and with the approval of the Food and Drug Administration. We have used hpGRF-40 which was synthesized and formulated as previously described.[2,9] Our initial studies were performed in normal subjects, in adults with acromegaly, in adults with GH deficiency, and in children with short stature.

2. *Investigations Using hpGRF*

The ability to use synthetic GRF in man has allowed us and others to investigate some aspects of the physiology of GH secretion. The combination of human investigations with basic *in vivo* and *in vitro* studies may yield new and effective therapies including those for GH-deficient children and possibly for the prevention of those long-term complications of diabetes mellitus thought to be due to GH excess. Moreover, new approaches to the diagnosis and treatment of acromegaly may be forthcoming.

2.1. *Studies in Normal Men*

Our first studies in humans consisted of the administration of 1 μg/kg body weight hpGRF-40 given as an intravenous (i.v.) bolus.[9] hpGRF-40 selectively stimulated GH secretion (Fig. 1) with serum GH concentrations

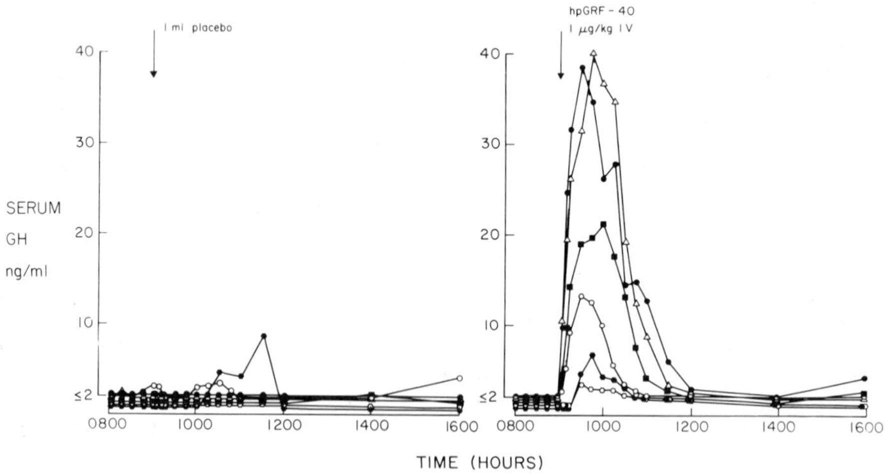

Figure 1 Serum growth hormone before and after placebo (left panel) and hpGRF (right panel) in six normal men. Note the variability of response to hpGRF among the subjects. Reproduced with permission from Thorner *et al.*[9]

increasing within 5 min and reaching a peak between 30 and 60 min (20.4 ± 6.5 ng/ml compared with 2.1 ± 0.1 ng/ml after placebo). Serum levels of prolactin, thyrotropin (TSH), luteinizing hormone (LH), and corticotropin (measured indirectly through plasma cortisol) were not increased after administration of hpGRF-40. Similarly, the concentrations of blood glucose, plasma insulin, glucagon, pancreatic polypeptide, cholecystokinin, gastrin, gastric inhibitory peptide, motilin, and somatostatin were unaffected by hpGRF-40. There were no changes in blood pressure, pulse rate, or body temperature, and no side effects were noted. These studies were repeated using hpGRF-44 by Wood *et al.*,[10] Rosenthal *et al.*,[11] and Gelato *et al.*[12]

A dose–response relationship was next determined by documenting the effect of graded doses of hpGRF-40 on GH release in 12 healthy men.[13] Mean (± SEM) peak GH levels were 2.56 ± 1.00, 12.94 ± 2.54, 17.45 ± 1.90, 17.68 ± 2.97, 18.04 ± 5.44, and 21.33 ± 3.86 ng/ml after vehicle and 0.1, 0.33, 1.0, 3.3, and 10 μg/kg hpGRF-40 given as an i.v. bolus, respectively. Peak values were observed 30–60 min after administration of hpGRF-40. There was considerable variability in response among individual subjects, and no dose–response relationship between the doses and maximal GH levels achieved was observed. However, the higher doses of 1.0, 3.3, and 10.0 μg/kg were associated with a more prolonged, biphasic pattern of GH release. Side effects of facial flushing of less than 5-min duration occurred in four of six subjects who received 3.3 μg/kg and in all six who received 10 μg/kg of hpGRF-40. No changes in serum glucose, LH, TSH, prolactin, plasma cortisol, or eight enteropancreatic hormones were observed following hpGRF-40. Serum somatomedin C levels (mean ± SEM) increased over basal levels by 0.173 ± 0.07 U/ml (p = 0.0435) 24 hr after hpGRF-40 in eight subjects. Plasma immunoreactive GRF levels (mean ± SEM) measured 5 min after injection were 4.86 ± 0.87, 23.90 ± 4.1, and 66.6 ± 17.6 ng/ml after 1.0, 3.3, and 10 μg/kg of hpGRF-40, respectively. Serum GH responses after insulin-induced hypoglycemia were compared to GH responses following hpGRF-40. Comparable GH stimulation occurred with both provocative tests. Mean (± SEM) peak GH after insulin was 20.2 ± 1.04 ng/ml and 20.9 ± 3.16 following hpGRF-40. Rosenthal *et al.*,[11] using 0.5, 5, and 10 μg/kg hpGRF-44, failed to demonstrate a dose-related increase or a prolonged duration of action with the 5 or 10 μg/kg doses.

2.2. Studies in Normal Women

Having shown in normal men that hpGRF-40 could stimulate GH secretion, we next examined its effects in normal women.[14] We administered hpGRF-40 (3.33 μg/kg; test day) or an equivalent volume of vehicle (control day) as an i.v. bolus at 9 A.M. to five normal subjects. All women were studied during the early follicular, late follicular, and midluteal phases of the menstrual cycle. Serum concentrations of GH, prolactin, LH, and follicle-stimulating hormone (FSH) were measured at intervals between 8 A.M. and 12 P.M. Serum somatomedin C concentrations were obtained prior to and 24 hr following administration of vehicle or hpGRF-40. Within 1–3 min following injection of hpGRF-

40 all women described warmth which was localized to the head and neck and exhibited facial flushing. No changes in pulse rate or blood pressure were noted. Median maximal levels of serum GH (ng/ml; control versus test day) were higher following hpGRF-40 during the early follicular (6.2 versus 23.1; $p = 0.0216$), late follicular (5.4 versus 22.7; $p = 0.0216$), and luteal (3.2 versus 16.8; $p = 0.0122$) phases of the menstrual cycle. When expressed as change from baseline, median integrated serum GH levels (ng/ml per hr) were higher following hpGRF-40 than control values during the early follicular (1.18 versus 13.5; control versus test day) and luteal (−2.22 versus 5.16) phases of the menstrual cycle ($p = 0.0122$). Although higher in four of five subjects, median integrated serum GH levels were not increased following hpGRF-40 during the late follicular phase (1.44 versus 4.53; $p = 0.17$). Serum somatomedin C values 24 hr after hpGRF-40 were higher compared to those 24 hr after vehicle at all stages of the menstrual cycle. hpGRF-40 did not stimulate the release of prolactin, LH, or FSH. These studies suggest that hpGRF-40 selectively stimulates the release of biologically active GH in women, but that, at the dose tested, hpGRF-40-stimulated GH release does not vary as a function of the gonadal hormone environment.

2.3. Routes of Administration

2.3.1. Intranasal hpGRF-40

Since GRF will probably require administration several times each day, parenteral administration may interfere with its utility as a therapeutic agent. We thus undertook a preliminary investigation of the effects of intranasal administration of hpGRF-40.[15] Six normal men were given hpGRF-40 (30 μg/kg; test day) or an equivalent volume of vehicle alone (control day) at 9 A.M. Immunoreactive GH was measured in serum obtained at intervals between 8 A.M. and 12 P.M. Mean (± SEM) integrated serum levels of GH (ng/ml per hr) prior to and following administration of vehicle were not different (1.27 ± 0.57 versus 0.87 ± 0.28; $p = 0.54$). However, following hpGRF-40 administration, GH levels increased significantly (0.53 ± 0.03 versus 2.88 ± 0.75; $p = 0.022$). Peak levels of serum GH were detected within 30 min following hpGRF-40. Except for mild burning of the nasal mucosa reported by one subject, no side effects were noted. Subsequently we have investigated the dose–response relationship of intranasal hpGRF-40 administration (3–100 μg/kg) and its effects on GH secretion. Only the 30-μg/kg and 100-μg/kg doses of hpGRF-40 stimulated GH secretion when administered by this route.

The dose required to stimulate GH secretion by the intranasal route was extremely large thus precluding use of the peptide by this route using this formulation. However, alternative formulations or the development of shorter sequence and more potent analogues may allow this method of administration to be practicable.

2.3.2. *Subcutaneous hpGRF-40*

We have therefore administered hpGRF-40 (1–10 μg/kg) subcutaneously. Our results demonstrate a dose–response relationship between hpGRF-40 and GH release over this dose range. Although 1 μg/kg hpGRF-40 administered subcutaneously appears to stimulate GH release, the 3.3- and 10-μg/kg doses were associated with greater and more prolonged increases.

2.4. Pharmacokinetics of hpGRF Administration

Most hypothalamic hormones are small peptides with very short half-lives. It was therefore important to define the pharmacokinetics of GRF in the human. Using two different techniques we investigated the metabolic clearance rate and plasma disappearance rate in normal men by single injection and constant infusion techniques.[16] Immunoreactive GRF levels were measured by radioimmunoassay using an extracted plasma specimen with hpGRF-40 as radioiodionated tracer, an antibody raised against hpGRF(1–20), and a second antibody separation method.

The metabolic clearance rate (MCR) and plasma disappearance rate ($t_{\frac{1}{2}}$) of hpGRF-40 were determined in normal adult male subjects. Single i.v. injections of 1, 3.3, and 10 μg/kg hpGRF-40 were administered. Plasma immunoreactive (IR) GRF levels were measured during the subsequent 180 min, and biexponential curve analysis was performed. Graded dose constant infusions of hpGRF-40 at rates of 1, 3.3, 10, and 33 ng/kg per min were administered and the MCR calculated from measurement of steady-state plasma IR-GRF levels at each infusion rate. The postinfusion disappearance rate was determined by linear regression analysis of plasma IR-GRF levels during the 120-min period after cessation of the infusion.

The calculated MCR (l/m^2 per day) during the single injection study was 194 ± 17.5 and was not significantly different from the calculated value during the constant infusion study (202 ± 16). The disappearance rate during the single injection study was subdivided into two linear phases: an initial equilibration phase (7.6 ± 1.2 min) and a subsequent elimination phase (51.8 ± 5.4 min). The latter was similar to the linear disappearance rate observed (41.3 ± 3.0 min) after cessation of the constant infusion. The chromatographic and biological characteristics of plasma IR-GRF, 30 min after injection, were similar to those of synthetic hpGRF-40.

2.5. Studies in Acromegalic Patients

We next addressed the question of whether GH secretion in acromegalic patients was autonomous or remained responsive to hpGRF stimulation, and whether these responses might correlate with other clinical or chemical features of their condition. Twenty-nine patients aged 20–72 were tested; of these, seven were untreated, and the others had received various combinations of surgical, radiation, or bromocriptine therapy. No patient had ectopic hpGRF secretion. Twenty-five of the twenty-nine had significant GH responses to hpGRF-40 i.v. bolus in doses of 1 (21 patients) or 3.3 μg/kg (8 patients). The magnitude of the responses was often much higher than normal (e.g., in one patient GH rose from 8 to 502 ng/ml), but was highly variable. There was no significant correlation of hpGRF-40 responses with the patients' age, sex, duration of disease, history of prior therapy, basal GH level, somatomedin C, PRL, or the GH response to thyroid-releasing hormone (TRH). In contrast to the study of Wood and colleagues,[10] we found no significant correlation between the magnitude of the GH responses to GRF stimulation and to glucose suppression. Two patients who did not respond to hpGRF-40 at 1 μg/kg remained unresponsive when retested at 3.3 μg/kg. One patient was studied before, and again 5 and 40 days following, successful adenomectomy. She had a supranormal response preoperatively, a blunted response at 5 days, and a normal response at 40 days postoperatively. This suggests that there is no long-term pituitary refractoriness to hpGRF after tumor removal. Frohman and colleagues[17] found wide variability in the responsiveness of somatotroph adenomas to partially purified GRF *in vitro*, and it may be that part of the reason for the great heterogeneity of responses in these patients is that the test doses may have fallen at different points on their dose–response curves.

Thus, the majority of patients with acromegaly remain responsive to hpGRF, but a standard single-bolus hpGRF test appears to be of limited utility in assessing the status of the disease in any given patient.

2.6. Studies in GH-Deficient Adults

In our initial study hpGRF-40, 10 μg/kg, was administered intravenously to six normal young men and 12 adult patients who had presented in childhood with GH deficiency [seven patients had isolated GH deficiency, four had multiple anterior pituitary hormone deficiencies, and one had Hand-Schuller-Christian (HSC) disease].[18] hpGRF-40 administration increased serum GH concentrations in all normal subjects and in three of seven patients with isolated GH deficiency and in the one with HSC disease; however, the mean serum GH concentration in the patients who responded was less than that of the normal subjects (Fig. 2). Serum somatomedin C concentrations were increased 24 hr after a single dose of hpGRF-40 in 8 of 10 patients with GH deficiency (Fig. 3). All subjects experienced flushing in response to hpGRF-40. One patient with isolated GH deficiency received 0.33 μg/kg hpGRF-40 every 3 hr for 5 days. Despite the modest increase in GH in response to a subsequent dose of 10 μg/kg hpGRF-40, serum somatomedin C levels increased within 12 hr from 0.06 to 0.1 U/ml and peaked at 0.36 U/ml at 72 hr (Fig. 4); in addition, the patient with HSC disease, treated with hpGRF-40 daily for 5 days, demonstrated an increase in serum somatomedin C from 0.4 to 0.58 U/ml. The increase after hpGRF-40 in serum GH levels in this patient and the similar or greater responses in three of seven patients suggest that at least some of these patients may have hypothalamic GHRH deficiency and that hpGRF-40 may be useful in distinguishing pituitary disease from hypothalamic disease. Similar results

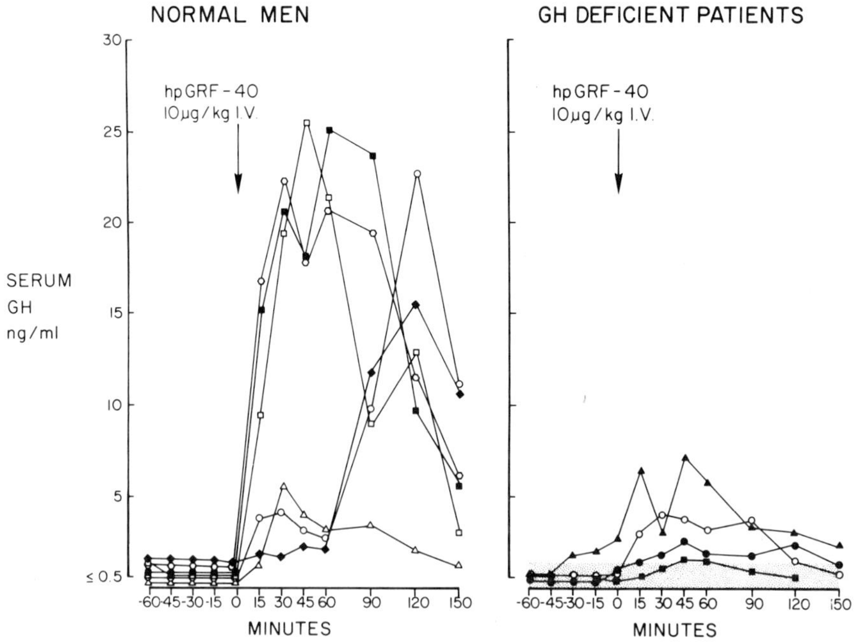

Figure 2 Serum GH concentrations in six normal men and 12 patients with GH deficiency in response to hpGRF-40 administered intravenously at time 0 (9 A.M). Shaded area represents no response in eight patients. Reproduced with permission from Borges *et al.*[18]

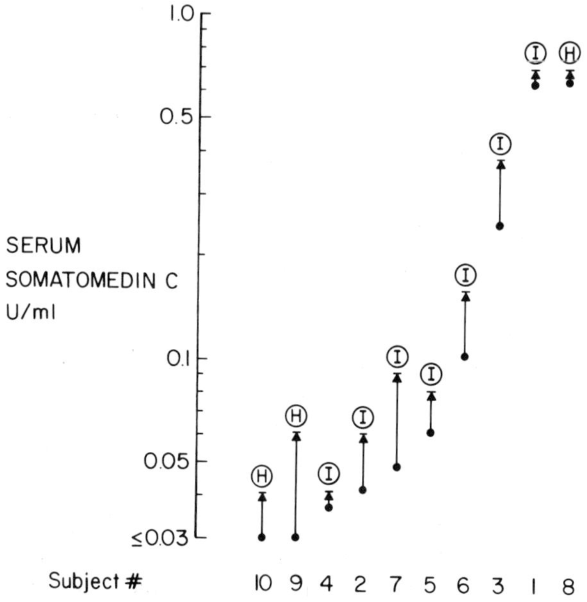

Figure 3 Serum somatomedin C concentrations in seven patients with isolated GH deficiency (I) and three with idiopathic hypopituitarism (H) before (●) and 24 hr after (▲) administration of hpGRF-40 (10 μg/kg). Reproduced with permission from Borges *et al.*[18]

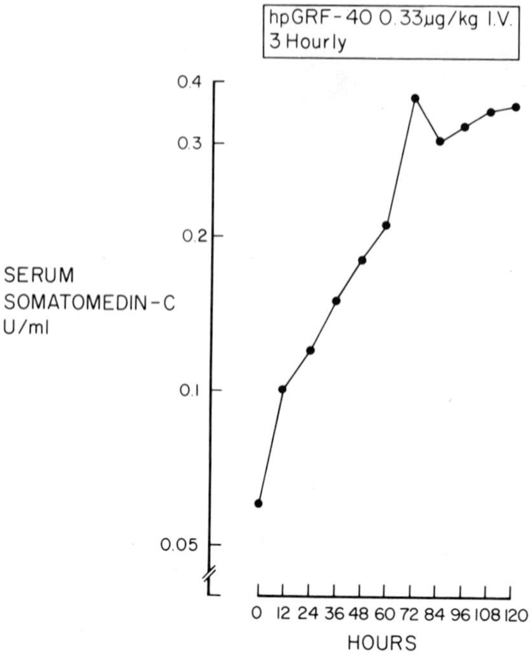

Figure 4 Serum somatomedin C concentrations in patient with isolated GH deficiency during 5 days of treatment with hpGRF-40. Reproduced with permission from Borges *et al.*[18]

with both hpGRF-44 and hpGRF-40 have been reported respectively by Wood *et al.*[10] and by Grossman *et al.*[19]

In addition, we have studied seven adult subjects who had presented in childhood with idiopathic GH deficiency by challenging them with 10 μg/kg hpGRF-40 prior to and following 5 days of treatment with 0.33 μg/kg hpGRF-40 administered i.v. every 3 hr.[20] Serum GH levels were monitored daily for 90 min after the 8 A.M. doses of 0.33 μg/kg hpGRF-40, and serum somatomedin C was measured at 8 A.M. and 2 P.M. In addition, plasma levels of cholesterol, high-density-lipoprotein (HDL) cholesterol, and triglycerides were measured daily at 8 A.M. Three hours after the last 0.33-μg/kg dose, all subjects were rechallenged with 10 μg/kg hpGRF-40. In response to the initial 10-μg/kg challenge with hpGRF-40, and although serum GH levels rose in four of seven subjects, the mean maximum GH level was no different from that after vehicle alone. Within 12 hr after initiation of the intermittent administration of hpGRF-40, mean (± SEM) serum somatomedin C had risen by 0.1 ± 0.04 U/ml and, at the end of the 5-day period, had increased from 0.27 ± 0.14 to 0.89 ± 0.29 U/ml (normal, 0.94 ± 0.30; mean ± 1 SD). In response to the second challenge with 10 μg/kg hpGRF-40, serum GH levels rose in two of the three subjects who initially failed to respond, and the increase in serum GH was greater in three of the four subjects who had responded to the first dose. In addition, unlike the first dose, the mean maximal serum GH level achieved in response to the second 10-μg/kg dose of hpGRF-40 was higher when compared to vehicle (p = 0.014). Although there was no change during the 5-day period in plasma cholesterol or HDL cholesterol, plasma triglycerides increased from 85.0 ± 17.9 to 169.7 ± 26.1 mg/dl (mean ± SEM; basal versus day 5). These data demonstrate that 5 days of intermittent hpGRF-40 administration augments GH secretion in some adults with GH deficiency, suggesting that the somatotropes that are present in idiopathic GH deficiency may be primed by hpGRF-40. The rise in serum somatomedin C to normal levels after multiple injections of hpGRF-40 is encouraging, since circulating levels of somatomedin C may be more important than the increase in immunoreactive GH levels as an index of response for induction of linear growth. The demonstration of biological effects of hpGRF-40 in all seven subjects, without any serious adverse effects, suggests that hpGRF-40 has promise in the treatment of GH deficiency.

2.7. Studies in Children with Short Stature

Forty children with short stature were evaluated for GH reserve following pharmacological tests and after a single intravenous bolus injection of hpGRF-40. These children were grouped into four diagnostic categories: (1) idiopathic GH deficiency (IGHD); (2) organic hypopituitarism; (3) intrauterine growth retardation (IUGR); and (4) constitutional delay of growth (CD), by standard clinical criteria and physiological and pharmacological tests of GH reserve. A fifth category, unclassified by these criteria, was also included.

Subjects were tested on two consecutive days: (1) following the intravenous administration of L-arginine (0.5 mg/kg over 30 min) and the oral administration of L-dihydroxyphenylalanine (L-dopa, 9 mg/kg) and (2) following the administration of hpGRF-40, 3.3 μg/kg, as an intravenous bolus.

The GH-deficient patients in categories 1 and 2 failed to raise their circulating GH concentrations to more than 7 ng/ml following the arginine–L-dopa test, but some showed marked GH responses following hpGRF-40 administration. As a group, however, the lowest responses (mean ± SE) to the releasing factor were noted in the organic hypopituitary (3.4 ± 1.1 ng/ml) and in the IGHD (8.2 ± 2.4 ng/ml) categories. All children in the IUGR, CD, and unclassified groups responded briskly to hpGRF-40, although there was wide variation of the peak GH level (5 to more than 25 ng/ml).

Circulating concentrations of somatomedin C did not differ in any category 24 hr following hpGRF-40 injection when compared to preinfusion values.

These data indicate that hpGRF-40 can be employed to evaluate GH reserve in short children and may be useful in the diagnosis of hypothalamic–pituitary disorders. The ultimate test of the utility of hpGRF-40 in children with GH deficiency or short stature is the therapeutic response (growth) when compared to the presently available natural or synthetic GH preparations.

3. Conclusions

In the short time that synthetic hpGRF has been available we have determined that it is a specific secretagogue (except in unusual circumstances) for GH secretion; i.e., it fails to affect the secretion of the other anterior pituitary hormones or of at least eight enteropancreatic hormones. The peptide is effective in stimulating GH secretion when administered intravenously, subcutaneously, or intranasally although at least 100-fold the effective intravenous dose is required for the intranasal route. Doses as low as 0.1 μg/kg i.v. produce maximal GH secretion although higher doses have a more prolonged duration of action.

The pharmacokinetics of GRF are unlike those of TRH, gonadotropin-releasing hormone (GnRH), or somatostatin, but similar to those of corticotropin-releasing factor, which is of similar molecular size having 41 amino acids.

The ability of hpGRF to stimulate GH secretion in only a minority of GH-deficient adults, but somatomedin C in a majority, and the high proportion of children of short stature who respond to hpGRF make it likely that hpGRF could have a role in therapy of GH deficiency. The role of hpGRF in other human disease is more speculative but may include the treatment of negative nitrogen balance seen in the aging process, and of wound healing and burns and possibly of short stature in the absence of GH deficiency, to mention a few. hpGRF also has a potential role in diabetes mellitus in that GH may be important in the development of complications of the disorder.

In the next year we hope to be able to determine whether hpGRF therapy is effective in stimulating linear growth in GH-deficient children.

ACKNOWLEDGMENTS. We thank Dr. Michael J. Cronin for helpful assistance; Mrs. Sandra W. Jackson, Ms. Fotini Beziriannidis, Mrs. Laura Dolan, Mrs. Jean Chitwood, Pattie Hellmann, Ms. Kathryn Wolf, and the staff of the Clinical Research Center for their help; the staff of the laboratories at our respective institutions for technical assistance; Mrs. Joyce Paciolla for assistance in the preparation of the manuscript; and the Pituitary Hormone Distribution Program of the National Institute of Arthritis, Diabetes, Digestive and Kidney Diseases for radioimmunoassay reagents for measurements of human GH and LH. These studies were supported in part by US Public Health Service research grants: General Clinical Research Grant RR-847; HD-13197 (M.O.T.); AM-32632 (M.O.T.); CIA 1-K03-HD-00439 (W.S.E.); AM-26741 (P.B.L); AM-209117 (P.B.L.); AA-03504 (P.B.L.); HD-13527 (P.B.L.); and AM-30667 (L.A.F).

References

1. M. O. Thorner, R. L. Perryman, M. J. Cronin, A. D. Rogol, M. Drazinin, A. Johanson, W. Vale, E. Horvath, and K. Kovacs, Somatotroph hyperplasia: Successful treatment of acromegaly by removal of a pancreatic islet tumor secreting a growth hormone releasing factor, *J. Clin. Invest.* **70:**965–977, 1982.
2. J. Rivier, J. Spiess, M. O. Thorner and W. Vale, Characterization of growth hormone-releasing factor from a human pancreatic islet tumor, *Nature* **300:**276–278, 1982.
3. J. Spiess, J. Rivier, M. O. Thorner and W. Vale, Sequence analysis of a growth hormone releasing factor from a human pancreatic islet tumor, *Biochemistry* **21:**6037–6040, 1982.
4. R. Guillemin, P. Brazeau, P. Bohlen, F. Esch, N. Ling and W. Wehrenberg, Growth hormone-

releasing factor from a human pancreatic tumor that caused acromegaly, *Science* **218:**585–587, 1982.

5. J. Rivier, J. Spiess and W. Vale, Human and rat hypothalamic growth hormone releasing factors (GRF), in: *Proceedings of Eighth American Peptide Symposium* (Victor J. Hruby and Daniel H. Rich, eds.), Pierce Chemical Company, 1983, pp. 853–856.
6. U. Gubler, J. J. Monahan, P. T. Lomedico, R. S. Bhatt, K. J. Collier, B. J. Hoffman, P. Bohlen, F. Esch, N. Ling, F. Zeytin, P. Brazeau, M. S. Poonian and L. P. Gage, Cloning and sequence analysis of cDNA for the precursor of human growth hormone-releasing factor, somatocrinin, *Proc. Natl. Acad. Sci. USA* **80:**4311–4314, 1983.
7. K. E. Mayo, W. Vale, J. Rivier, G. Rosenfeld and R. M. Evans, Expression cloning and sequencing of a cDNA encoding human growth hormone releasing factor, *Nature* **306:**86–88, 1983.
8. J. Spiess, J. Rivier and W. Vale, Characterization of rat hypothalamic growth hormone-releasing factor, *Nature* **303:**532–535, 1983.
9. M. O. Thorner, J Rivier, J. Spiess, J. L. C. Borges, M. L. Vance, S. R. Bloom, A. D. Rogol, M. J. Cronin, D. L. Kaiser, W. S. Evans, J. D. Webster, R. M. MacLeod and W. Vale, Human Pancreatic growth-hormone releasing factor selectively stimulates growth-hormone secretion in man, *Lancet* **1:**24–28, 1983.
10. S. M. Wood, J. L. C. Ch'ng, E. F. Adams, J. D. Webster, G. F. Joplin, K. Maschiter, and S. R. Bloom, Abnormalities of growth hormone release in response to human pancreatic growth hormone releasing factor (GRF-(1-44)) in acromegaly and hypopituitarism, *Br. Med. J.* **286:**1687–1691, 1983.
11. S. M. Rosenthal, E. A. Schriock, S. L. Kaplan, R. Guillemin and M. M. Grumbach, Synthetic human pancreatic growth hormone-releasing factor ($hpGRF_{1\text{-}44}\text{-}NH_2$) stimulates growth hormone secretion in normal man, *J. Clin. Endocrinol. Metab.* **57:**677–679, 1983.
12. M. C. Gelato, O. Pescovits, F. Cassorla, D. L. Loriaux and G. R. Merriam, Effects of a growth hormone releasing factor in man, *J. Clin. Endocrinol. Metab.* **57:**674–676, 1983.
13. M. L. Vance, J. L. C. Borges, D. L. Kaiser, W. S. Evans, R. Furlanetto, J. L. Thominet, L. A. Frohman, A. D. Rogol, R. M. MacLeod, S. Bloom, J. Rivier, W. Vale, M. O. Thorner, Human pancreatic tumor growth hormone releasing factor (hpGRF-40): Dose response relationships in normal man, *J. Clin. Endocrinol. Metab.* **58:**838–844, 1984.
14. W. S. Evans, J. L. C. Borges, M. L. Vance, D. L. Kaiser, A. D. Rogol, R. Furlanetto, J. Rivier, W. Vale, M. O. Thorner, Effects of human pancreatic growth hormone releasing factor-40 on serum growth hormone, prolactin, luteinizing hormone, follicle-stimulating hormone and somatomedin C concentrations in normal women throughout the menstrual cycle, *J. Clin. Endocrinol. Metab.* **59:**1006–1010, 1984.
15. W. S. Evans, J. L. C. Borges, D. L. Kaiser, M. L. Vance, R. P. Sellers, R. M. MacLeod, W. Vale, J. Rivier, and M. O. Thorner, Intranasal administration of human pancreatic tumor GH-releasing factor-40 stimulates GH release in normal men, *J. Clin. Endocrinol. Metab.* **57:**1081–1083, 1983.
16. L. A. Frohman, J. L. Thominet, C. B. Webb, M. L. Vance, H. Uderman, J. Rivier, W. Vale, and M. O. Thorner, Metabolic clearance and plasma disappearance rates of human pancreatic tumor growth hormone releasing factor in man, *J. Clin. Invest.* **73:**1304–1311, 1984.
17. C. B. Webb, J. L. Thominet and L. A. Frohman, Ectopic growth hormone releasing factor stimulates growth hormone release from human somatotroph adenomas in vitro, *J. Clin. Endocrinol. Metab.* **56:**417–419, 1983.
18. J. L. C. Borges, R. M. Blizzard, M. C. Gelato, A. D. Rogol, M. L. Vance, R. M. MacLeod, D. L. Loriaux, J. Rivier, R. M. Blizzard, R. Furlanetto, W. S. Evans, D. L. Kaiser, G. R. Merriam, J. Spiess, W. Vale and M. O. Thorner, Effects of human pancreatic tumour growth hormone releasing factor on growth hormone and somatomedin C levels in patients with idiopathic growth hormone deficiency, *Lancet* **2:**119–124, 1983.
19. A. Grossman, M. O. Savage, J. A. H. Wass, N. Lytras, J. Sueiras-Diaz, D. H. Coy, and G. M. Besser, Growth-hormone-releasing factor in growth hormone deficiency: Demonstration of a hypothalamic defect in growth hormone release, *Lancet* **2:**137–138, 1983.
20. J. L. C. Borges, R. M. Blizzard, W. S. Evans, R. Furlanetto, A. D. Rogol, D. L. Kaiser, J. Rivier, W. Vale and M. O. Thorner, Stimulation of growth hormone (GH) and somatomedin C in idiopathic GH-deficient subjects by intermittent pulsatile administration of synthetic human pancreatic tumor GH-releasing factor, *J. Clin. Endocrinol. Metab.* **59:**1–6, 1984.

28

Acromegaly Related to Tumoral Secretion of Growth Hormone-Releasing Factor

Geneviève Sassolas and Pierre Chatelain

1. *Introduction*

The clinical evidence of acromegaly in a patient with a past history of treated and cured thymic tumor and on-going hyperparathyroidism, in contrast with an unmodified sella turcica, led us to suspect a putative growth hormone (GH) or growth hormone-releasing factor (GRF) secreting tumor. The discovery of high plasma levels of somatostatin, pancreatic polypeptide (PP), and glucagon suggested a tumor of the pancreas. A body scan documented the existence of a large pancreatic tumor.

Three molecular forms of GRF's were isolated, characterized, and sequenced (hpGRF 1-44 NH_2, hpGRF 1-40, hpGRF 1-37) from that tumor by R. Guillemin *et al.*[1]

By many criteria, hpGRF 1-44-NH_2 was rapidly shown to be closely related to human hypothalamic GRF.[2,3] We report the results of clinical studies performed with the synthetic replicate of hpGRF 1-44-NH_2 (referred to as GRF).

2. *Clinical Investigation of a Patient with a GRF- and Somatostatin-Secreting Tumor*

2.1. *Case History*[4]

A 55-year-old man was referred to the Centre de Médecine Nucléaire of Lyon, France, in July 1981 for acromegaly. The family history was negative for

Geneviève Sassolas • Centre de Médecine Nucléaire, Hôpital Neuro-Cardiologique, 69394 Lyon Cedex 3, France. ***Pierre Chatelain*** • INSERM U 34, Hôpital Debrousse, 69005 Lyon, France.

endocrine disorders. He had complained of nephrolithiasis for 17 years. An asymptomatic 70-g, well-circumscribed thymic tumor recognized on routine chest X ray had been surgically removed in 1975. It was diagnosed as a carcinoid tumor of the thymus when we reviewed the case. This tumor was treated by radiotherapy when it recurred in 1978. Radiological examination did not show any relapse in the mediastinum.

The progressive appearance of the characteristic morphological features of acromegaly had been noted by the patient since 1974. Non-insulin-dependent diabetes mellitus was diagnosed in 1975. During the same period, the patient, who had previously fathered two children, reported sexual impotence and progressive testicular atrophy. On admission, he weighed 78 kg and measured 178 cm in height and presented all the typical features of acromegaly. Blood pressure was 130/80 mm Hg. The only clinical sign that could have been attributed to pituitary deficiency was testicular atrophy. The diagnosis of acromegaly was fully confirmed by laboratory investigations. A chronic and slight hypercalcemia with hyperparathyroidism was found. The sella turcica was slightly enlarged, with a calculated volume of 1500 mm^3 (normal range: 242–1042 mm^3), according to the method of Di Chiro and Nelson; there had been no modification in the volume of the sella between 1975 and 1982. Head computerized tomography scan was normal. On the basis of these data, we decided not to submit the patient to surgical treatment for acromegaly, but to undertake a medical treatment with bromocriptine, which was carried over a 4-month period at doses of 10–20 mg daily, without clinical or biological improvement of GH secretion but with total suppression of prolactin (PRL).

The progressive worsening of the patient's general condition and the elevation of the erythrocyte sedimentation rate (120 mm) without evidence of infection led us to hypothesize the existence of a tumorous process. Among the tumorous markers measured, marked increase in the plasma levels of somatostatin, pancreatic polypeptide (PP), and glucagon, alone, suggested the endocrine nature of this tumor and pointed to the pancreas. A body scan demonstrated the presence of a voluminous pancreatic tumor without hepatic metastases. A puncture under echography led to a cytological study suggestive of an endocrine tumor. Surgery was performed on April 16, 1982. The pancreas was found to be totally invaded by a large tumor, measuring 25 × 12 cm, responsible for portal hypertension. Sequential removal of the pancreas, the spleen, the stomach, and the lymph nodes was performed. The tumor consisted of two contiguous masses of equal size; one, located in the tail, was yellowish, partially calcified, and cystic; the other, located in the head, was soft and reddish. Lymph node metastases were present in the hilum of the liver and of the spleen. A small tumor, 3 cm in diameter, was found in the duodenal submucosa, 5 cm inferior to the papilla of Vater.

Following surgery, the patient recovered uneventfully. Diabetes has been subsequently controlled by insulin therapy. A good nutritional state has been maintained with supplemental pancreatic enzymes, and the patient has been gaining weight. GH levels were undetectable on the 10th day after surgery. Calcemia remained elevated. The patient underwent follow-up examinations

Table 1 Pre- and Postoperative Plasma Levels of Somatomedin C

	Preoperative	Postoperative 6.82	Postoperative 9.82	Normal range
Plasma Somatedin C	6	0.2	0.6	0.2–2 IU/ml

in June and September 1982, and in January, June, and September 1983. The recurrence of the tumorous process was suspected in January 1983 (9 months postsurgery) because of a rise in plasma levels of GH, GRF, and somatostatin. Hepatic echography performed at the same time revealed the existence of small, diffuse hepatic nodules. In August and October 1983, the patient underwent treatment with streptozotocin and 5-fluorouracil.

2.2. Laboratory Investigations

The materials and the methods used for these investigations have previously been reported in details.[4] Pre- and postoperative data are given in Tables 1–4 and in Figs. 1–5.

In May 1983, when GRF radioimmunoassay (RIA) became available, plasma GRF was evaluated a posteriori on frozen preoperative plasma samples and revealed levels consistently higher than 50 μg/liter (GRF is not detectable in normal plasma with our technique). Plasma GRF level (μg/liter) was 2 in May 1982 and between 5 and 7 in September; the 24-hr sampling in June 1983 showed the following levels: 12.5 at 8 A.M., 13.7 at 2 P.M., 23.5 at 10 P.M., and 18.2 at 12 P.M.

2.3. Pathology

A complete report of the studies performed on this tumor, including histological methods, immunocytochemistry, and electron microscopy, has been

Table 2 Plasma Levels of Various Substances Considered as Tumor Markers[a]

	Preoperative 1.82	Postoperative 6.82	Postoperative 9.82	Postoperative 5.83	Postoperative 9.83	Normal range
Calcitonin	<150	<150	ND	ND	ND	<150 pg/ml
HCG	<0.5	<0.5	ND	<0.5	ND	<1 ng/ml
Substance P	12	14	8	34	ND	<20 pg/ml
VIP	16	19	7	44	19	<50 pg/ml
Gastrin	34	4	8	8	62	<80 pg/ml
Pancreatic PP	4000	<10	17	29	55	<500 pg/ml
Somatostatin	690	20	69	110	180	<50 pg/ml
Glucagon	475	265	220	210	ND	<200 pg/ml

[a] VIP = vasointestinal peptide. Note the decrease in pancreatic peptide, somatostatin, and glucagon concentration in early postoperative period.

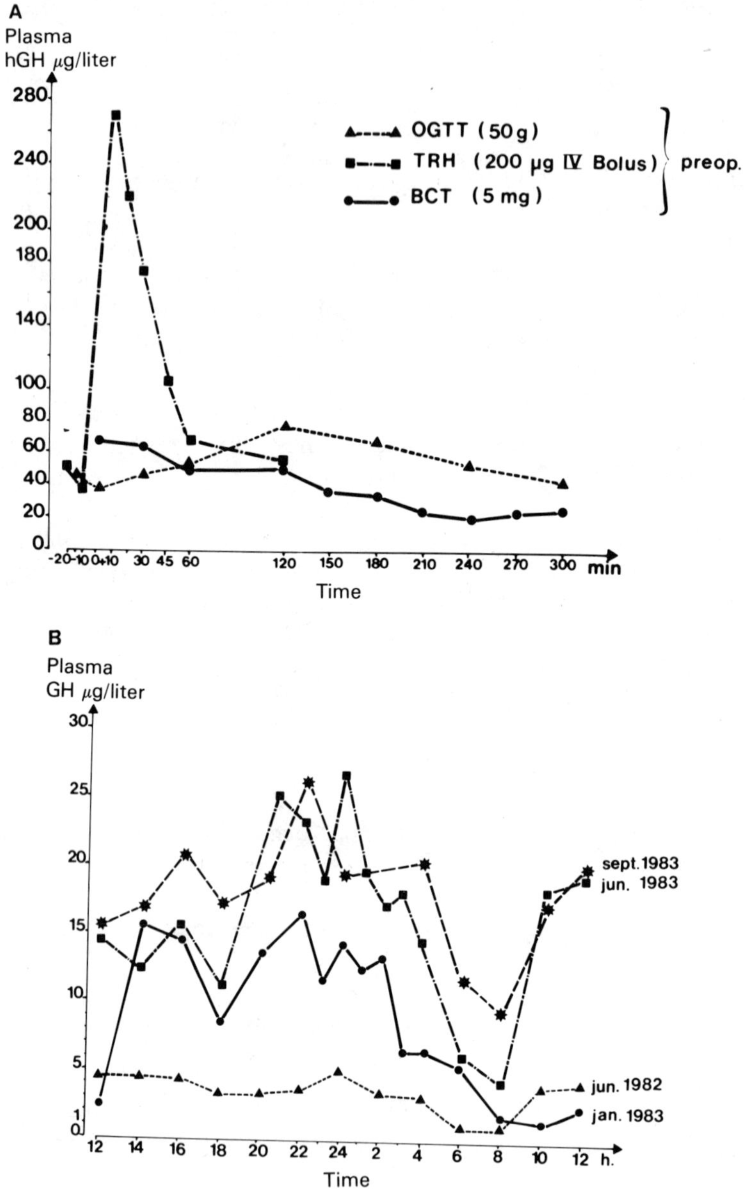

Figure 1 Pre- and postoperative plasma growth hormone (GH) investigations. (A) Preoperative GH investigations with glucose load, thyroid-releasing hormone (TRH), and bromocriptine (BCT). (B) Early and late postoperative plasma GH levels on 24 hr samplings.

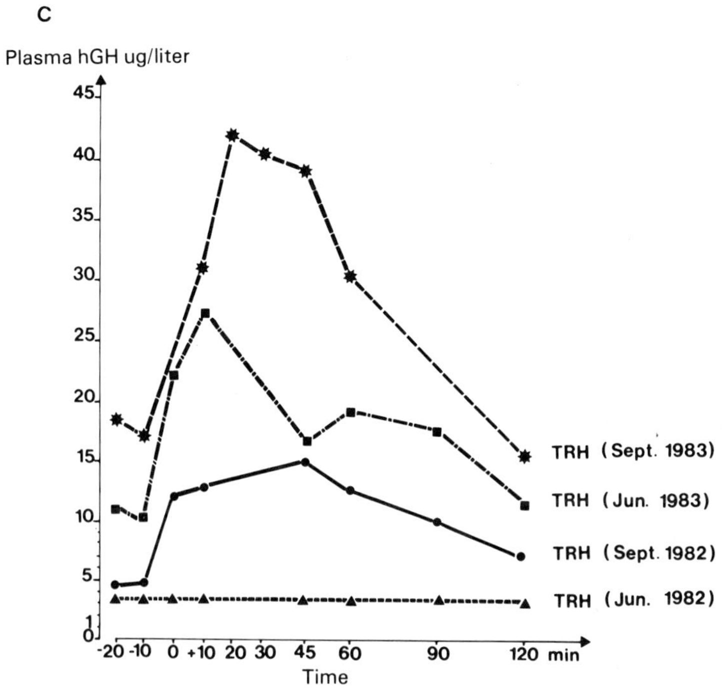

Figure 1 (C) Early and late postoperative plasma GH levels under TRH.

published by Berger *et al.*[5] These studies revealed a malignant endocrine pancreatic tumor with varying types of growth.

2.3.1. *The Cephalic Mass*

The cephalic mass, from which were extracted peptides displaying a GH-releasing activity, consisted of cells arranged in medullary masses. Amyloid deposits were found between tumor cells. The Grimelius stain showed the

Table 3 Plasma Levels of Ca^{2+}, P, and Parathormone at Various Pre- and Postoperative Periods

		Postoperative			
	Preoperative	6.82	6.83	9.83	Normal range
Plasma Ca^{2+}	2.9–3.10	2.80	2.85	2.9	2.30–2.60
(mmole/liter)			1.00	1.10	0.80–1.50
Plasma P	0.75–0.80	1.11			
(mmole/liter)					
Plasma PTH (ng/liter)	2480	2294	2183	2300	450–1450

Table 4 Hormonal Preoperative Investigations

		Preoperative	Normal range
T_4		98	55–150 mmole/liter
TRH test	Basal TSH	3.8	<7 mUI/liter
	Maximum TSH	11.3	5–20 mUI/liter
Metyrapone			
	Basal ACTH	20	35 ± 15 ng/liter
	Maximum ACTH	228	150 ng/liter
	Basal 11-deoxycortisol	0.1	0.1 μg/100 ml
	Maximum 11-deoxycortisol	14.5	>12 μg/100 ml

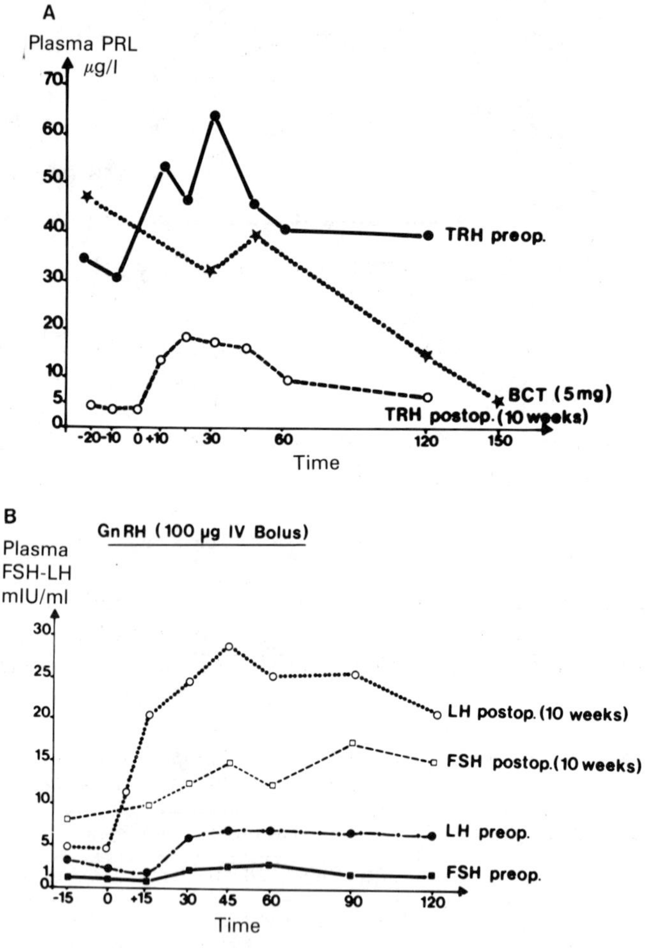

Figure 2 Pre- and postoperative investigations of plasma prolactin and gonadotropins. (A) Preoperative levels of prolactin under thyroid-releasing hormone (TRH) and bromocriptine; postoperative levels of prolactin under TRH. (B) Plasma follicle-stimulating hormone (FSH) and luteinizing hormone (LH) under gonadotropin-releasing hormone (GnRH)I in pre- and postoperative periods.

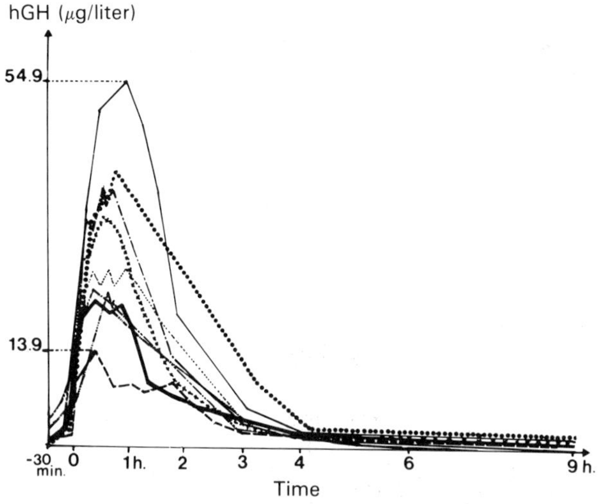

Figure 3 Ten individual plasma GH values before and after 150-μg i.v. bolus GRF, in normal adults.

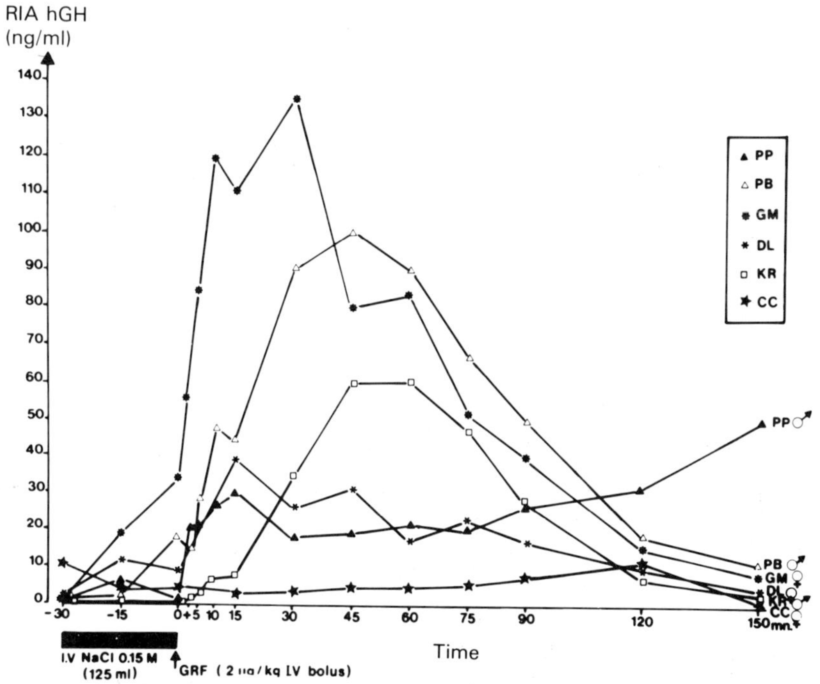

Figure 4 Plasma growth hormone (GH) values before and after 2-μg/kg i.v. bolus of growth hormone-releasing hormone (GRF) in six normal prepubertal children (7–9 years old).

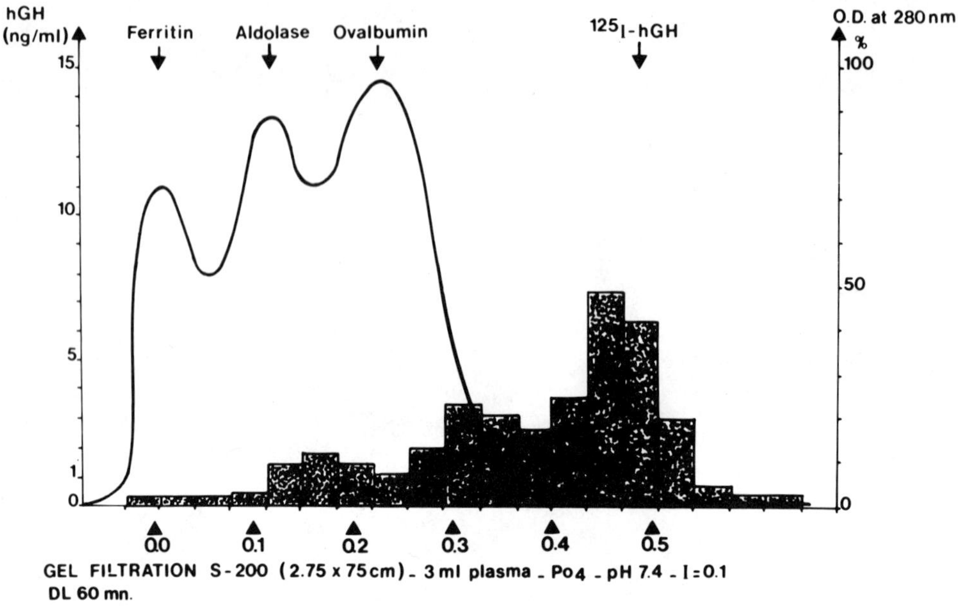

Figure 5 Gel filtration chromatography of 3 ml of heparinized plasma 60 min after growth hormone-releasing factor Subject D.L. (Fig. 4), Sephacryl S 300. 2.75 × 75 column. Buffer PO_4, pH 7.4, ionic strength = 0.1. Flow rate 1 ml/mn 4 ml/fraction at 4°C. Growth hormone radioimmunoassay (ng/ml of pool) was evaluated from five pooled fractions.

presence of numerous weakly reactive cells. Most of the tumor cells reacted strongly with the antiserum to neuron-specific enolase. Ten percent of the cell population was stained with antisomatostatin antiserum, although a large number reacted with antiserum raised against anti hpGRF 1-40. Control for inhibition showed a total disappearance of this staining by addition of GRF 1-44 NH_2 to the antiserum.

2.3.2. The Caudal Mass

The caudal mass consisted of a zone of gyriform nodules, with strong diffuse argyrophilia, and of a zone of trabecular pattern areas, with moderate argyrophilia. Immunochemistry revealed in the gyriform nodules three types of immunoreactive cells, the majority stained with antiglucagon constituting entire nodules, some PP-reacting cells constituting a few ribbons within the nodules, and some somatostatin-storing cells irregularly scattered.

In the areas presenting a microlobular pattern, some cells reacted weakly with anti-ACTH 17-39, anti-ACTH 1-24, anti-βLPH, anti-α-endorphin, and anti-β-endorphin antisera. Total extinction of the stainings was obtained adding each of the pancreatic peptides and β-endorphin to the corresponding antisera.

2.3.3. Lymph Node Metastases

The metastatic lesions removed from the mesentery and the hilum of the liver showed a pattern of growth similar to that of the cephalic mass. Most of the cells were strongly stained by anti-GRF 1-40 antiserum. The lymph node adjacent to the spleen contained a ribbon-patterned tumor mainly composed of PP-storing cells.

2.3.4. Duodenal Tumor

The microscopic image was characteristic of a foregut carcinoid tumor. Immunocytochemistry showed that it consisted mainly of somatostatin-reacting cells.

2.4. Isolation and Characterization of Peptides Displaying GH-Releasing Activity and Somatostatin

Out of the cephalic mass of that tumor, R. Guillemin *et al*, isolated three peptides with GH-releasing activity which were sequenced and referred to as hpGRF 1-44 NH_2, 1-40, and 1-37.(1) Two peptides with GH-inhibiting activity were also characterized from that tumor, one being identical to the ovine somatostatin-14 and the other to somatostatin-28.(6) Hormonal RIA of scattered tumorous fragments did not demonstrate any significant concentration of GH, PRL, luteinizing hormone, follicle-stimulating hormone, thyroid-stimulating hormone, parathyroid hormone (PTH).

2.5. Discussion

As hypothesized, this case of acromegaly was the clinical expression of a GRF-secreting tumor. In agreement with the others reported, it shows that GRF-producing tumors may be responsible for a long-standing acromegaly, similar in many respects to those due to pituitary somatotropic adenomas. Basal plasma GH levels and pattern of GH response to provocative tests are similar and fail to differentiate between the two types of acromegaly. Of special interest is the suggestion that chronic GRF stimulation of somatotrophs induces responsiveness to thyrotropin-releasing hormone (TRH). This is based on the relationship between the variations in plasma GH patterns after TRH and the variations in plasma GRF levels assessed in the pre-, early, and late postoperative phases.

The unchanged subnormal sella turcica between 1975 and 1982 suggests the absence of an evolutive pituitary adenoma. Somatotroph hyperplasia induced by chronic GRF hyperproduction has been confirmed by the studies of Kovacs and Horvath reported by Thorner(7) and can explain the minimal sellar enlargement seen here and in similar cases.

Hyperprolactinemia is commonly reported in acromegaly with pituitary adenoma and is most often explained by adenomatous PRL production. In the ectopic acromegalic syndrome, however, the pathophysiology of hyperprolactinemia remains unclear. In previously published cases, amenorrhea with galactorrhea or isolated galactorrhea[8] was described, and hyperprolactinemia was noted in three instances.[7,9,10] In the case reported by Thorner *et al.*[7] there was no change in plasma PRL levels after removal of the GRF-producing tumor. In our case, hyperprolactinemia disappeared following tumor removal. The absence of PRL in the tumor itself led us to hypothesize that pancreas tumor might have been producing a lactotroph-stimulating agent. The direct GRF effect on PRL secretion under chronic exposure to GRF remains to be evaluated.

In each of the three GRF-secreting tumors reported by Frohman[11] significant amounts of somatostatinlike immunoreactivity were detected. Our observation confirms that in the acromegalic syndrome, it may be worthwhile to evaluate plasma somatostatin even in the absence of the reported features of somatostatinoma syndrome as a marker of a digestive tract endocrine tumor.

Resection of the pancreatic tumor was not followed by any change in the hypercalcemia or elevated plasma PTH levels, and no immunoreactive PTH was detected in the tumor. It is likely, therefore, that the hypercalcemia is due to parathyroid adenoma(s) or hyperplasia, as has previously been reported in cases of intestinal and ovarian carcinoid tumors[12] and in familial cases of multiple endocrine neoplasia including thymic carcinoid tumors.[13,14]

This case led to the discovery of three peptides with GRF activity, among which is hpGRF 1-44 NH_2. The synthetic replicate of this peptide is highly active *in vitro,*[1] *in vivo,* in animals,[15,16] and in human beings.[17–20]

3. Clinical Results with Synthetic Replicate of GRF 1-44 N H_2 Isolated from That Tumor

Studies reported were performed after approval of the protocols by the Ethical Committee of Université Claude Bernard and Hospices Civils de Lyon.

3.1. Clinical Results in Normal Adult Men

Synthetic replicate of hpGRF 1-44 NH_2 was given by i.v. bolus to 10 normal young men at doses of 75, 150, 300, and 600 μg and placebo in a random order. Clinical and biological tolerance were excellent. At every dose and in every subject, the plasma GH rise induced by GRF is early, dramatic, and prolonged (Fig. 3). The means for the GH peaks are between 22 and 28 ng/ml. There is a significant difference in GH patterns between the individuals. Plasma PRL patterns show an early slight and dose-dependent rise after GRF. Details on this study are being reported elsewhere.

3.2. Clinical Results in Normal Prepubertal Children

A 2 μg/kg dose of GRF given intravenously to six normal prepubertal children (7–9 years old) induced a heterogenous plasma GH rise (Fig. 4).

3.3. Bioactivity and Circulating Forms of hGH Released under GRF

In normal adults and children hGF released under GRF is biologically active in the NB_2 lymphoma cell system; circulating plasma hGF is found under three molecular forms at 60 min after GRF (Fig. 5).

3.4. Clinical Results in GH-Deficient Patients (Table 5)

3.4.1. GH-Deficient Adults

A 100-μg i.v. bolus GRF injection was given to eight GH-deficient adults (Table 5). In three idiopathic GH-deficient patients, GRF induced a slight GH rise. In five cases, secondary to craniopharyngioma treated by surgery, GRF induced a significant response only in the patient who had most recently been submitted for surgery.

Table 5 GRF in GH Deficiency: Adults and Children[a]

No.	Age	Diagnosis	Pituitary hormone deficiency	Peak GH under GRF (μg/liter)
1	46	Idiopathic	Global deficiency	4.4
2	19	Idiopathic	GH + FSH + LH	8
3	21	Idiopathic	GH + FSH + LH	3.3
4	33	Craniopharyngioma 20 years postsurgery	Global deficiency	<1
5	23	Craniopharyngioma 10 years post XRT	Global deficiency	1.3
6	21	Craniopharyngioma 7 years postsurgery and XRT	Global deficiency	<1
7	28	Craniopharyngioma 1 year postsurgery	GH + FSH + LH	16.4
8	11	Hypothalamic germinoma	GH + AVP	
		PreXRT		13.1
		Post XRT		15.8
9	14	Hypothalamic germinoma 1 year post XRT	GH + TSH	9.0
10	18	Supra sellar tumor	GH + ACTH + TSH + FSH + LH + AVP	14.2
11	9	4 years post XRT for brain tumor	GH	11
12	10	2 years post XRT for brain tumor	GH	9.5
13	18	12 years post head trauma	GH + ACTH + FSH + LH + AVP	4.5
14	14	Idiopathic	GH	3.0
15	2.6	Idiopathic	GH	7.2
16	6	Idiopathic	GH + TSH	88
17	8	Idiopathic	GH	30

[a] XRT = radiotherapy; GH = growth hormone; FSH = follicle-stimulating hormone; LH = luteinizing hormone; AVP = arginine vasopressive; TSH = thyroid-stimulating hormone.

3.4.2. GH-Deficient Children

GH response to hpGRF was tested in 10 GH-deficient children, whose clinical history and features are summarized in Table 5. Four children presented with idiopathic GH deficiency, the six others with GH deficiency secondary to a central nervous system (CNS) tumor with or without radiotherapy or to head trauma. Criteria used to define GH deficiency included growth failure and growth velocity 2 SD below the normal for bone age, peak hGH < 5 ng/ml, after the administration of two or more standard pharmacological stimuli (glucagon–propranolol, arginine–insulin, or insulin alone).

Of 10 children, 7 had a significant increase in serum hGH level (above 8 ng/ml) after an i.v. injection of 2 μg/kg hpGRF. Five of the seven had GH deficiency secondary to a CNS tumor. Their response to GRF was comparable to that of control subjects, but the magnitude of the peaks (9–15.8 ng/ml) was smaller. In the two others with idiopathic GH deficiency, GRF injection induced a marked increase in plasma GH level. This suggests a hypothalamic defect.

The three remaining patients did not show any response. One of them had complete hypopituitarism secondary to head trauma that occurred 12 years previously. The two others had idiopathic deficiency, and further studies, including repeated GRF infusion, are needed for a better understanding of their defect.

ACKNOWLEDGMENTS. These studies were performed with the collaboration of R. Guillemin, P. Brazeau, N. Ling, J. A. Chayvialle, C. Partensky, R. Cohen, B. Claustrat, G. Berger, J. Trouillas, B. Bloch, J. P. Boissel, S. Biot-Laporte, S. Ferry, A. Elm Charfi, F. Borson, R. François, L. David, M. David, A. Ruitton, H. Cohen, F. Frederich, and C. Metral, who constituted a team involved in GRF studies in Lyon, France. They were supported in part by the following administrations: Hospices Civils de Lyon, INSERM, Faculté de Médecine Alexis Carrel and SANOFI, and Fondation pour la Recherche Médicale Française.

References

1. R. Guillemin, P. Brazeau, P. Böhlen, F. Esch, N. Ling, and W. B. Wehrenberg, Growth hormone-releasing factor from a human pancreatic tumor that caused acromegaly, *Science* **218:**585, 1982.
2. B. Bloch, P. Brazeau, N. C. Ling, *et al.*, Immunohistochemical detection of growth hormone-releasing factor in brain, *Nature* **301:**607, 1983.
3. P. Böhlen, P. Brazeau, B. Bloch, N. Ling, R. Gaillard, and R. Guillemin, Human hypothalamic growth hormone-releasing factor (GRF): Evidence for two forms identical to tumor-derived GRF-44-NH_2 and GRF-40, *Biochem. Biophys. Res. Commun.* **114:**930, 1983.
4. G. Sassolas, J. A. Chayvialle, C. Partensky, G. Berger, J. Trouillas, F. Berger, B. Claustrat, R. Cohen, C. Girod, and R. Guillemin, Acromegalie, expression clinique de la production de facteurs de libération de l'hormone de croissance (GRF) par une tumeur pancréatique, *Ann. Endocrinol. (Paris)* **44:**343, 1983.
5. G. Berger, J. Trouillas, B. Bloch, G. Sassolas, F. Berger, C. Partensky, J. A. Chayvialle, P. Brazeau, B. Claustrat, F. Lesbros, and C. Girod, Multihormonal carcinoid tumor of the pancreas secreting growth hormone-releasing factor as a cause of acromegaly, *Cancer* **54:**2097, 1984.

6. P. Böhlen, P. Brazeau, F. Esch, N. Ling, and W. B. Wehrenberg, Human growth hormone releasing factor and somatostatin from two pancreatic tumors: Isolation and characterization, *Regulatory Peptides* **6:**343, 1983.
7. M. O. Thorner, R. L. Perryman, M. J. Cronin, A. D. Rogol, M. Draznin, A. Johanson, W. Vale, E. Horvath, and K. Kovacs, Somatotroph hyperplasia. Successful treatment of acromegaly by removal of a pancreatic islet tumor secreting a growth hormone-releasing factor, *J. Clin. Invest.* **82:**965, 1982.
8. R. H. Caplan, L. Koobb, R. M. Abellera, A. S. Pagliara, K. Kovacs, and R. Randall, Cure of acromegaly by operative removal of an islet cell tumor of the pancreas, *Am. J. Med.* **64:**874, 1978.
9. A. Gomez-Pan, M. F. Scanlon, M. O. Thorner, L. H. Rees, A. V. Schally, R. Hall, and G. M. Besser, Effect of somatostatin on abnormal growth hormone and prolactin secretion in patients with the carcinoid syndrome, *Clin. Endocrinol.* **10:**575, 1979.
10. S. A. Leveston, D. W. Mc Keel, P. J. Buckley, K. Deschryver, M. H. Greider, B. M. Jaffe, and W. H. Daughaday, Acromegaly and Cushing's syndrome associated with a foregut carcinoid tumor, *J. Clin. Endocrinol. Metab.* **53:**682, 1981.
11. L. A. Frohman, M. Szabo, and M. Berelowitz, Partial purification and characterization of a peptide with growth hormone-releasing activity from extrapituitary tumors in patients with acromegaly, *J. Clin. Invest.* **65:**43, 1980.
12. N. A. Samaan, R. C. Hickey, T. D. Bedner, and M. L. Ibanez, Hyper parathyroidism and carcinoid tumor, *Ann. Intern. Med.* **82:**205, 1975.
13. J. Rosai, E. Higa, and J. Davie, Mediastinal endocrine neoplasm in patients with multiple endocrine adenomatosis. A previously unrecognized association, *Cancer* **29:**1075, 1972.
14. J. L. Manes and H. B. Taylor, Thymic carcinoid in familial multiple endocrine adenomatosis, *Arch. Pathol.* **95:**252, 1973.
15. W. B. Wehrenberg, N. Ling, P. Brazeau, F. Esch, P. Böhlen, A. Baird, S. Ying, and R. Guillemin, Somatocrinin, growth hormone-releasing factor, stimulates secretion of growth hormone in anesthetized rats, *Biochem. Biophys. Res. Commun.* **109:**38, 1982.
16. W. B. Wehrenberg, N. Ling, P. Böhlen, F. Esch, P. Brazeau, and R. Guillemin, Physiological role of somatocrinin and somatostatin in the regulation of growth hormone secretion, *Biochem. Biophys. Res. Commun.* **109:**562, 1982.
17. G. Sassolas, H. Rousset, R. Cohen, P. Chatelain, B. Claustrat, S. Laporte, B. Laferrere, A. El Charfi, and S. Ferry, La somatocrinine provoque la libération de somathormone dans un cas de déficit en somathormone d'origine hypothalamique chez l'enfant, *C.R. Acad. Sci. Paris*, Série III **296:**527, 1983.
18. G. Sassolas, J. P. Boissel, R. Cohen, P. Chatelain, S. Laporte, J. Galleyrand, H. Cohen, S. Ferry, and P. Maire, La somatocrinine (GRF) provoque la libération de somathormone (GH) biologiquement active chez l'homme adulte, *C.R. Acad. Sci. Paris*, Série III **296:**1113, 1983.
19. S. M. Rosenthal, E. A. Schriock, S. L. Kaplan, R. Guillemin, and M. M. Grumbach, Synthetic human pancreas growth hormone-releasing factor (hpGRF 1-44 NH_2) stimulates growth hormone secretion in normal men, *J. Clin. Endocrinol. Metab.* **57:**677, 1983.
20. M. C. Gelato, O. Pescovitz, F. Cassorla, D. Lynn Loriaux, and G. R. Merriam, Effects of a growth hormone releasing factor in man, *J. Clin. Endocrinol. Metab.* **57:**674, 1983.
21. S. M. Wood, J. Ching, E. F. Adams, J. Webster, G. G. Joplin, K. Mashiter, and S. R. Bloom, *In vivo* and *in vitro* effects of hpGRF (1-44)-NH_2 on GH secretion from normal and tumorous human pituitary, The 65th Annual Meeting of the Endocrine Society, San Antonio, 1983, Abstr. 204.
22. K. Von Werder, O. A. Muller, R. Hartl, M. Losa, and G. K. Stalla, Growth hormone-releasing factor. Stimulation test in normal controls and acromegalic patients, *J. Endocrinol. Invest* **7:**185, 1984.

29

Studies with Growth Hormone-Releasing Factor (GRF) in the Human

Effect of a Single Pulse, Continuous Infusion, or Multiple Pulses of GRF on Growth Hormone (GH) Release in Normal and GH-Deficient Children and Adults

Elizabeth A. Schriock, Stephen M. Rosenthal, Catherine A. Egli, David A. Harris, Berthold P. Hauffa, J. Anthony Hulse, Robert H. Lustig, Selna L. Kaplan, and Melvin M. Grumbach

In 1982 Guillemin *et al.*[1] isolated, characterized, and synthesized an amidated, 44-amino-acid growth hormone-releasing factor (GRF_{44}) from the pancreatic tumor of a patient with acromegaly. GRF_{44} and two shorter, homologous peptides that lack a *C*-terminal amide, GRF_{37} and GRF_{40}, stimulated GH release in cultured pituitary cells and *in vivo*.[1] An identical 40-residue peptide was isolated by Rivier *et al.*[2] from the tumor of a second acromegalic patient. There is increasing evidence that these tumor-derived growth hormone-releasing factors (GRF) are identical to human hypothalamic GRF and are encoded by a single GRF gene.[3–6] The availability of synthetic GRF provides an opportunity to define further the control and dynamics of GH secretion in the human in health and disease.

Elizabeth A. Schriock, Stephen M. Rosenthal, Catherine A. Egli, David A. Harris, Berthold P. Hauffa, J. Anthony Hulse, Robert H. Lustig, Selna L. Kaplan, and Melvin M. Grumbach • Department of Pediatrics, University of California–San Francisco, San Francisco, California 94143.

1. Subjects and Methods

1.1. GRF

The $GRF_{1\text{-}44}$-NH_2 (Lot No. 92-117-3G-34-40) and $GRF_{1\text{-}40}$-OH (Lot No. 92-112-3G-76-112) were synthesized by Dr. Nicholas Ling at the Salk Institute for Biological Studies and kindly provided by Dr. Roger Guillemin. The peptides were dissolved in 0.45% physiological saline to give a final concentration of 1 mg/ml with 0.5% human serum albumin added.

1.2. Subjects and Study

The GRF test protocol as previously described[7–9] was undertaken in the Pediatric and General Clinical Research Centers at the University of California San Francisco with approval of the UCSF Committee on Human Research and the U.S. Food and Drug Administration. GRF was administered in various doses with different frequencies to non-GH-deficient- and GH-deficient individuals.

A. Studies with a single intravenous pulse of GRF involved:
   1. Seven normal young men aged 19.5–33 years given placebo (physiological saline) and 0.5 μg/kg, 5 μg/kg, and/or 10μg/kg of GRF_{44} at intervals of at least 1 week.[7] Five of these men were also given 0.5 μg/kg of GRF_{40}.
   2. Seven normal women aged 21–33 years given 1 μg/kg of GRF_{44} at early follicular, mid-, and luteal phases of the menstrual cycle.
   3. Seven non-GH-deficient children and adolescents aged 4–16 years given 5 μg/kg of GRF_{44}.
   4. Administration of 5 μg/kg of GRF_{44} to 27 patients 4.75–24.5 years of age and one additional patient, aged 40 years, with severe GH deficiency as defined by GH responses <3 ng/ml to at least two pharmacological stimuli of GH release, i.e., arginine, insulin, or L-dopa.[8] GH deficiency was diagnosed 13–16 years prior to GRF_{44} testing in all of the adult patients with GH deficiency except one whose presentation was consistent with long-standing GH deficiency. Distribution of these patients according to etiology of GH deficiency and the presence of multiple hypothalamic–pituitary hormone deficiencies is given in Table 1. Twelve patients had thyroid-stimulating hormone responses to synthetic thyroid hormone-releasing factor (TRF) consistent with hypothalamic TRF deficiency and/or elevated basal prolactin concentrations consistent with a deficiency of prolactin release-inhibiting factor. The same dose of GRF_{44} was also given to seven patients 5–14 years of age with isolated, partial GH deficiency; GH responses to pharmacological provocative tests were between 3 and 7 ng/ml. Five of these patients had idiopathic GH deficiency, and two had midline developmental defects.

Table 1 Distribution of Patients with Severe GH Deficiency by Etiology

	Isolated GH deficiency (n = 12)	Multiple hypothalamic–pituitary hormone deficiencies (n = 16)
Idiopathic	5	11
Familial[a]	2	—
Organic		
Postirradiation[b]	4	1
Midline defects	1	2
Craniopharyngioma	—	1
TB meningitis	—	1

[a] Autosomal dominant; mother and daughter.
[b] Cranial irradiation for leukemia or non-hypothalamic–pituitary tumors.

B. A study of the effects of a continuous infusion of GRF_{44} involved eight normal men 24–39 years of age given a 3-μg/kg i.v. pulse of GRF_{44} followed 30 min later by a 2-hr infusion of 3 μg/kg per hr of GRF_{44}.[9]

C. Stimulation of GH release by multiple pulses of GRF_{44} was evaluated in six GH-deficient patients (9.42–18.67 years old) given an intravenous pulse of GRF_{44} every 2 hr beginning at 8 A.M. for a total of 6–12 doses. The first and last pulses were 5 μg/kg in all individuals except one whose study was discontinued after the 10th pulse; all other pulses were 1 μg/kg i.v. of GRF_{44}.

1.3. Methods

Blood samples were obtained at −30, 0, 10, 15, 20, 30, 45, 60, 90, 120, 150, 180, 210, and 240 min and 24 hr after GRF administration. Measurement of plasma growth hormone (GH), luteinizing hormone (LH), follicle-stimulating hormone (FSH), thyroid-stimulating hormone (TSH), prolactin (PRL) insulin, and cortisol was by radioimmunoassay (RIA) as previously described.[7] Sensitivity of the GH assay was 0.5 ng/ml. Serum somatomedin C (SMC) was measured after acid–ethanol extraction by RIA[10] using as tracer synthetic SMC (Li, lot No. 4378B), the Van Wyk rabbit antiserum distributed by NIADDK, and an Ortho adult serum standard (No. 1778-5, Ortho Diagnostics, Raritan, NJ); sensitivity was 0.1 U/ml. Plasma somatostatin concentrations were determined by Dr. Arthur Vinik (University of Michigan) by RIA.[11] Plasma catecholamines were measured at −15, 0, 5, 10, and 15 min post-GRF by a modified Peuler–Johnson radioenzymatic assay[12,13] by Dr. Michael Roizen (University of California, San Francisco).

1.4. Statistics

Statistical analyses were done using unpaired (p) or paired (p^*) t tests unless otherwise indicated. Data are expressed as mean ± SEM.

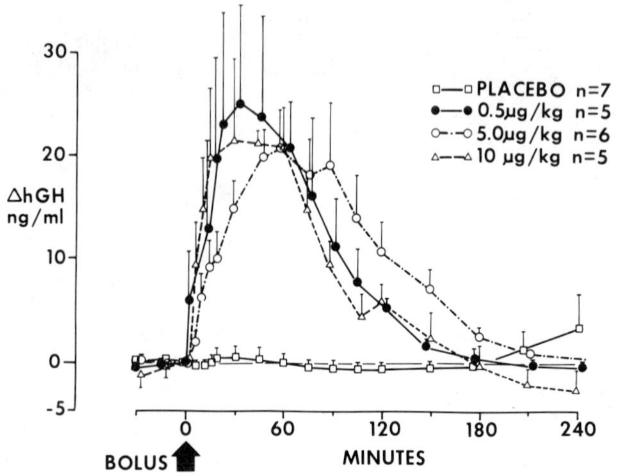

Figure 1 Increment above baseline (Δ) of plasma growth hormone (GH) (mean ± SEM) in normal men after a single i.v. pulse of GRF_{44}. From Rosenthal *et al.*[7]

2. Results

2.1. Single-Pulse Studies

Plasma GH concentration increased in seven normal young men after a single i.v. pulse of GRF_{44} (Fig. 1). Peak GH concentrations after 0.5, 5, or 10 μg/kg i.v. of GRF_{44} (27.5 ± 11.3, 24.4 ± 4.5, 30.4 ± 8.0 ng/ml) were comparable. The responses to the 5- and 10-μg/kg doses were greater than after placebo (6.3 ± 2.8 ng/ml; $p < 0.005$ and $p < 0.01$, respectively). The integrated GH response areas after 0.5, 5, and 10 μg/kg GRF were also similar to one another (2283 ± 1055, 2364 ± 452, 2756 ± 656 ng/min per ml^{-1}) and greater than after placebo (284 ± 103 ng/min per ml^{-1}; $p < 0.05$, $p = 0.0005$, $p < 0.002$). An increase in plasma GH concentration was detected 5–10 min after GRF_{44} administration, and peak levels were observed by 30–120 min (mean, 53.4 ± 6.7 min).

The peak GH responses of any single individual to the three doses were variable with coefficients of variation ranging from 14.9 to 42.3%. In the normal young men aged 19.5–33 years there was a negative correlation between chronological age and peak GH concentration after 0.5, 5, and 10 μg/kg of GRF_{44} (Fig. 2), $r = -0.558$, $p < 0.05$).

In the five men given 0.5 μg/kg i.v. of GRF_{40}, peak GH concentrations (23.6 ± 6.9 ng/ml) were comparable to those after the equivalent dose of GRF_{44}. Figure 3 illustrates the similarity of the integrated GH response areas evoked by GRF_{40} (1665.8 ± 665.0 ng/min per ml^{-1}) and GRF_{44} (2283 ± 1055 ng/min per ml^{-1}).

The GH responses to GRF_{44} of young women studied at three different phases of the menstrual cycle were lower during the luteal phase (10.4 ± 1.2

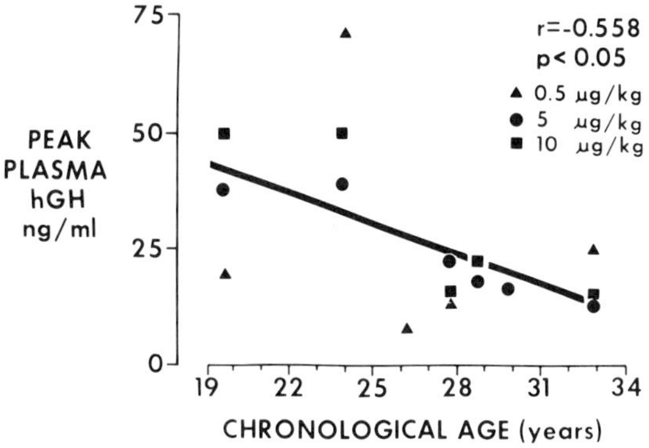

Figure 2 Correlation between chronological age and peak plasma growth hormone (GH) response to GRF_{44} in seven normal men 19.5–33 years of age. From Schriock *et al.*[8]

ng/ml) than at midcycle (22.3 ± 6.2 ng/ml; $p < 0.05$, Wilcoxon) or during the follicular phase (20.9 ± 5.6 ng/ml; $p = 0.05$, Wilcoxon). The peak GH levels after GRF_{44} in the midcycle and follicular phases were comparable to those of the men.

The seven non-GH-deficient children had an immediate increase in plasma GH after 5 μg/kg of GRF_{44} with most subjects attaining peak values between 10 and 45 min. The magnitude of the peak GH response in the normal children (24.4 ± 5.6 ng/ml) was similar to that of the young adult men and follicular-phase women. GH responses of the three pubertal adolescents tended to be lower (16.1 ± 4.1, range 8.0–20.8 ng/ml) than those of the prepubertal children (30.5 ± 8.5, range 10.0–51.1 ng/ml).

Twenty-three of the twenty-eight patients with severe GH deficiency had an increase in plasma GH concentration after GRF_{44}. The mean peak concentration of GH attained by these 23 patients was 5.1 ± 0.9 ng/ml, which was significantly less than the 25.1 ± 2.9 ng/ml attained by the non-GH-deficient men, women, and children (Fig. 4, $p < 0.0001$).

As shown in Fig. 5, patients with idiopathic, postirradiation, or other organic causes of GH deficiency had similar peak GH responses to GRF_{44} (4.5 ± 1.3, 6.4 ± 1.3, and 3.2 ± 1.4 ng/ml, respectively). Comparison of the plasma GH concentrations after GRF_{44} of patients with multiple hypothalamic–pituitary

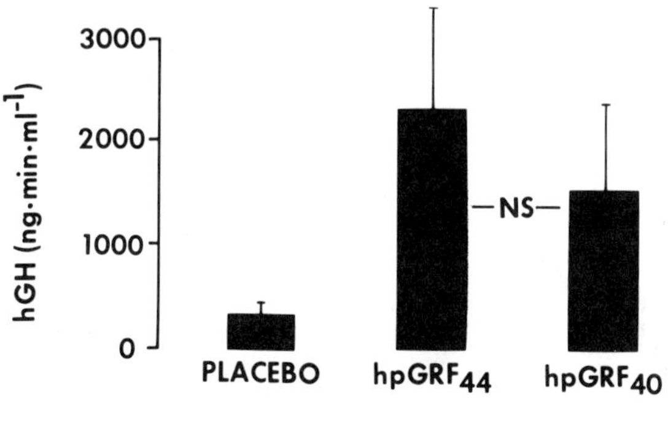

Figure 3 Integrated area of growth hormone (GH) response after GRF_{44} and GRF_{40} (0.5 μg/kg i.v.) in five normal men.

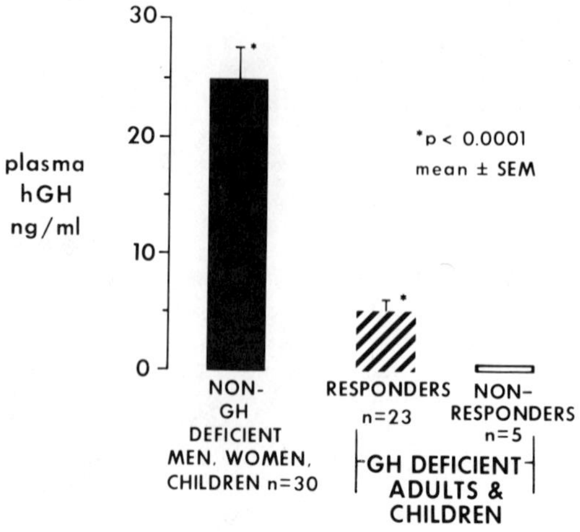

Figure 4 Peak concentration of plasma growth hormone (GH) after GRF_{44} in 28 patients with severe GH deficiency compared to non-GH-deficient individuals.

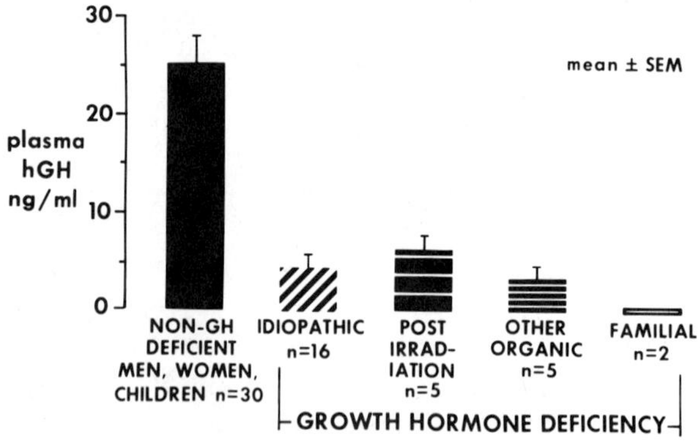

Figure 5 Peak growth hormone (GH) responses to GRF_{44} in patients grouped according to etiology of severe GH deficiency.

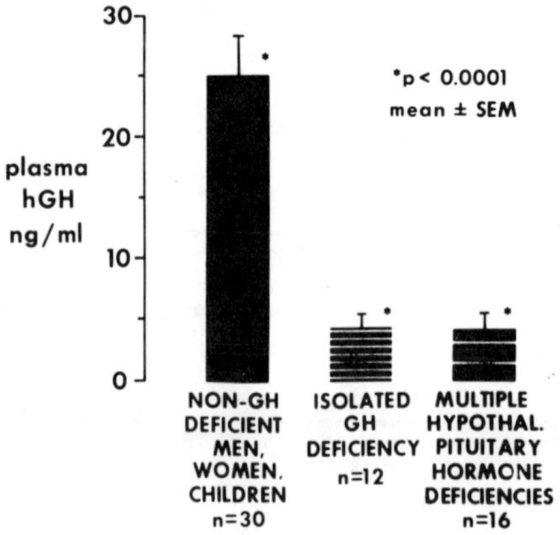

Figure 6 Maximum concentrations of growth hormone (GH) stimulated by GRF_{44} in patients with severe, isolated GH deficiency or multiple hypothalamic–pituitary hormone deficiencies. The significant difference between the GH-deficient patients and the non-GH-deficient individuals is indicated.

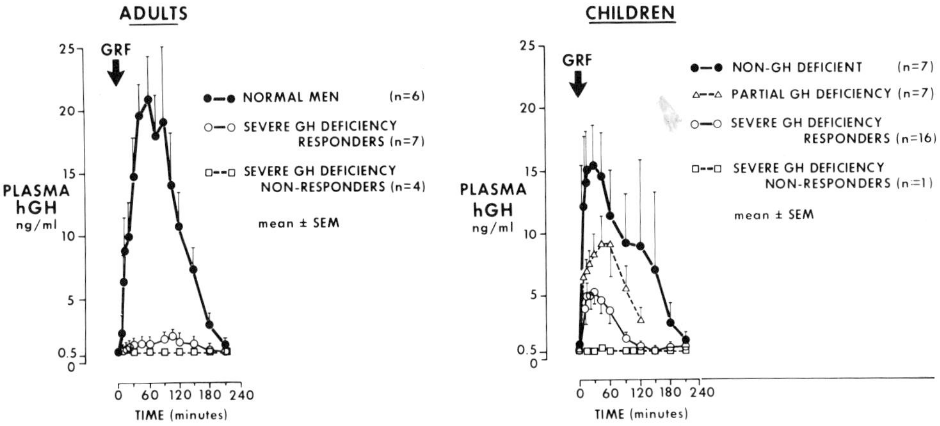

Figure 7 Mean plasma growth hormone (GH) response (± SEM) to GRF_{44} (5 μg/kg) in normal young adults and GH-deficient adults (left panel) and non-GH-deficient and GH-deficient children (right panel). Adapted with additions from Schriock *et al.*[7]

hormone deficiencies and those with isolated GH deficiency (Fig. 6) revealed no difference (4.3 ± 1.2 and 4.3 ± 1.1 ng/ml, respectively), and both groups responded less than the non-GH-deficient patients ($p < 0.0001$).

Patients younger than 18 years of age with severe GH deficiency tended to have greater GH responses to GRF_{44} than patients older than 18 years of age (Fig. 7). Sixteen of the seventeen children responded to GRF_{44} with a mean peak plasma GH concentration of 6.4 ± 1.2 ng/ml, which was greater than the 2.4 ± 0.3 ng/ml achieved by the seven adults who responded (Fig. 8; $p < 0.05$). Though the mean peak GH response of the GH-deficient children was less than that of the non-GH-deficient children (24.4 ± 5.6 ng/ml; $p = 0.0002$), there was overlap in the range of responses of the two groups (1.1–16.5 and 8.0–51.1 ng/ml, respectively).

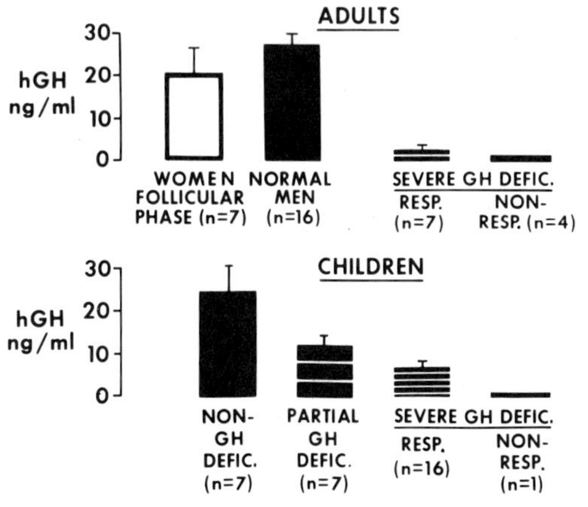

Figure 8 Growth hormone (GH) response to GRF_{44} by age of GH-deficient patients (adults 18–40 years; children and adolescents 4.75–16.3 years) and severity of GH deficiency (non-GH-deficient adults versus GH-deficient adult responders, $p < 0.002$; non-GH-deficient children versus children with severe GH deficiency who respond to GRF_{44}, $p < 0.0002$; GH-deficient adult responders versus children with severe GH deficiency who respond, $p < 0.05$; non-GH-deficient children versus children with partial GH deficiency, not significant; children with partial GH deficiency versus children with severe GH deficiency who respond to GRF, $p < 0.02$).

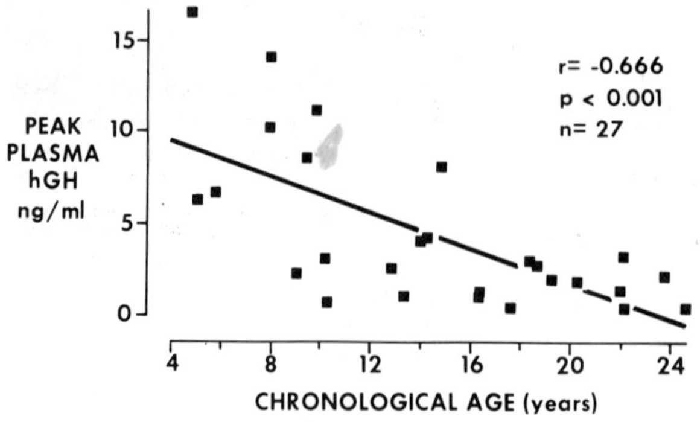

Figure 9 Correlation of chronological age with peak plasma growth hormone (GH) response to GRF_{44} (5 μg/kg) in patients 4.75–24.5 years of age with severe GH deficiency. Modified with additions from Schriock *et al.*[8]

Two of the four adults who did not respond to GRF_{44} were a mother and daughter with familial GH deficiency transmitted as an autosomal dominant trait; neither had a deletion of the GH gene. The other two adults and the one child who failed to respond had other evidence of hypothalamic dysfunction, i.e., a deficiency of thyrotropin-releasing factor or of prolactin release-inhibiting factor.

In the 27 severely GH-deficient children, adolescents, and young adults, peak GH response to GRF_{44} correlated negatively with chronological age by linear regression analysis (Fig. 9, $r = -0.666$, $p < 0.001$). Divided by age groups, the responses of the 17 GH-deficient children and adolescents correlated significantly with chronological age ($r = -0.605$, $p < 0.02$), but those of the GH-deficient adults did not ($r = -0.219$, $p > 0.50$).

In comparison to the children with severe GH deficiency (Fig. 8), the seven children with partial GH deficiency achieved a greater mean peak concentration of plasma GH after GRF_{44} administration (6.4 ± 1.2 versus 12.0 ± 1.9 ng/ml, respectively; $p < 0.02$). This peak value for the children with partial GH deficiency was not statistically different from that of the non-GH-deficient children (24.5 ± 5.6 ng/ml), though the highest response of the children with partial GH deficiency (17.3 ng/ml) was considerably lower than that of the non-GH-deficient children (51.1 ng/ml). The peak GH response to GRF_{44} in these GH-deficient children did not correlate with chronological age.

All patients with severe or partial GH deficiency who responded to GRF_{44} had greater responses to GRF_{44} than to standard pharmacological stimuli (Fig. 10) (6.7 ± 1.0 versus 1.8 ± 0.3 ng/ml, $p^* < 0.0001$); 50% of patients less than 15 years of age achieved plasma GH concentrations greater than 7 ng/ml after GRF_{44}. In the GH-deficient children and adolescents, the maximum GH response to any one of the pharmacological tests correlated positively with the peak response to GRF ($r = 0.537$, $p < 0.01$).

Plasma PRL, LH, FSH, TSH, cortisol, insulin, catecholamine, and somatostatin concentrations were unaffected by administration of a single pulse of GRF_{44} to the normal men. GH-deficient patients with multiple hypothalamic–pituitary hormone deficiencies had a small but significant increase in plasma

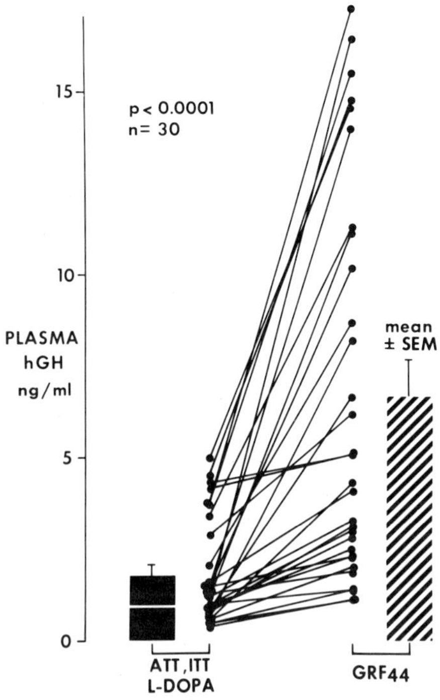

Figure 10 Peak concentration of plasma growth hormone (GH) after pharmacological tests of GH release and after GRF_{44} (5 μg/kg i.v.) in 30 GH-deficient patients responsive to GRF_{44}. Indicated is the highest GH response to arginine (ATT), insulin (ITT), or L-dopa of each individual connected by a solid line to his peak GH response to GRF_{44}. Bars represent mean peak GH responses for the group to pharmacological tests or GRF_{44}.

PRL concentration (1.6 ± 0.5 ng/ml, $p < 0.02$) 30 min after GRF_{44}. The PRL concentration in patients with isolated GH deficiency was unaffected.

In the normal men the mean concentration of serum SMC increased from 0.83 ± 0.06 to 1.13 ± 0.09 IU/ml ($p^* < 0.002$) 24 hr after 5 μg/kg IV GRF_{44} and from 0.96 ± 0.06 to 1.18 ± 0.06 IU/ml ($p^* < 0.03$) after 10 μg/kg of GRF_{44}. Although not statistically significant, five of six men also had an increase in SMC concentration after placebo (0.96 ± 0.03 to 1.07 ± 0.09 IU/ml, n = 6). SMC only increased a mean of 0.03 ± 0.01 IU/ml 24 hr after a single pulse of GRF_{44} in 15 patients with severe GH deficiency.

2.2. *Continuous Infusion Study*

Six of the eight young men who received a 3-μg/kg pulse of GRF_{44} followed 30 min later by a 2-hr GRF_{44} infusion (3 μg/kg per hr) had striking increases in plasma GH concentrations (Fig. 11). In these six men the pulse/infusion regimen evoked peak plasma GH concentrations similar to those elicited by a single 5-μg/kg i.v. pulse of GRF_{44} (31.2 ± 6.0 versus 24.4 ± 4.5 ng/ml), and these peaks occurred at similar times (52.5 ± 10.8 versus 65 ± 8.4 min). The mean integrated GH response area of the pulse/infusion study (3542 ± 794.9 ng/min per ml^{-1}) was similar to that of the single-pulse study (2364.2 ± 452 ng/min per ml^{-1}). Plasma GH concentrations declined in all six men prior to

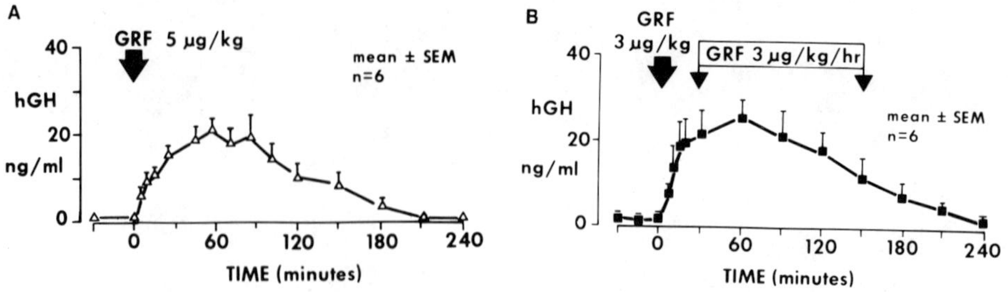

Figure 11 Effect of a single pulse of GRF_{44} (5 μg/kg) on plasma GH (A) compared to a single pulse of GRF_{44} (3 μg/kg i.v.) followed 30 min later by a 2-hr continuous infusion (3 μg/kg per hr) of GRF_{44} in six normal young men (B).

termination of the infusion. The two men with blunted responses had spontaneous GH pulses just prior to receiving GRF_{44}.

2.3. *Multiple-Pulse Study*

Higher GH responses were achieved in all six patients after multiple pulses of GRF_{44} than after the initial, single GRF_{44} pulse. The pulse number that elicited the highest plasma concentration of GH varied from patient to patient, and the magnitude of this response was greater than 7 ng/ml in all patients. Not all GRF_{44} pulses were followed by a GH response, though a majority were. The GH response of one GH-deficient child to 12 hr of pulsing with GRF_{44} is illustrated in Fig. 12.

None of the adults or children experienced any untoward side effects from the GRF pulse or infusion. Transient facial flushing beginning 30–90 sec after GRF injection was observed in a majority of the adult subjects but in only a

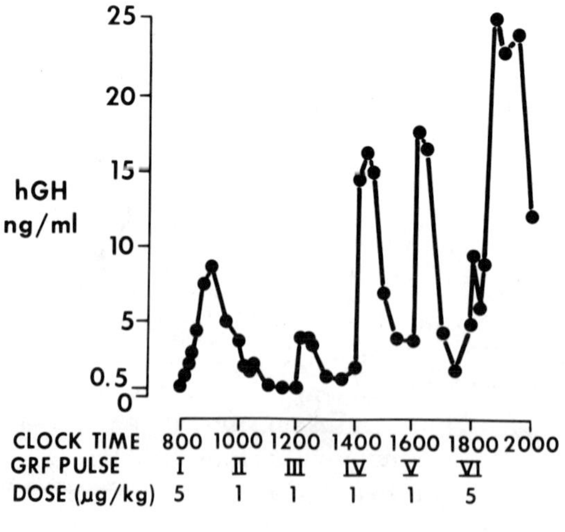

Figure 12 Plasma growth hormone (GH) response to six pulses of GRF_{44} in a child with severe GH deficiency.

few of the children. Small, transient decreases in blood pressure and increases in pulse rate were observed in the young adult men[7] but not in the children. No significant changes in blood chemistry profiles including liver enzymes and renal function tests occurred after GRF.

3. Discussion

3.1. Comparison of GH Response to GRF_{44} and GRF_{40}

Synthetic GRF_{44} and GRF_{40} have been previously reported by our group and others to be potent and specific stimulants of GH secretion in normal young men.[7,9,14–17] *In vitro* studies[1] indicated GRF_{44} to be a more potent stimulant of GH release than GRF_{40}, but subsequent *in vivo* studies in the rat[18] failed to demonstrate a difference in potency between the two. Similarly, our data indicate that in humans comparable GH responses are stimulated by the two GRF peptides. The 0.5-, 5-, and 10-μg/kg i.v. doses of GRF_{44} evoked similar GH responses suggesting that 0.5 μg/kg of GRF_{44} is a maximally stimulating dose in normal young men. Detection of a difference in the specific activities of GRF_{40} and GRF_{44} *in vivo* may require the use of a smaller dose.

Considerable variation, unexplained by differences in dose, occurred in the individual GH responses to GRF_{44}. This has been observed in rhesus monkeys and unanesthesized rats as well as man.[14,19,20] Fluctuations in hypothalamic somatostatin or induction of somatotrope refractoriness by a recent pulse of endogenous GRF, though the two are not mutually exclusive, may explain the variability of responses. Consistency of GH responses to synthetic GRF greatly improved in rats after the administration of antisera to somatostatin[20]; this finding favors the former possibility.

3.2. Variations in Responsiveness during the Menstrual Cycle

Fluctuations in sex steroid levels may also modulate somatotrope responsiveness to GRF as suggested by the lower GH responses to GRF of women in the luteal phase as compared to other phases of the menstrual cycle. This is consistent with earlier observations of changes in GH concentration in the ambulatory state and in response to pharmacological stimuli at different phases of the menstrual cycle.[21–23]

3.3. Change in GH Response to GRF with Age

Pubertal adolescents and adults have greater GH responses to pharmacological stimuli than prepubertal children[24–28] just as estrogen- or testosterone-treated subjects have augmented responses to these tests.[23,29,30] Though our study group was small, the prepubertal, non-GH-deficient children did not have lower GH responses to GRF testing than the adolescent or young adult subjects. However, GH responses to GRF_{44} decreased with age in the normal young men suggesting that some factor other than sexual maturation was influencing GH response to GRF. Shizume *et al.* (personal communication) have also observed a decrease in responsivity to GRF with age in adults, and this may be related at least in part to changes in the release of hypothalamic somatostatin as suggested by *in vivo* and *in vitro* rat studies.[31,32]

3.4. *Systemic Effects*

Transient vasodilation as evidenced by facial flushing, minimal blood pressure decreases, and pulse rate increases in the men[7] was not associated with changes in peripheral plasma epinephrine, norepinephrine, or dopamine concentration. Only a few of the children had facial flushing, and this was not accompanied by significant changes in vital signs. The flushing that occurred by 30 sec post-GRF injection in some patients could be due to a direct effect of GRF on the vascular system. Other members of the glucagon–secretin family to which GRF is related have vasodilatory activity.[33]

3.5. *Continuous versus Pulsatile Modes of GRF Administration*

An alternate pattern of administration of GRF_{44}, a single pulse followed 30 min later by a 2-hr infusion, resulted in peak GH concentrations and integrated GH responses similar to those elicited by a single 5 μg/kg i.v. pulse of GRF_{44} in normal young men. Failure of the GRF_{44} infusion to enhance or sustain GH release suggests that the prior GRF_{44} pulse may have led to refractoriness to GRF or induced somatostatin release, thereby blocking any further GRF effect. Anesthetized rats given a 500 ng/min GRF_{40} infusion for 30 min also did not sustain elevated GH levels for the duration of the infusion.[18] Rats infused with 20 ng/min of GRF_{40} for 30 min maintained increased GH concentrations until the infusion was terminated, but the GH levels achieved were much lower than with the larger dose of GRF. Pituitary depletion of GH may have occurred in the rats receiving the larger dose as the amount of GH secreted was similar to the total GH content of the rat pituitary.[18] On the contrary, pituitary GH depletion is unlikely in the men since the human pituitary contains 4–8 mg of GH,[34] and the men secreted less than this by the end of the infusion. Furthermore, patients with functional GH deficiency were able to respond to multiple, consecutive pulses of GRF.

3.6. *GRF in GH Deficiency*

The majority of GH-deficient patients, including those with postirradiation GH deficiency,[35] responded to a single pulse of GRF_{44} with increases in plasma GH concentration. This indicates the presence of functional somatotropes in these patients and implicates an abnormality in the synthesis, release, or transport of endogenous GRF as the cause of their GH deficiency. Many of the patients who responded had other evidence of hypothalamic dysfunction.

The magnitude of the peak plasma GH concentrations after GRF_{44} in the GH-deficient patients varied; some responses were quite blunted, and others were within the range of the non-GH-deficient individuals. All responses to GRF_{44} were greater than those elicited by the pharmacological tests; this reflects the direct stimulating effect of GRF_{44} on the somatotrope and indicates the potential usefulness of GRF in the assessment of readily releasable GH reserve.

Some of the heterogeneity of the GH responses to GRF_{44} in the GH-deficient patients is related to the negative correlation between chronological age and GH response to GRF_{44}. This inverse relationship implies that chronic deficiency of endogenous GRF results in decreased responsivity to a single pulse of GRF_{44}. Consequently, a negative GH response to a single pulse of GRF_{44} does not exclude a hypothalamic abnormality. Patients with partial GRF deficiency had higher GH responses to GRF_{44} than those with severe deficiency. These findings suggest that priming of the somatotrope with GRF could enhance GH release to GRF_{44}. Indeed, the higher GH peaks achieved in GH-deficient patients with pulsatile administration of GRF_{44} support this hypothesis. Thus, the identification of hypothalamic GRF deficiency in some patients may require more than one GRF_{44} pulse. These preliminary results also suggest that GRF-deficient patients should benefit from treatment with GRF or with an analogue.

A desirable effect of treatment with synthetic GRF would be an increase in the serum concentration of the GH-dependent factor SMC. An appreciable rise in SMC concentration in GH-deficient patients would be expected after multiple pulses of GRF_{44} but not necessarily after a single pulse. Indeed the minimal increase in SMC concentration after a single pulse of GRF_{44} is of questionable significance; we have observed a larger increment in SMC after multiple pulses of GRF_{44}. Similarly, Borges *et al.*[36] have reported a comparable, small rise in SMC (from 0.19 ± 0.08 to 0.23 ± 0.08 IU/ml) for a group of hypopituitary adults after a single 10 μg/kg i.v. injection of GRF_{40} and a marked increase in SMC from 0.06 IU/ml to 0.36 IU/ml in a GH-deficient patient after 72 hr of receiving 0.33 μg/kg of GRF_{40} every 3 hr.

3.7. Limitations of the GRF Test

Synthetic GRF_{44} is useful for assessing readily releasable GH reserve and diagnosing hypothalamic GRF deficiency. However, the failure to elicit a rise in the concentration of GH after a single-pulse GRF test does not exclude a deficiency of hypothalamic GRF as the defect in patients with GH deficiency. Although the GRF test is safe and simple to administer, it does have important limitations.

The decline in responsiveness to GRF with increasing age in normal adults (ref. 8 and Shizume *et al.*, personal communication) must be considered in interpreting GRF tests in older patients. The adolescents and young adults with severe, chronic GH deficiency had uniformly lower responses than the normal subjects. However, the GH responses of patients with the recent onset of GH deficiency or partial GH deficiency due to GRF deficiency may be in the range of normal.[8,16] In children the single-pulse GRF test does not discriminate between normal children and those with hypothalamic GRF deficiency as the cause of the GH deficiency. Accumulation of additional age-related data in normal children and the investigation of GH responses to submaximal doses of GRF may refine the test. At present the GRF test is a useful adjunct to the traditional pharmacological and physiological tests of GH release and, with limitations, for the diagnosis of GRF deficiency.

4. Summary

Synthetic GRF_{44} is a safe and potent secretagogue for GH in men, women, and children. GRF_{40} and GRF_{44} appear to be equipotent in stimulating GH release in normal young men.

Considerable variation in the individual GH responses to GRF_{44} of normal young men probably reflects fluctuations in hypothalamic somatostatin release. Responsivity to GRF_{44} decreases with chronological age in normal men.

Somatotrope responsiveness to GRF_{44} in normal, young women is diminished during the luteal phase of the menstrual cycle.

GH secretion is not sustained during a 2-hr GRF_{44} infusion preceded by a pulse of GRF_{44} in normal young men. The GRF test is useful in distinguishing hypothalamic GRF deficiency from primary pituitary GH deficiency. The GRF-induced rise in plasma GH in a majority of GH-deficient patients suggests that a hypothalamic abnormality is the most common cause of GH deficiency.

Lack of response in GH-deficient patients to a single pulse of GRF_{44} does not exclude GRF deficiency since unresponsiveness may reflect chronic, severe lack of GRF stimulation to the somatotropes. Readily releasable GH reserve may be assessed by the GRF test.

The higher concentrations of GH obtained after multiple pulses of GRF_{44} than after a single pulse of GRF_{44} in GH-deficient patients are consistent with a priming effect of GRF on the somatotrope. These studies support the usefulness of GRF_{44} or an analogue as a potential therapeutic agent for the treatment of GRF deficiency.

ACKNOWLEDGMENTS. The authors thank Dr. Nicholas Ling and Dr. Roger Guillemin for providing the synthetic GRF_{44}; Dr. Arthur I. Vinik for the assay of somatostatin levels; Bryan Frazer and Dr. Michael Roizen for performing catecholamine assays; the nurses of the Pediatric and General Clinical Research Centers for their assistance; and Christine Stav for secretarial assistance.

E. A. S., C. A. E., D. A. H., and R. H. L. are Trainees in Pediatric Endocrinology under a program sponsored by The National Institute of Arthritis, Diabetes, Digestive and Kidney Diseases, NIH, USPHS. B. P. H. is a Trainee in Pediatric Endocrinology, sponsored by Deutsche Forschungsgemeinschaft (DGF - grant Ha 1268/1-1). J. A. H. is a Trainee in Pediatric

Endocrinology, sponsored by the King's Fund and St. Thomas Hospital, England.

This work was supported in part by a grant from the National Institute of Child Health and Human Development, NIH, USPHS, and was carried out in the Pediatric Clinical Research Center (Grant RR 01271-01) of the University of California, San Francisco.

References

1. R. Guillemin, P. Brazeau, P. Bohlen, F. Esch, N. Ling, and W. B. Wehrenberg, Growth hormone-releasing factor from a human pancreatic tumor that caused acromegaly, *Science* **218:**585, 1982.
2. J. Rivier, J. Spiess, M. Thorner, and W. Vale, Characterization of a growth hormone-releasing factor from a human pancreatic islet tumour, *Nature* **300:**276, 1982.
3. B. Bloch, P. Brazeau, N. Ling, P. Bohlen, F. Esch, W. B. Wehrenberg, R. Benoit, F. Bloom, and R. Guillemin, Immunohistochemical detection of growth hormone-releasing factor in brain, *Nature* **301:**607, 1983.
4. J. Spiess, J. Rivier, and W. Vale, Characterization of rat hypothalamic growth hormone-releasing factor, *Nature* **303:**532, 1983.
5. P. Bohlen, P. Brazeau, B. Bloch, N. Ling, R. Gaillard, and R. Guillemin, Human hypothalamic growth hormone releasing factor (GRF): Evidence for two forms identical to tumor derived GRF-44-NH_2 and GRF-40, *Biochem. Biophys. Res. Commun.* **114:**930, 1983.
6. K. E. Mayo, W. Vale, J. Rivier, M. G. Rosenfeld, and R. M. Evans, Expression-cloning and sequence of a cDNA encoding human growth hormone-releasing factor, *Nature* **306:**86 1983.
7. S. M. Rosenthal, E. A. Schriock, S. L. Kaplan, R. Guillemin, and M. M. Grumbach, Synthetic human pancreas growth hormone-releasing factor ($hpGRF_{1\text{-}44}$-NH_2) stimulates growth hormone secretion in normal men, *J. Clin. Endocrinol. Metab.* **57:**677, 1983.
8. E. A. Schriock, R. H. Lustig, S. M. Rosenthal, S. L. Kaplan, and M. M. Grumbach, Effect of growth hormone (GH)-releasing hormone (GRH) on plasma GH in relation to magnitude and duration of GH deficiency in 26 children and adults with isolated GH deficiency or multiple pituitary hormone deficiencies: evidence for hypothalamic GRH deficiency, *J. Clin. Endocrinol. Metab.* **58:**1043, 1984.
9. S. M. Rosenthal, E. A. Schriock, S. L. Kaplan, R. Guillemin, and M. M. Grumbach, Synthetic human pancreatic growth hormone releasing factor (hpGRF(1-44)-NH_2) stimulates growth hormone (GH) secretion in normal men, *Endocrinology* **112:**83A, 1983.
10. G. Van Vliet, D. M. Styne, S. L. Kaplan, and M. M. Grumbach, Growth hormone treatment for short stature, *N. Engl. J. Med.* **309:**1016, 1983.
11. A. I. Vinik, N. S. Levitt, B. L. Pimstone, and L. Wagner, Peripheral plasma somatostatin-like immunoreactive responses to insulin hypoglycemia and a mixed meal in healthy subjects and in noninsulin-dependent maturity-onset diabetes, *J. Clin. Endocrinol. Metab.* **52:**330, 1981.
12. M. F. Roizen, J. Moss, D. P. Henry, *et al.*, Effects of halothane on plasma catecholamines, *Anesthesiology* **41:**432, 1974.
13. J. D. Peuler and G. A. Johnson, Simultaneous single isotope radioenzymatic assay of plasma norepinephrine, epinephrine and dopamine, *Life Sci.* **21:**625, 1977.
14. M. O. Thorner, J. Rivier, J. Spiess, J. L. Borger, M. L. Vance, S. R. Bloom, A. D. Rogol, M. J. Cronin, D. L. Kaiser, W. S. Evans, J. D. Webster, R. M. MacLeod, and W. Vale, Human pancreatic growth-hormone-releasing factor selectively stimulates growth-hormone secretion in man, *Lancet* **1:**24, 1983.
15. S. M. Wood, J. L. C. Ch'ng, E. F. Adams, J. D. Webster, G. F. Joplin, K. Mashiter, and S. R.

Bloom, Abnormalities of growth hormone release in response to human pancreatic growth hormone releasing factor (GRF (1-44)) in acromegaly and hypopituitarism, *Br. Med. J.* **286:**1687, 1983.

16. A. Grossman, M. O. Savage, J. A. H. Wass, N. Lytras, J. Sueiras-Diaz, D. H. Coy, and G. M. Besser, Growth-hormone-releasing factor in growth hormone deficiency: Demonstration of a hypothalamic defect in growth hormone release, *Lancet* **2:**137, 1983.
17. M. C. Gelato, O. Pescovitz, F. Cassorla, D. L. Loriaux, and G. R. Merriam, Effects of a growth hormone releasing factor in man, *J. Clin. Endocrinol. Metab.* **57:**674, 1983.
18. W. B. Wehrenberg, N. Ling, P. Brazeau, F. Esch, P. Bohlen, A. Baird, S. Ying, and R. Guillemin, Somatocrinin, growth hormone releasing factor, stimulates secretion of growth hormone in anesthetized rats, *Biochem. Biophys. Res. Commun.* **109:**382, 1982.
19. O. F. X. Almeida, R. S. Rittmaster, and G. R. Merriam, Potency and specificity of a growth hormone releasing factor in a primate and *in vitro*, Program of the 65th Annual Meeting of The Endocrine Society, San Antonio, TX, Williams and Wilkins, Baltimore, 1983, p. 153 (abstr.).
20. W. B. Wehrenberg, N. Ling, P. Bohlen, F. Esch, P. Brazeau, and R. Guillemin, Physiological roles of somatocrinin and somatostatin in the regulation of growth hormone secretion, *Biochem. Biophys. Res. Commun.* **109:**562, 1982.
21. A. G. Frantz and M. T. Rabkin, Effects of estrogen and sex difference on secretion of human growth hormone, *J. Clin. Endocrinol.* **25:**1470, 1965.
22. T. J. Merimee, S. E. Fineberg and J. E. Tyson, Fluctuations of human growth hormone secretion during menstrual cycle: Response to arginine, *Metabolism* **18:**606, 1969.
23. T. J. Merimee and S. E. Fineberg, Studies of the sex based variation of human growth hormone secretion, *J. Clin. Endocrinol.* **33:**896, 1971.
24. K. Shizume, M. Irie, F. Matsuzaki, Y. Shishiba, and T. Saito, Growth hormone response to hypoglycemia in puberty and adolescence, *Endocrinol. Jpn.* **16:**459, 1969.
25. M. A. Sperling, F. M. Kenny, and A. L. Drash, Arginine-induced growth hormone responses in children: Effect of age and puberty, *J. Pediatr.* **77:**462, 1970.
26. S. D. Frasier, J. M. Hilburn, and F. G. Smith, Effect of adolescence on the serum growth hormone response to hypoglycemia, *J. Pediatr.* **77:**465, 1970.
27. S. L. Kaplan, and M. M. Grumbach, The ontogenesis of hypothalamic–hypophysiotropic releasing factor regulation of hGH secretion, Excerpta Medica International Congress Series No. 244 "Growth and Growth Hormone," Proc. 2nd Intl. Symp, Milan, Excerpta Medica, Amsterdam-Oxford, 1971, pp. 382.
28. F. Minuto, A. Barreca, S. Ferrini, G. Mazzocchi, P. Del Monte and G. Giordano, Growth hormone secretion in pubertal and adult subjects, *Acta Endocrinol.* **99:**161, 1982.
29. R. Youlton, S. L. Kaplan and M. M. Grumbach, Growth and growth hormone. IV. Limitations of the growth hormone response to insulin and arginine and of the immunoreactive insulin response to arginine in the assessment of growth hormone deficiency in children, *Pediatrics* **43:**989, 1969.
30. L. G. Martin, J. W. Clark and T. B. Connor, Growth hormone secretion enhanced by androgens, *J. Clin. Endocrinol.* **28:**425, 1968.
31. W. E. Sonntag, V. W. Hylka and J. Meites, Impaired ability of old male rats to secrete growth hormone *in vivo* but not *in vitro* in response to hpGRF (1-44), *Endocrinology* **113:**2305, 1983.
32. W. E. Sonntag, L. J. Forman, N. Miki, R. W. Steger, T. Ramos, A. Arimura, and J. Meites, Effects of CNS active drugs and somatostatin antiserum on growth hormone release in young and old male rats, *Neuroendocrinology* **33:**73, 1981.
33. B. M. Jaffe, Hormones of the gastrointestinal tract, in: *Endocrinology* (L. J. DeGroot, G. F. Cahill, W. D. Odell, L. Martini, J. T. Potts, D. H. Nelson, E. Steinberger, and A. I. Winegrad, eds.), Grune & Stratton, New York, 1979, p. 1669.
34. W. H. Daughaday, The adenohypophysis, in: *Textbook of Endocrinology* (R. H. Williams, ed.), W. B. Saunders Company, Philadelphia, 1981, p. 90.
35. R. H. Lustig, E. A. Schriock, S. L. Kaplan, and M. M. Grumbach, Effect of growth hormone-releasing factor on growth hormone release in children with radiation-induced growth hormone deficiency, *Pediatrics* **76:**274, 1985.

36. J. L. Borges, R. M. Blizzard, M. C. Gelato, R. Furlanetto, A. D. Rogol, W. S. Evans, M. L. Vance, D. L. Kaiser, R. M. MacLeod, G. R. Merriam, D. L. Loriaux, J. Spiess, J. Rivier, W. Vale, and M. O. Thorner, Effects of human pancreatic tumour growth hormone releasing factor on growth hormone and somatomedin C levels in patients with idiopathic growth hormone deficiency, *Lancet* **2:**119, 1983.

30

Somatostatin

Biosynthesis, Regulation of Secretion, and Role in Growth Hormone Regulation

Seymour Reichlin

1. Introduction

Growth hormone (GH) release inhibitory factor(s) were discovered by Krulich and collaborators during the course of efforts to assay GH-releasing factor from rat hypothalamus.[1] These authors demonstrated that hypothalamic extracts contained regionally localized GH-releasing and inhibitory factors and hypothesized that the secretion of GH was regulated by the interaction of the two kinds of chemical factors, each under the control of the nervous system. This hypothesis has since been well substantiated by more specific methods. For this reason, understanding the mechanism of biosynthesis and regulation of secretion of somatostatin is relevant to understanding GH regulation. The somatostatinergic system has also been important to study because it is one of the best models of neuropeptide biology.

GH release inhibitory activity was rediscovered and chemically characterized by Brazeau *et al.*[2] The term somatostatin (somatotropin release inhibitory factor, SRIF) was applied to the 14-amino-acid cyclic peptide they finally were able to isolate using as a bioassay dispersed pituitary cell cultures.

Subsequent work has considerably expanded the initially simply concept of somatostatin as a tetradecapeptide (14-amino-acid-containing peptide) primarily involved in the regulation of GH secretion. Somatostatinlike peptides constitute a family of related molecules including the originally identified somatostatin (designated somatostatin 14), amino-terminal extended somatostatin (somatostatin 28),[3] somatostatin 28 (1-12),[4] and still larger forms, which

Seymour Reichlin • Endocrine Division, Department of Medicine, Tufts University School of Medicine, New England Medical Center, Inc., Boston, Massachusetts 02111.

Table 1 Established Structure of Several Secreted Forms of Somatostatin and Somatostatin-Related Peptides in Mammals

Ser	Ala	Asn	Ser	Asn	Pro	Ala	Met	Ala	Pro	Arg	Glu	Arg	Lys	Ala	Gly	Cys	Lys	Asn	Phe	Phe	Trp	Lys	Thr	Phe	Thr	Ser	Cys
1	2	2	4	4	6	7	8	9	10	11	12	13	14	15	16	17	18	19	20	21	22	23	24	25	26	27	28
Somatostain 28(1–12)																Somatostatin 14											
Somatostatin 28																											

range in molecular weight according to location and species from 11.5 to 15.7K daltons (Table 1). When the larger forms of somatostatin were first recognized, it was reasonable to consider them to be prohormones, but more recent studies have shown that under some circumstances somatostatin 28 and even the larger forms can be secreted as such. All possess biological activity both in regulating GH secretion and in other functions.[(5)] These larger forms can therefore be considered to be both prohormones and hormones. Because somatostatin is distributed widely in cells throughout the nervous system and in many extraneural tissues where it exerts effects on a wide range of structures, including neurons, epithelia, endocrine, and exocrine glands, its first designation as "somatostatin" is seen to be too restrictive a term.

My purpose in this chapter is to summarize current knowledge of the anatomy of somatostatin containing hypothalamic neurons of the biosynthesis and secretion of this peptide and its physiological role in regulation of GH secretion. Several comprehensive reviews have been published.[(6–10)]

2. Anatomical Distribution

Somatostatin is not limited to the hypothalamus or to higher vertebrates. It is distributed widely but selectively in cells throughout the body of vertebrates. The peptide is found in most, but not all, organs and displays specific and selective functions depending on its location.

Within the vertebrate nervous system, somatostatinergic neurons form part of the tuberoinfundibular complex that terminates in the median eminence. In this region they function in anterior pituitary regulation. Somatostatinergic terminals also are found in the neurohypophysis, where they may modulate secretion of vasopressin and oxytocin. Other long tracts are present, particularly in the limbic system, and some project from the hypothalamus caudally to brain stem and spinal cord. Somatostatin fibers make up one or more classes of sensory neurons located in spinal, sympathetic, and vagus nerves, retina, and auditory nerves. In addition, local short-range control neurons occur especially in the cerebral cortex, hippocampus, brain stem, and spinal cord.

Outside the nervous system, somatostatin is distributed in specific secretory cells of the gut, in pancreatic islets, in the salivary glands, and in the urinary excretory system of some species.

Somatostatin has also been found in all phyla including the protozoa.[(11–13)] In all peptidergic neuronal systems that have been studied, somatostatin has been shown to be localized to nerve endings in secretory granules, to be released in response to depolarizing stimuli (chemical or electrical), to be degraded by locally present enzymes, and to bind to specific receptors, both pituitary and neural. In the case of pituitary regulation it thus qualifies as a neurohormone, i.e., a secretion of nerve cells that is released into the blood; in the nervous system it qualifies as a neurotransmitter or neuromodulator. Outside the brain, somatostatin can serve as a paracrine regulator (side-by-side secretion), as an autocrine regulator (self-regulator), in some instances as a true hormone, and

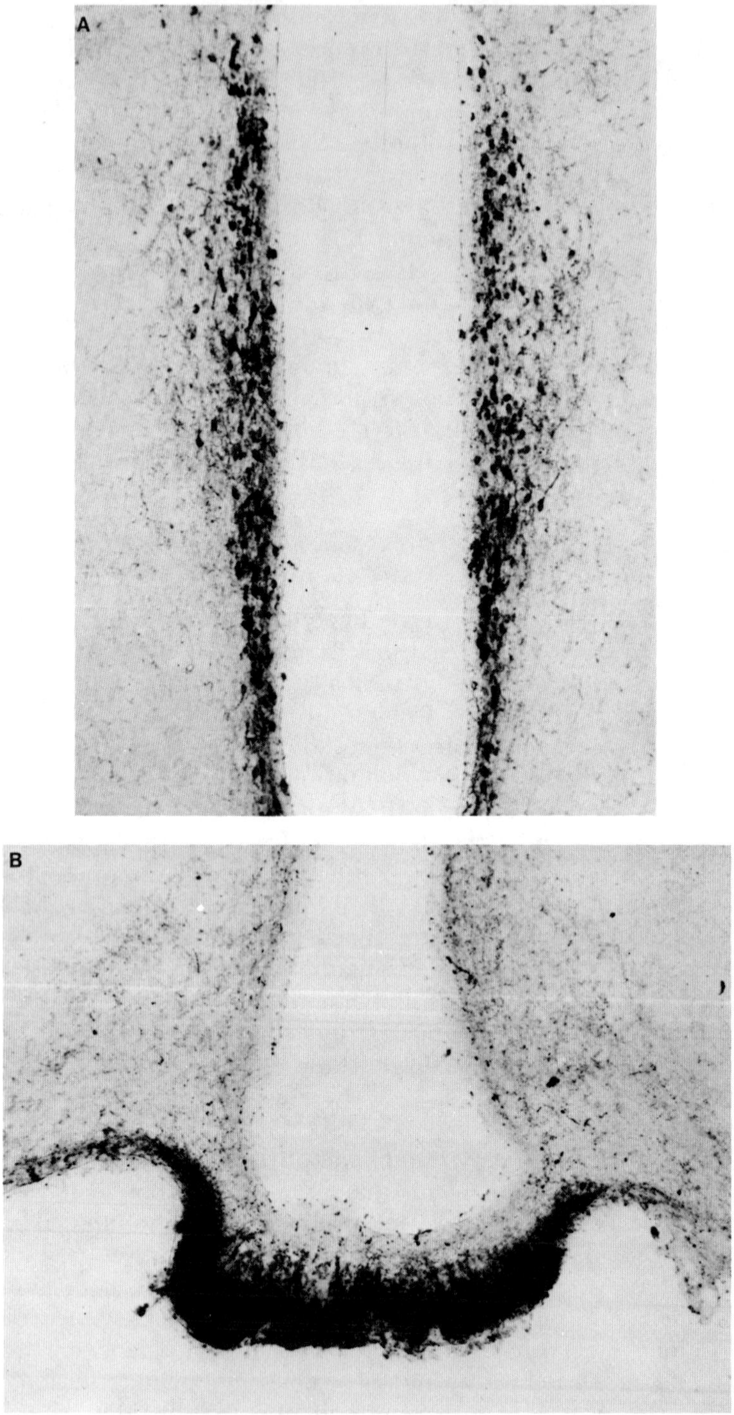

as a "lumone," which is a regulatory factor acting on the gut after secretion into the lumen of the gut. In relation to regulation of GH secretion, the distribution in the tuberoinfundibular system is most pertinent.

Somatostatin cells and fibers are found in a bewildering distribution in many parts of the hypothalamus and limbic system, with many terminating in the median eminence which is the site of origin of the hypophysial–portal vessels that are destined to perfuse the anterior pituitary. The principal source of fibers projecting to the median eminence is a layer several cells thick located in the anterior periventricular region of the hypothalamus in a position immediately under the ependymal lining (Fig. 1). Analogous periventricular cells also form the medial division (parvocellular division) of the paraventricular nucleus. Both regions are rich in somatostatin-containing nerve cells. From the anterior periventricular nucleus somatostatin-containing fibers sweep laterally through the hypothalamus at the level of the suprachiasmatic nucleus and are reassembled posteriorly into a condensed bundle that enters the median eminence through its anterior margin. Small contributions to the median eminence come from the dorsal arcuate nucleus (Fig. 1).

Immunohistochemical findings of peptide distribution in the brain are fully confirmed by immunoassays of hypothalamic nuclei punched out by the Palkovitz technique[14,15] and by study of changes in concentration of somatostatin in the median eminence after selective lesions of the anterior periventricular–median eminence projections.[16–19]

Somatostatinergic fibers originating in the periventricular region (in addition to their projections to the median eminence) also project in short-distance trajectories to several hypothalamic nuclei that are considered to be part of the limbic system (visceral brain).[20–22]

Periventricular hypothalamic somatostatinergic fibers also project caudally into the brain stem and spinal cord.[23]

The system of ependyma-derived neurohemal organs arranged around the 3rd ventricle (subfornical organ, pineal, subcommissural organ, median eminence, and organum vasculosum of the lamina terminals) and in the floor of the 4th ventricle (area postrema) sometimes referred to as the "windows to the brain" all contain nerve terminals of somatostatinergic neurons.[24] Their function is unknown, but most likely they contribute peptide to the intraventricular cerebrospinal fluid which conceivably could influence tuberoinfundibular function.

As a general rule, it can be stated that the brain somatostatin neural system has no direct effect on the somatostatin system outside of the central nervous system. Lesions of the hypothalamus have no effect on pancreatic somatostatin[25] or on somatostatin content of the cerebral cortex.[17]

Figure 1 Anatomy of somatostatinergic fibers in the rat related to anterior pituitary regulation. (A) The periventricular plexus located in the anterior hypothalamus. (B) The median eminence. These figures were prepared from thick frozen sections stained with the perioxidase-antiperoxidase (PAP) technique and were provided through the courtesy of Dr. Ronald Lechan.

3. Biosynthesis, Processing, and Transport in Neurons

The amino acid sequence of human[26] and rat[27] preprosomatostatin have been elucidated by recombinant techniques using mRNA extracted from somatostatin-secreting tumors (Fig. 2) The entire gene sequence of brain somatostatin has been determined by Montminy *et al.*[28] (Fig. 3). Sequential processing of the peptide to smaller forms [corresponding to somatostatin 14, somatostatin 28, somatostatin 28 (1-12)], and larger molecular forms (prohormone, preprohormone) takes place within the endoplasmic reticulum and is translocated in the Golgi apparatus for processing and assembly into granule form (see refs. 29 and 30). Somatostatin 14 conceivably could be cleaved from somatostatin 28, but in our studies,[31] the S14 and S28 forms appear to be separately cleaved from the prohormone. In epithelial cells the secretory granules are translocated along the cytoskeleton to the plasma membrane for

TGCGGACCTGCGTCTAGACTGACCCACCGCGCTCAAGCTCGGCTGTCTGAGGCAGGGGAG

ATG	CTG	TCC	TGC	CGT	CTC	CAG	TGC	GCG	CTG	GCC	GCG	CTC	TGC	ATC	
Met	*Leu*	*Ser*	*Cys*	*Arg*	*Leu*	*Gln*	*Cys*	*Ala*	*Leu*	*Ala*	*Ala*	*Leu*	*Cys*	*Ile*	15
GTC	CTG	GCT	TTG	GGC	GGT	GTC	ACC	GGG	GCG	CCC	TCG	GAC	CCC	AGA	
Val	*Leu*	*Ala*	*Leu*	*Gly*	*Gly*	*Val*	*Thr*	*Gly*	*Ala*	*Pro*	*Ser*	*Asp*	*Pro*	*Arg*	30
CTC	CGT	CAG	TTT	CTG	CAG	AAG	TCT	CTG	GCG	GCT	GCC	ACC	GGG	AAA	
Leu	*Arg*	*Gln*	*Phe*	*Leu*	*Gln*	*Lys*	*Ser*	*Leu*	*Ala*	*Ala*	*Ala*	*Thr*	*Gly*	*Lys*	45
CAG	GAA	CTG	GCC	AAG	TAC	TTC	TTG	GCA	GAA	CTG	CTG	TCT	GAG	CCC	
Gln	*Glu*	*Leu*	*Ala*	*Lys*	*Tyr*	*Phe*	*Leu*	*Ala*	*Glu*	*Leu*	*Leu*	*Ser*	*Glu*	*Pro*	60
AAC	CAG	ACA	GAG	AAC	GAT	GCC	CTG	GAG	CCT	GAG	GAT	TTG	CCC	CAG	
Asn	*Gln*	*Thr*	*Glu*	*Asn*	*Asp*	*Ala*	*Leu*	*Glu*	*Pro*	*Glu*	*Asp*	*Leu*	*Pro*	*Gln*	75
GCA	GCT	GAG	CAG	GAC	GAG	ATG	AGG	CTG	GAG	CTG	CAG	AGG	TCT	GCC	
Ala	*Ala*	*Glu*	*Gln*	*Asp*	*Glu*	*Met*	*Arg*	*Leu*	*Glu*	*Leu*	*Gln*	*Arg*	*Ser*	*Ala*	90
AAC	TCG	AAC	CCA	GCC	ATG	GCA	CCC	CGG	GAA	CGC	AAA	GCT	GGC	TGC	
Asn	*Ser*	*Asn*	*Pro*	*Ala*	*Met*	*Ala*	*Pro*	*Arg*	*Glu*	*Arg*	*Lys*	*Ala*	*Gly*	*Cys*	105
AAG	AAC	TTC	TTC	TGG	AAG	ACA	TTC	ACA	TCC	TGT	TAG	CTTTAATATT			
Lys	*Asn*	*Phe*	*Phe*	*Trp*	*Lys*	*Thr*	*Phe*	*Thr*	*Ser*	*Cys*	*Stop*				116

GTTGTCTCAGCCAGACCTCTGATCCCTCTCCTCCAAATCCCATATCTCTTCCTTAACTCC

CAGCCCCCCCCCCCAATGCTCAACTAGACCCTGCGTTAGAAATTGAAGACTGTAAATACAA

AATAAAATTATGGTGAAATTATGpoly(A)

Figure 2 Structure of preprosomatostatin in the rat deduced by recombinant genetic techniques from rat medullary thyroid carcinoma. From Goodman *et al.*[27]

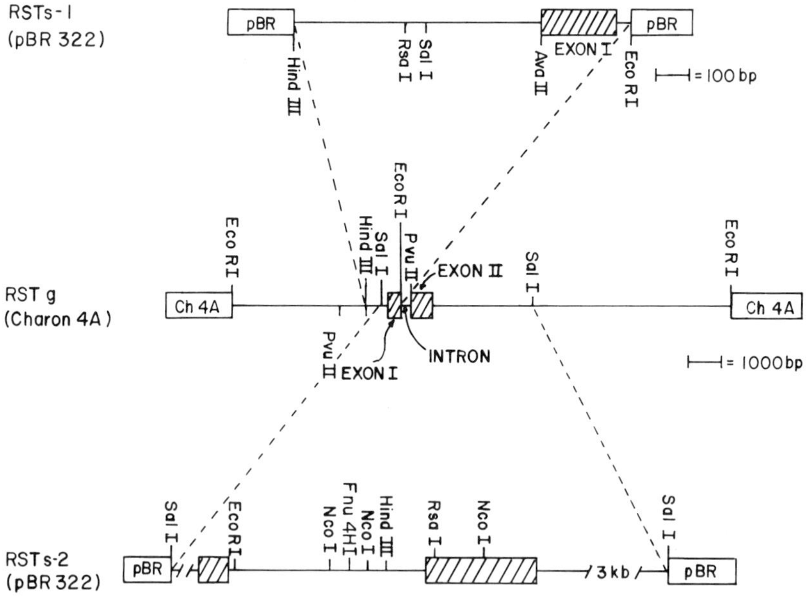

Figure 3 Gene sequence of somatostatin in brain deduced by genetic recombinant techniques. From Montminy *et al.*[28]

exocytic secretion: in neurons the granules move from cell bodies to nerve terminals by axoplasmic transport. The transport of somatostatin in granules within neurons has been well documented in studies of effects of section of the axonal projections to the median eminence (which depletes median eminence somatostatin),[16–18] by intraventricular injections of colchicine which poisons microtubular action (Lechan, unpublished observations), and in ligation studies of the sciatic[32] and vagus nerve[33] of the rat. Transport rate of the mobile fraction in sciatic nerve is approximately 400 mm/24 hr, thus corresponding to the fast transport rate characteristic of secretion granules in neurosecretory fibers of the neurohypophysis and in the sympathetic nerves. Only 20% of the somatostatin in sciatic nerve appears to be in the rapidly mobile fraction. Content of somatostatin in sensory ganglia of the sciatic nerve is estimated to turn over more than nine times in 24 hr.[32]

In some tissues, like the sciatic nerve, somatostatin 14 predominates,[32] while in gut mucosa somatostatin 28 predominates.[34] Cerebral cortical cells grown in tissue culture secrete appreciable amounts of somatostatin 28, somatostatin 11.5 Kd, as well as somatostatin 14,[35] and these ratios correspond to the proportions of larger and smaller forms extractable from brain and found in human cerebrospinal fluid[14] (Fig. 4). Somatostatin 28 (1-12) is also secreted as such.[4] Differences in proportionate distribution reflect cellular differences in posttranslational processing of somatostatin and may have functional importance. Somatostatin 14, 28, and 11.5 K all inhibit GH secretion.[5] Somatostatin 28 is actually more potent than somatostatin 14 in eliciting some biological

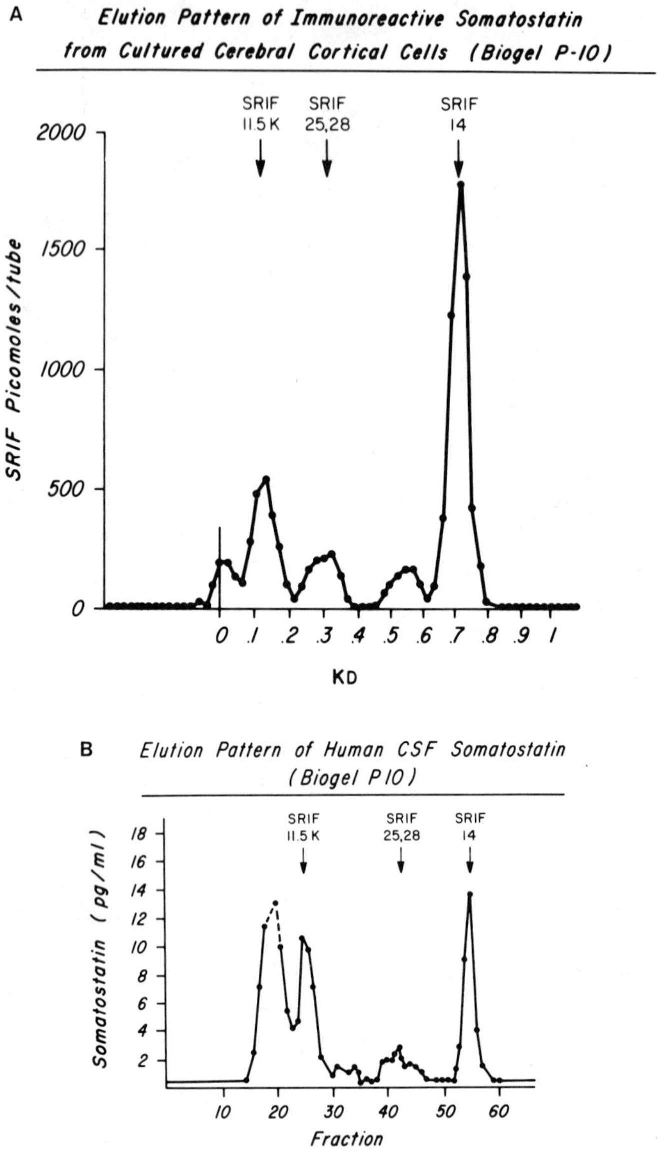

Figure 4 Chromatographically separated forms of immunoreactive somatostatin found in extracts of cerebral cortical cells and cerebrospinal fluid. (A) Present in cerebral cortical cells, and (B) found in human cerebrospinal fluid. In all preparations it can be shown that forms corresponding to somatostatin 14, 28, and 11.5K-daltons are present. Hence, the prohormone as well as the hormones themselves are secreted.

effects, notably the suppression of pancreatic insulin secretion,[36] and somatostatin 28 suppresses release of serum gastrin, insulin, and pancreatic polypeptide, just as does somatostatin 14. However, unlike somatostatin 14, the larger form does not affect mesenteric circulation, oxygen uptake, or intestinal motility.[37] By at least one criterion, somatostatin 28 is more potent as a neuromodulator than somatostatin 14.[38] Each form can be regarded as a hormone, and the larger forms may be prohormones as well. In this regard the situation with somatostatin is similar to that for gastrin. "Large" gastrin consists of 34 and small gastrin of 17 amino acids. Both forms occur in and are secreted by gastrin-secreting cells, and both have biological activity.[39]

4. Function of Somatostatin in Regulation of GH Secretion

The influence of the central nervous system, of central neurotransmitters, and of central neuropeptides on GH secretion has been well documented.[40] Much study has been devoted therefore to the elucidation of the role that altered somatostatin release plays in mediating these neural effect.

Lesions of the somatostatinergic tuberoinfundular tract in the base of the hypothalamus lead to depletion of median eminence somatostatin and to increased basal levels of GH[18] (and TSH[14]). Contrariwise, electrical stimulation of the anterior periventricular area in the rat (corresponding to the cells of origin of the principal hypothalamic–median eminence projection) inhibits GH secretion[40,41] and, as anticipated, increases the release of somatostatin into the hypophysial–portal vessels.[42] That somatostatin is involved in regulation of GH release under normal circumstances is shown by studies in rats that utilize antisera to the peptide. Acute release of GH[43,44] and reversal of the characteristic inhibition of GH release caused by physical stress[45] and by starvation[46] are observed to follow antiserum injection.

Somatostatin secretion also appears to be at least one of the factors mediating "short-loop" feedback control of GH secretion. GH deficiency (as produced by hypophysectomy) leads to decrease in somatostatin content of the median eminence, a change reversed by GH administration[47] (see refs 4 and 48 reviews). Exposure of the hypothalamus to GH in the living animal brings about suppression of GH release[49,50] from the pituitary and increases release of somatostatin into the hypophysial–portal vessels.[51] GH appears to influence somatostatin secretion in two different ways, one as a direct regulator, and the second through somatomedin C. Exposure of hypothalamic blocks to GH[47,52] or to somatomedin C *in vitro*[53] stimulates somatostatin release; intracerebroventricular injection of somatomedin C in the rat inhibits GH secretion.[50] These relations are summarized in Fig. 5.

5. Neurotransmitter Modulation of Somatostatin Secretion

Depolarization of somatostatinergic nerve cells is followed by secretion of somatostatin (for reviews, see refs. 7, 9, 35, and 36). A secretory response

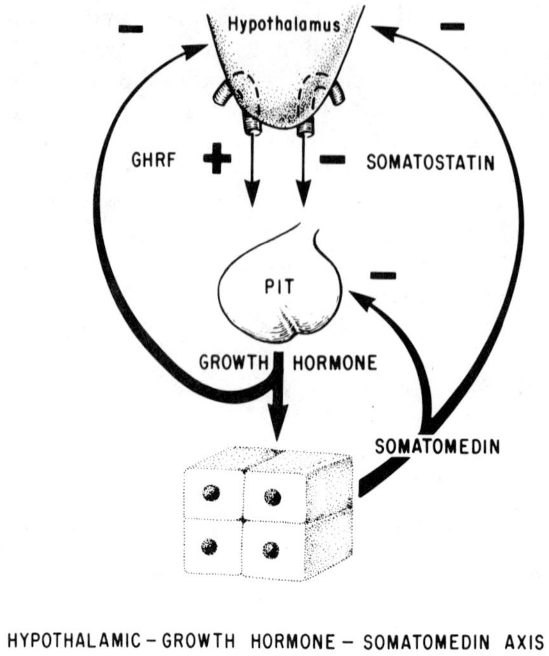

Figure 5 Role of somatostatin in regulation of GH secretion. Growth hormone release is stimulated by growth hormone-releasing hormone and inhibited by somatostatin. Growth hormone exerts negative feedback on its own secretion by several mechanisms (indicated by a − in the diagram). These include stimulation of release of somatostatin from the hypothalamus and stimulation of release of somatomedin C from the liver. Somatomedin C, in turn, stimulates hypothalamic somatostatin release and also acts directly on the pituitary (PIT) to inhibit growth hormone secretion, including that brought about by growth hormone-releasing hormone. The effects of somatomedin C and of growth hormone on the secretion of growth hormone-releasing hormone have not been defined. Figure reproduced from Martin and Reichlin,(90) with permission of the publisher. Legend reproduced from Reichlin.(10)

occurs if Ca^{2+} is present in the media and is blocked by agents that interfere with calcium entry, including Ca^{2+} and verapamil.(54) It is clear that membrane control of somatostatin release is mediated in the same way as other neurotransmitter release processes.(55) Release of somatostatin is induced by veratridine, an agent that holds Na^+ channels in an open configuration and is blocked by the sodium channel blocker tetrodotoxin.(54)

Electrical stimulation of neurohypophysis *in vitro*(56) and stimulation of the preoptic area *in vivo*(42) leads to somatostatin release, presumably owing to neuronal membrane depolarization.

The neurotransmitters that will activate somatostatin release from various neural preparations are listed in Table 2 (summarized in ref. 9). In our own studies of cerebral cell regulation by neurotransmitters acetylcholine showed a dose-related activation of somatostatin release, an effect blocked by addition of atropine,(57) indicating that it is the muscarinic form of the receptor that is being activated. On the other hand, Hollander and colleagues report that acetylcholine inhibits somatostatin release.(58) Robbins and Landon report that glutamate and aspartate also stimulate somatostatin secretion.(59) The most convincing inhibitory effects in cortex were exerted by GABA, which suppressed somatostatin release.(57) Picrotoxin, a GABA antagonist, reversed GABA effects and, in fact, by itself produced a marked increase in somatostatin release. These cell preparations contain GABAergic cells. Gamse *et al.* have reported similar findings in hypothalamic cell cultures treated with GABA.(60)

Influence of neuropeptides on somatostatin release is also summarized in

Table 2 Factors Reported to Influence Somatostatin Secretion from Hypothalamus and Cerebral Cortical Preparations[a]

Hormonal changes	
GH deficiency	
↓	Hoffman and Baker, 1977, immunohistochemistry median eminence
↓	Fernandez-Durango *et al.*, 1978, hypothalamic content
↓	Patel, 1979, hypothalamic content
GH	
↑	Hoffman and Baker, 1977, immunohistochemistry median eminence
↑	Sheppard *et al.*, 1978, hypothalamic fragment release
↑	Patel, 1979, hypothalamic content
↑	Pimstone *et al.*, 1979, hypothalamic fragment release
↑	Molitch and Hlivyak, 1980, content of hypothalamus
↑	Chihara *et al.*, 1981, portal vessels
↑	Berelowitz *et al.*, 1981, content and release from MBH fragment
Somatomedin C	
↑	Berelowitz *et al.*, 1981, release from MBH fragment
Thyroidectomy	
↔	Fernandez-Durango *et al.*, 1978, hypothalamic content
↔	Gillioz *et al.*, 1979, portal blood
↓	Berelowitz *et al.*, 1980, content and release from MBH fragment
Thyroid hormone	
↔	Fernandez-Durango *et al.*, 1978, hypothalamic content
↔	Gillioz *et al.*, 1979, portal blood
↑	Berelowitz *et al.*, 1980, content and release from MBH
Neurotransmitters	
Catechol depletion	
↓	Turkelson *et al.*, 1979, portal vessel
Dopamine	
↑	Wakabayashi *et al.*, 1977, synaptosome release
↑	Negro-Villar *et al.*, 1978
↑	Chihara *et al.*, 1979, portal vessel
↑ cortex ↓ hypo	Bennet *et al.*, 1979, synapotosome release
↑	Maeda and Frohman, 1980, hypothalamic fragment release
↑	Ojeda *et al.*, 1980, hypothalamic fragment release
↔	Terry *et al.*, 1980, hypothalamic fragment release
↑	Berelowitz *et al.*, 1980, hypothalamic fragment release
Norepinephrine	
↑	Negro-Villar *et al.*, 1978, hypothalamic fragment release
↑	Chihara *et al.*, 1979, portal
↑	Turkelson *et al.*, 1979, portal blood
↑ cortex ↓ hypo	Bennet *et al.*, 1979, synaptosome release
↔	Terry *et al.*, 1980, hypothalamic fragment release
↔	Maeda and Frohman, 1980, hypothalamic fragment release
↔	Robbins *et al.*, 1982, cortical neuron release
Serotonin	
↔	Chihara *et al.*, 1979, portal blood
↔	Turkelson *et al.*, 1979, portal blood
↑ cortex ↓ hypo	Bennet *et al.*, 1979
↔	Maeda and Frohman, 1980, hypothalamic fragment release
↓	Richardson *et al.*, 1981, hypothalamic fragment release
Acetylcholine	
↑	Chihara *et al.*, 1979, portal blood
↔	Terry *et al.*, 1980, hypothalamic fragment release
↓	Richardson *et al.*, 1980, hypothalamic fragment release
↑	Robbins *et al.*, 1982, cortical neuron release

(*continued*)

Table 2 (*continued*)

GABA	
↑	Turkelson *et al.*, 1979, portal blood
↓	Gamse *et al.*, 1980, hypothalamic cells release
↔	Terry *et al.*, 1980, hypothalamic fragments
↓	Robbins *et al.*, 1982, cortical neuron release
Peptides	
Glucagon	
↑	Abe *et al.*, 1978, portal (249).
↔	Epelbaum *et al.*, 1979, MBH fragment release
↑	Shimatsu *et al.*, 1981, hypothalamic fragment release
VIP	
↓	Epelbaum *et al.*, 1979, MBH fragment release
↓	Epelbaum *et al.*, 1980, MBH fragments
↓	Shimatsu *et al.*, 1981, hypothalamic fragment release
↑	Hermansen, 1980 (253)
Secretin	
↓	Epelbaum *et al.*, 1979, hypothalamic fragment release
Substance P	
↑	Sheppard *et al.*, 1978, hypothalamic fragment release
↔	Epelbaum *et al.*, 1979, hypothalamic fragments
↔	Maeda and Frohman, 1980, hypothalamic fragment release
↑	Hermansen, 1980
↔	Abe *et al.*, 1981, portal blood
Opiates	
↔	Sheppard *et al.*, 1979, hypothalamic fragment release
↓	Epelbaum *et al.*, 1980, hypothalamic fragments
↔	Abe *et al.*, 1981, portal blood
↓	Drouva *et al.*, 1981, hypothalamic fragments
Neurotensin	
↑	Sheppard *et al.*, 1979, hypothalamic fragments
↑	Maeda and Frohman, 1980, hypothalamic fragments
↑	Berelowitz *et al.*, 1981, hypothalamic fragments
↑	Abe *et al.*, 1981, portal blood
CCK-8 and CCK-4	
↔	Shimatsu *et al.*, 1981, hypothalamic fragment release
↑	Hermansen, 1980
Bombesin	
↑	Abe *et al.*, 1981, portal blood
↔	Hermansen, 1980
Metabolic factors	
Glucose	
↑	Negro-Villar *et al.*, 1980, MBH fragment release
Cyclic AMP	
↔	Maeda and Frohman, 1980, hypothalamic fragment release
↑	Robbins *et al.*, 1982, cortical neuron release
Prostaglandins	
↔	Ojeda *et al.*, 1980, hypothalamic fragment release
↔	Terry *et al.*, 1980, hypothalamic fragment release
Miscellaneous	
Melatonin	
↔	Terry *et al.*, 1980, hypothalamic fragments
↑	Richardson *et al.*, 1980

[a] References cited showing effect on content only give no direct evidence of modified release. ↑, Increased release; ↓, decreased release; ↔, unaltered release. Table adapted from Reichlin.[9] Reproduced from Reichlin.[(10)] In addition to the papers cited, the following have been reported to stimulate somatostatin secretion: the amino acids glutamate and aspartate,[(59)] corticotropin release hormone,[(91)] acetylcholine,[(92)] VIP.[(61)]

Table 2. Most work has been carried out on hypothalamic preparations. The most consistent changes reported are increased somatostatin release after GH exposure, glucagon, and neurotensin, while inhibition of release has been reported to occur after exposure to vasoactive intestinal peptide (VIP) and secretin. Recent work carried out in my laboratory indicates that VIP stimulates somatostatin release rather than inhibiting it.[(61)] The analogue peptide histidine isoleucine (PHI) was approximately 1/15 as potent a releaser of somatostatin, while other related peptides including secretin, GHRH 1-40-OH had no effect on somatostatin secretion. The response to this agent is very similar to the known effects of VIP in stimulating the formation of cyclic 3′,5′AMP (cAMP) in rat cerebral cortical slices.[(62)] Since high concentrations of cAMP analogues were reported to increase somatostatin release in this preparation (Robbins *et al.*[(57)]), I think it likely that VIP may act through the mediation of cAMP, as has been reported for many other secretory systems.[(63)] On the other hand, Maeda and Frohman[(64)] reported that dibutyryl cAMP was without effect on somatostatin secretion. Nevertheless, I believe that these data suggest that there are two mechanisms by which somatostatin release from neurons is brought about, one by calcium channel and sodium channel mediated reactions, and the other by one or more second messengers, including cAMP.

The known effects of neurotransmitters and neuropeptides on growth hormone secretion in the intact animal can only partially be explained by the actions of these agents on somatostatin secretion (Table 3). Perhaps the most relevant are the effect of GABA and of β-endorphin, both of which inhibit somatostatin secretion and stimulate GH secretion. The action of neurotensin, which inhibits GH secretion (acting centrally) and stimulates somatostatin release, is also of possible relevance to physiological control of GH. On the other hand, several neurotransmitters and neuropeptides known to stimulate GH release when applied to the hypothalamus (dopamine, norepinephrine, glucagon) either stimulate the release of somatostatin or have no effect, depending on the study, thus suggesting that these agents must be acting through the growth hormone-releasing hormone (GHRH) system. Serotonin, a hypothalamically mediated GH releaser, apparently has no effect on the secretion of somatostatin. Acetylcholine, which also releases GH by acting at the hypothalamic level (see ref. 65), as noted previously, was found to stimulate somatostatin secretion in most, but not all studies. This neurotransmitter, in common with several others cited previously, probably mediates its effects through GHRH release.

6. Somatostatin Regulation of Thyroid-Stimulating Hormone Secretion

The fact that somatostatin is a link in the negative feedback control of thyroid-stimulating hormone (TSH) secretion means that thyroid effects on the hypothalamus must also be considered as part of the somatostatin–GH regulatory system. Elucidation of the role of somatostatin in regulation of TSH release has followed much the same logical sequence of development as that of

Table 3 Comparison of Factors influencing Somatostatin, GH Release, and TSH Release[a]

	GH secretion	TSH secretion
Increased somatostatin secretion		
Anterior periventricular stimulation	↓	↓
Hormonal changes		
GH administration	↓	↓
Somatomedin C	↓	?
Thyroid hormone excess	↑	↓
Neurotransmitters		
Dopamine	↑	↓
Norepinephrine	↑	↑
Melatonin	↓	↓ ↑
Acetylcholine	↑	?
Peptides		
Glucagon	↑	?
Neurotensin	↓	?
Bombesin		?
Decreased somatostatin secretion		
Hormonal changes		
Thyroid hormone deficiency	↓	↑
Neurotransmitters		
Catechol depletion	↓	↓
Serotonin	↑	↓ ↑
GABA	↑	↓
Peptides		
VIP	↑	↔
Secretin	↑	?
Unaltered somatostatin secretion		
Neurotransmitters		
Dopamine	↑	↓
Norepinephrine	↑	↑
Serotonin	↑	↓ ↑
Acetylcholine	↑	?
Melatonin	↓	↓
Peptides		
Glucagon	↑	?
Substance P	?	↔
Endorphins	↑	↓
TRH	↓	?
Cholecystokinin	?	?

[a] ↑, Elevated; ↓, decreased; ↔, no change; ↓ ↑, both increases and decreases have been reported; ?, no data available, or inadequate studies. Reproduced from ref. 9. The table should be modified to show that VIP, reported to inhibit somatostatin secretion in some studies, is reported in others to increase somatostatin secretion. Since VIP injected into the third ventricle of the rat reportedly increases GH secretion, it seems likely that the major effect of this peptide on GH secretion is mediated by increased GRH secretion.

GH regulation. Somatostatin inhibits release of TSH from isolated pituitary preparations[66] and reduces basal TSH levels in rats[67,68] and man.[69] The effect is readily observed in hypothyroid man[70] and after thyroid-releasing hormone (TRH) stimulation.[71,72] Administration of antisomatostatin antibody potentiates the TSH response to cold exposure[73] or to TRH administration

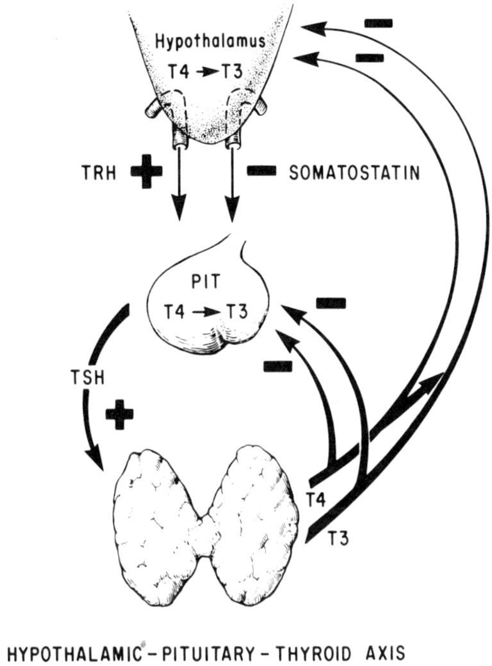

Figure 6 Role of somatostatin in regulation of TSH secretion. Secretion of thyrotropin is regulated by the interaction of a releasing factor (thyrotropin-releasing hormone, TRH) and an inhibitory factor (somatostatin). Thyroid hormones, triodothyronine and thyroxine (T_4 and T_3), act directly on the pituitary to inhibit thyrotropin secretion. Thyroid hormone also acts at the hypothalamic level to stimulate somatostatin release. The effect of thyroid hormone on secretion of thyrotropin-releasing hormone has not been established with certainty. Finally, T_4 is converted to T_3 in the periphery, the pituitary, and the hypothalamus: in all sites, T_3 is more potent than T_4. Figure reproduced from Martin and Reichlin,[(90)] with permission of the publisher. Legend reproduced from Reichlin,[(10)] with permission of the publisher.

(in the rat).[(74)] These results clearly indicate that at the level of the pituitary, secretion of TSH is regulated by the interplay of TRH and somatostatin as well as by thyroid hormone (see refs. 9, 10, and 75 for reviews). In addition, thyroid hormone appears to modulate somatostatin secretion by the hypothalamus. Berelowitz and co-workers report that hypothyroid rats show reduced hypothalamic somatostatin content, and that release of the peptide *in vitro* is reduced and is stimulated from hypothalami of normal rats *in vitro* by exposure to T_3.[(76)] Portal vessel somatostatin concentration is less than normal in hypothyroid rats.[(77)] Somatostatin acting on the hypothalamus also inhibits TRH release.[(78)] Thus, somatostatin, which inhibits both TSH and GH and is part of the physiological regulatory mechanism (in the rat), provides at lease one common pathway by which thyroid hormone can modify GH secretion and vice versa (Fig. 6).

7. Mechanisms of Action of Somatostatin

7.1. Somatostatin Receptors

Somatostatin-binding receptors have been identified by competitive binding assays in rat pituitary cells,[(79–81)] in plasma membranes isolated from bovine pituitaries,[(82)] in whole brain synaptosome membranes,[(83,84)] and in isolated rat islet cells.[(85)] Binding affinities and binding kinetics have been defined.[(79–83)]

Both pituitary and brain possess a single high-affinity site. Alteration in somatostatin receptor population is emerging as a possible mode of cell control by hormones and substrates. For example, Schonbrunn and Tashjian have shown that exposure of a GH-secreting rat pituitary tumor line to TRH acutely increases the number, but not the affinity, of the somatostatin receptors, but that chronic exposure for 4–72 hr causes a decrease in the number of somatostatin receptors.[80] Exposure to cortisol also causes a decrease in somatostatin receptor number but by different mechanisms.[86] In a detailed analysis of structure–function relationships, Srikant and Patel[81] find that pituitary somatostatin receptors behave differently toward several analogues from brain receptors. They have inferred, therefore, that there may be more than one class of somatostatin receptor. In contrast to several other peptide systems studied previously, such as the endogenous opioids, the concentration of somatostatin in various parts of the brain is not closely correlated with receptor concentrations in the same areas.[83]

7.2. Action of Somatostatin

Most studies of somatostatin effects have been directed at pancreatic islet and pituitary somatotrope cells (see refs 6 and 9 for review), virtually nothing being known about the biochemical–physiological basis of somatostatin action on neurons. Although the concentration of cAMP is reduced in both pituitary and pancreas by exposure to somatostatin, its inhibitory effect on secretion is not mediated through changes in cyclic nucleotides. cAMP-induced secretion is also reduced by somatostatin. It has been concluded, therefore, that somatostatin acts "downstream" from cAMP to inhibit secretion.

Several reports suggest an important role of Ca^{2+} in somatostatin effects. Somatostatin reduces membrane permeability to calcium, and the inhibitory actions of somatostatin on hormone release are partially or completely reversed by exposure to A23187, an ionophore that opens calcium channels in both the external membrane and membranes of intracellular organelles. These findings have generally been interpreted to mean that somatostatin acts to block calcium-mediated intracellular activation.

A study by Pace and Tarvin has provided insight into the way that somatostatin may act to inhibit pancreatic islet cell function,[87] which may be a model of action in pituitary cells. The crucial new finding in their work is that somatostatin, in addition to inhibiting Ca^{2+} uptake by islet cells, also increases the permeability of cells to K^+ (as inferred from ^{86}Rubidium washout). Since chemical agents that interfere with outward K^+ flux from islet cells (quinine and tetraethylamonium) block somatostatin effects on both insulin secretory response and the electrophysiological effects of high glucose on islet cells, these workers postulate that somatostatin acts to block K^+ outflux, which in turn reduces the accumulation of intracellular Ca^{2+} thereby disrupting glucose-induced stimulus–secretion coupling.

Mechanisms (including possible second messengers) by which somatostatin release is modified by neurotransmitters and neuropeptides have not been elucidated. We have found that relatively high concentrations of dibutyryl cAMP will release somatostatin from cerebral cortical cells in culture,[57] but Maeda and Frohman[64] reported no changes in release from hypothalamic fragments, and prostaglandins E_2, E_{2a} have been reportedly ineffective in modifying hypothalamic somatostatin release.[88,89]

ACKNOWLEDGMENT. Work from the author's laboratory cited in this chapter was supported by U.S.P.H.S. Grants AM16684 and T32AM07039.

References

1. L. Krulich, A. P. S. Dhariwal, and S. M. McCann, Stimulatory and inhibitory effects of purified hypothalamic extracts on growth hormone release from rat pituitary *in vitro, Endocrinology* **83:**783–790, 1968.
2. P. Brazeau, W. Vale, R. Burgus, N. Ling, M. Butcher, J. Rivier, and R. Guillemin, Hypothalamic peptide that inhibits the secretion of immunoreactive growth hormone, *Science* **179:**77–79, 1973.
3. L. Pradayrol, H. Jörnvall, V. Mutt, and A. Ribet, *N*-terminally extended somatostatin: The primary structure of somatostatin −28, *FEBS Lett.* **109:**55–58, 1980.
4. C. Bakhit and F. E. Bloom, Release of somatostatin-28 (1-12) from rat hypothalamus *in vitro, Nature* **301:**524–526, 1983.
5. O. P. Rorstad, J. Epelbaum, P. Brazeau, and J. B. Martin, Chromatographic and biological properties of immunoreactive somatostatin in hypothalamic and extrahypothalamic brain regions of the rat, *Endocrinology* **105:**1083–1092, 1979.
6. I. S. Gottesman, L. J. Mandarino, and J. E. Gerich, Somatostatin, in: *Special Topics in Endocrinology and Metabolism,* Vol. 4, M. Cohen and P. Foa, eds., Alan R. Liss, New York, 1982, pp. 177–243.
7. S. Reichlin, S. Robbins, and R. Lechan, Somatostatin and the nervous system, in: *Proceedings of 2nd International Somatostatin Symposium, Athens, 1981* (J. Rosenthal, S. Raptis, and J. Gerich, eds.), Attempto Publishing House, University Press, Tubingen, W. Germany, 1984, pp. 21–37.
8. S. M. McCann, L. Krulich, A. Negro-Vilar, S. R. Ojeda, and E. Vijayan, Regulation and function of panhibin (somatostatin), *Adv. Biochem. Psychopharmacol.* **22:**131–143, 1980.
9. S. Reichlin, Somatostatin, in: *Brain Peptides* (D. T. Krieger, M. Brownstein, and J. B. Martin, eds.), John Wiley & Sons, New York, 1982, pp. 712–752.
10. S. Reichlin, Somatostatin, *N. Engl. J. Med.* **309:**1495–1501, 1556–1563, 1983.
11. S. Falkmer, R. Elde, C. Hellerstrom, and B. Petersson, Phylogenetic aspects of somatostatin in the gastroenteropancreatic (GEP) endocrine system, *Metabolism* **27**(Suppl. 1)**:**1193–1196, 1978.
12. I. M. D. Jackson, Phylogenetic distribution and function of the hypophysiotropic hormones of the hypothalmus, *Am. Zoologist* **18:**385–399, 1978.
13. M. Berelowitz, D. LeRoith, H. Von Schenk, C. Newgard, M. Szabo, L. A. Frohman, J. Shiloach, and J. Roth, Somatostatin-like immunoreactivity and biologic activity is present in tetrahymena-pyriformis, a ciliated protozoan, *Endocrinology* **110:**1939–1944, 1982.
14. M. Palkovits, M. J. Brownstein, A. Arimura, H. Sato, A. V. Schally, and J. S. Kizer, Somatostatin content of the hypothalamic ventromedial and arcuate nuclei and the circumventricular organs in the rat, *Brain Res.* **109:**430–434, 1976.
15. M. Brownstein, A. Arimura, H. Sato, A. V. Schally, and J. S. Kizer, The regional distribution of somatostatin in the rat brain, *Endocrinology* **96:**1456–1461, 1975.
16. M. J. Brownstein, A. Arimura, R. Fernandez-Durango, A. V. Schally, M. Palkovits, and J. S. Kizer, The effect of hypothalamic deafferentation on somatostatin-like activity in the rat brain, *Endocrinology* **100:**246–249, 1977.
17. J. Epelbaum, J. O. Willoughby, P. Brazeau, and J. B. Martin, Effects of brain lesions and hypothalamic deafferentation on somatostatin distribution in the rat brain, *Endocrinology* **101:**1495–1502, 1977.

18. V. Critchlow, R. W. Rice, K. Abe, and W. Vale, Somatostatin content of the median eminence in female rats with lesion-induced disruption of the inhibitory control of growth hormone secretion, *Endocrinology* **103:**817–825, 1978.
19. S. M. Urman and V. Critchlow, Long term elevations in plasma thyrotropin (TSH) but not growth hormone (GH) concentrations associated with lesion-induced depletion of median eminence (ME) somatostatin (SRIF), *Endocrinology* **42:**659–664, 1983.
20. T. Hökfelt, R. Elde, K. Fuxe, O. Johansson, H. Ljungdahl, M. Goldstein, R. Luft, S. Efendic, G. Perez de la Mora, L. Possani, R. Tapia, L. Teran and R. Palacios, Aminergic and peptidergic pathways in the nervous system with special reference to the hypothalamus, in: *The Hypothalamus* S. Reichlin, R. J. Baldessarini and J. B. Martin, eds., Raven Press, New York, 1978, pp. 69–135.
21. B. Krisch, Hypothalamic and extrahypothalamic distribution of somatostatin-immunoreactive elements in the rat brain, *Cell Tissue Res.* **195:**499–513, 1978.
22. C. Bennett-Clarke, M. A. Romagnano, and S. A. Joseph, Distribution of somatostatin in the rat brain: telencephalon and diencephalon, *Brain Res.* **188:**473–486, 1980.
23. B. Krisch, Somatostatin-immunoreactive fiber projections into the brain stem and the spinal cord of the rat, *Cell Tissue Res.* **217:**531–552, 1981.
24. G. Pelletier, Immunohistochemical localization of somatostatin, *Prog. Histochem. Cytochem.* **12:**1–41, 1980.
25. Y. C. Patel, K. Hoyte, and J. B. Martin, Effect of anterior hypothalamic lesions on neurohypophysial and peripheral tissue concentrations of somatostatin in the rat, *Endocrinology* **105:**712–715, 1979.
26. L-P. Shen, R. L. Pictet, and W. J. Rutter, Human somatostatin 1: Sequence of the cDNA, *Proc. Natl. Acad. Sci. USA* **79:**4574–4579, 1982.
27. R. H. Goodman, D. C. Aron, and B. A. Roos, Rat pre-prosomatostatin, *J. Biol. Chem.* **258:**5570–5573, 1983.
28. M. R. Montminy, R. H. Goodman, S. J. Horovitch, and J. F. Habener, Primary structure of the gene encoding rat pre-prosomatostatin, *Proc. Natl. Acad. Sci. USA* **81:**3337–3340, 1984.
29. J. F. Habener, Principles of peptide-hormone biosynthesis, *Adv. Biochem. Psychopharmacol.* **28:**21–43, 1981.
30. T. G. Warren and D. Shields, Cell-free biosynthesis of somatostatin precursors: evidence for multiple forms of pre-prosomatostatin, *Proc. Natl. Acad. Sci. USA* **79:**3729–3733, 1982.
31. R. J. Robbins, S. Mothon, and S. Reichlin, Somatostatin-28 is not the precursor of somatostatin-14 in cerebral cortical cells, Program of the Endocrine Society 65th Annual Meeting, San Antonio, June 8–10, Williams and Wilkins, Baltimore, 1983, pp. 85.
32. C. G. Rasool, A. L. Schwartz, J. A. Bollinger, S. Reichlin, and W. G. Bradley, Somatostatin distribution and axoplasmic transport in rat peripheral nerve, *Endocrinology* **108:**996–1001, 1981.
33. D. B. MacLean, S. F. Lewis, and D. Engler, Studies of the axoplasmic transport of somatostatin in the vagus nerve of the rat, *Endocrinology* **114:**1145–1154, 1984.
34. Y. C. Patel, T. Wheatley and C. Ning, Multiple forms of immunoreactive somatostatin: Comparison of distribution in neural and nonneural tissues and portal plasma of the rat, *Endocrinology* **109:**1943–1949, 1981.
35. S. Reichlin, Somatostatin in the nervous system, in: *Molecular Genetic Neuroscience* (F. E. Bloom, S. Bird, and F. O. Schmitt, eds.), Raven Press, New York, 1982, pp. 359–372.
36. C. A. Meyers, W. A. Murphy, T. W. Redding, D. H. Coy, and A. V. Schally, Synthesis and biological actions of prosomatostatin, *Proc. Natl. Acad. Sci. USA* **77:**6171–6174, 1980.
37. S. J. Konturek, J. Tasler, J. Jaworek, W. Pawlik, K. M. Walus, V. Schusdziarra, C. A. Meyers, D. H. Coy, and A. V. Schally, Gastrointestinal secretory, motor, circulatory, and metabolic effects of prosomatostatin, *Proc. Natl. Acad. Sci. USA* **78:**1967–1971, 1981.
38. M. Brown, J. Speiss, J. C. Reubi, M. Perrin, and W. Vale, Somatostatin-28 (SS 28), somatostatin-14 (SS 14) and their analogs: Comparative biological actions, Program of the 63rd Annual Meeting the Endocrine Society Cincinnati, Williams and Wilkins, Baltimore, 1981, p. 164a.
39. G. J. Dockray, Gastrin overview, in: *Gut Hormones* (S. R. Bloom ed.), Churchill Livingstone, Edinburgh, 1978, pp. 129–139.

40. J. B. Martin, P. Brazeau, G. S. Tannenbaum, J. O. Willoughby, J. Epelbaum, L. C. Terry, and D. Durand, Neuroendocrine organization of growth hormone regulation, in: *The Hypothalamus* (S. Reichlin, R. J. Baldessarini, and J. B. Martin, eds.), Raven Press, New York, 1978, pp. 329–351.
41. L. C. Terry and J. B. Martin, The effects of lateral hypothalamic–medial forebrain stimulation and somatostatin antiserum on pulsatile growth hormone secretion in freely behaving rats: Evidence for a dual regulatory mechanism, *Endocrinology* **109:**622–627, 1981.
42. K. Chih, C. Arimura, C. Kubli-Garfias, and A. V. Schally, Enhancement of immunoreactive somatostatin release into hypophysial portal blood by electrical stimulation of the preoptic area in the rat, *Endocrinology* **105:**1416–1418, 1979.
43. A. Arimura, W. Smith, and A. V. Schally, Blockade of the stress-induced decrease in blood GH by anti-somatostatin serum in rats, *Endocrinology* **98:**540–543, 1976.
44. L. Ferland, F. Labrie, M. Jobin, A. Arimura, and A. V. Schally, Physiologic role of somatostatin in the control of growth hormone and thyrotropin secretion, *Biochem. Biophys. Res. Commun.* **68:**149–156, 1976.
45. L. C. Terry, J. O. Willoughby, P. Brazeau, J. B. Martin, and Y. Patel, Antiserum to somatostatin prevents stress-induced inhibition of growth hormone secretion in the rat, *Science* **192:**565–566, 1976.
46. G. S. Tannenbaum, J. Epelbaum, E. Colle, P. Brazeau, and J. B. Martin, Antiserum to somatostatin reverses starvation-induced inhibition of growth hormone but not insulin secretion, *Endocrinology* **102:**1909–1914, 1978.
47. M. Berelowitz, S. L. Firestone, and L. A. Frohman, Effects of growth hormone excess and deficiency on hypothalamic somatostatin content and release and on tissue somatostatin distribution, *Endocrinology* **109:**714–719, 1981.
48. S. Reichlin, Systems for the study of regulation of neuropeptide secretion, *Adv. Biochem. Psychopharmacol.* **28:**573–597, 1981.
49. G. S. Tannebaum, Evidence for autoregulation of growth hormone secretion via the central nervous system, *Endocrinology* **107:**2117–2120, 1980.
50. H. Abe, M. E. Molitch, J. J. Van Wyk, and L. E. Underwood, Human growth hormone and somatomedin C suppress the spontaneous release of growth hormone in anesthestized rats, *Endocrinology* **113:**1319–1324, 1983.
51. K. Chihara, N. Minamitani, H. Kaji, A. Arimura, and T. Fujita, Intraventricularly injected growth hormone stimulates somatostatin release into rat hypophysial portal blood, *Endocrinology* **109:**2279–2281, 1981.
52. M. C. Sheppard, S. Kronheim, and B. L. Pimstone, Stimulation by growth hormone of somatostatin release from the rat hypothalamus *in vitro, Clin. Endocrinol.* **9:**583–586, 1978.
53. M. Berelowitz, M. Szabo, L. A. Frohman, S. Firestone, L. Chu, and R. L. Hintz, Somatomedin-C mediates growth hormone negative feedbank by effects on both the hypothalamus and the pituitary, *Science* **212:**1279–1281, 1981.
54. R. J. Robbins, R. E. Sutton, and S. Reichlin, Sodium-and calcium-dependent somatostatin release from dissociated cerebral cortical cells in culture, *Endocrinology* **110:**496–499, 1982.
55. L. L. Iverson, C. M. Lee, R. F. Gilbert, S. Hunt, and P. C. Emson, Regulation of neuropeptide release, *Proc. R. Soc. Lond. B* **210:**91–111, 1980.
56. Y. C. Patel, H. H. Zingg, and J. J. Dreifuss, Somatostatin secretion from the rat neurohypophysis and stalk median eminence (SME) *in vitro*: Calcium-dependent release by high potassium and electrical stimulation, *Metabolism* **27:**1243–1245, 1978.
57. R. Robbins, R. Sutton, and S. Reichlin, Effects of neurotransmitters and cyclic AMP on somatostatin release from cultured cerebral cortical cells, *Brain Res.* **234:**377–386, 1982.
58. S. B. Richardson, C. S. Hollander, R. D'Eletto, P. W. Greenleaf, and C. Thaw, Acetylcholine inhibits the release of somatostatin from rat hypothalamus *in vitro, Endocrinology* **107:**112–129, 1980.
59. R. J. Robbins and R. M. Landon, Somatostatin release from cerebral cortical cells: influence of amino acid neurotransmitters, *Brain Res.* **273:**374–378, 1983.
60. R. Gamse, D. Vaccaro, G. Gamse, M. DuPage, T. O. Fox, and S. F. Leeman, Release of

immunoreactive somatostatin from hypothalamic cells in culture: Inhibition by γ-aminobutyric acid, *Proc. Natl. Acad. Sci. USA* **77:**5552–5556, 1980.
61. L. Tapia-Arancibia and S. Reichlin, Vasoactive intestinal peptide and PHI stimulate somatostatin release from rat cerebral cortical and diencephalic cells in dispersed cell culture, *Brain Res.* **336:**67–72, 1985.
62. M. Quik, L. Iverson, and S. Bloom, Effect of vasoactive intestinal peptide (VIP) and other peptides on cAMP accumulation in rat brain, *Biochem. Pharmacol.* **27:**2209, 1978.
63. B. Amiranoff and G. Rosselin, VIP receptors and control of cyclic AMP production, in: *Vasoactive Intestinal Peptide* (S. Said, ed.), Raven Press, New York, 1982, pp. 307–322.
64. K. Maeda and L. A. Frohman, Release of somatostatin and thyrotropin releasing hormone from rat hypothalamic fragments *in vitro, Endocrinology* **106:**1837–1842, 1980.
65. L. C. Terry, Neuropharmacologic regulation of anterior pituitary hormone secretion in man, in: *Hormone-Secreting Pituitary Tumors* (JR Givens, ed.), Year Book Medical Publishers, Chicago, 1982, pp. 27–44.
66. W. Vale, C. Rivier, P. Brazeau, and R. Guillemin, Effects of somatostatin on the secretion of thyrotropin and prolactin, *Endocrinology* **95:**968–977, 1975.
67. P. Tanjasiri, X. Kozbur, and W. Florsheim, Somatostatin in the physiologic feedback control of thyrotropin secretion, *Life Sci.* **19:**657–660, 1976.
68. K. Chihara, A. Arimura, M. Chihara, and A. V. Schally, Studies on the mechanism of growth hormone and thyrotropin responses to somatostatin antiserum in anesthetized rats, *Endocrinology* **103:**1916–1922, 1978.
69. J. Weeke, A. Hansen, and K. Lundbaek, Inhibition by somatostatin of basal levels of serum thyrotropin (TSH) in normal man, *J. Clin. Endocrinol. Metab.* **41:**168–171, 1975.
70. C. Lucke, B. Hoffken, and A. von zur Muhlen, The effect of somatostatin on TSH levels in patients with primary hypothyroidism, *J. Clin. Endocrinol. Metab.* **41:**1082–1084, 1975.
71. T. Siler, S. Yen, W. Vale, and R. Guillemin, Inhibition by somatostatin of the release of TSH induced in man by throtropin-releasing factor, *J. Clin. Endocrinol. Metab.* **38:**742–745, 1974.
72. J. Weeke, A. Hansen, and K. Lundbaek, The inhibition by somatostatin of the thyrotropin response to thyrotropin-releasing hormone in normal subjects, *Scand. J. Clin. Lab. Invest.* **33:**101–103, 1974.
73. A. Gordin, A. Arimura, and A. V. Schally, Effect of thyroid hormone excess and deficiency on serum thyrotropin in rats immunized passively with antiserum to somatostatin, *Proc. Soc. Exp. Biol. Med.* **153:**319–323, 1976.
74. A. Arimura and A. V. Schally, Increases in basal and thyrotropin releasing hormone (TRH)-stimulated secretion of thyrotropin (TSH) by passive immunization with antiserum to somatostatin in rats, *Endocrinology* **98:**1069–1072, 1976.
75. S. Reichlin, J. B. Martin, and I. M. D. Jackson, Regulation of thyroid-stimulating hormone (TSH) secretion, in: *The Endocrine Hypothalamus* (S. L. Jeffcoate and J. S. M. Hutchinson, eds.), Academic Press, London, 1978, pp. 230–269.
76. M. Berelowitz, K. Maeda, S. Harris, and L. A. Frohman, The effect of alterations in the pituitary–thyroid axis on hypothalamic content and *in vitro* release of somatostatin-like immunoreactivity, *Endocrinology* **107:**24–29, 1980.
77. H. Abe, R. Robbins, and S. Reichlin, Somatostatin 14 and 28 are secreted by tuberoinfundibular neurons, Program of the 64th Annual Meeting of the Endocrine Society, San Francisco, June 16–18, Williams and Wilkins, Baltimore, 1982, p. 309 (abstr.).
78. Y. Hirooka, C. S. Hollander, S. Suzuki, P. Ferdinand, and S. I. Juan, Somatostatin inhibits release of thyrotropin releasing factor from organ cultures of rat hypothalamus, *Proc. Natl. Acad. Sci. USA* **75:**4509–4513, 1978.
79. A. Schonbrunn and A. H. Tashjian, Jr., Characterization of functional receptors for somatostatin in rat pituitary cells in culture, *J. Biol. Chem.* **253:**6473–6483, 1978.
80. A. Schonbrunn and A. H. Tashjian, Jr., Modulation of somatostatin receptors by thyrotropin-releasing hormone in a clonal pituitary cell strain, *J. Biol. Chem.* **255:**190–198, 1980.
81. C. B. Srikant and Y. C. Patel, Somatostatin analogs. Dissociation of brain receptor binding affinities and pituitary affinities and pituitary actions in the rat, *Endocrinology* **108:**341–343, 1981.

82. J. W. Leitner, R. M. Rifkin, A. Maman, and K. E. Sussman, Somatostatin binding to pituitary plasma membranes, *Biochem. Biophys. Res. Commun.* **87:**919–927, 1979.
83. C. B. Srikant and Y. C. Patel, Somatostatin receptors: Identification and characterization in rat brain membranes, *Proc. Natl. Acad. Sci USA* **78:**3930–3934, 1981.
84. J. C. Reubi, M. H. Perrin, J. E. Rivier, and W. Vale, High affinity binding sites for a somatostatin-28 analog in rat brain, *Life Sci.* **28:**2191–2198, 1981.
85. P. S. Mehler, A. L. Sussman, A. Maman, J. W. Leitner, and K. E. Sussman, Role of insulin secretogogues in the regulation of somatostatin binding by isolated rat islets, *J. Clin. Invest.* **66:**1338, 1980.
86. A. Schonbrunn, Glucocorticoids down-regulate somatostatin receptors on pituitary cells in culture, *Endocrinology* **110:**1147–1154, 1982.
87. C. S. Pace and J. T. Tarvin, Somatostatin: Mechanism of action in pancreatic islet β-cells, *Diabetes* **30:**836–842, 1981.
88. S. R. Ojeda, A. Negro-Vilar, A. Arimura, and S. M. McCann, On the hypothalamic mechanism by which prostaglandin E_2 stimulates growth hormone release. *In vivo* and *in vitro* studies, *Neuroendocrinology* **31:**1–7, 1980.
89. L. C. Terry, O. P. Rorstad, and J. B. Martin, The release of biologically and immunologically reactive somatostatin from perifused hypothalamic fragments, *Endocrinology* **107:**784–800, 1980.
90. J. B. Martin and S. Reichlin, *Clinical Neuroendocrinology*, 2nd ed., F. A. Davis, Philadelphia, 1986 (in press).
91. R. A. Peterfreund and W. W. Vale, Ovine corticotropin-releasing factor stimulates somatostatin secretion from cultured brain cells, *Endocrinology* **112:**1275–1278, 1983.
92. R. A. Peterfreund and W. W. Vale, Muscarinic cholinergic stimulation of somatostatin secretion from long term dispersed cell cultures of fetal rat hypothalamus: Inhibition of γ-aminobutyric acid and serotonin, *Endocrinology* **112:**526–534, 1983.

V

CHEMISTRY OF HUMAN GROWTH HORMONE

Physiological Studies

31

Conformational Comparison of Human Pituitary Growth Hormone and Human Chorionic Somatomammotropin (Human Placental Lactogen) by Second-Order Absorption Spectroscopy

Thomas A. Bewley

Human pituitary growth hormone (hGH) and the placental hormone HCS (hPL) are both potent lactogens,[1–3] but they differ markedly in growth-promoting potency.[3–6] The primary structures[7,8] of these two proteins have been shown[9,10] to be very closely related, with 28 replacements out of 191 residues, only seven of which can be considered nonconservative. Many laboratories have reported both similarities and differences in secondary and tertiary structures of these proteins (for a review see ref. 11).

It is quite probable that the similarities in primary, secondary, and tertiary structures are responsible for the nearly equivalent lactogenic activities of these molecules. It is equally probable that one or more of the few differences in primary structure, acting through subtle alterations in secondary and/or tertiary structure, give rise to the quantitative disparity in growth-promoting potencies. Much might be learned about how hGH functions to promote growth by understanding why HCS (hPL) does not.

One of the major differences noted in previous conformational studies [11,13–16] involves the local microenvironment of the single Trp residue, in both hormones. This difference, whose precise nature has never been elucidated,

Abbreviations used: hGH, human pituitary growth hormone; HCS(hPL), human chorionic somatomammotropin (human placental lactogen); CD, circular dichroism; E_M, molar extinction coefficient; A, absorptivity; Abs., absorbance.

Thomas A. Bewley • Laboratory of Molecular Endocrinology, University of California, San Francisco, California 94143.

gives rise to inequalities in ultraviolet (UV) absorption[11,12] and significant differences in circular dichroism (CD) spectra,[11,13–16] although it has been shown that in both proteins, the Trp side chain is buried within the hydrophobic interior of the molecule. The differences in UV absorption are particularly noteworthy in view of the fact that all chromophoric residues, including the single Trp, all eight Tyr, both disulfide bonds, and 11 of the 13 Phe residues in hGH occur in identical sequence positions in HCS(hPL) and are flanked by identical or highly homologous peptide segments.

In this report, the UV absorption spectra of hGH and HCS(hPL) are further compared by second-order absorption spectroscopy. In addition, the molar extinction coefficients (E_M) of both hormones have been reevaluated in 0.1 M Tris-Cl (pH 8.2).

1. Materials and Methods

The hGH and HCS(hPL) were prepared as described by Li *et al.*[17] and Neri *et al.*,[18] respectively. Thermolysin, lot 73326, was obtained from Calbiochem. All UV spectra were taken at 25°C on a Perkin–Elmer Model 552 double-beam, recording spectrophotometer. The spectrophotometer was also equipped with the second-derivative accessory supplied by the manufacturer. The wavelength positions of all second-order minima were corrected as described previously.[19] Molar extinction coefficients were determined from zero-order spectra of native and thermolysin digested proteins as described elsewhere.[19] Two different preparations of each hormone were used, with three to five digestions of each preparation.

2. Results and Discussion

The zero-order UV absorption spectra of native hGH and HCS(hPL) at pH 8.2 are presented in Fig. 1. Both spectra display a very weak shoulder between 310 and 300 nm, a weak shoulder near 290 nm, a distinct shoulder near 285 nm, and an absorption maxium at 276.6 ± 0.1 nm for hGH and 276.4 ± 0.1 nm for HCS(hPL). In addition, there is a series of closely spaced, weak shoulders between 270 and 245 nm. Although there is considerable apparent similarity in these two spectra, the E_M for hGH was found to be 20, 350 ± 100 ($M^{-1}cm^{-1}$) at the absorption maximum, while that for HCS(hPL) was found to be only 19,200 ± 180 ($M^{-1}cm^{-1}$). These E_M values correspond to absorptivities of

$$A_{1.0\ \mathrm{cm}}^{0.1\%} = 0.920 \pm 0.004 \text{ for hGH}$$

and

$$A_{1.0\ \mathrm{cm}}^{0.1\%} = 0.863 \pm 0.008 \text{ for HCS(hPL)}.$$

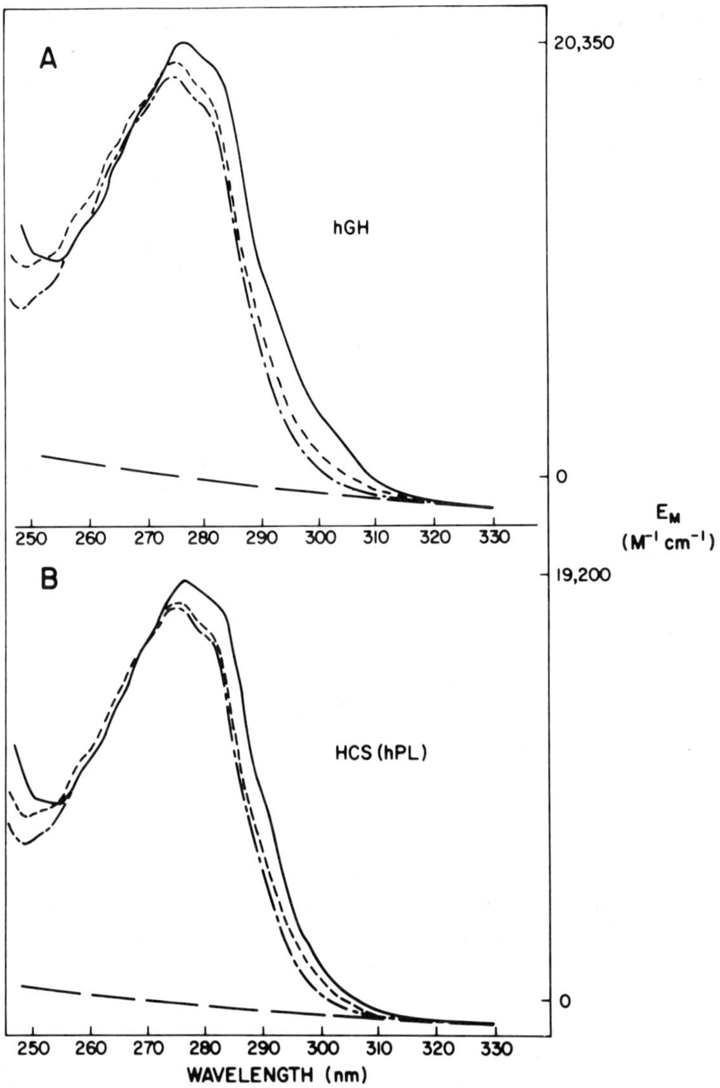

Figure 1 Zero-order absorption spectra of (A) hGH and (B) HCS(hPL), including the native protein (——), a 24-hr thermolysin digest at pH 8.2 (– – –), and the digest after titration to pH 1.5 (–·–). The heavier dashed line near the bottom of each set of spectra is the light-scattering correction. The zero position on the ordinate scale has been placed at the height of the light-scattering correction under the wavelength of maximum absorption.

Both are within 4% of previously published values.[12,20,21] This difference in the E_M of the two proteins is significant and clearly implies that the conformations of the molecules have produced unequal hyperchromic effects on one or more of the major chromophoric groups.[19,23] Since hyperchromic effects are generally accompanied by red shifting of the absorption bands toward higher wavelengths,[23] we might also expect to see some absorption features of hGH at somewhat higher wavelengths than in HCS(hPL). Indeed, the absorption maximum of hGH is slightly "red" of that of HCS (hPL), but we cannot readily assign this to any particular chromophore type, nor does it indicate what type of structural difference might be involved.

A more concrete feeling for the degree of hyperchromicity and red shifting that the conformation of a globular protein can produce may be obtained by comparing the native spectra with those taken after exhaustive digestion with a proteolytic enzyme such as thermolysin.[19,22] As shown in Fig. 1, both digested proteins (pH 8.2) display a considerable loss in extinction (hypochromicity) and a general blue shifting of the entire spectrum toward lower wavelengths. The enormous alterations in the hGH spectrum between 310 and 290 nm are most noteworthy.

Although the hypo- and hyperchromic shifts can be readily evaluated from zero-order spectra in terms of changes in E_M, the exact degree of red or blue shifting of the many weak shoulders is not at all clear. This problem can be largely circumvented by employing second-order absorption spectroscopy in which the second derivative of absorption ($d^2Abs./d\lambda^2$) is plotted against wavelength. In this mode, the shoulders appearing in Fig. 1 will all become distinct negative maxima. These second-order peaks indicate the wavelength positions of the absorption band centers in the various shoulders to about ± 0.1 nm.

Figure 2 presents the second-order absorption spectra of native hGH and HCS(hPL). The weak shoulders in the zero-order spectra between 270 and 245 nm appear as six sharp, negative peaks in the second-order spectra. The corrected wavelength positions of these maxima in both hGH and HCS(hPL) are listed in Table 1. By comparison with second-order spectra of model compounds (Table 1), all six of these absorption bands may be assigned to the Phe ^1L_b [0-0] transition and its five major, higher vibrational modes.[23,24,25] It appears that the average Phe residue in HCS(hPL) may absorb at a slightly higher wavelength than in hGH. This indicates a slightly more hydrophobic (nonpolar) environment for the average HCS(hPL) Phe residue.[23] Following digestion and titration to pH 1.5, all six Phe absorption bands, in both proteins, blue-shift to wavelength positions equivalent to the model compound (Table 1). This proves that the digestion and titration have successfully removed all conformational effects from the Phe absorption (normalization). The data in Table 1 further indicate that the Tyr and Trp absorptions are also essentially fully normalized in the titrated digests. As discussed elsewhere,[19] this point is crucial to the validity of the E_M determinations reported above.

By further reference to model compound spectra,[23,24,25] the absorption bands arising from the Tyr ^1L_b [0-0] transition and one of its higher vibrational modes may also be identified in the second-order spectra of the native hormones. The Tyr bands in hGH appear at essentially the same wavelengths as those in HCS(hPL) (Table 1), suggesting a similar nonpolarity for the average environment of these chromophores.

In this region of the spectrum, two different electrons in the indole ring

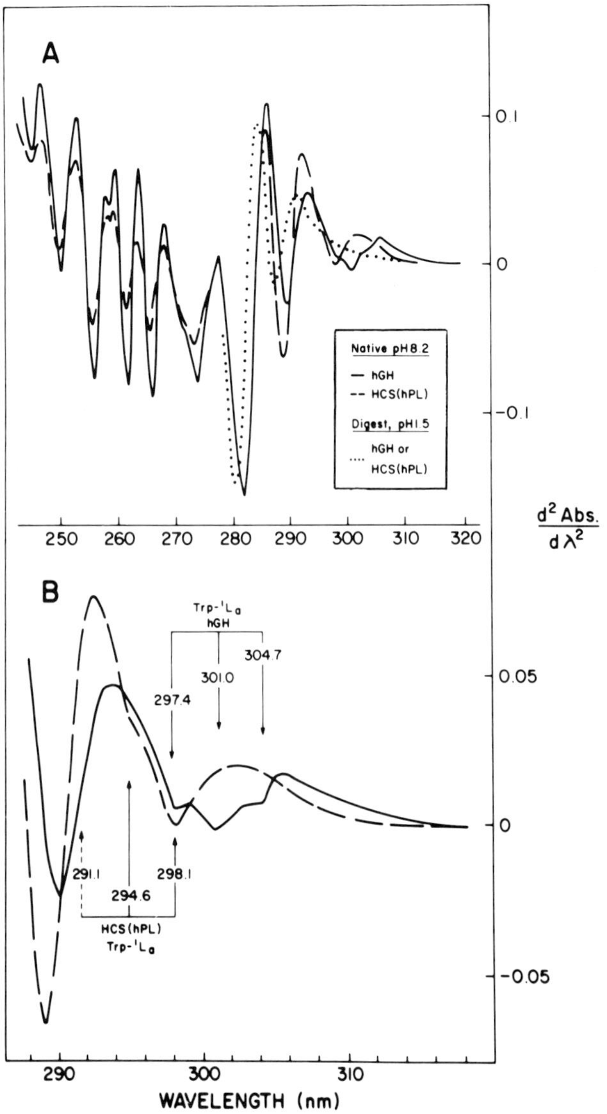

Figure 2 Second-order absorption spectra of native hGH at pH 8.2 (——), native HCS(hPL) at pH 8.2 (– – –), and the 24-hr thermolysin digest of either protein after titration to pH 1.5 (····). The pH 1.5 digests of hGH and HCS(hPL) provide essentially equivalent spectra. (A) The entire spectra of the native proteins, along with a partial spectrum of the digests. (B) An expanded view of the two native molecules in the region of Trp absorption. The wavelength positions of the Trp ^1L_a absorption bands have been calculated from the known[27] spacing of the vibronic modes after assigning the positions of the [0-0] bands. These calculated positions compare very favorably with the positions of actual second-order peaks (Table 1).

Table 1 Second-Order Spectral Bands of Native and Digested hGH and HCS(hPL)[a,b]

hGH			HCS(hPL)			Model compounds		
Native		Digest	Native		Digest	Trp[c]	Tyr[c]	Phe[c]
304.2	Trp^1L_a*	—	—	—	—	—	—	—
301.0	Trp^1L_a	—	298.1*	Trp^1L_a*	—	—	—	—
298.2 (s)	Trp^1L_a	—	~295 (ws)	Trp^1L_a	—	—	—	—
290.9	Trp^1L_b*	287.8*	290.0*	Trp^1L_b*	288.0	287.7*	—	—
283.2	Tyr^1L_b*	281.5*	283.4*	Tyr^1L_b*	281.6	281.1	281.9*	—
276.2	Tyr^1L_b	274.5	276.0	Tyr^1L_b	274.9	—	274.8	—
~273 (vws)	Trp^1L_b	~272 (ws)	~273 (vws)	Trp^1L_b	~272 (ws)	271.6	—	—
268.3	Phe^1L_b*	267.1*	268.6*	Phe^1L_b*	267.1	—	266.7	267.3*
264.5	Phe^1L_b	263.2	264.7	Phe^1L_b	263.5	—	—	263.5
261.5 (w)	Phe^1L_b	260.3	263.1 (w)	Phe^1L_b	260.8 (w)	—	—	260.0 (w)
258.6	Phe^1L_b	257.3	258.9	Phe^1L_b	257.5	—	—	257.4
252.9	Phe^1L_b	251.3	253.2	Phe^1L_b	251.6	—	—	251.7
247.8	Phe^1L_b	246.4	247.9	Phe^1L_b	246.7	—	—	246.7

[a] Corrected as described elsewhere. (19). Native at pH 8.2, digests and models at pH 1.5. w = Weak peak; s = shoulder; ws = weak shoulder; vws = very weak shoulder.
[b] The * denotes the [0–0] band of each chromophoric electron.
[c] *N*-acetyl-L-Trp-amide, *N*-acetyl-L-Tyr-amide, and *N*-acetyl-L-Phe-ethyl ester.

of Trp may simultaneously produce both ^1L_a and ^1L_b [0-0] transitions along with their associated higher vibronic modes.[(26,27)] The ^1L_b [0-0] transition is easily identified[(19,24–27)] as the relatively intense second-order peak at 290.0 nm in HCS(hPL) and 290.9 nm in hGH (Table 1). This nearly 1.0-nm difference is highly significant and readily visible in the second-order spectra (Fig. 2B). In the absence of additional information, it seems that the microenvironment of the hGH Trp residue is more hydrophobic than that of HCS(hPL). However, as explained later, this is probably not the case.

In the absorption spectra of most Trp model compounds dissolved in H_2O, the ^1L_a [0-0] transition is not visible owing to overlapping with the more powerful ^1L_b band. Addition of nonpolar solvents, or reagents that form hydrogen bonds with the indole ring > NH group, usually produces unequal red shifting of the ^1L_a and ^1L_b bands.[(26,27)] The ^1L_a bands red-shift much more strongly, pulling them out from under the more intense ^1L_b bands. The presence of absorption and CD bands above ~293 nm in the spectra of many native, unconjugated, globular proteins is due to this phenomenon.[(23,26)] Similarly, it is the lack of conformation which makes it impossible to identify the Trp ^1L_a [0-0] transition in the spectra of either the model compound or the titrated digests (Table 1).

In Fig. 1, the very broad, weak shoulders appearing above 300 nm in the zero-order spectra of native hGh and HCS(hPL) can arise only from weak ^1L_a absorption bands.[(26,27)] It is impossible to give a detailed description of this absorption from the zero-order spectra. However, the second-order spectra (Fig. 2 and Table 1) depict it very clearly. In HCS(hPL), the Trp ^1L_a [0-0] band lies at 298.1 nm with one higher vibronic mode being visible as a shoulder near 295 nm. The second higher vibronic mode, which should appear at 291.1 nm [calculated as described elsewhere[(26,27)] from the position of the [0-0] band], is still hidden by the powerful Trp ^1L_a [0-0] absorption at 290 nm. In native hGH, the Trp ^1L_a bands are so strongly red-shifted that the [0-0] transition and both higher vibronic modes are visible well above the Trp ^1L_b absorption.

The amount of red shift existing in both Trp transitions, as well as in the Tyr and Phe transitions, can be obtained by simply subtracting the wavelength positions of the various [0-0] bands of appropriate model compounds dissolved in H_2O from the wavelengths of the corresponding [0-0] bands in the native proteins. The Trp ^1L_a [0-0] position in H_2O is taken to be 292.0 nm.[(27)] These average red-shift values are listed in Table 2 for both hormones. The shifts shown for Tyr, Phe, and Trp in HCS(hPL) are typical of many folded, globular proteins[26,27] and are due to these residues being buried (on the average) within the nonpolar interiors of these molecules. The same is true of the Phe and Tyr red shifts in hGH. In contrast, the enormous, 12.2-nm red shift in the Trp ^1L_a [0-0] absorption of hGH is not typical. It cannot be explained as being due to an unusually hydrophobic environment.[(27)] Fortunately, Trp ^1L_a red shifts of this magnitude, although rare, are not unknown. As reported by Strickland *et al.*,[(27)] Trp ^1L_a red shifts of precisely this magnitude (12–13 nm) occur when the indole ring >NH group is simultaneously surrounded by a nonpolar environment *and* is hydrogen bonded to a negatively charged carboxylate ion.

Table 2 Average Red Shifting of Chromophores in Native hGH and HCS(hPL)

Protein	Chromophore transition	Red shift (nm)
hGH	Trp ^1L_a	12.2
	Trp ^1L_b	2.9
	Tyr ^1L_b	1.3
	Phe ^1L_b	1.0
HCS(hPL)	Trp ^1L_a	6.1
	Trp ^1L_b	2.0
	Tyr ^1L_b	1.5
	Phe ^1L_b	1.3

Hydrogen bonding to other groups such as protonated carboxyls, amide carbonyl, and amino groups, or imidazole side chains, even in nonpolar solvents, will produce red shifts of only 8–9 nm at most.[27] Hydrogen bonding of the indole ring >NH group to a carboxylate ion also very conveniently explains the extra red shift in the hGH Trp ^1L_b absorption, beyond that produced by the nonpolar environment as seen in HCS(hPL).[27]

Thus, a Trp-carboxylate hydrogen-bonded complex, buried in a hydrophobic environment, may be the specific structural detail that gives rise to the long-known differences in the UV absorption and CD properties of the Trp residues in hGH and HCS(hPL). Additional experimentation will be required to confirm the presence of such a complex and, if confirmed, to identify the residue that provides the carboxylate ion. The relationship between this structural detail and the growth-promoting potencies of these hormones will also be of interest.

3. Summary

Examination of the second-order absorption spectrum of native hGH strongly suggests the presence of a hydrogen-bonded complex between the single Trp side chain and a carboxylate ion. The complex is buried within a nonpolar region of the hGH molecule. Although the homologous Trp in HCS(hPL) is also buried within a nonpolar region, there is no evidence that it is hydrogen bonded. The molar extinction coefficients of hGH and HCS(hPL) have been refined.

ACKNOWLEDGMENT. This work was supported in part by a grant (AM-18677) from the National Institutes of Health.

References

1. W. R. Lyons, C. H. Li, and R. E. Johnson, Mammary-stimulating activities of human pituitary hormones, *Acta Endocrinol.* **51:**(Suppl.) 1145, 1960.

2. A. Chadwick, S. J. Folley, and C. A. Gemzell, Lactogenic activity of human pituitary growth hormone, *Lancet* **2**:241, 1961.
3. J. B. Josimovich and J. S. MacLaren, Presence in the human placenta and term serum of a highly lactogenic substance immunologically related to pituitary growth hormone, *Endocrinology* **71**:209, 1962.
4. J. R. Florini, G. Tonelli, C. B. Breuer, J. Coppola, J. Ringler, and P. H. Bell, Characterization and biological effects of purified placental protein (human), *Endocrinology* **79**:692, 1966.
5. S. L. Kaplan and M. M. Grumbach, Studies of a human and simian placental hormone with growth hormone-like and prolactin-like activities, *J. Clin. Endocrinol. Metab.* **24**:80, 1964.
6. C. H. Li, On the characterization of human chorionic somatomammotropin, *Ann. Sclavo Inst. Siena* **12**:651, 1970.
7. C. H. Li, Hormones of the adenohypophysis, *Proc. Am. Phil. Soc.* **116**:365, 1972.
8. C. H. Li, J. S. Dixon, and D. Chung, Amino acid sequence of human chorionic somatomammotropin, *Arch. Biochem. Biophys.* **155**:95, 1973.
9. T. A. Bewley, J. S. Dixon, and C. H. Li, Sequence comparison of human pituitary growth hormone, human chorionic somatomammotropin, and ovine pituitary growth and lactogenic hormones, *Int. J. Peptide Protein Res.* **4**:281, 1972.
10. H. D. Niall, M. L. Hogan, R. Sauer, I. Y. Rosenblum, and F. C. Greenwood, Sequences of pituitary and placental lactogenic and growth hormones: Evolution from a primordial peptide by gene reduplication, *Proc. Natl. Acad. Sci. USA* **68**:866, 1971.
11. T. A. Bewley, The chemistry of human chorionic somatomammotropin, in: *Hormonal Proteins and Peptides*, Vol. 4 (C. H. Li, ed.), Academic Press, New York, 1977, pp. 62–95.
12. B. Shome and H. G. Friesen, Purification and characterization of monkey placental lactogen, *Endocrinology* **89**:631, 1971.
13. T. A. Bewley, H. Kawauchi, and C. H. Li, Comparative studies of the single tryptophan residue in human chorionic somatomammotropin and human pituitary growth hormone, *Biochemistry* **11**:4179, 1972.
14. S. Aloj and H. Edelhoch, The molecular properties of human chorionic somatomammotropin, *J. Biol. Chem.* **246**:5047, 1971.
15. S. Aloj and H. Edelhoch, The molecular properties of human growth hormone, *J. Biol. Chem.* **247**:1146, 1972.
16. J. Russell, L. M. Sherwood, K. Kowalski, and A. B. Schneider, Recombinant hormones from fragments of human growth hormone and human placental lactogen, *J. Biol. Chem.* **256**:296, 1981.
17. C. H. Li, W. K. Liu, and J. S. Dixon, Human pituitary growth hormone VI. Modified procedure of isolation and NH_2-terminal amino acid sequence, *Arch. Biochem. Biophys.*, **51**:(Suppl.) 327, 1962.
18. P. Neri, P. Tarli, and F. Cocola, Effects of pH and protein concentration on the molecular weight of human chorionic somatomammotropin, *Ital. J. Biochem.* **19**:111, 1972.
19. T. A. Bewley, A novel procedure for determining protein concentrations from absorption spectra of enzyme digests, *Anal. Biochem.* **123**:55, 1982.
20. T. A. Bewley and C. H. Li, Circular dichroism studies on human chorionic somatomammotropin, *Arch. Biochem. Biophys.* **144**:589, 1971.
21. T. A. Bewley and C. H. Li, Human pituitary growth hormone XV. The optical rotatory dispersion in acidic, neutral and basic solutions, *Biochimica et Biophysica Acta* **140**:201, 1967.
22. T. A. Bewley, P. Colosi, and F. Talamantes, Conformational studies of secreted mouse pituitary prolactin, *Biochemistry* **21**:4238, 1982.
23. D. B. Wetlaufer, Ultraviolet spectra of proteins and amino acids, *Adv. Protein Chem.* **17**:303, 1962.
24. C. Balestrieri, G. Colonna, A. Giovane, G. Irace, and L. Servillo, Second-derivative spectroscopy of proteins. A method for the quantitative determination of aromatic amino acids in proteins, *Eur. J. Biochem.* **90**:433, 1978.
25. C. Balestrieri, G. Colonna, A. Giovane, G. Irace, and L. Sevillo, Second-derivative spectroscopy of proteins: Studies on tyrosyl residues, *Anal. Biochem.* **106**:49, 1980.
26. E. H. Strickland, Aromatic contributions to circular dichroism spectra of proteins, *C.R.C. Crit. Rev. Biochem.* **2**:113, 1974.
27. E. H. Strickland, C. Billups, and E. Kay, Effects of hydrogen bonding and solvents upon the tryptophanyl ^1L_a absorption band. Studies using 2,3-dimethylindole, *Biochemistry* **11**:3657, 1972.

32

Multiple Forms of Human Growth Hormone

Urban J. Lewis, Luciano G. Frigeri, Morton B. Sigel, Gene F. Tutwiler, and Willard P. Vanderlaan

1. Introduction

Production of multiple forms of human growth hormone (hGH) by the pituitary gland may provide a mechanism for controlling posttranslational proteolytic modification of the hormone. Double-base sequences (adjacent arginine and/or lysine residues), the ones most often encountered in proteolytic processing of precursors, are few in the growth hormones. For the protein to serve as a prohormone, then, there must be a process whereby the peptide chain can be cleaved to produce specific, metabolically active fragments. Our hypothesis is that each of the multiple forms (different gene products, products of alternative mRNA processing, and posttranslational modifications) has a characteristic conformation that exposes different yet specific regions to proteolytic attack. Spatial structures, then, could direct cleavage at specific bonds. Although positive evidence for the hypothesis is lacking, there are a growing number of observations that support the idea. This paper will review those data, interpreting them as supportive of the hypothesis. Figure 1 provides a summary of the relationships between the known multiple forms of hGH.

2. Gene Products

The entire DNA sequence, including the introns, of the gene that is responsible for the major 22,000-dalton form of hGH has been reported.[1,2]

Urban J. Lewis, Luciano G. Frigeri, Morton B. Sigel, and Willard P. Vanderlaan • The Whittier Institute for Diabetes and Endocrinology, La Jolla, California 92037. *Gene F. Tutwiler* • McNeil Pharmaceutical, Spring House, Pennsylvania 08540.

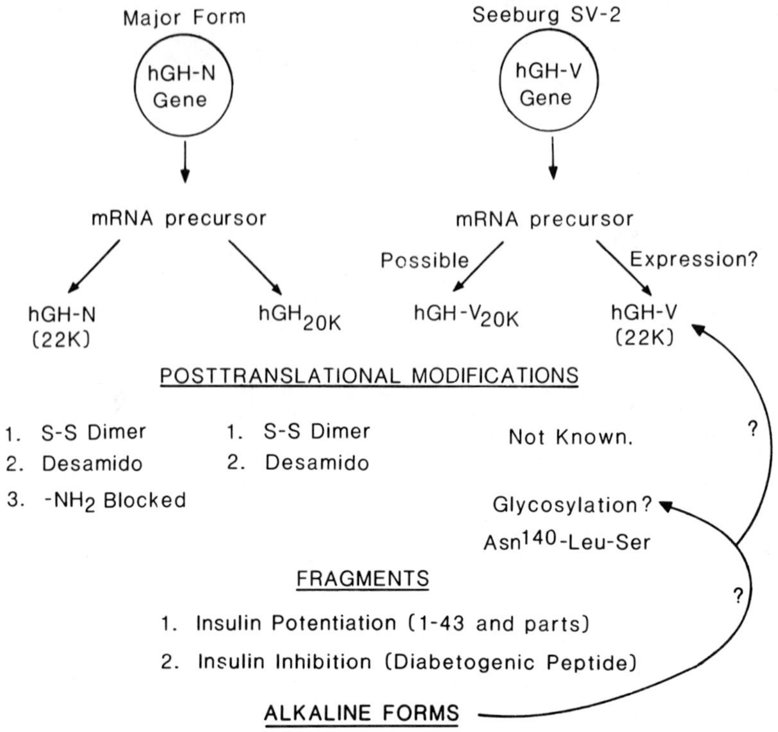

Figure 1 Interrelationships of the multiple forms of human growth hormone. The mRNA transcript of the gene for the major form can be processed so that either a 22K-dalton or a 20K-dalton form is translated. Expression of the hGH-V gene in the pituitary gland has not been demonstrated, but the gene has all the structural requirements necessary for expression. The mRNA precursor could be processed to form a 20K-dalton variant. Posttranslational modifications of hGH-V are as yet unknown. Placement of the alkaline forms in a specific category cannot be made. They may be related to the hGH-V gene products.

The pituitary gland produces a remarkable amount of the hormone. As much as 5–10 mg/gland can be isolated.[3] The wet weight of a single gland is about 400 mg, which means that nearly 10% of the dry weight is growth hormone. On the other hand, only about 10 μg of prolactin per gland was estimated by Chapman *et al.*[3] This disproportionate amount of hGH is not reflected in serum values, which are 2–5 ng/ml for hGH[4] and 10–20 ng/ml for prolactin.[5] Why such an enormous amount of growth hormone is produced but not secreted is still an unanswered question. The prohormone hypothesis for hGH accommodates these facts by assigning a substrate role to the hormone and its posttranslational modifications for production of fragments none of which is detected by current radioimmunoassays.

In addition to the gene for the major form of hGH, Seeburg[6] has reported a second gene for the hormone. Although there is no direct evidence that the gene is expressed, Seeburg discusses the likelihood that this does occur. From a biochemical point of view, there is no reason why the gene could not be expressed, and Seeburg points out that the pituitary gland may not be the tissue where this takes place. The gene has been inserted into a simian virus and expressed in monkey kidney.[7] The gene product is designated hGH-V. The DNA sequence of the gene indicates 13 amino acid substitutions distributed throughout the hGH sequence, which imparts a more alkaline isoelectric point with little change in molecular weight. Nine of the substitutions are not conservative and might be expected to change significantly the properties of the hormone. Nothing is known of its biological properties although immunoreactivity and receptor binding activity have been studied.[8] Binding to hGH receptors was only moderately affected but reactivity toward hGH antiserum was markedly decreased. There are three noteworthy amino acid substitutions in the sequence of hGH-V. The first is introduction of an arginine at position 18, which generates an Arg–Arg sequence, a common structure for proteolytic processing. The second is the substitution of a leucine for the proline at 37, which eliminates a tripeptide sequence common to GH of several species and may be important in insulin potentiation activity of GH (see Section 6.1). The third interesting substitution in hGH-V is the introduction of an Asn–Gln–Ser sequence when asparagine replaces lysine at position 140. This sequence is a prerequisite for glycosylation[9] and indicates that a glycosylated form may be found, as has recently been done for prolactin[10] (see Section 4.4). The significance of these amino acid substitutions is that they not only change overall conformation but introduce specific sequences that could be involved in directing proteolytic processing. Geisow[11] has written an excellent article on this topic.

3. mRNA Processing

Wallis[12] was the first to propose a mechanism for formation of the 20,000-dalton variant of hGH (hGH_{20K}). DeNoto *et al.*[13] later provided experimental information in support of the idea. Because no separate gene had been found for the variant, the deletion of the 15 amino acids (residues 32–46) must take place during synthesis. At present all indications are that the coding sequences for amino acids 32–46 are treated as part of intron B of the hGH gene and are thus eliminated during the splicing process of the precursor mRNA. The resulting mRNA transcript is translated into the smaller hGH_{20K} protein. Not only conformational changes[14] but also alteration of immuno-[15] and receptor-binding activities[16] results from the deletion.

Seeburg[6] points out that the hGH-V gene could also produce a 20,000-dalton form because it has all the necessary coding sequences. This, together with the known formation of hGH_{20K}, suggests that alternative mRNA processing is a control mechanism for production of a particular form having special biological properties. The control processes involved in directing synthesis of one or the other of the possible mRNA transcripts are completely unknown.

That the hGH_{20K} variant serves a special function is suggested by its altered biological activities.[17] The ability to rapidly lower blood glucose and free fatty acids is greatly diminished, as is the ability to stimulate a prolonged rise in the serum free fatty acids. As discussed later, these properties are dependent on the 32-46 sequence deleted from hGH_{20K}. The question we are seeking to answer is does the deletion produce a form of hGH that by itself participates in a different set of cellular actions or does the alteration in structure permit proteolytic production of biologically active peptides that function in energy metabolism?

4. Posttranslational Modifications

At this time the significance of the posttranslational modifications of hGH is as little understood as is the physiological role of the different gene products and alternate mRNA-splicing products. We will attempt to give an explanation based on the prohormone hypothesis.

4.1. Disulfide Dimer

An interchain disulfide of hGH has been isolated from the pituitary gland.[18] The location of the disulfide bridge was not identified, but because no unusual tryptic peptides were observed by peptide mapping, an antiparallel linkage was suggested. Schneider *et al.*[19] showed that the disulfide dimer of human placental lactogen resulted from a nonparallel orientation of the two monomers held by a linkage at the COOH terminal disulfide. The S-S dimer of hGH probably has the same type of bridge.

Formation of the disulfide bond between two molecules of hGH essentially eliminates growth-promoting activity but has little effect on the lactogenic activity as measured by the pigeon crop sac assay.[18] The importance of the lactogenic activity of hGH is unknown, but if it has biological significance, the S-S dimer would be a form where growth activity is eliminated when only lactogenic activity is to be expressed. Another possibility is that dimer formation changes the conformation of the monomer so that a different surface of the molecule is exposed to proteolytic attack.

4.2. Desamido Forms

The ease with which hGH deamidates complicates purification of its variants. For example, the alteration in charge that occurs during deamidation of hGH_{20K} can nullify ion exchange separation. The first desamido form of hGH_{20K} has almost the same charge as undeamidated hGH making separation extremely difficult. Because deamidation occurs nonenzymatically during isolation, the forms have been considered to be artifacts. However, when human pituitary tissue was examined by electrophoresis within an hour after surgical removal,[20] a component with the mobility of desamido hGH was observed.

Conclusive identification of the deamidated forms, however, needs a new procedural approach.

If the desamido forms of hGH are shown to exist in the pituitary gland or the circulation, one possible function of deamidation may be to direct proteolytic cleavage. Evidence for this came from studies that compared the cleavage products produced by limited hydrolysis of hGH and desamido hGH with subtilisin.(21) The results indicated that when asparagine^{152} is deamidated to produce the first desamido form, subtilisin no longer cleaved the chain at position 159. Deamidation did then influence how the hGH was cleaved. A point not widely discussed is that asparagine may deamidate through formation and subsequent opening of a succinimide ring.(22) When the ring is opened, either an α- or β-aspartyl form is produced. The β form has an entirely different structure from the α form (aspartyl form), and this alteration could alter attack by a proteinase.

4.3. *Acylation of the NH_2 Terminus*

Electrophoresis of partially purified preparations of hGH at pH 7.8 indicated that some samples contained a slightly more acidic form of the hormone.(23) It was more acidic than the major form but not as acidic as the first deamidated modification. Sequence analysis indicated a blocked amino terminus, but identification of the blocking group was unsuccessful. Additional quantities of this material must be isolated to continue these studies.

Acylation of the NH_2 terminus is a likely explanation, and if this is found to be the case, one would expect that acetylation, for example, of the NH_2 terminus would make that portion of the molecule even more hydrophobic. This could alter folding properties and susceptibility to proteolysis in a region that we now know is involved in the insulin-potentiating properties of hGH (see Section 6.1).

4.4. *Glycosylation*

A glycosylated form of either GH or prolactin had not been identified until such a modification of ovine prolactin was reported.(10) The carbohydrate unit is linked to asparagine.(31) The prerequisite structure for such a glycosylation is Asn–X–Ser(Thr),(9) and although it is present in ovine human and porcine prolactin, hGH does not have this sequence. However, the hGH-V gene codes for Asn^{140}–Leu–Ser, which suggests that a glycosylated form of this variant may be produced. The function of the carbohydrate in glycoprotein hormones is unknown, but Loh and Grainer(24) reported that glycosylation of proopiomelanocortin is necessary to prevent rapid proteolytic degradation. In addition Eggo and Burrow(25) suggest that glycosylation is necessary for the correct processing of thyroglobulin. Glycosylation of growth hormone and prolactin may serve similar functions. Beeley(26) has pointed out that glycosylation frequently occurs at β turns where the surface of the protein is exposed. He suggests that attachment of carbohydrate may protect such regions from proteolytic attack.

5. Alkaline Forms

Two alkaline forms of hGH (as judged by electrophoretic mobility) can be detected in various pituitary fractions if the electrophoresis is carried out in urea.[23] Insufficient quantities of these forms have prevented complete structure analysis; so we do not know how they differ from hGH. One intriguing possibility is that they are related to hGH-V. Seeburg[6] predicted a more alkaline isoelectric point for hGH-V, which is true for the unidentified forms. The slower mobility of the more alkaline of the two forms may be due not to charge but rather to a carbohydrate side chain which retards its passage through the polyacrylamide.

6. Biologically Active Fragments

6.1. Insulin-Potentiating Peptides

The altered biological activities of hGH_{20K} suggested that the 32-46 region of the molecule is involved in carbohydrate metabolism (Section 3). Frigeri *et al.*,[27] Dryburgh *et al.*,[28] and Rudman *et al.*[29] have demonstrated this. Synthetic hGH (32–46) and rat GH (32–45) enhanced glucose incorporation into mouse liver and muscle *in vivo*. In addition, the peptides improved insulin-stimulated clearance of blood glucose. This activity was demonstrated in rodents, dogs, and monkeys. There is no indication that the peptides have insulinlike activity themselves, but they do potentiate the action of insulin. Having noted this property for the 32–46 region, we found it most interesting that we were able to isolate hGH (1–43) from pituitary extracts. As yet we have not found hGH (32–46) in the pituitary gland. Noteworthy is the fact that the 1–43 peptide also contains the 1–15 sequence reported by Bornstein *et al.*[30] to be insulin potentiating. The relationship of the 32–46 and 1–15 regions has recently been discussed.[31]

Frigeri *et al.*[27] observed an effect of hGH (1–43) on adipose tissue, which was not seen with hGH (32–46). Sigel *et al.* (unpublished observations), in work carried out in our laboratories, found that hGH (1–43) enhanced the binding of insulin to receptors prepared from liver of a female rat. Binding was improved by 30% during a 4-hr incubation. The 32–46 sequence and hGH caused about a 10% increase in binding. Frigeri *et al.* (unpublished observations) noted a similar difference in reactivity of hGH (1–43) and hGH (32–46) in glucose incorporation into liver. The larger peptide caused a 101% enhancement; the shorter one produced only a 23% improvement.

An important point to make is that the insulin-potentiating action of the hGH peptides represents the first time a fragment has been shown to have equal or greater activity than the intact hormone. The biological activities of various portions of the hGH molecule have been examined,[(32)] but on a molar basis they have never approached the potency of hGH.

Comparison of the amino acid sequence of the 32–46 region of all the growth hormones whose structures are known[(33)] shows that the Tyr^{35}–Ile–Pro sequence is invariable. Two recent papers make this even more noteworthy. In the first report, by Chothia *et al.*,[(34)] crystallographic analysis of the insulin molecule showed that two of the areas of the A and B chains that are brought together by folding involve Tyr(A19), Ile(A2), and Pro(B28). This, of course, is the hGH sequence in the insulin-potentiating region (Tyr^{35}-Ile–Pro) of hGH. The full significance of this is still to be elucidated. In the second report, by Baldwin *et al.*,[(35)] phosphorylation of hGH by a tyrosine kinase was studied. $Tyrosine^{35}$ was a major phosphorylation site. Whether or not this observation is related to the tyrosine kinase activity of the insulin receptor must be determined.

6.2. *Insulin Antagonist*

The diabetogenic activity of the pituitary gland is attributed mainly to GH. Various portions of the molecule have been reported to exhibit this activity, but there is no agreement on the active region. This has been discussed in a recent review.[(31)] Our work supports the view that intact GH does indeed produce glucose intolerance but that there is a low-molecular-weight peptide, still unidentified, which is 50–100 times more active on a molar basis. Because limited proteolysis of hGH with subtilisin increases the diabetogenic activity of hGH,[(36)] we think the low-molecular-weight peptide we are trying to purify from pituitary extracts is probably a fragment of hGH. Although we do not know the structure of the peptide, we do know some of its biological properties. A single dose of our most highly purified diabetogenic material produced glucose intolerance in a dog at a dosage of 5 μg/kg body weight.[(37)] The sample was known to be free of intact hGH. From 1 to 2 mg/kg was required to produce the same intolerance with hGH.

Using a less purified preparation of the pituitary diabetogenic peptide, Frigeri *et al.*[(38)] noted that the fraction was at least 100 times more potent than hGH in causing glucose intolerance in obese mice. The material also inhibited insulin-stimulated glucose oxidation by adipose tissue *in vitro*. As yet we do not know whether the glucose intolerance or inhibition of glucose utilization can be reversed with the insulin-potentiating peptides.

Bornstein *et al.*[(30)] found insulin-inhibiting activity in the carboxyl terminal portion of hGH. We have tested 177 to 191 (disulfide bridge intact) in the dog *in vivo* to see whether it produced glucose intolerance and on adipose tissue *in vitro* for its ability to inhibit insulin action (CO_2 production). Activity was not found in either test, which leads us to believe that the low-molecular-weight peptide we are trying to isolate from pituitary extracts is different from the Bornstein peptide.

7. Summary

The hypothesis that GH serves as a prohormone is at this time no more than a framework for designing experiments. No matter how many fragments are isolated from pituitary extracts, the idea will be considered an important control mechanism only when each fragment is shown to have a special function and is found to circulate. Sufficient data are accumulating to encourage additional work to examine these possibilities.

References

1. H. M. Goodman, F. DeNoto, J. C. Fiddes, R. A. Hallewell, G. S. Page, S. Smith, and E. Tischer, Structure and evolution of growth hormone related genes, in: *Miami Winter Symposium*, Vol. 17 (W. A. Scott, R. Werner, D. R. Joseph, and J. Schultz, eds.), Academic Press, New York, 1980, pp. 155–178.
2. D. D. Moore, M. D. Walder, D. J. Diamond, M. A. Conkling, and H. M. Goodman, Structure, expression and evolution of growth hormone genes, *Rec. Prog. Hormone Res.* **38:**197, 1982.
3. G. E. Chapman, A. G. C. Renwick, and J. N. Livesey, The isolation of human pituitary hormones from frozen glands, *J. Clin. Endocrinol. Metab.* **53:**1008, 1981.
4. D. C. Parker, Arginine infusion and growth hormone secretion, *Lancet* **2:**1036, 1968.
5. Y. Ehara, T. Siler, G. Van den Berg, Y. N. Sinha, and S. S. C. Yen, Circulating prolactin levels during the menstrual cycle: Episodic release and diurnal variation, *Am. J. Obstet. Gynecol.* **117:**962, 1973.
6. P. H. Seeburg, The human growth hormone gene family: Nucleotide sequences show recent divergence and predict a new polypeptide hormone, *DNA* **1:**239, 1982.
7. G. N. Pavlakis, N. Hizuka, P. Gorden, P. Seeburg, and D. H. Hamer, Expression of two human growth hormone genes in monkey cells by cloning in simian virus 40, *Proc. Natl. Acad. Sci. USA* **78:**7398, 1981.
8. N. Hizuka, C. M. Hendricks, G. N. Pavlakis, D. H. Hamer, and P. Gorden, Properties of human growth hormone polypeptides: Purified from pituitary extracts and synthesized in monkey kidney cells and bacteria, *J. Clin. Endocrinol. Metab.* **55:**545, 1982.
9. O. P. Bahl and R. H. Shah, Glycoenzymes and glycohormones, in: *The Glycoconjugates*, Vol. 1. *Mammalian glycoproteins and glycolipids* (H. I. Horowitz and W. Pigman, eds.), Academic Press, New York, 1977, pp. 385–422.
10. U. J. Lewis, R. N. P. Singh, L. J. Lewis, B. K. Seavey, and Y. N. Sinha, Glycosylated ovine prolactin, *Proc. Natl. Acad. Sci. USA* **811:**385–389, 1984.
11. M. J. Geisow, Polypeptide secondary structure may direct the specificity of prohormone conversion, *FEBS Lett.* **87:**111, 1978.
12. M. Wallis, Growth hormone: Deletions in the protein and introns in the gene, *Nature* **284:**512, 1980.
13. F. M. DeNoto, D. D. Moore, and H. M. Goodman, Human growth hormone DNA sequence and mRNA structure: Possible alternative splicing, *Nucleic Acid Res.* **9:**3719, 1981.
14. G. E. Chapman, K. M. Rogers, T. Brittain, R. A. Bradshaw, O. J. Bates, C. Turner, P. D. Cary, and C. Crane-Robinson, The 20,000 molecular weight variant of human growth hormone. Preparation and some physical and chemical properties, *J. Biol. Chem.* **256:**2395, 1981.
15. M. B. Sigel, W. P. VanderLaan, E. F. Vander Laan, and U. J. Lewis, Measurement of multiple forms of hGH: Cross-reactivities in conventional and 2-chain radioimmunoassays, *Endocrinology* **106:**92, 1980.
16. M. B. Sigel, N. A. Thorpe, M. S. Kobrin, U. J. Lewis, and W. P. VanderLaan, Binding characteristics of a biologically active variant of human growth hormone (20K) to growth hormone and lactogen receptors, *Endocrinology* **108:**1600, 1981.

17. L. G. Frigeri, S. M. Peterson, and U. J. Lewis, The 20,000 dalton structural variant of human growth hormone: Lack of some early insulinlike effects, *Biochem. Biophys. Res. Commun.* **91:**778, 1979.
18. U. J. Lewis, S. M. Peterson, L. F. Bonewald, B. K. Seavey, and W. P. VanderLaan, An interchain disulfide dimer of human growth hormone, *J. Biol. Chem.* **252:**3697, 1977.
19. A. B. Schneider, K. Kowalski, J. Russell, and L. M. Sherwood, Identification of the interchain disulfide bonds of dimeric human placental lactogen, *J. Biol. Chem.* **254:**3782, 1979.
20. F. Talamantes, J. Lopez, U. J. Lewis, and C. B. Wilson, Multiple forms of growth hormone: Detection in medium from cultured pituitary adenoma explants, *Acta Endocrinol.* **98:**8, 1981.
21. U. J. Lewis, R. N. P. Singh, L. F. Bonewald, and B. K. Seavey, Altered proteolytic cleavage of human growth hormone as a result of deamidation, *J. Biol. Chem.* **256:**11645, 1981.
22. L. Graf, S. Bajusz, A. Patthy, E. Barat, and G. Cseh, Revised amide location for porcine and human adrenocorticotropic hormone, *Acta Biochim. Biophys. Acad. Sci. Hung.* **6:**415, 1971.
23. U. J. Lewis, R. N. P. Singh, L. F. Bonewald, L. J. Lewis, and W. P. VanderLaan, Human growth hormone: Additional members of the complex, *Endocrinology* **104:**1256, 1979.
24. Y. P. Loh and H. Grainer, The role of the carbohydrate in the stabilization processing, and packaging of the glycosylated adrenocorticotropin-endorphin common precursor in toad pituitaries, *Endocrinology* **105:**474, 1979.
25. M. C. Eggo and G. N. Burrow, Glycosylation of thyroglobulin. Its role in secretion, iodination and stability, *Endocrinology* **113:**1655, 1983.
26. J. G. Beeley, Peptide chain conformation and the glycosylation of glycoproteins, *Biochem. Biophys. Res. Commun.* **76:**1051, 1977.
27. L. G. Frigeri, N. Ling, C. Rudman, W. P. VanderLaan, and U. J. Lewis, Increased *in vivo* glucose utilization by amino terminal peptides of growth hormone, *1983 Research Symposium on Diabetes and Exercise*, July 27–29, Burlington, Vermont, 1983 (abstr.).
28. J. R. Dryburgh, C. G. Rudman, and N. Stebbing, The effect of a fragment of human growth hormone on induced hyperglycemia in normal and glucose intolerant animals, *Endocrinology* **112**(Suppl.):379, 1983 (abstr.).
29. C. G. Rudman, N. Stebbing, L. G. Frigeri, and U. J. Lewis, Some effects of synthetic fragments of human growth hormone on glucose metabolism, *19th Annual Meeting of the European Association for Study of Diabetes*, Oslo, Norway, September 14–17, 1983.
30. J. Bornstein, J. McD. Armstrong, H. P. Taft, F. M. Ng, and M. K. Gould, The mechanism of the diabetogenic effects of pituitary growth hormone, *Postgrad. Med. J.* **49**(Suppl.):219, 1973.
31. U. J. Lewis, Variants of growth hormone and prolactin and their posttranslational modifications, *Ann. Rev. Physiol.* **46:**33–42, 1984.
32. C. R. Reagan, J. L. Kostyo, J. B. Mills, M. H. Moseley, and A. E. Wilhelmi, Isolation and characterization of fragments of reduced and S-carbamidomethylated human growth hormone produced by plasmin digestion. II. Biological and immunological activities, *Endocrinology* **102:**1377, 1978.
33. R. N. P. Singh, B. K. Seavey, L. J. Lewis, and U. J. Lewis, Human growth hormone peptide 1–43: Isolation from pituitary glands, *J. Protein Chem.* 1984 **2:**425–436, 1983.
34. C. Chothia, A. M. Lesk, G. G. Dodson, and D. C. Hodgkin, Transmission of conformational change in insulin, *Nature* **302:**500, 1983.
35. G. S. Baldwin, B. Grego, M. T. W. Hearn, J. A. Knesel, F. J. Morgan, and R. J. Simpson, Phosphorylation of human growth hormone by epidermal growth factor-stimulated tyrosine kinase, *Proc. Natl. Acad. Sci. USA* **80:**5276, 1983.
36. U. J. Lewis, R. N. P. Singh, W. P. VanderLaan, and G. F. Tutwiler, Enhancement of the hyperglycemic activity of human growth hormone by enzymic modification, *Endocrinology* **101:**1587, 1977.
37. R. N. P. Singh, L. J. Lewis, R. O'Brien, U. J. Lewis, and G. F. Tutwiler, Characterization of the pituitary hyperglycemic factor as a low molecular weight peptide, *Endocrinology* **110**(Suppl.):102, 1982 (abstr.)
38. L. G. Frigeri, G. L. Wolff, and G. Robel, Impairment of glucose tolerance in yellow (A^{vy}/A)(BALB/c x VY)F-1 hybrid mice by hyperglycemic peptide(s) from human pituitary glands, *Endocrinology*, **113:**2097–2105, 1983.

33

The Multivalent Nature of Growth Hormone

Jack L. Kostyo

It has been known for many years that pituitary growth hormone (GH) preparations isolated from a variety of mammalian species exhibit multiple biological properties that are difficult to reconcile with one common mechanism of action of the hormone. Most GH preparations are not only anabolic but can be diabetogenic, producing hyperglycemia and glucose intolerance in certain species such as the dog and man and causing hyperinsulinemia and insulin resistance in many species. Moreover, in GH-deficient subjects, these GH preparations produce transient insulinlike effects, including the induction of hypoglycemia, resulting from the stimulation of glucose uptake and utilization by peripheral tissues and the inhibition of lipolysis. Finally, primate GH preparations are also lactogenic, whereas those of infraprimates are not.

A question that has been raised repeatedly over the years[1,2] is whether these diverse biological activities are intrinsic properties of the GH molecule itself or whether they are due to the presence of contaminants or proteolytic degradation products in GH preparations. It is certainly well known that even purified preparations of GH are quite heterogeneous,[2] containing charge isomers, enzymatically cleaved forms of the hormone, and, in the case of human GH (hGH) preparations, a genetic variant (20K hGH). In the case of the lactogenic activity of the primate GHs, there is general agreement that this is an intrinsic property of the hormones themselves, due to their structural similarity to the prolactins. Some question has remained, however, about the diabetogenic and insulinlike properties of the hormone. For example, Lewis and his colleagues[2] have suggested that the diabetogenic activity of hGH preparations is due primarily to contamination with low-molecular-weight (5000–6000) pituitary peptides, and that the hGH molecule itself has little

Jack L. Kostyo • Department of Physiology, University of Michigan Medical School, Ann Arbor, Michigan 48109.

intrinsic diabetogenic activity. Whether these low-molecular-weight peptides are proteolytic degradation products of the hGH molecule has not been disclosed. On the other hand, substantial evidence has been obtained in our laboratory, among others, indicating that diabetogenic and insulinlike activities are intrinsic properties of the GH molecule and not a reflection of contamination of the growth-promoting principal with other pituitary substances or proteolytic degradation products. This chapter will be a summary of this evidence, as well as a brief review of structure–function data suggesting that there are multiple active sites in the GH molecule accounting for its multivalent character.

Submitting crude native hGH to ion exchange chromatography[3] and gel filtration results in the separation of the native hormone (A-type hGH) from aggregates, from more acidic forms of the hormone, and from most of the enzymatically cleaved forms of the hormone (e.g., C-type hGH[4]). We have found that native A-type hGH isolated in this manner is not only growth-promoting (ca. 2 IU/mg) but possesses diabetogenic activity,[4–6] as assessed by its ability to increase fasting hyperglycemia and worsen glucose tolerance in the obese (ob/ob) mouse when administered daily for three days.[7] In this assay, threshold diabetogenic doses of the hormone range from 10 to 50 μg/day. We have also found A-type hGH to have insulinlike activity,[4,5,8] when tested for the *in vitro* ability to stimulate [^{14}C]glucose oxidation to $^{14}CO_2$ by isolated adipose tissue of hypophysectomized rats. In this assay, threshold responses are obtained with concentrations of the hormone ranging from 1 to 5 nM.

In an attempt to remove any low-molecular-weight peptides that might be adsorbed noncovalently to A-type hGH, the hormone was submitted to gel filtration on a column of Sephacryl S-200 in the presence of 6 M guanidine-HCl.[6] The elution pattern obtained suggested that the A-type hGH applied to the column was essentially free of low-molecular-weight peptides adsorbed noncovalently to the hormone. The "guanidine-treated hGH" recovered from the column was found to retain substantial growth-promoting activity in the 9-day weight gain test in hypophysectomized rats (1.5 IU/mg). Moreover, it was found to possess diabetogenic activity in the ob/ob mouse equivalent to that of the A-type hGH from which it was derived and to retain full *in vitro* insulinlike activity on isolated rat adipose tissue.[6] Thus, it seems unlikely that the diabetogenic and insulinlike activities of native A-type hGH can be attributed to contamination of the hormone with low-molecular-weight peptides or proteolytic degradation products. Of course, it is conceivable that proteolytic processing of the GH molecule within the organism may be required for the expression of certain activities of the hormone, as has been suggested for both the diabetogenic[9] and insulinlike[10] activities.

Direct evidence supporting the view that diabetogenic and insulinlike activities are intrinsic properties of the GH molecule itself was obtained in experiments with biosynthetic methionyl-hGH. Biosynthetic methionyl-hGH, which had been produced by recombinant DNA methodology,[11] was obtained from KabiVitrum AB (Stockholm, Sweden). This material is reported to have growth-promoting activity in hypophysectomized rats equivalent to that of purified native hGH.[11] It produced a marked increase in fasting blood glucose concentration and an impairment in glucose intolerance when administered at doses of 50 and 100 μg/day for 3 days to ob/ob mice, responses comparable to those produced with native A-type hGH.[6] Moreover, the biosynthetic hGH had *in vitro* insulinlike activity equivalent to that of native A-type hGH.[6]

Finally, the biological characterization of GH peptide complexes produced by noncovalent complementation of large fragments of the hormone provided additional strong support for the conclusion that diabetogenic and insulinlike activities are intrinsic properties of GH. A complex produced by noncovalent complementation of hGH peptide 1–134 with hGH peptide 135–191 had considerably more diabetogenic activity in the ob/ob mouse than the constituent peptides themselves.[4] This was also true for the *in vitro* insulinlike activity of the complex. Also, a hybrid complex produced by noncovalent complementation of porcine GH peptide 1–133 with hGH peptide 135–191 had substantial diabetogenic activity,[12] whereas the two peptides from which the complex was produced had no activity themselves at the doses tested. Thus, it seems highly unlikely that the diabetogenic and insulinlike activities of these complexes, which are generated when weakly active or inert GH fragments are combined, can be explained on the basis of contamination with other substances.

Preliminary experiments conducted recently in our laboratory on highly purified GH's of inframammalian origin, which were prepared and provided by Dr. Harold Papkoff and his colleagues,[13] suggest that diabetogenic activity is also a property of the GH's of lower vertebrates. The GH's of the ostrich, sea turtle, snapping turtle, bullfrog, and a teleost, *Tilapia*, were all found to have the ability to increase fasting blood glucose concentration and worsen glucose tolerance in the ob/ob mouse. Considering that these hormones possess significant growth-promoting activity in the rat, perhaps with the exception of *Tilapia* GH, it would appear that the structural determinants responsible for growth-promoting activity and diabetogenic activity have not changed appreciably during the course of lower vertebrate evolution, despite substantial evolution of the overall structure of the GH molecule.[13] Whether these nonmammalian GH's possess insulinlike activity has not yet been determined.

From the foregoing, it seems reasonable to conclude that the diabetogenic and insulinlike activities of GH are intrinsic properties of the hormone molecule, and that in this sense it is multivalent. Given that, one may ask whether the growth-promoting, diabetogenic, and insulinlike activities are produced by the same active site in the molecule, or whether the hormone possesses multiple active sites, one for each of its respective activities. Structure–function studies of the hGH and pGH molecules conducted by our group over the past decade have provided indirect evidence supporting the latter concept. Studies of various S-alkylated derivatives of plasmin-[8] and thrombin-digested[5] hGH and, more recently, of noncovalent recombinants of fragments of hGH[4] and pGH[12] have indicated that structural modification of the molecule can produce marked changes in the activity profile of the hormone. For example, the recombinant molecule formed by the noncovalent complementation of hGH peptides 1–134 and 135–191 had 35% the growth-promoting activity, 100% the diabetogenic activity, and 20% the insulinlike activity of native hGH.[4] Quite recently, we have found[14] that S-carboxymethylated hGH lacks significant growth-promoting and insulinlike activities, and yet it retains about 50% the diabetogenic activity of native hGH. It has been our general experience with hGH that most modifications in the molecule result in a marked diminution in insulinlike activity, whereas diabetogenic activity remains high.

Such nonparallel shifts in the profile of biological activities may indicate that there are separate active sites and hence separate receptors for each of the activities of GH, and that structural modifications made in the molecule can interfere in a differential manner with the various active site–receptor interactions. In agreement with this view is recent work[15] on the 20K structural variant of hGH showing that, whereas it has full growth-promoting activity, its insulinlike activity appears to be significantly attenuated. There is no precise knowledge of the location of these active sites in the GH molecule, other than that they must reside within the *N*-terminal two-thirds of the molecule. Extensive biological characterization (see Mills *et al.*[5] for review) of hGH peptides 1–134 and 135–191 has indicated that peptide 1–134 has significant but markedly attenuated growth-promoting, diabetogenic, and insulinlike activities, whereas peptide 135–191 by itself is biologically inert. Recent studies[16] with a small synthetic fragment (residues 31–44) of the hGH molecule suggest that the active site for the insulinlike activity may reside in the region that is deleted (residues 32–46) in the 20K variant of hGH, perhaps accounting for the low insulinlike activity of the latter. Clearly, further work will be necessary to determine the domains responsible for the various biological activities of the GH molecule and to establish whether or not separate receptors exist for these active sites. The failure to date to obtain evidence for the existence of individual receptors for GH's anabolic, diabetogenic, and insulinlike activities (i.e., heterogeneity of GH binding sites) may be a reflection of the possibility that receptors for only one of these activities of the hormone may predominate on a given target cell.

References

1. E. B. Astwood, Relation of the metabolic effects of corticotropin concentrates to growth hormone, in: *The Hypophyseal Growth Hormone, Nature and Actions* (R. W. Smith, Jr., O. H. Gaebler, and C. N. H. Long, eds.), McGraw-Hill, New York, 1955, pp. 286–292.
2. U. J. Lewis, R. N. P. Singh, G. F. Tutwiler, M. B. Sigel, E. F. VanderLaan, and W. P. VanderLaan, Human growth hormone: A complex of proteins, *Recent Prog. Horm. Res.* **36:**477, 1980.
3. J. B. Mills, R. B. Ashworth, A. E. Wilhelmi, and A. S. Hartree, Improved method for the extraction and purification of human growth hormone, *J. Clin. Endocrinol. Metab.* **29:**1456, 1969.
4. C. R. Reagan, J. L. Kostyo, J. B. Mills, S. E. Gennick, J. L. Messina, S. A. Wagner, and A. E. Wilhelmi, Recombination of fragments of human growth hormone: Altered activity of the recombinant molecule, *Endocrinology* **109:**1663, 1981.
5. J. B. Mills, J. L. Kostyo, C. R. Reagan, S. A. Wagner, M. H. Moseley, and A. E. Wilhelmi, Fragments of human growth hormone produced by digestion with thrombin: chemistry and biological properties, *Endocrinology* **107:**391, 1980.

6. S. E. Gennick, J. L. Kostyo, and S. E. Sauder, Diabetogenic effects of purified native and biosynthetic human growth hormones in the ob/ob mouse, *Fed. Proc.* **42:**972, 1983.
7. C. R. Reagan, The diabetogenic activity of fragments of human growth hormone in obese (ob/ob) mice, *Diabetes* **27:**883, 1978.
8. C. R. Reagan, J. L. Kostyo, J. B. Mills, M. H. Moseley, and A. E. Wilhelmi, Isolation and characterization of fragments of reduced and S-carbamidomethylated human growth hormone produced by plasmin digestion. II. Biological and immunological activities, *Endocrinology* **102:** 1377, 1978.
9. A. J. Lostroh and M. E. Krahl, A hyperglycemic peptide from pituitary growth hormone: preparation with pepsin and assay in ob/ob mice, *Proc. Natl. Acad. Sci. USA* **71:**1244, 1974.
10. J. McD. Armstrong, J. Bornstein, J. O. Bromley, S. L. Macaulay, and F. M. Ng, Parallel insulin-like actions of human growth hormone and its part sequence hGH7–13, *Acta Endocrinol.* **102:**492, 1983.
11. A. Skottner, K. Fhölenhag, L. Fryklund, H. Flodh, B. Holnström, A. Wichman, and K.-G. Thorngren, Biosynthetic human somatotropin and pituitary somatotropin: A comparison, in: *Abstracts of the 64th Annual Meeting of the Endocrine Society*, San Francisco, CA, 1982 (abstr. 85).
12. S. E. Gennick, J. L. Kostyo, J. B. Mills, and S. Eden, A hybrid noncovalent complex of fragments of porcine and human growth hormones with diabetogenic activity, *Endocrinology* **112:**2069, 1983.
13. H. Papkoff, S. W. Farmer, and T. Hayashida, Non-mammalian growth hormones: naturally occurring analogs, in: *Growth Hormone and Other Biologically Active Peptides* (A. Pecile and E. E. Müller, eds.), Excerpta Medica, Amsterdam, 1979, pp. 45–51.
14. C. M. Cameron, J. L. Kostyo, J. A. Rillema, and S. E. Gennick, Reduced and S-carboxymethylated human growth hormone: A probe for diabetogenic action, *Am. J. Physiol.* **247:**E639, 1984.
15. L. G. Frigeri, S. M. Peterson, and U. J. Lewis, The 20,000-dalton structural variant of human growth hormone: Lack of some early insulin-like effects, *Biochem. Biophys. Res. Commun.* **91:**778, 1979.
16. N. A. Yudaev, Yu. A. Pankov, Yu. M. Keda, E. T. Sazina, T. A. Osipova, Yu. P. Shwachkin, and M. N. Ryabtsev, The effect of synthetic fragment 31–44 of human growth hormone on glucose uptake by isolated adipose tissue, *Biochem. Biophys. Res. Commun.* **110:**866, 1983.

34

Isolation of Growth Hormone Receptor

James P. Hughes, Lisa L. Wheeler, Timothy J. Fields, Jeanette M. Schepper, and Henry G. Friesen

1. Introduction

Although a number of methodologies have been used to isolate growth hormone (GH) receptors, affinity chromatography, using either GH[1,2] or concanavalin A[3] covalently coupled to a gel matrix, has proven to be the most effective technique for large-scale purification of the receptor. Despite the efficacy of affinity chromatography, the most highly purified preparations of GH receptor, for which purity has been assessed, have contained only 6–12% receptor.[2,3] One factor that has greatly hindered isolation of more highly purified preparations has been the limited resolution obtained using separation techniques based on molecular charge. Only marginal increases in purity have been effected using ion exchange chromatography[3] or isoelectric focusing.[2] Thus, further purification of the GH receptor has necessitated application/development of new separation methodologies. One new technique showing promise is adsorption of receptor to organomercurial matrices. Preliminary data show that these matrices, which form covalent linkages with proteins bearing free sulfhydryl groups, bind the rabbit GH receptor. The results further show that receptor adsorbed to organomercurial agarose is readily recovered and suggest, therefore, that this technique may constitute an important high-yield step in purification of the GH receptor. The recent development of a monoclonal antibody specific for the GH receptor in rabbit liver[4] provides yet another means of exploiting affinity chromatography for purification of this receptor. Initial results[4] indicate that antibody coupled to a solid support is a highly selective tool with which to further purify the GH receptor.

James P. Hughes, Lisa L. Wheeler, Timothy J. Fields, and Jeanette M. Schepper • Department of Life Sciences, Indiana State University, Terre Haute, Indiana 47809. *Henry G. Friesen* • Departments of Physiology and Pediatrics, University of Manitoba, Winnipeg, Manitoba, Canada R3E OW3.

2. Solubilization/Initial Isolation

2.1. Detergents

Rabbit liver contains a relatively high concentration of GH receptors; thus, liver from this species has been used almost exclusively as the source of receptors in purification studies. Solubilization of liver membranes (i.e., GH receptor) generally requires the use of detergents. The nonionic detergents Triton X-100 and X-305 have been most widely used for this purpose; however, ionic detergents,[5] low-ionic-strength buffers,[6] and even water[7] have been used with some success. Moreover, a recent report[5] showed that GH receptors solubilized in denaturing conditions using sodium dodecyl sulfate (SDS) recovered some GH-binding activity. Thus, there is as yet no consensus on optimal conditions for solubilization of the GH receptor.

2.2. Affinity Chromatography

Following solubilization, initial isolation of the GH receptor in large-scale purification studies has been achieved using affinity chromatography in which either hGH[2] or concanavalin A[3] has been coupled to agarose. Human GH selects for GH receptors on the basis of a specific hormone–receptor interaction, while concanavalin A is less specific in that it recognizes carbohydrate moieties associated with many glycoproteins, including the GH receptor. Whereas affinity chromatography based on either hGH[2] or concanavalin A[3] allows separation of GH receptors from prolactin (PRL) receptors, coupled GH offers a greater one-step purification (~70-fold) of the receptor. More recently, the highly specific interaction between GH and its receptor has been exploited using a biotin–streptavidin system.[5] Biotinylated hGH was incubated with receptor solubilized from rabbit liver, and the hormone–receptor complex was adsorbed to agarose-coupled streptavidin. Receptor was separated from hormone by treatment with 5 M $MgCl_2$ or 2% SDS. Although not assessed quantitatively, the purity of the receptor preparation was analyzed by SDS electrophoresis. With the exception of contaminating streptavidin, the major component identified by electrophoresis was a 67,000-dalton protein. This protein was identified as a subunit of the GH receptor on the basis of GH-binding studies. These studies further demonstrate that hGH-affinity chromatography is a powerful tool for isolation of GH receptor.

3. Separation Based on Molecular Size and Charge

3.1. Gel Filtration/SDS Electrophoresis

Using gel filtration on Sepharose 6B, Waters and Friesen[2] achieved an eightfold further purification of GH receptor isolated by hGH-affinity chromatography. Moreover, several other studies[3,5,7,8] also have successfully used gel filtration for purification of the GH receptor. A recent study by Haeuptle *et al.*[5] has suggested that SDS electrophoresis also might be useful in purification of the GH receptor. These investigators reported a 60% recovery of binding activity after treatment of receptor with 2% SDS and showed that following electrophoresis in the presence of SDS, GH receptors retained some binding activity. Further characterization of SDS-treated GH receptor is required, however, before the value of SDS electrophoresis in purification of the receptor can be assessed. It is apparent, however, that fractionation of the GH receptor on the basis of molecular size constitutes an important step in purification of the receptor.

3.2. DEAE Cellulose/Isoelectric Focusing

Fractionation of the GH receptor on the basis of molecular charge (i.e., isoelectric focusing and ion exchange chromatography) has been of limited value in purifying the receptor. Using DEAE cellulose, Tsushima *et al.*[3] achieved only a modest threefold increase in purification. Ion exchange chromatography was of some use, however, as a means of decreasing high concentrations of detergent.[3] Similar results were obtained using isoelectric focusing,[2] but this technique had a further drawback in that it added unwanted ampholytes to the preparation. Hence, techniques based on molecular charge, which generally are powerful fractionation tools, appear to be of limited value in purification of the GH receptor.

4. Adsorption to Organomercurial Agarose

Proteins that contain free sulfhydryl groups are adsorbed to organomercurial agarose (i.e., Affi Gel 501) by formation of a covalent mercaptide linkage. The linkage can be broken, and the protein eluted, by addition of a competing sulfhydryl (e.g., mercaptoethanol). Hence, affinity chromatography using organomercurial agarose selectively purifies sulfhydryl-containing proteins.

Data in our laboratory (unpublished results) suggested that subunits of the rabbit GH receptor in microsomal membrane preparations contained free sulfhydryl groups; thus, binding of receptor to organomercurial agarose was examined. GH receptor (~100 pmole), purified using hGH-affinity chromatography as outlined by Waters and Friesen,[2] was incubated for 15 min with 0.1 ml of packed organomercurial agarose beads in phosphate buffer (0.05 M, pH 7.2) containing 1% Triton X-100. Less than 10% of the original (Fig. 1, control) binding activity remained in solution (Fig. 1, unbound) after incubation. Moreover, even after the agarose beads were washed five times with the phosphate–Triton buffer, treatment of the organomercurial agarose with 10 mM mercaptoethanol resulted in complete recovery of GH-binding activity (Fig. 1, eluate). The covalent interaction between the rabbit GH receptor and agarose-bound mercury is resistant to dissociation under stringent wash conditions (i.e., high salt, low pH), allowing for removal of a high proportion of the contaminants

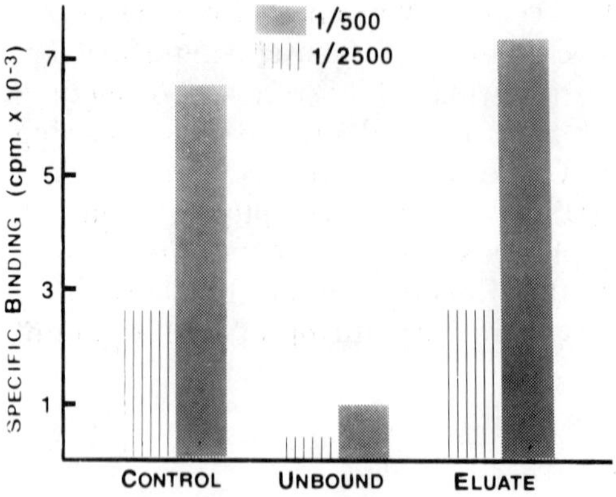

Figure 1 Adsorption of GH-binding sites, purified from rabbit liver by hGH-affinity chromatography, to organomercurial agarose. Specific binding of ^{125}I[hGH] to hGH-purified sites in a fraction not exposed to organomercurial agarose beads (control), to sites remaining in an identical fraction after treatment with 0.1 ml of packed beads (unbound), and to sites dissociated from the beads using 10 mM 2-mercaptoethanol (eluate). In binding assays, fractions were used at final dilutions of 1/500 and 1/2500.

that adsorb nonspecifically to the beads. Recovery of binding activity actually exceeded 100%; however, it is unknown whether this represents experimental error or whether, in some manner, the chromatographic procedure "unmasks" GH-binding sites. The data clearly show (1) that the major GH-binding protein in microsomal membranes prepared from rabbit liver adsorbs rapidly to organomercurial agarose, (2) that adsorbtion is not inhibited by a high concentration (1%) of Triton, and (3) that the GH-binding activity is readily eluted from the agarose beads. Thus, these preliminary results strongly suggest that affinity chromatography using organomercurial agarose may be an effective technique for purification of the rabbit liver GH receptor.

5. Monoclonal Antibody

5.1. Binding Characteristics

Simpson *et al.*[4] have recently reported production of a monoclonal antibody specific for the rabbit GH receptor. The antibody blocks binding to the rabbit GH receptor; however, it does not block binding to the rabbit PRL receptor or to GH receptors in human lymphocytes or in rat liver. Recent studies in our laboratory (data not shown) indicate that the converse is also true or that GH can compete with antibody for binding to the receptor. Thus, the monoclonal antibody is highly specific for the rabbit GH receptor, and the antibody appears to recognize a region of the receptor close to the hormone-binding site.

5.2. Affinity Chromatography

The potential usefulness of the antibody in purification of the GH receptor was examined by Simpson *et al.*[4] using affinity chromatography. Antibody was coupled to Sepharose 4B and incubated with solubilized liver membranes. After incubation, the antibody–agarose beads were washed three times to remove contaminating proteins and treated with 0.1 M glycine/HCl (pH 2) to dissociate

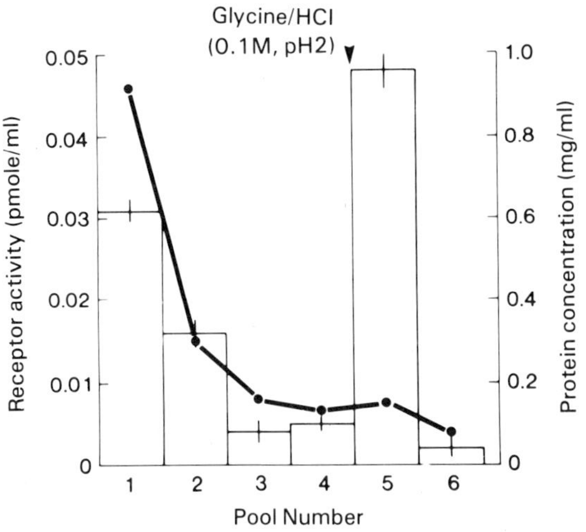

Figure 2 The vertical bars show receptor activity in pmoles of hGH per milliliter of each pool after dialysis. Error is expressed as the SD for three determinations. Pool 1 is the unbound material (250 ml); pool 2 is a Tris buffer (0.025 M Tris-HCl, 0.01 M $MgCl_2$, 0.1% BSA, pH 7.6) wash (2.3 ml); pool 3 is a second Tris buffer wash (2.4 ml); pool 4 is a 2 M $MgCl_2$ wash (6.0 ml); pool 5 is a 0.1 M glycine, pH 2.0 wash (9.8 ml); and pool 6 is the final Tris buffer wash (3.0 ml). ●—●, protein concentration in each pool. Reproduced from Simpson *et al.*,[4] with permission.

receptor/antibody complexes (Fig. 2). When expressed per milligrams of protein, the GH-binding activity in the 0.1 M glycine fraction was 9.2-fold higher than that in the original solubilized-membrane fraction. The investigators established, therefore, that antibody immobilized on agarose retained binding activity and that antibody–agarose was an effective technique for use in purification of the rabbit GH receptor.

Although the increase in GH binding was substantial, it was somewhat disappointing considering the high specificity of the antibody. However, no attempt was made in the study[4] to optimize the conditions for affinity chromatography. In particular, the protein coupled to Sepharose was a sodium sulfate precipitate from ascitic fluid rather than purified antibody, and proteinase inhibitors were not added to the solubilized membrane preparation. Hence, an even higher degree of purification might be expected when affinity chromatography using antibody is performed under optimal conditions.

5.3. Immunoprecipitation of Receptor

Precipitation of the rabbit GH receptor using the monoclonal antibody has been examined in preliminary studies. A double-antibody system was employed because the anti-GH-receptor antibody is of a class (IgE) not bound by Protein A. Rabbit antimouse IgE was generated in our laboratory using a monoclonal IgE generously provided by Dr. R. J. Schwenk. A GH-receptor preparation (~0.5% receptor) purified using hGH-affinity chromatography was iodinated by the lactoperoxidase method of Thorell and Johansson.[9] The iodinated preparation (1×10^7 cpm) ± 20 μg of unlabeled hGH was incubated with antibody-containing ascitic fluid (5 μl) for 16 hr at 21°C. Antimouse IgE serum was added to a final dilution of 1/25, and the incubation was continued for another 16 hr. Specific binding (i.e., binding inhibited by hGH) was 2×10^4 cpm, or approximately 0.2% of the total counts. When the immunoprecipitates were examined by SDS electrophoresis, several radioactive bands were observed; however, hGH only decreased the intensity of a broad band at 55,000 daltons. Hence, preliminary data suggest that the antibody recognizes a subunit of the rabbit GH receptor similar in size to the subunits identified in cross-linking studies.[10,11]

6. Summary

Progress in isolation of the GH receptor has been relatively limited in the 4 years subsequent to the initial large-scale purification reported by Waters and Friesen.[2] Although affinity chromatography has proven effective for isolation of the GH receptor, conventional separation methodologies have been ineffective in this regard. Accordingly, purification schemes using these methodologies have not yielded highly purified preparations of the GH receptor. Two new techniques have been developed, however, which may be useful for further purification of the GH receptor. Adsorption of GH receptor to organomercurial agarose shows promise as a purification step and as a means of eliminating excess detergent. Moreover, the covalent nature of the interaction between receptor and organomercurial matrix allows for use of more stringent conditions for removal of contaminants. The development of a monoclonal antibody specific for the rabbit GH receptor provides another highly selective ligand with which to exploit affinity chromatography. Preliminary data[4] already have demonstrated the value of the antibody in isolating the GH receptor and in eliminating contaminating PRL receptors. A simple purification scheme suggested by these data is (1) solubilization of the GH receptor using Triton X-100, (2) initial isolation of the receptor using GH-affinity chromatography, (3) further purification of the receptor using antibody-affinity chromatography, (4) adsorption of receptor to organomercurial agarose, and (5) final purification of the receptor using gel filtration. Hence, in the next 2 years, rapid progress should be made toward final purification of the GH receptor.

ACKNOWLEDGMENTS. The authors thank Dr. R. J. Schwenk for the generous gift of mouse IgE. The authors also gratefully acknowledge the excellent secretarial assistance of Anita Kritz.

References

1. C. J. McIntosh, J. Warnecke, M. Nieger, A. Barner, and J. Köbberling, Solubilization and partial purification of a growth hormone receptor from rabbit liver, *FEBS Lett.* **66:**149, 1976.
2. M. J. Waters and H. G. Friesen, Purification and partial characterization of a nonprimate growth hormone receptor, *J. Biol. Chem.* **254:**6815, 1979.
3. T. Tsushima, N. Sasaki, Y. Imai, F. Matsuzaki, and H. G. Friesen, Characteristics of solubilized human-somatotropin-binding protein from the liver of pregnant rabbits, *Biochem. J.* **187:**479, 1980.
4. J. S. A. Simpson, J. P. Hughes, and H. G. Friesen, A monoclonal antibody to the growth hormone receptor of rabbit liver membranes, *Endocrinology* **112:**2137, 1983.

5. M-T Haeuptle, M. L. Aubert, J. Djiane, and J-P. Kraehenbuhl, Binding sites for lactogenic and somatogenic hormones from rabbit mammary gland and liver, *J. Biol. Chem.* **258:**305, 1983.
6. A. C. Herington, S. D. Elson, and S. I. Ymer, Water soluble receptors for human growth hormone from rabbit liver, *J. Recept. Res.* **2:**203, 1981.
7. M. Gottsmann and K. V. Werder, hGH-receptor protein in rabbit liver microsomes, *Acta Endocrinol.* **82**(suppl. 202)**:**40, 1976.
8. H. Chr. Blossey, Studies on the molecular architecture and composition of the GH receptor from rabbit liver, *Horm. Metab. Res.* **11:**616, 1979.
9. J. I. Thorell and B. G. Johansson, Enzymatic iodination of polypeptides with ^{125}I to high specific activity, *Biochim. Biophys. Acta* **251:**363, 1971.
10. T. Tsushima, H. Murakami, K. Wakai, O. Isozaki, Y. Sato, and K. Shizume, Analysis of hepatic growth hormone binding sites of pregnant rabbit crosslinked to ^{125}I-labelled human growth hormone, *FEBS Lett.* **147:**49, 1982.
11. J. P. Hughes, J. S. A. Simpson, and H. G. Friesen, Analysis of growth hormone and lactogenic binding sites cross-linked to iodinated human growth hormone, *Endocrinology* **112:**1980, 1983.

35

Structural Studies of the Growth Hormone Receptor by Affinity Labeling

David B. Donner, Kazuyo Yamada, Kenneth E. Lipson, and Andrea Dorato

1. Introduction

Detailed information of receptor structure can be obtained by covalently coupling radioactively labeled peptide hormones to membrane binding sites with chemical cross-linking reagents.[1] Disuccinimidyl esters have been particularly useful for such affinity labeling as they react rapidly with free amino groups under physiological conditions, thereby minimizing distortion of receptor structure.[2] Using such reagents, the receptors for insulin[3,4] and the insulinlike growth factors[5,6] have been characterized.

Binding sites for human growth hormone (hGH) have been identified on hepatocytes isolated from normal male rats.[7] GH interacts with such receptors specifically and with high affinity.[8,9] These properties have been advantageously used to couple ^{125}I-labeled hGH to sites on rat hepatic tissue. This has permitted a description of the subunit structure of the hepatic GH receptor.

2. The Experimental Procedure

^{125}I-labeled hGH and hepatocytes were prepared as previously described.[10,11] Hepatocytes ($\sim 10^6$ cells/ml) were incubated with [^{125}I]hGH (5

Abbreviations used: HBSS, Hank's balanced salt solution; hGH, human growth hormone; bGH, bovine growth hormone; PRL, ovine prolactin; [^{125}I]hGH, iodine-125-labeled human growth hormone.

David B. Donner, Kazuyo Yamada, Kenneth E. Lipson, and Andrea Dorato • Memorial Sloan-Kettering Cancer Center and The Sloan-Kettering Division, Graduate School of Medical Sciences, Cornell University, New York, New York 10021.

nM) in HBSS[1]/1% bovine serum albumin at 23°C and centrifuged (1500 × g, 3 min). The cell pellet was resuspended in ice-cold HBSS and recentrifuged. This process was repeated twice to remove albumin and unbound [^{125}I]hGH from the hepatocytes. The hepatocytes were suspended in HBSS at 15°C, and freshly prepared disuccinimidyl suberate dissolved in dimethyl sulfoxide was added to a final concentration of 1 mM. After 15 min the reaction was terminated by addition of 10 volumes of ice-cold 10 mM Tris, 1 mM EDTA, pH 7.4. The mixture was centrifuged, and the cell pellet was washed with 5 ml of ice-cold HBSS. The hepatocytes were suspended in 0.25 M sucrose containing phenylmethylsulfonyl fluoride (2 mM), disrupted with 30 strokes in a tight-fitting dounce homogenizer and drawn through a 22-gauge needle 10 times. The homogenate thus produced was centrifuged at 800 × g for 10 min, and the supernatant was centrifuged at 100,000 × g for 15 min to yield a membrane pellet to which [^{125}I]hGH was covalently bound. The pellet was solubilized in boiling 3% sodium dodecyl sulfate, 10% glycerol, 6 M urea, 0.001% bromphenol blue, 0.06 M Tris-HCl (3 min).

Samples were analyzed on 1.5 mm sodium dodecyl sulfate polyacrylamide gels using the discontinuous system described by Laemmli[(12)] at 24-mA constant current for 4–5 hr. Gels were stained for 1 hr in 0.5% Coomasie blue R in 25% isopropyl alcohol, 7% acetic acid, destained, and dried. The molecular weight standards were myosin (200,000), B-galactosidase (116,250), phosphorylase B (92,500), bovine serum albumin (66,200), and ovalbumin (45,000). The positions of these standards are shown to the left of each autoradiograph. The apparent molecular weights of cross-linked species are to the right. Autoradiography was carried out in cassettes with Kodak X-Omat XAR-2 film using a Dupont Cronex Lightning Plus enhancing screen. Lead plates (1/8 in. thick) were inserted between cassettes to prevent fogging of films. Films were exposed from 7 to 21 days at −80°C prior to development.

3. Results and Discussion

The properties of GH receptors on a number of target cells including the hepatocyte have been characterized by equilibrium and kinetic binding measurements.[(13)] In studies with hepatocytes from normal male rats, [^{125}I]hGH uptake increased with the applied hormone concentration. Whereas the specific binding of GH to hepatocytes is saturable,[(8)] the total radioactive uptake is not (Fig. 1, bottom). The nonlinearity of the total uptake confirms the presence of a specific component of GH binding. The linear increase of [^{125}I]hGH binding at high hormone concentrations demonstrates that nonspecific uptake must be considered in the interpretation of both radioreceptor and affinity labeling experiments. For this reason, specific hormone binding (Fig. 1) and the incorporation of [^{125}I]hGH into covalent complexes (Figs. 2–4) have been directly compared in parallel experiments. The effects of time, hormone concentration, and various competitors on the incorporation of [^{125}I]hGH into

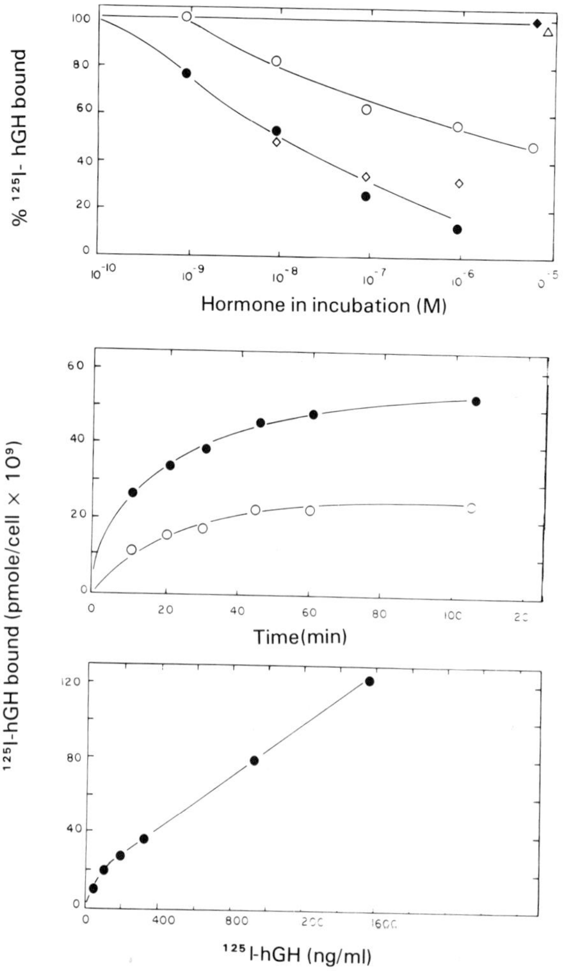

Figure 1 Characterization of [^{125}I]hGH binding to hepatocytes (Top) Hepatocytes were incubated with [^{125}I]hGH (5 nM) and various concentrations of hGH (●), bGH (◇), ovine prolactin (○), insulin (◆), or glucagon (△) for 1 hr at 23°C. Binding of [^{125}I]hGH was then assayed by microcentrifugation. Each point is the mean of triplicate determinations. (Middle) Hepatocytes were incubated with [^{125}I]hGH (5 nM) or with [^{125}]hGH + hGH (10μM) at 23°C in HBSS/1% albumin. At the indicated times triplicate aliquots from each incubate were assayed for [^{125}I]hGH binding by microcentrifugation. Total (●) and specific (○) binding curves are shown. (Bottom) Hepatocytes were incubated with various concentrations of [^{125}I]hGH for 1 hr at 23°C. Triplicate aliquots from each incubate were then assayed for total [^{125}I]hGH uptake by microcentrifugation.

covalent complexes have permitted identification of the components of the GH receptor with some degree of assurance.

During affinity labeling experiments, [^{125}I]hGH was incorporated into a predominant 130,000-dalton complex as well as into a number of lower-molecular-weight complexes (Fig. 2A). The uptake of [^{125}I]hGH into the 130,000-dalton complex was saturable with respect to hormone concentration. Half-saturation of the formation of the 130,000-dalton species occurred in the presence of about 0.8 nM [^{125}I]hGH, a concentration close to the dissociation constant (K_D = 2 nM) for the GH receptor on freshly isolated liver cells.[(8)] The incorporation of [^{125}I]hGH into the 130,000-dalton complex was specifically inhibited by native hGH (Fig. 2B). The amount of the 130,000-dalton complex formed (Fig. 2B) closely paralleled the increase of specific GH binding during an association reaction (Fig. 1, middle).

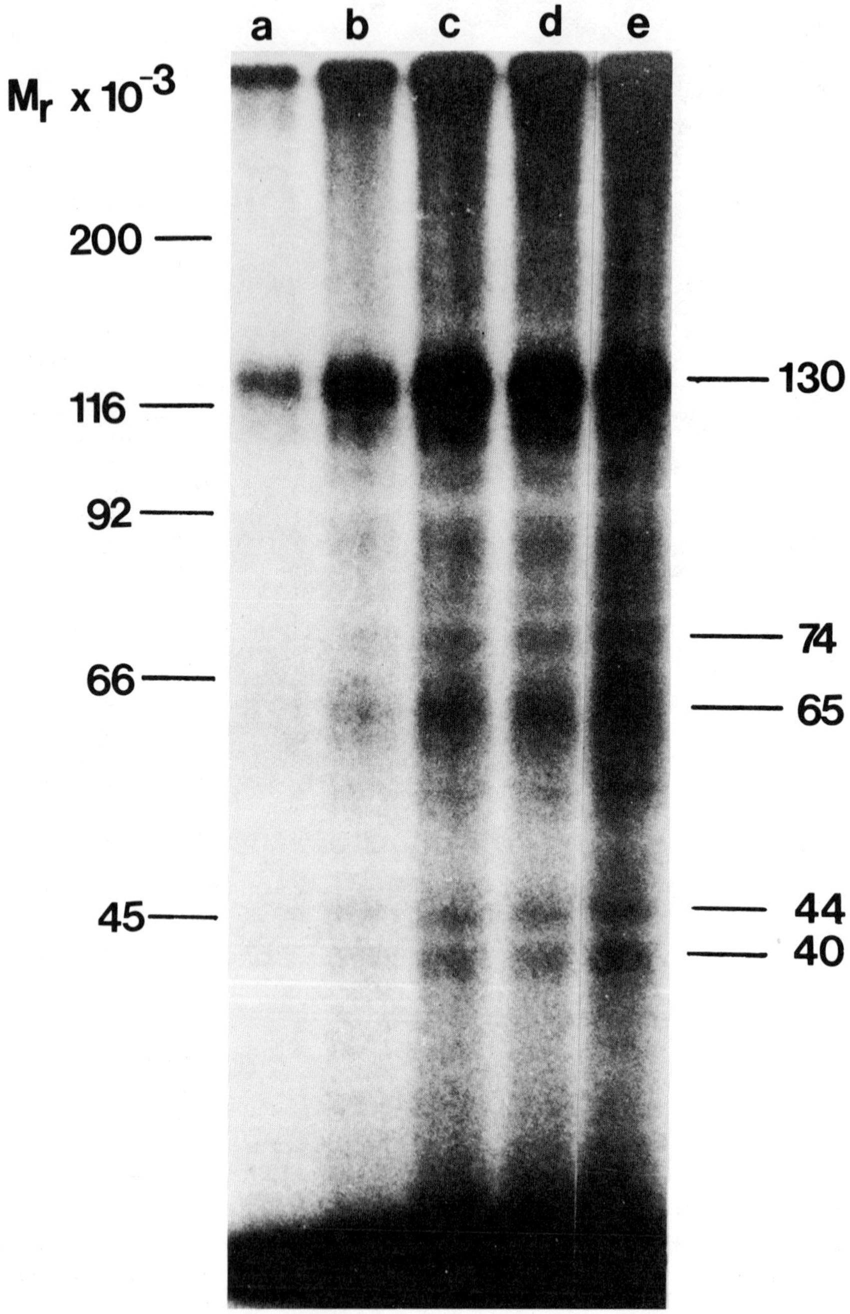

Figure 2 (left) Effect of cross-linking various concentrations of [^{125}I]hGH to hepatocytes. Hepatocytes were incubated with various concentrations of [^{125}I]hGH: lane a, 0.4 nM; lane b, 0.8 nM; lane c, 2 nM; lane d, 4 nM; lane e, 20 nM for 1 hr at 23°C. After cross-linking, a crude membrane fraction was prepared from each incubate. Membranes were solubilized in the presence of 100 mM dithiothreitol and characterized on a 7.5% polyacrylamide gel (acrylamide:bisacrylamide 37.5 to 1).

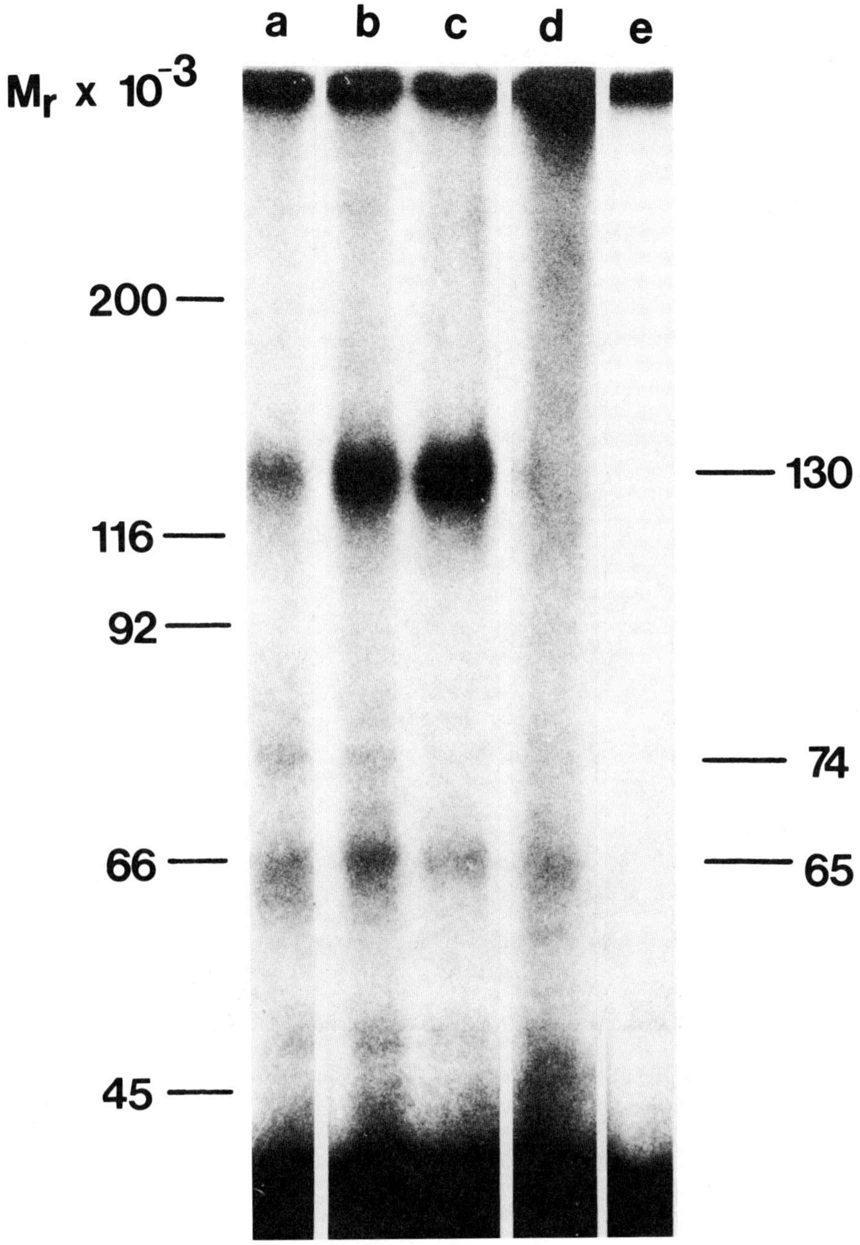

Figure 2 (right) Crosslinking of [^{125}I]hGH to hepatocytes after incubations of various times. Hepatocytes were incubated with [^{125}I]hGH (5 nM) or [^{125}I]hGH + hGH (10 μM) at 23°C. Crosslinking of bound [^{125}I]hGH was initiated after various times of incubation. Membranes isolated from each incubate were solubilized in the presence of 100 mM dithiothreitol and characterized on a 7.5% polyacrylamide gel. Lane a, 10 min [^{125}I]hGH association; lane b, 45-min [^{125}I]hGH association; lane c, 105 min [^{125}I]hGH association; lane d, 105 min [^{125}I]hGH association in which membranes were solubilized in the absence of dithiothreitol; lane e, 105 min association in which hGH was present.

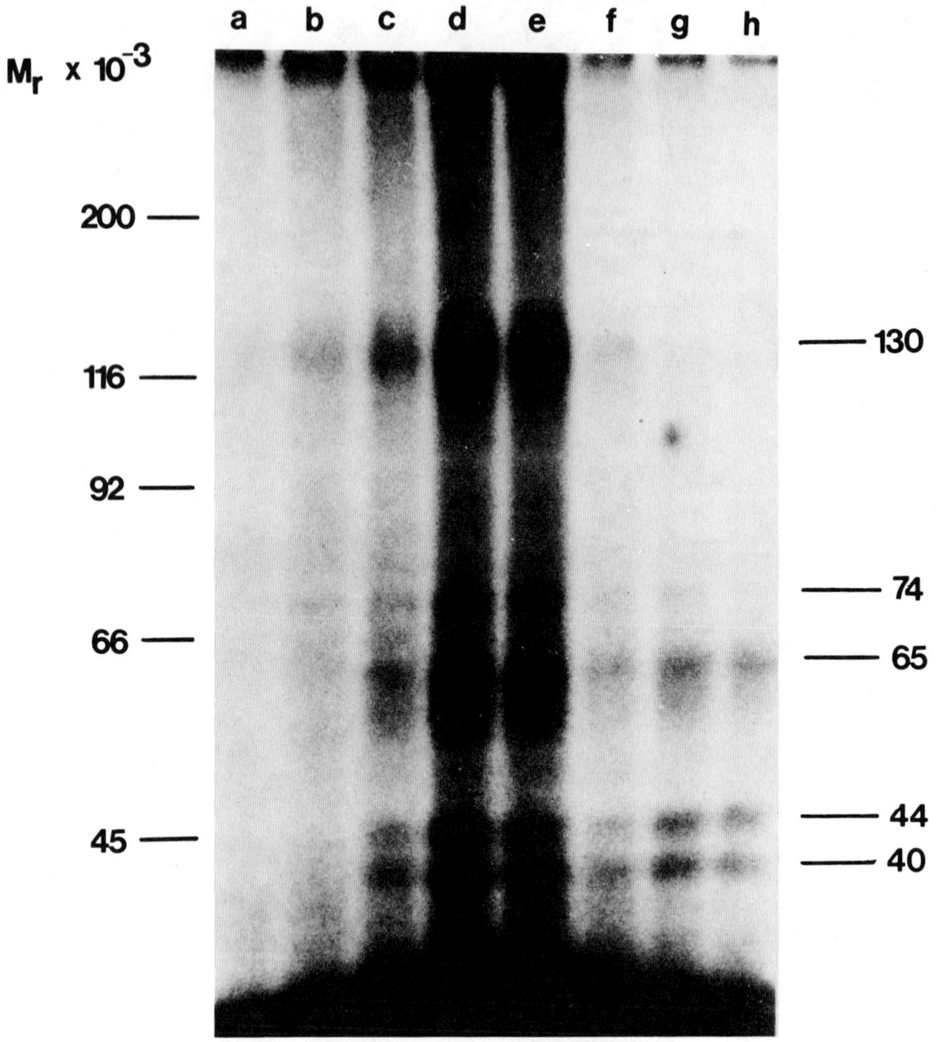

Figure 3 (left) Competition of human growth hormone and bovine growth hormone for [^{125}I] hGH cross-linking. Hepatocytes were incubated with [^{125}I]hGH (5 nM) for 1 hr at 23°C in the presence of: lane a, 900 nM hGH; lane b, 90 nM hGH; lane c, 9 nM hGH; lane d, 0.9 nM hGH; lane e, control; lane f, 9 nM bGh; lane g, 90 nM bGH; lane h, 900 nM bGH. Cross-linking of hormone was carried out, and a membrane fraction prepared from each incubate was solubilized in the presence of 100 mM dithiothreitol and characterized on a 7.5% polyacrylamide gel.

hGH has weak lactogenic(14) as well as growth-promoting activity(15) and binds to GH as well as prolactin (PRL) receptors.(7,16,17) Either bovine growth hormone or PRL inhibited a component of the specific uptake of [^{125}I]hGH by hepatocytes (Fig. 1, top). hGH and PRL, but not bovine GH, compete for [^{125}I]PRL binding to hepatocytes from male or female rats. PRL, however, does

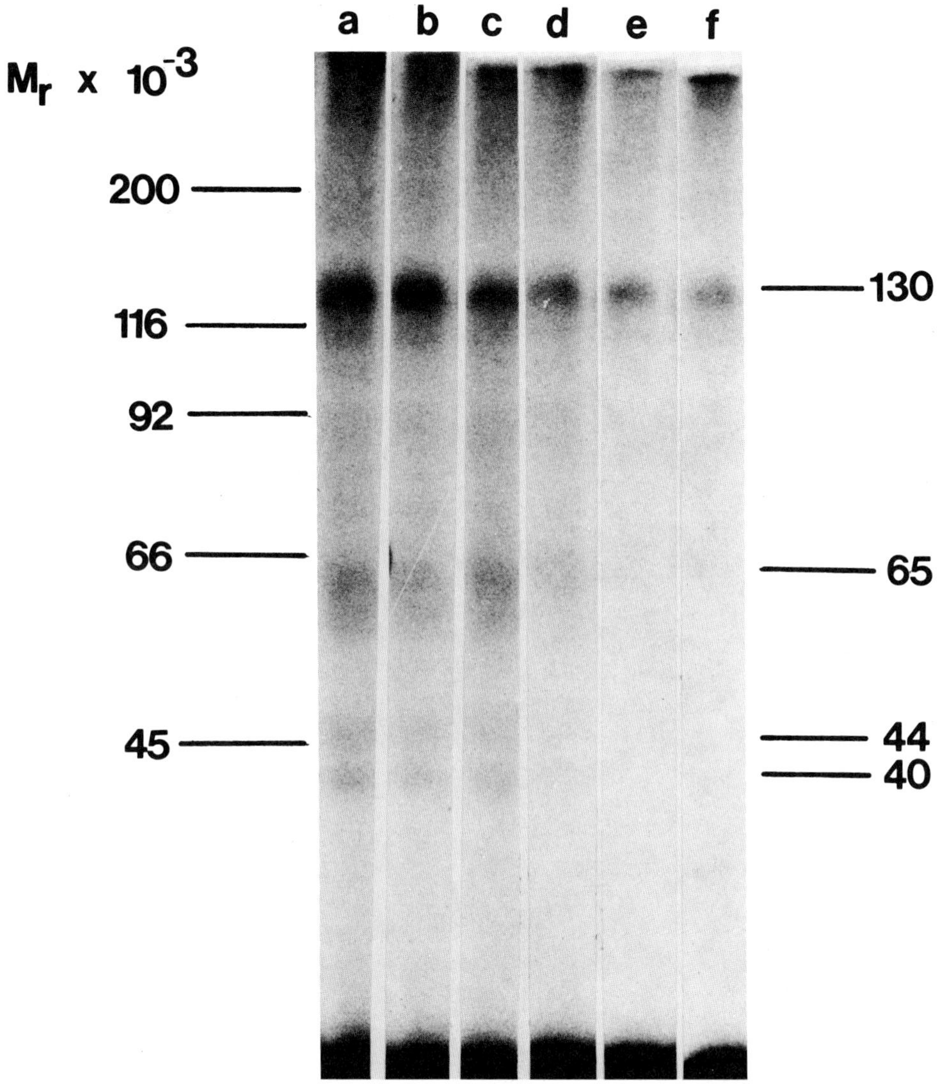

Figure 3 (right) Competition of prolactin (PRL) for [^{125}I]hGH cross-linking. Hepatocytes were incubated with [^{125}I]hGH (5 nM) in the presence of: lane a, control lane b, 0.9 nM PRL; lane c, 9 nM PRL; lane d, 90 nM PRL; lane e, 900 nM PRL; lane f, 9000 nM PRL. Cross-linking was carried out, and a membrane fraction prepared from each incubate was solubilized in the presence of 100 mM dithiothreitol and characterized on a 7.5% polyacrylamide gel.

not compete for [^{125}I]bovine GH uptake, whereas bovine GH and hGH are equally effective in this regard. These observations suggest that bGH and PRL bind to somatogenic and lactogenic receptors, respectively.

The different specificities of hGH, bGH, and PRL for binding were used to distinguish somatogenic from lactogenic complexes into which [^{125}I]hGH

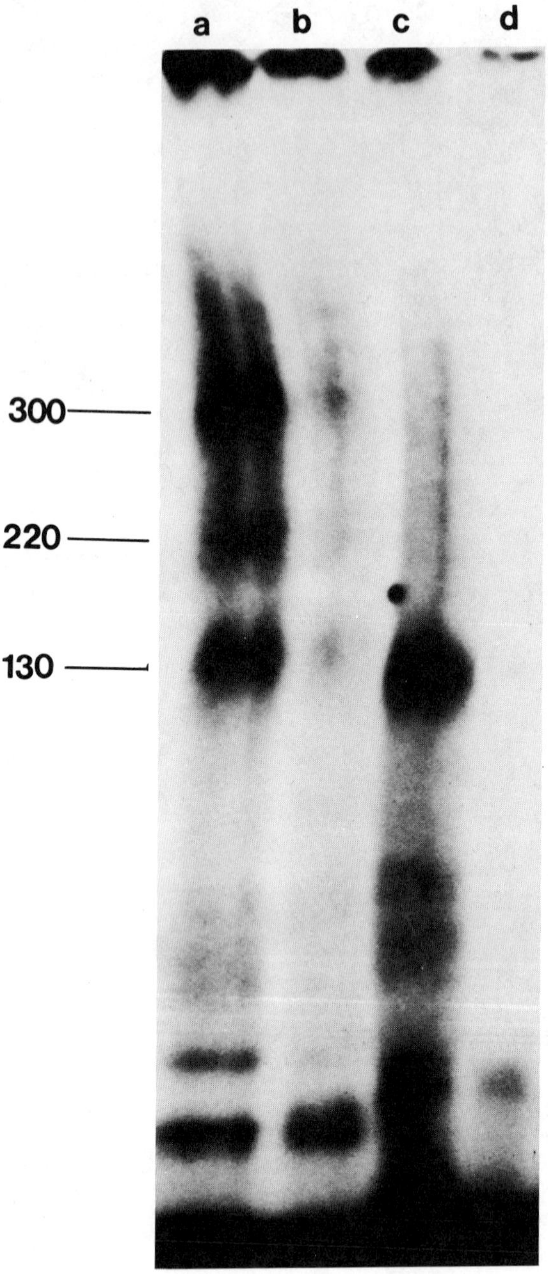

Figure 4 Cross-linking [^{125}I]hGH to hepatocytes and solubilization in the absence of dithiothreitol. Hepatocytes were incubated with [^{125}I]hGH (5 nM) at 23°C for 1 hr in the presence or absence of 10 μM hGH. Cross-linking was terminated by shaking each incubate with 10 mM Tris, 1 mM EDTA, 5 mM *N*-ethylmaleimide (pH 7.4) for 15 min. Membranes prepared from each incubate were solubilized in extraction medium containing 5 mM *N*-ethylmaleimide with or without 100 mM dithiothreitol. Equal amounts of radioactivity were applied to a linear 4–20% polyacrylamide gel (acrylamide:bisacrylamide 100:1): lane a, [^{125}I]hGH unreduced; lane b, [^{125}I]hGH + hGH unreduced; lane c, [^{125}I]hGH reduced; lane d, [^{125}I]hGH + hGH reduced.

was coupled. bGH inhibited the incorporation of [^{125}I]hGH into the 130,000-dalton complex (Fig. 3A) without significantly affecting the isolation of the lower-molecular-weight complexes. hGH competed for uptake of [^{125}I]hGH into all saturable species. Low amounts of PRL competed for [^{125}I]hGH incorporation into lower-molecular-weight complexes (Fig. 3B). Much higher concentrations of PRL inhibited [^{125}I]hGH uptake into the 130,000-dalton species. This may have been due to trace contamination of this pituitary hormone with GH. These experiments suggest that [^{125}I]hGH receptor complexes can be segregated into a somatogenic complex of M_r 130,000 and lactogenic complexes of lower apparent molecular weight.

The results summarized in the preceding paragraphs support the view that the 130,000-dalton complex is formed by interaction of [^{125}I]hGH with a component of the GH receptor. Disulfide bonds may stabilize receptor structure, and the 130,000-dalton complex had been isolated from membranes solubilized in the presence of the reducing agent dithiothreitol. In order to determine whether such covalent bonds affect the structure of the receptor for GH, [^{125}I]hGH was covalently coupled to hepatic membrane receptors which were then solubilized in the absence or presence of dithiothreitol. In this experiment, *N*-ethylmaleimide was present in the solubilization medium to prevent disulfide bond interchange between free sulfhydryl groups. In the absence of reductant, complexes of M_r 300,000, 220,000, and 130,000 were isolated on a porous polyacrylamide gel (Fig. 4, lane a). Native hGH (lane b) and bGH (not shown) completely inhibited the incorporation of [^{125}I]hGH into the 300,000-, 220,000-, and 130,000-dalton species suggesting that these are saturable components of the GH receptor. The 300,000- and 220,000-dalton complexes were not isolated from membranes solubilized in the presence of reductant (lane c) whereas the amount of the 130,000-dalton species was increased. These observations suggest that the GH receptor is an aggregate in which some subunits are retained in a larger complex by interchain disulfide bonds. As with insulin receptors,[18] monomeric components of the GH receptor were isolated from membranes in the absence of reductant. Subtracting the molecular weight of growth hormone (~22,000 daltons) from somatogenic complexes suggests that [^{125}I]hGH had bound to species of M_r 280,000, 200,000, and 100,000. The molecular weights of the species to which GH bound have suggested a simple model by which to describe this receptor: a trimeric aggregate of 100,000-dalton subunits held together by covalent and noncovalent forces.

[^{125}I]hGH has been covalently incorporated into a complex of M_r 130,000 on IM-9 lymphocytes.[19] A GH receptor of 300,000 daltons from pregnant rabbit liver membranes contained a subunit of M_r 75,000 as well as other minor components.[20] A number of complexes of M_r less than 130,000 were isolated after covalently coupling hormone to receptors on microsomal fractions of rabbit liver,[19,21] rabbit mammary gland, and rat liver.[19] The different molecular weights for the GH receptor may result from different methods and tissues used in various studies. In addition, in attempting to propose a model by which to describe the GH receptor, it is possible that not all of the components of this species have been identified. The β subunit of the insulin receptor, for example, was not readily labeled using disuccinimidyl suberate but was identified after metabolic labeling and immunoprecipitation.[18] In view of these facts, the trimeric structure proposed to describe the GH receptor should be considered an attractive model upon which to base further study.[22]

ACKNOWLEDGMENTS. D.B.D. is the recipient of Research Career Development Award AM 01045 from NIH. K.Y. is the recipient of a fellowship from the Ministry of Education, Science and Culture of Japan, on leave from the Faculty of Pharmaceutical Science, Nagoya City University, Nagoya, Japan. K.E.L. is the recipient of National Research Service Award Am 07112 from NIH. This research was supported by grants AM 30788, AM 22121, and CA 08748 from NIH.

References

1. P. S. Linsley, M. Das, and C. F. Fox, Affinity labeling of hormone receptors and other ligand binding sites, in: *Receptors and Recognition* (S. Jacobs and P. Cuatrecasas, eds), Chapman and Hall, London, 1981, series B, volume 11, pp. 89–113.
2. P. F. Pilch and M. P. Czech, The subunit structure of the high affinity insulin receptor, *J. Biol. Chem.* **255:**1722, 1980.
3. S. Jacobs and P. Cuatrecasas, Insulin receptor: Structure and function, *Endocrine Rev.* **2:**251, 1981.
4. M. P. Czech and J. Massague, Subunit structure and dynamics of the insulin receptor, *Fed. Proc.* **41:**2719, 1982.
5. S. D. Chernausek, S. Jacobs, and J. J. Van Wyk, Structural similarities between human receptors for somatomedin C and insulin: analysis by affinity labeling, *Biochemistry* **20:**7345, 1981.
6. J. Massague and M. P. Czech, The subunit structures of two distinct receptors for insulin-like growth factors I and II and their relationship to the insulin receptor, *J. Biol. Chem.* **257:**5038, 1982.
7. D. B. Donner, Covalent coupling of human growth hormone to its receptor on rat hepatocytes, *J. Biol. Chem.* **258:**2736, 1983.
8. D. B. Donner, J. Casadei, L. Hartstein, D. Martin, and M. Sonenberg, Characterization of the slowly dissociable human growth hormone binding component of isolated rat hepatocytes, *Biochemistry* **19:**3293, 1980.
9. D. B. Donner, Interconversion between different states of affinity of the human growth hormone receptor on rat hepatocytes: effects of fractional site occupancy on receptor availability, *Biochemistry* **10:**3300, 1980.
10. D. B. Donner, D. W. Martin, and M. Sonenberg, Accumulation of a slowly dissociable peptide hormone binding component by isolated target cells, *Proc. Natl. Acad. Sci. USA* **75:**672, 1978.
11. D. B. Donner, K. Nakayama, S. Tani, U. Lutz, and M. Sonenberg, The uptake and degradation of growth hormone fragments by isolated rat hepatocytes, *J. Biol. Chem.* **253:**6717, 1978.
12. U. K. Laemmli, Cleavage of structural proteins during the assembly of the head of bacteriophage T_4, *Nature (London)* **227:**680, 1970.
13. M. Wallis, Receptors for growth hormone, prolactin and the somatomedins, in: *Cellular Receptors for Hormones and Neurotransmitters* (D. Schluster and A. Levitski, eds.), John Wiley & Sons, New York, 1980, pp. 163–183.
14. D. L. Kleinberg and A. G. Frantz, Human prolactin: Measurement in plasma by *in vitro* bioassay, *J. Clin. Invest.* **50:**1557, 1971.
15. H. M. Evans, M. E. Simpson, W. Marx, and E. Kibrick, Bioassay of the pituitary growth hormone. Width of the proximal epiphysial cartilage of the tibia in hypophysectomized rats, *Endocrinology* **32:**13, 1943.

16. P. A. Kelly, B. I. Posner, T. Tsushima, and H. G. Friesen, Studies of insulin, growth hormone and prolactin binding: Ontogenesis, effects of sex and pregnancy, *Endocrinology* **95:**532, 1974.
17. M. B. Ranke, C. A. Stanley, D. Rodbard, L. Baker, A. Bongiovanni, and J. S. Parks, Sex differences in binding human growth hormone to isolated rat hepatocytes, *Proc. Natl. Acad. Sci. USA* **73:**847, 1976.
18. M. Kasuga, J. A. Hedo, K. M. Yamada, and C. R. Kahn, The structure of insulin receptor and its subunits, *J. Biol. Chem.* **257:**10392, 1982.
19. J. P. Hughes, J. S. A. Simpson, and H. G. Friesen, Analysis of growth hormone and lactogenic binding sites cross-linked to iodinated growth hormone, *Endocrinology* **112:**1980, 1983.
20. M. J. Waters and J. G. Friesen, Purification and partial characterization of a nonprimate growth hormone receptor, *J. Biol. Chem.* **254:**6815, 1979.
21. M-T. Haeuptle, M. L. Aubert, J. Dijiane, and J-P. Kraehenbuhl, Binding sites for lactogenic and somatogenic hormones from rabbit mammary gland and liver, *J. Biol. Chem.* **258:**305, 1983.
22. K. Yamada and D. B. Donner, Structures of the somatotropin receptor and prolactin receptor on rat hepatocytes characterized by affinity labelling, *Biochem. J.* **220:**361, 1984.

36

Binding of Growth Hormone to Hepatic Receptors

Wayne V. Moore, Karen Kover, and Chung Ho Hung

Most polypeptide hormones are thought to exert an intracellular effect by binding to specific cell membrane receptors with subsequent stimulation of adenylate cyclase and a resultant elevation of intracellular cyclic AMP levels. A major exception to this schema appears to be growth hormone (GH) and its effect on the liver. Even though GH does bind to cell membrane receptors in hepatic tissue,[1–12] a stimulatory effect of GH on adenylate cyclase in liver has not been observed. These observations make GH–receptor interactions basically interesting. The liver is of special interest in GH binding since it is a putative target organ for GH in the production of somatomedin C or IGF-I. GH also binds to other tissue, notably renal and adipose,[13–15] but this chapter will focus on GH binding to hepatic tissue.

Human GH (hGH) has been the commonly used ligand for investigating GH interactions with membrane receptors probably because hGH exhibits significant binding to liver receptors from a variety of different species. Figure 1 illustrates the total and nonspecific binding for [^{125}I]hGH to liver membranes isolated from the liver of different adult female mammals by the method of Neville.[16] The binding incubation medium included 10 mM calcium with 50 mM Tris-HCl, pH 7.4, in 0.1% albumin (Fig. 1). Membranes obtained from the livers of chicken and fish do not bind appreciable quantities of hGH while liver membranes from adult female human, rabbit, rat, hamster, mouse, cow, sheep, goat, dog, and cat do exhibit appreciable binding of [^{125}I]hGH. The phylogenetic orders represented by the animals listed on Fig. 1 are listed in Table 1. The percentage bound was determined at the smallest concentration of [^{125}I]hGH and may be representative of the relative capacity of the different membranes for hGH assuming that the individual membranes and individual

Wayne V. Moore, Karen Kover, and Chung Ho Hung • Section of Pediatric Endocrinology, Department of Pediatrics, University of Kansas Medical Center, Kansas City, Kansas 66103.

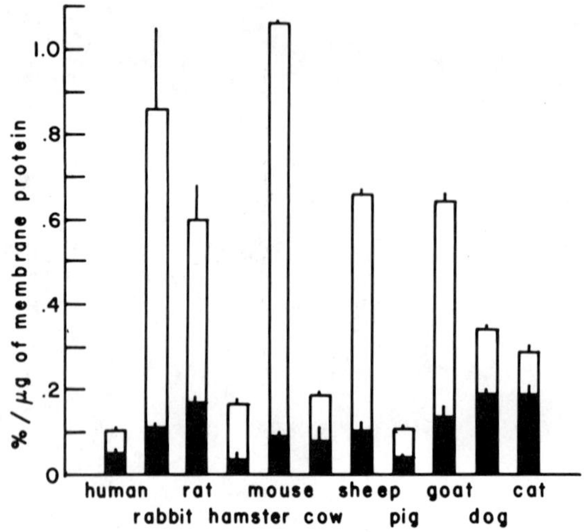

Figure 1 The binding of [^{125}I]hGH to liver membranes from several different species of the class Mammalia. The membranes were incubated with 1 ng/ml of [^{125}I]hGH in 50 mM Tris-HCl (pH 7.4) with 10 mM calcium and 0.1% albumin for 2 hr at 30°C. The bound [^{125}I]hGH was separated from unbound hGH by sedimenting the membranes through a layer of 10% sucrose in a microfuge tube. The percentage of [^{125}I]hGH in the membrane pellet per µg of membrane protein plus 1 SD is depicted (open bars). The amount of [^{125}I]hGH bound in the presence of excess unlabeled hGH (nonspecific binding) is depicted in the solid bars.

binding conditions are comparable. In fact, conditions of optimal binding of hGH to the rabbit, rat, hamster, and mouse liver membranes include not only divalent cation but lower pH in the range of 5–6. Therefore the percentage of [^{125}I]hGH bound by the different species may be only qualitatively comparable.

The competitive inhibition of [^{125}I]hGH binding to the rodent membranes indicates the typical lactogenic hormone nature of the receptors for hGH in the rat, hamster, and mouse liver membranes based on the cross-reactivity of oPrl and hGH (hGH P-17, 2 IU/mg; bGH B-18, 0.81 IU/mg; oPrl-NIH, 21 IU/mg; hPrl-NIH, 20 IU/mg; hPl 3000A; rGH B-7, 1.6 IU/mg from the National Hormone and Pituitary Program) (Fig. 2). The adult female animal was chosen for the source of liver in each of the orders since receptors for hGH and lactogenic hormones are most likely to be present in the adult female liver.[1–12] The binding of hGH to male human, cow, pig, and goat liver membranes has also been determined and does not differ significantly from that of the adult

Table 1 Classification of Animals Used to Determine Binding of [^{125}I]hGH to Liver Membranes

Class
Mammalia
Subclass
Theria
Order
Primates—man
Lagomorpha—rabbit
Rodentia—mouse, rat, hamster
Carnivora—dog, cat
Artiodactyla—cow, sheep, pig, goat

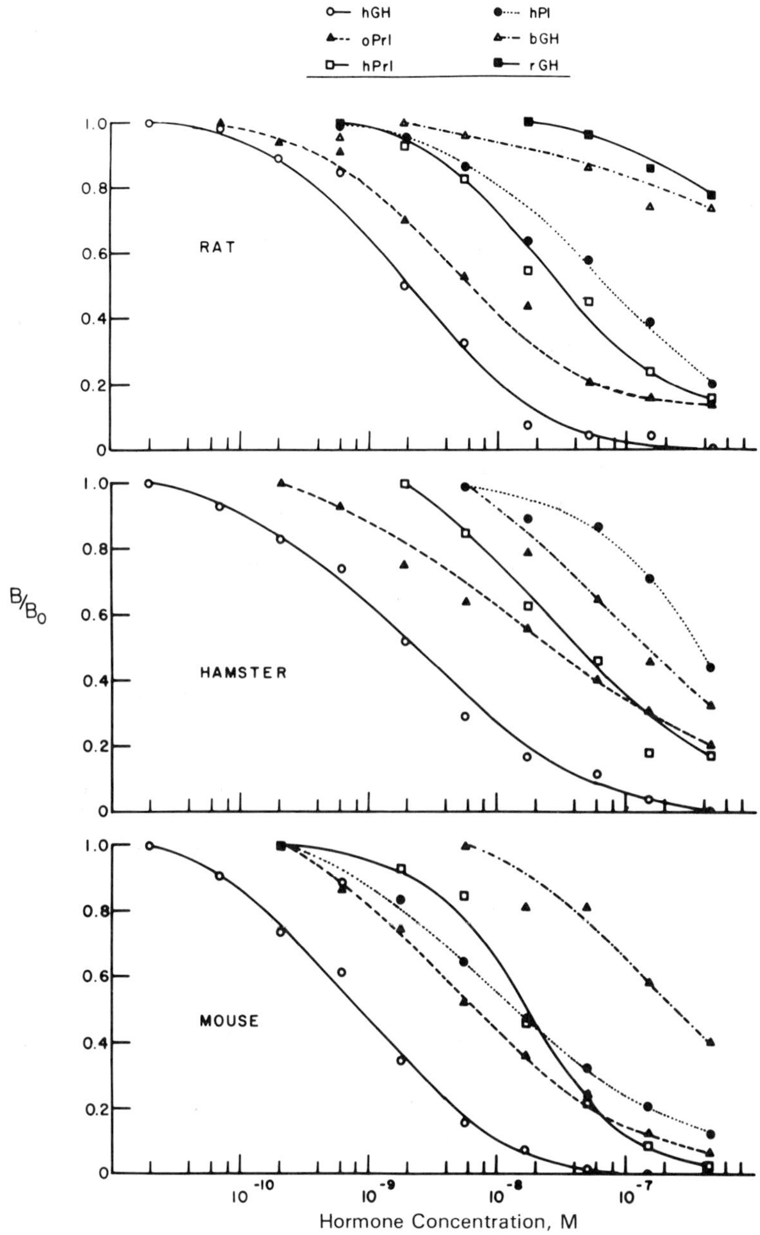

Figure 2 Competitive binding profiles of [^{125}I]hGH binding to rodent liver membranes. The ratio (B/B_o) of the amount of [^{125}I]hGH bound at the smallest concentration of [^{125}I]hGH (B_o) to the amount bound at each concentration of competing hormone is plotted versus the concentration of competing hormone. The nonspecific binding of [^{125}I]hGH is subtracted from each value to normalize B/B_o for hGH from 1 to 0. hGH = human growth hormone; oPRL = ovine prolactin; hPRL = human prolactin; hPl = human placental lactogen; bGH = bovine growth hormone; rGH = rat growth hormone.

female animal. Rat (rGH) and bovine GH (bGH) are not effective competitors for hGH binding to adult female rodent liver membranes. This probably results from a preponderance of the lactogenic hormone receptors in the female rat liver.[9]

The competitive inhibition of hGH binding to rabbit liver plasma membranes indicates that only hGH is effective in competing for [^{125}I]hGH binding (Fig. 3). These receptor sites for hGH have been designated as "somatotropic" or "somatogenic" since ovine prolactin (PRL) does not interact significantly with [^{125}I]hGH binding to these receptors but neither does bGH. Hughes[17] has reported competition between rabbit GH and [^{125}I]hGH binding to rabbit liver membranes.

In human liver membranes, only hGH is effective in competing for [^{125}I]hGH binding (Fig. 4). It is interesting to note that human placental lactogen, which has the highest degree of primary sequence homology with hGH (Table 2), does not cause any displacement of [^{125}I]hGH from these receptors. Carr and Friesen[18] have reported competition between [^{125}I]hGH and ovine and bovine placental lactogens in human liver membranes. This has also been observed in IM-9 lymphoblastoid cell lines.[19]

In the order of Artiodactyla, which includes sheep, pig, cow, and goat, hGH is again the most effective competitor for [^{125}I]hGH binding (Fig. 5). In this order, there appears to be a significant interaction of bGH with the [^{125}I]hGH binding since bGH is equipotent to hGH binding in sheep liver membranes and exhibits significant competition in the pig and cow liver membranes for [^{125}I]hGH binding.

The [^{125}I]hGH binding to dog and cat liver membranes exhibits relatively high nonspecific binding, and these membranes provide an instance in which hGH is not the most effective competitor for [^{125}I]hGH binding (Fig. 6) since oPrl and hPl are more effective than hGH in competing for [^{125}I]hGH binding.

This survey of hGH binding to liver membranes from several mammalian species indicates that the "lactogenic" nature of the hGH receptors resides primarily in liver membranes from the order of Rodentia (Fig. 7). This cross-reactivity of hGH and ovine PRL (oPRL) is curious since oPRL does not share a high degree of primary structural homology with hGH (Table 2). Human placental lactogen exhibits a high degree of structural homology with hGH but does not interact significantly with the [^{125}I]hGH binding. Since regions of primary structural homology are not readily apparent between oPrl and hGH, the binding site of hGH must occupy a small portion of the hGH molecule. Even though hGH binds to liver membranes from many mammalian species, the characteristics of binding are reasonably consistent within a phylogenetic order. The broad range of hGH binding could be explained by a multipotent binding site or multiple sites that have been maintained during evolutionary development. The latter would be consistent with the suggestion of Niall *et al.*[20] that hGH has resulted from a repeated duplication of a short sequence in its phylogenetic evolution.

Mode *et al.*[21] have identified rGH as the pituitary factor responsible for the induction of the PRL receptors in female rat liver. Concomitant with the

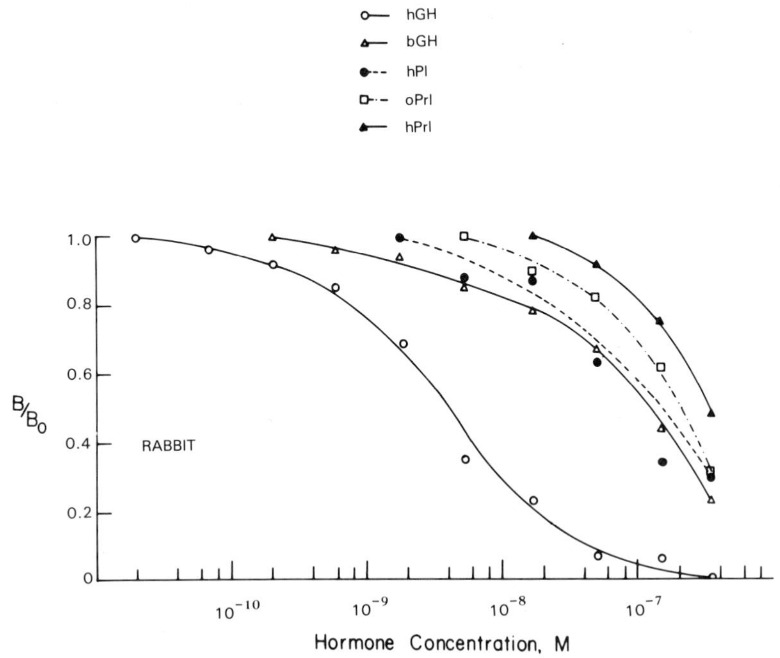

Figure 3 Competitive binding profiles of [^{125}I]hGH binding to rabbit liver plasma membranes. The methods are described in the legend to Fig. 2.

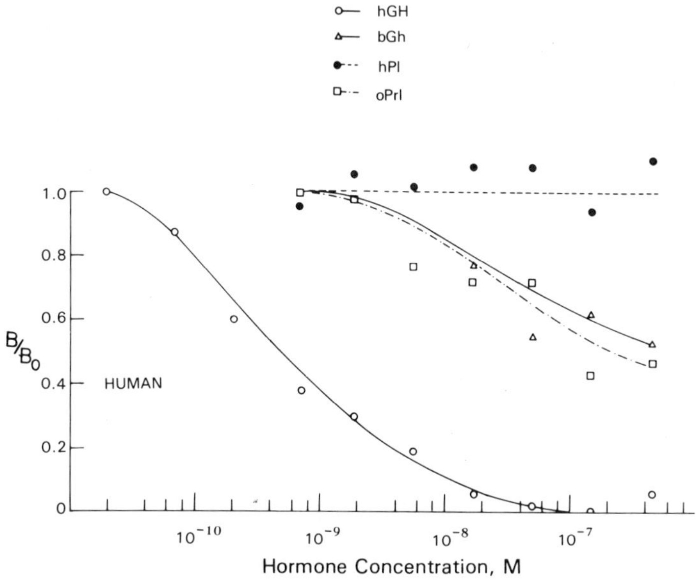

Figure 4 Competitive binding profiles of [^{125}I]hGH binding to human liver membranes. The methods are described in the legend to Fig. 2.

Table 2 Comparison of the Difference in Primary Sequence of Growth Hormones, Ovine Prolactin, and Human Placental Lactogen[15–17] [a]

	Number of differences					
	1	2	3	4	5	6
1 oPRL		143	142	144	145	143
2 hPRL	72		29	77	79	
3 hGH	72	15		67	69	65
4 bGH	73	73	35		2	23
5 oGH	73	73	36	1		23
6 rGH	75		34	12	12	
	Percent difference					

[a] oPRL = ovine prolactin; hPRL = human prolactin; hGH = human growth hormone; bGH = bovine growth hormone; oGH = ovine growth hormone; rGH = rat growth hormone.

induction of PRL receptors, they observe a "feminization" of liver metabolism of androstenedione and hypothesize that the sex hormones control hepatic steroid metabolism and PRL receptors indirectly through a hypothalamopituitary feedback mechanism.[22] The secretory pattern of rGH in male versus female animals may be responsible for the appearance of the PRL receptors since they appear following continuous GH treatment mimicking the female secretory pattern but not the episodic GH secretory pattern of male rats.

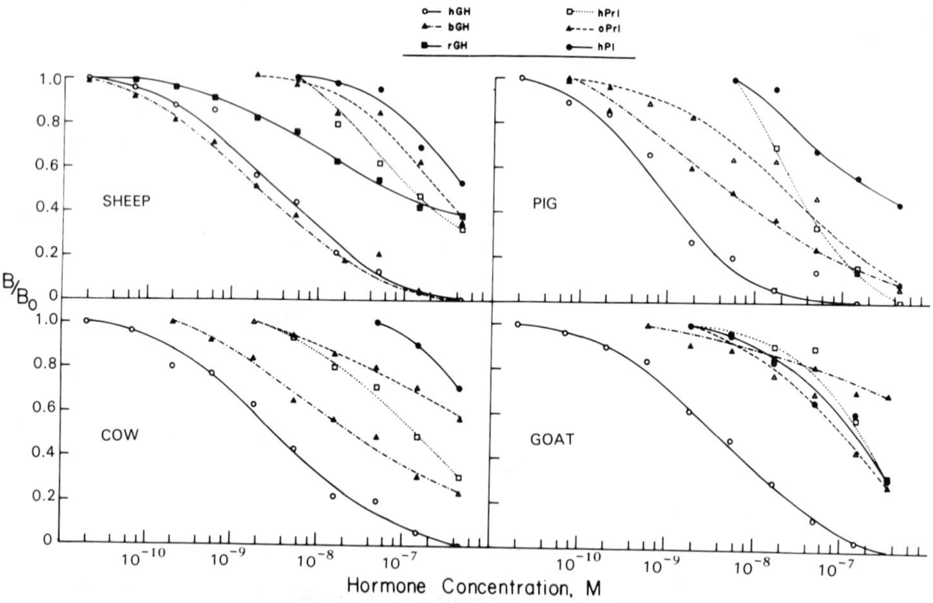

Figure 5 Competitive binding profiles of [^{125}I]hGH binding to liver membranes from animals of the order Artiodactyla. The methods are described in the legend to Fig. 2.

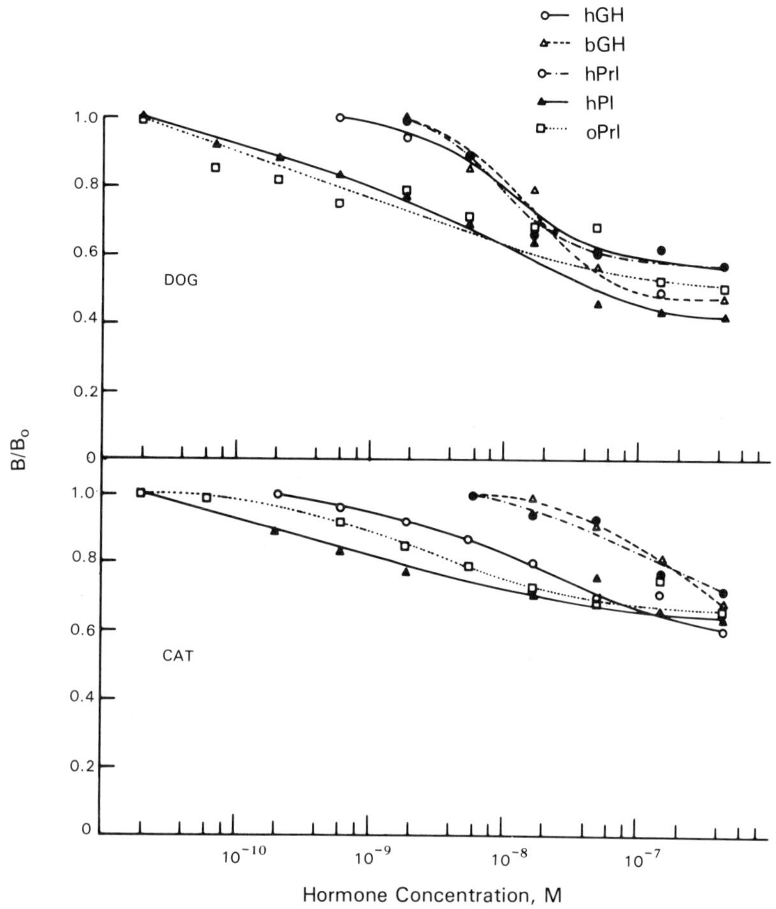

Figure 6 Competitive binding profiles of [^{125}I]hGH binding to liver membranes from adult female cat and dog. The [^{125}I]hGH binding was not normalized owing to the relatively high nonspecific binding and to hormones other than hGH competing more effectively for hGH binding. Abbreviations as in the legend to Fig. 2.

Consequently, hGH may be interacting with physiologically meaningful PRL receptors on liver from female rats due to its "lactogenic" potential. The presence of PRL receptors in liver of other species cannot be excluded on the basis of lack of competitive interaction between oPRL and hGH.[23] In this regard, we have observed [^{125}I]oPrl binding in rabbit liver membranes but minimal competition between [^{125}I]hGH binding and oPrl (Fig. 8).[24] This supports the presence of separate "somatotropic" and "lactogenic" hormone receptors not only in female rodents but in adult females of other species as well.

In any case, the cross-reactivity of [^{125}I]hGH and oPRL in the adult female or pregnant rodent liver plasma membranes is interesting.[23,24]

hGH binding to rat liver plasma membranes exhibits a significant pH

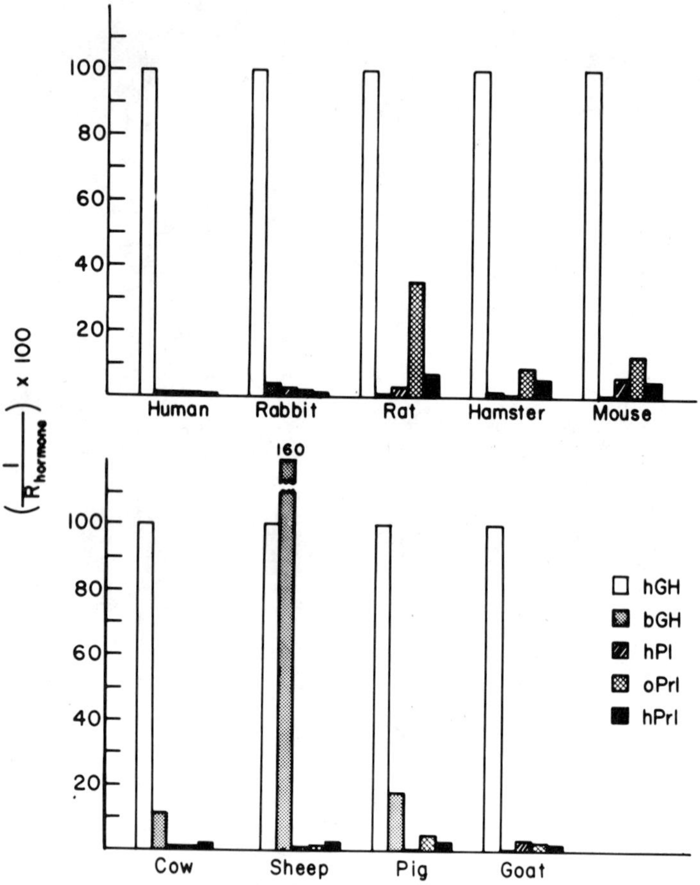

Figure 7 A comparison of the competitive properties of various hormones for [^{125}I]hGH binding. $R_{hormone}$ is defined as the ratio of the ID_{50} (the concentration of hormone producing half-maximal displacement of [^{125}I]hGH for a given hormone and the ID_{50} of hGH. ($^{1/}R_{hormone}$) × 100 depicts the ID_{50} of a hormone in competing for [^{125}I]hGH binding as a percent of the ID_{50} of hGH. Abbreviations as in the legend to Fig. 2.

optimum with maximum binding between pH 5 and 6 and binding at physiological pH that is several-fold less than optimum (Fig. 9). Cations also result in significant enhancement of hGH binding to these liver plasma membranes with the divalent cations calcium and magnesium being most effective in enhancing the hGH binding (Fig. 10).[24,25] The maximum binding appears to occur at approximately 10 mM divalent cation concentration. In contrast, maximum binding of [^{125}I]hGH occurs at approximately 150 mM concentration of monovalent cation. The effect of cations on hGH binding may be related to increased structural rigidity of the hGH molecule at higher cation concentrations. By polarization of fluorescence, we have determined that hGH is a

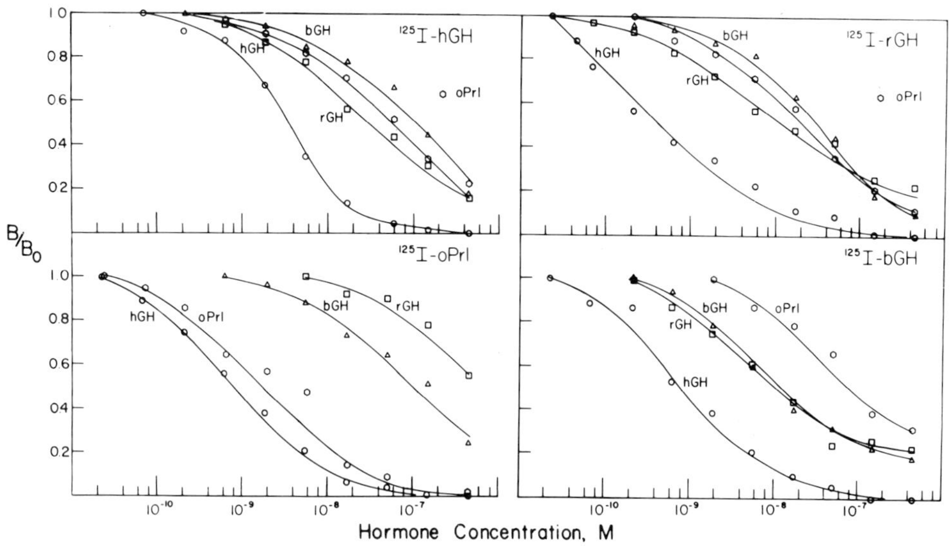

Figure 8 Competitive binding profiles of [^{125}I]hGH, -rGH, -oPRL, and -bGH binding to pregnant rabbit liver membranes. A constant concentration of iodinated hormone (0.72 ng/ml of [^{125}I]hGH, 0.61 ng/ml of [^{125}I]oPRL, 0.61 ng/ml of [^{125}I]bGH, or 0.74 ng/ml of [^{125}I]rGH) was incubated with increasing concentration of unlabeled hormone. B/B_o is normalized as in Fig. 2. Abbreviations as in the legend to Fig. 2.

relatively flexible molecule in solution and that divalent and monovalent cations result in increases in structural rigidity that may be related to its enhanced binding.[(26)]

The binding of hGH in isolated rat liver hepatocytes exhibits an irreversibility such that only a small fraction of the bound hGH is displaceable with the addition of unlabeled hGH (Fig. 11)[(27–29)] This irreversibility of binding in the

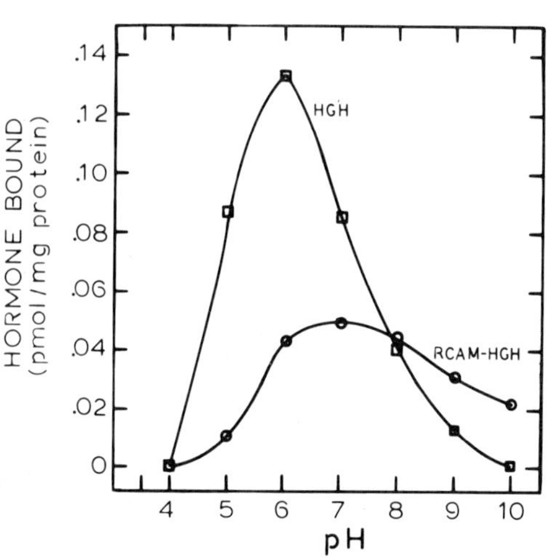

Figure 9 The effect of pH on the binding of [^{125}I]hGH and reduced and carbamidomethylated human growth hormone (RCAM-HGH) to female rat liver plasma membranes. The incubation medium contained 0.05 M buffer (acetate, pH 4 and 5; Hepes, pH 5–8; Tris, ph 8–10) in 0.1% albumin.

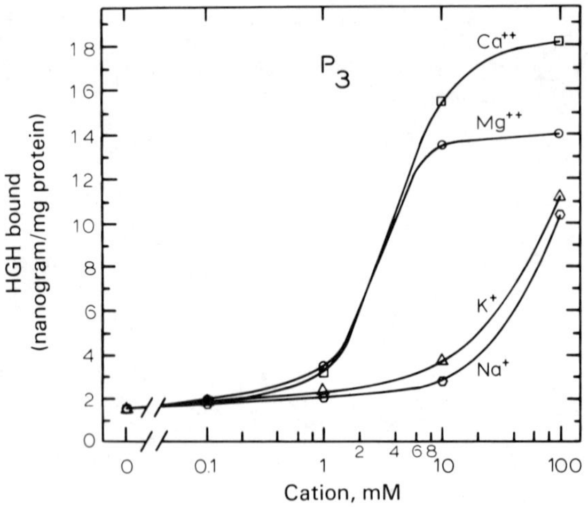

Figure 10 The effect of electrolytes on [125I]hGH binding to rat liver plasma membranes. The amount of [125I]hGH bound at each cation concentration was determined by incubating the membranes with [125I]hGH in 10 mM Hepes (pH 7.5), 0.1% albumin, and the indicated cation concentration for 2 hr at 30°C. The binding in the presence of monovalent cations plateaued at 150 mM cation concentration.

hepatocytes may be related to internalization of the bound GH [4,30,31] that has been observed in other cell preparations such as the IM-9 lymphocytes.[32] However, the irreversibility of binding is not dependent on an intact cell structure since the same irreversibility of binding is observed in isolated plasma membranes (Fig. 12).[27] Analysis of the bound [125I]hGH by Sephadex gel filtration suggests that there is no degradation of the bound hGH and internalization is not interrupted by lysosomotropic agents.[33] Recently, Hughes and Hughes[34] have reported that proteolysis of the bound GH can be detected if reduction of the GH molecule precedes polyacrylamide gel electrophoresis.

Binding of hGH to the membranes or isolated hepatocytes is stable by

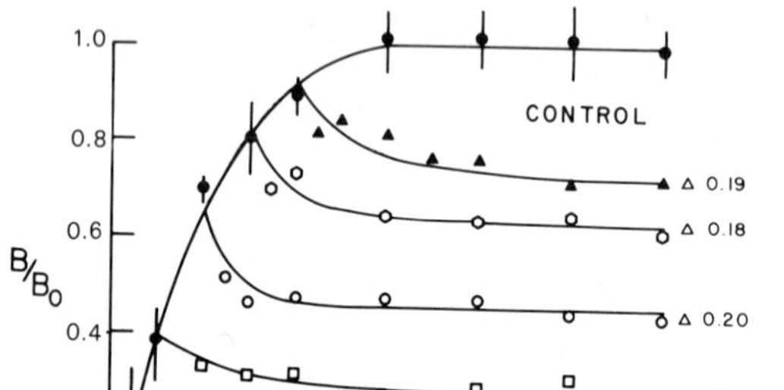

Figure 11 The time course of binding and displacement of [125I]hGH from isolated rat hepatocytes. Binding was determined by incubating the cells in Tris-buffered saline (25 mM Tris-HCl, 10 mM glucose, 5 mM KCl, 135 mM NaCl, 1.36 mM $CaCl_2$, 1.2 mM $MgSO_4$, and 15 mM Na acetate, pH 7.4) with [125I]hGH at 30°C. The relative amount of [125I]hGH bound (B/B_o) is plotted versus the time of incubation where B equaled the amount bound at each time and B_o equaled the total amount bound. The bound [125I]hGH was displaced at different time intervals of binding by adding 10 μg/ml of unlabeled hGH to the incubation medium. The number at the right of each dissociation curve represents the maximum decrease (△) in B/B_o after adding unlabeled hGH. (●) Total binding, (△) nonspecific binding, (⬢) hGH added at 15 min, (□) hGH added at 30 min, (○) hGH added at 60 min, (**O**) hGH added at 90 min, (▲) hGH added at 120 min.

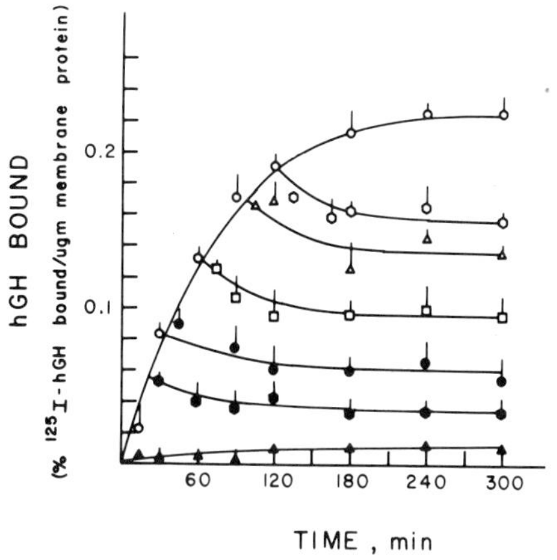

Figure 12 The time course of binding and displacement of [^{125}I]hGH in female rat liver plasma membranes. The incubation medium is described in the legend to Fig. 1. The bound [^{125}I]hGH was displaced at specified time intervals by adding 10 μg/ml of unlabeled human growth hormone (hGH) to the incubation medium. Symbols as in Fig. 11.

approximately 180 min of incubation and remains stable for at least 2–3 hr at 30°C (Fig. 11).[27,35] The binding does exhibit a temperature dependence so that maximum binding at 0–4°C is not obtained until 24 hr of incubation. Binding performed at temperatures above 37°C exhibits significant degradation of either hormone or receptor as indicated by decreased total binding. The internalized GH appears to localize to the Golgi area,[30,31] and Golgi membranes from rat liver exhibit binding that is similar to plasma membrane binding with the exception of a higher apparent affinity of the Golgi membranes for hGH (Fig. 13).[28]

Preliminary evaluations of the soluble receptor do not indicate a disulfide-linked subunit structure as is present in the insulin receptor. Regardless, the binding of [^{125}I]hGH to the rat liver plasma membrane or hepatocytes is affected by disulfide reduction.[27] Binding of the GH is completely interrupted by adding 10 mM dithiothreitol or 1% mercaptoethanol, which leads to a three- to fourfold increase in sulfhydryl concentration of the membranes (Fig. 14). The displacement of [^{125}I]hGH is greater in the presence of unlabeled hormone and reducing agents suggesting that the disulfide bonds may have some role in the irreversible binding of GH. Alkylation of the free sulfhydryls with *N*-ethylmaleimide does not affect specific growth hormone binding (Fig. 14). The effect of disulfide reduction on GH binding resides with the membranes as indicated by the following: (1) GH is not reduced under the conditions of binding, (2) pretreatment of the membranes with mercaptoethanol results in destruction of the binding capabilities of the membranes, and (3) preincubation of GH with mercaptoethanol does not significantly alter its binding ability (Fig. 15). The effect of the reducers is specific since this effect on binding can be prevented or reversed with oxidizers such as NAD and oxidized glutathione. Reduction and carbamidomethylation of hGH does alter its binding characteristics, but this effect is separate from the effect of disulfide reduction in the membranes.

We have recently turned our attention to the binding characteristics of the

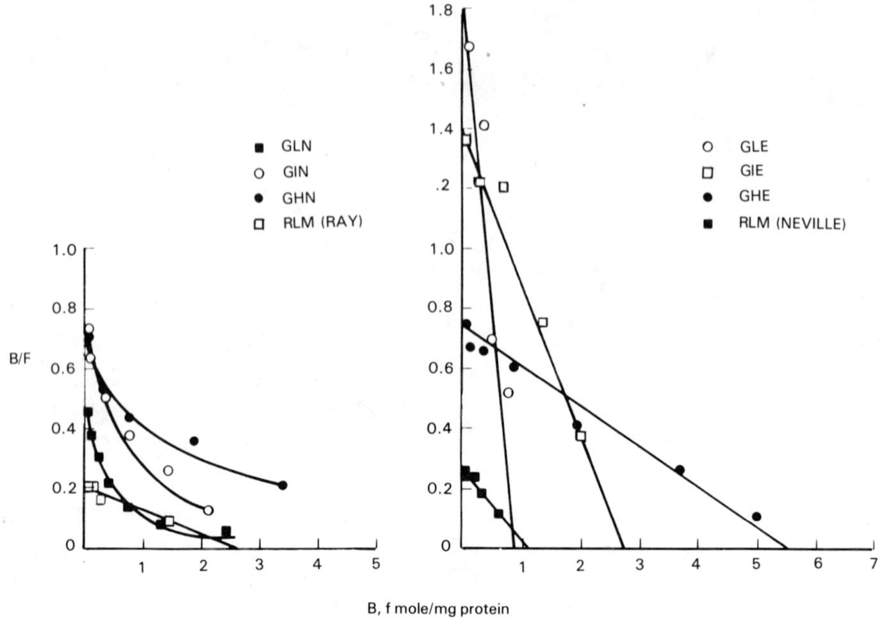

Figure 13 Scatchard plots for [^{125}I]hGH binding to plasma and Golgi membrane fractions from rat liver in normal and ethanol-treated rats. The incubation mixture contained 50 mM Tris-HCl (pH 7.5), 10 mM $CaCl_2$, 0.1% albumin, 0.13 ng/ml [^{125}I]hGH, 37–501 μg/ml of membrane protein. The results for a set of Golgi membranes (GLN, GIN, and GHN) and plasma membranes isolated by Neville's [RLM (Neville)] and Ray's [RLM (Ray)] method are presented as representative of the Scatchard plots obtained for the membranes. The ratio of the amount bound to the concentration of free hormone (B/F) is plotted versus the amount bound.

somatotropic GHs using a homologous receptor system. Our initial attempts to detect rGH binding to rat liver plasma membranes were not successful, but more recent preparations of rGH do bind to rat liver plasma membranes. The characteristics of rGH binding to rat liver plasma membranes are typical of GH binding in other homologous systems such as bGH binding to bovine liver membranes, hGH binding to human liver membranes, and ovine GH binding to ovine liver membranes. The binding of [^{125}I]bGH to bovine liver membranes is presented as an example of somatotropic GH binding. Binding of ^{125}I[bGH] is complete within 120–180 min of incubation (Fig. 16) in contrast to 240–300 min for hGH binding to the same membranes. The total binding of [^{125}I]bGH remains stable for an additional 2–3 hr of incubation at 30°C, but continued increases in nonspecific binding result in decreased specific binding after 180 min of the binding incubation. This binding is characterized by relatively high nonspecific binding and low total binding compared to hGH binding to the same membranes. The same irreversibility of bGH and hGH binding is observed to the bovine liver membrane.

The binding of bGH to bovine liver membranes does not exhibit the same

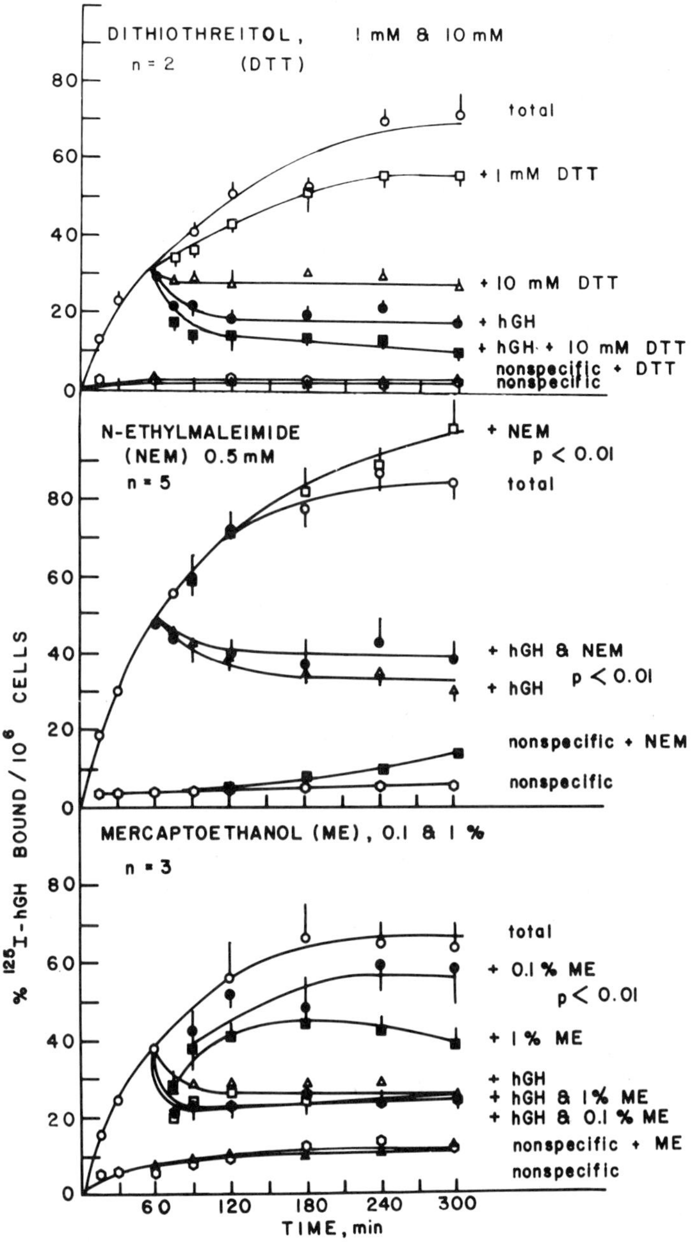

Figure 14 The effect of disulfide reduction and sulfhydryl alkylation on the time course and displacement of [^{125}I]hGH binding to isolated rat hepatocytes. The incubation conditions are described in the legend to Fig. 10. The specified concentration of the disulfide reducer or alkylater was added to the incubation medium at 60 min with or without 10 μg/ml unlabeled human growth hormone (hGH) to determine the effect of the agent on binding of displacement of [^{125}I]hGH.

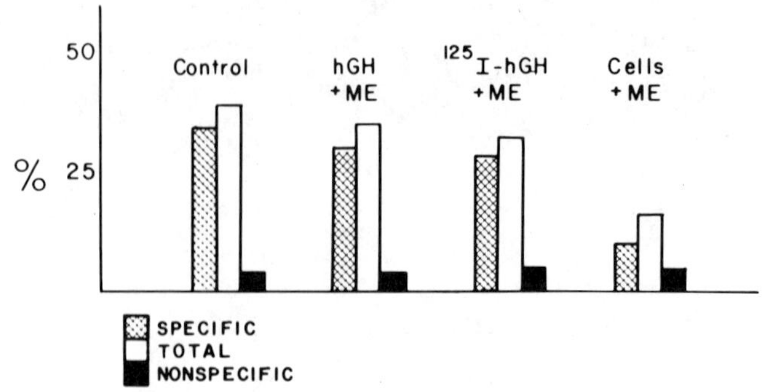

	Specific	Total	Non Specific
CONTROL	34.3	38.7	4.4
125I-hGH + ME	27.7	32.3	4.6
hGH + ME	30.3	34.6	4.3
CELLS + ME	10.0	15.8	5.8

Figure 15 HGH binding in response to selective pretreatment of unlabeled human growth hormone (hGH), [^{125}I]hGH, or hepatocytes with mercaptoethanol (ME). Isolated hepatocytes, [^{125}I]hGH, or unlabeled hGH was preincubated in the presence of 1% ME for 1 hr at 30°C (+ME). The treated component (hormone and cells) was then used in subsequent assay of [^{125}I]hGH binding. Hepatocytes (0.22×10^6) were incubated at 30°C for 2 hr with 1.36 ng [^{125}I]hGH in the presence (nonspecific) or absence (total) of 10 μg/ml unlabeled hGH. The ME concentration in the binding incubation medium was 0.1%.

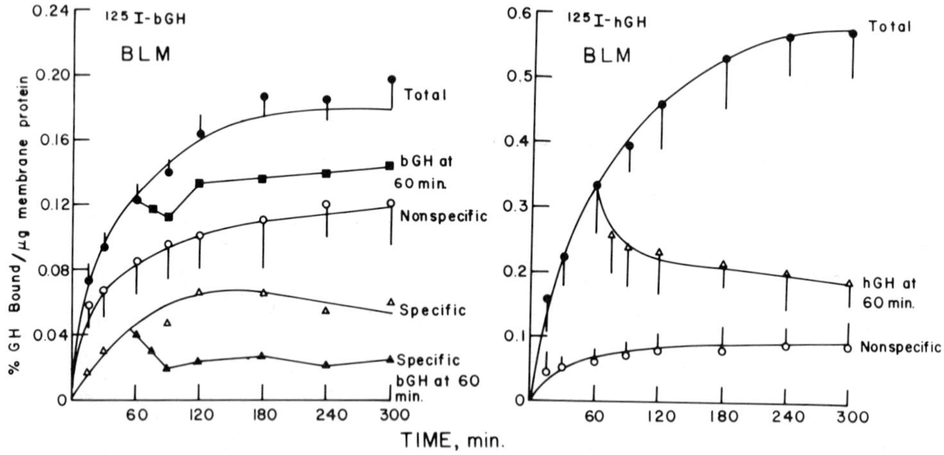

Figure 16 The binding of [^{125}]bGH and -hGH to bovine liver membranes (BLM). The amount of [^{125}I]GH bound to the membranes was determined in 50 mM Tris-HCl (pH 7.2), 10 mM $CaCl_2$, and 0.1% albumin at 30°C. Displaceability of the bound hGH was ascertained by adding 10 μg/μl of the respective unlabeled growth hormone (GH) at 60 min of the binding incubation.

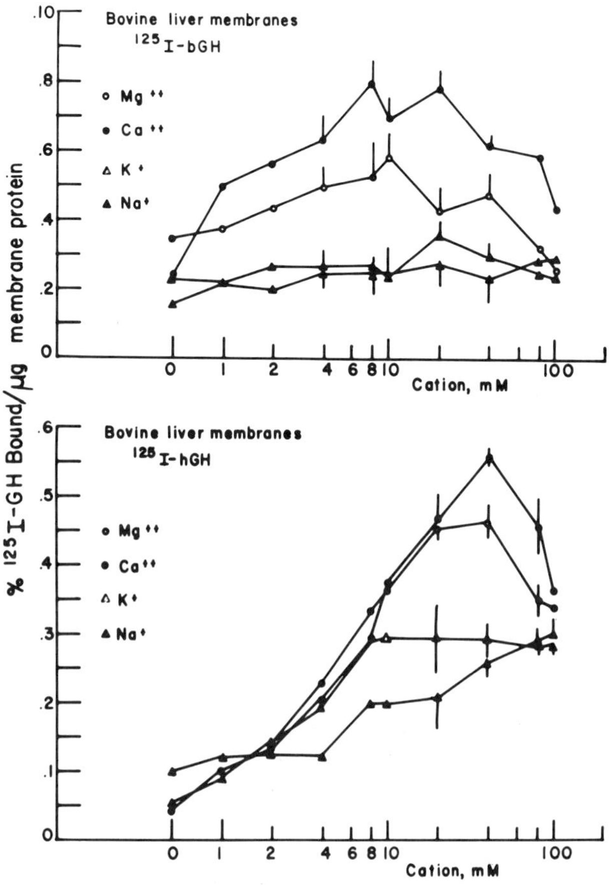

Figure 17 The effect of cations on [^{125}I]bGH and -hGH binding to bovine liver membranes. The amount of [^{125}I]GH specifically bound at each cation concentration was determined in 50 mM Tris-HCl (pH 7.2), 0.1% albumin, and the specified cation concentration at 30°C.

cation effect as hGH binding to either bovine, rabbit, or rat liver plasma membranes (Fig. 17).[25,36] There is no obvious peak of [^{125}I]bGH binding with the divalent cations, and the monovalent cations are generally ineffective in enhancing bGH binding. The bGH binding does exhibit a definite dependence on cations for binding as is illustrated in the effect of pH on the bGH binding (Fig. 18). Minimal binding is observed at physiological pH in the absence of cations; however, with cations a definite peak of binding is detected at approximately pH 6–6.5 with significant binding at physiological pH. bGH, rGH, and hGH are similarly effective competitors for [^{125}I]bGH binding, but only hGH competes effectively for [^{125}I]hGH binding (Fig. 19).

It appears that the characteristics of binding of the somatotropic growth hormones to homologous receptors differ appreciably from hGH or oPRL binding to the adult female rat liver membranes or hepatocytes and rabbit liver membranes. In comparison, the characteristics of hGH binding to bovine liver membranes are reminiscent of hGH binding to female rat or rabbit liver plasma membranes, while hGH binding to human liver membranes has characteristics

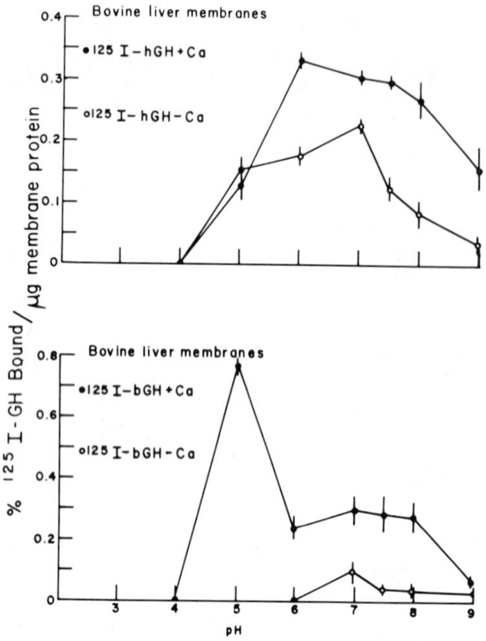

Figure 18 The effect of pH on [^{125}I]bGH and -hGH binding to bovine liver membranes in the presence (+Ca) and absence (−Ca) of 10 mM $CaCl_2$. The following buffers (50 mM concentrations) were used: acetate pH 4–5; MES, pH 5–6; PIPES, pH 6–7; and Tris, pH 7–9. No significant buffer effect was observed at the overlapping pH.

of bGH binding to bovine liver membranes. This supports the concept that the receptors for lactogenic and somatotropic hormones in the female liver are distinct and separate entities. Furthermore, the receptor for hGH in the different species may be distinct and separate from the receptor for homologous hormones in those systems.

With recent reports that fibroblasts in culture produce somatomedin C or IGF-I and exhibit a degree of responsiveness to GH,[37,38] we began investigating which cell type in the liver is responsible for GH binding. Approximately 60–70% of the volume of the liver is composed of hepatocytes; however, the lining of the sinusoids includes fixed macrophages (sharing mesenchymal origin with fibroblasts) and nonphagocytic endothelial cells. The fixed macrophage or Kupffer cell was studied as a candidate for a GH-responsive cell. The Kupffer cells can be isolated on the basis of their phagocytic ability. Prior to the perfusion of the liver for cell isolation, the rats are injected with colloidal iron which is phagocytized by the Kupffer cells.[39] Following the isolation of the cells from the liver, the Kupffer cells with the phagocytized iron can be separated from the hepatocytes by attraction to a magnet. The macrophages isolated from adult female rat liver do bind hGH with characteristics similar to the hepatocytes (Fig. 20).[40] The percent of [^{125}I]hGH bound to the Kupffer cells is less than to hepatocytes, but owing to differences in cell surface area, the concentration of surface receptors for hGH would be greater on the Kupffer cells. However, the hGH receptors in macrophages isolated from male rat and hypophysectomized rat liver do not exhibit significant binding of hGH (Fig. 21). The competitive binding properties of hGH in male and hypox female rat liver macrophages

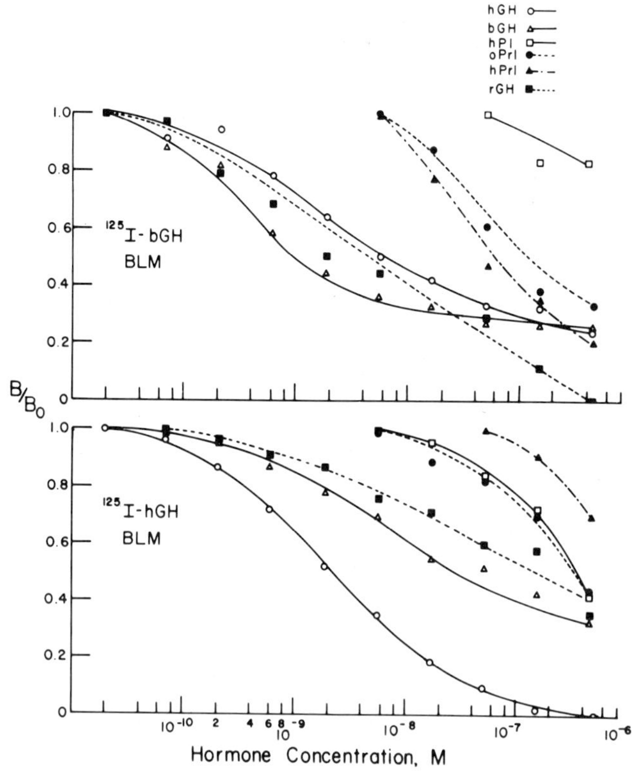

Figure 19 Competitive binding profiles of [^{125}I]bGH and -hGH binding to bovine plasma membranes. The incubation medium contained 50 mM Tris-HCl (pH 7.2), 0.1% albumin with 10 mM $CaCl_2$. The relative amount of bound B/B_o was determined at each specified concentration of unlabeled hormone. Abbreviations as in the legend to Fig. 2.

suggest that the hGH receptors on the female rat macrophages are for lactogenic hormones (Fig. 22). Therefore, it appears that the receptors for hGH in the macrophages from adult female rat liver recognize primarily the lactogenic portion of hGH. In addition, we have not detected significant binding of rGH to the Kupffer cells. Therefore, the liver macrophages are not a likely candidate as the GH but possibly a PRL-responsive cell in the rat liver. In contrast, Murphy *et al.*[41] have reported binding of [^{125}I]hGH and [^{125}I]hPRL to a cultured fibroblast cell line. It is notable that the effect of pH on hGH binding to the fibroblast is reminiscent of somatogenic GH binding to its receptor. We might speculate on a relationship between the binding of hGH to cultured fibroblasts and the observation that somatomedin C is produced by multiple tissues in the fetus,[42] or it may be possible that the binding of hGH to fibroblasts in culture represents binding secondary to transformation of the cell surface properties of the cell in culture. Another example of the appearance of hGH receptors in undifferentiated cells might be the presence of significant

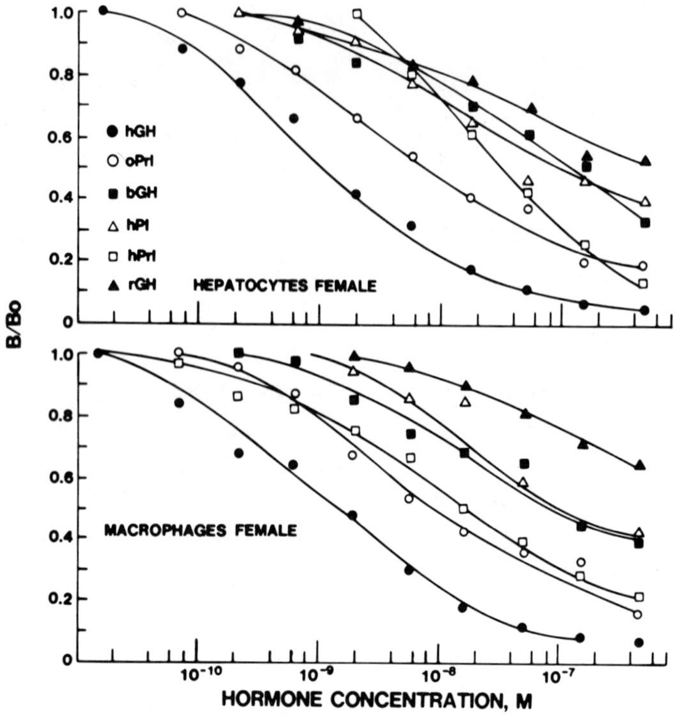

Figure 20 Competitive binding profiles of [^{125}I]hGH binding to isolated female rat hepatocytes and macrophages. The incubation medium is described in the legend to Fig. 10. Abbreviations as in the legend to Fig. 2.

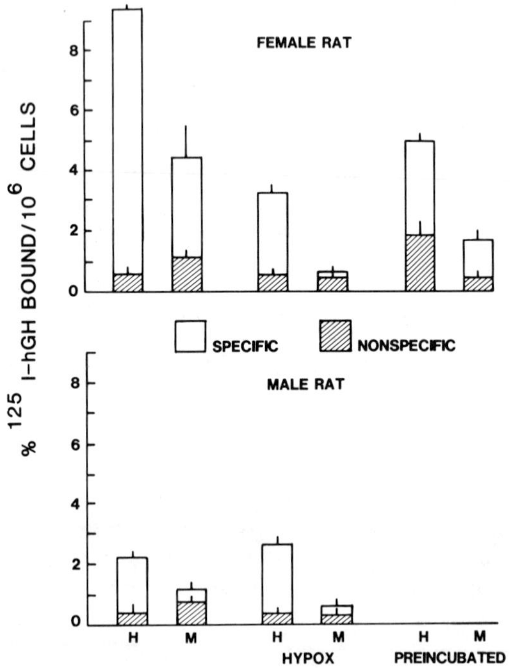

Figure 21 The effect of hypophysectomy and sex of the rat on the total and nonspecific binding of [^{125}I]hGH to rat liver hepatocytes (H) and macrophages (M). The incubation medium is described in the legend to Fig. 10.

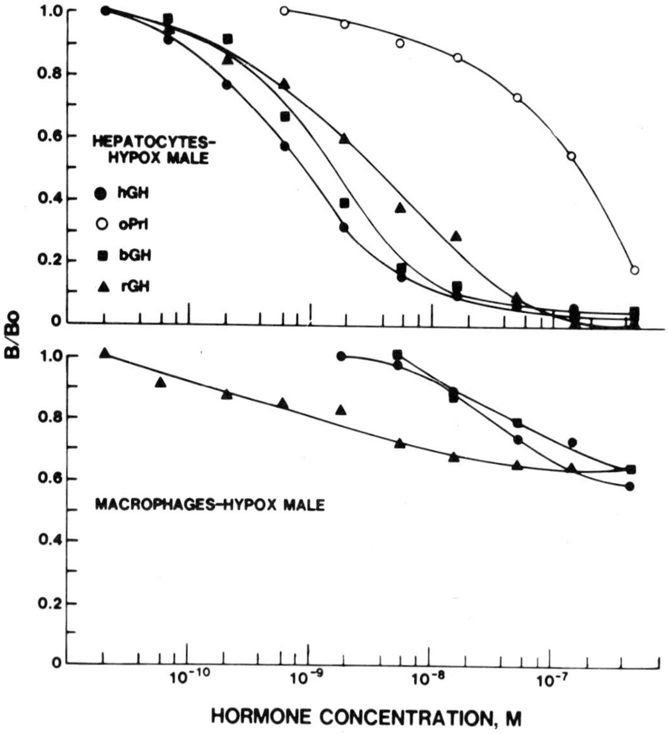

Figure 22 Competitive binding properties of [^{125}I]hGH binding to isolated hypophysectomized male rat hepatocytes and macrophages. The incubation medium is described in the legend to Fig. 10. Abbreviations as in the legend to Fig. 2.

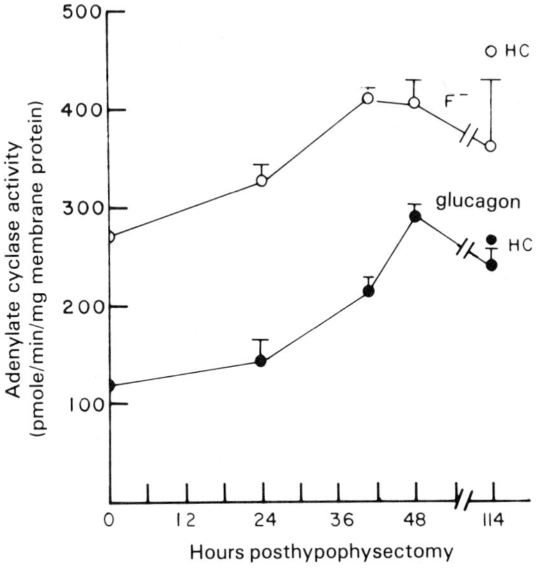

Figure 23 Time course of increase in adenylate cyclase activity of rat liver plasma membranes following hypophysectomy of adult female rats (Holtzman). The rats were hypophysectomized and then sacrificed at the specified time intervals following hypox for the isolation of liver plasma membranes. The activity at the time intervals is plotted for fluoride- (F) and glucagon-stimulated activities and is compared to the activity of a control group of rats that had been hypophysectomized (HC) at least 2 weeks previously. Each point represents the mean ± SD for six individual rats. The glucagon- and fluoride-stimulated activity was determined in the membranes obtained from six separate isolations. The incubation medium for the adenylate cyclase assay contained 0.33 mM cAMP, 4 mM $MgCl_2$, 25 mM Tris-HCl (pH 7.7), 20 mM creatine phosphate, 1 mM ATP, 0.1% albumin, 5 × 10^{-5}M GTP, 0.3 mg/ml creatine phosphokinase, and 40–150 μg of membrane protein.

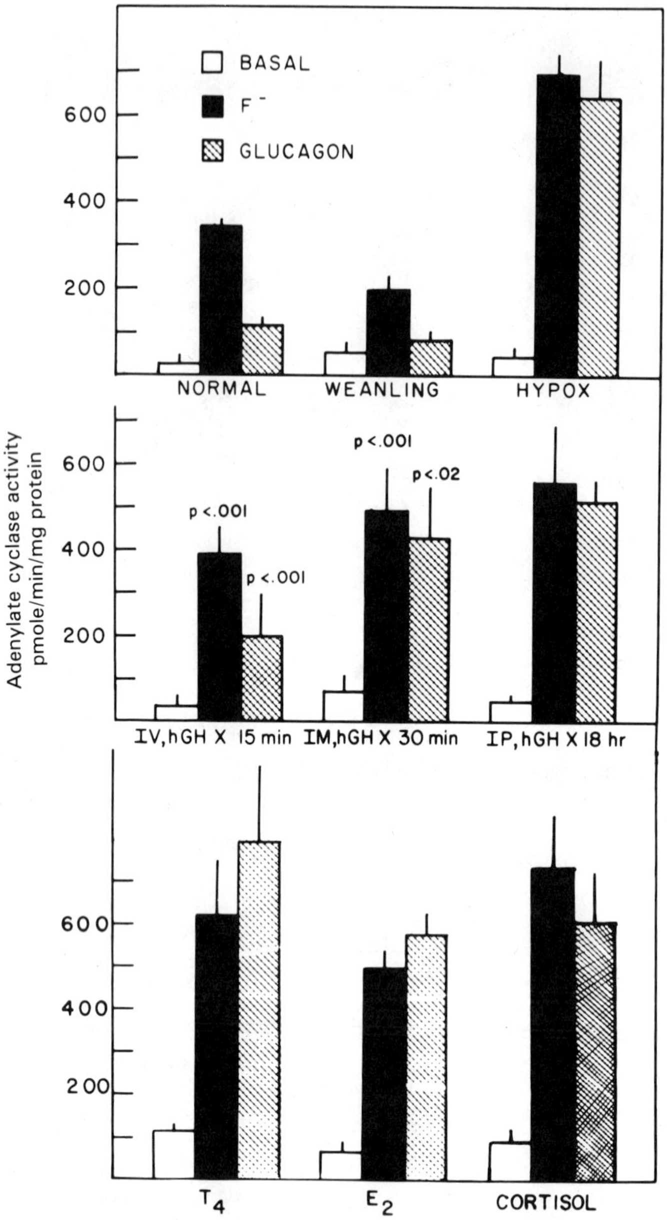

Figure 24 The effect of hypophysectomy and hormone replacement on the adenylate cyclase activity of rat liver plasma membranes. Rats were hypophysectomized for 2 weeks prior to starting hormone replacement with human growth hormone (hGH) (dosage indicated), T_4 (2 μg/100 g body weight i.m.), E_2 (50 μg/day in oil i.m.), or cortisol (0.1 ng/100 g body weight and daily i.m.).

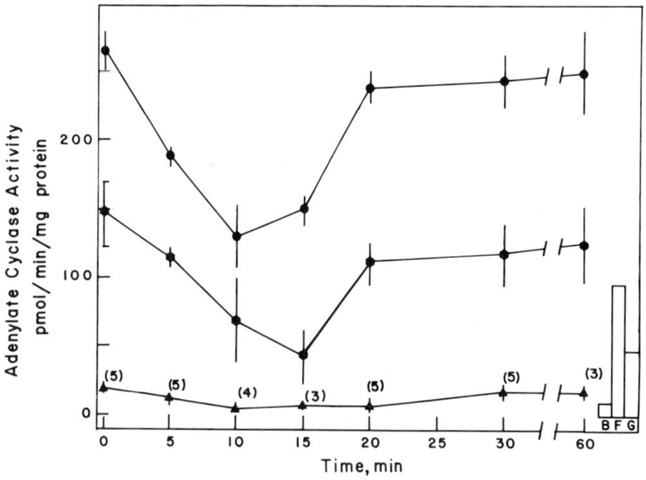

Figure 25 The time course of decrease in adenylate cyclase activity in hypophysectomized rat liver plasma membrane following intravenous injection of human growth hormone (hGH). One international unit of hGH was injected in the tail vein, and the animals were sacrificed at the time intervals listed following injection of hGH. Plasma levels of hGH at the time of sacrifice were 18–20 ng/ml. The membranes were isolated from the livers according to the Neville (16) procedure.

[^{125}I]hGH binding in IM-9 lymphocytes representing a lymphoblastoid cell line, while appreciable binding has not been detected in circulating monocytes or lymphocytes in human. The etiology of the differences in binding between *in vivo* and *in vitro* macrophages or fibroblasts may be important in understanding the control not only of GH receptors, but also of cell growth.

Resolution of many of the questions regarding the nature of the GH–

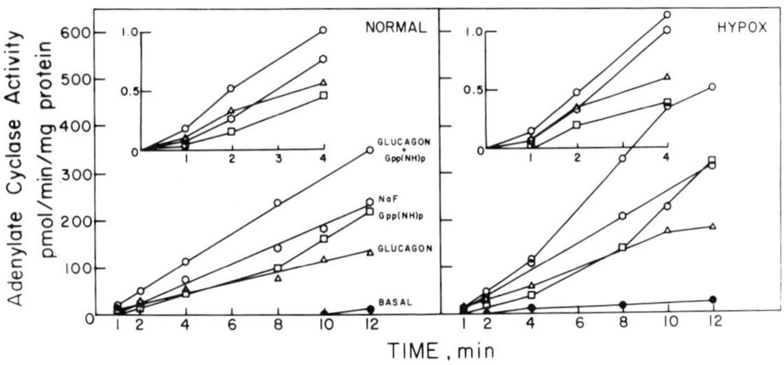

Figure 26 The time course of stimulation of adenylate cyclase in normal and hypox rat liver plasma membranes with glucagon, NaF, Gpp(NH)p, and glucagon plus Gpp(NH)p. The incubation medium is described in the legend to Fig. 23 except GTP was omitted. The amount of cAMP formed at each time interval was determined by stopping the reaction at the specified time with "stopping" solution. The detail of the first 4 min of the reaction is presented in the inset.

receptor interactions will require study of the isolated receptor. These studies are not beyond the descriptive stage, but there does appear to be variability in the GH receptor from different tissues. For example, the hGH receptor identified by cross-linking has an apparent molecular weight of 50–54K, 109K, 45K, and 40K from rabbit liver, IM-9 cells, rat liver, and rabbit mammary gland, respectively.[43] Monoclonal antibodies against the soluble receptor will probably be required for isolation of sufficient quantity of the receptor for in-depth study.[44]

Finally, we have explored the relationship of GH binding to adenylate cyclase activity of the liver plasma membranes. Initially, we observed that the glucagon- and fluoride-enhanced adenylate cyclase activity of the liver plasma membranes from hypophysectomized adult female rats is several-fold greater than activity in membranes obtained from normal adult female rats.[45] The increase in activity is observed within several hours following hypophysectomy and reaches a maximum within 24–48 hr posthypophysectomy (Fig. 23). Epinephrine- and fluoride-stimulated activity of diaphragm homogenate is not significantly different in the normal and hypophysectomized rats. The increased glucagon- and fluoride-stimulated adenylate cyclase activity returned to normal levels in rats treated with GH but not estradiol, cortisol, or thyroxine (Fig. 24). The time course of the decrease in activity following injection of hGH is consistent with the time course of internalization of hGH to the Golgi area in normal rat liver (Fig. 25). Rosberg *et al.*[46] have observed a suppressive effect of injected GH on epinephrine and Gpp(NH)p-stimulated adenylate cyclase activity in diaphragm muscle of normal young rats. The suppressive effect of GH on adenylate cyclase activity, of the muscle *in vitro* was observed only with the Gpp(NH)p-enhanced activity, suggesting a locus of effect at the regulatory protein.[47] In contrast, the characteristics of the adenylate cyclase activity in hypox rat liver plasma membranes suggest that the alteration in cyclase resides in the catalytic subunit (Fig. 26).

In summary, GH does bind to hepatocyte receptors from a variety of mammals. The characteristics of the GH binding and internalization suggest that the binding process is unique, but its uniqueness suggests biological relevance in the binding. The curious species specificity and cross-reactivity of the GH's and PRL's has resulted in some confusion regarding the nature of the GH receptor. The relationship of the somatotropic GH, hGH, and lactogenic hormone receptor is not entirely clear, but they may represent separate receptors or be the result of curious phylogenetic similarities. Direct evidence now supports a suppressive or desensitizing effect of GH on hepatic and muscle adenylate cyclase. This effect is consistent with the effects of anabolic hormones and with indirect evidence accumulated through the measurement of the effect of GH on cAMP production and the effect of phosphodiesterase inhibitors on GH action. We might speculate that GH binding either leads to internalization of the GH receptor and attendant adenylate cyclase with a resultant reduction in plasma membrane adenylate cyclase activity or occurs to an inhibitory hormone receptor with subsequent inhibition of adenylate cyclase by the recently described inhibitory GTP binding proteins (G_i). The relationship of GH's modulation of adenylate cyclase and the production of somatomedin C or IGF-I is not known but will be the subject of future research.

ACKNOWLEDGMENT. This work was supported by NIH, NIAMDD Grant AM 19544, and Research Career Development Award AM 00309 (to WVM).

References

1. B. I. Posner, P. A. Kelly, R. P. Shiu, and H. G. Friesen, Studies of insulin, growth hormone and prolactin binding: Tissue distribution, species variation and characterization, *Endocrinology* **95:**521–531, 1974.
2. B. I. Posner, P. A. Kelly, and H. G. Friesen, Induction of a lactogenic receptor in rat liver: Influence of estrogen and the pituitary, *Proc. Natl. Acad. Sci. USA* **71:**2407–2410, 1974.
3. B. I. Posner, Regulation of lactogen specific binding sites in rat liver: Studies on the role of lactogens and estrogens, *Endocrinology* **99:**1168–1177, 1976.
4. Z. Josefsberg, B. I. Posner, B. Patel, and J. J. M. Bergeron, The uptake of prolactin into female rat liver, *J. Biol. Chem.* **254:**209–214, 1979.
5. M. Postel-Vinay, and B. Desbuquois, Interaction of hGH with isolated rat liver cells, *Endocrinology* **100:**209–215, 1977.
6. P. A. Kelly, B. I. Posner, and H. G. Friesen, Effects of hypophysectomy, ovariectomy and cycloheximide on specific binding sites for lactogenic hormones in rat liver, *Endocrinology* **97:**1408–1415, 1975.
7. B. I. Posner, P. A. Kelly, and H. G. Friesen, Prolactin receptors in rat liver: Possible induction by prolactin, *Science* **188:**57–59, 1975.
8. M. Costlow, R. Buschow, and W. McGuire, Prolactin stimulation of prolactin receptors in rat liver, *Life Sci.* **17:**1457–1466, 1975.
9. M. B. Ranke, C. A. Stanley, D. Rodbard, L. Baker, A. Bongiovanni, and J. S. Parks, Sex differences in binding of human growth hormone to isolated rat hepatocytes, *Proc. Natl. Acad. Sci. USA* **130:**847–851, 1976.
10. A. C. Herington, H. G. Burger, and N. M. Veith, Binding to hGH to hepatic lactogenic binding sites: Regulation by oestrogens and androgens, *J. Endocrinol.* **70:**473–484, 1976.
11. A. C. Herington and N. M. Veith, The presence of lactogen but not growth hormone binding sites in the isolated rat hepatocyte, *J. Endocrinol.* **74:**323–334, 1977.
12. R. Knazek, S. Liu, and P. Gullino, Induction of lactogenic binding sites in the liver of the Snell dwarf mouse, *Endocrinology* **101:**50–58, 1977.
13. R. Gavin, R. J. Saltman, and S. E. Tollefsen, Growth hormone receptors in isolated rat adipocytes, *Endocrinology* **110:**637–643, 1982.
14. G. Grichting, L. K. Levy, and H. M. Goodman, Relationship between binding and biological effects of hGH in rat adipocytes, *Endocrinology* **113:**1111–1120, 1983.
15. L. P. Roquin, S. H. Sanchez, J. S. Bonifacino, and A. C. Paladini, Human somatotrophic binding to rabbit kidney microsomal membranes, *Biochem. J.* **200:**257–263, 1981.
16. D. M. Neville, Isolation of an organ specific protein antigen from cell-surface membrane of rat liver, *Biochem. Biophys. Acta* **154:**540–552, 1968.
17. J. P. Hughes, Identification and characterization of high and low affinity binding sites for growth hormone in rabbit liver, *Endocrinology* **105:**414–420, 1979.
18. D. Carr and H. G. Friesen, Growth hormone and insulin binding to human liver, *J. Clin. Endocrinol. Metab.* **42:**484–493, 1976.
19. M. A. Lesniak, P. Gorden, and J. Roth, Reactivity of non-primate growth hormones and prolactins with hGH receptors on cultured human lymphocytes, *J. Clin. Endocrinol. Metab.* **44:**838–849, 1977.
20. H. D. Niall, M. L. Hogan, R. Sauer, I. Y. Rosenblum, and F. C. Greenwood, Sequence of pituitary and placental lactogenic and growth hormones: Evolution from a premordial peptide by gene replication, *Proc. Natl. Acad. Sci. USA* **68:**866–869, 1971.
21. A. Mode, G. Norstedt, P. Eneroth, and J. Gustafsson, Purification of liver feminizing factor from rat pituitaries and demonstration of its identity with growth hormone, *Endocrinology* **113:**1250–1260, 1983.
22. G. Norstedt, P. Eneroth, J. Gustafsson, T. Hokfelt, and P. Skett, Hypothalamo–pituitary regulation of hepatic prolactin receptors in the rat, *Brain Res.* **192:**77–78, 1980.

23. J. A. Fix and W. V. Moore, Human growth hormone binding to rat hepatocytes: Effects of selected hormones on binding, *Biochem. Med.* **25:**1–14, 1981.
24. J. A. Fix, P. Leppert, and W. V. Moore, Growth hormone and prolactin binding to rabbit liver plasma membranes, *Horm. Metab. Res.* **13:**508–515, 1981.
25. W. V. Moore and D. Jin, Polymeric and monomeric hGH binding to rat liver plasma membranes, *J. Clin. Endocrinol. Metab.* **46:**374–380, 1978.
26. W. V. Moore, K. P. Wong, and A. B. Rawitch, Conformational effects of cations and pH on hGH in solution, *Endocrinology* **107:**822–829, 1980.
27. W. V. Moore, L. P. Wohnlich, and J. A. Fix, Role of disulfide bonds in human growth hormone binding and dissociation in silated rat hepatocytes and liver plasma membranes, *Endocrinology* **112:**2152–2158, 1983.
28. W. V. Moore, P. Leppert, and J. A. Fix, Comparison of human growth hormone binding to rat liver plasma and golgi membranes, *Horm. Metab. Res.* **13:**621–626, 1981.
29. D. B. Donner, K. Nakayama, S. Tani, U. Lutz, and M. Sonenberg, The uptake and degradation of growth hormone fragments of isolated rat hepatocytes, *J. Biol. Chem.* **253:**6717–6723, 1978.
30. B. I. Posner, Z. Josefsberg, and J. M. Bergeron, Intracellular polypeptide hormone receptors, *J. Biol. Chem.* **254:**12494–12499, 1979.
31. M. Postel-Vinay, C. Kayser, and B. Desbuquois, Fate of injected hGH in the female rat liver *in vivo*, *Endocrinology* **111:**244–251, 1982.
32. P. Barazzone, M. A. Lesniak, P. Gorden, E. V. Van Ohberghen, J. L. Carpentier, and L. Orci, Binding, internalization and lysosomol association of ^{125}I-hGH in cultured human lymphocytes, *J. Cell Biol.* **87:**360–369, 1980.
33. N. Hizuka, P. Gorden, M. Lesniak, J-L. Carpentier, and L. Orci, Effect of pH and lysosomotropic agents on membrane-associated and internalized ^{125}I-hGH in cultured human lymphocytes: A quantitative biochemical and electron microscopic study, *Endocrinology* **111:**1576–1580, 1982.
34. J. P. Hughes and E. F. Hughes, Cleavage of growth hormone by rabbit liver plasmalemma, *Proceedings of the 65th Annual Meeting of the Endocrine Society*, San Francisco, 1983, #546.
35. D. B. Donner, D. W. Martin, and M. Sonenberg, Accumulation of a slowly dissociable peptide hormone binding component by isolated target cells, *Proc. Natl. Acad. Sci. USA* **75:**672–676, 1978.
36. J. A. Fix, P. Leppert, and W. V. Moore, Growth hormone and prolactin binding to rabbit liver plasma membranes, *Horm. Metab. Res.* **13:**510–514, 1981.
37. D. R. Clemmons, L. E. Underwood, and J. J. VanWyk, Hormonal control of immunoreactive somatomedin production by cultured human fibroblast, *J. Clin. Invest.* **67:**10–16, 1981.
38. P. R. Atkison, E. R. Weidman, B. Bhaumick, and R. M. Bala, Release of somatomedin-like activity by cultured WI-38 human fibroblasts, *Endocrinology* **106:**2006–2012, 1980.
39. T. J. Wincek, A. L. Hupka, and F. W. Sweat, Stimulation of adenylate cyclase from isolated hepatocytes and Kupffer cells, *J. Biol. Chem.* **250:**8863–8873, 1975.
40. K. Kover and W. Moore, Comparison of hGH binding to isolated rat liver macrophages and hepatocytes, *Horm. Metab. Res.* **16:**193–197, 1984.
41. L. J. Murphy, E. Vrhovsek, and L. Lazarus, Characterization of specific growth hormone binding sites in mouse fibroblasts, *Endocrinology* **113:**750–757, 1983.
42. A. J. D'Ercole, G. T. Applewhite, and L. E. Underwood, Evidence that somatomedin is synthesized by multiple tissues in the fetus, *Develop. Biol.* **75:**315–328, 1980.
43. J. P. Hughes, J. S. A. Simpson, and H. G. Friesen, Analysis of growth hormone and lactogenic binding sites cross-linked to iodinated human growth hormone, *Endocrinology* **112:**1908–1985, 1983.
44. J. S. A. Simpson, J. P. Hughes, and H. G. Friesen, A monoclonal antibody to the growth hormone receptor of rabbit liver membranes, *Endocrinology* **112:**2137–2141, 1983.
45. P. Leppert, V. J. Guillory, L. J. Russo, and W. V. Moore, *In vivo* effect of hGH on hepatic adenylate cyclase activity, *Endocrinology* **109:**990–992, 1981.
46. S. Rosberg, K. Albertsson-Wikland, S. Eden, and O. Isaksson, Rapid effect of growth hormone or adenylate cyclase activity in diaphragm muscle of young rats, *Acta Endocr.* **225:**(Suppl.) 299, 1979.
47. K. Albertsson-Wikland, and S. Rosberg, Inhibition of adenylate cyclase activity in muscles by growth hormone, *Endocrinology* **111:**1855–1861, 1982.

37

Growth Hormone Action on Adipocytes

H. Maurice Goodman, Genevieve Grichting, and Vittorio Coiro

Adipose tissue appears to be one of the physiologically important targets for direct actions of growth hormone. Many of the metabolic responses to growth hormone that can be elicited in appropriately prepared experimental animals *in vivo* can also be shown *in vitro* when physiologically relevant concentrations of growth hormone are added to segments of epididymal fat or isolated fat cells. Metabolic responses to growth hormone, which are demonstrable either *in vivo* or *in vitro*, however, are complex and time-dependent. The initial responses are insulinlike, but these are short-lived and within a few hours give way to the more typical counterinsulin responses that are generally recognized as the physiological consequences of chronic treatment with growth hormone.[1,2]

Some of the effects of growth hormone that can be demonstrated *in vitro* in segments of rat epididymal adipose tissue or in isolated fat cells aare shown in Table 1. Among the early insulinlike effects are events that occur at the cell membrane, in the cytosolic portion of the cell, and in the mitochrondria. Stimulation of glucose utilization results from an increase in the transport of the sugar across the adipocyte membrane.[3] Increased influx of glucose results, in turn, in increased oxidation to CO_2, and the incorporation of glucose carbons into both the fatty acid and glycerol components of triglyceride. In addition, there is a disproportionate increase in the incorporation of glucose into glycogen as a consequence of activation of glycogen synthetase.[4] Another cytosolic enzyme, cyclic nucleotide phosophodiesterase, is also stimulated by growth hormone.[5] At least two intramitochondrial enzyme complexes, pyruvate dehydrogenase and branched-chain α-keto acid dehydrogenase, which catalyze the decarboxylation of pyruvate or the keto analogues of the branched-chain

H. Maurice Goodman, Genevieve Grichting, and Vittorio Coiro • Department of Physiology, University of Massachusetts Medical School, Worcester, Massachusetts 01605.

Table 1 *In Vitro Effects of Growth Hormone in Adipose Tissue*

Early effects (insulinlike)	
Increases:	1. Glucose transport and utilization[1,2,3]
	2. Decarboxylation of leucine[7]
	3. Cyclic nucleotide phosphodiesterase[5]
	4. Pyruvate dehydrogenase[6]
	5. Glycogen synthetase[4]
Decreases:	1. Lipolysis[5,7]
	2. Glycogenolysis[4]
	3. Cyclic AMP accumulation[5]
	4. Protein kinase activity[5]
Delayed effects	
	1. Induces refractoriness[1,2]
	2. Increases lipolysis[9,10,11,12]
	3. Increases glycogenolysis[13]
	4. Decreases glucose utilization[14]

amino acids, are also activated by growth hormone.[6,7] At the same time, growth hormone antagonizes the lipolytic[5,8] and glycogenolytic actions of catecholamines[4] by decreasing cyclic AMP accumulation in the cytosol[5] and somehow antagonizes the activation of protein kinase by cyclic AMP.[5] All of these effects are also produced by insulin, and the actions of the two hormones are qualitatively and, in some instances, even quantitatively indistinguishable.

With this passage of time, growth hormone produces other effects which are quite different from those of insulin. Among these are increases in lipolysis, which can be produced in the presence of theophylline[9] or glucocorticoid,[10–12] and in glycogenolyosis.[4,13] Decreased glucose utilization can be demonstrated either in adipocytes of normal rats[14] or in segments of adipose tissue obtained from hypophysectomized rats which were treated with thyroid hormone (Coiro and Goodman, unpublished observations). In addition, growth hormone induces refractoriness or insensitivity to its own insulinlike effects.[2,15] This action is unique and highly specific in two respects: Only growth hormone produces refractoriness to its own insulinlike effects, and only the insulinlike response to growth hormone and not to insulin is affected. The phenomenon of refractoriness will be considered further subsequently.

This confusing array of responses has been the subject of much discussion in the past. Various investigators have dismissed some aspects of growth hormone action as either pharmacological phenomena or artifacts attributable to other biologically active pituitary peptides which copurify with growth hormone. However, all of these responses can be produced with concentrations of growth hormone in the physiological range (1–300 ng/ml) under physiologically relevant conditions. Furthermore, for all the responses thus far examined, human growth hormone synthesized in bacteria and, therefore, free of contaminating pituitary peptides has reproduced the effects of the hormone isolated

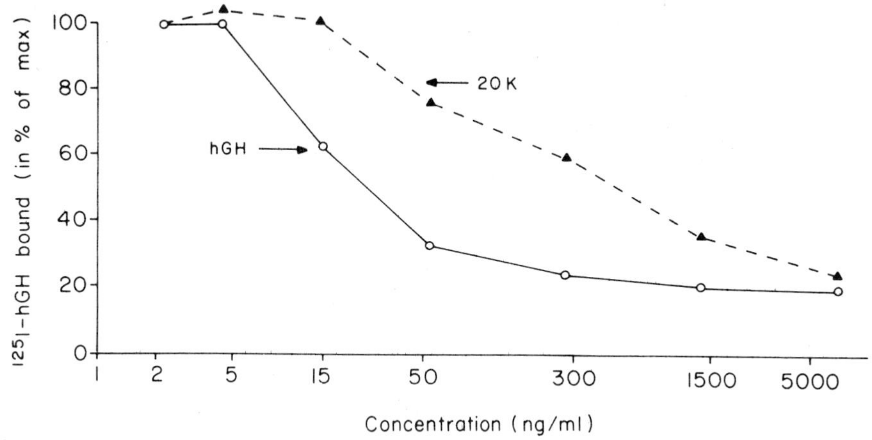

Figure 1 Displacement of [^{125}I]hGH bound to adipocytes by the 20K variant of human growth hormone (hGH). Adipocytes (10^6/ml) of normal rats were incubated with 2 ng/ml [^{125}I]hGH (HS 2223 OD, Kostyo; SA 35 μCi/μg) and various concentrations of unlabeled hGH (305-24-3, Lewis) or 20K (305-42-3, Lewis) for 2 hr. Each point is the mean of triplicate measurements.

from pituitary glands.[16] In addition, the naturally occurring 20K variant of growth hormone[17] also duplicates each of the effects of growth hormone so far tested. Although it has only 10–30% of the ability of human growth hormone to compete for binding sites on fat cells (Fig. 1), or to stimulate glucose oxidation (Fig. 2) and lipolysis, or to induce refractoriness, we conclude that production of both early and late effects is an intrinsic property of growth hormone.

Whether or not insulinlike effects are seen, however, is determined by physiological circumstances and by growth hormone itself. The responses listed in Table 1 are most readily demonstrated in tissues of rats made deficient in growth hormone by hypophysectomy. Under usual circumstances, insulinlike responses are not seen in tissues of normal rats, presumably because endogenously secreted hormone renders them refractory. Normal tissues, however, become fully capable of an insulinlike response if they are preincubated *in vitro* for 2–3 hr before they are exposed to growth hormone[18] (Fig. 3). Here and in subsequent figures, increased oxidation of uniformly labeled glucose to radioactive CO_2 will serve a simple and convenient index for insulinlike responses. In this experiment, 12 segments of epididymal fat were obtained from each of eight rats: three were tested immediately for responsiveness to growth hormone and insulin, three were allowed to preincubate for 1 hr in the absence of hormone and then tested in the 2nd hour, three were tested in the 3rd hour, and the final three in the 4th hour. In the 3rd and 4th hours, growth hormone significantly increased glucose oxidation. In the experiment shown, the responses to insulin were also enhanced in the 3rd and 4th hours, but such enhancement of the response to insulin was seen in only about half of the experiments. The appearance of insulinlike responses was completely blocked

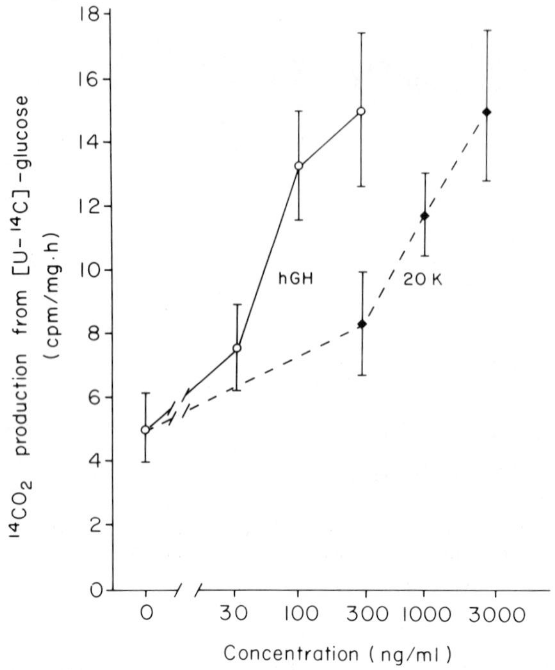

Figure 2 Effects of human growth hormone (hGH) and the 20K variant of hGH on the oxidation of [U^{14}C]glucose by segments of adipose tissue of hypophysectomized rats. Each point is the mean of 8–24 observations. The vertical brackets indicate SEMs. The hGH (305-24-3) and 20K (305-42-3) were generously provided by Dr. U. J. Lewis. Similar results were obtained with another preparation (305-75-3) of 20K.

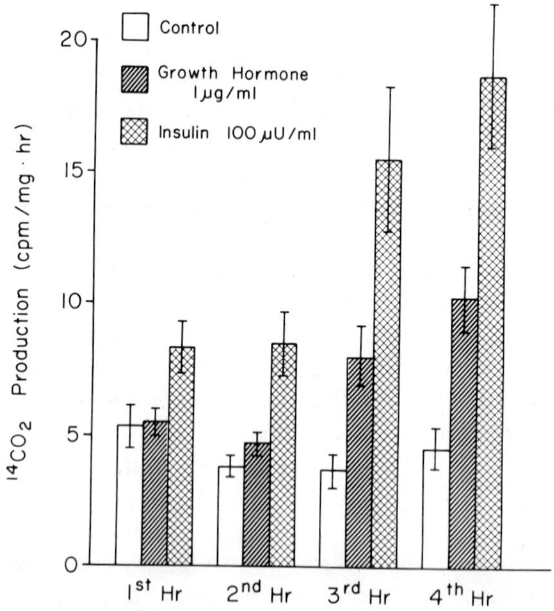

Figure 3 Effects of preincubation for various times on the response of adipose tissue from normal rats to growth hormone (GH) or insulin. Triplets of segments of epididymal fat from nine normal 58-day-old rats were incubated for 0, 1, 2, or 3 hr in Krebs–Ringer bicarbonate buffer containing 1 mg/ml glucose and 0.5% bovine serum albumin. The tissues were then transferred to medium containing [U^{14}C]glucose in the presence or absence of hormone for the final hour of incubation, and $^{14}CO_2$ was collected. Each bar represents the mean of nine observations; the SE's are indicated by the brackets. The response to GH was statistically significant ($P < 0.01$) in the 3rd and 4th hours. The increment produced by insulin in the 3rd and 4th hours was significantly ($P < 0.01$) greater than that seen in the 1st hour. Reprinted by permission from Goodman and Coiro.[17]

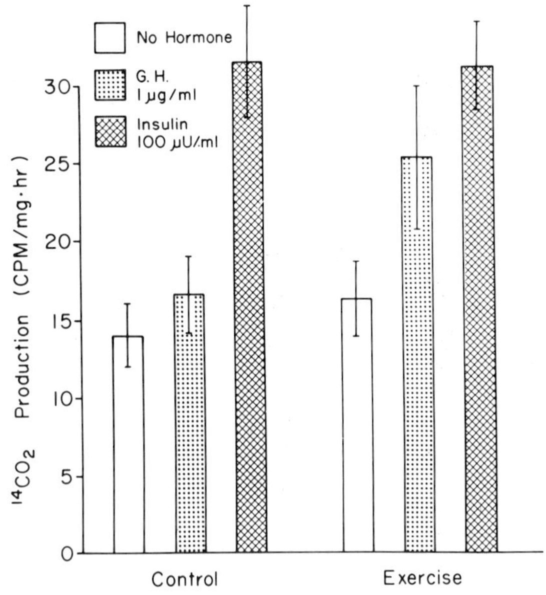

Figure 4 The effects of exercise on the response of adipose tissue to growth hormone (GH) *in vitro*. Exercised animals were made to swim for 15 min. Exercised and control rats were killed by cervical dislocation, and the tissues were studied immediately. Each bar represents the mean of eight observations. The vertical brackets indicate the SE's. The effects of insulin were significant ($P < 0.01$) in both groups, and those of GH only in the exercised group ($P < 0.01$). Reprinted by permission from Coiro *et al.*[20]

by a brief exposure to 300 ng/ml of growth hormone for the initial 15 min of incubation.[18]

Preincubation is not the only means of demonstrating an insulinlike response in normal adipose tissue. Insulinlike responses to growth hormone were also seen in tissues of weanling rats without a preincubation period[19] and in adipose tissue freshly excised from exercised rats[20] (Fig. 4). Exercise, in this case, consisted of swimming in a bucket of water for 15 min prior to sacrifice. Acute surgical stress produced by midline abdominal incision rapidly induced sensitivity in adipose tissue of normal rats[20] (Fig. 5). These tissues were excised within 1–2 min of the initiation of the surgical procedure. The pair of bars shown on the left of the figure indicate again that tissues freshly excised from normal nonstressed animals were refractory. The pair of bars in the middle of the figure indicate a significant insulinlike response in tissues excised from surgically stressed animals, and the pair of bars on the right indicate the lack of response to growth hormone in animals injected with 250 µg of the opioid receptor blocker naloxone 15 min prior to surgery. These data indicate that refractoriness can be quickly reversed and suggest that secretion of endogenous opioids may produce sensitivity, but we were only partially successful in reproducing the effects of stress by injection of β-endorphin.[20] The data in Fig. 6 suggest that a pituitary agent related to β-endorphin might be involved. Segments of adipose tissue were incubated in medium in which pituitary glands had previously incubated for 1 hr. Medium in which pituitary glands were incubated along with corticotropin releasing hormone (CRH) and vasopressin contained something which, when added to segments of adipose tissue, rendered them sensitive to the insulinlike effects of growth hormone.

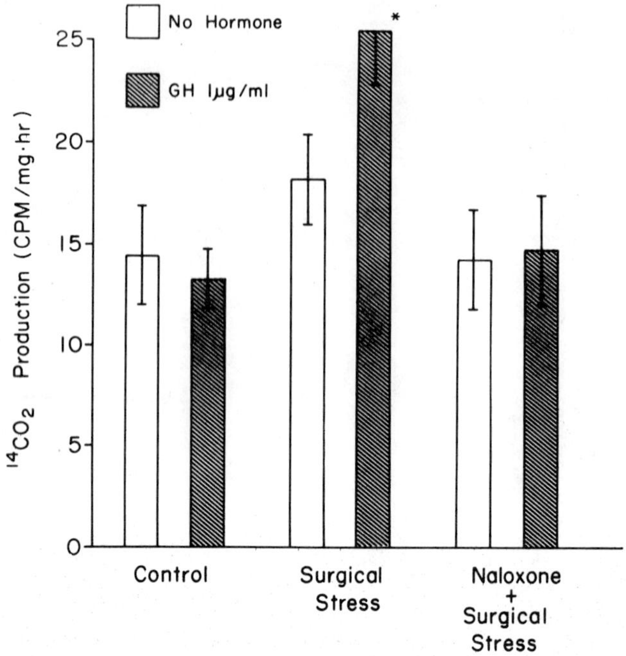

Figure 5 Inhibition by naloxone of the stress-induced insulinlike response to growth hormone (GH). Rats were injected i.p. with 250 μg naloxone 15 min before laparatomy. Tissues were excised from control animals after cervical dislocation and from experimental animals during laparatomy under ether anesthesia. Each bar represents the mean of eight observations; the vertical brackets indicate the SE's. The asterisk indicates a statistically significant ($p < 0.05$) effect of GH. Reprinted by permission from Coiro *et al.*.[20]

The effects of this pituitary secretion were blocked by naloxone. No insulinlike effect was seen in tissues that were incubated in medium from unstimulated pituitary glands, or when tissues were incubated with just CRH and vasopressin. It obviously would not have been possible to perform this experiment if the pituitary glands also released growth hormone into the medium. To perform these studies, therefore, we used pituitary glands of rats that had been thyroidectomized at least 2 weeks earlier, since such glands contain virtually no growth hormone.[21] These data indicate that a pituitary product secreted in response to CRH and blocked by naloxone is capable of inducing sensitivity to the insulinlike effects of growth hormone in tissues that would otherwise be refractory. We suggest that one or more components of the ACTH family which are secreted in response to stress may reverse refractoriness to the insulinlike effects of growth hormone in adipose tissue.

From the foregoing discussion, it is evident that sensitivity to insulinlike stimulation by growth hormone can be turned on and off physiologically. We consider next what is known about the termination of the insulinlike response and the induction of refractoriness. In Fig. 7 is shown, in schematic form, a summary of a large number of experiments that demonstrate the time course

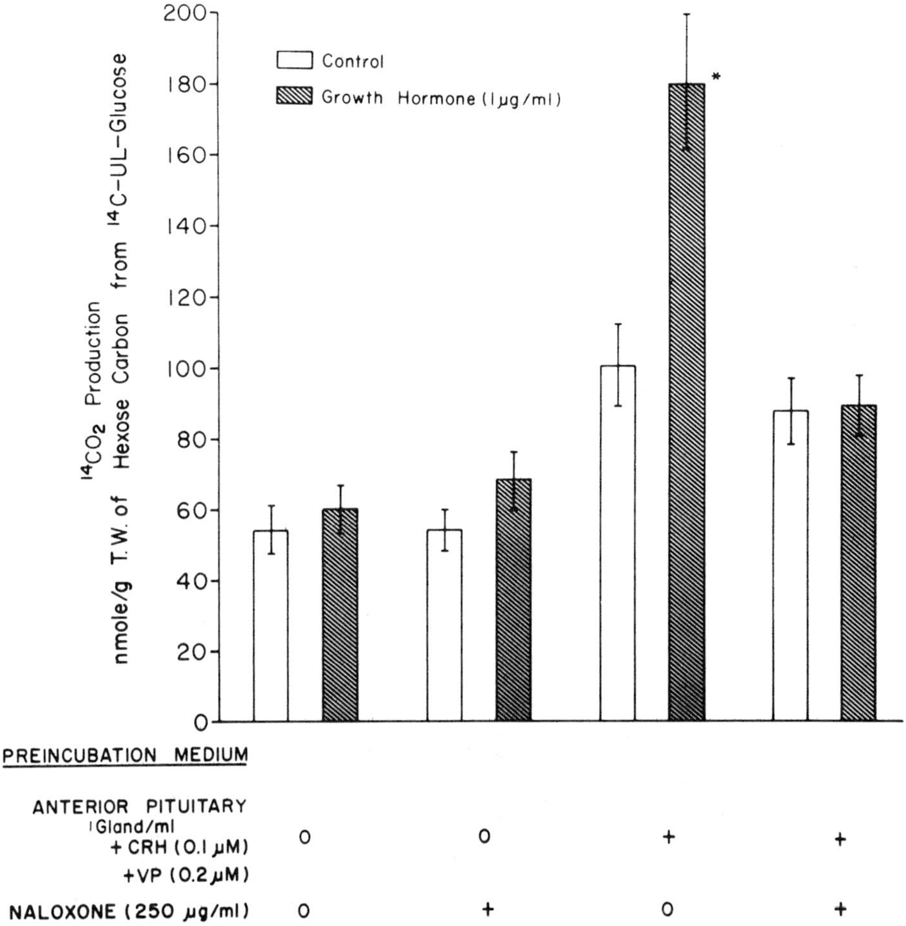

Figure 6 Induction of sensitivity to the insulinlike effects of growth hormone (GH) by preincubation of segments of adipose tissue with medium in which anterior pituitary glands had previously incubated for 1 hr in the presence of CRH and vasopressin. Pituitaries were obtained from rats that has been thryoidectomized 2 weeks earlier. Tissues were preincubated in the pituitary medium for 15 min in the absence or presence of naloxone and then reincubated for 1 hr along with [$U^{14}C$]glucose and GH. Each bar represents the mean of eight observations. The vertical brackets indicate the SE's. The asterisk indicates a statistically significant ($p < 0.01$) effect of GH.

for the effects of growth hormone on glucose oxidation and the action of agents that affect that process.[2,15] After a lag period of 20–30 min,[22] growth hormone stimulates glucose oxidation, which reaches a maximum after 1–2 hr and then gradually decays. By 3 hr, the effect is no longer significant despite the continued presence of growth hormone in the incubation medium. In contrast, stimulation of glucose oxidation by insulin persists as long as the hormone is present. The difference between insulin and growth hormone may be accounted for by an action of growth hormone to promote the synthesis of

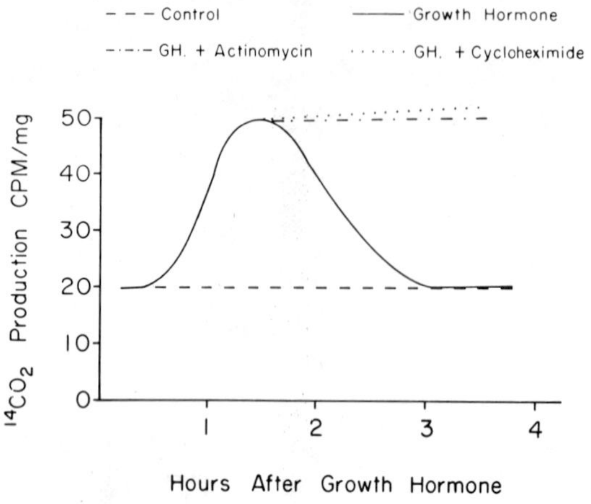

Figure 7 Schematic summary of the time course of the insulinlike response to growth hormone (GH) and its prolongation by inhibitors of RNA and protein synthesis.

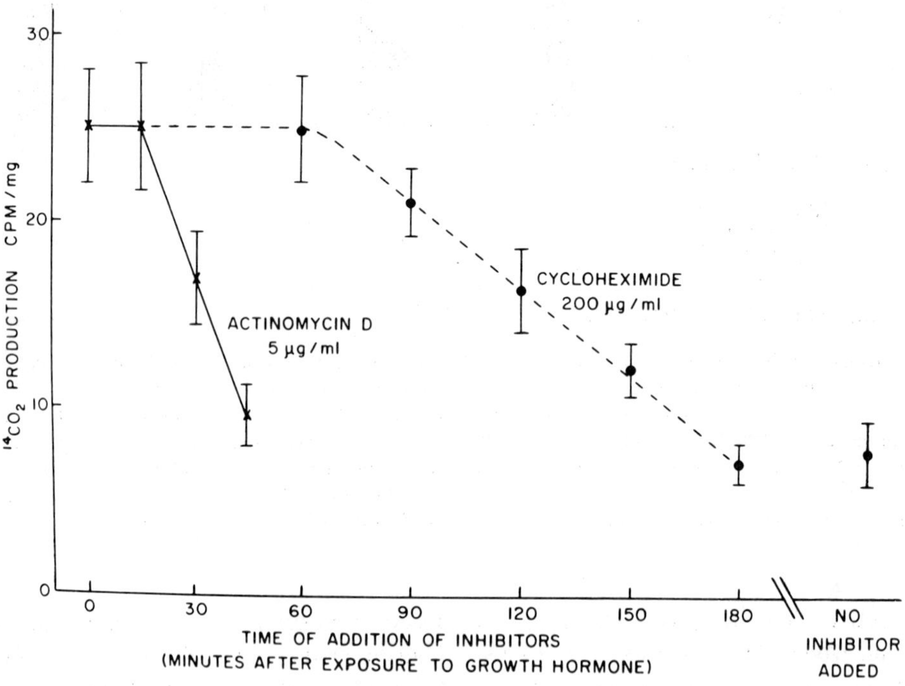

Figure 8 The effects of actinomycin D and cycloheximide on $^{14}CO_2$ production in the 4th hour after exposure of segments of epididymal fat from hypophysectomized rats to growth hormone. Each point represents the mean of eight observations. The vertical brackets give the SE's. Growth hormone (NIH-GH-S-7) was present at a concentration of 1 µg/ml throughout the 3-hr preincubation period and the 1-hr period (4th hour) in which $^{14}CO_2$ production was measured. Reprinted by permission from Goodman.(15)

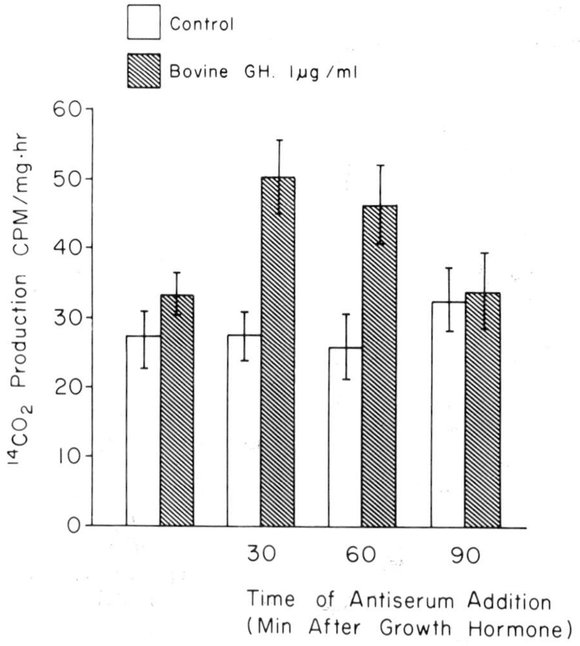

Figure 9 The effects of antiserum on the production of refractoriness by rat growth hormone (GH). Tissues were incubated with rat GH (1 μg/ml) for 60–120 min. Antiserum (5 μl) was added after 30, 60, or 90 min, and the tissues were transferred to fresh medium in the absence of antiserum or hormone until 180 min. The tissues treated with rat GH and no antiserum were transferred to hormone-free medium at 60 min. At 180 min, all tissues were transferred for the final hour of incubation to fresh medium containing [U-^{14}C]glucose in the presence or absence of 1 μg/ml bovine GH. Each bar represents the mean rate of $^{14}CO_2$ production by tissues of nine rats. The SE's are indicated by the vertical brackets. The effects of bovine GH were significant ($p < 0.01$) in tissues to which antiserum was added 30–60 min after rat GH.

new species of protein, which, in effect, shuts off the insulinlike response. When either actinomycin D, an inhibitor of RNA synthesis, or cycloheximide, an inhibitor of protein synthesis, is added along with growth hormone, the stimulation of glucose oxidation persists for at least 4 hr just as the response to insulin persists.

RNA synthesis appears to occur between 15 and 45 min after growth hormone is added.[15] To demonstrate this, we used an experimental protocol in which persistence of elevated rates of CO_2 production in the 4th hour was examined in tissues that were treated with growth hormone at time 0 followed at various times thereafter by actinomycin D (Fig. 8). When actinomycin D was added simultaneously with growth hormone, or 15 min later, increased oxidation of glucose persisted into the 4th hour. When added at 30 min, actinomycin D was only partially effective in sustaining elevated rates of glucose oxidation, and when it was added at 45 min, the rate of glucose oxidation in the 4th hour was identical to that observed when no inhibitor was added. We infer that the transcriptional event triggered by growth hormone was already complete by 45 min and, therefore, no longer subject to inhibition. Cycloheximide added as late as an hour after growth hormone was fully effective in maintaining the elevated rate of glucose oxidation in the 4th hour of incubation suggesting translation had not yet begun. When added 90, 120, or 150 min after growth hormone, cycloheximide was progressively less effective in sustaining the elevated rates of glucose oxidation suggesting that the induced proteins are

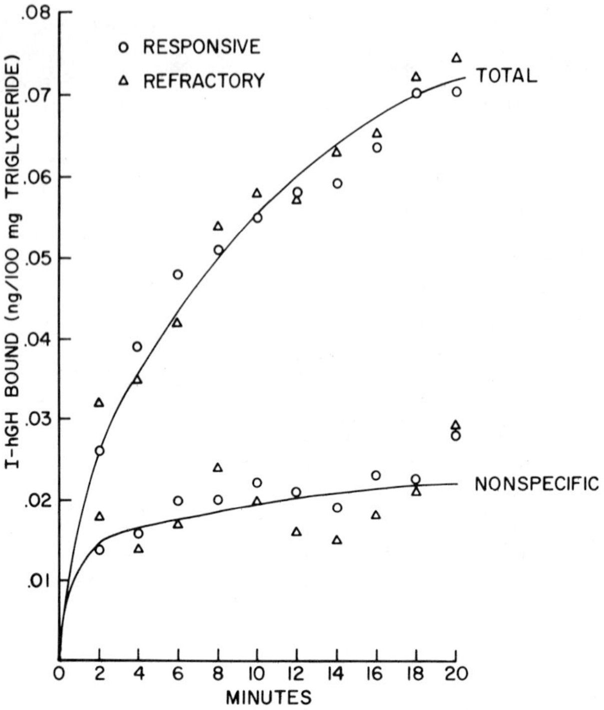

Figure 10 Specific binding of human growth hormone (hGH) to adipocytes responsive or refractory to the insulinlike actions of GH. One epididymal fat pad from each of 12 normal rats was cut into small segments and incubated for 15 min in the presence of 1 μg/ml ovine GH. The contralateral fat pads were preincubated in the absence of ovine GH. Isolated adipocytes were then prepared from each of the tissue pools, and the cells were allowed to incubate for an additional 2 hr in the absence of hormone. The cells were then collected by centrifugation and reincubated in the presence of a trace amount (2 ng/ml) of [^{125}I]hGH. For determination of nonspecific binding excess (5 μg/ml) of unlabeled hGH was added. Each point is the mean of triplicate measurements. Refractoriness and sensitivity were verified by measuring $^{14}CO_2$ production by aliquots of cells from each pool incubated in the 4th hour in the presence or absence of 1 μg/ml of ovine GH.

synthesized in this interval, coincident with the decline in the insulinlike response.

It is unlikely that the proteins responsible for terminating the insulinlike effect are also responsible for refractoriness. Figure 9 shows the response of tissues to bovine growth hormone added in the 4th hour of incubation. All of the tissues were exposed to rat growth hormone at 0 time. As already mentioned, fresh hormone added 3 hr after an initial exposure of tissues to growth hormone fails to induce a 2nd insulinlike response. Addition of excess antiserum to rat growth hormone at 30 or 60 min prevented refractoriness, as indicated by the substantial insulinlike response in the 4th hour.(23) When antiserum was added as late as 90 min, however, it was ineffective in this regard. From these data, we conclude that the cells were not committed to a refractory response to

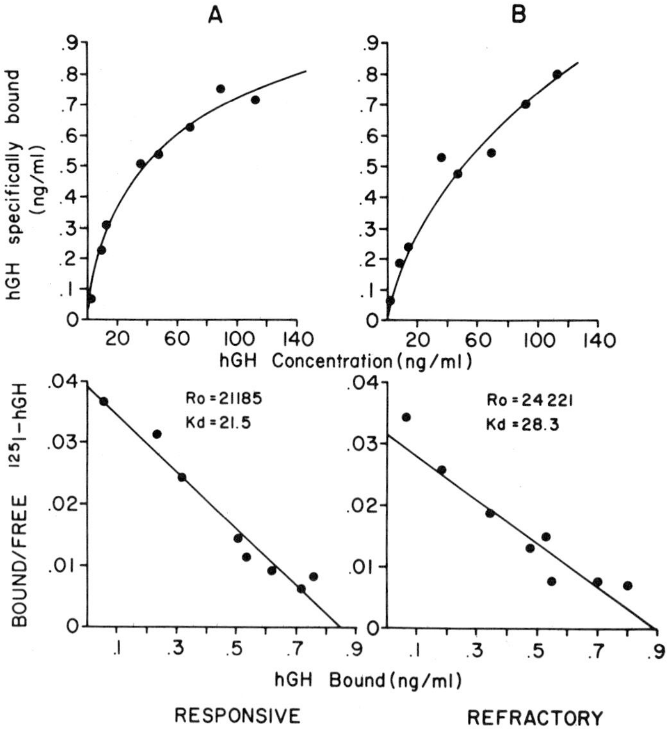

Figure 11 Specific binding of human growth hormone (hGH) to adipocytes responsive or refractory to the insulinlike action of GH. Cells were prepared as in Fig. 10 and incubated with 2 ng/ml of [^{125}I]hGH and the indicated amounts of unlabeled hGH. The upper panels indicate the concentration dependence of specific binding. The same data are replotted in the lower panels in Scatchard plots. The data presented in A were obtained with cells capable of responding to the insulinlike actions of hGH, while those in B were obtained with refractory cells, i.e., derived from those tissues preincubated with ovine GH. Reprinted by permission from Grichting *et al.*[24]

growth hormone until at least 60 min had elapsed. At this time, the RNA coding for the protein which terminates the insulinlike response had already been synthesized, as shown in the previous figure. Thus the irreversible step in the production of refractoriness does not occur until well after the fat cells are committed to terminate the insulinlike response. In fact, regardless of the time the antiserum was added, the insulinlike response was terminated as judged by restoration of the basal rate of glucose oxidation in the 4th hour. Furthermore, since a 2nd insulinlike response could be induced in the 4th hour in the tissues treated with antiserum within 60 min the protein(s) that terminated the first insulinlike response must have a very short half-life.

We still do not know the molecular mechanisms by which insulinlike responses are terminated or cells are made refractory to them. We observed no differences in the initial rates of binding of [^{125}I]human growth hormone to adipocytes which are responsive or refractory to the insulinlike effects of

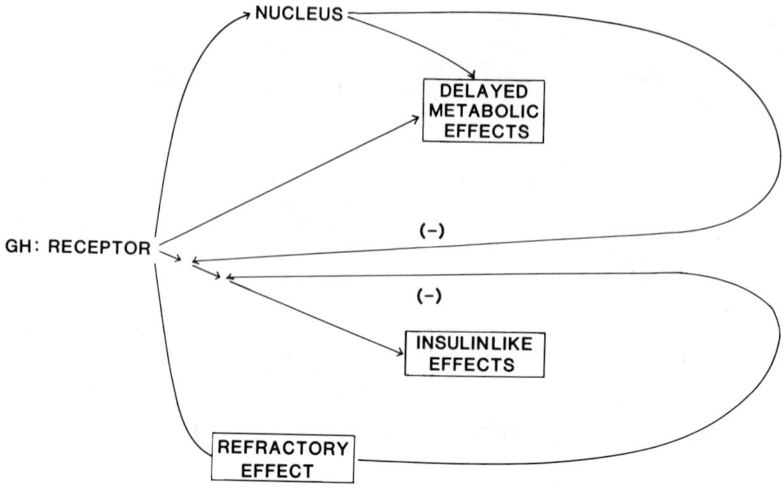

Figure 12 Hypothetical model for the actions of growth hormone (GH) on adipocytes.

growth hormone (Fig. 10). Each point is the mean of triplicate observations. The circles represent responsive cells and the triangles refractory cells. The upper curve illustrates the total amount of [^{125}I]growth hormone bound to the cells, and the lower curve indicates that amount of hormone which bound nonspecifically. By nonspecific binding, we mean that amount of binding which cannot be displaced by the addition of a large excess of unlabeled hormone. To obtain these data, 2 ng/ml of [^{125}I]human growth hormone (S.A. 1.4 mCi/nmole) was added to a suspension containing 1,000,000 fat cells/ml.

Additional binding studies(21) (Fig. 11) confirmed the observation that binding of labeled hormone is unchanged in cells that are either refractory to the insulinlike effects of growth hormone, shown in Fig. 11B, or cells that are responsive to the insulinlike effects of growth hormone, shown in Fig. 11A. The curves are no more different from each other than curves obtained in duplicate experiments on a single pool of cells. Scatchard analysis of the data from both groups of cells gave linear plots suggestive of a single class of about 20,000 receptors/cell, which appeared to be half-saturated at a concentration of around 20 ng of growth hormone/ml. These binding constants were confirmed in a large number of additional experiments which also revealed no difference in the binding characteristics of growth hormone in cells that are responsive or refractory to the insulinlike effects of growth hormone. It is noteworthy that the concentration of growth hormone needed to half-saturate binding sites very nearly approximates the concentration needed to produce half-maximal insulin-like effects.(24) Conclusions drawn from Scatchard analysis, however, must be tempered with caution, since growth hormone binding to fat cells does not satisfy all the conditions needed for equilibrium analysis.

Figure 12 shows our present working model for the actions of growth hormone on fat cells. Though not conclusive, the weight of available evidence suggests that growth hormone interacts with a single class of surface binding sites whose availability is unaltered by refractoriness. Growth hormone: receptor interaction triggers what appear to be four categories of responses. The insulinlike effects may be mediated by a second messenger akin to or identical with that described for insulin.[25–28] The refractory effect, which is triggered after a delay of 1 hr, prevents the appearance of insulinlike effects without modifying responses to insulin suggesting an action at an early site in the insulinlike pathway. Shutting off the insulinlike response appears to involve the genome, and since responses to insulin are not altered, an early site of action in the insulinlike pathway again seems likely. Finally, at least some aspects of the delayed metabolic responses, e.g., stimulation of lipolysis,[9] do not require synthesis of RNA or protein.

In summary, growth hormone produces multiple effects in adipose tissue. Responsiveness to the insulinlike effects is not always evident and can be turned on or off under different physiological conditions.

References

1. H. M. Goodman, *In vitro* actions of growth hormone on glucose metabolism in adipose tissue, *Endocrinology* **76:**216, 1965.
2. H. M. Goodman, Early and late effects of growth hormone on the metabolism of glucose in adipose tissue, *Endocrinology* **76:**1134, 1965.
3. H. M. Goodman, Effects of growth hormone on the penetration of L-arabinose in adipose tissue, *Endocrinology* **78:**819, 1966.
4. T. H. Honeyman and H. M. Goodman, Effects of growth hormone on glycogen metabolism in adipose tissue of hypophysectomized rats, *Am. J. Physiol.* **238:**389, 1980.
5. R. S. Birnbaum and H. M. Goodman, Studies on the mechanism of the antilipolytic effect of growth hormone, *Endocrinology* **99,**1336, 1976.
6. H. M. Goodman and G. P. Frick, Effects of growth hormone on the oxidation of [1-^{14}C]-pyruvate in adipose tissue of hypophysectomized rats, *Hormone Metab. Res.* **13:**351, 1981.
7. H. M. Goodman, Effects of growth hormone on the utilization of L-leucine in adipose tissue, *Endocrinology* **102:**210, 1978.
8. H. M. Goodman, Antilipolytic effects of growth hormone, *Metabolism* **19:**849, 1970.
9. H. M. Goodman, The effects of growth hormone on the lipolytic response of adipose tissue to theophylline, *Endocrinology* **82:**1027, 1968.
10. J. N. Fain, V. P. Kovacev, and R. O. Scow, Effect of growth hormone and duxamethasone on lipolysis and metabolism in isolated fat cells of the rat, *J. Biol. Chem.* **240:**2522, 1965.
11. H. M. Goodman, Multiple effects of growth hormone on lipolysis, *Endocrinology* **83:**300, 1968.
12. H. M. Goodman and G. Grichting, Growth hormone and lipolysis. A reevaluation, *Endocrinology* **113:**1697, 1983.
13. H. M. Goodman and H. J. Eisen, Growth hormone and phosphorylase activity in adipose tissue, *Endocrinology* **84:**414, 1969.
14. B. L. Maloff, J. H. Levine, and D. H. Lockwood, Direct effects of growth hormone on insulin action in rat adipose tissue maintained *in vitro*, *Endocrinology* **107:**538, 1980.
15. H. M. Goodman. Growth hormone and the metabolism of carbohydrate and lipid in adipose tissue, *Ann. N.Y. Acad. Sci.* **148:**419, 1968.
16. H. M. Goodman, Biological activity of bacterial derived human growth hormone in adipose tissue of hypophysectomized rats, *Endocrinology* **114:**131, 1984.
17. H. M. Goodman and V. Coiro, Induction of sensitivity to the insulin-like action of growth hormone in normal rat adipose tissue, *Endocrinology* **108:**113, 1981.

18. U. J. Lewis, J. T. Dunn, L. F. Bonewald, B. K. Seavey, and W. P. VanderLaan, A naturally occurring structural varient of human growth hormone, *J. Biol. Chem.* **253:**2679, 1978.
19. H. M. Goodman and V. Coiro, Effects of growth hormone on adipose tissue of weanling rats, *Endocrinology* **109:**2046, 1981.
20. V. Coiro, G. Grichting, and H. M. Goodman, Induction of insulin-like responses to growth hormone by stress, *Endocrinology* **109:**2213, 1981.
21. V. Coiro, L. E. Braverman, D. Christianson, S. L. Fang, and H. M. Goodman, Effect of hypopthyroidism and thyroxine replacement on growth hormone in the rat, *Endocrinology* **105:**641–646, 1979.
22. R. S. Birnbaum and H. M. Goodman, Comparison of several insulin-like effects of growth hormone, *Hormone Metab. Res.* **11:**136, 1979.
23. H. M. Goodman, Separation of early and late responses of adipose tissue to growth hormone, *Endocrinology* **109:**120, 1981.
24. G. Grichting, L. K. Levy, and H. M. Goodman, Relationship between binding and biological effects of hGH in hGH in rat and adipocytes, *Endocrinology* **113:**111, 1983.
25. J. R. Seals and L. Jarett, Activation of pyruvate dehydrogenase by direct addition of insulin to an isolated plasma membrane–mitochondria mixture: Evidence of insulin's second messener in a subcellular system. *Proc. Natl. Acad. Sci. USA* **77:**77, 1980.
26. J. R. Seals and M. P. Czech, Characterization of a pyruvate dehydrogenase activator released by adipocyte plasma membrane in response to insulin, *J. Biol. Chem.* **256:**2894, 1981.
27. J. Larner, K. Cheng, C. Schwartz, R. Dubler, S. Creacy, K. Kikuchi, S. Tamura, G. Galasko, C. Pullin, and M. Katz, Chemical mechanism of insulin action via proteolytic formation of mediator peptides, *Mol. Cell. Biochem.* **40:**155, 1981.
28. H. Begun, H. M. Tepperman, and J. Tepperman, Effect of high fat and high carbohydrate diets on adipose tissue pyruvate dehydrogenase and its activation by a plasma membrane-enriched fraction and insulin, *Endocrinology* **110:**1904, 1982.

VI

SOMATOMEDIN/ INSULINLIKE GROWTH FACTOR

38

Chemistry of Somatomedins

René E. Humbel

Somatomedins (SM) are defined as growth hormone (GH) dependent hormonal peptides mediating GH action.[1] The original observations that led to the postulate of such a growth factor in serum were made by Salmon and Daughaday in 1957.[2] Later, the originally proposed term "sulfation factor" was changed to "somatomedin."[1] Three different SM fractions from human serum differing in isoelectric point (pI) have been described as SMA, B, and C, respectively. At present, SMA is a questionable entity of unknown structure, SMB a nonentity of known structure,[3] and SMC has been shown to be identical with insulinlike growth factor I (IGF-I).[4] IGF I is one of the components of the serum insulinlike activities originally described as nonsuppressible insulinlike activity (NSILA).[5] The insulinlike effects of NSILA detracted at first from its main function as a growth factor. We now know that NSILA has definitely no blood-sugar-lowering effect under physiological conditions, but the *in vitro* insulinlike effects of NSILA led us to apply purification procedures that have been used for the isolation of insulin.[5] The basic strategy in isolating SM and NSILA was indeed very similar in the different laboratories as soon as the possible identity of SM and NSILA was suspected.[6,7] When the homology of the *N*-terminal sequence of NSILA to the one of insulin B chain had been established,[8] NSILA was renamed insulinlike growth factor to signal the structural kinship to insulin and to emphasize its tentative physiological significance as a growth factor.

Whereas the metamorphosis of the former NSILA to a serum growth factor occurred more by serendipity than by following a cleverly designed approach, the latter was definitely the case in the research on multiplication-stimulating activity (MSA). There, the "hypothesis that animal cells which multiply *in vitro* in the absence of serum produce MSA"[9] was the origin of the demonstration of this serum growth factor and eventually of its successful isolation and sequence determination.[10]

René E. Humbel • Biochemisches Institut, Universität Zürich, Switzerland.

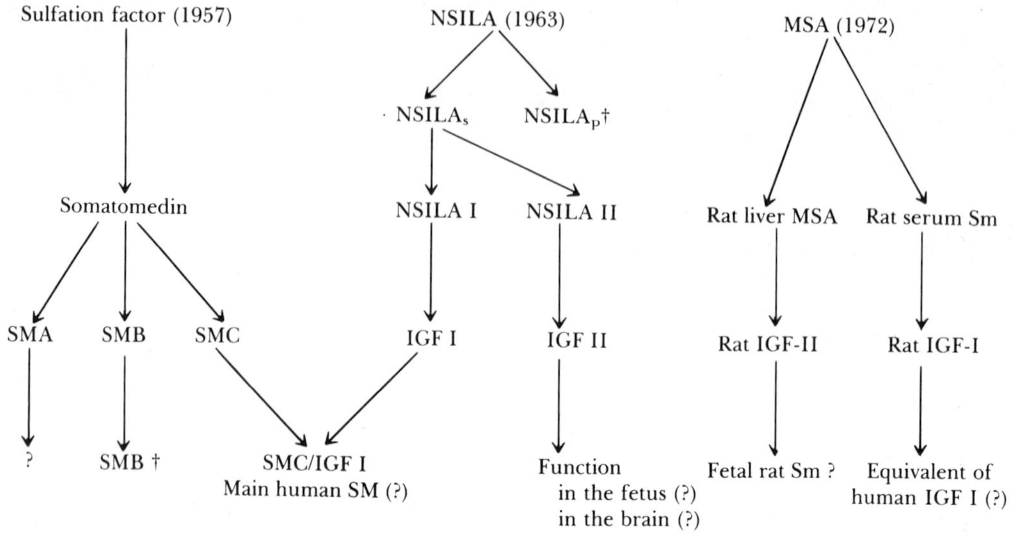

Figure 1 The discovery of somatomedins: A simplified scheme of routes, detours, and dead ends (†).

This convergence of three lines of research—SM, NSILA, and MSA—has led to a fruitful rallying of research efforts and at the same time to a deplorable semantic confusion. Now that the identity of SMC and of IGF-I and the homology of rat SM and rat MSA to human IGF-I and -II are established, the terms somatomedins, IGF's, and MSA are used interchangeably by most researchers. The Zurich group still has some reservations about equating human IGF-II with SM for reasons outlined later. A simplified scheme of the historical routes leading to the present-day concepts is proposed in Fig. 1.

The amino acid sequences of IGF-I and -II were published in 1978.[11,12] IGF-1 and -II are both single-chain peptides of M_r 7649 and 7471, respectively. IGF-I consists of 70 and IGF-II of 67 amino acid residues. The primary structure of IGF is similar to that of proinsulin: the single-chain peptide is cross-linked by three disulfide bridges and consists of a B-chain-like *N*-terminal part, a connecting peptide (C domain), and an A-chain-like part. In addition, a C-terminal extension not found in proinsulin is present (D domain). The C domain of IGF is considerably shorter than the C peptide of proinsulin and shows no apparent homology. The absence of pairs of basic residues at the B–C and C–A junction is the structural basis for the nonconvertibility of IGF to a two-chain structure analogous to the processing of proinsulin to insulin.

The amino acid sequences of IGF-I and -II are shown in Fig. 2. In the A and B domains, the sequence identity between insulin and IGF-I and -II is 49 and 47%, respectively; in the C domain it is 25%. The overall sequence identity between IGF-I and -II is 62%. Significantly, most of the hydrophobic residues, all cystine, and all glycine residues are identical in the three peptides. These residues are also invariant in all other insulins and form the hydrophobic core

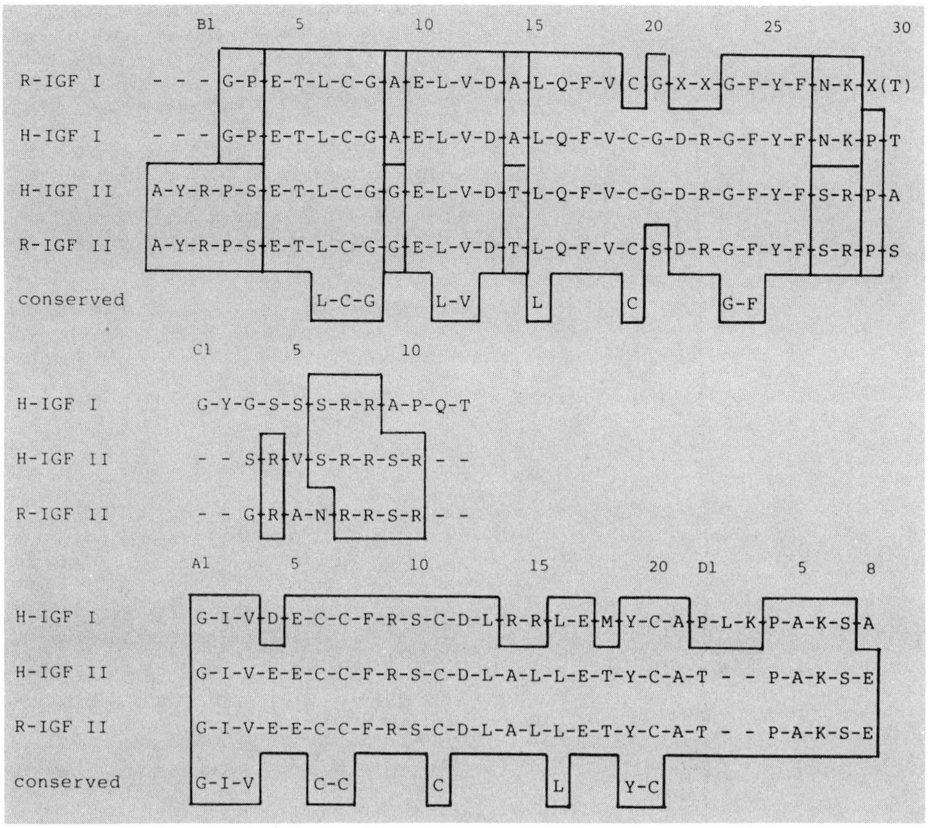

Figure 2 The amino acid sequences of human and rat IGF-I and -II. The residues concerned in all insulins and IGF's are indicated in the bottom line.

mainly responsible for the three-dimensional structure of insulin. On the assumption that the hydrophobic core of IGF has the same conformation as in insulin, models of IGF-I and -II could be constructed by model building[13] and interactive computer graphics.[14] Figure 3 shows a schematic representation of the conformational similarity of insulin, proinsulin, and IGF's. These models add considerable weight to the supposed homology of insulin and IGF's. More important, this model forms the structural basis for the partial cross-reactivity of IGF's with the insulin receptor and vice versa (for a more detailed discussion, see ref. 14).

Recently, we have worked out a simplified procedure to purify IGF from human serum.[15] During this work it became apparent that IGF-I with a pI of 8.25 and IGF-II with a pI of 6.5 are just the main components of a whole family of IGF's. Subforms of IGF-I with pI's of 8.7–9, 8.5, 7.9, and 7.5 and subforms of IGF-II with pI's of 7–7.5, 3.7–4, 4.8–5.1, and 8 could be separated from the main forms by chromatofocusing. The amino acid sequences of two

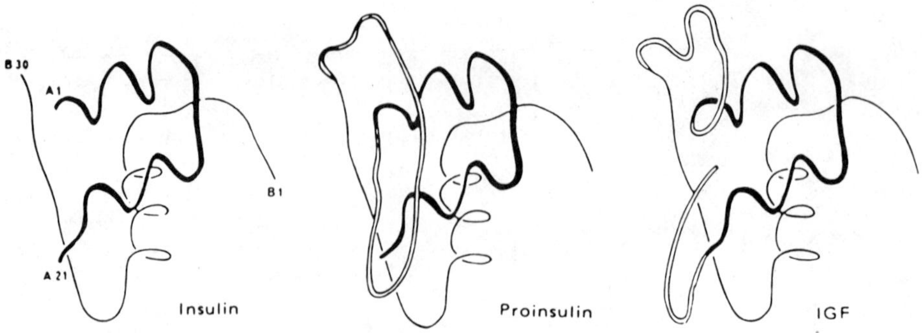

Figure 3 Schematic representation of the conformation similarity of insulin, proinsulin, and IGF. From Blundell *et al.*[13]

subforms has been recently determined.[16] The presence of additional forms of IGF besides the classical 8.3 and 6.5 forms has also been noted by other workers.[17,18]

As already mentioned, the amino acid sequence of SMC has been found to be completely identical to that of IGF-I.[4] The *N*-terminal sequence of rat SM isolated from serum of rats bearing a growth-hormone-secreting tumor is in all likelihood identical to that of human IGF-I.[19] The complete amino acid sequence of MSA from rat liver cell cultures is surprisingly similar to that of human IGF-II.[10] In fact, only five amino acid substitutions out of a total of 67 residues are present.

So far, human and rat IGF's are the only ones with a determined amino acid sequence. Judging from radioimmunoassays, it appears that other mammalian species and birds produce immunologically cross-reacting IGF's.[20]

A switch from IGF-II to IGF-I around birth has been observed in rat fibroblasts *in vitro*.[21] This is in contrast to the situation in man, where plasma IGF-II persists into adulthood in concentrations that are about three times higher than those of IGF I.[22] IGF-II in man is under less strict control of growth hormone,[22] is much less active on growth *in vivo*,[23] and is less active as feedback control on pituitary GH secretion.[24] This prompted us to search for physiological functions of IGF-II different from a classical SM. The recently demonstrated occurrence of IGF-II in granular form in human brain at widely varying concentrations in different brain regions[25] may point to a regulatory role in the central nervous system. The highest concentrations of apparently authentic IGF occur in the anterior pituitary, whereas other regions, notably thalamus, hypothalamus, and mammary bodies, contain relatively large amounts of higher-molecular-weight immunoreactivity. A 38K and 26K component seem to contain, in addition to IGF, an IGF-binding protein which might constitute a part of the common precursor. Similar high-molecular-weight IGF compounds have been described that were derived from liver.[26,27]

These facts, taken together with the mentioned heterogeneity of IGF in human serum, suggest to us that IGF-I and -II may be just the two main processing products from a large precursor, and that, at least in man, IGF-II may play a role not properly circumscribed by the term "somatomedin."

ACKNOWLEDGMENT. In the author's own research, the collaboration of E. Rinderknecht, G. Haselbacher, A. M. Honegger, U. Läubli, C. Lüthy, and P. Zumstein and the financial support by the Swiss National Science Foundation (grant 3.328.82) are gratefully acknowledged.

References

1. W. H. Daughaday, K. Hall, M. S. Raben, W. D. Salmon, Jr., J. L. Van den Brande, and J. J. Van Wyk, Somatomedin: Proposed designation for sulphation factor, *Nature* **235:**107, 1972.
2. W. D. Salmon, and W. H. Daughaday, A hormonally controlled serum factor which stimulates sulfate incorporation by cartilage in vitro, *J. Lab. Clin. Med.* **49:**825–836, 1957.
3. C-H. Heldin, Å. Wasteson, L. Fryklund, and B. Westermark, Somatomedin B: Mitogenic activity derived from contaminant epidermal growth factor, *Science* **213:**1122–1123, 1981.
4. D. G. Klapper, M. E. Svoboda, and J. J. Van Wyk, Sequence analysis of somatomedin C: Confirmation of identity with insulin-like growth factor I, *Endocrinology* **112:**2215–2217, 1983.
5. J. Zapf, E. R. Froesch, and R. E. Humbel, The insulin-like growth factors (IGF) of human serum: chemical and biological characterization and aspects of their possible physiological role, *Curr. Top. Cell. Reg.* **19:**257–309, 1981.
6. K. Hall, Human somatomedin. Determination, occurrence, biological activity and purification, *Acta endocrinol. (Kbh.)* **163** Suppl.: 1–52, 1973.
7. B. Morell, and E. R. Froesch, Fibroblasts as an experimental tool in metabolic and hormone studies. II. Effects of insulin and nonsuppressible insulin-like activity (NSILA-s) on fibroblasts in culture, *Eur. J. Clin. Invest.* **3:**119–123, 1973.
8. E. Rinderknecht and R. E. Humbel, Amino-terminal sequences of two polypeptides from human serum with nonsuppressible insulinlike and cell-growth-promoting activities: Evidence for structural homology with insulin B chain, *Proc. Natl. Acad. Sci. USA* **73:**4379–4381, 1976.
9. N. C. Dulak and H. M. Temin, A partially purified polypeptide fraction from rat liver cell conditioned medium with multiplication-stimulating activity for embryo-fibroblasts, *J. Cell Physiol.* **81:**153–160, 1973.
10. H. Marquardt, G. J. Todaro, L. E. Henderson, and S. Oroszlan, Purification and primary structure of a polypeptide with multiplication-stimulating activity from rat liver cell cultures, *J. Biol. Chem.* **256:**6859–6865, 1981.
11. E. Rinderknecht and R. E. Humbel, the amino acid sequence of human insulin-like growth factor I and its structural homology with proinsulin, *J. Biol. Chem.* **253:**2769–2776, 1978.
12. E. Rinderknecht and R. E. Humbel, Primary structure of human insulin-like growth factor II, *FEBS Lett.* **89:**283–286, 1978.
13. T. L. Blundell, S. Bedarkar, E. Rinderknecht, and R. E. Humbel, Insulin-like growth factor: A model for tertiary sturcture accounting for immunoreactivity and receptor binding, *Proc. Natl. Acad. Sci. USA* **75:**180–184, 1978.
14. T. L. Blundell, S. Bedarkar, and R. E. Humbel, Tertiary structures, receptor binding, and antigenicity of insulinlike growth factors, *Fed. Proc.* **42:**2592–2597, 1983.
15. P. P. Zumstein and R. E. Humbel, Purification of human insulin-like growth factors I and II, *Methods Enzymol.*, **109:**782–798, 1985.
16. P. P. Zumstein, C. Luthy, and R. E. Humbel, Amino acid sequence of a variant pro-form of insulin-like growth factor II, *Proc. Natl. Acad. Sci. USA* **82:**3169–3172, 1985.
17. M. E. Svoboda, J. J. Van Wyk, D. G. Klapper, R. E. Fellows, F. E. Grisson, and R. J. Schlueter, Purification of somatomedin C from human plasma: Chemical and biological properties, partial sequence analysis, and relationship to other somatomedins, *Biochemistry* **19:**790–797, 1980.

18. R. M. Bala and B. Bhaumick, Purification of a basic somatomedin, from human plasma Cohn fraction IV-1, with physicochemical and radioimmunoassay similarity to somatomedin-C and insulin-like growth factor, *Can J. Biochem.* **57:**1289–1298, 1979.
19. J. S. Rubin, I. Mariz, J. W. Jacobs, W. H. Daughaday, and R. A. Bradshaw, Isolation and partial sequence analysis of rat basic somatomedin, *Endocrinology* **110:**734–740, 1982.
20. D. M. Wilson and R. L. Hintz, Inter-species comparison of somatomedin structure using immunological probes, *J. Endocrinol.* **95:**59–64, 1982.
21. S. O. Adams, S. P. Nissley, S. Handwerger, and M. M. Rechler, Developmental patterns of insulin-like growth factor-I and -II synthesis and regulation in rat fibroblasts, *Nature* **302:**150–153, 1983.
22. J. Zapf, H. Walter, and E. R. Froesch, Radioimmunological determination of insulin-like growth factors I and II in normal subjects and in patients wtih growth disorders and extrapancreatic tumor hypoglycemia, *J. Clin. Invest.* **68,**1321–1330, 1981.
23. E. Schoenle, E. Hauri, T. Steiner, J. Zapf, and E. R. Froesch, Comparison of *in vivo* effects of insulin-like growth factors I and II and of growth hormone in hypophysectomized rats, *Acta Endocrinol.* **108:**167–174, 1985.
24. P. Brazeau, R. Guillemin, N. Ling, J. J. Van Wyk, and R. E. Humbel, Inhibition by somatomedins of growth hormone secretion stimulated by hypothalamic growth hormone releasing factor (somatocrinin, GRF) or the synthetic peptide hpGRF, *CR Acad. Sci. Paris t 295*, Série **III:**651–654, 1982.
25. G. K. Haselbacher, M. E. Schwab, A. Pasi, and R. E. Humbel, Insulin-like growth factor II (IGF II) in human brain: Regional distribution of IGF II and of higher molecular mass forms, *Proc. Natl. Acad. Sci. USA* **82:**2153–2157 (1985).
26. M. A. Aquaviva, C. B. Bruni, S. P. Nissley, and M. M. Rechler, Cell-free synthesis of rat insulin-like growth factor II, *Diabetes* **31:**656–658, 1982.
27. R. Vassilopoulou-Sellin and L. S. Phillips, Extraction of somatomedin activity from rat liver, *Endocrinology* **110:**582–589, 1982.

39

Synthetic Somatomedin C/Insulinlike Growth Factor I

Choh Hao Li

Insulinlike growth factors (IGF's) and somatomedin C (SMC) have been isolated in pure form from human plasma.[1,2] Complete amino acid sequences for IGF-I,[3] IGF-II,[4] and SMC[5] have been determined. From these structural data and various biological studies,[6] SMC and IGF-I are identical peptides consisting of 70 amino acids with three disulfide bridges as shown in Fig. 1.

1. Synthesis

Synthesis of protected peptide corresponding to SMC was performed by the solid-phase method[7] with use of aminomethyl resin to which Bocalanyl-4-(oxymethyl)-phenylacetic acid is attached.[8] Assembly of the required sequence was accomplished by procedures generally used in this laboratory.[9,10] The synthetic scheme[11] is shown in Fig. 2. From 1.60 g (0.60 mmole) starting resin, the yield of fully protected peptide resin was 7.92 g (weight gain was 87% of theory). A crude reduced form of SMC after treatment of protected peptide resin with HF and dithiothreitol in 6 M guanidine·HCl was subjected to purification on CM-cellulose in 8 M urea with a gradient of 0.01 to 0.2 M NH_4OAc. The purified reduced form of SMC was oxidized by stirring for 6 hr at 24°C in 2 M guanidine·HCl. After gel filtration on Sephadex G-50, the crude SMC monomer was purified by partition chromatography and chromatofocusing. From 1.34 g (0.1 mmole) of fully protected SMC, 6.8 mg of synthetic peptide was obtained with ca. 1% yield based on starting load of alanine to the resin.

Choh Hao Li • Laboratory of Molecular Endocrinology, University of California, San Francisco, California 94143.

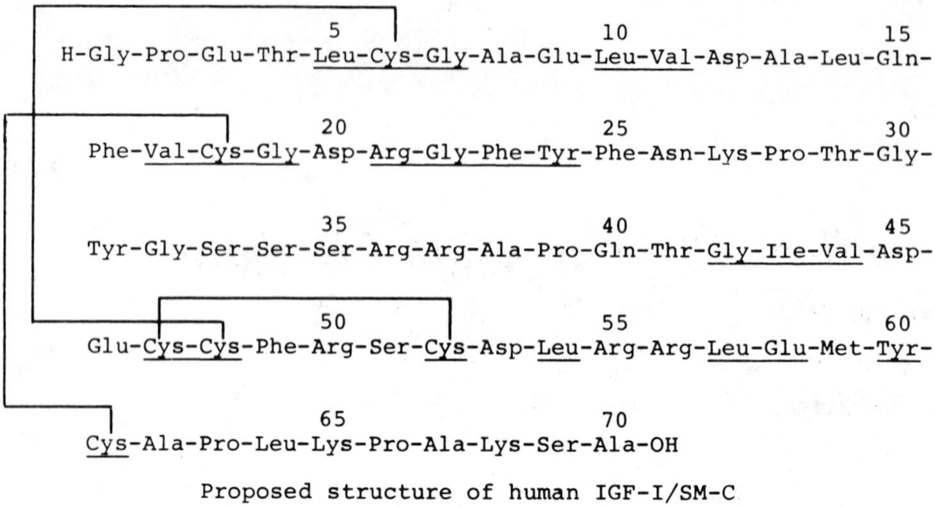

Figure 1 Proposed structure of human IGF-I/SMC.

(A) Boc-Ala-4-(oxymethyl)-phenylacetamidomethyl-resin

1. TFA-CH_2Cl_2, 15 min.
2. Diisopropylethylamine (DIEA)
3. Preformed symmetrical anhydride of Boc-Ser

↓

Boc-Ser-Ala-resin

1. TFA-CH_2Cl_2, 15 min.
2. DIEA
3. Preformed symmetrical anhydride of Boc-AA

↓

Fully protected SMC/IGF-I

1. TFA-CH_2Cl_2, 15 min
2. CH_2Cl_2, EtOH

↓

Protected SMC with free α-NH_2 group

(B) Protected SMC with free α-NH_2 group

1. HF, anisol/2-mercaptopyridine
2. Dithiothreitol (DTT) in 6 M guanidine-HCl
3. Dialysis in 0.1 M HOAc

↓

Crude reduced form of SMC

1. Purified on CM-cellulose in 8 M urea with 0.01–0.2 M NH_4OAc
2. Dialysis in 0.1 M HOAc

↓

Purified reduced form of SMC

1. 1 mg/ml in 2 M guanidine-HCl, pH 8.4
2. Stirred for 6 hr at 24°C
3. Dialysis in 0.1 M HOAc, 4°C, 2 days
4. Gel filtration on Sephadex G-50 in 0.1 M HOAc

↓

Crude SMC monomer

Figure 2 Scheme for the synthesis of SMC/IGF-I by the solid-phase method. Side-chain protecting group: Asp, cyclopentyl; Thr, Ser, and Glu, benzyl; Cys, 3,4-dimethylbenzyl; Met, sulfoxide; Tyr and Lys, 2-bromobenzyloxycarbonyl; Arg, tosyl.

(C) Crude SMC monomer

↓ Partition chromatography in a Sephadex G-50 column in the solvent system: 1-butanol/pyridine/0.1% HOAc 5:3:10 (volume/volume)

Purified SM-C (R_f0.13)

↓ 1. Chromatofocusing in a Polybuffer exchanger; starting buffer, pH 9.4 0.025 M ethanolamine-HCl; developed with pH 7.0 Polybuffer 96·HCl
2. The desired product mixed with butanol/pyridine and submitted to partition chromatography as above
3. Gel filtration in Sephadex G-50 in 1 M HOAc

Synthetic SMC/IGF-I

Figure 2 *(continued)*

2. *Homogeneity*

The purity of the synthetic product was established by isoelectric focusing, paper electrophoresis, high-performance liquid chromatography (HPLC), and amino acid analysis.[11] Table 1 presents the amino acid composition of synthetic SMC/IGF-I as compared with the theoretical values.

3. *Identity to the Natural Product*

In paper electrophoresis (20-μg sample) on Whatman 3MM paper at pH 6.7 (collidine acetate buffer) for 4 hr at 400 V, both natural and synthetic SMC/

Table 1 Amino Acid Analysis of Synthetic SMC/IGF-I

	Acid hydrolysate	
Amino acid	Experimental	Theoretical
Aspartic acid	4.82	5
Theonine	2.89	3
Serine	4.21	5
Glutamic acid	6.08	6
Proline	5.20	5
Half-cystine	5.28	6
Glycine	6.73	7
Alanine	6.19	6
Valine	2.72[a]	3
Methionine	0.95	1
Isoleucine	0.69[a]	1
Leucine	5.90	6
Tyrosine	2.73	3
Phenylalanine	4.12	4
Lysine	3.04	3
Arginine	5.97	6

[a] Low value for valine and isoleucine because of the acid resistance of Ile–Val sequence.

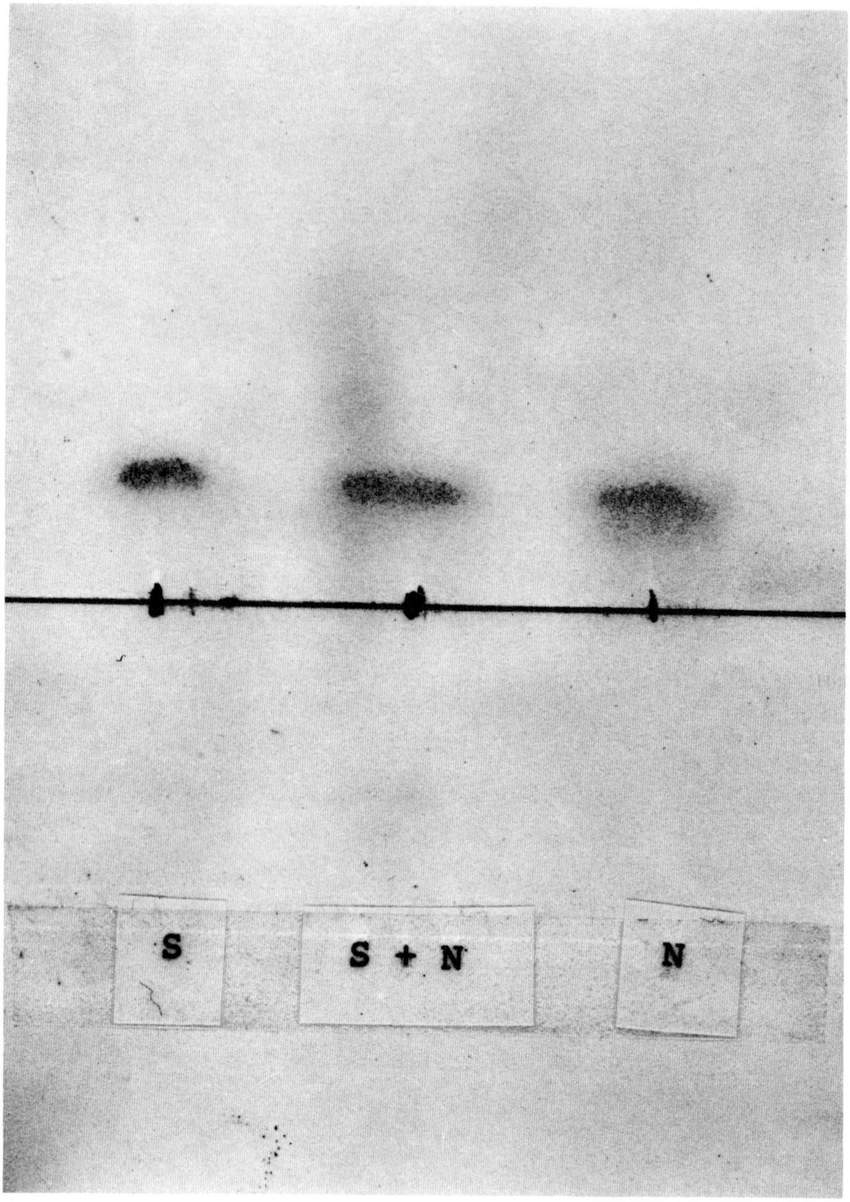

Figure 3 Paper electrophoresis of 20-μg samples of natural IGF-I/SMC (N), synthetic IGF-I/SMC (S), and a 1:1 mixture (S + N) on Whatman 3MM paper at pH 6.7 for 4 hr at 400 V.

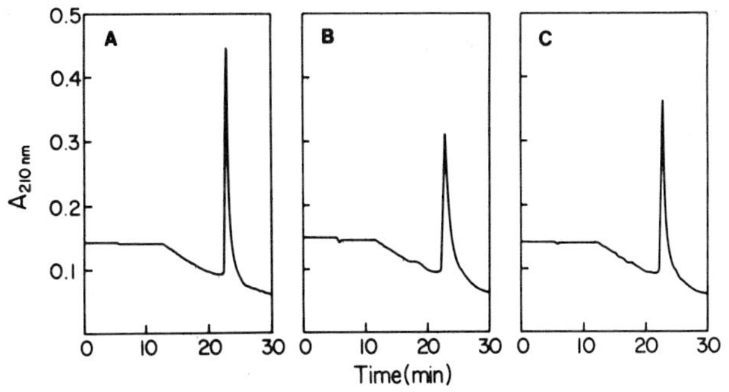

Figure 4 HPLC of 10-μg samples of synthetic IGF-I/SMC (A), natural IGF-I/SMC (B), and a 1:1 mixture (C) on an Alltech Vydac 201-TP column (4.5 × 250 mm). Elution by a linear gradient of 10–50% isopropanol in 0.1% TFA over 30 min.

IGF-I gave a single spot (ninhydrin detection) with identical mobility (Fig. 3). HPLC of a 1:1 mixture of synthetic and natural peptides showed a single peak (Fig. 4). Moreover, HPLC patterns of the peptide maps for thermolysin digests of natural and synthetic IGF-I are very similar (Fig. 5).

In collaboration with Drs. J. J. Van Wyk and W. E. Russell,[12] the biological and immunological activities of synthetic SMC were found to be identical to those of the natural product. In both radioimmunoassay[13] and radioreceptor assay[14] the displacement curves obtained by natural and synthetic SMC are identical. In addition, Van Wyk *et al.*[12] compared the ability of natural and synthetic SMC to induce DNA synthesis in G_o arrested confluent BALB c/3T3 cells. In this study, the two peptides at concentration of 16 ng/ml were indistinguishable in their stimulation of DNA synthesis.

The growth-promoting activity of synthetic SMC on circulating human erythroid progenitor cells has been investigated by Dr. D. W. Golde and Ms. N. Bersch (unpublished experiments). Using the same procedure as described,[15] natural and synthetic SMC exhibited nearly identical stimulating activity on human erythroid colony formation (Fig. 6). As in the case of insulin, peak activity was seen at 8–9 ng/ml, resulting in about a two-fold increase in burst-forming units-erythroid.

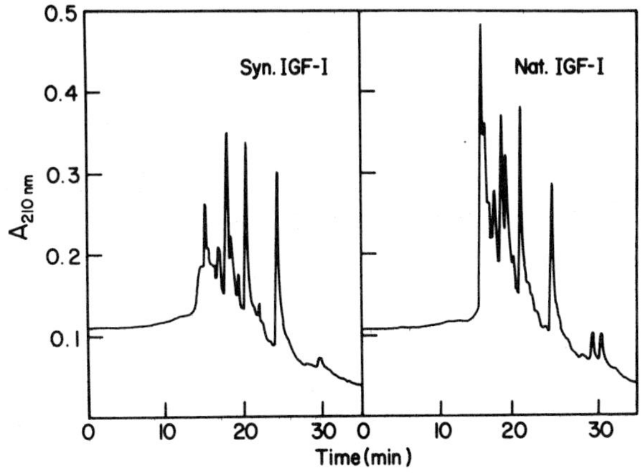

Figure 5 HPLC of thermolysin digests of synthetic and natural IGF-I/SMC. The peptide (50 μg) dissolved in 50 μl 0.2 M NH_4Ac of pH 8.2 containing 2.5 μg crystalline thermolysin (Serva) at 37°C for 16 hr. HPLC was performed as outlined in Fig. 4.

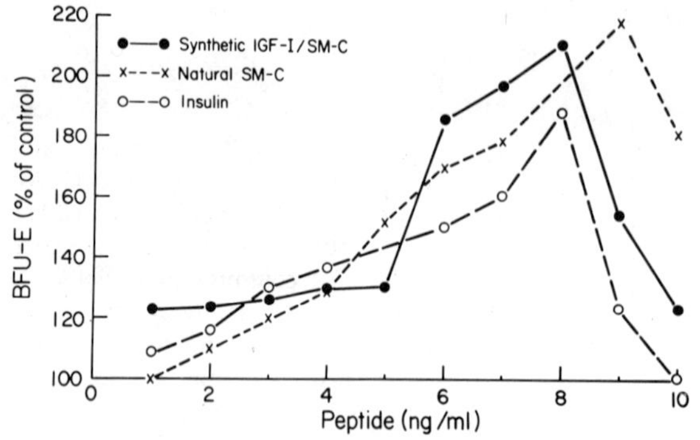

Figure 6 Effect of synthetic and natural SMC on the growth of human peripheral blood red cell colonies *in vitro* known as burst-forming units-erythroid.

4. Summary

A short account for the total synthesis of IGF-I/SMC has been presented. The synthetic product was found to be homogenous on paper electrophoresis, isoelectric focusing, HPLC, and amino acid composition. In paper electrophoresis, HPLC, and thermolysin peptide maps, synthetic and natural IGF-I/SMC behave similarly. In radioimmunoassay, radioreceptor assay, ability to induce DNA synthesis in 3T3 cells as well as growth-promoting activity on circulating human erythroid progenitor cells *in vitro*, synthetic and natural IGF-I/SMC have comparable activities.

ACKNOWLEDGMENTS. I thank Dr. R. E. Humbel for the natural IGF-I, for HPLC, paper electrophoresis, and peptide map experiments. I also thank Dr. Golde and Ms. N. Bersch for permission to quote their unpublished results. The assistance of David Chung is gratefully acknowledged. This work was supported in part by the National Institutes of Health (GM-2907 and AM-18677) and the Hormone Research Foundation.

References

1. E. Rinderknecht and R. E. Humbel, Polypeptides with nonsuppressible insulin-like and cell-growth promoting activities in human serum: Isolation, chemical characterization, and some biological properties of forms I and II, *Proc. Natl. Acad. Sci. USA* **73:**2365, 1976.
2. M. E. Svoboda, J. J. Van Wyk, D. G. Klapper, R. E. Fellow, F. E. Grissom, and R. J. Schlueter, Purification of somatomedin-C from human plasma: Chemical and biological properties, partial sequence analysis, and relationship to other somatomedins, *Biochemistry* **19:**790, 1980.
3. E. Rinderknecht and R. E. Humbel, The amino acid sequence of human insulin-like growth factor I and its structural homology with proinsulin, *J. Biol. Chem.* **253:**2769, 1978.
4. E. Rinderknecht and R. E. Humbel, Primary structure of human insulin-like growth factor II, *FEBS Lett.* **89:**283, 1978.
5. D. G. Klapper, M. E. Svoboda, and J. J. Van Wyk, Sequence analysis of somatomedin-C: Confirmation of identity with insulin-like growth factor I, *Endocrinology* **112:**2215, 1983.
6. J. J. Van Wyk, The somatomedins: Biological actions and physiologic control mechanisms, in *Hormonal Protein and Peptides, Vol. 12* (C. H. Li, ed.) Academic Press, New York 1984, pp. 81–125.

7. R. B. Merrifield, Solid phase peptide synthesis. I The synthesis of a tetrapeptide, *J. Am. Chem. Soc.* **85:**2149, 1963.
8. A. R. Mitchell, S. B. H. Kent, M. Engelhard, and R. B. Merrifield, A new synthetic route to tert-butyloxycarbonylaminoaceyl-4-(oxymethyl)phenylacetamidomethylresin, an improved support for solid-phase peptide synthesis, *J. Org. Chem.* **43:**2845, 1978.
9. C. H. Li, D. Yamashiro, L-F. Tseng, and H. H. Loh, Synthesis and analgesic activity of human β-endorphin, *J. Med. Chem.* **20:**325, 1977.
10. J. Blake, L-F. Tseng, and C. H. Li, Synthesis and analgesic activity of human β-endorphin analogs substituted at positions 17, 18, or 19, *Int. J. Pept. Prot. Res.* **15:**167, 1980.
11. C. H. Li, D. Yamashiro, D. Gospodarowicz, S. L. Kaplan, and G. Van Vliet, Total synthesis of insulin-like growth factor I (somatomedin C), *Proc. Natl. Acad. Sci. USA* **80:**2216, 1983.
12. J. J. Van Wyk, W. E. Russell, and C. H. Li, Synthetic somatomedin-C: Comparison with the natural hormone isolated from human plasma, *Proc. Natl. Acad. Sci. USA* **81:**740–742, 1984.
13. R. W. Furlanetto, L. E. Underwood, J. J. Van Wyk, and A. J. D'Ercole, Estimation of somatomedin-C levels in normals and patients with pituitary disease by radioimmunoassay, *J. Clin. Invest.* **60:**648, 1977.
14. A. J. D'Ercole, L. E. Underwood, J. J. Van Wyk, C. J. Decedue, and D. B. Foushee, Specificity, topography, and ontogeny of the somatomedin C receptor in mammalian tissues, in: *Growth Hormone and Related Peptides* (A. Pecile and E. E. Muller, eds.), Excerpta Medica, Amsterdam, 1976, p. 190.
15. N. Bersch, J. E. Groopman, and D. W. Golde, Jr., Natural and biosynthetic insulin stimulates the growth of human erythroid progenitors *in vitro, Clin. Endocrinol. Metab.* **55:**1209, 1982.

40

Biosynthesis of Rat Insulinlike Growth Factor II in Intact Cells and Cell-Free Translation

Matthew M. Rechler, Yvonne W.-H. Yang, Jeffrey E. Terrell, Angela M. Acquaviva, Harvey J. Whitfield, Joyce A. Romanus, C. Bruno Bruni, and S. Peter Nissley

1. MSA is the Rat Homologue of Human IGF-II

Dulak and Temin[1] first reported that the BRL-3A cell line established from normal rat liver secreted a family of polypeptides, which they termed MSA, that had multiplication-stimulating activity for chick embryo fibroblasts. MSA was purified by Moses and colleagues from conditioned media using Dowex chromatography, Sephadex G-75 gel filtration in 1 M acetic acid, and preparative gel electrophoresis, and shown to appear in multiple forms of Mr 16.3K, 8.7K, and 7.1K.[2] Marquardt *et al.*[3] purified MSA by a different purification scheme (i.e., Bio-Gel P-10 in 1 M acetic acid and high-performance liquid chromatography), and determined the amino acid sequence of a Mr 7484-dalton form. Mr 7484 MSA appears to correspond to our Mr 7.1K species. It is identical to human IGF-II at 62 of 67 amino acid loci, establishing that BRL-MSA represents the rat homologue of IGF-II.[3]

Abbreviations used: IGF, insulinlike growth factor; MSA, multiplication-stimulating activity; BRL, Buffalo rat liver; SDS–PAGE, sodium dodecyl sulfate–polyacrylamide gel electrophoresis.

Matthew M. Rechler, Yvonne W.-H. Yang, Jeffrey E. Terrell, Angela M. Acquaviva, Harvey J. Whitfield, Joyce A. Romanus, C. Bruno Bruni • Section on Growth and Development, Molecular, Cellular, and Nutritional Endocrinology Branch, National Institute of Arthritis, Diabetes, and Digestive and Kidney Diseases, National Institutes of Health, Bethesda, Maryland 20205. ***S. Peter Nissley*** • Endocrine Section, Metabolism Branch, National Cancer Institute, National Institutes of Health, Bethesda, Maryland 20205.

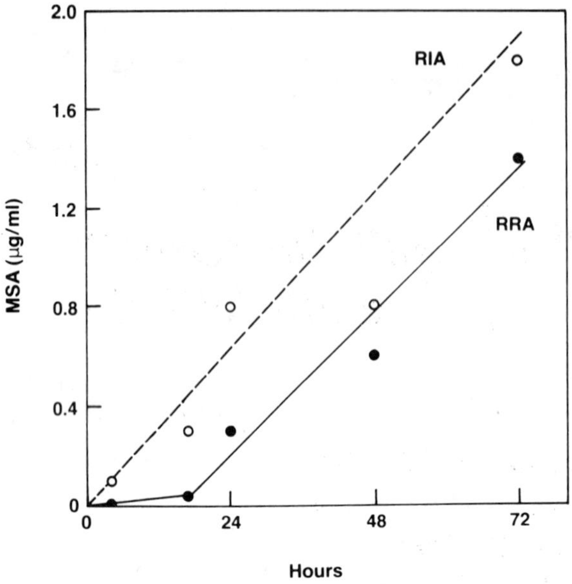

Figure 1 Time course of appearance of MSA (rat IGF-II) in BRL-3A conditioned media. Serum-free media was incubated with confluent BRL-3A cells for the indicated times, harvested, and fractionated on Sephadex G-75 in 1 M acetic acid as previously described.[4] The low-molecular-weight fractions containing MSA were pooled, and MSA content was determined in an MSA radioimmunoassay (○——○) and MSA radioreceptor assay (●——●) with Mr 8.7K MSA reference standard as previously described.[5,14] This experiment suggests that during the first 24 hr after media change, conditioned culture medium may contain immunoreactive MSA species that do not possess full receptor reactivity (presumably reflecting biological activity).

2. *Indirect Evidence Suggesting a Precursor for Mr 7.1K MSA*

The relationship of the higher-molecular weight MSA species to Mr 7.1K MSA was less clear-cut. The three forms possessed similar biological and binding properties, with the relative reactivity 7.1K > 8.7K > 16.3K.[2,4,5] Moreover, they had full and parallel cross-reactivity in an MSA radioimmunoassay,[5] suggesting the presence of common antigenic determinants and a possible precursor–intermediate-product relationship. Further indirect evidence consistent with processing and biological activation of a precursor came from comparison of the time course of appearance of MSA in BRL-3A medium using bioassay, receptor assay, and radioimmunoassay. Receptor-reactive[4] (Fig. 1) and biologically active MSA (unpublished results) is present at low levels for 17–24 hr after media change. By contrast, accumulation of immunoreactive MSA increases linearly without a lag period (Fig. 1). This suggested that at early times, conditioned media contains immunoreactive MSA that is not fully active, and that requires processing to obtain the mature active forms.

3. *Demonstration of Mr ~ 22K pre-pro-MSA and Mr ~ 20K pro-MSA by Cell-Free Translation and Biosynthetic Labeling in Intact Cells*

We have used two approaches to demonstrate directly higher-molecular weight precursor forms of MSA. In the 1st, intact BRL-3A cells were metabolically labeled with radioactive amino acids. Cell lysates were incubated with rabbit antiserum to Mr 8.7K MSA, and the immune complexes precipitated

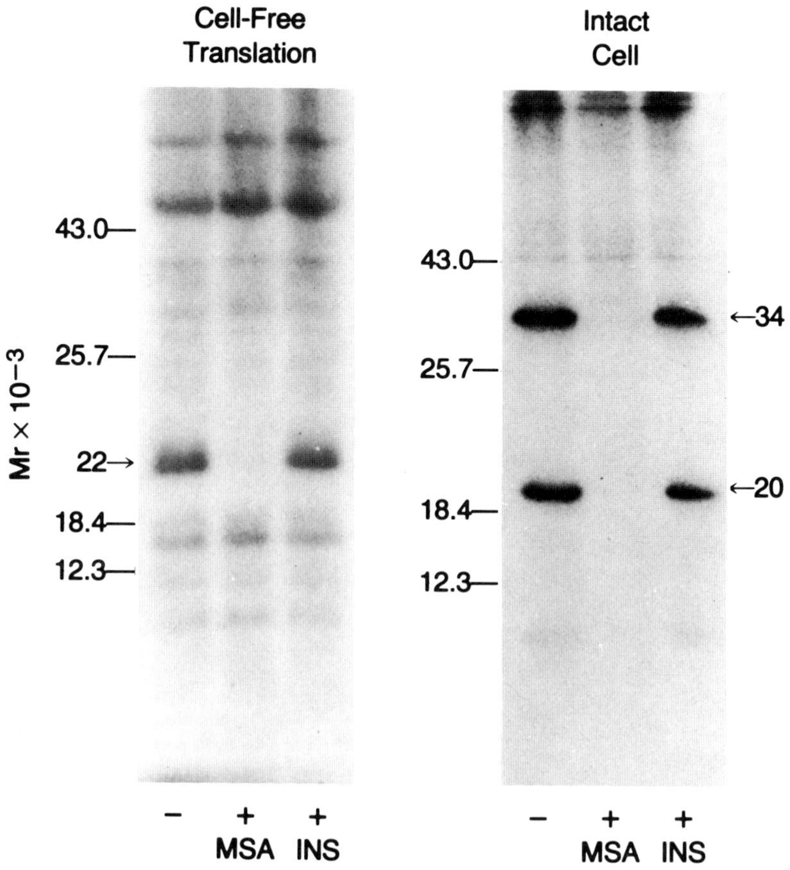

Figure 2 Specificity of immunoprecipitation of Mr ~ 22K and Mr ~ 20K proteins. (Left) Immunoprecipitation of [^{35}S]cysteine- and [^{35}S]methionine-labeled BRL-3A RNA-directed translation products by anti-MSA serum in the absence of unlabeled proteins (left), and in the presence of 15 μg of unlabeled MSA (center) or 15 μg of unlabeled mouse insulin (right). Immunoprecipitation of the Mr ~ 22K protein is inhibited by unlabeled MSA but not by insulin. Redrawn from Acquaviva *et al.*[7] (Right) Immunoprecipitation of [^{35}S]cysteine-labeled BRL-3A cell lysates by anti-MSA serum in the absence of unlabeled protein (left), and in the presence of 20 μg of unlabeled MSA (center) or 20 μg of unlabeled rat insulin (right). Immunoprecipitation of the Mr ~ 20K protein is inhibited by unlabeled MSA, but not by insulin. The Mr ~ 34K protein corresponds to IGF carrier protein (ref. 10 and our unpublished results). It is precipitated by antibodies to carrier protein present in this antiserum. Immunoprecipitation of the Mr ~ 34K protein is inhibited because the MSA preparation used in this experiment contains small amounts of carrier protein.

with *Staphylococcus aureus* strains containing protein A, and examined by SDS–PAGE (15% acrylamide, Laemmli system) and fluorography.[6] In the 2nd, high-molecular-weight RNA was extracted from BRL-3A cells with guanidine hydrochloride, purified by successive salt and alcohol precipitations, and used to direct protein synthesis in a rabbit reticulocyte lysate translation system.

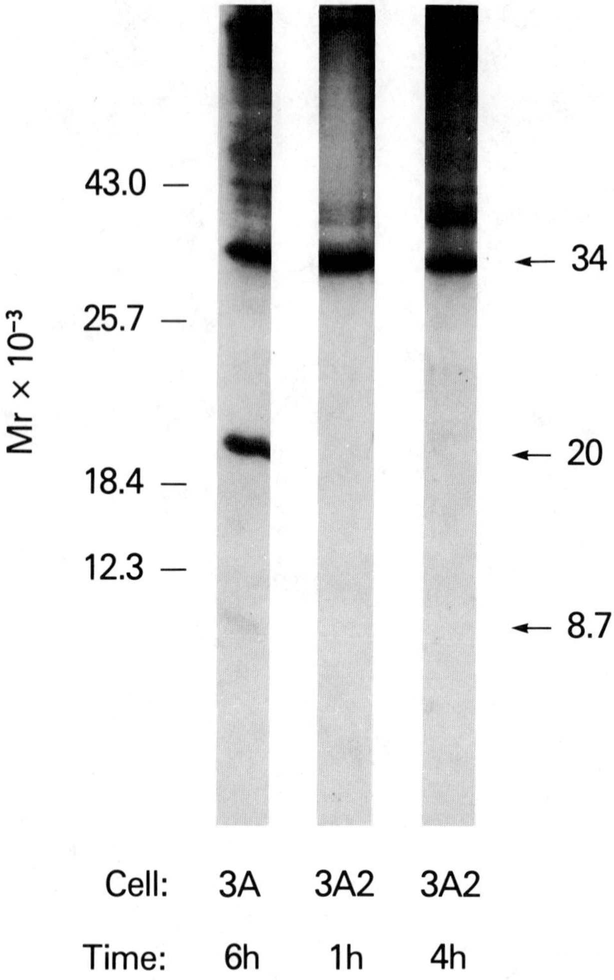

Figure 3 Immunoprecipitation of BRL-3A and BRL-3A2 lysates with anti-MSA serum. BRL-3A and BRL-3A2 cells were labeled with [^{35}S]cysteine (100 μCi/ml) for the indicated times. Cells were lysed and the lysates immunoprecipitated with anti-MSA serum. Immunoprecipitates were examined by SDS–PAGE and fluorography. BRL-3A, left lane; BRL-3A2, center and right lanes. The Mr ~ 20K protein is immunoprecipitated from BRL-3A but not BRL-3A2, lysates.

Immunoprecipitation, electrophoresis, and fluorography were performed as described previously.(7)

From both experimental approaches, the following picture emerges: (1) a Mr~ 20K protein, presumably pro-MSA, is rapidly labeled in BRL-3A cells and specifically immunoprecipitated from BRL-3A cell lysates; (2) a Mr ~22K protein, presumably pre-pro-MSA, is specifically immunoprecipitated from proteins whose translation is directed by BRL-3A RNA.

The specificity of immunoprecipitation was established by the following

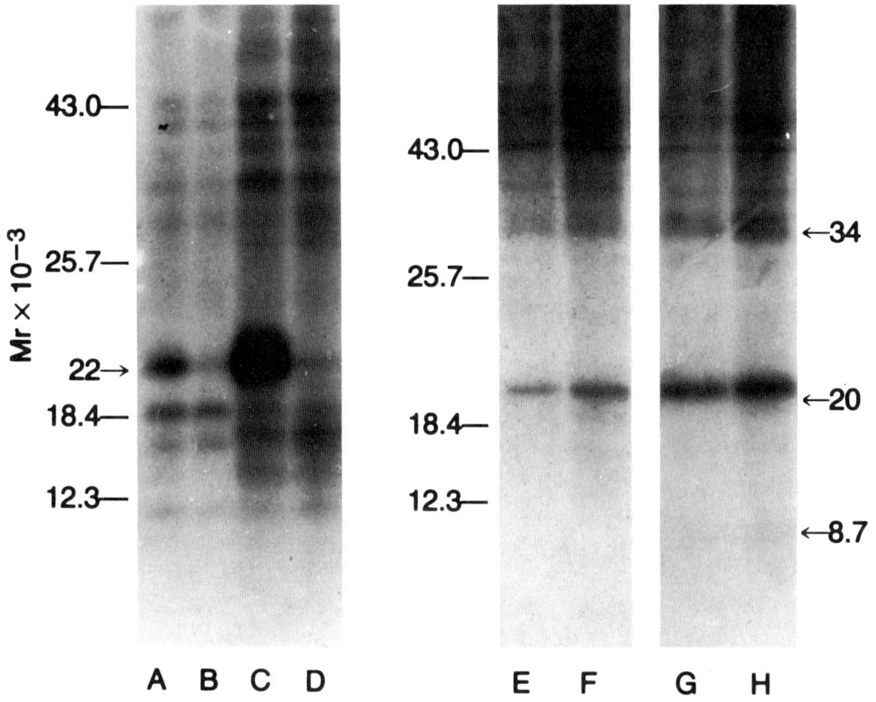

Figure 4 Labeling of Mr ~ 22K and ~ 20K MSA precursors with [^{35}S]cysteine and [^{35}S]methionine in cell-free translation (lanes A–D) and in intact cells (lanes E–H). (Lanes A–D) Immunoprecipitates of [^{35}S]proteins labeled in cell-free translation directed by BRL-3A RNA in the presence of 35 μCi [^{35}S]methionine (1215 Ci/mmole) (lanes A and B), or 35 μCi [^{35}S]methionine plus 15 μCi [^{35}S]cysteine (866 Ci/mmole) (lanes C and D). Immunoprecipitations were performed in the absence (lanes A and C) and presence (lanes B and D) of 8 μg of unlabeled MSA. (Lanes E–H) BRL-3A cells were incubated in methionine- and cysteine-deficient media for 2 hr, and labeled with 100 μCi/ml of [^{35}S]methionine (1053 Ci/mmole) (lanes E and F) and [^{35}S]cysteine (937 Ci/mmole) (lanes G and H) for 6 hr, respectively. Aliquots of the [^{35}S]-labeled cell lysates (50 μl, lanes E and G; 100 μl, lanes F and H) were immunoprecipitated with anti-MSA serum and compared on 15% polyacrylamide gel.

criteria: (1) immunoprecipitation by immune but not by nonimmune sera[7,8]; (2) inhibition of immunoprecipitation by excess unlabeled MSA but not by the homologous proteins, rat or mouse insulin (Fig. 2); (3) absence of the Mr ~ 20K and ~22K bands in immunoprecipitates of cell lysates and translated proteins using a subclone of BRL-3A cells, BRL-3A2, that does not secrete MSA into the culture medium, as determined by biological assay, receptor assay, and immunoassay (ref. 9 and unpublished results). Mr ~ 22K pre-pro-MSA was not detected in immunoprecipitates of [^{35}S]proteins whose translation was directed by BRL-3A2 RNA.[7] Similarly, Mr~20K pro-MSA was not immunoprecipitated from lysates of radiolabeled BRL-3A2 cells (Fig. 3). These results suggest that BRL-3A2 cells lack translationally competent RNA for MSA, or that the MSA synthesized is a variant form that is not recognized by MSA antibodies.

Although Mr 7484 MSA contains cysteine but lacks methionine residues,[3] the Mr ~ 22K and ~20K precursors were biosynthetically labeled using [^{35}S]methionine as precursor in both intact cells and in the cell-free system (Fig. 4). Incorporation with [^{35}S]methionine was lower than with [^{35}S]cysteine, however, in both systems.

4. *The Kinetics of Labeling and Processing Mr ~ 20K pro-MSA Are Consistent with Its Being a Prohormone*

One prerequisite for the Mr ~ 20K protein to be considered a prohormone would be its appearance in biosynthetic-labeling experiments prior to the appearance of lower-molecular-weight MSA species. In continuous labeling experiments, [^{35}S] radioactivity appears in the Mr ~ 20K protein at the earliest time examined (10 mins), is maximal between 40 and 60 min, and subsequently declines.[8,11] The Mr ~ 20K protein is not observed in the culture media. In pulse-chase experiments, as illustrated in Fig. 5, the Mr ~ 20K and Mr ~ 16K proteins are rapidly labeled, followed by the appearance of Mr ~ 8.7K MSA intracellularly.[11] This suggests that at least part of the processing of the Mr ~ 20K protein occurs prior to secretion. Specifically immunoprecipitated Mr ~ 19K, ~16K, 8.7K, ~8.0K, and ~ 7.1K proteins appear in media at later times.[11]

Various protease inhibitors were added to the culture medium during the biosynthetic labeling to determine whether they affected processing of Mr ~ 20K pro-MSA. Leupeptin, antipain, TPCK, TLCK, PMSF, aprotinin, and soybean trypsin inhibitor did not significantly affect the processing of Mr ~ 20K pro-MSA to Mr ~ 7.1K MSA.

5. *Mr ~ 22K pre-pro-MSA Can Be Converted to a Mr ~ 20K Protein Indistinguishable from pro-MSA by Incubation with Microsomal Membranes*

Secreted proteins typically are translated with an additional hydrophobic signal peptide, which is cleaved by a signal peptidase associated with microsomal membranes coincident with the translocation of the nascent protein to the interior of the rough endoplasmic reticulum.[12] The Mr ~ 22K product observed in the rabbit reticulocyte lysate translation system was postulated to represent pre-pro-MSA, since MSA is a secreted protein and since the signal peptidase is lacking from this preparation.[7] The size of the intracellular pro-MSA (Mr ~ 20K) was appropriate for removal of a prepeptide (signal peptide) of Mr ~ 2000.[12] Recently, we have demonstrated that incubation of BRL-3A RNA-directed translation products with microsomal membranes converts the

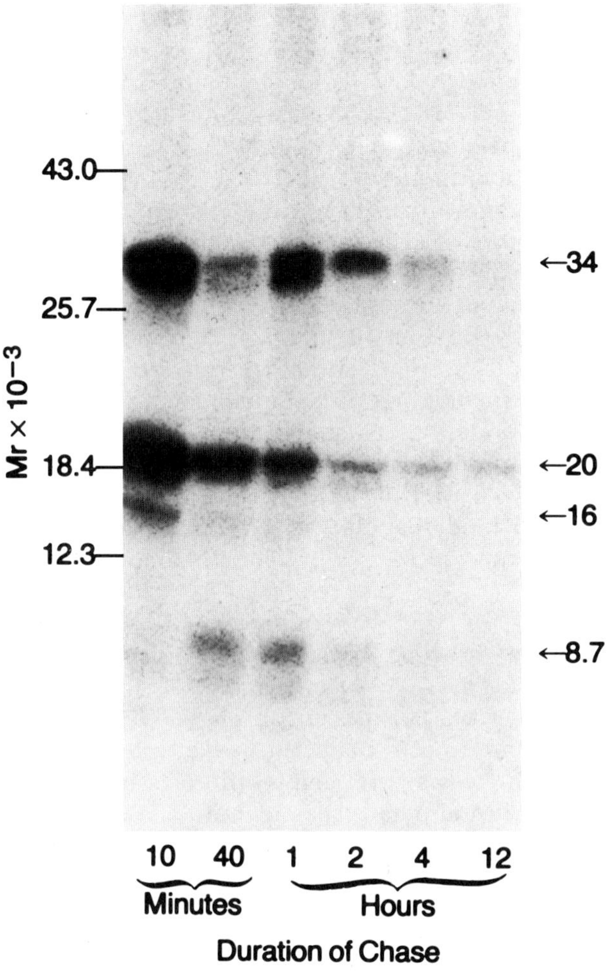

Figure 5 Turnover of [^{35}S]cysteine-labeled Mr ~ 20K pro-MSA in BRL-3A cells. BRL-3A cells were starved for 2 hr, and then incubated with 100 μCi/ml of [^{35}S]cysteine for 10 min. Then, fresh media containing 0.4 mM unlabeled cysteine was added for the indicated times (chase). Aliquots of cell lysates were immunoprecipitated with anti-MSA serum and compared on a 15% polyacrylamide gel. Mr ~ 34K carrier protein, Mr ~ 20K pro-MSA, and Mr ~ 16K MSA were labeled after 10 min incubation with [^{35}S]cysteine. Mr 8.7K MSA was labeled after 40–60 min of chase, during which time labeling of Mr ~ 20K and ~ 16K forms was decreasing.

Mr ~ 22K protein to a Mr ~ 20K protein that is immunoprecipitated by anti-MSA serum and is electrophoretically indistinguishable from the Mr ~ 20K protein (pro-MSA) identified in radiolabeled intact BRL-3A cells.

6. Conclusions

We have identified Mr ~ 22K pre-pro-MSA and Mr ~ 20K pro-MSA in immunoprecipitates of [^{35}S]proteins labeled in cell-free translation and in intact BRL-3A cells, respectively. The specificity of immunoprecipitation was established by: (1) inhibition by unlabeled MSA but not by insulin; (2) requirement for immune serum; (3) inability to detect these proteins in a nonproducer subclone (BRL-3A2). The Mr ~ 20K protein is labeled rapidly, is observed intracellularly but not in media, and is converted with time to lower-molecular-weight MSA species. Treatment of Mr ~ 22K pre-pro-MSA with microsomal membrane preparations that contain signal peptidase generates a Mr ~ 20K protein that is indistinguishable from Mr ~ 20K pro-MSA. Together these results provide consistent and compelling evidence that MSA is synthesized in BRL-3A cells as a Mr ~ 22K preprohormone.

Our results in BRL-3A cells suggest that MSA and its carrier protein are synthesized as separate proteins, rather than as part of a single large precursor polyprotein. This was suggested by the results of cell-free translation, since the carrier protein (Mr ~ 34K) is larger than the observed prepro-MSA.[7] Although it was possible that the Mr ~ 22K protein did not represent a full-length protein precursor from an intact mRNA, the fact that treatment of the Mr ~ 22K protein with a microsomal membrane preparation containing signal peptidase generated a Mr ~ 20K protein that was indistinguishable from the Mr ~ 20K protein labeled in intact cells makes this extremely unlikely. The concordance of these results suggests that the mRNA and Mr ~22K initial translation product are full-length, and that the Mr ~ 20K protein that is rapidly labeled intracellularly does not represent a fragment of a large, rapidly degraded precursor.

We propose that the pathway for MSA biosynthesis in BRL-3A cells, a continuous cell line, reflects the biosynthetic pathway for IGF-II, and possibly IGF-I, in normal tissues *in vivo* and *in vitro*. Not only is Mr 7484 BRL-MSA 93% homologous to IGF-II isolated from normal human plasma, but peptides indistinguishable from BRL-MSA have been identified in fetal rat serum,[13] explants of fetal rat liver,[14] and 3rd passage rat embryo fibroblasts.[15] Mr ~ 22K and ~ 20K pre-pro- and pro-MSA also have been identified in rat embryo fibroblasts suggesting that precursor and processing mechanisms in BRL-3A cells may be present in embryonic fibroblasts as well. In addition, cultured human fibroblasts synthesize peptides immunologically related to IGF-I[16,17] that Clemmons and Shaw (personal communication) have purified and shown to have a Mr ~ 20K. Higher-molecular-weight IGF's also have been described in human plasma and cerebrospinal fluid (IGF-II),[18] in extracts of human brain (IGF-II),[19] and in media conditioned by human fibrosarcoma 8387.[20]

References

1. N. C. Dulak and H. M. Temin, A partially purified polypeptide fraction from rat liver cell conditioned medium with multiplication-stimulating activity for embryo fibroblasts, *J. Cell Physiol* **81**:153, 1973.
2. A. C. Moses, S. P. Nissley, P. A. Short, M. M. Rechler, and J. M. Podskalny, Purification and characterization of multiplication-stimulating activity. Insulin-like growth factors purified from rat-liver-cell-conditioned medium, *Eur. J. Biochem.* **103**:387, 1980.
3. H. Marquardt, G. J. Todaro, L. E. Hendeson, and S. Oroszlan, Purification and primary structure of a polypeptide with multiplication-stimulating activity from rat liver cell cultures: Homology with human insulin-like growth factor II, *J. Biol. Chem.* **256**:6859, 1981.
4. S. P. Nissley, M. M. Rechler, A. C. Moses, H. J. Eisen, O. Z. Higa, P. A. Short, I. Fennoy, C.

B. Bruni, R. M. White, Evidence that multiplication stimulating activity (MSA) purified from the BRL-3A rat liver cell line is found in rat serum and fetal liver organ cultures, in: *Hormones and Cell Culture*, Cold Spring Harbor Laboratory Press, Cold Spring Harbor, New York, Vol. 6., 1979, pp. 79–94.

5. A. C. Moses, S. P. Nissley, P. A. Short, and M. M. Rechler, Immunological cross-reactivity of multiplication stimulating activity polypeptides, *Eur. J. Biochem.* **103**:401, 1980.
6. Y. W.-H. Yang, A. M. Acquaviva, C. B. Bruni, J. A. Romanus, S. P. Nissley, and M. M. Rechler, Biosynthesis of multiplication stimulating activity (MSA) in rat liver cells: Demonstration of pre-pro-MSA and pro-MSA, in: *Insulin-Like Growth Factors/Somatomedins: Basic Chemistry, Biology, and Clinical Importance* (E. M. Spencer, ed.), Walter de Gruyter Co., New York, Berlin, 1983, pp. 603–610.
7. A. M. Acquaviva, C. B. Bruni, S. P. Nissley, and M. M. Rechler, Cell-free synthesis of rat-insulin-like growth factor II, *Diabetes*, **31**:656, 1982.
8. Y. W.-H. Yang, J. A. Romanus, A. M. Acquaviva, T.-Y. Liu, S. P. Nissley, and M. M. Rechler, Intracellular biosynthetic precursor of multiplication-stimulating activity (MSA), in a rat liver cell line, *J. Cell Biol.* **95**:196a, 1982.
9. S. P. Nissley, P. A. Short, M. M. Rechler, J. M. Podskanly, and H. G. Coon, Proliferation of Buffalo rat liver cells in serum-free medium does not depend upon multiplication stimulating activity (MSA), *Cell* **11**:441, 1977.
10. D. J. Knauer, F. W. Wagner, and G. L. Smith, Purification and characterization of multiplication-stimulating activity (MSA) carrier protein, *J. Supramol. Structure* **15**:177, 1981.
11. Y. W.-H. Yang, J. A. Romanus, T.-Y. Liu, S. P. Nissley, and M. M. Rechler, Biosynthesis of rat insulin-like growth factor II (rIGF-II). I. Immunochemical demonstration of a ~20,000 dalton biosynthetic precursor of rat IGF-II in metabolically-labeled BRL-3A rat liver cells, *J. Biol. Chem.* **260**:2570–2577, 1985.
12. G. Kreil, Transfer of proteins across membranes, *Annu. Rev. Biochem.* **50**:317, 1981.
13. A. C. Moses, S. P. Nissley, P. A. Short, M. M. Rechler, R. M. White, A. B. Knight, and O. Z. Higa, Increased levels of multiplication-stimulating activity, an insulin-like growth factor, in fetal rat serum, *Proc. Natl. Acad. Sci. USA* **77**:3649, 1980.
14. M. M. Rechler, H. J. Eisen, O. Z. Higa, S. P. Nissley, A. C. Moses, E. E. Schilling, I. Fennoy, C. B. Bruni, L. S. Phillips, and K. L. Baird, Characterization of a somatomedin (insulin-like growth factor) synthesized by fetal rat liver organ cultures, *J. Biol. Chem.* **254**:7942, 1979.
15. S. O. Adams, S. P. Nissley, L. A. Greenstein, Y. W-H. Yang, and M. M. Rechler, Synthesis of multiplication-stimulating activity (rat insulin-like growth factor II) by rat embryo fibroblasts, *Endocrinology* **112**:979, 1983.
16. D. R. Clemmons, L. E. Underwood, and J. J. Van Wyk, Hormonal control of immunoreactive somatomedin production by cultured human fibroblasts, *J. Clin. Invest.* **67**:10, 1981.
17. P. R. Atkison, E. R. Weidman, B. Bhaumick, and R. M. Bala, Release of somatomedin-like activity by cultured WI-38 human fibroblasts, *Endocrinology* **106**:2006, 1980.
18. G. Haselbacher and R. Humbel, Evidence for two species of insulin-like growth factor II (IGF II and "big" IGF II) in human spinal fluid, *Endocrinology* **110**:1822, 1982.
19. G. K. Haselbacher, M. E. Schwab, A. Pasi, and R. E. Humbel, Insulin-like growth factor II (IGF-II) in human brain: Regional distribution of IGF II and of higher molecular mass forms, *Proc. Natl. Acad. Sci. USA* **82**:2153–2157, 1985.
20. H. Marquardt, G. L. Wilson, and G. J. Todaro, Isolation and characterization of a multiplication-stimulating activity (MSA)-like polypeptide produced by a human fibrosoma cell line, *J. Biol. Chem.* **255**:9177, 1980.

41

The Insulinlike Growth Factor Receptors

Michael P. Czech, Cristina Mottola, Kin-Tak Yu, and Yoshitomo Oka

1. Introduction

Early studies on the binding kinetics of receptors that exhibit relatively high affinity for [^{125}I]insulinlike growth factors[1] indicated that heterogeneous receptor populations resided in many cell types.[1–3] These receptor populations appeared to differ in their exact affinity for [^{125}I]IGF-I versus [^{125}I]IGF-II, and marked differences in the abundance of these receptor populations could be observed among various tissues. It was also apparent from binding kinetic measurements that [^{125}I]IGF-I and [^{125}I]IGF-II exhibited low but significant affinity for insulin receptors[4] and that conversely [^{125}I]insulin could bind with low affinity to at least one population of IGF receptors. From these binding studies, it became clear that a family of receptors with cross-reacting affinities for insulin and the IGF's existed in many cells, but the number of receptor types and their structural identity were unknown.

Development of affinity-labeling techniques for peptide hormone receptors only a few years ago has allowed access to the identities and structural aspects of the IGF receptors. The first successful IGF receptor identification was performed in our laboratory[5] using in combination the cross-linker disuccinimidyl suberate and [^{125}I]multiplication-stimulating activity, which appears to be equivalent to rat IGF-II.[6] We had earlier synthesized this cross-linker as an affinity-labeling reagent for the insulin receptor,[7–9] and it has been useful for other receptor systems as well.[10,11] More recently, photoactive analogues of

Abbreviations used: IGF, insulinlike growth factor; Ig, immunoglobulin.

Michael P. Czech, Cristina Mottola, Kin-Tak Yu, and Yoshitomo Oka • Department of Biochemistry, University of Massachusetts Medical School, Worcester, Massachusetts 01605.

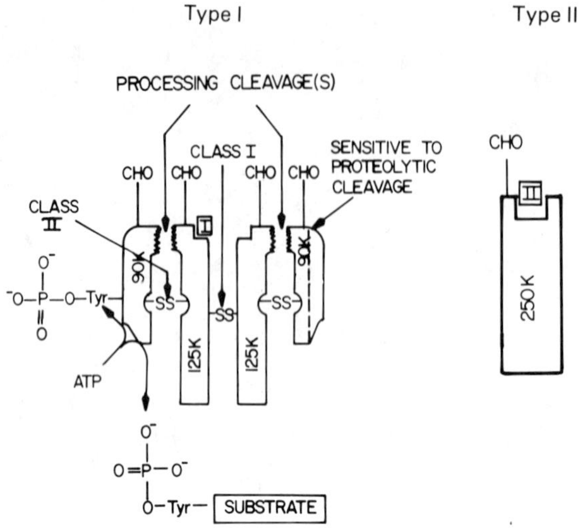

Figure 1 Postulated subunit structures of the insulin and IGF-I receptors (type I) and IGF-II receptors (type II). See text for detailed description.

the IGF's have been utilized to affinity-label a 2nd IGF receptor system.[12] Affinity cross-linking techniques also succeeded in visualizing this 2nd IGF receptor structure.[13–15] Other powerful methodologies have now been applied to one or both of these two IGF receptor systems, including affinity purification by our laboratory[16] and others[17] and specific antibody production.[18,19] The purpose of this chapter is to summarize the insights that have been gained about these IGF receptor systems using these biochemical techniques.

2. *Structure of the IGF Receptors*

The impact of a number of important studies on the structural characterization of the IGF receptors has led to the concept that two structural classes of receptor systems reside on target cell surfaces.[15,20,21] It is possible that other receptor components not yet identified are also involved in IGF actions, but the data now available appear to be accounted for quite well by these two receptor classes. The proposed subunit compositions and notable structural elements are depicted schematically in Fig. 1. The receptors for insulin and the IGF peptides can be classified into two groups, which we denote as type I and type II. The receptor structures for insulin and IGF-I are classified as type I and appear to consist of heterotetrameric subunit compositions.[8,12–15,18] Each of these receptors seems to be composed of two subunits in the 125K–135K range, denoted as α subunits, and two subunits of approximately 90K, denoted as β subunits.

For both the insulin and IGF-I receptors, all α and β subunits within the receptor complex are disulfide-linked.[8,12–15,18] The disulfide linkage or linkage(s) between α and β subunits in the type I receptors are relatively resistant to reduction by exogeneous reductants.[8,15,22] Thus, these receptors in the native membrane or in nondenaturing detergent solution (Triton X-100) are not cleaved by 10 mM dithiothreitol.[8,15,22] In contrast, the disulfide linkage or linkages that connect the two α-S-S-β receptor halves are exquisitely sensitive to reductants. These disulfides are denoted as class I linkages and in Fig. 1 are proposed to link two α subunits. The data available have not unequivocally eliminated the possibility that the two receptor halves are linked by disulfides between α and β or β and β subunits.

The detailed methodology and experiments that led our laboratory group to propose this heterotetrameric structure for the insulin receptor has been previously reviewed.[23,24] Although the data relating to the IGF-I receptor structure are not as extensive as those for the insulin receptor, the results strongly support the concept that the IGF-I receptor shares the same general subunit composition and overall design as that for the insulin receptor.[13–15,18,25] Another interesting feature of the type I receptor structures is the likelihood that α and β subunits are initially synthesized as a continuous polypeptide chain. Results from three independent laboratory groups have demonstrated that an apparent precursor subunit of about 200,000 daltons can be immunoprecipitated with antisera against IGF-I or insulin receptors.[26–28] Monensin, known to inhibit the processing of other receptors, causes the accumulation of this biosynthetically labeled precursor subunit. Furthermore, peptide mapping of this precursor reveals the presence of peptides similar to both α and β subunits.[29] These data support the model that type I receptor processing during the transition from endoplasmic reticulum to plasma membrane involves one or more proteolytic cleavages in a precursor polypeptide that yields one α and one β subunit.

In addition to the above-noted structural similarities between the insulin and IGF-I receptor complexes, several other notable apparent homologies have been demonstrated. For example, two different patients found to exhibit alterations in insulin receptor affinity or number have been shown to exhibit parallel alterations in IGF-I receptors.[30–32] These results suggest the possibility that common genes or gene segments may actually code for the two receptors, although definitive proof for this concept will require more detailed structural analysis. Consistent with this interpretation are results demonstrating significant similarities in the peptide maps of the insulin versus IGF-I receptors in human placenta.[13] Furthermore, crossreactivity of antiinsulin receptor antibody with IGF-I receptors has been documented.[18,31,33] Taken together, these data strongly indicate that at least partial regions of significant sequence homologies exist between insulin and IGF-I receptors.

Another important similarity between the insulin and IGF-I receptors is that they both appear to contain intrinsic tyrosine kinase activity *in vitro*. Experiments first reported by Kasuga, Kahn, and colleagues[34,35] first demonstrated that insulin addition to intact cells labeled with ^{32}P resulted in phosphorylation of the insulin receptor β subunit. More recent experiments have shown that insulin added to detergent-solubilized insulin receptors activates a tyrosine kinase activity that catalyzes autophosphorylation of the insulin receptor β subunit and tyrosine phosphorylation of certain added substrates such as histone proteins.[36–40] The tyrosine kinase activity associated with the insulin receptor copurifies with the insulin receptor after affinity–purification techniques.[41] Interestingly, phosphorylation of the insulin receptor in intact cells results in predominantly phosphoserine phosphorylation, whereas in cell-free systems, tyrosine phosphate residues predominate in response to insulin.[35,36,41]

Recently, Jacobs and colleagues[42] have demonstrated that similar phosphorylation reactions occur in respect to the IGF-I receptor complex. While the latter system has not been as carefully studied, the data available indicate very similar characteristics of the associate kinase activity for these two type I receptors. Recent results also suggest interaction of the insulin and IGF-I receptors in that IGF-I receptor stimulation apparently leads to insulin receptor phosphorylation.[43] The physiological role of the associated tyrosine kinase activities of these receptors has yet to be determined.

In striking contrast to the above-noted structural characteristics of the type I receptors, the type II IGF receptor exhibits a completely different spectrum of properties. While both categories of receptors are apparent glycoproteins, the type II IGF receptor exhibits no disulfide-linked subunits.[14–17] In human mutants that exhibit parallel alterations in insulin and IGF-I receptors, the IGF-II receptor is normal, at least by affinity-labeling analysis.[39] Furthermore, recent studies in our laboratory with purified IGF-II receptor preparations demonstrate that this receptor is devoid of associated kinase activity when assayed under conditions identical to those employed for detection of insulin receptor kinase activity (unpublished observations). These marked differences in properties of the type II IGF receptor and the type I IGF receptor suggest that the physiological roles of these receptor types might be quite different.

3. Biological Actions of the IGF Receptors

An important question related to the two IGF receptor types is their specific physiological role in mediating biological responses. A convenient way to analyze this question is to consider the various biological responses initiated by insulin and the IGF's in two broad groups. One class of responses consist of the rapid, acute metabolic actions that are classically associated with insulin action, while the other class of responses can be considered to include more long-term cellular growth effects (e.g., DNA synthesis). A variety of approaches have been taken to examine this issue: (1) analysis of exact dose–response relationships between growth effects and the concentrations of insulin, IGF-I, or IGF-II required to stimulate such responses; (2) specific blockade of one of the receptor types with antireceptor immunoglobulin and analyzing cellular responses to the peptides under these conditions; and (3) utilization of cell types that express only one or two of the possible three receptor structures for insulin and the IGF's in order to examine the function of each receptor independent of the presence of the other receptors.

Using these strategies, it has been possible to make certain definitive

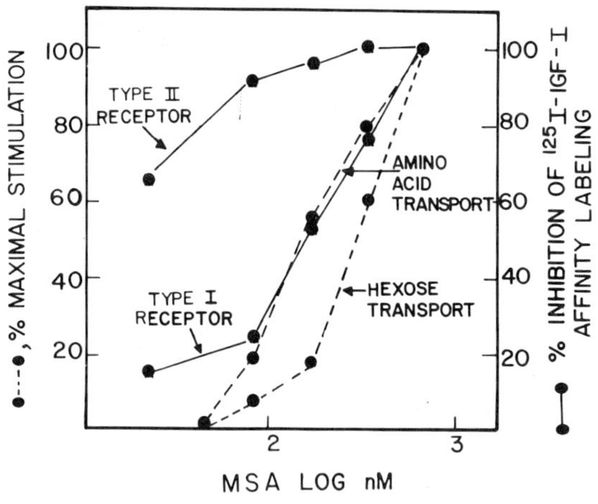

Figure 2 Comparison of the stimulatory effects of MSA on xylose and AIB transport with its inhibitory effects on the [^{125}I]affinity labeling of the type I and type II IGF receptors. The stimulatory effects of MSA at different concentrations is plotted as percentage of maximal stimulation after subtraction of the corresponding basal values. The curves for the percentage of inhibition by unlabeled MSA of the [^{125}I]-affinity labeling of the type I and type II IGF receptor were derived from parallel incubations. Reproduced from Yu and Czech,[49] with permission.

conclusions regarding the biological actions of each of the insulin and IGF receptor types. Insulin receptor is well known to cause rapid effects on a large number of transport and enzymatic steps in metabolic pathways (for review, see refs. 44 and 45). These effects include modulation of hexose transport activity, glycogen synthase, pyruvate dehydrogenase, and hormone-sensitive lipase. However, the ability of insulin receptor to mediate DNA synthesis and cell growth is a more complicated issue. In some cell types (e.g., the H-35 hepatoma cell in culture) there is no doubt that insulin receptor can stimulate these activities because insulin at concentrations as low as 10^{-10} M activates H-35 cell proliferation.[46–47] This fact, in combination with the finding that H-35 cells have no IGF-I receptor and the IGF-II receptor that is present does not bind insulin,[15] strongly implicates the insulin receptor in the growth response of these cells.

On the other hand, human fibroblasts contain insulin receptor that appears incapable of increasing cell growth and DNA synthesis.[4,48] Thus, insulin has a very low potency in stimulating this response in human fibroblasts, and blockade of the insulin receptor with antireceptor antibody does not inhibit this response.[4] That the IGF-I receptor is indeed mediating insulin and IGF-I effects on cell growth in human fibroblasts is strongly supported by the data. Furthermore, in strain E carcinoma cells that contain only IGF-I receptor and no insulin or IGF-II receptor, IGF-I is capable of stimulating cell division with a dose–response curve compatible with its action through the IGF-I receptor (unpublished observations).

Because the structural homology between insulin and IGF-I receptors is strong, we recently tested the hypothesis that IGF-I receptor mimics insulin receptor actions on hexose and amino acid transport, rapid effects characteristic of insulin action. For these experiments, we employed isolated soleus muscle from rat and determined the dose–response relationships between the growth factors and these transport processes.[49] One of the key findings is shown in Fig. 2. As the concentration of IGF-II in the incubation medium is raised, activation of both hexose and amino acid transport occurs in a dose–response relationship where the half-maximal response is about 10^{-7} M. In parallel incubations, the ability of IGF-II to compete for [^{125}I]IGF-I binding to the IGF-I receptor and [^{125}I]IGF-II binding to the IGF-II receptor was assessed. As shown in Fig. 2, the IGF-II receptor is almost fully saturated with IGF-II under the conditions where only a small degree of biological activity is observed. On the other hand, the biological responses to IGF-II parallel its occupancy of the IGF-I receptor. In experiments not illustrated, IGF-II binding to insulin receptor was assessed under identical conditions. Binding to the insulin receptor was too low to account for the biological activities of IGF-II in these studies.[(49)] Thus, we can conclude that the IGF-I receptor is functionally homologous to the insulin receptor, at least with regard to modulating hexose and amino acid transport in muscle.

The role of the IGF-II receptor is unresolved. It seems established that it is incapable of modulating hexose transport and other rapid actions that are ascribed to insulin action, because the potency of IGF-II on these responses is low, indicating it acts through the insulin receptor. The role of the IGF-II receptor in the control of cell proliferation, if any, is unclear. Experiments with H-35 cells, which contain only insulin and IGF-II receptors and not IGF-I receptors, indicate that IGF-II modulates DNA synthesis with a dose–response relationship that approximately parallels binding of this ligand to the IGF-II receptor.[(47–50)] However, insulin receptor is a potent stimulator of DNA synthesis as well, and the concentrations of IGF-II that stimulate this biological response are at levels which allow its significant binding to the insulin receptor. Thus, from these studies it cannot be decided whether insulin receptor or IGF-II receptor or both are involved in mediating such effects of IGF-II.[(47,50)] In other cell types, this issue is even more complicated to unravel owing to the presence of IGF-I receptor. Thus the role of the IGF-II receptor is an important question for further investigation.

4. Immunology of the IGF Receptors

Recently a good deal of progress has been made in obtaining antibody preparations directed against either the type I or type II IGF receptor structures. Two approaches have succeeded in providing anti-type I IGF receptor immunoglobulins.[(1)] One of these involves antiserum from patients with acanthosis nigricans and insulin resistance, previously found to have circulating antiinsulin receptor Ig.[(51,52)] The structural homologies between the insulin and IGF-I receptor structures suggest that cross-reactivity between such antisera and the IGF-I receptor might be observed. This has indeed found to be the case.[(31)] The degree of cross-reactivity for IGF-I receptor antiserum from various patients would be expected to vary significantly. A major problem with the use of these antisera for the specific immunoprecipitation of IGF-I receptors is the cross-reactivity with the insulin receptor. Nonetheless, such antisera may be useful when using cell systems that express IGF-I receptor but not insulin receptor.

Recently, Jacobs and colleagues have been able to produce monoclonal antibodies against IGF-I receptor.(18) In these experiments, mice were injected with partially purified insulin receptor, and hybridoma cell lines were produced and screened for antiinsulin receptor and anti-IGF-I receptor activity. Antiinsulin receptor Ig and and anti-IGF-I receptor Ig populations could be identified, and cross-reactivity for the two receptors varies substantially among different preparations. These data again suggest that certain regions of the two receptors may be very similar while other regions may vary in specific structure. These antibody populations generated by such hybridoma techniques have been useful in specifically immunoprecipitating IGF-I receptor in the absence of substantial insulin receptor. Using such preparations of anti-IGF-I receptor Ig, Jacobs and colleagues have been able to demonstrate tyrosine kinase activity associated with the IGF-I receptor.(42)

Specific antisera against the IGF-II receptor have recently been produced in our laboratory. In these experiments, IGF-II receptor was purified by affinity chromatography on immobilized rat IGF-II(16) and purified receptor injected into rabbits. Rabbits were injected with 50 μg of purified IGF-II receptor from rat placental membranes in Freund's complete adjuvant and then boosted 8 weeks later by injection of 12 μg of purified IGF-II receptor. The Ig fraction from blood drawn 4 weeks after the boost was prepared by ammonium sulfate precipitation. The anti-IGF-II receptor Ig is capable of forming immunoprecipitation lines on agar immunodiffusion plates against purified IGF-II receptor.(19) Furthermore, the enzyme-linked immunosorbent assay ELISA showed a positive reaction with this anti-IGF-II receptor antiserum at dilutions up to 5000. These data indicate that a potent rabbit antiserum against IGF-II receptor is in hand.

We have recently tested the ability of the anti-IGF-II receptor antibody to effect [^{125}I]IGF-II binding to its receptor. In a number of plasma membrane preparations or intact cell systems the anti-IGF-II receptor Ig is a potent inhibitor of [^{125}I]IGF-II binding.(19) Figure 3 shows a typical experiment performed with isolated adipocyte plasma membranes. The cells were incubated with 3 μg of control or anti-IGF-II receptor Ig in the presence of [^{125}I]IGF-II or [^{125}I]insulin, and association of labeled ligands with cells was allowed to proceed for 45 min at room temperature. Binding of [^{125}I]IGF-II to the membranes was inhibited about 75% under the conditions of these experiments, whereas insulin binding was not significantly affected. These results are representative of those obtained with several other systems. Solubilization of IGF-II receptor with Triton X-100 leads to even more dramatic inhibition of [^{125}I]IGF-II binding to its receptor by the antireceptor Ig preparation. Under these conditions virtually 100% inhibition of [^{125}I]IGF-II binding can be observed owing to very low concentrations of antireceptor Ig. This anti-IGF-II receptor antibody preparation should be very useful in further assessing the biological role of the IGF-II receptor because this receptor can now be specifically blocked with the antireceptor Ig in intact cells.

5. Insulin Action on the IGF-II Receptor

Experiments from several laboratories have now demonstrated that [^{125}I]IGF-II binding to fat cells or H-35 hepatoma cells is markedly stimulated by physiological doses of insulin.(4,53–55) Affinity labeling of IGF-II receptor by [^{125}I]IGF-II and the cross-linker disuccinimidyl suberate in the presence and absence of insulin shows that indeed this increased [^{125}I]IGF-II binding reflects increased association of the peptide with the Mr = 250,000 IGF-II receptor structure.(55) Recent experiments in our laboratory demonstrated that plasma membranes derived from insulin-treated adipocytes exhibited higher levels of [^{125}I]IGF-II binding compared to controls, whereas low-density microsomes prepared from insulin-treated adipocytes exhibited decreased [^{125}I]IGF-II binding.(55) Scatchard analysis of [^{125}I]IGF-II binding to these membrane fractions indicated that the effect of insulin is to modulate the number of apparent receptor sites in these fractions and indicated a redistribution of receptors from one membrane fraction (low-density microsomes) to the other (plasma membranes). In contrast, Scatchard analysis of [^{125}I]IGF-II binding to intact cells in the presence and absence of insulin indicated that the increased [^{125}I]IGF-II binding in response to insulin was due to an apparent affinity increase in the IGF-II receptor for [^{125}I]IGF-II.(54,55) Thus, the data available have been ambiguous in terms of the mechanism whereby insulin action increases [^{125}I]IGF-II binding to cells.

Recent work in our laboratory has been able to clarify this problem. Our strategy has been to circumvent the need to use [^{125}I]IGF-II binding as a probe for an affinity versus a number change in IGF-II receptors by utilizing anti-IGF-II receptor Ig binding to intact cells as an estimate of the number of IGF-II receptors present. The binding of antireceptor antibody to IGF-II receptor was previously shown to be independent of the ligand binding site because immunoprecipitation lines between antisera and purified IGF-II receptor could easily be generated even under conditions where the receptor was saturated with IGF-II.(19) We thus reasoned that measurement of antireceptor Ig binding to the cell surface might adequately reflect the number of IGF-II receptors present on the cell surface. In order to monitor antireceptor Ig binding to intact cells, isolated fat cells were treated with or without insulin and incubated in the presence of anti-IGF-II receptor Ig for 30 min at 21°C. The cells were then washed free of antireceptor Ig and incubated with [^{125}I]goat antirabbit IgG for 30 min at 21°C. The [^{125}I]goat antirabbit IgG was composed of the IgG fraction of serum from goat raised against rabbit IgG. The cells were again washed and centrifuged through silicone fluid to assess the label that bound to cells. We found that the amount of anti-IGF-II receptor Ig that bound to intact fat cells in the presence of insulin, as reflected by the [^{125}I]goat antirabbit IgG binding, was increased by insulin action to the same degree as was [^{125}I]IGF-II binding itself.(19)

This parallel increase in the binding of anti-IGF-II receptor Ig and [^{125}I]IGF-II strongly suggests that an increase in the number of IGF-II receptors

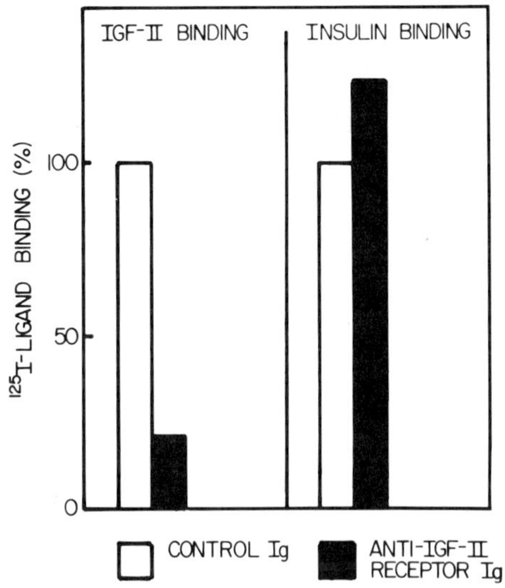

Figure 3 Inhibitory action of anti-IGF-II receptor Ig on [^{125}I]IGF-II binding to rat adipocytes plasma membrane. Purified rat adipocytes plasma membranes (69 μg) were suspended in 100 μl of Krebs–Ringer phosphate buffer, pH 7.4, containing 1% bovine serum albumin and 3 μg of control (empty bars) or anti-IGF-II receptor Ig (solid bars). After 45 min at room temperature of end-over-end mixing, [^{125}I]IGF-II (final concentration 1 nM) was added in 100 μl of the same buffer, and the samples were further incubated for 30 min at room temperature. The reaction mixture was diluted with 1 ml of cold Krebs–Ringer phosphate buffer, pH 7.4, and the membranes centrifuged for 7 min at 12,000 × g. The pellets were washed once; then the bottom of the incubation tubes, containing the pelleted membranes, was cut and counted for radioactivity. The characterization of the antibodies and further experimental details are reported by Oka *et al.*[19] The values obtained in the presence of control Ig are expressed as 100% binding and are, respectively, 3.2 fmoles of [^{125}]IGF-II and 0.8 fmoles of [^{125}I]insulin bound per tube. The nonspecific binding values were measured in the presence of 250 nM unlabeled hormone and subtracted from experimental values. The average values of two determinations from representative experiment are shown.

is expressed in response to insulin on the intact fat cell surface. That the affinity of the IGF-II receptor for antireceptor IGF-II receptor Ig is modulated by insulin is most unlikely because in these experiments we employed concentrations of anti-IGF-II receptor that were well above those required to saturate binding sites on the IGF-II receptor. It is also unlikely that insulin action acts to expose new antigenic sites on the IGF-II receptor as a consequence of a conformation change of the receptor. Control experiments that we performed indicated that receptor on control cells was able to adsorb all of the anti-IGF-II receptor Ig from the anti-IGF-II receptor preparation, indicating that control IGF-II receptor contains all the antigenic sites that the receptors on insulin-treated cells possess. These data then demonstrate that insulin acts rapidly to mobilize receptors in fat cells from a location not exposed to a cell surface location that is exposed to the extracellular medium. Further investigations underway are designed to determine the precise mechanism(s) that are involved in this action of insulin.

What could be the physiological importance of increased IGF-II receptors in response to insulin? The answer to this question is clouded by the lack of knowledge we have about the physiological role of the IGF-II receptor itself. If this receptor were involved in modulating cell DNA synthesis or other biological effects, the ability of insulin to increase the activity of the IGF-II receptor by increasing its presence on the cell surface might lead to potentiation of IGF-II action. However, it is possible that the IGF-II receptor does not signal biological responses but rather serves another function(s). We can only guess what these might be, but one possibility is that it serves the role of a transport protein. That is, perhaps this receptor acts to transport IGF-II from one location to another. In the case of adipocytes, the transport would presumably be from the extracellular medium to the cellular compartment. However, in the case of certain cells such as endothelial or epithelial cells, this receptor might serve a transcellular transport role for IGF-II. At present, however, these possible roles of the IGF-II receptor must be considered unsupported speculation.

6. Future Directions

It is clear that over the past several years important new advances have been made in our understanding of the receptor systems for insulin and the IGF's. The general subunit structures postulated for the type I and type II receptor systems provide a useful framework for future studies. A most important objective will be to obtain at least partial primary amino acid sequences of these receptors in order to refine our understanding about the homologies between the insulin and IGF-I receptors as well as possible relationships to the IGF-II receptor. We should be prepared for the possibility that structural microheterogeneity exists among the type I receptors in different tissues such that the type I receptors actually represent a spectrum of closely related structures. Also, the fact that the IGF-II receptor has affinity for IGF-I as well as IGF-II suggests possible partial sequence homology with the IGF-I receptor. There is little doubt that significant insight will be achieved when sequence analysis of these receptors is available.

Other important questions also await future investigations. For example, the recent demonstration in our laboratory that the type I IGF receptor mediates hexose and amino acid transport activation and therefore probably other acute metabolic effects similar to insulin action suggests a potentially important role for IGF-I in muscle physiology. These results also suggest a possible important role for the IGF-I receptor in modulating metabolic activities of other cell types. Another important physiological question relates to the role of IGF-II receptor. Its function as a signaling receptor is in question in our opinion, and other possible roles for this protein should be considered and investigated. Also, the functional relevance of insulin action to increase the expression of IGF-II receptors on the cell surface needs to be elucidated. Clearly, this question is closely related to the question of IGF-II receptor function itself. The fact that we now have significant biochemical information about the insulin and IGF receptor family, yet so many important questions remain, indicates that this field will be a very fertile one for future research.

References

1. K. Megyesi, C. R. Kahn, J. Roth, E. R. Froesch, R. E. Humbel, J. Zapf, and D. M. Neville, Jr., Insulin and nonsuppressible insulin-like activity (NSILA's): Evidence for separate plasma membrane receptor sites, *Biochem. Biophys. Res. Commun.* **52:**307, 1974.
2. J. Zapf, E. Schoenle, and E. R. Froesch, Insulin-like growth factors I and II: Some biological actions and receptor binding characteristics of two purified constituents of nonsuppressible insulin-like activity of human serum, *Eur. J. Biochem.* **87:**285, 1978.
3. G. L. King, C. R. Kahn, M. M. Rechler, and S. P. Nissley, Direct demonstration of separate receptors for growth and metabolic activities of insulin and multiplication-stimulating activity (an insulin-like growth factor) using antibodies to the insulin receptor, *J. Clin. Invest.* **66:**130, 1980.
4. R. N. Marshall, L. E. Underwood, S. J. Voina, D. B. Foushee, and J. J. Wan Wyk, Characterization of the insulin and somatomedin C receptors in human placental membranes, *J. Clin. Endocrinol. Metab.* **39:**284, 1974.
5. J. Massague, B. J. Guillette, and M. P. Czech, Affinity labeling of multiplication stimulating activity receptors in membranes from rat and human tissues, *J. Biol. Chem.* **256:**2122, 1981.
6. H. Marquardt, G. J. Todaro, L. E. Henderson, and S. Oroszlan, Purification and primary structure of a polypeptide with multiplication-stimulating activity from rat liver cell cultures, *J. Biol. Chem.* **256:**6859, 1981.
7. P. F. Pilch and M. P. Czech, Interaction of crosslinking agents with the insulin effector system of isolated fat cells, *J. Biol. Chem.* **254:**3375, 1979.
8. P. F. Pilch and M. P. Czech, The subunit structure of the high affinity insulin receptor, *J. Biol. Chem.* **255:**1722, 1980.
9. J. Massague, P. F. Pilch, and M. P. Czech, Electrophoretic resolution of three major insulin receptor structures with unique subunit stoichiometries, *Proc. Natl. Acad. Sci. USA* **77:**7137, 1980.
10. J. Massague, B. J. Guillette, M. P. Czech, C. J. Morgan, and R. A. Bradshaw, Identification of a nerve growth factor receptor protein in sympathetic ganglia membranes by affinity labeling, *J. Biol. Chem.* **256:**9419, 1981.
11. K. Glenn, D. F. Bowen-Pope, and R. Ross, Platelet-derived growth factor. III. Identification of a platelet derived growth factor receptor by affinity labeling, *J. Biol. Chem.* **257:**5172, 1982.
12. B. Bhaumick, R. M. Bala, and M. D. Hollenberg, Somatomedin receptor of human placenta: solubilization, photolabeling, partial purification and comparison with insulin receptor, *Proc. Natl. Acad. Sci. USA* **78:**4279, 1981.
13. S. D. Chernausek, S. Jacobs, and J. J. Van Wyk, Structural similarities between human receptors for somatomedin C and insulin: analysis by affinity labeling, *Biochemistry* **20:**7345, 1981.
14. M. Kasuga, E. Van Obberghen, S. P. Nissley, and M. M. Rechler, Demonstration of two subtypes of insulin-like growth factor receptors by affinity crosslinking, *J. Biol. Chem.* **256:**5305, 1981.
15. J. Massague and M. P. Czech, The subunit structures of two distinct receptors for insulin-like growth factors I and II and their relationship to the insulin receptor, *J. Biol. Chem.* **257:**5038, 1982.
16. C. L. Oppenheimer and M. P. Czech, Purification of the type II insulin-like growth factor receptor from rat placenta, *J. Biol. Chem.* **258:**8539, 1983.
17. G. P. August, S. P. Nissley, M. Kasuga, L. Lee, L. Greenstein, and M. M. Rechler, Purification of an insulin-like growth factor II receptor from rat chondrosarcoma cells, *J. Biol. Chem.* **258:**9033, 1983.
18. F. C. Kull, Jr., S. Jacobs, Y.-F. Su, M. E. Svoboda, J. J. Van Wyk, and P. Cuatrecasas, Monoclonal antibodies to receptors for insulin and somatomedin C, *J. Biol. Chem.* **258:**6561, 1983.
19. Y. Oka, C. Mottola, C. L. Oppenheimer, and M. P. Czech, Insulin activates the appearance of IGF-II receptors on the adipocyte cell surface, *Proc. Natl. Acad. Sci. USA* **81:**4028, 1984.
20. M. P. Czech, Structural and functional homologies in the receptors for insulin and the insulin-like growth factors, *Cell* **31:**8, 1982.

21. M. P. Czech, C. L. Oppenheimer, and J. Massague, Interrelationships among receptor structures for insulin and peptide growth factors, *Fed. Proc.* **42:**2598, 1983.
22. J. Massague and M. P. Czech, Role of disulfides in the subunit structure of the insulin receptor, *J. Biol. Chem.* **257:**6729, 1982.
23. M. P. Czech, J. Massague, and P. F. Pilch, The insulin receptor: structural features, *Trends Biochem. Sci.* **6:**222, 1981.
24. M. P. Czech, Insulin action, *Am. J. Med.* **70:**142, 1981.
25. M. Kasuga, E. Van Obberghen, S. P. Nissley, and M. M. Rechler, Structure of the insulin-like growth factor receptor in chicken embryo fibroblasts, *Proc. Natl. Acad. Sci. USA* **79:**1864, 1982.
26. J. A. Hedo, C. R. Kahn, M. Hayashi, K. M. Yamada, and M. Kasuga, Biosynthesis and glycosylation of the insulin receptor, *J. Biol. Chem.* **258:**10020, 1983.
27. P. J. Deutsch, F. W. Wan, O. M. Rosen, and C. S. Rubin, Latent insulin receptors and possible receptor precursors in 3T3-L1 adipocytes, *Proc. Natl. Acad. Sci. USA* **80:**133, 1983.
28. S. Jacobs, F. C. Kull, Jr., and P. Cuatrecasas, Monensin blocks the maturation of receptors for insulin and somatomedin C: Identifiation of receptor precursors, *Proc. Natl. Acad. Sci. USA* **80:**1228, 1983.
29. M. Kasuga, J. A. Hedo, K. M. Yamada, and C. R. Kahn, The structure of insulin receptor and its subunits, *J. Biol. Chem.* **257:**10392, 1982.
30. E. E. Van Obberghen-Schilling, M. M. Rechler, J. A. Romanus, A. B. Knight, S. P. Nissley, and R. E. Humbel, Receptors for insulin-like growth factor I are defective in fibroblasts cultured from a patient with leprechaunism, *J. Clin. Invest.* **68:**1356, 1981.
31. R. G. Rosenfeld, D. Baldwin, Jr., L. A. Dollar, R. L. Hintz, J. M. Olefsky, and A. Rubenstein, Simultaneous inhibition of insulin and somatomedin C binding to cultured IM-9 lymphocytes by naturally occurring anti-receptor antibodies, *Diabetes* **30:**979, 1981.
32. J. Massague, G. F. Freidenberg, J. M. Olefsky, and M. P. Czech, Parallel decreases in the expression of receptors for insulin and insulin-like growth factor I in a mutant human fibroblast line, *Diabetes* **32:**541, 1983.
33. R. A. Roth, B. Maddux, K. Y. Wong, D. M. Styne, G. Van Vliet, R. E. Humbel, and I. D. Goldfine, Interactions of a monoclonal antibody to the insulin receptor with receptors for insulin-like growth factors, *Endocrinology* **112:**1865, 1983.
34. M. Kasuga, F. A. Karlsson, and C. R. Kahn, Insulin stimulates the phosphorylation of the 95,000 dalton subunit of its own receptor, *Science* **215:**185, 1982.
35. M. Kasuga, Y. Zick, D. L. Blith, F. A. Karlsson, H. U. Häring, and C. R. Kahn, Insulin stimulation of phosphorylation of the β subunit of the insulin receptor, *J. Biol. Chem.* **257:**9891, 1982.
36. M. Kasuga, Y. Zick, D. L. Blithe, M. Crettaz, and C. R. Kahn, Insulin stimulates tyrosine phosphorylation of the insulin receptor in a cell-free system, *Nature* **298:**667, 1982.
37. F. Machicao, T. Urumow, and O. H. Wieland, Phosphorylation–dephosphorylation of purified insulin receptor from human placenta, *FEBS Lett.* **149:**96, 1982.
38. E. Van Obberghen, B. Rossi, A. Kowalski, H. Gazzano, and G. Ponzio, Receptor-mediated phosphorylation of the hepatic insulin receptor: evidence that the Mr 95,000 receptor subunit is its own kinase, *Proc. Natl. Acad. Sci. USA* **80:**945, 1983.
39. Y. Zick, J. Whittaker, and J. Roth, Insulin stimulated phosphorylation of its own receptor, *J. Biol. Chem.* **258:**3431, 1983.
40. L. A. Stadtmauer and O. M. Rosen, Phosphorylation of exogenous substrates by the insulin receptor-associated protein kinase, *J. Biol. Chem.* **258:**6682, 1983.
41. M. Kasuga, Y. Fujita-Yamaguchi, D. L. Blithe, and C. R. Kahn, Tyrosine-specific protein kinase activity is associated with the purified insulin receptor, *Proc. Natl. Acad. Sci. USA* **80:**2137, 1983.
42. S. Jacobs, F. C. Kull, Jr., H. S. Earp, M. E. Svoboda, J. J. Van Wyk, and P. Cuatrecasas, Somatomedin C stimulates the phosphorylation of the beta subunit of its own receptor, *J. Biol. Chem.* **258:**9581, 1983.
43. J. B. Rubin, M. A. Shia, and P. F. Pilch, Stimulation of tyrosine-specific phosphorylation in vitro by insulin-like growth factor I, *Nature* **305:**438, 1983.
44. M. P. Czech, Molecular basis of insulin action, *Annu. Rev. Biochem.* **46:**359, 1977.

45. M. P. Czech, Insulin action: Second messengers, in: *Handbook of Diabetes Mellitus* (M. Brownlee, ed.), Garland Press, New York, 1981, pp. 117–149.
46. J. W. Koontz and M. Iwahashi, Insulin as a potent, specific growth factor in a rat hepatoma cell line, *Science* **211:**947, 1981.
47. J. Massague, L. A. Blinderman, and M. P. Czech, The high affinity insulin receptor mediates growth stimulation in rat hepatoma cells, *J. Biol. Chem.* **257:**13958, 1982.
48. M. M. Rechler, J. M. Podskalny, I. D. Goldfine, and C. A. Wells, DNA Synthesis in human fibroblasts: Stimulation by insulin and by nonsuppressible insulin-like activity (NSILA's), *J. Clin. Endocrinol. Metab.* **39:**512, 1974.
49. K.-T. Yu and M. P. Czech, The type I insulin-like growth factor receptor mediates the rapid effects of multiplication-stimulating activity on membrane transport systems in rat soleus muscle, *J. Biol. Chem.*, **259:**3090, 1984.
50. J. W. Koontz, The role of the insulin receptor in mediating the insulin-stimulated growth response in Reuber H-35 cells, *Mol. Cell Biochem.* **58:**139, 1984.
51. J. S. Flier, C. R. Kahn, J. Roth, and R. S. Bar, Antibodies that impair insulin receptor binding in an unusual diabetic syndrome with severe insulin resistance, *Science* **190:**63, 1975.
52. C. R. Kahn, J. S. Flier, R. S. Bar, J. A. Archer, P. Gorden, M. D. Martin, and J. Roth, The syndromes of insulin resistance and acanthosis nigricans: insulin receptor disorders in man, *N. Engl. J. Med.* **294:**739, 1976.
53. E. Schoenle, J. Zapf, and E. R. Froesch, Effects of insulin and NSILA on adipocytes of normal and diabetic rats: Receptor binding, glucose transport, and glucose metabolism, *Diabetologia* **13:**243, 1977.
54. G. L. King, M. M. Rechler, and C. R. Kahn, Interactions between the receptors for insulin and the insulin-like growth factors on adipocytes, *J. Biol. Chem.* **257:**10001, 1982.
55. C. L. Oppenheimer, J. E. Pessin, J. Massague, W. Gitomer, and M. P. Czech, Insulin action rapidly modulates the apparent affinity of the insulin-like growth factor II receptor, *J. Biol. Chem.* **248:**4824, 1983.

42

The Somatomedin Binding Proteins

Raymond L. Hintz

1. Introduction

The majority of steroidal and thyroid hormones are carried in plasma on specific macromolecular binding proteins. Peptide hormones, such as growth hormone and insulin, circulate free in plasma rather than in macromolecular complexes and do not have specific plasma binding proteins. An exception to this generalization is the somatomedin/insulinlike growth factor (SM/IGF) group of hormonal peptides. From the early observations of Daughaday and Kipnis on "sulfation factor"[1] it was apparent that the bulk of this bioactivity was a nondialyzable macromolocule. Similarly, it was clear from early parallel work on the "nonsuppressible insulinlike activity" (NSILA)[2] that this biological activity was also carried as a macromolecule in native plasma. However, a portion of the NSILA was soluble in acid–ethanol and became a small peptide of 5000–10,000 daltons under these conditions (NSILA-S). It was independently shown that essentially all of the sulfation factor bioactivity was converted under acid conditions to a 5000- to 10,000-dalton peptide.[3] This similarity in the behavior of sulfation factor and NSILA-S was one of the pieces of evidence that led to the realization that these two bioactivities were the same substance, and the general designation somatomedin (SM)[4] was proposed. Since then, two human SM peptides have been sequenced by Rinderknecht and Humbel; insulinlike growth factor I (IGF-I)[5] and insulinlike growth factor II (IGF-II).[6] IGF-I is identical or nearly so with SMC.[7] In the past decade, there have been many studies on these hormonal peptides and how they are carried in plasma.

2. Plasma Forms of the Somatomedins

The early observations about the macromolecular character of plasma SM have been confirmed and extended using a variety of techniques. The SM's

Raymond L. Hintz • Department of Pediatrics, Stanford University, Stanford, California 94305.

are macromolecular in plasma, but under acid conditions the 5000- to 10,000-dalton SM peptides are released. SM exists in at least two forms in plasma. The majority (approximately 80%) of somatomedins in normal plasma is carried as part of a 150,000-mol. wt. complex. This 150K complex is under growth hormone control, just as is the total plasma content of SM. The remainder of plasma SM activity is carried as a somewhat smaller complex which has a molecular weight variously estimated as between 30,000 and 50,000 daltons. The amount of this "45K" SM complex in plasma is less firmly linked to GH than is the 150K form, and the proportion of 45K complex has been demonstrated to be increased in plasma both from hypopituitary patients[(8)] and in the newborn.[(9)] In addition, some workers have estimated that perhaps 1% or more of the SM in plasma may exist as "free" SM peptide.[(10,11)] However, recent estimates, using highly sensitive radioimmunoassays, have placed the amount of "free" SM peptide plasma at much lower than 1%.[(12,13)]

The general outline of the SM complexes appear to be similar in other mammalian species. The best studied species other than the human is the rat. In the rat, like the human, the predominant plasma SM is the 150,000-daltons mol. wt. complex.[(14)] In addition to the 150,000-dalton complex a smaller, approximately 45,000-mol. wt., complex is found. As is true in the human, it appears to be the 150,000-dalton complex form that is most exquisitely under growth hormone control. Thus, the proportion of 45,000-dalton-mol. wt. complex increases in situations in which the stimulation by growth hormone is reduced. By using partially purified rat SM complex reinjected into rats[(15)] and radioactive SM peptide preequillibrated with plasma and injected into rats,[(16)] it has been clearly shown that the two complex forms have quite different half-lives in plasma. In both these studies the 150,000-mol. wt. complex has a much longer half-life than the 45,000-mol. wt. complex, and both complexes have a significantly longer half-life than that of any unbound SM peptide injected. As is true in the human, fetal rats have a higher proportion of the smaller complex than do adults, and the proportion of the 150,000-dalton SM complex increases rapidly after birth.

The information on the plasma forms of SM in other species is limited. However, it is clear from the work of Liberti[(17)] that bovine SM also exists in plasma almost totally as a macromolecular complex. In screening a wide range of species for SM, Wilson and Hintz[(18)] demonstrated that a number of mammalian species are similar to human SM's since they are macromolecular in plasma but dissociable in acid conditions. Thus, it appears that the broad outlines of SM plasma forms are similar over a wide variety of species.

3. Plasma Binding Proteins for the SM's

A partial explanation of the observations on the plasma forms of SM evolved with the demonstration of specific, reversible association of SM peptides with higher-molecular-weight plasma proteins. This was initially demonstrated using partially purified SM peptides.[(19)] When purified SM peptide is combined with acid-stripped macromolecules, the SM peptide binds tightly to the macromoleular fraction on chromatography under neutral conditions. This was confirmed using radioactively labeled pure SM peptides.[20,21] Using the radioactive SM peptides, it was demonstrated that the binding to plasma proteins had high affinity, with an estimated equilibrium constant of 0.1–1 nm, similar to the tissue receptors for SM peptides. In addition, the binding to these plasma proteins was very specific for SM/IGF peptides, although most workers have found an affinity preference for IGF-II compared to SMC/IGF-I. Unlike the SMC/IGF-I tissue receptors, insulin was unable to compete with the SM/IGF peptides for the plasma binding protein even at high concentrations. These observations led to the concept that unlike other peptide hormones, the SM's had specific plasma binding proteins. Just as there are at least two forms of SM complex, there are at least two forms of the SM binding protein in plasma. When radioactive SM peptides are added to whole plasma and allowed to equilibrate, the vast majority of the radioactive SM associates with a specific binding protein of approximately 45,000 mol. wt. A smaller proportion associates with a specific binding protein equivalent to 150,000 mol. wt. Both these binding proteins have high affinities for SM peptides, with an estimated affinity constant in the range of 0.1–1 nm. Most of the published data are consistent with a single affinity class, without evidence of negative cooperativity. Like the tissue receptors for SM, both the binding proteins are quite specific for SM peptides. However, insulin does not seem to compete at all for either SM binding protein.

The kinetics of the complex forms of SM are unusual. No matter how long the incubation, the radioactive SM peptides do not completely equilibrate with the SM peptides in the complex forms. This makes the measurement of SM binding proteins a complicated task. The most direct way of measuring SM binding protein has been to strip the SM peptide from the complex forms in strong acid conditions. Under these conditions, both complexes break down to SM peptides and a binding protein component that has an approximate molecular weight of 40,000 daltons. The binding protein and the SM peptide can be easily separated by chromatography and then measured separately. By using this approach, a number of groups have estimated the amounts of "total" SM binding protein in plasma. The amount of SM binding protein measured by these techniques clearly varies with the growth hormone status of the individual. The amount of binding protein is low in hypopituitary individuals and high in acromegalic patients when compared to normal.[11,22] Thus, the total amount of SM binding protein appears to reflect the amount of SM peptide present in plasma. However, there is some question as to the absolute quantitation of these measurements of binding protein, since kinetic evidence[12] and other indirect observations suggest that the characteristics of the binding proteins may well be changed by the acid treatment.

A different method of measuring the binding protein in the 150,000-mol. wt. SM complex has been suggested by White and his co-workers.[23] This method is based on the observation that when the 150K and 45K SM complexes are separated, radioactive SM peptide can equilibrate with the 150,000 complex without the complication of the presence of the smaller binding protein form. Using this technique, White and his co-workers have confirmed that the amount of SM binding protein in the 150,000-mol. wt. binding protein is under growth hormone control. Thus, they obtained high values in acromegalic patients and low values in pituitary patients.

Another approach to measuring binding protein has been developed.[24] By keeping the temperature of incubation low, the bulk of the radioactive SM is bound by an "unsaturated" binding protein component of approximately 40,000 mol. wt. The hormonal control of this form of binding protein is markedly different from the control of the bulk of the binding protein which is part of the 150,000-mol. wt. SM complex. The estimated number of binding sites for this unsaturated binding protein in plasma does not appear to be different in hypopituitary patients from the number found in normal plasma. However, the unsaturated binding protein from hypopituitary patients has a higher affinity for SMC/IGF-I. It also has a somewhat smaller apparent molecular weight.[8]

In addition to the binding proteins characterized in human plasma, another SM binding protein from human sources has been demonstrated in amniotic fluid.[25] Unlike the plasma SM/IGF binding proteins, all of the binding protein found in human amniotic fluid appears to be in a small form of approximately 35,000 molecular weight. The amount of this binding protein is maximum in midtrimester and decreases toward term. What role this binding protein plays a fetal physiology and what relationship, if any, this binding protein has to the binding plasma proteins remain to be elucidated.

Specific binding proteins for SM peptides have been demonstrated in other species as well. Again, the best studied species, in addition to the human, is the rat. The binding proteins in the rat appear to be similar to those in the human. Binding proteins for SM/IGF peptides have been demonstrated in association with both the large SM complex and the smaller SM complex form. The amount of SM binding protein in rat plasma has been clearly demonstrated to be under growth hormone control.[14] As hypox rats are treated with growth hormone, the amount of binding protein in the 45,000-mol. wt. complex form decreases, and the amount of binding protein associated with the larger SM complex increases. SM binding protein activity has been demonstrated in a number of other mammalian species including rabbits, dogs, and cattle.[21] In addition to this, binding proteins have been demonstrated in cultures of fetal mouse limb bud[26] and from rat brain slices *in vitro*.[27] Although the data are still preliminary, it appears likely that the binding proteins and the SM peptides are closely associated throughout evolution.

4. Characterization of the SM Binding Proteins

A number of attempts have been made to further characterize and purify the SM binding proteins from plasma. The majority of the workers have used acid to split the complex and binding protein apart and then separate the SM peptides and their binding proteins by column chromatography. After this type of partial purification, the reassociation of SM peptide and binding protein component does not lead to reconstruction of the 150,000-mol. wt. complex, but to a form of roughly 45,000 molecular weight. On the basis of this and kinetic data, it has been proposed that the 150,000-mol. wt. complex is actually made up of three subunits: the SM peptide, the binding protein component, and an acid-labile component which is denatured under acid conditions.(12,28) The simplest explanation of the 45K complex is that it represents the association of the SM peptide and the binding protein subunit. Using an approach not involving acid for characterizing and partially purifying the binding protein, Furlanetto has developed a similar model.(29) Using ion exchange chromatography as the initial step to separate the binding proteins from the SM peptides and mixing experiments with these fractions, he also proposed that the 150,000-molecular-weight SM complex has a three-subunit structure, with one subunit being the SM peptide, a 2nd subunit containing the binding protein activity, and a 3rd subunit which is acid labile. Homogeneity of the human plasma binding protein has not been achieved because as purification attempts have proceeded past the initial steps, the recovery of binding protein has decreased rapidly. It is unclear whether these poor yields are due to the adsorption of binding protein to surfaces after it is removed from the complex or due to denaturation of the binding protein so that it can no longer be detected. The binding protein in human amniotic fluid has been characterized in greater detail.(25,30) It appears to be a 30,000- to 33,000-mol. wt. protein with an acidic isoelectric point of between 4.5 and 5.0. Recently Drop and Kortleve(31) have reported further purification of the amniotic fluid binding protein and preliminary findings on a radioimmunoassay for the amniotic fluid binding protein. Both the purified plasma binding protein(32) and the amniotic fluid binding protein(25,30) have been shown to inhibit the *in vitro* biological action of the SM peptides.

The binding protein from rat sources has been characterized by a number of laboratories. When the approach of separating the binding protein from the SM peptide with acid is used, the rat plasma binding protein appears to have a 40,000 mol. wt., similar to the partially purified human binding protein.(14) Knauer, Wagner, and Smith,(33) working on the multiplication-stimulating activity binding protein from rat liver cell cultures, have achieved more extensive purification and characterization of the binding protein subunit. Using affinity chromatography, they purified to apparent homogeneity two binding proteins with molecular weights of 30,000 and 31,500 daltons. These are either nonidentical subunits or secondary modifications introduced by purification. The binding proteins from rat liver purfusions have been characterized by Binoux *et al.*(34) They appear to be similar in their characteristics to those from human and rat plasma, with a molecular weight of approximately 40,000. More recent work by this group suggests that rat liver cultures are actually producing two binding proteins.(35) One of these binding proteins has an affinity preference for IGF-I, while the other binding protein has a higher affinity for IGF-II.

Table 1 Models of SM/IGF Bioactivity

1. "Free" SM/IGF peptide active
   BP-SM → BP + "free" SM → SM-receptor → action
2. SM complex active
   BP-SM → BP-SM-receptor → action

5. Physiological Significance of the Binding Protein Phenomenon

While the broad outline of the plasma forms of SM's and the binding protein phenomenon seems clear, the physiological significance of this unique phenomenon in peptide endocrinology is unsettled. A number of theories have been proposed that attempt to put this phenomenon in perspective. A potential physiological role for the binding protein relates to the longer half-life of SM peptide–binding protein complex. When SM peptide, either radioactive or nonradioactive, is injected into an experimental animal, the initial half-life is very similar to that of other peptide hormones (10–20 min). However, the 150K form has a much longer half-life, with estimates ranging from 3 to 18 hr rather than the few minutes typical of peptide hormones. This long half-life explains the remarkable constancy of plasma levels of SM/IGF throughout a 24-hr period when compared to other peptide hormones. Teleologically, the long-term anabolic actions that the SM's perform may require the presence of SM's at constant levels, while those of other peptide hormones such as insulin and growth hormone need to change rapidly to regulate acute metabolic responses. Another physiological role of binding proteins/complex forms of SM could be to protect the body from excess insulinlike action. Since the original description of NSILA in plasma, it has been paradoxical that this potent insulinlike bioactivity in plasma was not more evident *in vivo*. The amount of insulinlike activity exhibited by the SM peptides in plasma is clearly modified by the amount bound in the protein complex.[36] Thus, in the complex form the SM peptides appear to be unable to gain full access to the insulin receptor. Another potential role of the SM complex form is the possibility that only certain tissues would be able to remove the peptide from the binding protein and thus have biologically potent free SM available to their receptor sites. It is also possible that the SM binding protein complex might actually be a vehicle by which the somatomedins can be internalized into cells where growth-promoting actions may take place.

Table 2 Reasons for Thinking Only the "Free" SM Peptide Is Active

1. Analogy to TBG, SBG, etc.
2. K equilibration of SM $\frac{11}{44}$ total concentration of SM complex
3. Limited diffusion of the SM complex
4. Purified BP receptor binding and biological activity

Table 3 Reasons for Thinking the SM Complex Is the Active Form

1.	Complex equipotent with SM peptide in bioassay and RRA
2.	Concentration of free SM $\frac{11}{44}$ K equilibration
3.	Incomplete equilibration of [^{125}I]SM with complex
4.	Excess of unsaturated BP in plasma

It is unclear whether the complex forms actually have complete access to the SM receptor sites at the tissue level and lead to biological action. Two models of the role of the SM/IGF complexes can be proposed (Table 1). In the 1st model, only the "free" SM/IGF can gain access to the tissue receptor and is the active form. The 2nd general model proposes that the SM peptide in the macromolecular complex can interact with the tissue receptor and is the major metabolically active form. The reasons for thinking that only the free SM/IGF is active (Table 2) include the analogy to steroidal and thyroid hormones, the observation that the equilibrium constant of SM/IGF is an order of magnitude less than the measured circulating concentration of SM/IGF, and *in vivo* experiments indicating that the bound forms of SM do not diffuse readily across capillary walls. Thus, the amount of SM in plasma may not reflect the amount at the tissue level. Finally, experiments have shown that adding purified SM binding protein to SM peptides in *in vitro* system inhibits not only receptor binding, but biological action of SM peptides. However, there seems to be almost as much reason to favor the concept that the complex is the active form (Table 3). In at least some experimental settings the SM complex can compete equally for the receptor site with the equivalent amount of free peptide. The estimated concentration of free SM/IGF is an order of magnitude less than the apparent equilibrium constant. The lack of complete equilibration of radioiodinated SM/IGF with the 150K complex also argues against this being a kinetically simple situation. Since the SM complex formed by the combination of SM/IGF peptide and purified binding protein is not the major physiological form in plasma, it is unclear whether the inhibition seen in the mixing experiments is biologically relevant. Finally, the presence of an excess of free binding sites in plasma suggests that any free SM/IGF peptide would be quickly removed. Thus, it not yet possible to choose definitively between these models of SM action. The role of SM peptides as mediators of growth hormone action appears to be on strong experimental footing, but the physiological roles of the binding protein and the SM complex forms in plasma are still at an early stage of study.

6. Summary and Conclusions

SM peptides are bound tightly and specifically to a plasma macromolecular complex *in vivo*. There are at least two macromolecular forms of SM complexes, 150,000 daltons and 45,000 daltons. These macromolecular complexes clearly increase the half-life of SM's in plasma. The 150,000-dalton-macromol. wt. SM complex appears to have a three-subunit structure consisting of the SM peptide, a specific SM/IGF binding protein, and an acid-labile component. In addition to the binding proteins in plasma contained in the 150K and 45K complexes, there are unsaturated SM/IGF binding proteins in plasma of approximately 40,000 mol. wt. The SM complex forms not only have longer half-lives, but in this form their biological actions (both insulinlike and growth-promoting) are clearly modified. It is possible that the complex forms play a role in tissue selectivity and/or delivery to the intracellular space of active SM peptide. The understanding of the SM complexes, plasma complexes, and binding proteins is necessary before the physiological roles of the SM peptides can be clearly defined.

References

1. W. H. Daughaday and D. M. Kipnis, The growth promoting and anti-insulin actions of somatotropin, *Rec. Prog. Hormone Res.* **22:**49–93, 1966.
2. E. R. Froesch, H. Burgi, W. A. Muller, R. E. Humbel, A. Jacob, and A. Labhart, Nonsupressible insulinlike activity of human serum: Purification, physicchemical and biological properties and its relation to total serum ILA, *Rec. Prog. Hormone Res.* **23:**565–605, 1967.
3. J. J. Van Wyk, K. Hall, and R. P. Weaver, Partial purification of sulfation factor and thymidine factor from plasma, *Biochim. Biophys. Acta* **192:**560–562, 1969.
4. W. H. Daughaday, K. Hall, M. S. Raben, W. D. Salmon, J. L. van den Brande, and J. J. Van Wyk, Somatomedin: Proposed designation for sulfation factor, *Nature* **235:**197, 1972.
5. E. Rinderknecht and R. E. Humbel, The amino acid sequence of human insulin-like growth factor I and its structural homology with proinsulin, *J. Biol. Chem.* **253:**2769–2776, 1978a.
6. E. Rinderknecht and R. E. Humbel, Primary structure of human insulin-like growth factor II, *FEBS Lett.* **89:**283–296, 1978b.
7. M. E. Svoboda, J. J. Van Wyk, D. G. Klapper, R. E. Fellows, F. E. Grisson, and R. J. Schlueter, Purification of somatomedin-C from human plasma: Chemical and biological properties, partial sequence analysis, and relationship to other somatomedins, *Biochemistry* **19:**790–797, 1980.
8. L. Borsi, R. G. Rosenfeld, F. Liu, and R. L. Hintz, Somatomedin peptide distribution and somatomedin binding protein content in cord plasma: Comparison to normal and hypopituitary plasma, *J. Clin. Endocrinol. Metab.* **54:**223–228, 1982.
9. A. J. D'Ercole, D. F. Willson, and L. E. Underwood, Changes in the circulating form of serum somatomedin-C during fetal life, *J. Clin. Endocrinol. Metab.* **51:**674–676, 1980.
10. K. Megyesi, C. R. Kahn, J. Roth, and P. Gorden, Circulating NSILA-S in man: Preliminary studies of stimuli in vivo and of binding to plasma components, *J. Clin. Endocrinol. Metab.* **41:**475–484, 1975.
11. D. S. Schalch, U. E. Heinrich, J. G. Koch, C. J. Johnson, and R. J. Schlueter, Non-suppressible insulin like activity: Development of a new sensitive competitive protein binding assay for the determination of serum levels, *J. Clin. Endocrinol. Metab.* **46:**664–671, 1978.
12. R. L. Hintz and F. Liu, Somatomedin plasma binding proteins, in: *Growth Hormone and Other Biologically Active Peptides* (A. Pecile and E. E. Muller, eds.), Excerpta Medica, Amsterdam, 1980, pp. 133–143.
13. W. H. Daughaday, A. P. Ward, A. C. Goldberg, B. Trevedi, and M. Kapadia, Characterization of somatomedin binding in human serum by ultracentrifugation and gel filtration, *J. Clin. Endocrinol. Metab.* **55:**916–921, 1982.
14. A. C. Moses, S. P. Nissley, K. L. Cohen, and M. M. Rechler, Specific binding of a somatomedin-like polypeptide in rat serum depends on growth hormone, *Nature* **263:**137–140, 1976.

15. K. L. Cohen and S. P. Nissley, The serum half-life of somatomedin activity: evidence for growth hormone dependence, *Acta Endocrinol.* **83:**243–258, 1976.
16. U. Kaufmann, J. Zapf, B. Torretti, and E. R. Froesch, Demonstration of a specific serum carrier protein of nonsuppressible insulin-like activity *in vivo*, *J. Clin. Endocrinol. Metab.* **44:**160–166, 1977.
17. J. Liberti, Purification of bovine somatomedin, *Biochem. Biophys. Res. Commun.* **67:**1226–1233, 1975.
18. D. M. Wilson and R. L. Hintz, Interspecies comparison of somatomedin structure using immunological probes, *J. Endocrinol.* **95:**59–64, 1982.
19. R. L. Hintz, E. M. Orsini, and M. G. Van Camp, Demonstration of reversible binding of somatomedin to plasma proteins, *Endocrinology* **94**(suppl):A-71, 1974.
20. J. Zapf, M. Waldvogel, and E. R. Froesch, Binding of NSILA to human serum: Evidence for a carrier protein, *Arch. Biochem. Biophys.* **168:**638–645, 1975.
21. R. L. Hintz and F. Liu, Demonstration of specific plasma protein binding sites for somatomedin, *J. Clin. Endocrinol. Metab.* **45:**988–995, 1977.
22. J. Zapf, B. Morrell, H. Walter, Z. Laron, and E. R. Froesch, Serum levels of insulin-like growth factor and its carrier protein in various metabolic disorders, *Acta Endocrinol.* **95:**505–517, 1980.
23. R. M. White, S. P. Nissley, A. C. Moses, M. M. Rechler, and R. E. Johnsonbaugh, The growth hormone dependence of a somatomedin binding protein in human serum, *J. Clin. Endocrinol. Metab.* **53:**49–57, 1981.
24. R. L. Hintz, F. Liu, R. G. Rosenfeld, and S. F. Kemp, Plasma somatomedin binding proteins in hypopituitarism: Changes during growth hormone therapy, *J. Clin. Endocrinol. Metab.* **53:**100–104, 1981.
25. R. H. Chochinov, I. K. Mariz, A. S. Hajek, and W. H. Daughaday, Characterization of a protein in mid-term human amniotic fluid which reacts in the SM-C radioreceptor assay, *J. Clin. Endocrinol. Metab.* **44:**902–908, 1977.
26. A. J. D'Ercole, G. T. Applewhite, and L. E. Underwood, Evidence that somatomedinis synthesized by multiple tissues in the fetus, *Dev. Biol.* **75:**315–328, 1980.
27. M. Binoux, L. Hossenlopp, L. Lassarre, and N. Hardouin, Production of insulin-like growth factors and their carriers by rat pituitary gland and brain explants in culture, *FEBS Lett.* **124:**178–184, 1981.
28. R. L. Hintz and F. Liu, Human somatomedin plasma binding proteins, in: *Somatomedin and Related Peptides* (G. Giordano, ed., Excerpta Medica, Amersterdam, 1979, pp. 143–147.
29. R. W. Furlanetto, The somatomedin C binding protein: Evidence for a heterologous subunit structure, *J. Clin. Endocrinol. Metab.* **51:**12–19, 1980.
30. S. L. S. Drop, G. Valiquette, H. J. Guyda, M. T. Corvol, and B. I. Posner, Partial purification and characterization of a binding protein for insulin-like activity inhuman amniotic fluid: A possible inhibitor of insulin-like activity, *Acta Endocrinol.* **90:**505–518, 1979.
31. S. L. S. Drop, D. J. Kortleve, and H. J. Guyda, Isolation of a somatomedia binding protein from preterm amniotic fluid. Development of a radio immunoassay, *J. Clin. Endocrinol. Metab.* **59:**899–907, 1984.
32. J. Zapf, E. Schoenle, G. Jagars, I. Sand, J. Grunwald, and E. R. Froesch, Inhibition of the action of non-suppressible insulin-like activity on isolated fat cells by binding to its carrier protein, *J. Clin. Invest.* **63:**1077–1084, 1979.
33. D. J. Knauer, F. W. Wagner, and G. L. Smith, Purification and characterization of multiplication stimulating activity carrier protein, *J. Supramol. Struct. Cell Biochem.* **15:**177–191, 1981.
34. M. Binoux, C. Lassarre, S. Hardouin, Somatomedin production by rat liver in organ culture, *Acta Endocrinol.* **99:**422–430, 1982.
35. M. Binoux, S. Hardouin, C. Lassarre, and P. Hassenlopp, Evidence for production by the liver of two IGF binding proteins with similar molecular weights but different affinities for IGF I and IGF II. Their relations with serum and cerebrospinal fluid IGF binding proteins, *J. Clin. Endocrinol. Metab.* **55:**600–603, 1982.
36. C. Meuli, J. Zapf, and E. R. Froesch, NSILA carrier protein abolishes the action of NSILA-S on perfused rat heart, *Diabetologia* **14:**255–259, 1978.

43

Direct Action of Growth Hormone on Cartilage Growth

Olle G. P. Isaksson, Staffan Edén, Kerstin Albertsson-Wikland, John-Olov Jansson, Ulf Friberg, and Kjell Madsen

1. Introduction

It is generally considered that the stimulatory effect of growth hormone (GH) on skeletal growth and development is mediated by GH-dependent plasma factors—somatomedin(s)—that are produced in the liver in response to GH.[1,2] This theory is mainly based on the fact that it has been difficult to demonstrate stimulatory effects of GH *in vitro* on cartilage metabolism, whereas somatomedin(s), when added *in vitro*, stimulate a number of metabolic and cellular processes in cartilage tissue.[1,2] Recently, Schoenle *et al.*[3] demonstrated that continuous administration of highly purified IGF-I (somatomedin C) to hypophysectomized rats, by means of osmotic minipumps, stimulated epiphyseal cartilage growth, supporting this hypothesis. The effect observed was comparable to that of small doses of continuously infused GH (12.5–25 mU hGH/day).

Administration of GH to hypophysectomized rats has been shown to stimulate cartilage metabolism, as measured by the uptake of radioactive sulfate, before any measurable increase in the plasma level of somatomedins. This finding may suggest that GH *in vivo* stimulates cartilage metabolism without increasing the level of circulating somatomedins.[4] Moreover, several studies have demonstrated direct effects of GH on the metabolism of adipose tissue, liver, and skeletal muscle and direct effects on the cellular growth of hemato-

Olle G. P. Isaksson, Staffan Edén, Kerstin Albertsson-Wikland, and John-Olov Jansson • Department of Physiology, University of Göteborg, S-400 33 Göteborg, Sweden. ***Ulf Friberg and Kjell Madsen*** • Department of Histology, Karolinska Institutet, S-10401 Stockholm, Sweden.

poietic cells.[5–7] Since it has been difficult to achieve stimulation of skeletal growth by somatomedins of the same magnitude as by GH, we have explored the possibility of direct effects of GH on growth and development of cartilage.

2. Results

2.1. Local Administration of GH in Vivo

In the first series of experiments the effect of local administration of GH on longitudinal bone growth and tibial width was studied. In brief, hypophysectomized rats were anesthetized, and a small incision was made in the skin at the medial part of the knee joint. The cartilage growth plate was identified under a stereomicroscope. Subsequently, saline or various doses of GH or prolactin were injected (injection volume 2 μl) into the growth plate or subperiosteally at the level of the growth plate. While the animals were still anesthetized, they were also given a single intraperitoneal injection of oxytetracycline (10 mg/kg; Terramycin, Pfizer). In most experiments hGH injections were given one to four times at 1- to 3-day intervals. One to five days after the last injection, the rats were killed by decapitation, and the accumulated bone growth of the proximal tibia of both sides was determined by measuring the distance between the cartilage and the tetracycline band, as previously described.[8,9] In one experiment (Fig. 4) the width of the growth plate of the proximal tibia was determined.

Administration of human GH (hGH) into the growth plate of hypophysectomized rats stimulated longitudinal bone growth, as shown in Fig. 1.[10] The growth response of the left (saline-injected) tibia of the GH-treated animals was not increased compared to rats receiving saline only, suggesting that the total dose of hGH used was too low to stimulate the linear growth of the whole animal in this experiment. Injection of an equal dose of ovine prolactin (oPRL) according to the same protocol had no stimulatory effect on longitudinal bone growth, as shown in Fig. 2, indicating a specific effect of GH on longitudinal bone growth. Figure 3 shows that administration of hGH subperiosteally at the level of the growth plate also stimulated longitudinal bone growth. This finding suggests that GH can penetrate the cartilage of the growth plate by diffusion or that GH is absorbed by blood vessels supplying the growth plate.

As shown in Fig. 4, local administration of GH also increased tibial width. A stimulatory effect of GH was apparent 1 day after the start of hGH injections and persisted over the 4-day injection period. After 4 days of GH treatment, there was a slight increase in the width of the contralateral epiphyseal growth plate, suggesting that repeated local administration of GH had systemic effects. From the results presented in Figs. 1–4 it is apparent, that local injection of GH into the growth plate, or subperiosteally adjacent to the growth plate, can stimulate longitudinal bone growth as well as tibial width of the leg injected, suggesting a direct effect of GH on the chondrocytes in the growth plate. To further explore this possibility, experiments were undertaken to investigate

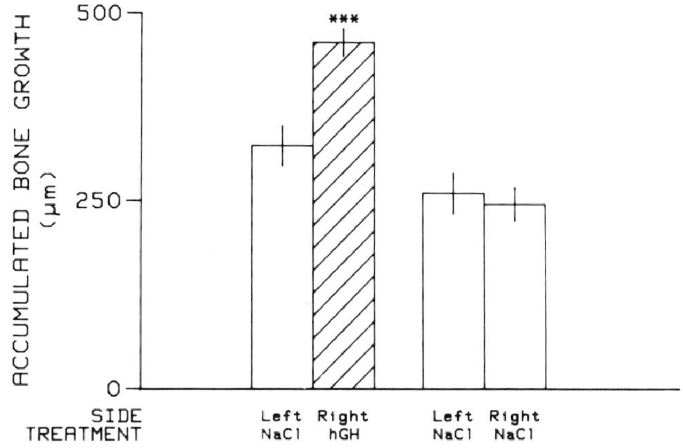

Figure 1 Effect of local administration of human growth hormone (hGH) on longitudinal bone growth. Male rats (36 days old) were hypophysectomized, and 14, 16, and 19 days later 10 μg of hGH was injected into the proximal growth plate of the right tibia of six animals. The contralateral leg received the same volume of saline (first bar on the left). The two bars on the right represent three rats that received saline in both the left and right proximal growth plate of the tibia. Longitudinal bone growth was measured by the tetracycline method as described in the text. Values are means ± SE. The effect of hGH was highly significant ($p < 0.001$ by paired t test). Reprinted by permission from Isaksson *et al.* [10] Copyright © 1982 AAAS.

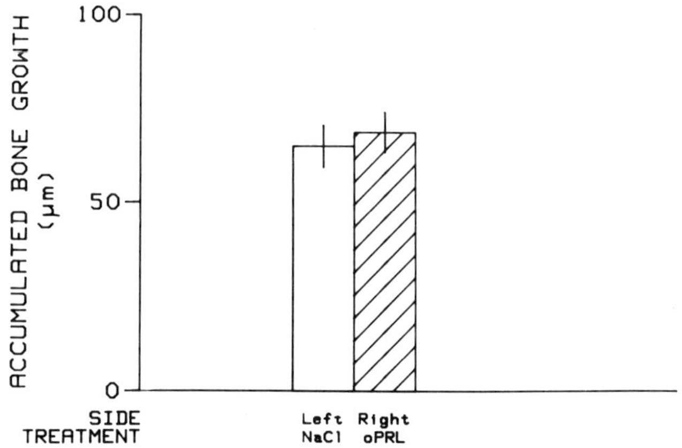

Figure 2 Effect of local administration of ovine prolactin (oPRL-P-B4NIH) on longitudinal bone growth. Six hypophysectomized rats received three injections of prolactin (10 μg/injection) or saline in the proximal growth plate of the right and left tibia, respectively, according to the same protocol as that in Fig. 1, except that injections started 34 days after hypophysectomy. Values are means ± SE. Reprinted by permission from Isaksson *et al.*[(10)] Copyright © 1982 AAAS.

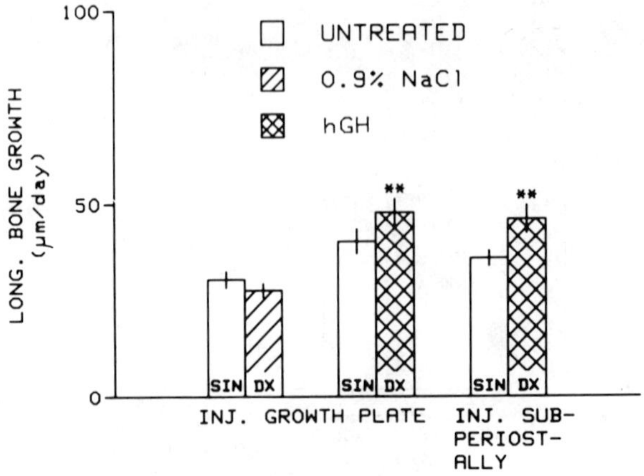

Figure 3 Effect of local administration of human growth hormone (hGH) on longitudinal bone growth. Male rats (35 days old) were hypophysectomized, and 24 and 27 days later 5 µg of hGH was injected into the growth plate, or subperiosteally at the level of the growth plate, of the right proximal tibia. Saline was injected into the growth plate of the right tibia in a control group of animals (first two bars on the left). Values are means ± SE. There were three to five animals in each group. $^{**}p < 0.01$ by paired *t* test.

whether chondrocytes have specific binding sites for GH, and whether GH *in vitro* stimulates the growth of cultured chondrocytes.

2.2. *Specific Binding and Proliferative Effects of GH in Cultured Chondrocytes*

Chondrocytes were isolated from rabbit ear of 4-week-old male New Zealand white rabbits as described previously[11] and from rat rib growth plates of 5-week-old male rats of the Sprague–Dawley strain according to Shimomura *et al.*[12] In brief, the cartilage was cleaned mechanically from skin, muscle, and perichondrium. Subsequently, the cartilage was cut into small pieces and digested overnight at 37°C in 0.1% (weight/volume) clostridial collagenase (Type IA, Sigma) in complete medium. The next day, the cells were washed twice in serum-free medium, counted in a hemocytometer, and plated in tissue culture plates or flasks. The cells were plated at a density of approximately 3×10^5 cells/cm^2 in F12 medium supplemented with 10% heat-inactivated newborn calf

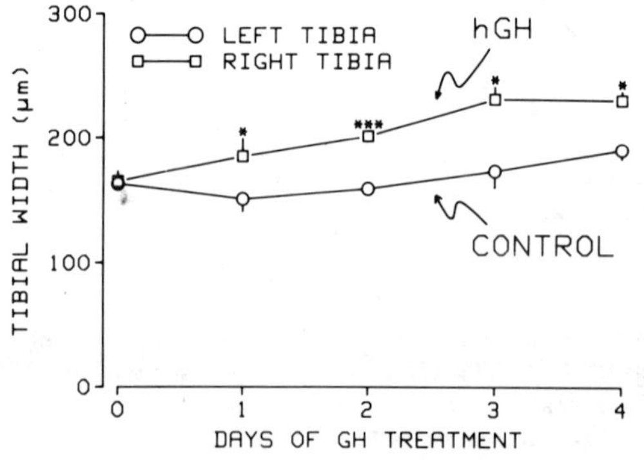

Figure 4 Time course effect of local administration of human growth hormone (hGH) on tibial width. Two weeks after hypophysectomy of 35-day-old male rats, 10 µg of hGH was injected daily into the growth plate of the right proximal tibia for 4 days. At various time intervals after the start of GH treatment, the width of the growth plate of the proximal tibia of both legs was measured in undecalcified longitudinal and frontal sections. Values are means ± SE. There were three to five animals at each time point. $^{*}p < 0.05$ by paired *t* test. $^{***}p < 0.001$ by paired *t* test.

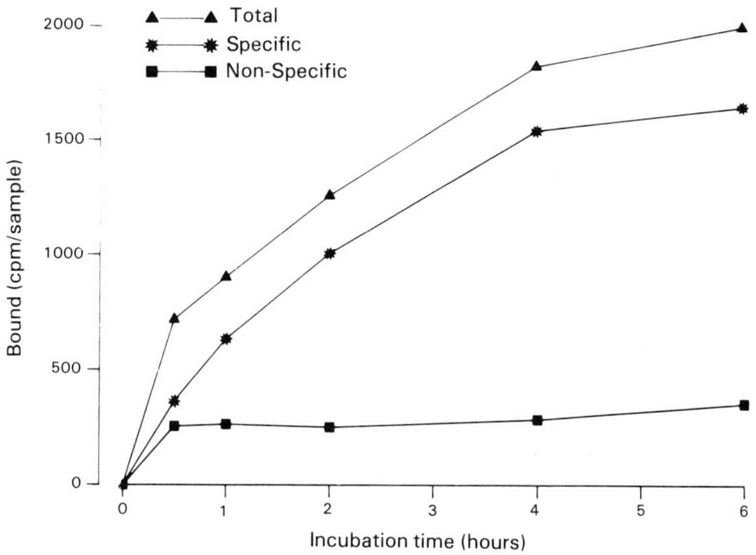

Figure 5 Binding of [^{125}I]hGH to ear chondrocytes from 4-week-old rabbits. Nonspecific binding is binding of [^{125}I]hGH in the presence of an excess (1 μg/ml) of unlabeled human growth hormone. Reprinted by permission from Edén *et al.*[14] Copyright © 1983 by The Endocrine Society.

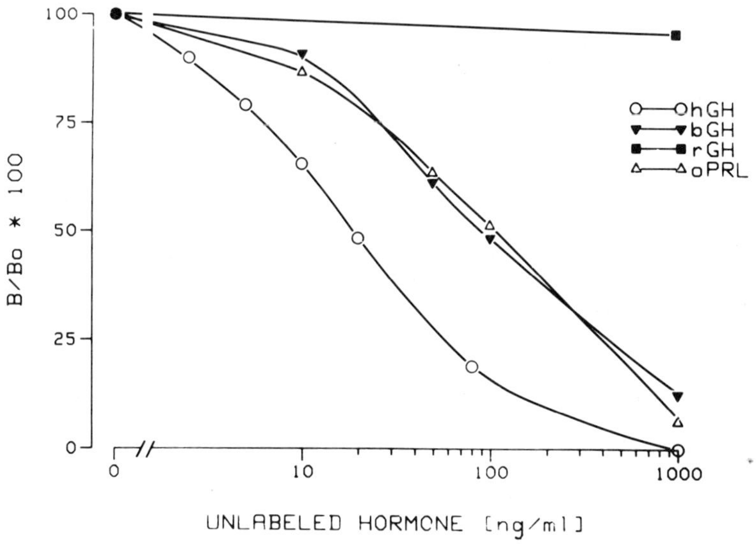

Figure 6 Displacement of [^{125}I]hGH binding to ear chondrocytes from 4-week-old rabbits by unlabeled human growth hormone (hGH), bovine growth hormone (bGH), rat growth hormone (rGH), and ovine prolactin (oPRL). Reprinted by permission from Edén et al.[14] Copyright © 1983 by The Endocrine Society.

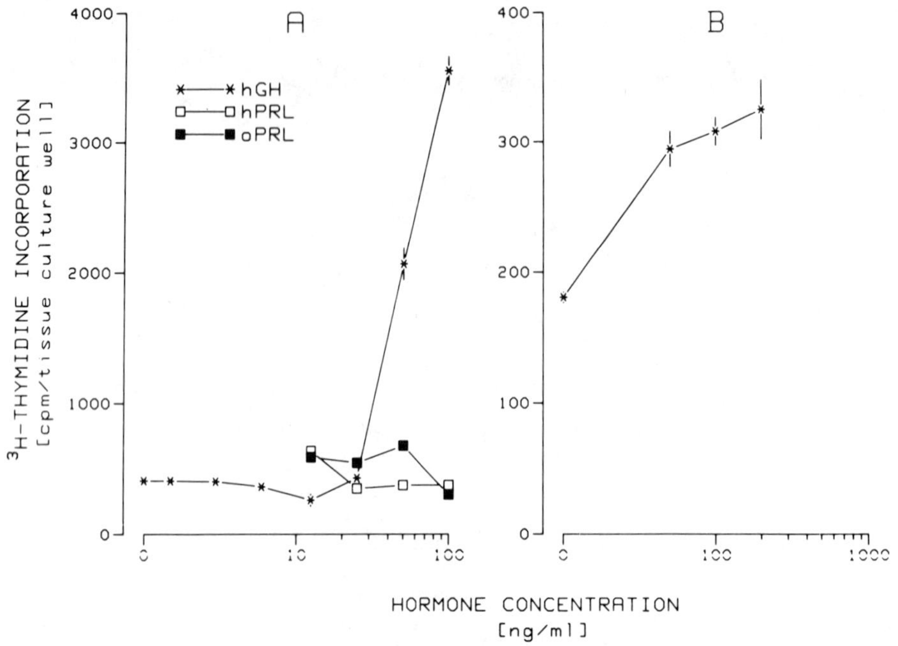

Figure 7 Effect of human growth hormone (hGH), ovine prolactin (oPRL), and human prolactin (hPRL) on thymidine incorporation in chondrocytes from rabbit ear (A) and the effect of hGH on thymidine incorporation in chondrocytes from rat rib growth plate (B). After collagenase digestion, the cells were plated in 96-well tissue culture plates in F-12 containing 10% NCS (0.2 ml/well) and were allowed to attach overnight. The medium was then changed to MCDB-104, and the cells were preincubated for 24 hr. The cells were then incubated in fresh MCDB-104 for 24 hr in the absence and presence of various concentrations of hGH, oPRL, and hPRL. After another 24 hr of incubation, the medium was changed to fresh MCDB-104 with the same addition plus [^{3}H]thymidine (5 μCi/ml; specific activity 9.9 Ci/ml). The incubation was terminated by three washes of the cells in phosphate-buffered saline (PBS) before treatment with 0.5% Triton X-100 (100 μl/well). The chondrocytes were collected in a multiple-cell harvester (Skatron) on glass-fiber filters. The filters were dried at 70°C and then counted in 2.5 ml of Lipoluma:Lumasolve:water, 10:1:0.2 (Lumac) in a Packard liquid scintillation spectrometer. Each point is the mean of six determinations ± SE. Data are shown from an experiment using ear chondrocytes from one rabbit (A) or chondrocytes from the pooled rib growth cartilage of 10 rats (B). The experiment presented in A was performed three times, and the experiment presented in B twice, with similar results. Reprinted by permission from Madsen *et al.*[15] Copyright © 1983 Macmillan Journals Ltd.

serum (Flow Labs.). The medium was further supplemented with Hepes (10 mM), Tes (10 mM), gentamycin (100 μg/ml), and L-ascorbic acid (50 μg/ml). The cells were cultured at 37°C in an atmosphere of air containing 5% CO_2. Ninety percent of the cells obtained were chondrocytes showing different degrees of maturation, and less than 10% of the cells were small cells from the peripheral part of the cartilage, presumably chondroblasts. These latter cells are not fibroblasts since they produce cartilagelike proteoglycans and do not synthetize large amounts of dermatan sulfate or type I collagen.[13]

After 24–72 hr of culture, the cells to be used in the binding assay were

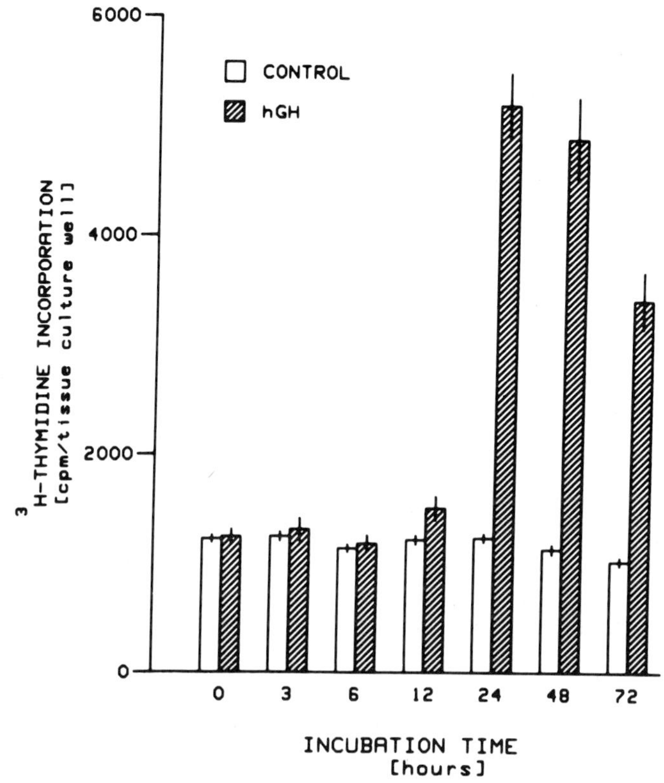

Figure 8 Time course for the stimulatory effect of human growth hormone (hGH) on [^{3}H]thymidine incorporation in chondrocytes from rabbit ear. The cells were plated in 96-well tissue culture plates in F-12 containing 10% NCS (0.2 ml/well) and allowed to attach overnight. They were then preincubated for 48 hr in MCDB-104. The cultures were then incubated for various periods in the absence or presence of hGH (100 ng/ml). For cultures incubated for 48 or 72 hr, the medium was changed every 24 hr. During the last 6 hr of the incubation, the cells were pulse-labeled with [^{3}H]thymidine (8 μCi/ml). The cultures were collected and counted as described in the legend to Fig. 1. Each point is the mean of 24 cultures ± SE. Data from an experiment using chondrocytes from one rabbit are shown. The experiment was repeated twice using different rabbits and resulted in similar findings. Reprinted by permission from Madsen *et al.*[(15)] Copyright © 1983 Macmillan Journals Ltd.

exposed to 0.5% collagenase for 45 min at 37°C. The digestion was terminated by adding EDTA to the medium (final concentration 50 mM). This procedure also released the cells from the walls of the culture flask. The suspended cells were then washed three times in Hepes buffer with 1% BSA, pH 7.5. After the washings, the cells were resuspended to a final concentration of about 1×10^6 cells/ml, and aliquots of the cell suspension were incubated with [^{125}I]hGH (specific activity 100–150 μCi/μg) and various concentrations of unlabeled hormones. The cells were incubated for 0–6 hr at 24°C in a water bath.

Figure 5 shows the time course of [^{125}I]hGH binding to chondrocytes isolated from rabbit ear.[(14)] It can be seen that total binding increased with time and that an apparent steady state was reached after 4–6 hr. Total binding was 1–3% of total counts added, and nonspecific binding was 10–20% of total binding. The ability of hGH, bovine GH (bGH), rat GH (rGH), and oPRL to compete with [^{125}I]hGH binding to rabbit ear chondrocytes is shown in Fig. 6. All added hormones, except rGH, competed with [^{125}I]hGH binding in a dose-dependent manner. Half-maximal binding for hGH was observed at a concentration of 10–20 ng/ml and for bGH and oPRL of about 100 ng/ml. The number of hGH molecules specifically bound to each cell was calculated to about 9000.

The effect of hGH, oPRL, and hPRL on the rate of cell division, as measured by thymidine incorporation, in chondrocytes from rabbit ear and rat rib growth plate was next investigated. As shown in Fig. 7A, hGH caused a dose-dependent increase in thymidine incorporation, whereas oPRL and hPRL had no effect.(15) The minimal effective concentration of hGH was 50 ng/ml. This concentration of hGH evoked an apparent maximal effect in chondrocytes from rat rib growth plate, as shown in Fig. 7B.

To study the time course effect of GH on thymidine incorporation, cultured chondrocytes from rabbit ear were pulse-labeled with [^{3}H]thymidine for 6 hr at various times after the start of incubation. Figure 8 shows that a stimulatory effect of hGH became apparent between 12 and 24 hr after the start of culture. The effect persisted for 72 hr with an apparent decrease between 48 and 72 hr.

The results from these experiments demonstrate that chondrocytes isolated from rabbit ear and epiphyseal cartilage have specific binding sites for GH. The results further demonstrate that physiological concentrations of GH, i.e., concentrations of GH that frequently can be recorded in the plasma during episodes of GH secretion both in the rabbit and in the rat, stimulate the rate of proliferation of cultured chondrocytes.

3. Discussion

The results from the present experiments demonstrate that local administration of GH into, or adjacent to, the growth plate of the proximal tibia of hypophysectomized rats stimulates longitudinal bone growth and increases the width of the growth plate on the side injected with the hormone. Furthermore, cultured chondrocytes from rabbit ear show specific binding sites of hGH, and GH stimulates DNA synthesis of cultured chondrocytes from rabbit ear and rat rib growth cartilage indicating that GH directly interacts with chondrocytes in the growth plate. This effect of GH cannot be attributed to an alteration in the circulating level of somatomedin(s). However, it is possible that the chondrocytes themselves, in response to GH, produce somatomedin(s) and perhaps other growth factors that are important for the full expression of the stimulatory effect of GH. In support of this theory is the recent demonstration that the stimulatory effect of hGH on the growth of cultured human fibroblast is not expressed when the incubation medium is changed frequently.(16) Thus, fibro-

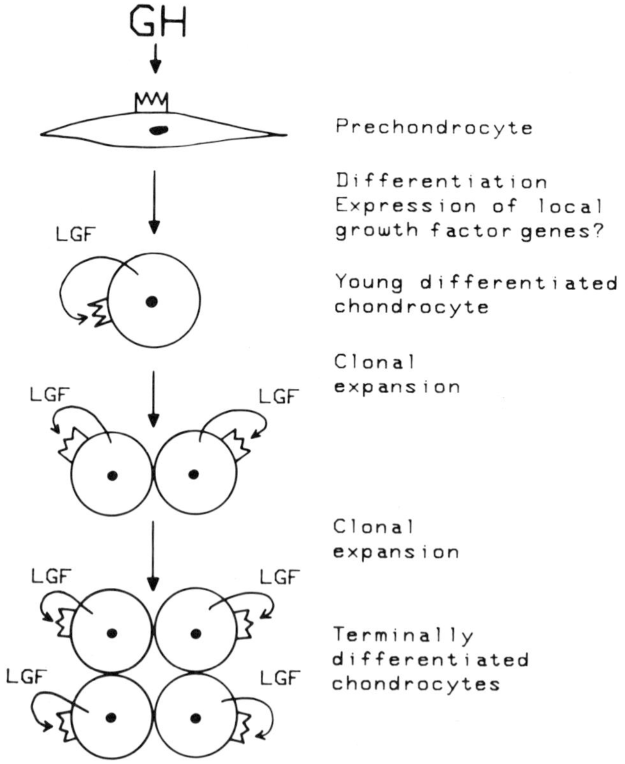

Figure 9 Hypothetical mechanism for the stimulatory effect of growth hormone on longitudinal bone growth at the level of the epiphyseal growth plate. Reprinted by permission from Isaksson *et al.*[20]

blasts may synthesize and release growth factors that are important for the expression of the growth response to GH. Indeed, it has been shown that cultured fibroblasts do synthesize somatomedinlike growth factors.[17]

4. *Hypothetical Mechanism for the Stimulatory Effect of GH on Longitudinal Bone Growth*

The discovery by Green and co-workers that GH stimulates the differentiation of preadipose cells from a cloned 3T3 preadipose line and multinucleated muscle cells from myoblasts of a cloned muscle line[18,19] suggests that stem cell chondrocytes might be the target cells for GH in the growth plate and that GH stimulates the differentiation of these cells. We suggest the following hypothesis for the stimulatory effect of GH on longitudinal bone growth (Fig. 9): During the process of cell differentiation, genes that code for growth factors of the somatomedin class such as IGF-1, are expressed. This gene activation results in an increased local production of IGF-1 that promotes the clonal expansion of chondrocytes through paracrine or autocrine mechanisms. According to this theory, longitudinal bone growth will be the result of both cell differentiation—directly stimulated by GH—and clonal expansion of cells in the proliferative layer of the growth plate due to the local production of growth factors.

ACKNOWLEDGMENTS. The authors are indebted to Dr Jack L. Kostyo for providing highly purified bovine GH (bGH; B 1132-2) and ovine PRL (oPRL; 7M 1251). Human GH (Crescormon) was a gift from Kabi AB, Stockholm, Sweden. Rat GH (rGH; NIAMDD rat GH-B1) and ovine prolactin (oPRL-P-B4-NIH) were kindly provided by the NIADDK. This study was supported by the Swedish Medical Research Council (Proj. no. 03355, 04250), by the Magnus Bergvall Foundation, and by the Funds of Karolinska Institutet and Faculty of Medicine, University of Göteborg, Sweden.

References

1. L. S. Phillips and R. Vassilopoulou-Sellin, Somatomedins, *N. Engl. J. Med.* **302:**371, 1980.
2. W. H. Daughaday, Growth hormone and the somatomedins, in: *Endocrine Control of Growth* (W. H. Daughaday, ed.), Elsevier, New York, 1981, pp. 1–24.
3. E. Schoenle, J. Zapf, R. E. Humbel, and E. R. Froesch, Insulin-like growth factor I stimulates growth in hypophysectomized rats, *Nature* **296:**252, 1982.
4. L. S. Phillips, A. C. Herington, and W. H. Daughaday, Hormone effects on somatomedin action and somatomedin generation, in: *Advances in Human Growth Hormone Research* (S. Raiti, ed.), DHEW Publication No. (NIH) 74-612, Washington, D.C., 1974, pp. 50–67.
5. H. M. Goodman and J. Schwartz, Growth hormone and lipid metabolism, in: *Handbook of Physiology,* Vol. IV, Part 2 (E. Knobil and W. H. Sawyer, eds.), American Physiology Society, Washington, D.C., 1974, pp. 211–232.
6. J. L. Kostyo and O. Isaksson, Growth hormone and regulation of somatic growth, in: *International Review of Physiology: Reproductive Physiology II,* Vol. 13 (R. O. Greep, ed.), University Park Press, Baltimore, 1977, pp. 255–274.
7. C. Gauwerky, D. W. Golde, and C. H. Li, Growth hormone polypeptides stimulate proliferation of K562 human erythroleukemia cells, *J. Clin. Endocrinol. Metab.* **51:**1208, 1980.
8. L. I. Hansson, Daily growth in length of diaphysis measured by oxytetracycline in rabbit normally and after medullary plugging, *Acta Orthop. Scand.* **101**(Suppl.)**:**34, 1967.
9. L. I. Hansson, K. Menander-Sellman, A. Stenström, and K-G. Thorngren, Rate of normal longitudinal bone growth in the rat, *Calc. Tiss. Res.* **10:**238, 1972.
10. O. G. P. Isaksson, J-O. Jansson, and I. A. M. Gause, Growth hormone stimulates longitudinal bone growth directly, *Science* **216:**1237, 1982.
11. K. Madsen and S. Lohmander, Production of cartilage-typic proteoglycans in cultures of chondrocytes from elastic cartilage, *Arch. Biochem. Biophys.* **196:**192, 1979.
12. Y. Shimomura, T. Yoneda, and F. Suzuki, Osteogenesis by chondrocytes from growth cartilage of rat rib, *Calc. Tiss. Res.* **19:**179, 1975.
13. K. Madsen, S. Moskalweski, K. von der Mark, and U. Friberg, Synthesis of proteoglycans, collagen, and elastin by cultures of rabbit auricular chondrocytes—Relation to age of the donor, *Dev. Biol.* **96:**63, 1983.
14. S. Edén, O. G. P. Isaksson, K. Madsen, and U. Friberg, Specific binding of growth hormone to isolated chondrocytes from rabbit ear and epiphyseal plate, *Endocrinology* **112:**1127, 1983.
15. K. Madsen, U. Friberg, P. Roos, S. Edén, and O. Isaksson, Growth hormone stimulates the proliferation of cultured chondrocytes from rabbit ear and rat rib growth cartilage, *Nature* **304:**545, 1983.

16. D. R. Clemmons and J. J. Van Wyk, Somatomedin-C and platelet-derived growth factor stimulate human fibroblast replication, *J. Cell. Physiol.* **106:**361, 1981.
17. D. R. Clemmons, L. E. Underwood, and J. J. Van Wyk, Hormonal control of immunoreactive somatomedin production by cultured human fibroblasts, *J. Clin Invest.* **67:**10, 1981.
18. M. Morikawa, T. Nixon, and H. Green, Growth hormone and the adipose conversion of 3T3 cells, *Cell* **29:**783, 1982.
19. B. T. Nixon and H. Green, Growth hormone promotes the differentiation of myoblasts and preadipocytes generated by azacytidine treatment of 10T $\frac{1}{2}$ cells, *Proc. Natl. Acad. Sci. USA* **81:**3429, 1984.
20. O. Isaksson, S. Eden, J.-O. Jansson, A. Lindahl, J. Isgaard, and A. Nilsson, Sites of action of growth hormone on somatic growth, *J. Anim. Sci.* 1985, in press.

44

Increased Somatomedin Inhibitors in Renal Failure

L. S. Phillips and T. G. Unterman

1. Introduction

Poor growth is a common problem in uremic children. Broyer *et al.*[1] and others have demonstrated the decrease in growth velocity that occurs after the development of renal disease in childhood, and others[2,3] have shown that growth is poorly responsive to conservative therapy and conventional dialysis approaches. Ritz *et al.*[4] have shown that uremic children have diminished bone age for chronological age and diminished height age for bone age. As shown by Schiffrin *et al.*,[5] such growth impairment occurs despite normal levels of growth hormone (GH), prolactin, and insulin. The cause of growth failure is obscure and has been attributed to a variety of factors, including skeletal resistance to hormone action, abnormal mineral metabolism, impaired synthesis of active vitamin D metabolites, and poor nutrition. Although growth in uremic children appears to improve when their nutritional intake is increased,[6] uremic rats exhibit decreased gain in length when compared with pair-fed controls[4]; thus, defective growth involves factors beyond poor nutrition.

Since the skeletal growth-promoting effects of GH are attributed to the action of somatomedins,[7] there has been interest in the possibility that abnormal production and/or action of somatomedins could mediate the growth failure associated with uremia. Early studies with radioreceptor assays indicated that circulating somatomedins might be elevated in renal failure[5,8] and that somatomedins increased with progressive impairment in renal function. More recently, examinations with more highly specific radioimmunoassays and ra-

L. S. Phillips • Division of Endocrinology and Metabolism, Department of Medicine, Emory University School of Medicine, Atlanta, Georgia 30303. ***T. G. Unterman*** • Center for Endocrinology, Metabolism and Nutrition, Northwestern University Medical School, Chicago, Illinois 60611.

dioreceptor (RRA) assays have revealed a 50% decrease in IGF-I and a 200% increase in IGF-II[9]; since the cartilage-stimulating activity of IGF-II is approximately half that of IGF-I, such observations suggest that total biologically active somatomedins may be normal in renal failure.

2. Methods

Our laboratory has used biological assay systems to probe the role of somatomedins in uremic growth failure. Since skeletal elongation occurs via proliferation of cartilage at epiphyseal plates, cartilage has been used as a substrate for measuring circulating substances that may promote or inhibit growth. Bioassay systems employing cartilage from normal young pigs[10] or hypophysectomized rats[11] are sensitive to stimulation by somatomedins, but relatively insensitive to effects of other circulating anabolic hormones such as insulin, triiodothyronine, or testosterone. Stimulation has generally been assessed by uptake of $^{35}SO_4$. (Since inorganic sulfate is elevated in renal failure,[10] ^{35}S-specific activity must be corrected appropriately for each sample.) These systems measure *net* circulating somatomedin activity, reflecting both the cartilage-stimulating effects of somatomedins and the presence of inhibitory factors that may antagonize such stimulation[11]; inhibitory activity is best detected by the ability of samples to *decrease* cartilage stimulation produced by somatomedins (in whole serum or after partial purification[12]).

The sensitivity of the cartilage systems to both stimulatory and inhibitory factors was utilized to measure biologically active somatomedins and inhibitors after separation by column chromatography. One-milliliter serum samples were gel-filtered over 48 hr at 4°C on 1.2 × 100 cm columns, either Sephadex G-50 fine (1 M formic acid, pH 2.4) or Sephadex G-25 superfine (0.1 M ammonium formate, pH 7.0). Column effluent was pooled into eight equal fractions between the void volume and the salt peak, lyophilized twice, and reconstituted in 2 ml assay buffer. With this approach, somatomedins are found at K_d 0.4–0.8 on the Sephadex G-50 column, high-molecular-weight inhibitors at K_d 0.1–0.4 with Sephadex G-50 and K_d 0–0.1 with Sephadex G-25, and low-molecular-weight inhibitors at K_d 0.1–1.0 with Sephadex G-25. In the rat cartilage assay system, somatomedins were measured in incubations with 10% final concentration (volume/volume) of reconstituted fractions (stimulation of sulfate uptake expressed as percent above buffer level) and somatomedin inhibitors in incubations with 1% normal human serum and 8% added fractions (inhibition expressed as percent stimulation by serum inhibited).

3. Results

In preliminary studies, we examined net circulating somatomedin activity by porcine cartilage bioassay in 53 men with a spectrum of renal function.[13] Nine had normal renal function (creatinine 1.2 mg/dl or less), 37 had mild to

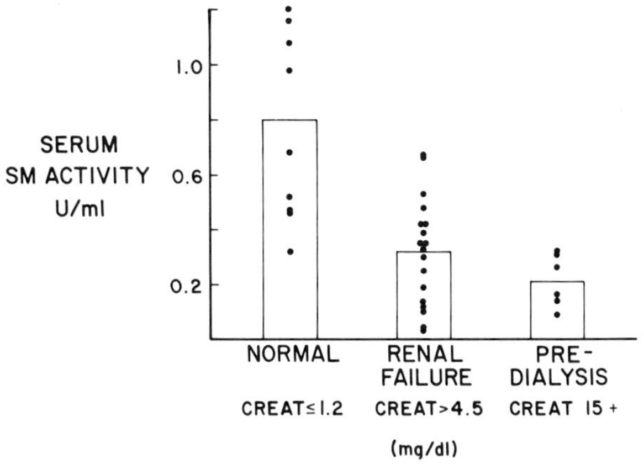

Figure 1 Somatomedin activity measured by porcine cartilage bioassay in 34 men with normal and impaired renal function (Nine normals, 18 subjects with creatinine > 4.5 mg/dl, and seven subjects prehemodialysis with creatinine greater than 15 mg/dl.)

advanced renal failure (creatinine 1.3 mg/dl or above, not requiring dialysis), and seven required chronic hemodialysis (mean creatinine 26 mg/dl predialysis). For each serum sample, inorganic sulfate was measured by $^{133}BaCl_2$ precipitation, allowing appropriate correction of assay results. As shown in Fig. 1, among these groups, progressive decrease in renal function was associated with a progressive decrease in serum somatomedin activity. Although there was overlap, mean somatomedin activity within each group differed significantly from that of the other groups ($p < 0.05$). Further studies (not shown) revealed that somatomedin activity was inversely correlated with both urea nitrogen and creatinine ($p < 0.001$ for both). Such findings contrasted with reports of elevated somatomedins as measured by RRA and suggested that decreased somatomedin activity in uremia might be due to an increase in a circulating inhibitor rather than a decrease in somatomedins.

The possibility that such putative inhibitors might be of low molecular weight was explored by examining somatomedin activity after dialysis *in vivo* and *in vitro*. As shown in Table 1, somatomedin activity rose with hemodialysis in six of seven subjects, and mean somatomedin activity doubled after a single 4- to 6-hr dialysis procedure ($p < 0.02$). When pooled uremic plasma was dialyzed against assay buffer for 24 hr, somatomedin activity rose 58%. In combination, such observations suggested that net circulating somatomedin activity might be decreased in renal failure owing to the presence of a factor (or factors) of low molecular weight.

We searched for such a uremic somatomedin inhibitor by examining the sera of eight normal volunteers and eight uremic subjects (prehemodialysis) after somatomedins and inhibitors were separated by column chromatography.[(14)] Figure 2 illustrates the profile of biologically active somatomedins found in normal and uremic serum. After gel filtration at acid pH to separate somatomedins from their carrier proteins, significant stimulatory activity was found only between K_d 0.38 and 0.75. In this bioassay system, which is sensitive to both IGF-I and IGF-II, there was no significant difference between somatomedins in the serum of normal subjects as compared to uremic subjects.

Table 1 Altered Somatomedin Activity with Dialysis

Patient	Hemodialysis, *in vivo* Pre	Post
1	0.67	0.76 U/ml
2	0.38	0.67
3	0.55	0.27
4	0.19	0.57
5	0.14	0.36
6	0.16	0.86
7	0.21	0.30
Mean	0.26 ± .04	0.54 ± .09
	Spectrapor-3 dialysis, *in vitro* Pre	Post
	0.43	0.68 U/ml

These observations confirm the analysis of the data of Goldberg *et al.*[9]; in combination, they indicate that decreased circulating somatomedin activity in uremia is not due to a decrease in somatomedins.

To examine the possibility that low somatomedin activity in uremia might be due to a high-molecular-weight inhibitor (as found in malnutrition[15] and

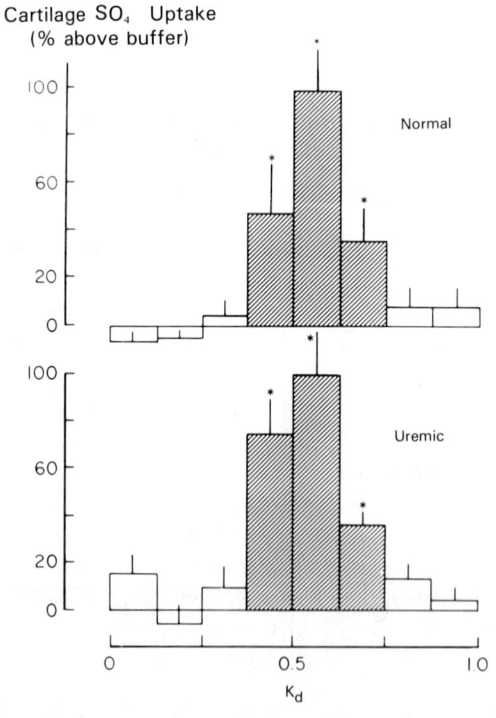

Figure 2 Somatomedins in normal and uremic serum. Samples from eight normals and eight uremics were chromatographed on Sephadex G-50, pH 2.4. Stimulation of SO_4 uptake by somatomedins is expressed as % above buffer, mean ± SEM. Hatching and asterisks show $p < 0.05$. From *J. Clin. Endocrinol. Metab.*, with permission.[21]

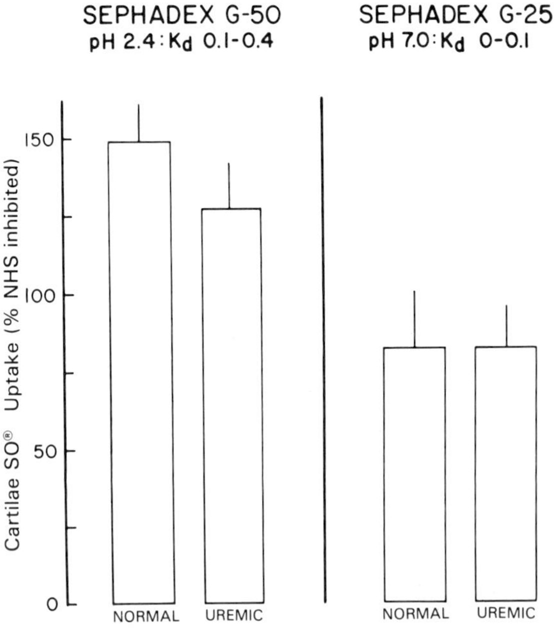

Figure 3 High-molecular-weight inhibitors in normal and uremic serum. High-molecular-weight inhibitors were measured at K_d 0.10–0.4 after chromatography of serum samples on Sephadex G-50, pH 2.4, and at K_d 0–0.1 after chromatography on Sephadex G-25, pH 7.0. Inhibition of cartilage sulfate uptake is expressed as % stimulation by normal human serum (NHS) inhibited, mean ± SEM. From *J. Clin. Endocrinol. Metab.*, with permission.[21]

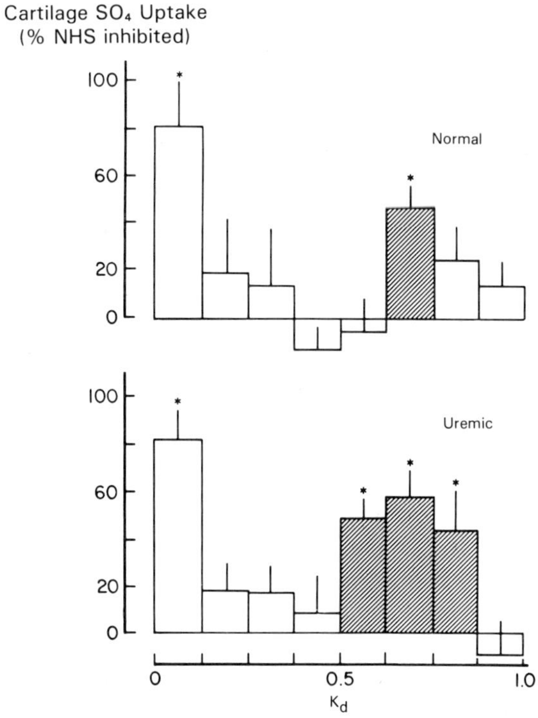

Figure 4 Low-molecular-weight somatomedin inhibitors in normal and uremic serum. Samples were gel-filtered on Sephadex G-25, pH 7.0, and inhibitory activity expressed as in Fig. 3. Asterisks show $p < 0.05$; hatching shows low-molecular-weight inhibition (K_d 0.5–0.88). From *J. Clin. Endocrinol. Metab.*, with permission.[21]

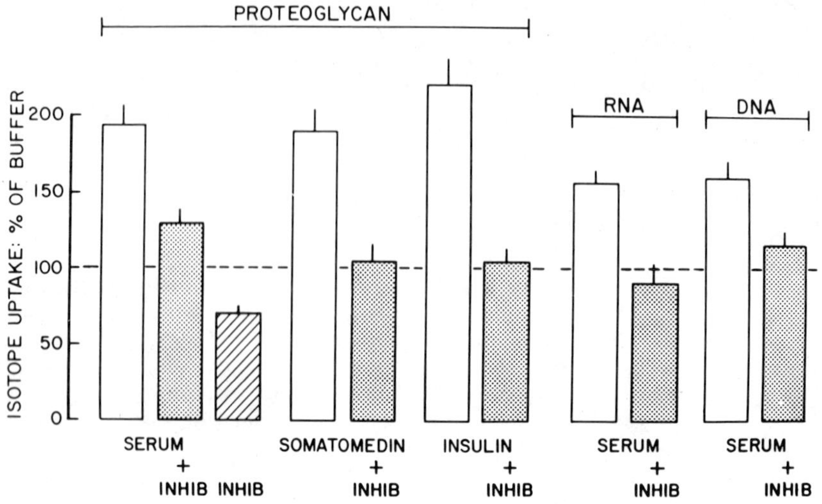

Figure 5 Cartilage effects of the uremic inhibitor. Hypophysectomized rat costal cartilage was incubated 48 hr with or without addition of stimulators [serum, somatomedins as found in Fig. 2 (K_d 0.5–0.62), and insulin 10 mU/ml], and the peak inhibitory fraction as found in uremic serum (from Fig. 4, K_d 0.62–0.75). Exposure to growth factors alone is shown by open bars, the peak inhibitory fraction alone by hatching, and growth factors with the inhibitory fraction by stippling. Isotope uptake ($^{35}SO_4$ into cartilage proteoglycan, [^{3}H]uridine into cartilage RNA, [^{3}H]thymidine into cartilage DNA) is expressed as % of buffer level, mean ± SEM, n = at least 4 for each determination. From *J. Clin. Endocrinol. Metab.*, with permission.[21]

diabetes[16]), such inhibitors were searched for in normal and uremic serum after chromatography at neutral and acid pH. As shown in Fig. 3, high-molecular-weight inhibitors were found in both sera, but there was no significant difference between levels of inhibitors under normal or uremic serum conditions. Thus, decreased somatomedin activity in uremia cannot be attributed to the presence of increased levels of high-molecular-weight inhibitors.

Low-molecular-weight somatomedin inhibitors were studied in uremic and normal serum after chromatograpy on Sephadex G-25, pH 7.0 (Fig. 4). In this system, inhibitors were found in both normal and uremic serum. However, the sum of significant inhibitory activity in the uremic sera (151 ± 23%) was significantly greater than total inhibitory activity in the normal subjects (47 ± 9%), $p < 0.001$. Based on comparison with elution of C-peptide, bacitracin, cyanocobalamin, and glutathione, peak inhibitory activity (K_d 0.62–0.75) had an apparent molecular size of ~1000. To examine the chemical nature of the inhibitory activity in the peak fraction, the fraction was exposed to proteolytic enzymes immobilized on carboxymethyl cellulose beads. Incubation with chymotrypsin or protease decreased inhibitory activity in this fraction, suggesting that the inhibitor is a peptide.

We next examined the effects of the low-molecular-weight inhibitor on different metabolic processes in cartilage. Segments of costal cartilage from hypophysectomized rats were incubated with or without different stimulators

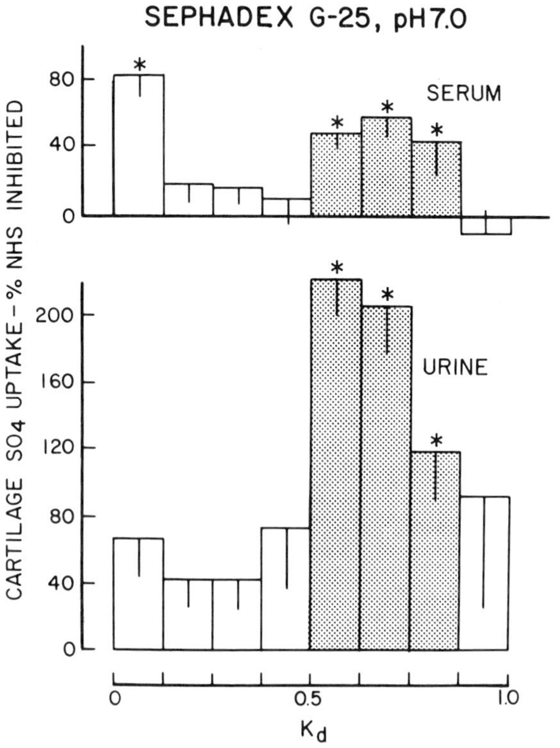

Figure 6 Somatomedin inhibitors in normal urine verus uremic serum. Pooled overnight urine from four normal subjects was chromatographed on Sephadex G-25, pH 7.0, as in Fig. 4. Shown for contrast is the profile of inhibition in uremic serum (upper panel), from Fig. 4. Asterisks show $p < 0.05$; hatching shows the low-molecular-weight inhibitory area. Mean ± SEM. From *J. Clin. Endocrinol. Metab.*, with permission.[21]

(whole normal human serum, a somatomedin fraction separated from normal serum by acid gel filtration, and insulin), with or without addition of the peak inhibitory fraction. As shown in Fig. 5, addition of the inhibitor to cartilage incubations containing normal human serum significantly decreased the stimulation of sulfate uptake into cartilage chondroitin sulfate. The inhibitor also blunted sulfate uptake as stimulated by somatomedins or insulin. Exposure of cartilage to inhibitor alone (without added stimulators) decreased cartilage sulfate uptake significantly below buffer levels. The inhibitor also blunted the ability of whole serum to stimulate uridine uptake into cartilage RNA and thymidine uptake into cartilage DNA. Somatomedin/inhibitor interactions on cartilage were further examined by exposing cartilage to increasing concentrations of normal human serum, with or without addition of the peak inhibitory fraction. When results were subjected to Lineweaver–Burk analysis, plotting inverse sulfate uptake versus inverse concentration, the findings indicated that interactions were noncompetitive (not shown). In combination, these observations provided evidence for broad antianabolic effects of the inhibitor on cartilage, including blunting of a variety of metabolic processes and direct effects on cartilage in the absence of added stimulators.

To explore other potential antianabolic effects of the uremic inhibitor, segments of rat epididymal fat pads were exposed to insulin with or without addition of the peak inhibitory fraction. Although insulin (100 uU/ml) provided substantial stimulation of glucose oxidation (1-[^{14}C]glucose conversion of CO_2 increased 400% above buffer level), addition of the peak inhibitory fraction did not blunt insulin action on this tissue. Thus, the actions of this uremic inhibitor may be specific for antagonism of growth-related processes (due to somatomedins) and may not include blunting of calorie storage (due to insulin).

As shown in Fig. 6, normal urine contains an inhibitor of apparent size similar to that found in uremic serum. After chromatography on Sephadex G-25, pH 7.0, pooled urine from four normal subjects had significant inhibitory activity in the same region, K_d 0.5–0.88, as inhibitory activity in the serum of eight uremic subjects (from Fig. 4). These observations are consistent with the hypothesis that the inhibitor found in uremic serum may represent a factor or factors that is normally cleared from the circulation by the kidneys.

4. Discussion

Studies of uremic children have consistently revealed impaired growth,[17] which may not be restored either by hemodialysis[2] or by continuous ambulatory peritoneal dialysis.[3] Poor growth in uremia occurs even after nutrient intake is taken into consideration[4] and cannot be attributed to low levels of GH, insulin, or somatomedins. Our studies indicate that uremic serum contains increased levels of a low-molecular-weight factor(s) that has broad antianabolic activity on cartilage. The inhibitor may normally be cleared from the circulation by the kidneys, inasmuch as lower levels were found in normal serum and an inhibitor of comparable size was found in normal urine.

The apparent size of the uremic inhibitor (mol. wt. ~1000) makes it likely that the inhibitor is a uremic "middle molecule." Such factors may contribute to the uremic syndrome in a variety of ways, including antagonism of insulin binding,[18] insulin resistance,[19] and inhibition of protein synthesis.[20] To the extent that such factors cause growth failure in uremia, their removal should improve growth. While conventional dialysis techniques may not restore the growth of uremic children to normal, such apparent lack of efficacy may result from dialytic approaches based on urea nitrogen and creatinine as markers. It is conceivable that the higher-molecular-weight uremic inhibitor identified in these studies, which has greater biological relevance and may be a better indicator of potential uremic toxicity, would be a more useful guide in determining therapeutic approaches.

5. Summary and Conclusion

Somatomedin activity is decreased in uremic serum, but rises with dialysis *in vivo* or *in vitro*. Biologically active somatomedins are comparable in normal and uremic serum. High-molecular-weight somatomedin inhibitors are comparable in normal and uremic serum.

Uremic serum contains significantly more low-molecular-weight somatomedin inhibitor(s) than normal serum. The uremic inhibitor(s) (mol. wt. ~1000) blunts cartilage stimulation by both somatomedins and insulin, but does not affect insulin action on adipose tissue. The uremic inhibitor(s) appears to be peptide in nature. An inhibitor(s) of size comparable to the uremic inhibitor(s) is found in normal urine.

Defective clearance of a circulating low-molecular-weight somatomedin inhibitor(s) may lead to decreased somatomedin activity and poor growth in uremia.

ACKNOWLEDGMENTS. We thank Ms. M. McGuire for assistance in preparation of the manuscript, Dr. Norbert Freinkel for encouragement and support, and Drs. J. D. Kopple (Renal Division, Department of Medicine, VA Wadsworth Medical Center and University of California Los Angeles) and F. del Greco (Renal Division, Department of Medicine, Northwestern University Medical School) for obtaining serum from uremic subjects. This work was supported in part by DHHS Training Grant AM-07169 and by Research Grants AM-10699, Am-21483, and AM-33475 from the National Institutes of Health.

References

1. M. Broyer, C. Kleinknecht, C. Loirat, C. Marti-henneberg, and M. P. Roy, Growth in children treated with long-term hemodialysis, *J. Pediat.* **84:**642, 1974.
2. O. Mehls, E. Ritz, G. Gilli, and W. Kreusser, Growth in renal failure, *Nephron* **21:**237, 1978.
3. I. B. Salusky, L. Cucullo, P. Nelson, and R. N. Fine, Continuous ambulatory peritoneal dialysis in children, *Pediatr. Clin North Am.* **29:**1005, 1982.
4. E. Ritz, O. Mehls, B. Krempien, G. Gilli, H. Udes, and W. Harendza, Skeletal growth in uremia, in: *Phosphate Metabolism* (S. G. Massry and E. Ritz, eds.), Plenum Press, New York, 1977, pp. 515–531.
5. A. Schiffrin, H. Guyda, P. Robitaille, and B. Posner, Increased plasma somatomedin reactivity in chronic renal failure as determined by acid gel filtration and radioreceptor assay, *J. Clin. Endocrinol. Metab.* **46:**511, 1977.
6. J. M. Simmons, C. J. Wilson, D. E. Potter, and M. A. Holliday, Relation of calorie deficiency to growth failure in children on hemodialysis and the growth response to calorie supplementation. *N. Engl. J. Med.* **285:**653, 1971.
7. L. S. Phillips and R. Vassilopoulou-Sellin, Somatomedins, *N. Engl. J. Med.* **302:**371, 438, 1980.
8. K. Takano, K. Hall, K. W. Kastrup, N. Hizuka, K. Shizume, K. Kawai, M. Akimoto, T. Takuma, and N. Sugino, Serum somatomedin A in chronic renal failure, *J. Clin. Endocrinol. Metab.* **48:**371, 1979.
9. A. C. Goldberg, B. Trivedi, J. A. Delmez, H. R. Harter, and W. H. Daughaday, Uremia reduces serum insulin-like growth factor I, increases insulin-like growth factor II, and modifies their serum protein binding, *J. Clin. Endocrinol. Metab.* **55:**1040, 1982.
10. L. S. Phillips, A. J. Pennisi, D. C. Belosky, C. Uittenbogaart, R. B. Ettenger, M. H. Malekzadeh, and R. N. Fine, Somatomedin activity and inorganic sulfate in children undergoing dialysis, *J. Clin. Endocrinol. Metab.* **46:**165, 1978.
11. L. S. Phillips and T. D. Scholz, Nutrition and somatomedin. IX. Blunting of insulin-like activity by inhibitor in diabetic rat serum, *Diabetes* **31:**97, 1982.
12. L. S. Phillips, D. C. Belosky, and L. A. Reichard, Nutrition and somatomedin. V. Inhibitor(s) in serum from diabetic rats, *Endocrinology 104:*1513, 1979.
13. L. S. Phillips and J. D. Kopple, Circulating somatomedin activity and sulfate levels in adults with normal and impaired kidney function. *Metabolism* **30:**1091, 1981.

14. L. S. Phillips, A. C. Fusco, T. Unterman, and F. del Greco, Increased somatomedin inhibitors in renal failure, *Clin. Res.* **31:**472A, 1983 (abstr.).
15. W. D. Salmon, Jr., Interaction of somatomedin and a peptide inhibitor in serum of hypophysectomized and starved, pituitary-intact rats, *Adv. Metab. Disord.* **8:**183, 1975.
16. L. S. Phillips, V. R. Bajaj, and A. C. Fusco, Nutrition and somatomedin. XI. Studies of somatomedin inhibitors in rats with streptozotocin-induced diabetes, *Diabetes* **32:**1117, 1983.
17. D. E. Potter and I. Greifer, Statural growth of children with renal disease, *Kidney Int.* **14:**334, 1978.
18. S. Milutinovic, D. Breyer, N. Jankovic, V. Molnar, A. Stefovic, and B. Rocic, Inhibitor of insulin binding to erythrocytes in plasma of uremic patients, *Nephron* **34:**99, 1983.
19. R. Dzurik, V. Hupkova, P. Cernacek, E. Valovicova, and T. R. Niederland, The isolation of an inhibitor of glucose utilization from the serum of uremic subjects, *Clin. Chim. Acta* **46:**77, 1973.
20. C. Delaporte and F. Gros, *In vitro* inhibition of protein synthesis by dialysates of plasma from uremic patients, *Eur. J. Clin. Invest.* **11:**139, 1981.
21. L. S. Phillips, Somatomedin inhibitor in uremia, *J. Clin. Endocrinol. Metab.* **59:**764–772, 1984.

45

Action of Somatomedins on Cell Growth

Effect of Selective Neutralization of Somatomedin C (Insulinlike Growth Factor I) with a Monoclonal Antibody

Judson J. Van Wyk, William E. Russell, Louis E. Underwood, Marjorie E. Svoboda, G. Yancey Gillespie, W. Jackson Pledger, Eli Y. Adashi, and Samuel D. Balk

1. Somatomedins Stimulate Both Differentiated Cell Functions and Cellular Proliferation

Research on growth factors wsa initiated in our laboratory with the erroneous expectation that the growth hormone-dependent factor that stimulated DNA synthesis in cartilage would prove different from the growth hormone-dependent "sulfation" factor that stimulated proteoglycan synthesis. The dual-isotope method we established to monitor the separation of these putative factors, however, quickly established that the same humoral signal that stimulated DNA synthesis also stimulated the differentiated functions characteristic of chondrocytes.[1] Since increased synthesis of proteins and proteoglycans could be demonstrated before cell division had occurred, expression of differentiated

Judson J. Van Wyk, William E. Russell, Louis E. Underwood, and Marjorie E. Svoboda • Department of Pediatrics, University of North Carolina, School of Medicine, Chapel Hill, North Carolina 27514. ***G. Yancey Gillespie*** • Department of Surgery and Cancer Research Center, University of North Carolina, School of Medicine, Chapel Hill, North Carolina 27514. ***W. Jackson Pledger*** • Department of Pharmacology and Cancer Research Center, University of North Carolina, School of Medicine, Chapel Hill, North Carolina 27514. ***Eli Y. Adashi*** • Department of Obstetrics and Gynecology, University of Maryland, Baltimore, Maryland 21201. ***Samuel D. Balk*** • Department of Pathology, New England Deaconess Hospital, Boston, Massachusetts 02116.

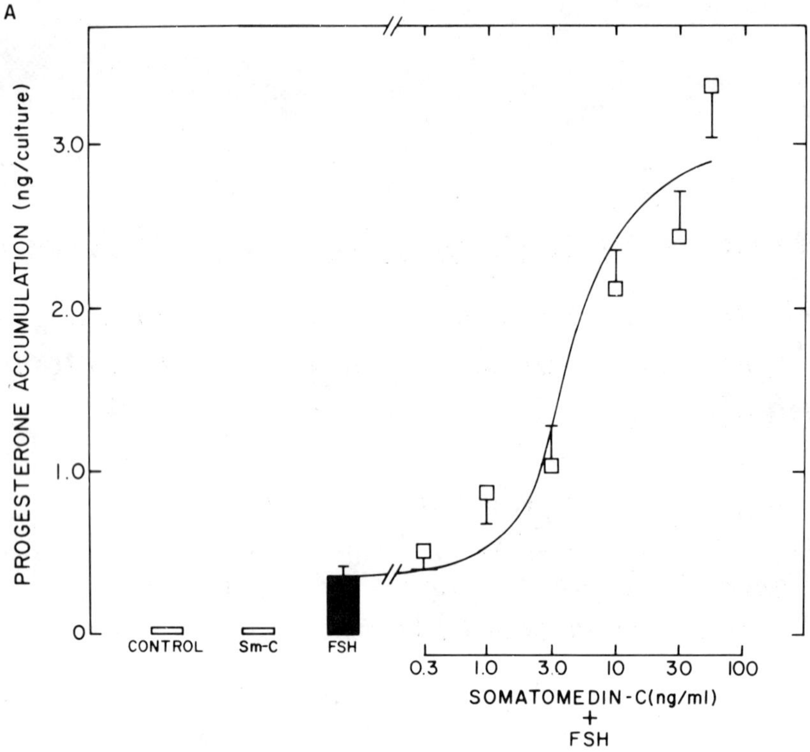

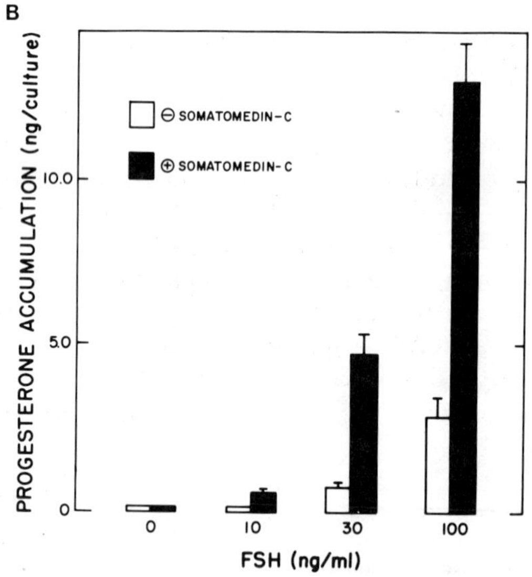

Figure 1 Effect of SMC/IGF-I on follicle-stimulating hormone (FSH)-stimulated progesterone accumulation by cultured rat granulosa cells. Granulosa cells from immature hypophysectomized diethylstilbesterol-treated rats were cultured under serum-free conditions for 72 hr after which progesterone in the media was measured by radioimmunoassay. (A) SMC dose–response relation-

cell functions was clearly a primary event and not the result of more highly differentiated cells arising as a consequence of cellular proliferation.

We have recently found that somatomedin C (SMC/IGF-I)* similarly stimulates both cell replication and highly differentiated responses in granulosa cells. Although it had previously been well documented that SMC/IGF-I stimulates the proliferation of bovine granulosa cells,[3] Adashi *et al.*[4] have now found that SMC acts synergistically with follicle-stimulating hormone (FSH) to stimulate progesterone synthesis in diethylstilbesterol-primed granulosa cells from immature hypophysectomized rats. Progesterone accumulation in the presence of a submaximal dose of FSH (20 ng/ml) was increased nearly tenfold in the presence of SMC/IGF-I, with an ED_{50} of 2.2 ng of SMC/IGF-I/ml (Fig. 1A). Furthermore, a steroidogenically inert dose of FSH (10 ng/ml) was rendered steroidogenically potent following the addition of SMC/IGF-I (50 nmg/ml) (Fig. 1B). This synergism was abolished when a specific monoclonal antibody to SMC/IGF-I was added to this medium. Although the significance of these striking effects remains to be clarified, the synergistic effect of SMC/IGF-I on FSH action might explain why growth hormone treatment of children with isolated growth hormone deficiency facilitates the development of pubertal changes as well as skeletal growth.

2. *In Vitro Growth-Promoting Actions of SMC*

2.1. *Range of Responsive Cell Types*

Although SM was first described as a skeletal growth factor, it has been found to stimulate DNA synthesis and cell proliferation in cell types of diverse embryological origin and derived from species ranging from amphibia to man. This is illustrated by the partial listing of responsive cell types shown in Table 1. In addition, Sato has shown that many different cell types fail to proliferate in defined, serum-free media unless insulin is present in pharmacological concentrations.[5] When careful comparisons have been made in such "insulin-dependent" cells, it has usually been found that the SM's produce similar effects but at molar concentrations far below those of insulin. At unphysiologically high concentrations insulin can function as an SM surrogate by virtue of its ability to interact with the Type I SMC/IGF-I receptor.[5–8] Although insulin does not cross-react with the Type II receptor for IGF-II,[8] it can increase

* Somatomedin C and insulinlike growth factor I are different names for the same peptide,[2] which will be referred to in the text as SMC/IGF-I.

ship. The study was carried out in the absence or presence of FSH (20 ng/ml) with or without increasing (0.3–50 ng/ml) concentrations of SMC. (B) FSH dose–response relationship. This study is similar to that in Fig. 1A except that it was carried out in the absence or presence of SMC (50 ng/ml) with or without increasing (10, 30, and 100 ng/ml) concentrations of FSH. The results represent the mean ± SEM of three determinations.

Table 1 Examples for Tissues Responsive to Somatomedin C

Fibroblasts	Muscle
Avian, murine, human	Bovine vascular smooth muscle
	Rat myoblasts
Gonadal	Hematopoietic
Rat Sertoli cells	Murine, proerythroid
Bovine granulosa	Freund's erythroleukemia
Rat ovarian tumor cells	Human lymphocytes
Hepatic	Pituitary tumor (GH_3)
Rat fetal hepatocytes	
Epithelial	Mesenchymal
Frog lens	Murine fetal limb buds
	Chicken heart mesenchymal cells
Chondrocytes	
Rat, rabbit, pig	

sensitivity to IGF-II by increasing the apparent concentration of Type II receptors on the outer surface of at least some cell types (see Czech *et al.*, this volume, Chapter 41).

2.2. Interaction of SM with Other Growth Factors

The effect of SM by itself on DNA synthesis and cell proliferation is usually quite small and well below the maximal effect of serum. This suggests that SM must act in concert with other growth factors and hormones to stimulate these effects. Such interactions of SMC/IGF-I with other growth factors have been most thoroughly explored in BALB/c 3T3 mouse fibroblasts. When these cells are grown to confluency in serum, they become arrested in G_O, but if they are then exposed to fresh serum derived from clotted whole blood, they will enter G_1 and, after a minimum lag period of 12 hr, reinitiate DNA synthesis. Serum prepared from platelet-poor plasma, however, will not elicit the same effect as serum prepared from whole blood unless the cells are first exposed for a few hours to the platelet-derived growth factor (PDGF).(10) This growth factor, which is normally released from α granules in platelets during blood coagulation, is greatly diminished in serum prepared from platelet-poor plasma (PPP), since PPP is prepared by removing formed elements of the blood before clotting is allowed to occur. Stiles *et al.* showed that after pretreatment of confluent BALB/c 3T3 cells with PDGF, entry into DNA synthesis was initiated 12 hr after the addition of PPP from normal rats, whereas SM-deficient plasma from hypophysectomized rats was nearly inactive in this respect unless fortified with exogenous SMC.(11)

Leof *et al.* showed that a combination of epidermal growth factor (EGF) plus SMC/IGF-I is just as effective as PPP in stimulating PDGF-treated BALB/c 3T3 cells to traverse G_1 and that high dosages of insulin can substitute for SMC/IGF-I in this regard.(12,13) In the study shown in Fig. 2, neither SMC/IGF-I nor EGF had much effect on [^{3}H]thymidine labeling of nuclear DNA in the

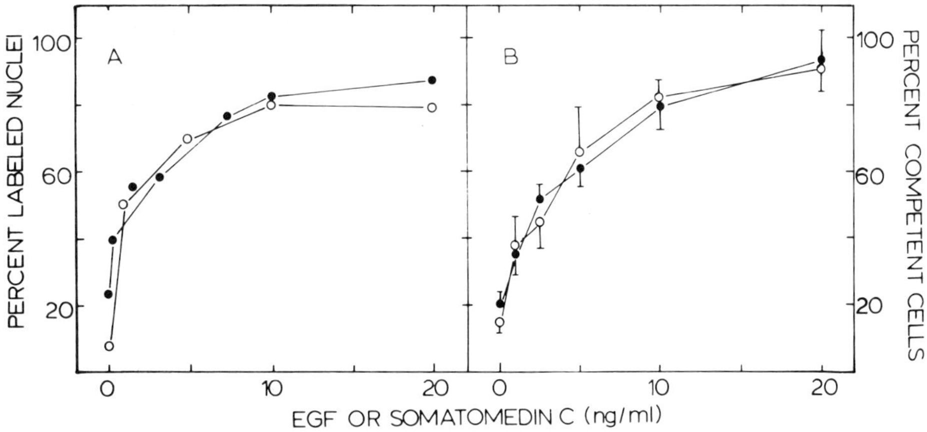

Figure 2 Dose–response curves of EGF and SMC/IGF-I on nuclear labeling of BALB/c 3T3 cells. (A) BALB/c 3T3 cells were grown to confluence in 96-well Nunc microtiter plates in 0.2 ml Dulbecco-Vogt Modified Eagle's Medium (DMEM) containing 10% calf serum. Two to three days after the cells reached confluence (4–5 days after initial plating) the spent medium was removed, and 0.2 ml DMEM containing 5 μg/well partially purified PDGF was added for 5 hr. The cultures were then washed, and DMEM containing 5 μCi/ml [^{3}H]thymidine plus 20 ng/ml EGF and varying concentrations of SMC (●—●) or media containing 20 ng/ml SMC and varying concentrations of EGF (○—○) was added. The cells were fixed 24 hr later and processed for autoradiography. In this experiment the addition of 5% PPP allowed 90% of the PDGF-treated cells to enter S phase. (B) Confluent cultures of BALB/c 3T3 cells were exposed to 0.7 μg/well PDGF further purified by Biogel P150 chromatography or 50ng/well fibroblast growth factor for 5-½ hr. The cultures were then washed and exposed to DMEM containing various concentrations of EGF or SMC as described for Fig. 2A. Individual points represent the mean ± SEM of four experiments (two using Biogel PDGF, one using CM-Sephadex PDGF, and one using FGF). The data are plotted as the percent of competent cells that entered S phase. In all cases 100% was defined as the percentage of PDGF- or FGF-treated cells that initiated DNA synthesis in DMEM supplemented with 5% PPP. Reproduced with permission from Leof *et al.*[12]

absence of the other, but similar dose–response curves were observed when the other hormone (either EGF or SMC/IGF-I) was present in a concentration of about 20 ng/ml. An unexpected finding from this study was that in plasma-free media, the minimum transit time across G_1 to the initiation of DNA synthesis was shortened from 12 hr to 10 hr. Addition of the SM-binding proteins present in plasma to media containing pure SMC/IGF-I and EGF caused reversion to the 12-hr transit time.[13]

The synergism between EGF and either high concentrations of insulin or nanomolar concentrations of SMC/IGF-I on DNA synthesis in confluent cultures of BALB/c 3T3 cells has also been observed on the rate of cell proliferation in rapidly growing cultures of other cell types.[14,15] In the study shown in Fig. 3, low-density cultures of chicken heart mesenchymal cells were plated out in increasing concentrations of SMC/IGF-I in the presence or absence of 100 ng/ml of EGF. In this system SMC/IGF-I has a modest effect on cell proliferation in the absence of EGF and a striking effect in the presence of EGF.

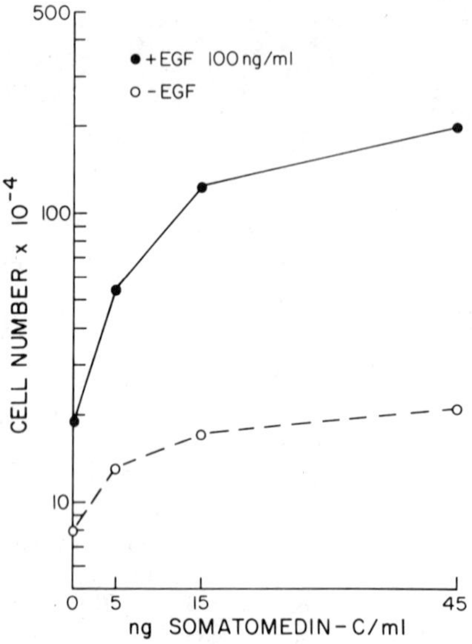

Figure 3 Proliferation of chicken heart mesenchymal cells after 4 days of incubation in media with graded concentrations of SMC in the absence or presence of 100 ng/ml EGF. Each point represents the mean of three 35-mm culture dishes, and no standard error exceeded half the height of the symbol used. At the beginning of the experiment (day 0) there were 7.0 ± 0 cells/dish. The culture media were changed daily. The culture conditions were identical to those described previously by Balk *et al.*[15] No EGF (○—○); EGF 100 ng/ml (●—●).

3. Immunoneutralization Studies with a Monoclonal Antibody to SMC

The fact that a mixture of purified growth factors can mimic the action of serum sheds no light on whether or not the ingredients of this mixture are the physiologically active substances in serum itself. A more elegant approach would be to determine the consequences of selectively removing the growth factor in question from plasma or serum. We have recently accomplished this by developing highly specific monoclonal antibodies to SMC/IGF-I and to the Type I receptor for SM-C/IGF-I.[9]

3.1. Development of Monoclonal Antibodies

BALB/c mice were immunized with a preparation of SMC/IGF-I which was isolated by the method of Svoboda *et al.*[16] and then coupled to mouse albumin in a final molar ratio of 6:1. Immunization was initiated by intramuscular injection of 100 μg of conjugated SMC/IGF-I and 37 μg of free SMC/IGF-I emulsified in Freund's complete adjuvant. Three subsequent challenges at 6-week intervals utilized i.m. injections with 80 μg of the conjugated and 30 μg of free SMC/IGF-I in incomplete Freund's adjuvant. Two weeks after the 4th i.m. injection, the mouse with the highest titer of serum antibodies was given 50 μg of conjugated SMC/IGF-I and 18 μg of free SMC/IGF-I intraperitoneally.

Four days after the i.p. injection, the mouse was killed, and 80% of the splenocytes were fused directly with an equal number of P3X63Ag8.653 myeloma cells[17] by a modification of the technique described by Galfre *et al.*[18] using 50% polyethylene glycol 1500. The remaining 20×10^6 splenocytes were

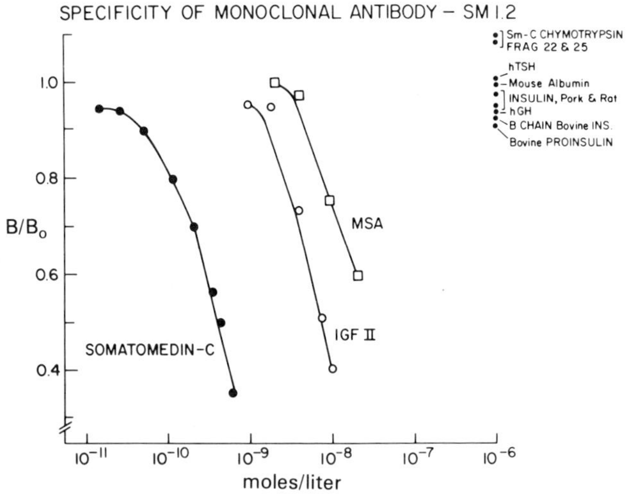

Figure 4 Specificity and affinity of monoclone sm 1.2 for the substances indicated was determined by preincubating the test substance with the antibody (4 hr, 22°C) and then adding 10,000 c/m of [^{125}I]SMC/IGF-I for an additional 15 hr before precipitating the immunoglobulins with PEG-6000 at a final concentration of 12.5%. B/B_o is the ratio of the counts specifically bound under the conditions shown to those in the absence of any unlabeled ligand.

cultured in thymocyte-conditioned medium in the presence of 2.4 μg of unconjugated SMC/IGF-I/ml before the survivors were fused with myeloma cells.[(19)] At the end of 5 days in culture, 8.9 × 10^6 viable cells were fused with 11.6 × 10^6 "653" myeloma cells. After each fusion the cells surviving overnight incubation in selective medium were dispensed into 96-well microculture trays.

Supernates containing colonial outgrowths were initially screened for evidence of mouse immunoglobulin (Ig) using a semiquantitative radioimmunoassay. Positive wells were then screened for anti-SMC/IGF-I activity by incubating supernates with [^{125}I]SMC/IGF-I and precipitating the complex with either 12.5% polyethylene glycol-6000 (PEG 6,000) or an appropriate dilution of rabbit antimouse Ig antiserum. Wells with anti-SMC/IGF-I activity were cloned by limiting dilution, and the monoclones were amplified by intraperitoneal injection of 3 × 10^6 viable cells into Pristane-primed BALB/c mice. Ascites fluids were collected and the immunoglobulin fraction purified by sequential precipitation in 18% and 15% sodium sulfate.

Only three of the wells derived from the 85 × 10^6 splenocytes that had been fused directly with myeloma cells were found to secrete anti-SMC/IGF-I antibodies, and none of these cultures could be successfully perpetuated and cloned. In striking contrast to this were the results from the 8.9 × 10^6 cells which had been boosted *in vitro*. Of the 80 wells that were plated from this fusion, 24 were found by both methods to secrete anti-SMC/IGF-I antibody. These were cloned by limiting dilution.

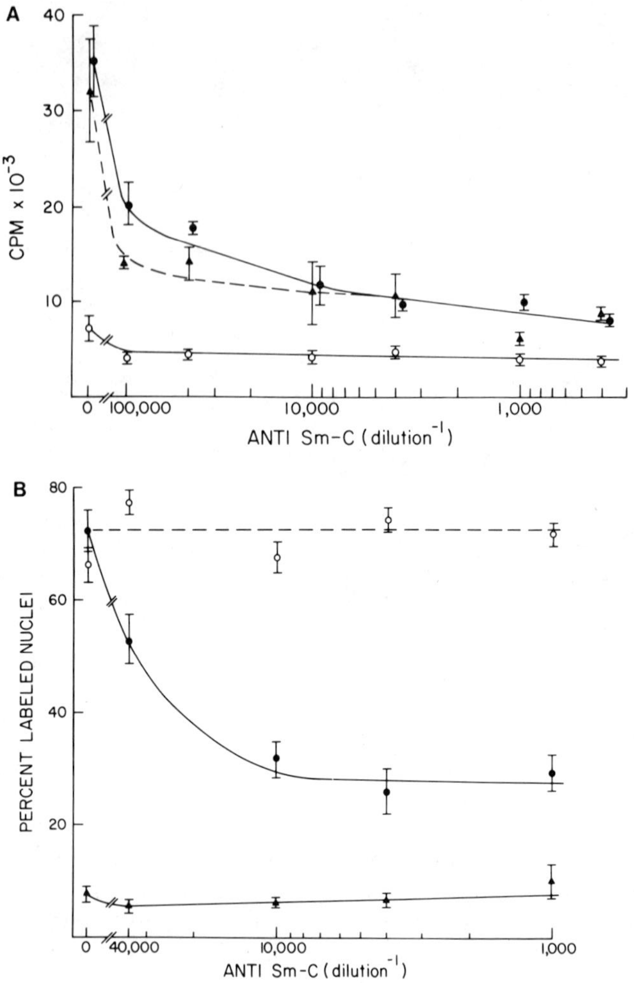

Figure 5 The effects of monoclonal antibody on DNA synthesis induced in PDGF-primed BALB/c 3T3 cells by plasma and by the combination of EGF with SMC or insulin. (A) Thymidine incorporation: 0.1% PPP, (○—○); SMC, 20 ng/ml + EGF, 10 ng/ml, (●—●); and 5% PPP, (▲—▲). All results are expressed as the mean of triplicate wells ± SEM. (B) Autoradiography: 0.1% PPP, (▲—▲); SMC, 20 ng/ml, + EGF, 10 ng/ml, (●—●); insulin, 1 μg/ml + EGF, 10 ng/ml, (○—○). Two predetermined fields were counted from each of two to four identical wells, and results expressed as percent labeled nuclei ± SEM. Reproduced with permission from Russell *et al.*[20]

3.2. Characterization of Anti-SMC/IGF-I Monoclones

Two monoclones (sm 1.2 and sm 5.1) were selected for further characterization. Ascitic fluid from both clones contained IgG_1, *kappa* immunoglobulins that bound 40–60% of trace quantities of [^{125}I]SMC/IGF-I at dilutions greater than 10^{-5} in the PEG radioimmunoassay (RIA). Affinity constants were determined to be 1.09 and 1.01×10^{10} 1/mole for sm 1.2 and sm 5.1, respectively.

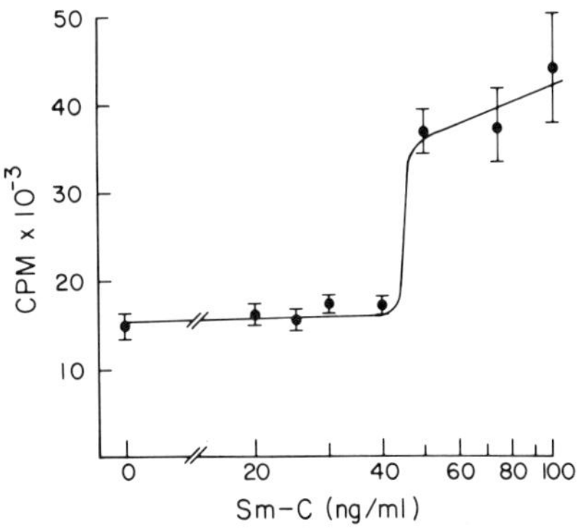

Figure 6 The effects of excess somatomedin C (SMC) on the inhibitory effect of monoclonal antibody. Thymidine incorporation was determined as described in the text. Following exposure to PDGF for 5 hr increasing amounts of SMC were added to cultures containing EGF, 10 ng/ml, and sm 1.2, 1:4000. Reproduced with permission from Russell *et al.*[20]

Both clones were quite specific for SMC/IGF-I. IGF-II was 5% as potent as SMC/IGF-I in competing with [^{125}I]SMC/IGF-I for binding to sm 1.2, and multiplication-stimulating activity was approximately 1% as active as SMC/IGF-I (Fig. 4). Neither chymotryptic fragments of SMC/IGF-I, various insulin preparations, human growth hormone, human thyroid-stimulating hormone mouse epidermal growth factor (mEGF), nor mouse albumin exerted activity at concentrations up to 10^{-6} molar.

3.3. In Vitro Immunoneutralization Studies with sm 1.2

The affinity of monoclonal antibody and sm 1.2 was sufficiently high to permit its use in selectively neutralizing SMC/IGF-I in various culture media.[20] Figure 5A shows that, following pretreatment with PDGF, the combination of EGF + SMC/IGF-I was as effective as 5% human PPP in stimulating [^{3}H]thymidine incorporation into trichloroacetic-acid-precipitable material in BALB/c 3T3 cells. Whereas the monoclonal antibody inhibited [^{3}H]thymidine uptake to a comparable extent whether induced by 5% PPP or by SMC/IGF-I + EGF, it had little influence on the baseline uptake observed in 0.1% PPP. The inhibitory effect of the antibody on the mitogenic effect produced by the EGF:SMC/IGF-I combination could be overcome by adding additional SMC/IGF-I to the mixture (Fig. 6). When nuclear labeling was used as an index of DNA synthesis (Fig. 5B), sm 1.2 produced a comparable inhibition of SMC/IGF-I-dependent labeling as that depicted in Fig. 5A. However, the antibody failed to influence the labeling when 1 μg/ml of insulin was used to supply the requirement for SMC/IGF-I.

These results were identical whether soluble antibody was incorporated directly into the incubation medium, or whether the experimental media were preincubated with sepharose-bound sm 1.2 to remove the growth factor prior to exposure of the cells. To further exclude the possibility that the inhibitory effect of the monoclonal antibody might be due to some factor in the ascites fluid other than the antibody, we tested the effect of a similarly processed IgG monoclone to horse apoferritin and observed no inhibition of the stimulatory effect of SMC/IGF-I + EGF or PPP (data not shown).

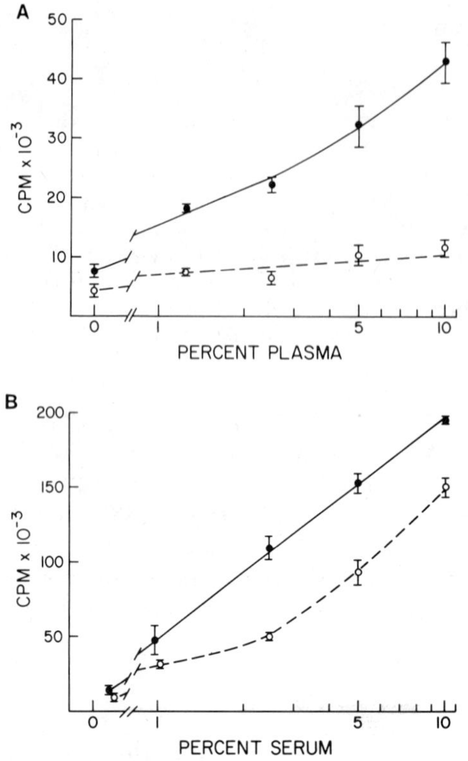

Figure 7 The effects of monoclonal antibody (1:4000) on the DNA synthesis induced by human platelet-poor plasma (A) and calf serum (B). Thymidine incorporation was determined in the presence of plasma or serum without antibody (○—○), or in the presence of sm 1.2 , 1:4000, (●—●). Reproduced with permission from Russell *et al.*[20]

Complete inhibition of the mitogenic effect of human plasma was evident up to a plasma concentration of 10% (Fig. 7A). The antibody also diminished the potency of calf serum, but this effect was less complete (Fig. 7B). We attribute these differences between human plasma and calf serum to higher concentrations of PDGF in calf serum since the requirement of BALB/c 3T3

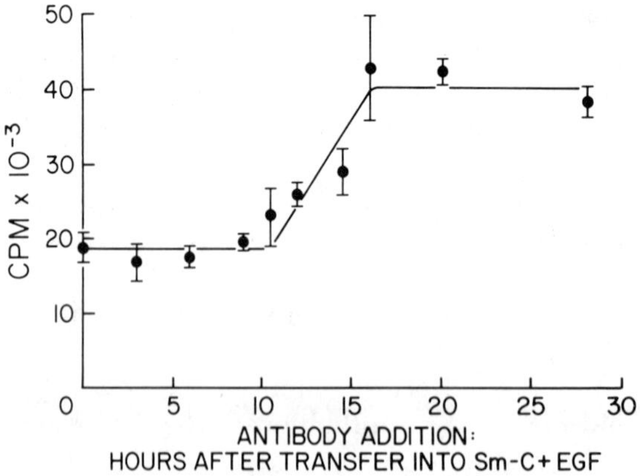

Figure 8 The time course of SMC dependent commitment to DNA synthesis. PDGF-primed cells were exposed to medium containing SMC, 20 ng/ml, + EGF, 10 ng/ml, and [^{3}H]thymidine as described. At the times indicated after the addition of growth factors, the media of triplicate wells were replaced with identical media containing sm 1.2, 1:4000. At 28 hr the incorporation of [^{3}H]thymidine was arrested by the addition of ascorbic acid and the cells harvested as described in the text. Reproduced with permission from Russell *et al.*[20]

Table 2 Inhibition of DNA Synthesis in BALB/c 3T3 Cells by sm 1.2[a]

Conditions	Antibody sm 1.2 (1:4000)	CPM ± SEM
0.1% PPP	0	33,609 ± 868
	+	22,270 ± 1087
Insulin (1 μg/ml)	0	61,119 ± 2728
	+	61,291 ± 2295
EGF (10 ng/ml)	0	82,476 ± 1724
	+	57,195 ± 2097
EGF (10 ng/ml) +	0	126,364 + 4494
insulin (1 μg/ml)	+	136,661 ± 2000

[a] BALB/c 3T3 cells (clone A-31) were cultured in a 1:1 mixture of Dulbecco's modified Eagle medium and Ham's F-12 (DME medium/F-12), supplemented with 10% calf serum. Cells were grown to confluence in 96-well microtiter plates in 0.2 ml of serum-containing culture medium. Within 48 hr of reaching confluence, the medium was replaced for 5 hr with serum-free DME medium/F-12 containing PDGF at 20 ng/well. The PDGF-containing medium was replaced with culture medium containing [^{3}H]methyl thymidine (6 Ci/mmol; 1 Ci = 37 GBq) at 1 μCi per well, and either plasma or growth factors with or without antibody. After 28 hr, the cells were fixed by the addition of 1 M ascorbic acid (0.06 ml per well), and the plates were processed for autoradiagraphy or thymidine incorporation.

Thymidine incorporation was determined after the monolayers were rinsed in serum-free DME medium/F-12, fixed and washed twice in 5% trichloroacetic acid at 4°C and then solubilized in 0.1 M NaOH with 1% sodium dodecyl sulfate for counting. All results are expressed as the mean of triplicate wells ± SEM. From Russell, *et al.*, (8) with permission.

cells for the growth factor present in plasma is progressively reduced with increasing concentrations of PDGF.[(21)]

To determine whether at some point in G_1 commitment to DNA synthesis no longer requires the presence of Sm-C/IGF-I, we added the monoclonal antibody at progressively later time intervals to the medium of PDGF-primed cells containing a combination of EGF and SMC/IGF-I and measured the rate of thymidine incorporation over 28 hr as described in Fig. 5A. As shown in Fig. 8, addition of the antibody at any time prior to the 9th hour after adding SMC/IGF-I and EGF completely prevented entry into DNA synthesis; thereafter an increasing percentage of the cells entered the S phase. Since the transit time through G_1 following the addition of EGF and SMC/IGF-I is 10 hr in the absence of plasma,[(13)] it can be concluded that irrevocable commitment to DNA synthesis requires SMC/IGF-I and occurs approximately at the G_1/S boundary. These findings agree well with the "W" point of irrevocable commitment described by Pledger *et al.*[(21)]

A consistent finding in our studies was that sm 1.2 lowered baseline DNA synthesis when only 0.1% PPP was present in the media and also produced a small inhibition of the DNA synthesis induced by EGF alone (Table 2). Since no exogenous SMC/IGF-I had been added to the media, we postulated that the small inhibition of DNA synthesis caused by the antibody when only EGF was added to the medium was due to an inhibitory effect of the antibody on small amounts of endogenous SMC/IGF-I produced by the cells themselves. This interpretation is supported by the failure of the antibody to block the mitogenic effect of EGF in the presence of insulin when added in sufficiently high concentrations to serve as a SMC/IGF-I surrogate (Table 2). This interpretation

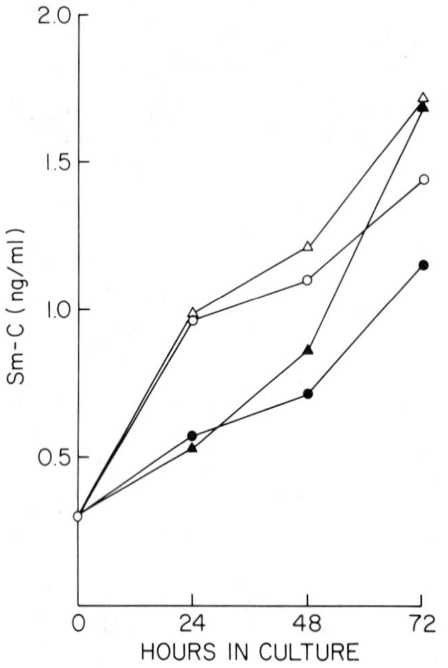

Figure 9 Production of immunoreactive somatomedin C (SMC) by confluent BALB/c 3T3 cells. Low-passage BALB/c 3T3 cells (Clone A-31) were grown to confluence in 10-cm-diameter tissue culture plates (GIBCO) in a medium made up of equal parts Dulbecco's Modified Eangle's Medium and Ham's F-12 (DMEM/F-12) supplemented with 10% calf serum (Hyclone, Logan UT). After reaching confluence one group of cells was exposed for 5 hr to serum-free DMEM/F-12 containing PDGF 20 ng/ml while the other group remained in DMEM/F-12 containing 10% calf serum. After 5 hr, the two groups were further subdivided with half of the plates in each group being refed with medium containing epidermal growth factor 10 ng/ml, and the other half receiving 0.1% PPP medium that was essentially serum or growth factor-free (DMEM/F-12 with 0.1% platelet poor plasma). At the times indicated the medium was aspirated and assayed for SMC by a radioimunoassay. Each point is the immunoreactive SMC content of one dish. Open symbols, no exposure to PDGF; O, cultured in 0.1% PPP, ▲, cultured in EGF 10 ng/ml. Closed symbols, 5-hr exposure to PDGF before culture in 0.1% PPP (●—●), EGF, 10 ng/ml (▲—▲).

was confirmed by finding that low, but significant, concentrations of immunoreactive SMC/IGF-I accumulate in the medium of density-inhibited BALB/c 3T3 cells cultured in the presence of PDGF and EGF, although neither growth factor appeared to stimulate the production of somatomedinlike activity above that produced in basal medium (Fig. 9).

The production by BALB/c 3T3 cells of substances that cross-react with our polyclonal antibody to SMC/IGF-I is consistent with our finding that, under

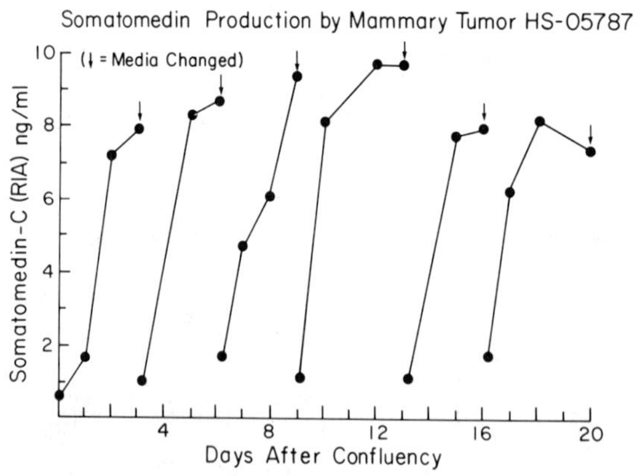

Figure 10 A human mammary carcinosarcoma cell line H50578T (courtesy of Dr. W. A. Nelson-Rees and the Cell Culture Department of the Naval Biosciences Laboratory of the School of Public Health, University of California, Berkeley) was grown to confluency in 100-mm Falcon dishes, and culture was continued in a medium consisting of 50% Dulbecco's Modified Eagle's Medium and 50% Ham's F-12 medium. The medium was enriched with 5 μg/ml transferrin. The media were changed at the points indicated and the spent media analyzed for SMC by radioimmunoassay.

appropriate circumstances, many normal as well as neoplastic cell lines secrete somatomedinlike substances into their media.[22] We found, for example, that media from five out of six human tumor cell lines cross-reacted in our SMC/IGF-I RIA (JJVW, unpublished data). In the study shown in Fig. 10, a mammary tumor cell line (H50578T) was grown to confluency in 10-cm dishes, and the SM in the serum-free media was measured by our RIA at the times shown. The conditioned media were harvested and replaced with new media every 3–4 days. Under these circumstances, the production of immunoreactive SMC/IGF-I continued unabated for at least 3 weeks. Unlike our findings with human fibroblasts,[23,24] the production of SMC/IGF-I-like substances by this tumor was not influenced by PDGF or human growth hormone, but did increase in response to cholera toxin, dibuteryl AMP, forskolin, and isobutyl methylxanthine.

4. Conclusion

The data presented here lend strong support to the hypothesis that SMC/IGF-I/IGF-I is a necessary component of plasma for the promotion of growth and/or differentiation in certain types of cultured cells. The role of IGF-II in the total mitogenicity of plasma cannot, however, be deduced from these studies. Although the affinity of sm 1.2 for IGF-II is only 5% as great as for SMC/IGF-I, we used relatively high concentrations of the antibody. Also, although IGF-II is not as potent a mitogen as SMC/IGF-I, the concentration of IGF-II in human plasma is three to four times higher than that of SMC/IGF-I. To resolve the relative roles of SMC/IGF-I and IGF-II it will be necessary to selectively neutralize each peptide and/or their respective receptors with antibodies of greater specificity.

The present studies reveal how powerful the subtractive approach can be in dissecting by means of specific antibodies the contribution of a single growth factor to the biological effects of plasma, serum, and other complex mixtures. Even more important may be the ability of this approach to demonstrate unambiguously the autocrine role of endogenously produced SM in the growth of cells and organs *in vitro* in experiments where no exogenous SM is present in the medium.

ACKNOWLEDGMENTS. Judson J. Van Wyk is a recipient of USPHS Research Career Award K06 AM14115. This work was also supported by USPHS Training Grant AM07129 and Research Grants AM01022 (JJVW), HD08299 (LEU), CA30479 (GYG), and CA24193 and CA16086 (WJP), and by grants from the Bressler Fund (EYA) and the Tobacco Research Council (SB).

References

1. J. J. Van Wyk, L. E. Underwood, R. L. Hintz, S. J. Voina, and R. P. Weaver, The somatomedins: A family of insulin-like peptides under growth hormone control, *Recent Prog. Hormone Res.* **30**:259, 1974.

2. D. G. Klapper, M. E. Svoboda, and J. J. Van Wyk, Sequence analysis of somatomedin-C: Confirmation of identify with insulin-like growth factor I, *Endocrinology* **112:**2215, 1983.
3. N. Savion, G. M. Lui, R. Laherty, and D. Gospodarowicz, Factors controlling proliferation and progesterone production by bovine granulosa cells in serum-free medium, *Endocrinology* **109:**409, 1981.
4. E. Adashi, C. E. Resnick, M. E. Svoboda, and J. J. Van Wyk, A novel role for somatomedin-C in the cytodifferentiation of the ovarian granulosa cell, *Endocrinology* **115:**1227, 1984.
5. J. Bottenstein, I. Hayashi, S. Hutchings, H. Masui, J. Mather, D. B. McClure, O. Sugayuki, A. Rizzino, G. Sato, G. Serrero, R. Wolfe, and R. Wu, The growth of cells in serum-free hormone-supplemented media, in: *Methods in Enzymology, LV III, Cell Culture* (W. B. Jacoby and I. H. Pastan, eds.), Academic Press, New York, 1979, pp. 94–109.
6. J. J. Van Wyk, L. E. Underwood, R. L. Hintz, D. R. Clemmons, and R. N. Marshall, Explorations of the insulin-like effects and growth promoting properties of somatomedin by membrane receptor assays, in: *Advances in Metabolic Disorders,* Vol. 8, Academic Press, New York, 1975, p. 127.
7. S. D. Chernausek, S. Jacobs, and J. J. Van Wyk, Structural similarities between human receptors for somatomedin-C and insulin: Analysis by affinity labeling, *Biochemistry* **20:**7345, 1981.
8. J. Massague and M. P. Czech, The subunit structures of two distinct receptors for insulin-like growth factors I and II and their relationship to the insulin receptor, *J. Biol. Chem.* **257**(9)**:**5038, 1982.
9. F. C. Kull, Jr., S. Jacobs, Y-F. Su, M. E. Svoboda, J. J. Van Wyk and P. Cuatrecasas, Monoclonal antibodies to receptors for insulin and somatomedin-C, *J. Biol. Chem.* **258**(10)**:**6561, 1983.
10. W. J. Pledger, C. D. Stiles, H. N. Antoniades, and C. H. Scher, Induction of DNA synthesis in BALB/c 3T3 cells of serum components: Reevaluation of the commitment process, *Proc. Natl. Acad. Sci. USA* **74:**4481, 1977.
11. C. D. Stiles, C. T. Capone, C. D. Scher, H. N. Antoniades, J. J. Van Wyk, and W. J. Pledger, Dual control of cell growth by somatomedins and platelet derived growth factor, *Proc. Natl. Acad. Sci. USA* **76:**1279, 1979.
12. E. B. Leof, W. Wharton, J. J. Van Wyk, and W. J. Pledger, Epidermal growth factor and somatomedin-C regulate G_1 progression in competent BALB/c 3T3 cells, *Exp. Cell Res.* **141:**107, 1982.
13. E. B. Leof, J. J. Van Wyk, E. J. O'Keefe, and W. J. Pledger, Epidermal growth factor (EGF) is required only during the traverse of early G1 in PDGF stimulated density-arrested BALB/c 3T3 cells, *Exp. Cell Res.* **147:**202, 1983.
14. S. D. Balk, A. Morisi, H. S. Gunther, M. E. Svoboda, J. J. Van Wyk, S. P. Nissley, and C. Scanes, Somatomedins (insulin-like growth factors), but not growth hormone, are mitogenic for chicken heart mesenchymal cells and act synergistically with epidermal growth factor and brain fibroblast growth factor, *Life Sci.* **35:**335, 1984.
15. S. D. Balk, R. P. C. Shiu, M. M. LaFleur, and L. L. Young, Epidermal growth factor and insulin cause normal chicken heart mesenchymal cells to proliferate like their Rous sarcoma virus-infected counterparts, *Proc. Natl. Acad. Sci. USA* **79:**1154, 1982.
16. M. E. Svoboda, J. J. Van Wyk, G. D. Klapper, R. E. Fellows, J. E. Grissom, and R. J. Schlueter, Purification of somatomedin-C from human plasma: Chemical and biological properties, partial sequence analysis and relationship to other somatomedins, *Biochemistry* **19:**790, 1980.
17. J. F. Kearney, A. Radbruch, B. Liesgang and K. Rajewsky. A new mouse myeloma cell line that has lost immunoglobulin expression but permits the construction of antibody-secretory hybrid cell lines, *J. Immunol.* **123:**1548, 1979.
18. G. Galfre, S. C. Howe, C. Milstein, G. W. Butcher, and J. C. Howard, Antibodies to major histocompatibility antigens produced by hybrid cell lines, *Nature* **266:**550, 1977.
19. C. L. Reading, Theory and methods for immunization in culture and monoclonal antibody production, *J. Immunol. Methods* **53:**261, 1982.
20. W. E. Russell, J. J. Van Wyk, and W. J. Pledger, Inhibition of the mitogenic effects of plasma by a monoclonal antibody to somatomedin-C, *Proc. Natl. Acad. Sci. USA* **81:**2389–2392, 1984.
21. W. J. Pledger, C. D. Stiles, H. N. Antoniades, and C. D. Scher, An ordered sequence of events

is required before BALB/c 3T3 cells become committed to DNA synthesis, *Proc. Natl. Acad. Sci. USA* **75:**2839, 1978.

22. J. J. Van Wyk, L. E. Underwood, A. J. D'Ercole, D. R. Clemmons, W. J. Pledger, W. R. Wharton, and E. B. Leof, Role of somatomedin in cellular proliferation, in: *Biology of Normal Human Growth* (E. R. Ritzen, ed.), Raven Press, New York, 1981, pp. 223–239.
23. D. R. Clemmons and J. J. Van Wyk, Somatomedin-C and platelet derived growth factor stimulate human fibroblast replication, *J. Cell Physiol.* **106:**361, 1981.
24. D. R. Clemmons, L. E. Underwood, and J. J. Van Wyk, Hormonal control of immunoreactive somatomedin production by cultured human fibroblasts, *J. Clin. Invest.* **67:**10, 1981.

46

Regulation of Serum Insulinlike Growth Factor II

William H. Daughaday

It is now generally accepted that there are two different insulinlike growth factors (IGF) of human serum: IGF-I/somatomedin C (SMC) and IGF-II. Although both peptides possess insulinlike and somatomedinlike properties, they are not functionally equivalent. Each possesses a preferential receptor which is variably expressed in different tissues. In this chapter I shall review some of the information about serum concentration of IGF-I/SMC and IGF-II which strongly suggests considerable independence of regulation of IGF-II.

Specific radioimmunoassay (RIA) of IGF I has been achieved with several different polyclonal[1–3] antibodies and monoclonal antibodies.[4,5] RIA's of IGF-II have been developed with antibodies raised against the intact peptide[3] and the C domain of the molecule.[6]

In our laboratory we have measured IGF-I/SMC with the SMC antibody provided by Underwood and Van Wyk and distributed by the National Pituitary Program. Purified IGF-I has been a gift of Rene Humbel of Zurich. This antibody has high specificity for IGF I with less than 1% cross-reaction from IGF-II. It also has proved cross-reactive with basic peptides of all the mammalian and nonmammalian species we have examined.

We measure IGF-II with a receptor assay that employs receptors present in rat placental membranes.[7] Rat placental membranes contain a high concentration of IGF-II receptors and a low concentration of IGF-I receptor sites. IGF-I has less than 1% of the potency of IGF-II to displace [^{125}I]IGF-II from the membrane in our hands although others have reported as much as 10% cross-reactivity in this assay. The reason for this discrepancy of results is not evident.

There is a continuing debate concerning the need to eliminate binding

William H. Daughaday • Metabolism Division, Washington University School of Medicine, St. Louis, Missouri 63110.

Table 1 Assays for IGF-I and IGF-II in Mammalian and Nonmammalian Species

	IGF-I RIA (*a*)	(ng/ml) (*b*)	IGF-II RRA (ng/ml)
Mammalian			
Rat	—	855	31
Guinea pig	459	615	3200
Rabbit	222	250	539
Man	218	250	830
Cow	—	130	788
Pig	111	—	900
Kangaroo	—	208[c]	120[c]
Oppossum	—	48	30
Nonmammalian			
Chicken	17	22	35
Turtle	None	5	100
Toad	None	11	<25
Trout	None	4	<25

[a] Results from Wilson and Hintz.[24]
[b] Present study.
[c] Kangaroo serum was assayed after acid–ethanol extraction.

protein before IGF-I assays. We found that only about one-third of human serum IGF-I was reactive in direct assays.[8] We described an acid ethanol method that excludes most human binding protein. The validity of this method has been confirmed by Zapf *et al.*[3] Clemmons *et al.*[9] have reported that temporary acidification of serum followed by heparin addition neutralizes binding protein. Our experience, however, has shown considerable residual binding of [^{125}I]IGF-I and [^{125}I]IGF-II after acidification and heparin when examined by the charcoal separation method.

While there is no agreement on the necessity of eliminating binding protein from IGF-I assays, it is essential for IGF-II assays because of the higher affinity of this peptide for serum binding proteins. Most of our IGF-II assays are done on acid–ethanol extracts of serum although acid gel filtration through Sephadex G-75 has been used for critical measurements of animal sera.

When the IGF-I RIA and IGF-II radio receptor assay (RRA) were applied to animal sera, we found considerable variation in the measured IGF-I and IGF-II contents (Table 1). The adult rat has a high concentration of IGF-I and a low concentration of IGF-II. In contrast, guinea pig serum, which has nearly as high a concentration of IGF-I, has a very high concentration of IGF-II. Marsupial serum has low concentrations of IGF-II as compared to human serum. We have been able to detect IGF-II receptor reactivity in chicken and turtle serum but not in toad and trout serum. A separate IGF-II may have been a reptilian acquisition.

There is evidence that IGF-II may be selectively increased in fetal development. This is most evident in the rat in which Moses *et al.*[10] found that rIGF-II was increased 20 to 200 times over adult rat serum. We have confirmed

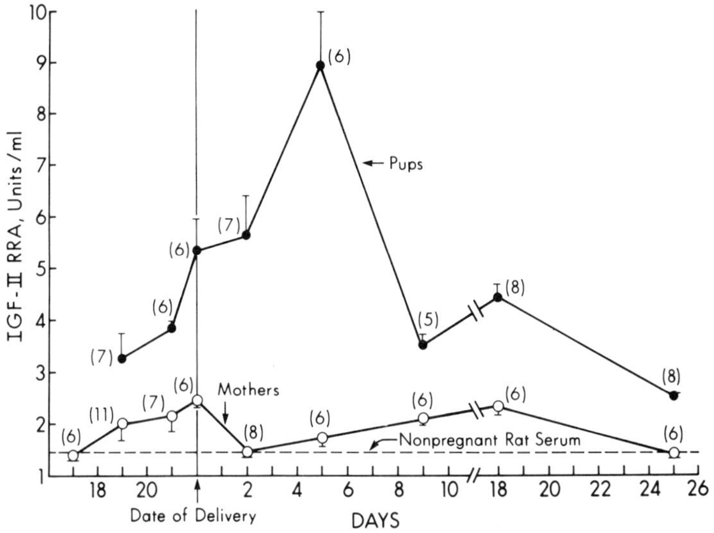

Figure 1 Measurement of IGF-II receptor activity in acid–ethanol extracts of serum from fetal, neonatal, and maternal rats. Reprinted by permission from Daughaday *et al.*[11]

an elevation of IGF-II in fetal and neonatal rats with our IGF-II RRA[11] (Fig. 1).

There have been four studies of human cord serum with simultaneous measurements of IGF-I and IGF-II. In Table 2 I have tabulated the results of these assays both as nanograms/milliliter and the ratio of cord serum to pooled adult normal serum indicated as units/milliliter. The IGF-I concentration was found to be from 27 to 60% of the normal adult value. Although the results calculated in nanograms varied greatly, cord serum IGF-II measured by IGF-II RIA by Zapf *et al.*[3] had about 50% of adult levels as estimated from his figure. Bennett *et al.*[12] measured IGF-II with an RIA using an antiserum developed against the C peptide domain and arrived at a similar IGF-II concentration. Gluckman *et al.*[13] found that cord serum IGF-II was 83% of

Table 2 IGF-I and IGF-II Peptides in Human Umbilical Cord Serum

Author	IGF-I	IGF-II
Zapf *et al.*[3]	ca. 0.3 IU/ml (RIA)[a] (ca. 50 ng/ml)	ca. 0.5 IU/ml (RIA)[a] (ca. 370 ng/ml)
Bennett *et al.*[12]	0.6 IU/ml (RIA) (113 ± 35 ng/ml)	0.41 IU/ml (RIA) (282 ± 84 ng/ml)
Gluckman *et al.*[13]	0.30 ± 0.13 IU/ml (RIA) (220 ng/ml)	0.83 ± 0.29 (RRA) (870 ng/ml)
Present study	0.27 ± 0.13 IU/ml (RIA) (68 ng/ml)	1.51 ± 0.18 (RRA) (1250 ng/ml)

[a] Estimated from Zapf *et al.*,[3] Fig. 5.

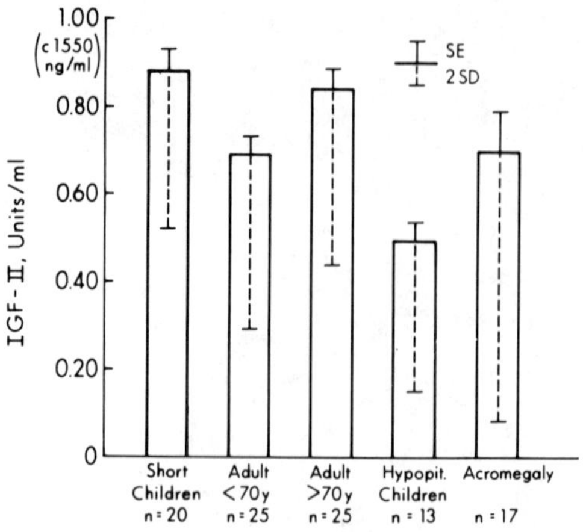

Figure 2 IGF-II levels measured by RRA on acid–ethenol extracts of sera from children and adult subjects, hypopituitary and acromegalic patients.

normal when measured on acid ethanol extract by our rat placental membrane RRA. Using the same method, we found that mean cord serum IGF-II was 151% of normal adult serum.

Because of the higher measured IGF-II content of cord sera with the rat placental membrane RRA than by RIA, we have initiated studies on molecular heterogeneity of IGF-II in adult and cord sera to determine whether the same peptides were present in both sera. We were mindful of earlier studies of Sara *et al.*[14] with a brain membrane RRA which suggested the presence of unique SM peptides in cord sera.

IGF-I and IGF-II differ in the degree of growth hormone (GH) dependence. As shown originally by Zapf *et al.*[3] and subsequently confirmed by us[15] (Fig. 2) and by Hintz and Liu,[6] IGF-II levels fall in hypopituitarism but not to the extent reached by IGF-I in this condition. GH treatment restores normal concentrations of IGF-II. Our results with short-term GH treatment are shown in Fig. 3. We, as well as others, have failed to find any elevation of IGF-II in acromegaly. The concentration of GH in normal individuals appears to be sufficient to maintain maximal IGF-II values.

This also seems to occur during puberty during which IGF-I levels have been observed to rise about twofold,[16,17] but no significant change of IGF II was observed by Luna *et al.*[18] The question is unresolved whether or not the pubertal rise in IGF-I can be accounted for by an increase in mean serum GH. It is possible that a synergism between sex steroids and GH is responsible for the remarkable rise in IGF-I.

In elderly individuals a decrease in serum IGF-I has been reported, and this has been attributed to a decrease in GH secretion.[19] A similar decrease in serum IGF-II probably does not occur. Widmer *et al.*[20] found that mean RIA IGF-II levels in individuals over the age of 50 were actually higher than in adults under 50 years of age. We reported that RRA IGF-II was modestly increased in individuals over the age of 70 years.[7]

There has been much interest in the effect of starvation and protein–

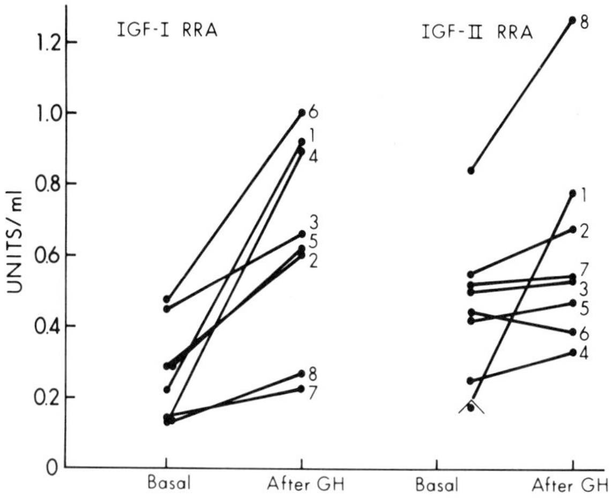

Figure 3 Measurement of serum IGF-I (RIA) and IGF-II (RRA) in hypopituitary children before and after GH 0.1 IU/kg every 12 hr for four doses. Reprinted by permission from Daughaday *et al.*[7]

calorie malnutrition on serum levels of bioassayable somatomedins and of IGF-I/SMC as measured by RIA. Clemmons *et al.*[21] have shown that levels fall in fasting and malnutrition on serum levels of bioassayable SM's and of IGF-I/SMC as measured by RIA. Clemmons *et al.*[21] have shown that levels fall in fasting and recovery is dependent on protein intake. Thus far, little attention has been directed toward nutritional factors affecting serum IGF-II levels. We have made measurements of IGF-I and IGF-II in sera of four young men fed isocaloric diets of varying protein content for 8 days at Massachusetts Institute of Technology by Yang and his co-workers. The sera were provided by Dr. D. E. Mathews of Washington University School of Medicine. The composition of the experimental diets is summarized in Table 3.

We found a moderate lowering of serum IGF-I on the 0.3 and 0.0 g/kg protein diets. There was no significant change in serum IGF-II on any of the experimental diets.

Table 3 Effect of Dietary Protein on Serum IGF-I and IGF-II Protein Intake[a]

Protein intake (g/kg)	1.5	0.6	0.3	0.0
IGF-I (U/ml)	1.33 ± 0.07	1.12 ± 0.53	1.04 ± 017	0.96 ± 0.14
IGF-II (U/ml)	0.98 ± 0.04	0.89 ± 0.08	0.96 ± 0.08	0.96 ± 0.12
	Composition of Experimental Diet[b]			
Protein (g/kg)	1.5	0.6	0.3	0.0
% Carbohydrate	47	56	56	63
% Fat	40	40	41	37
% Protein	13	5	3	0

[a] In grams per kilogram per day. Conducted at Massachusetts Institute of Technology by R. D. Yang, D. E. Mathews, Z. M. Wen, , D. M. Bier, and V. R. Young [25]. Sera provided for IGF-I and IGF-II by D. E. Mathews.
[b] Four young male volunteers, mean age 21 years, were fed isocaloric diets of varying protein content and a mean caloric content of 48kcal/kg. Two days of unrestricted diets preceded the two low-protein diets.

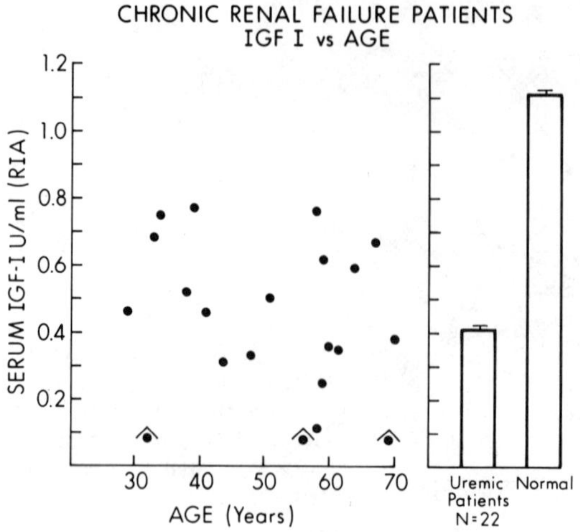

Figure 4 Serum IGF-I levels in adult patients on dialysis. Reprinted by permission from Goldberg *et al.*[22]

The only disease state we have studied in detail with respect to IGF-II has been uremia.[22] We found that serum IGF-I was lowered in uremic patients (Fig. 4) undergoing dialysis, and serum IGF-II was elevated three- to fourfold (Fig. 5). In addition, we observed an alteration of IGF binding protein pattern with increased total available binding and a shift to smaller-sized IGF binding protein complexes.

The possibility that certain large tumors secrete IGF-like peptides was first suggested by Megyesi *et al.*,[23] who reported that sera of such patients frequently had increased reactivity in an IGF radioreceptor assay using rat hepatic membranes. We reported that 10 to 14 similar patients had increased reactivity in our IGF-II RRA. In most cases the elevations were modest and unlikely to have been the cause of hypoglycemia. In these patients IGF-I levels were reduced.

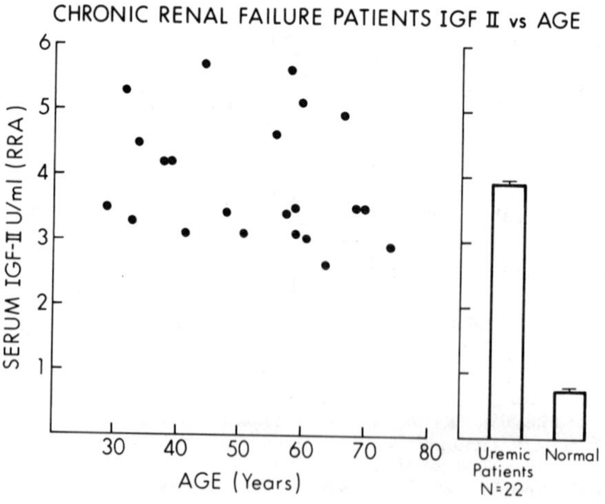

Figure 5 Serum IGF-II levels in adult patients on dialysis. Reprinted by permission from Goldberg *et al.*[22]

In an extensive investigation of sera of patients with tumor-associated hypoglycemia, Widmer *et al.*[20] were unable to document increased levels of hepatic receptor reactive IGF or immunoreactive IGF-II. The reason for the discrepancy of results is not evident.

ACKNOWLEDGMENTS. This work was supported by Research grants AM 01526 and AM 05105 from the National Institute of Arthritis, Diabetes, Digestion and Kidney Diseases, Bethesda, Maryland. The author acknowledges the participation of Dr. Anne C. Goldberg, Dr. Ann P. Ward, Ms. Ida Mariz, Ms. Bakula Trivedi, Ms. Milan Kapadia, and Ms. Bonnie Mills.

References

1. R. W. Furlanetto, L. Underwood, J. J. Van Wyk, and A. J. D'Ercole, Estimation of somatomedin-C levels in normals and patients with pituitary disease by radioimmunoassay, *J. Clin. Invest.* **60:**648, 1977.
2. R. M. Bala and B. Bhaumick, Radioimmunoassay of a basic somatomedin: Comparison of various assay techniques and somatomedin levels in various sera, *J. Clin. Endocrinol. Metab.* **49:**770, 1979.
3. J. Zapf, H. Walter, and E. R. Froesch, Radioimmunological determination of insulinlike growth factors I and II in normal subjects and patients with growth disorders and extrapancreatic tumor hypoglycemia, *J. Clin Invest.* **68:**1321, 1981.
4. R. C. Baxter, S. Axiak, and R. L. Raison, Monoclonal antibody against human somatomedin-C insulin-like growth factor-I, *J. Clin. Endocrinol. Metab.* **54:**474, 1982.
5. U. K. Laubli, W. Baier, H. Binz, M. R. Celio, and R. E. Humble, Monoclonal antibodies directed to human insulin-like growth factor I (IGF I). Use for radioimmunoassay and immunopurification of IGF, *FEBS Lett.* **149:**109, 1982.
6. R. L. Hintz and F. Liu, A radioimmunoassay for insulin-like growth factor-II specific for the C-peptide region, *J. Clin. Endocrinol. Metab.* **54:**442, 1982.
7. W. H. Daughaday, B. Trivedi, and M. Kapadia, Measurement of insulin-like growth factor II by a specific radioreceptor assay in serum of normal individuals, patients with abnormal growth hormone secretion, and patients with tumor-associated hypoglycemia, *J. Clin. Endocrinol. Metab.* **53:**289, 1981.
8. W. H. Daughaday, I. K. Mariz, and S. L. Blethen, Inhibition of access of bound somatomedin to membrane receptor and immunobinding sites—A comparison of radioreceptor and radioimmunoassay of somatomedin in native and acid-ethanol-extracted serum, *J. Clin. Endocrinol. Metab.* **51:**781, 1980.
9. D. R. Clemmons, L. E. Underwood, P. G. Chatelain, and J. J. Van Wyk, Liberation of immunoreactive somatomedin-C from its binding proteins by proteolytic enzymes and heparin, *J. Clin. Endocrinol. Metab.* **56:**384, 1983.
10. A. C. Moses, S. P. Nissley, P. A. Short, M. M. Rechler, R. M. White, A. B. Knight, and O. Z. Higa, Increased levels of multiplication-stimulating activity, an insulinlike growth factor, in fetal rat serum, *Proc. Natl. Acad. Sci. USA* **77:**3649, 1980.
11. W. H. Daughaday, K. A. Parker, S. Borowsky, B. Trivedi, and M. Kapadia, Measurement of somatomedin related peptides in fetal, neonatal and maternal rat serum by IGF-I RIA, IGF-II RRA and MSA RRA after acid-ethanol extraction, *Endocrinology* **110:**575, 1982.
12. A. Bennett, D. M. Wilson, F. Lui, R. Nagashima, R. G. Rosenfeld, and R. L. Hintz, Levels of insulin-like growth factors I and II in human cord blood, *J. Clin. Endocrinol. Metab.* **57:**609, 1983.
13. P. D. Gluckman, J. J. Johnson-Barrett, J. H. Butler, B. W. Edgar, and T. R. Gunn, Studies of insulin-like growth factor-I and -II by specific radioligand assays in umbilical cord blood, *Clin. Endocrinol.* **19:**405, 1983.

14. V. R. Sara, K. Hall, C. H. Rodeck, and L. Wetterberg, Human embryonic somatomedin, *Proc. Natl. Acad. Sci. USA-Biol. Sci.* **78:**3175, 1981.
15. J. Zapf, B. Morell, H. Walter, Z. Laron, and E. R. Froesch, Serum levels of insulinlike growth factor (IGF) and its carrier protein in various metabolic disorders, *Acta Endocrinol.* **95:**505, 1980.
16. K. Hall, G. Enberg, M. Ritzen, H. Svan, L. Fryklund, and K. Takano, Somatomedin-A levels in serum from healthy children and from children with growth hormone deficiency or delayed puberty, *Acta Endocrinol.* **94:**155, 1980.
17. R. M. Bala, J. Lopatka, A. Leung, E. McCoy, and R. G. McArthur, Serum immunoreactive somatomedin levels in normal adults, pregnant women at term, children at various ages, and children with constitutionally delayed growth, *J. Clin. Endocrinol. Metab.* **52:**508, 1981.
18. A. M. Luna, D. M. Wilson, C. J. Wibbelsman, R. C. Brown, R. J. Nagashima, R. L. Hintz, and R. G. Rosenfeld, Somatomedins in adolescence: a cross-sectional study of the effect of puberty on plasma insulin-like growth factor I and II levels, *J. Clin. Endocrinol. Metab.* **57:**268, 1983.
19. D. Rudman, M. H. Kutner, C. M. Rogers, M. F. Lubin, G. A. Fleming, and R. P. Bain, Impaired growth hormone secretion in the adult population, *J. Clin. Invest.* **67:**1361, 1981.
20. U. Widmer, J. Zapf, and E. R. Froesch, Is extrapancreatic tumor hypoglycemia associated with elevated levels of insulin-like growth factor II? *J. Clin. Endocrinol. Metab.* **55:**833, 1982.
21. D. R. Clemmons, A. Klibanski, L. E. Underwood, J. W. McArthur, E. C. Ridgway, I. Z. Beitins, and J. J. Van Wyk, Reduction of plasma immunoreactive somatomedin C during fasting in humans, *J. Clin. Endocrinol. Metab.* **53:**1247, 1981.
22. A. C. Goldberg, B. Trivedi, J. A. Delmez, H. R. Harter, and W. H. Daughaday, Uremia reduces serum insulin-like growth factor I, increases insulin-like growth factor II, and modifies their serum protein binding, *J. Clin. Endocrinol. Metab.* **55:**1040, 1982.
23. K. Megyesi, C. R. Kahn, J. Roth, and P. Gorden, Hypoglycemia in association with extrapancreatic tumors: demonstration of elevated plasma NSILA-s by a new radioreceptor assay, *J. Clin. Endocrinol. Metab.* **38:**931, 1974.
24. D. M. Wilson and R. L. Hinz, Interspecies comparison of somatomedin structure using immunological probes, *J. Endocrinol.* **95:**59–64, 1982.
25. R. D. Yang; D. E. Mathews, Z. M. Wen, D. M. Bier, and V. R. Young, Response of alanine metabolism in humans to manipulation of dietary protein and energy intakes, *Am. J. Physiol.*, 1986, in press.

47

Somatomedin C/Insulinlike Growth Factor I

Regulation and Clinical Applications

Louis E. Underwood, Eric P. Smith, Judson J. Van Wyk, David R. Clemmons, A. Joseph D'Ercole, M. R. Pandian, Michael A. Preece, and Wayne V. Moore

Plasma concentrations of somatomedin C/insulinlike growth factor I (SMC/IGF-I) are modulated by a variety of factors. Cardinal among these are growth hormone (GH) and nutritional status. In this chapter we review current understanding of the influence of GH and nutrition and SMC/IGF-I. We also present nomative data on SMC/IGF-I in over 800 normal infants and children and our assessment of the value of measuring this peptide in growth diagnosis and other conditions.

1. Materials and Methods

1.1. Normal Subjects

The study of SMC/IGF-I values in normal children includes 846 normal children (369 boys and 477 girls) ranging in age from 0.34 to 17.89 years. Criteria for inclusion in the study were supine length or standing height

Louis E. Underwood, Eric P. Smith, Judson J. Van Wyk, David R. Clemmons, and A. Joseph D'Ercole • Department of Pediatrics, University of North Carolina, School of Medicine, Chapel Hill, North Carolina 27514. ***M. R. Pandian*** • Nichols Institute, San Juan Capistrano, California 92675. ***Michael A. Preece*** • Department of Growth and Development, Institute of Child Health, University of London, London, WCIN 1EH England. ***Wayne V. Moore*** • Section of Pediatric Endocrinology, Department of Pediatrics, University of Kansas Medical Center, Kansas City, Kansas 66103.

between the 3rd and the 97th centiles on the growth charts of Tanner and Whitehouse[1] and no acute or known chronic illness. Informed consent for participation in the study was obtained prior to sampling. More than half the samples were obtained during well-child visits to an outpatient clinic in Kansas City, Kansas. In approximately 200 children between 12 and 14 years, samples were collected during the course of studies directed at assessing nutritional intake in healthy girls (Greensboro, NC) and studies of the biochemical correlates of pubertal development (London, England). Height and weight measurements were made by trained health workers during the same encounter in which blood was sampled. Breast, pubic hair, and genital development were determined by health workers according to the classification of Tanner.[2] The only exceptions to this were 140 teen-age girls who were asked to perform a self-assessment of breast and pubic hair development by comparing their developmental stage with photographs that were representative of each stage. Blood was collected in tubes containing sodium EDTA, promptly chilled, and centrifuged to separate formed elements, and the plasma was stored at −20°C prior to being shipped on dry ice to the laboratory.

1.2. Hypopituitary Children

The study also includes 53 children who were proven to be GH deficient by demonstrating subnormal growth rates for age and failure to raise serum GH concentrations above 4 ng/ml in response to two or more provocative stimuli. Most of these children also had undergone other pituitary function tests, and many had evidence of deficient secretion of other pituitary hormones. Those children who had received GH therapy previously were removed from treatment at least 2 weeks prior to sampling.

1.3. Laboratory Methods

The SMC/IGF-I content of the plasma samples from normal and hypopituitary children were assayed at Nichols Institute (San Juan Capistrano, CA), using the method detailed by Copeland *et al.*[3] The only modifications of the assay from that reported previously are that the SMC antibody is incubated with sample or standard for 1 hr prior to addition of tracer, and after addition of tracer the assay is incubated overnight. The standard was a pool of normal adult plasma (Ortho, lot #1778-5) in which the potency was designated to be 1.66 IU/ml.

2. Results and Discussion

2.1. Comments on Plasma Radioimmunoassay Methods

There are pitfalls in radioimmunoassay (RIA) measurements of SMC/IGF-I. These occur because of the involvement of the SM binding protein in plasma and because of the strong ionic properties of the peptide itself. We have chosen an assay system that uses polystyrene tubes and protamine-containing buffer.[3] Compared to the other systems we have tested, this method offers the advantage of lower nonspecific binding of tracer, better linearity and parallelity between plasma unknown and purified SMC/IGF-I, lower interassay variability, and the ability to measure levels in unextracted EDTA plasma.[4] It has the disadvantage, however, that about 60–70% of the activity in plasma is "covered up" and cannot be measured.[4] Despite this, plasma values obtained with this assay correlate extremely well with results obtained using other methods, and in extensive comparisons we have found that pretreatment of samples before assay, or use of different assay systems, does not change or improve the clinical interpretation of results (in preparation).

Some investigators have tried to minimize the effects of the binding protein by incubating plasma samples at acid pH before assay.[5,6] Others have reported the extraction of serum with acid ethanol[7] or have separated small-molecular-weight somatomedin from plasma binding proteins by chromatography in G50 Sephadex at acid pH.[8] The latter method has the disadvantage that the SMC/IGF-I may not be completely stripped from the binding protein. The most important factor determining whether these procedures are necessary appears to be the affinity between antibody and SMC/IGF-I. When the affinity of the antibody is sufficiently greater than the affinity of SMC/IGF-I for its binding protein, plasma sample preparatory procedures are not mandatory.

2.2. *Plasma SMC/IGF-I in Normal Infants and Children*

2.2.1. *Newborn and Early Childhood*

Using a variety of assay methods, the SM's have been reported to be low at birth and during the first few years of life.[9–12] These low values have been confirmed using the RIA for SMC/IGF-I,[13–15] and correlations have been drawn with various parameters of growth. In one study of cord sera from 145 full-term newborns,[15] SMC/IGF-I correlated with birth weight, birth length, and placental weight, and significantly lower levels were observed in small-for-gestational-age infants (<2500 g) than in appropriately sized infants. In another study, SMA also was found to correlate with birth weight.[11] Using RIA's specific for SMC/IGF-I and IGF-II and a radioreceptor assay for total SM, Bennett *et al.* confirmed that these peptides are relatively low in newborns and correlate with birth weight.[16]

SMC/IGF-I values during early childhood are also low.[13,14,17] In a study of 24 healthy infants from 6 weeks of age onward, Kaplowitz *et al.* observed that there was little change occurring during the first 15 months of life, and SMC/IGF-I did not correlate with either height centile or growth rate.[17]

The implication of these low levels is unclear in light of the rapid growth of infants. The question to be asked is whether SMC/IGF-I stimulates fetal or early childhood growth. It is possible that a multiplication-stimulating activity (MSA)-like peptide is more relevant to fetal growth than SMC/IGF-I, since MSA

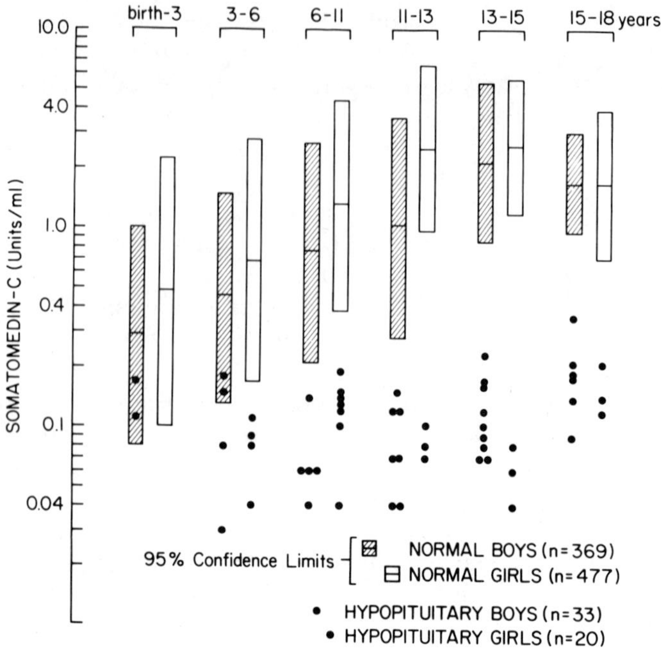

Figure 1 Plasma somatomedin C concentrations in 846 normal children between 0.34 and 17.89 years of age and in 53 children with growth hormone deficiency. Bars showing group values of normal children are plotted on a log scale as means and 95% confidence limits. Values for individual hypopituitary children are shown. (Reproduced with permission from J. D. Wilson and D. W. Foster (eds.), *Williams Textbook of Endocrinology*, 7th ed., W. B. Saunders, Philadelphia, copyright © 1985.)

values are high in fetal rats.[18] In humans, however, IGF-II (the human equivalent of rat MSA) is low.[16] Another possibility is that fetal tissues are especially sensitive to SM's.[10] A mechanism for such increased sensitivity is suggested by the finding that SM receptors on newborn circulating monocytes are increased.[19] It also is possible that SM in fetal serum is not a meaningful measure of SM activity in tissues, and that paracrine and autocrine growth regulation may dominate endocrine control mechanisms.[15]

2.2.2. *Values through Childhood and Puberty*

Plasma concentrations of SMC/IGF-I rise progressively from 1 year of age until puberty (Fig. 1). During this period the group mean values increase 2 $\frac{1}{2}$-fold, reaching 1 unit/ml at the onset of puberty. There is an additional dramatic increase during puberty so that at midpuberty the mean value is between 2.5 and 3 U/ml, with girls having a slightly higher value than boys. This late childhood increment in SMC/IGF-I may be linked to the events of puberty, since girls achieve peak values between 11 and 13 years, while boys have their

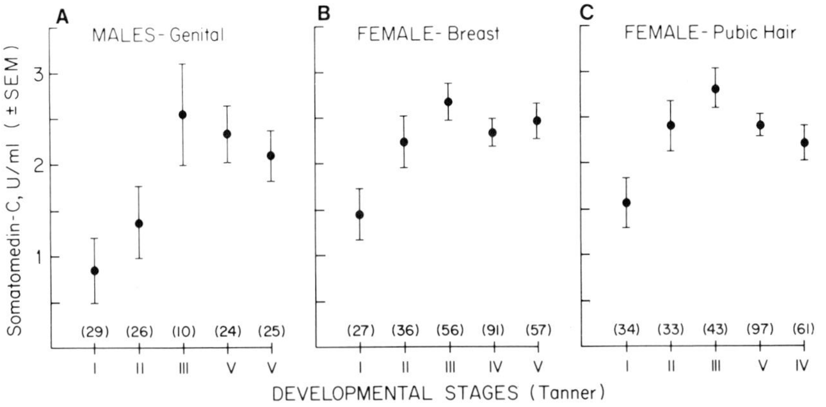

Figure 2 Plasma somatomedin C concentrations in normal children according to stages of pubertal development. The numbers of children in each group are given in parentheses. The girls included in the developmental stage I groups were 9 years of age or more. The boys included in the genital stage I group were 10 years of age or more. Statistically significant differences by Student's t test are as follows: (A) For male genital stages, stage I is significantly different from III, IV, and V ($p < 0.05$ to $p < 0.01$) (for male pubic hair, there were no significant differences, perhaps because the number within each group was 15 or less); (B) for female breast development, stage I was significantly different from III, IV, and V ($p < 0.05$ to $p < 0.01$); (C) for pubic hair in females, stage I is significantly different from stages III and IV ($p < 0.05$).

highest values between 13 and 15 years. When the pubertal rise in SMC/IGF-I is plotted as a function of developmental status rather than chronological age (Fig. 2), it is apparent that maximal values are achieved in the midpubertal groups (Tanner stage III pubic hair, breasts, and genital development). These data were collected cross-sectionally, and longitudinal studies are needed for confirmation. However, the results agree with those of other cross-sectional studies[13,20,21] and with small-scale longitudinal studies on normal children[22] and on boys with constitutional growth delay.[11] Copeland *et al.* have shown that baboons, primates that also have a pubertal growth spurt, also increase their SMC/IGF-I markedly during puberty.[23]

That GH causes the pubertal SMC/IGF-I rise is suggested by the observation that SMC/IGF-I responses to GH injections correlate with bone age[24,25] and that GH secretion is thought to be increased during puberty.[26,27] It seems less likely that androgens directly stimulate SM production since their administration along with GH to hypopituitary boys does not improve the SMC/IGF-I response to GH over that observed with GH alone.[28]

The possible role of estrogens in the pubertal rise in SMC/IGF-I is uncertain. Low doses of ethinyl estradiol administered to girls with Turner's syndrome stimulate linear growth without increasing SMC/IGF-I levels[29]; intermediate doses increase SMC/IGF-I, but do not stimulate growth[29]; and high-dose estrogen therapy arrests growth in tall girls and decreases the SMC levels, probably by direct inhibition of SM production.[30]

2.3. Value of SMC/IGF-I in Growth Diagnosis

In children with unequivocal GH deficiency, plasma SMC/IGF-I concentrations are consistently below 0.25 U/ml (Fig. 1). A normal value in a short child is strong evidence against the diagnosis of hypopituitarism, particularly when the patient is over 5–6 years of age where the normal range does not overlap with hypopituitary.(31) The SMC/IGF-I may be normal in the face of GH deficiency in some patients who have undergone surgery for craniopharyngioma. These children, who may have excessive or normal linear growth, frequently exhibit hyperphagia and excessive weight gain and sometimes have elevated prolactin levels.(32) Also, hypopituitary adults with markedly elevated prolactin levels secondary to pituitary tumors may have normal SMC/IGF-I levels despite GH deficiency.(33)

While a normal SMC/IGF-I value suggests that GH deficiency is not present, a low value in a growth-retarded child is not diagnostic of hypopituitarism. Of the factors that are associated with reduced SMC/IGF-I, undernutrition is perhaps the most prevalent. We believe that suboptimal nutrition may be the mechanism whereby SMC/IGF-I is reduced in many growth-retarding, systemic illnesses. Values are also reduced in hypothyroidism.(34,35) Another factor that complicates interpretation in older children is that it is not known whether the SMC/IGF-I level should be interpreted with respect to normative data for chronological age or for "developmental age" (skeletal age or stage of puberty). Our data suggest that at least in normal children, it may be better to use developmental age.

Reports differ in their estimates of the value of the SMC test in determining whether short children are GH deficient. In one study of 41 short children,(36) plasma SMC/IGF-I values confirmed or supported the clinical diagnosis and GH tests in 30 patients. In seven of the remaining 11 cases the SMC/IGF-I value was low and growth rate was slow despite apparently normal GH secretion. This suggests that the growth failure was due to some mechanism other than GH deficiency. In the other four, GH was low while SMC/IGF-I concentrations were normal. Explanation of these discordant results must await further clarification of the cause of the growth failure in these children.(37)

Moore *et al.*(38) measured SMC/IGF-I in 143 short children and performed GH-provocative tests on 78 patients who had SMC/IGF-I values below 0.5 IU/ml. They found that half of these patients (values below 0.5 units/ml) were GH deficient and concluded that SMC screening for GH deficiency is as efficient and more convenient than the exercise-induced GH screening test. On the other hand, Dean *et al.*(39) concluded that SMC/IGF-I measurements cannot be used to diagnose GH deficiency, since approximately 15% of their patients with presumed hypopituitarism had serum values within the normal range. Other recent reviews also have considered the value of SM measurements in growth diagnosis.(40,41)

Since concentrations are GH dependent, one might assume that the short-term plasma SMC/IGF-I responses to GH could be used to predict the growth responses to long-term therapy. While one study(42) suggests that this is indeed the case, our own experience and that of others do not bear this out. While chronic GH therapy raises plasma SMC/IGF-I in most hypopituitary subjects, we and Rosenfeld *et al.*[25] have observed low values in the face of an acceptable growth rate. Others have reported that SMC/IGF-I responses to GH in non-GH-deficient children do not correlate with growth during GH therapy.[43,44]

More experience is needed before the SMC/IGF-I test assumes its proper place in the assessment of short children. We believe that this test is but one more useful tool in the assessment of growth disorders, and that it must be used in conjunction with careful clinical assessment of the patient, as well as other appropriate laboratory studies.

2.4. Plasma SMC/IGF-I in Acromegaly

SMC/IGF-I concentrations are uniformly elevated in the plasma of patients with acromegaly and in children with gigantism due to pituitary GH excess. In our initial study of 57 acromegalic adults, the mean value was 10 times higher than in control subjects.[45] The SMC/IGF-I correlated with indices of clinical severity such as heel pad thickness (r = 0.73), fasting glucose (r = 0.74), and 1-hr postprandial glucose (r = 0.77). These correlations were substantially better than those observed between GH and indices of clinical severity. Furthermore, in patients with clinically active disease but low or marginal GH values, the SMC/IGF-I was consistently elevated. Since our initial report we have continued to use the test for diagnosis of acromegaly. In over 150 patients with clinically active disease, we have consistently found the plasma SMC/IGF-I to be elevated. Also, in four children with GH excess, heights above the 97th centile, rapid growth, and soft tissue signs of GH excess, the SMC/IGF-I values were strikingly elevated. Errors in interpretation of high values may occur with normal pubertal children, who may have values as high as 5 units/ml, and in pregnant women.[46]

Stonesifer *et al.*[47] reported that SMC/IGF-I did not correlate with clinical improvement any better than GH in 15 acromegalic adults treated 3–15 years earlier. This study is seriously weakened by the fact that no objective index of disease activity was monitored over time, and no assessment of symptoms was made prior to therapy. Rather, the patient's recall of symptoms was the primary source of clinical information.

We also have noted that although SMC/IGF-I levels fall dramatically in acromegalic patients who improve with treatment, concentrations may not fall to the normal range. We believe, therefore, that while a single SMC/IGF-I measurement provides the best available indicator of disease activity, it cannot be used as an index of how much treatment-related improvement has already occurred. It also should not be used as the sole means for determining whether additional therapy is needed. Serial measurements of SMC/IGF-I, however, are useful in monitoring the course of acromegaly and the response to therapy.[45,48]

2.5. Relationship between SMC/IGF-I and Nutritional Status

Many workers have shown that nutritional status and short-term dietary intake are important determinants of the SM activity measured in bioassays.

Using the RIA for SMC/IGF-I, we have shown that fasting for 10 days markedly reduces plasma levels, and that this reduction in SMC/IGF-I correlates with changes in nitrogen balance.[49] In another study, the dietary components responsible for regulation of SMC/IGF-I were assessed by fasting normal adult volunteers for 5 days and then refeeding with three diets of different composition for 5 days each.[50] Refeeding with the normal diet restored SMC/IGF-I nearly to normal. Refeeding with a diet that had approximately $\frac{1}{3}$ normal energy and $\frac{1}{3}$ normal protein resulted in a further fall in SMC/IGF-I. Refeeding with a diet of normal energy, but only $\frac{1}{3}$ normal protein, produced an intermediate response. From this study we conclude that both protein and energy intake are regulators of SMC/IGF-I in normal adult humans, and that energy intake may be of greater importance.

The importance of nutrient insufficiency in inhibiting SMC/IGF-I secretion has been confirmed by Merimee *et al.*,[51] who showed that fasted subjects are unable to raise their plasma SMC/IGF-I following the injection of GH. This observation mirrors, to some degree, the situation in chronic malnutrition, where GH is elevated and SMC/IGF-I is low.

Studies are not yet adequate to know the value of SMC/IGF-I in disorders of altered nutrition. The test shows promise, however, of being useful for assessing overall nutritional condition and the status of nitrogen balance. It seems probable that isolated measurements of SMC/IGF-I will have little value in assessing nutrition, because levels are reduced in a variety of inflammatory diseases, liver disease, poorly controlled insulin-dependent diabetes, and disorders associated with undernutrition. Since the mechanisms by which such diseases lower SMC/IGF-I are not certain, it may be impossible to distinguish between low SMC/IGF-I due to nutritional insufficiency and depression of peptide synthesis by an underlying disease. We believe that measurements of SMC/IGF-I during the course of therapy might be helpful in assessing adequacy of nutrient intake and progress of therapy. For example, in patients undergoing bowel surgery, decisions concerning adequacy of protein and calorie supplementation might be facilitated by knowing the presurgery SMC/IGF-I value and the results of serial measurements after surgery. Obviously, much work remains to be done before the test can be applied to such situations.

3. Conclusions

SMC/IGF-I is a growth factor whose role in promoting growth remains as unsettled as the mechanisms of its production and regulation. It is clearly GH dependent, but because of the many other factors impinging on its production, we are just beginning to appreciate the clinical information it provides. Future studies on the role of sex steroids, thyroid hormone, nutritional status, and underlying disease should further enhance its usefulness in clinical medicine.

ACKNOWLEDGMENTS. This work was supported by National Institute of Health Research Grants AM-01022 and HD-08299, NIH Training Grant AM 07129.

The Clinical Research Unit of the North Carolina Memorial Hospital is supported by grant RR-0046 from the General Clinical Research Centers Program of the Division of Research Resources, NIH. Judson J. Van Wyk is a recipient of a USPHS Research Career Award 5K06A AM 14115. A. Joseph D'Ercole is a USPHS Research Career Development Awardee (HD 00435). David R. Clemmons is a Jefferson-Pilot Fellow in Academic Medicine, University of North Carolina.

The authors express their appreciation to Joan Mueller, Esther Carlton, and Linda Jones of Nichols Institute, to Terry Bazzare (University of North Carolina, Greensboro) for providing samples on normal girls, and to Mary G. Murphy and Evyonne Bruton for technical assistance and Christine Silva for help in preparing this manuscript.

References

1. J. M. Tanner and R. H. Whitehouse, Clinical longitudinal standards for height, weight, height velocity, weight velocity and stages of puberty, *Arch. Dis. Child.* **51:**170, 1976.
2. J. M. Tanner, *Growth at Adolescence*, 2nd ed., Blackwell Scientific Publications, Oxford, 1962.
3. K. C. Copeland, L. E. Underwood, and J. J. Van Wyk, Induction of immunoreactive somatomedin-C in human serum by growth hormone: Dose response relationships and effect on chromatographic profiles, *J. Clin. Endocrinol. Metab.* **50:**690, 1980.
4. P. G. Chatelain, J. J. Van Wyk, K. C. Copeland, S. L. Blethen, and L. E. Underwood, Effect of *in vitro* action of serum protease or exposure to acid on measurable immunoreactive somatomedin-C in serum, *J. Clin. Endocrinol. Metab.* **56:**376, 1983.
5. R. M. Bala and B. Bhaumick, Radioimmunoassay of a basic somatomedin: Comparison of various assay techniques and somatomedin levels in various sera, *J. Clin. Endocrinol. Metab.* **49:**770, 1979.
6. L. E. Underwood, A. J. D'Ercole, K. C. Copeland, J. J. Van Wyk, T. Hurley, and S. Handwerger, Development of a heterologous radioimmunoassay for somatomedin-C in sheep blood, *J. Endocrinol.* **93:**31, 1982.
7. W. H. Daughaday, I. K. Mariz, and S. L. Blethen, Inhibition of access of bound somatomedin to membrane receptors and immunobinding sites: A comparison of radioreceptor and radioimmunoassay of somatomedin in native and acid-ethanol-extracted serum, *J. Clin. Endocrinol. Metab.* **51:**781, 1980.
8. J. Zapf, H. Walter, and E. R. Froesch, Radioimmunological determination of insulinlike growth factors I and II in normal subjects and in patients with growth disorders and extrapancreatic tumor hypoglycemia, *J. Clin. Invest.* **68:**1321, 1981.
9. T. P. Foley, R. DePhilip, A. Perricelli, and A. Miller, Low somatomedin activity in cord serum from infants with intrauterine growth retardation, *J. Pediatr.* **96:**605, 1980.
10. I. K. Ashton and J. Vesey, Somatomedin activity in human cord plasma and relationship to birth size, insulin, and growth hormone and prolactin, *Early Hum. Dev.* **2:**115, 1978.
11. K. Hall, G. Enberg, M. Ritźen, H. Svan, L. Fryklund, and K. Takano, Somatomedin A levels in serum from healthy children and from children with growth hormone deficiency or delayed puberty, *Acta Endocrinol.* **94:**155, 1980.
12. P. D. Gluckman and M. W. Brinsmead, Somatomedin in cord blood: Relationship to gestational age and birth size, *J. Clin. Endocrinol. Metab.* **43:**1378, 1976.
13. R. M. Bala, J. Lopatka, A. Leung, E. McCoy, and R. G. McArthur, Serum immunoreactive somatomedin levels in normal adults, pregnant women at term, children at various ages, and children with constitutionally delayed growth, *J. Clin. Endocrinol. Metab.* **52:**508, 1981.
14. R. W. Furlanetto, L. E. Underwood, J. J. Van Wyk, and A. J. D'Ercole, Estimation of

somatomedin-C levels in normals and patients with pituitary disease by radioimmunoassay, *J. Clin. Invest.* **60:**648, 1977.
15. A. J. D'Ercole and L. E. Underwood, Growth factors in fetal growth and development, in: *Fetal Endocrinology* (M. J. Novy and J. A. Resko, eds.), Academic Press, Inc., New York, 1981, pp. 155–182.
16. A. Bennett, D. M. Wilson, F. Liu, R. Nagashima, R. G. Rosenfeld, and R. L. Hintz, Levels of insulin-like growth factors I and II in human cord blood, *J. Clin. Endocrinol. Metab.* **57:**609, 1983.
17. P. B. Kaplowitz, A. J. D'Ercole, J. J. Van Wyk, and L. E. Underwood, Plasma somatomedin-C during the first year of life, *J. Pediatr.* **100:**932, 1982.
18. A. C. Moses, S. P. Nissley, P. A. Short, M. M. Rechler, R. M. White, A. B. Knight, and O. Z. Higa, Increased levels of multiplication-stimulating activity, an insulin-like growth factor, in fetal rat serum, *Proc. Natl. Acad. Sci. USA* **77:**3649, 1980.
19. R. Rosenfeld, A. V. Thorsson, and R. L. Hintz, Increased somatomedin receptor sites in newborn circulating mononuclear cells, *J. Clin. Endocrinol. Metab.* **48:**456, 1979.
20. S. L. Blethen, D. T. Sargent, M. G. Whitlow, and J. V. Santiago, Effect of pubertal stage and recent blood glucose control on plasma somatomedin-C in children with insulin-dependent diabetes mellitus, *Diabetes* **30:**868, 1981.
21. A. M. Luna, D. M. Wilson, C. J. Wibbelsman, R. C. Brown, R. J. Nagashima, R. L. Hintz, and R. G. Rosenfeld, Somatomedins in adolescence: A cross-sectional study of the effect of puberty on plasma insulin-like growth factor I and II levels, *J. Clin. Endocrinol. Metab.* **57:**268, 1983.
22. R. L. Rosenfield, R. Furlanetto, and D. Bock, Relationship of somatomedin-C concentrations to pubertal changes. *J. Pediatr.* **103:**723, 1983.
23. K. C. Copeland, T. J. Kuehl, and V. D. Castracane, Pubertal endocrinology of the baboon: Elevated somatomedin-C/Insulin-like growth factor I at puberty. *J. Clin. Endocrinol. Metab.* **55:**1198, 1982.
24. S. L. Blethen, W. H. Daughaday, and V. V. Weldon, Kinetics of the somatomedin-C/insulin-like growth factor-I: Response to exogenous growth hormone in GH-dependent children, *J. Clin. Endocrinol. Metab.* **54:**986, 1982.
25. R. G. Rosenfeld, S. F. Kemp, and R. L. Hintz, Constancy of somatomedin response to growth hormone treatment of hypopituitary dwarfism, and lack of correlation with growth rate, *J. Clin. Endocrinol. Metab.* **53:**611, 1981.
26. R. Illig, and H. Bucher, Testosterone priming of growth hormone release, evaluation of growth hormone secretion, *Pediatr. Adolesc. Endocrinol.* **12:**75, 1983.
27. J. D. Miller, G. S. Tannenbaum, E. Colle, and H. J. Guyda, Daytime pulsatile growth hormone secretion during childhood and adolescence, *J. Clin. Endocrinol. Metab.* **55:**989, 1982.
28. W. H. Craft and L. E. Underwood, Effect of androgens on plasma somatomedin-C responses to growth hormone, *Clin. Endocrinol.* **201:**549, 1984.
29. J. L. Ross, F. G. Cassorla, M. C. Skerda, I. M. Valk, D. L. Loriaux, and G. B. Cutler, A preliminary study of the effect of estrogen dose on growth in Turner's syndrome, *N. Engl. J. Med.* **309:**1104, 1983.
30. R. L. Rosenfield, Toward optimal estrogen replacement therapy, *N. Engl. J. Med.* **309:**1120, 1983.
31. L. E. Underwood, A. J. D'Ercole, and J. J. Van Wyk, Somatomedin-C and the assessment of growth, *Pediatr. Clin. North Am.* **27:**771, 1980.
32. H. Bucher, J. Zapf, T. Torresani, A. Prader, E. R. Froesch, and R. Illig, Insulin-like growth factors I and II, prolactin, and insulin in 19 growth hormone-deficient children with excessive, normal or decreased longitudinal growth after operation for craniopharyngioma, *N. Engl. J. Med.* **309:**1142, 1983.
33. D. R. Clemmons, L. E. Underwood, E. C. Ridgway, B. Kliman, and J. J. Van Wyk, Hyperprolactinemia is associated with increased immunoreactive somatomedin-C in hypopituitarism, *J. Clin. Endocrinol. Metab.* **52:**731, 1981.
34. B. Draznin, P. J. Burstein, U. E. Heinrich, C. J. Johnson, C. A. Emler, and D. S. Schalch, Insulin-like growth factor and its carrier protein in hypopituitary and hypothyroid children and adults, *Clin. Endocrinol.* **12:**137, 1980.

35. S. D. Chernausek, L. E. Underwood, R. D. Utiger, and J. J. Van Wyk, Growth hormone secretion and plasma somatomedin-C in primary hypothyroidism, *Clin. Endocrinol.* **19:**337, 1983.
36. E. O. Reiter and R. D. Lovinger, The use of a commercially available somatomedin-C radioimmunoassay in patients with disorders of growth, *J. Pediatr.* **99:**720, 1981.
37. A. J. D'Ercole, Commentary: Use of the RIA for somatomedin-C estimation, *J. Pediatr.* **99:**735, 1981.
38. D. C. Moore, H. A. Rogelio, E. K. Smith, and V. C. Kelly, Plasma somatomedin-C as a screening test for growth hormone deficiency in children and adolescents, *Hormone Res.* **16:**49, 1982.
39. H. J. Dean, J. G. Kellet, R. M. Bala, H. J. Guyda, B. Bhaumick, B. I. Posner, and H. G. Friesen, The effect of growth hormone treatment on somatomedin levels in growth hormone-deficient children, *J. Clin. Endocrinol. Metab.* **55:**1167, 1982.
40. U. Widmer, J. Zapf, and E. R. Froesch, Somatomedins and hGH activity, evaluation of growth hormone secretion, *Pediatr. Adolesc. Endocr.* **12:**133, 1983.
41. K. W. Kastrup, Can somatomedin stimulation by growth hormone be used as a test? Evaluation of growth hormone secretion, *Pediatr. Adolesc.* **12:**147, 1983.
42. D. Rudman, S. D. Moffitt, P. M. Fernkoff, W. J. McKenzie, J. M. Kenny, and R. P. Bain, The relationship between growth velocity and serum somatomedin-C concentrations. *J. Clin Endocrinol. Metab.* **52:**622, 1981.
43. G. Van Vleit, D. M. Styne, S. L. Kaplan, and M. M. Grumbach, Growth hormone treatment of short stature. *N. Engl. J. Med.* **309:**1016, 1983.
44. L. P. Plotnick, Q. L. Van Meter, and A. A. Kowarski, Human growth hormone treatment of children with growth failure and normal growth hormone levels by immunoassay: Lack of correlation with somatomedin generation, *Pediatrics* **71:**324, 1983.
45. D. R. Clemmons, J. J. Van Wyk, E. C. Ridgway, B. Kliman, R. N. Kjellberg, and L. E. Underwood, Evaluation of acromegaly by radioimmunoassay of somatomedin-C, *N. Engl. J. Med.* **301:**1138, 1979.
46. R. W. Furlanetto, L. E. Underwood, J. J. Van Wyk, and S. Handwerger, Serum immunoreactive somatomedin-C is elevated late in pregnancy. *J. Clin. Endocrinol. Metab.* **47:**695, 1978.
47. L. D. Stonesifer, R. M. Jordan, and P. O. Kohler, Somatomedin-C in treated acromegaly: Poor correlation with growth hormone and clinical response. *J. Clin. Endocrinol. Metab.* **53:**931, 1981.
48. J. A. H. Wass, D. R. Clemmons, L. E. Underwood, I. Barrow, G. M. Besser, and J. J. Van Wyk, Changes in circulating somatomedin-C levels in bromocriptine-treated acromegaly, *Clin. Endocrinol.* **17:**369, 1982.
49. D. R. Clemmons, A. Klibanski, L. E. Underwood, J. McArthur, E. C. Ridgway, I. Z. Beitins, and J. J. Van Wyk, Reduction of immunoreactive somatomedin-C during fasting in humans, *J. Clin. Endocrinol. Metab.* **53:**1247, 1981.
50. W. L. Isley, L. E. Underwood, and D. R. Clemmons, Dietary components that regulate serum somatomedin-C concentrations in humans, *J. Clin. Invest.* **71:**175, 1983.
51. T. J. Merimee, J. Zapf, and E. R. Froesch, Insulin-like growth factors in the fed and fasted states, *J. Clin. Endocrinol. Metab.* **55:**999, 1982.

48

Somatomedin/Insulinlike Growth Factor in the Human Fetus

S. Peter Nissley, Lynne A. Gaynes, and Robert M. White

1. Introduction

The hormonal regulation of human fetal growth is poorly understood. However, considerable evidence supports a role for insulin. Infants with pancreatic agenesis or transient diabetes mellitus have decreased birth lengths whereas fetal hyperinsulinemia accompanying maternal diabetes mellitus results in overgrowth.[1] Are the IGF's/SM's also important for fetal growth? What are the relative roles of IGF-I and IGF-II? Is there a distinct embryonic/fetal IGF? In this chapter we will briefly summarize the indirect evidence that points to a role for the IGF's in human fetal growth. We will not review data on IGF's in the maternal circulation or address the question of regulation of IGF levels in the mother and fetus. More general reviews of the regulation of fetal growth in other species as well as in humans have been provided by others.[2–6]

2. Evidence Supporting a Role for IGF's in Fetal Growth

2.1. Receptors for IGF's on Fetal Membranes and Cells

As discussed by Czech *et al.* (this volume, chapter 41), there are two classes of IGF receptors. The type I receptor generally binds IGF-I more avidly than

Abbreviations used: IGF, insulinlike growth factor; SM, somatomedin; ILAs, insulin-like activity; EGF, epidermal growth factor; MSA, multiplication-stimulating activity. We have used IGF/SM or simply IGF to refer to the entire family of insulinlike growth factors and somatomedins. See Humbel's paper in this volume (Chapter 38) for discussion of the terminology and relationship of the various IGF's and SM's.

S. Peter Nissley and Lynne A. Gaynes • Endocrine Section, Metabolism Branch, National Cancer Institute, National Institutes of Health, Bethesda, Maryland 20205. ***Robert M. White*** • Howard University Cancer Center, Departments of Oncology and Medicine, Howard University School of Medicine, Washington, D.C. 20059.

IGF-II and also recognizes insulin, whereas the type II receptor prefers IGF-II over IGF-I and does not bind insulin. The binding subunit of the type I receptor has a size similar to the subunit of the insulin receptor (mol.wt. = 130,000) in contrast to the much larger binding unit of the type II receptor (mol.wt. = 260,000). IGF receptors have been identified in membrane preparations from human placenta, liver, and brain, as well as on circulating mononuclear cells.

2.1.1. Placenta

An IGF receptor distinct from the insulin receptor was first identified by radiolabeled SMC binding to human placental membranes.[7] High concentrations of insulin completely inhibited the binding of SMC tracer, and later, affinity cross-linking experiments showed that the binding subunit had a mol.wt. = 130,000, identifying this receptor as a typical type I receptor.[8,9] Subsequently, Daughaday *et al.*[10] identified a binding site on human placental membranes which preferred IGF-II over IGF-I, and chemical cross-linking experiments of Massague and Czech[11] identified a type II receptor (mol.wt. = 260,000).

Deal *et al.*[12] compared the specific binding of IGF (ILAs), insulin, and EGF to membrane preparations from midterm (11–19 weeks) and term human placenta. Interestingly, radiolabeled IGF binding was higher than insulin and EGF tracer binding in midterm placental membranes. [^{125}I]IGF binding to midterm placenta was equal to the level of binding of IGF in term placenta. The IGF receptor in the early gestation placental membranes was not further characterized as a type I or type II receptor.

2.1.2. Liver

Sara *et al.*[13] recently examined the ontogeny of the specific binding of IGF-I and IGF-II to human liver membranes prepared from fetuses of <17 cm, 17–25 cm, 26–28 cm, and >28 cm. A high-affinity binding site for [^{125}I]IGF-II was found at all stages of development, and the concentration of the IGF-II receptor (mole/g) was also uniform throughout gestation. By contrast, the binding of [^{125}I]IGF-I was low, and insulin did not complete for the binding of [^{125}I]IGF-II, suggesting that the liver IGF binding site was a type II receptor.

2.1.3. Brain

Sara *et al.*[13] also examined the binding of IGF-I and IGF-II to human brain membranes from fetuses of different gestational ages. In contrast to the findings of liver, binding of [^{125}I]IGF-I was higher than binding of IGF-II at all gestational ages. With increasing gestational age there was a change in the binding characteristics of the receptor identified by [^{125}I]IGF-I binding (Fig. 1). For brain membranes from fetuses <17 cm, IGF-II was actually more effective than IGF-I in competing for binding of [^{125}I]IGF-I; for membranes

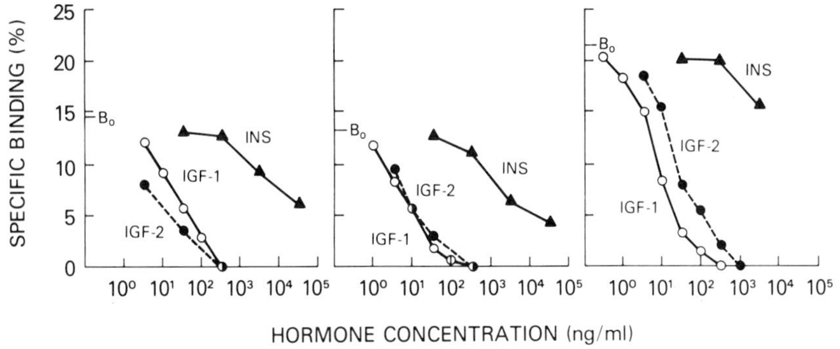

Figure 1 Specific binding of [^{125}I]IGF-I to human fetal brain plasma membranes (330 μg membrane protein/ml) at different developmental stages classified according to fetal length: <17 cm (left), 17–25 cm (middle), and >25 cm (right). Total specific binding is given as B_0. Displacement by different concentrations of unlabeled IGF-I (○——○), IGF-II (●---●), and insulin (▲——▲) is shown. Redrawn from Sara *et al.*,[13] reference with permission.

from 17- to 25-cm fetuses, IGF-I and IGF-II were equipotent; and IGF-I was more potent than IGF-II for binding to the membranes of fetuses >25 cm. Insulin competed for binding of IGF-I tracer at all gestational ages (Fig. 1) suggesting that this is a type I receptor. However, generally, IGF-I binds more avidly than IGF-II to the type I receptor. It is interesting that the one other exception to this rule is in cells of embryonic origin, chick embryo fibroblasts. In chick embryo fibroblasts, IGF-I and IGF-II are equipotent in competing for [^{125}I]IGF-I binding.[14] Kasuga *et al.*[15] have demonstrated by chemical cross-linking experiments that structurally the chick embryo fibroblast IGF receptor is a typical type I receptor. It would be of interest to perform chemical cross-linking of [^{125}I]IGF-I to the human brain membrane receptor at different gestational ages to see whether the change in binding characteristics is accompanied by a structural change.

2.1.4. *Mononuclear Cells from Cord Blood*

Rosenfeld *et al.*[16] reported that the binding of radiolabeled SMC to mononuclear cells from cord blood was 71% higher than the binding of [^{125}I]SMC to adult mononuclear cells (11.38 ± 0.74% versus 6.67 ± 0.51%). Scatchard analysis of the binding data showed that the increased binding was secondary to an increase in receptor number.

2.2. Biological Response to IGF's

There are relatively few reports in which the biological effects of highly purified IGF's have been examined in cells or tissues of human fetuses.

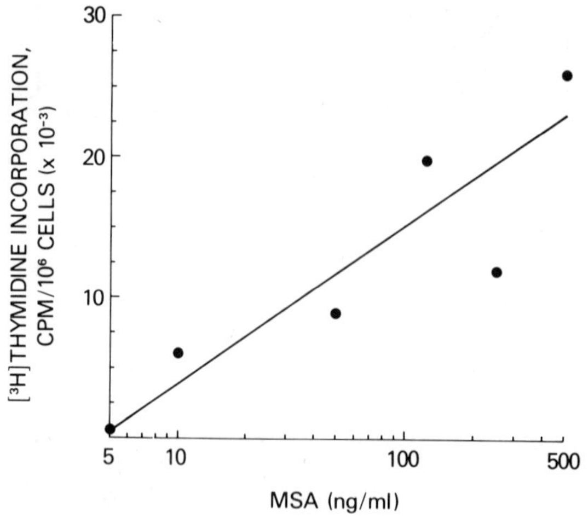

Figure 2 Effect of MSA on [^{3}H]thymidine incorporation into DNA in human fetal chondrocytes in the presence of 1.25% cord plasma. The data are plotted as stimulation (cpm) above the 1.25% plasma control (15,000 cpm/10^6 cells) versus MSA concentration (ng/ml) on a log scale. Adapted from Ashton and Pocock,[17] with permission.

2.2.1. *Fetal Chondrocytes*

Ashton and Pocock[17] measured the incorporation of [^{3}H]thymidine into the DNA of chondrocytes isolated from costal and articular cartilage of human fetuses of 18–22 weeks' gestation. When added to 1.25% plasma, MSA caused a significant increase in [^{3}H]thymidine incorporation with increasing dose (Fig. 2).

2.2.2. *Fibroblasts in Culture*

Weidman and Bala[18] demonstrated that a highly purified preparation of basic SM (probably identical to IGF-I/SMC) caused a small but significant increase in cell number of a strain of human fetal lung fibroblasts (WI 38). Similarly, MSA stimulated the multiplication of a different strain of human lung fibroblasts (HFL-1) (Table 1). The small response to both basic SM[18] and MSA (Table 1) compared to the response to 10% serum suggests that additional factors are required for a full mitogenic response, as has been shown for postnatal human skin fibroblasts.[19]

2.3. *Production of IGF's by Human Fetal Fibroblasts in Culture*

Atkinson *et al.*[20] demonstrated that cultured human fetal lung fibroblasts (WI 38) maintained in serum-free medium produced IGF's as measured by radioimmunoassay for basic SM (Fig. 3). Gel filtration under neutral conditions showed that the fibroblast IGF had a mol.wt. = 45,000, whereas gel filtration under acid conditions demonstrated a mol.wt. = 15,000 species, suggesting association of IGF with a binding protein under neutral conditions. Isoelectric focusing showed species with isoelectric points of 4.7 (mol.wt. = 29,000) and 6.1 (mol.wt. = 16,500).[21] The fibroblast IGF's also were active in a placental radioreceptor assay, stimulated $^{35}SO_4$ uptake in hypophysectomized rat carti-

Table 1 Effect of MSA on Multiplication of Human Fetal Lung Fibroblasts[a]

Addition	Cells/well (mean ± SD)
Medium	2.40 ± 0.05
MSA (100 ng/ml)	2.80 ± 0.13[b]
10% fetal calf serum	4.80 ± 0.58[b]

[a] Human fetal lung fibroblasts (HFL-1) were plated in 35-mm wells in Dulbecco's modified Eagles medium (DMEM) containing growth hormone and SMC-deficient human serum (1%). The following day the cells were counted in triplicate wells (2.37 ± 0.24) and medium changed to DMEM containing 0.25% BSA (medium), medium plus MSA (100 ng/ml), or DMEM containing 10% fetal calf serum. Sixty-six hours later the cells were counted again in triplicate wells.

[b] Different from medium control ($p < 0.005$).

lage, and stimulated multiplication of WI 38 cells. While the fibroblast IGF species were detected by a specific radioimmunoassay for basic SM (probably identical to IGF-I/SMC), there were several differences between the fibroblast IGF/SM and basic SM (IGF-I/SMC). The sizes of the fibroblast IGF's were

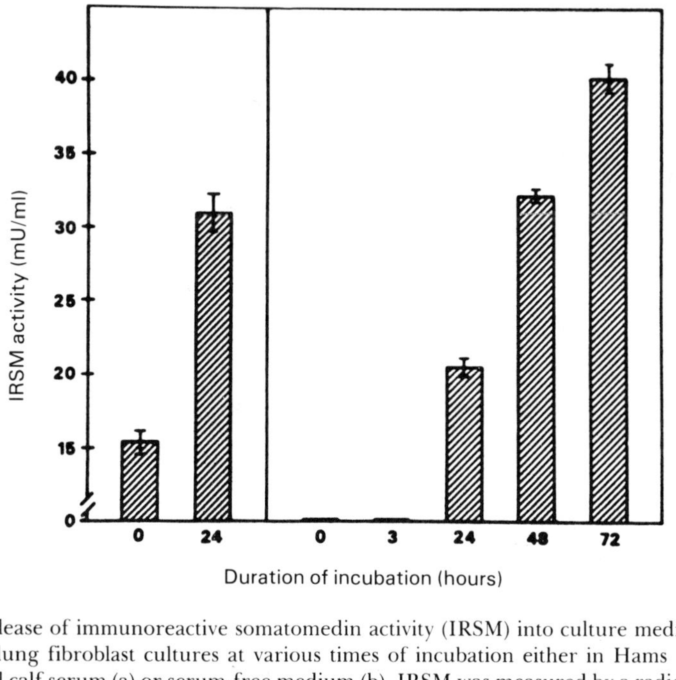

Figure 3 Release of immunoreactive somatomedin activity (IRSM) into culture medium of WI 38 human fetal lung fibroblast cultures at various times of incubation either in Hams F-12 medium plus 10% fetal calf serum (a) or serum-free medium (b). IRSM was measured by a radiommunoassay for basic somatomedin and expressed as mU/ml based on a reference serum standard. Reproduced from Atkison *et al.*[20] with permission.

larger than basic SM, displayed more acidic isoelectric points, and appeared to have higher intrinsic mitogenic activity for WI 38 fibroblasts.

We have also partially characterized IGF's in conditioned media from several strains of fibroblasts derived from skin and lung of early human embryos. Conditioned media were dialyzed against 2% acetic acid and then gel-filtered on Sephadex G-75 in 1 M acetic acid to dissociate and separate IGF's from binding proteins. IGF's were assayed in column fractions with a chick embryo fibroblast bioassay ([^{3}H]thymidine incorporation into DNA), a rat placental membrane radioreceptor assay, and specific radioimmunoassays for IGF-I/SMC and MSA. IGF binding proteins were detected in column fractions by ability to bind MSA tracer. Our preliminary findings indicate that in addition to immunoreactive SMC/IGF-I as described by Atkison *et al.*,[20] IGF-II-like activity, as measured by the rat placental radioreceptor assay, is also present in the fetal fibroblast conditioned media (Gaynes and Nissley, unpublished data).

2.4. Blood Levels of IGF's

2.4.1. Bioassays

Bioassays based on the measurement of the incorporation of $^{35}SO_4$ or [^{3}H]thymidine into cartilage from various sources have shown that cord blood values are lower than adult levels.[22–24] Using the porcine cartilage sulfation assay, Gluckman and Brinsmead[25] found a positive correlation between gestational age and SM levels.

2.4.2. Radioreceptor Assays

2.4.2a. Human Placental Membranes. The human placental membrane radioreceptor assay has usually been utilized with either radiolabeled SMC or SMA. In either case, the tracer is binding to the type I receptor and IGF-I/SMC is detected in preference to IGF-II. Nonetheless, this assay measures both IGF-I/SMC and IGF-II since IGF-II is 36–50% as potent as IGF-I.[10,26] Both D'Ercole *et al.*[27] and Hall *et al.*[28] have reported that SM levels in cord blood were lower than adult levels.

2.4.2b. Human Brain Membranes. As discussed in Section 2.1.3 (Fig. 1), Sara *et al.*[29] have developed a radioreceptor assay using human fetal brain membranes. IGF-I and IGF-II are equipotent in this assay, and insulin competes for binding. When this radioreceptor assay was used to make measurements in cord blood and early fetal samples (16–22 weeks) obtained by fetoscopy, high levels were found in the fetal samples (Fig. 4). In addition, the levels in cord blood were also twice as high as obtained by either the placental radioreceptor assay or SMA radioimmunoassay.[29] Sara and her colleagues proposed that the

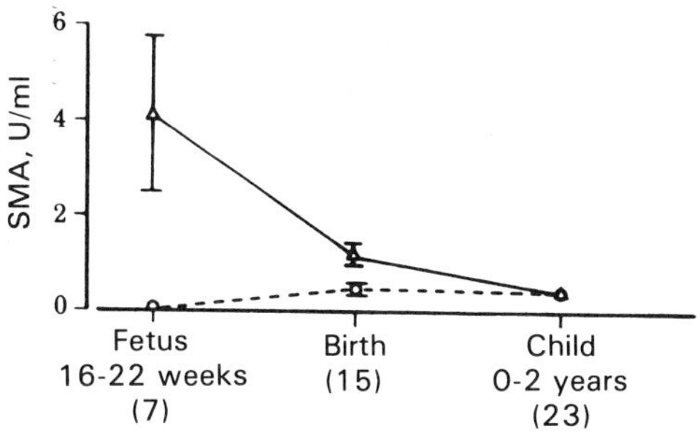

Figure 4 Levels of somatomedins throughout normal early development in humans determined by fetal brain radioreceptor assay (△——△) and SMA radioimmunoassay (○– – –○). The results are expressed as IU/ml based on a reference serum arbitrarily assigned a value of 1 IU/ml. IGF-I is 10 times more potent than and IGF-II 1/10th as potent as SMA in the SMA radioimmunoassay. Reproduced from Sara *et al.*,[29] with permission.

fetal brain membrane radioreceptor assay was able to detect a distinct embryonic IGF.

2.4.2c. Rat Placental Membranes. Daughaday *et al.*[10] reported that rat placental membranes have a preferential binding site for IGF-II. Chemical cross-linking experiments confirmed the presence of a typical type II receptor.[11] Using this radioreceptor assay, Daughaday *et al.*[30] measured levels in 39 cord sera. The mean concentration (1.55 ± 0.24 U/ml) was 2.5 times higher than the mean value (0.64 ± 0.04 U/ml) obtained for 32 adult sera. Recently Gluckman *et al.*[31] have utilized this same assay to measure IGF-II levels in cord blood of 177 normal infants. In contrast to the results of Daughaday *et al.*,[30] cord blood levels were not significantly higher than adult levels.

2.4.3. *Radioimmunoassays*

2.4.3a. IGF-I/SMC. IGF-I/SMC levels in full-term cord blood by radioimmunoassay have uniformly been reported to be less than adult levels.[28,31–35] Bennett *et al.*[35] have determined IGF-I levels in cord blood in normal term and preterm (22–37 weeks) infants (Fig. 5). There was a strong correlation between gestational age and IGF-I levels ($r = 0.63$).

2.4.3b. IGF-II. While there have been fewer reports of IGF-II levels by radioimmunoassay than for IGF-I/SMC, IGF-II concentrations in full-term cord blood have also been reported to be less than adult levels.[34,35] Bennett *et al.*[35] have measured IGF-II levels in cord blood of infants of various gestational ages (Fig. 6). There was a strong correlation with gestational age.

This pattern of IGF-II blood levels versus gestational age in human fetal blood is strikingly different than the developmental pattern we reported in the rat[36] using a radioimmunoassay for MSA, the rat homologue of human IGF-II. In the rat, MSA levels were high in fetal blood and declined postnatally to low levels, leading us to suggest that MSA is a fetal growth factor in the rat.

2.4.4. *IGF Binding Proteins in Fetal Blood*

D'Ercole *et al.*[37] reported that in the blood of seven of nine infants of 27 weeks or less, the circulating form of immunoreactive SMC had an apparent

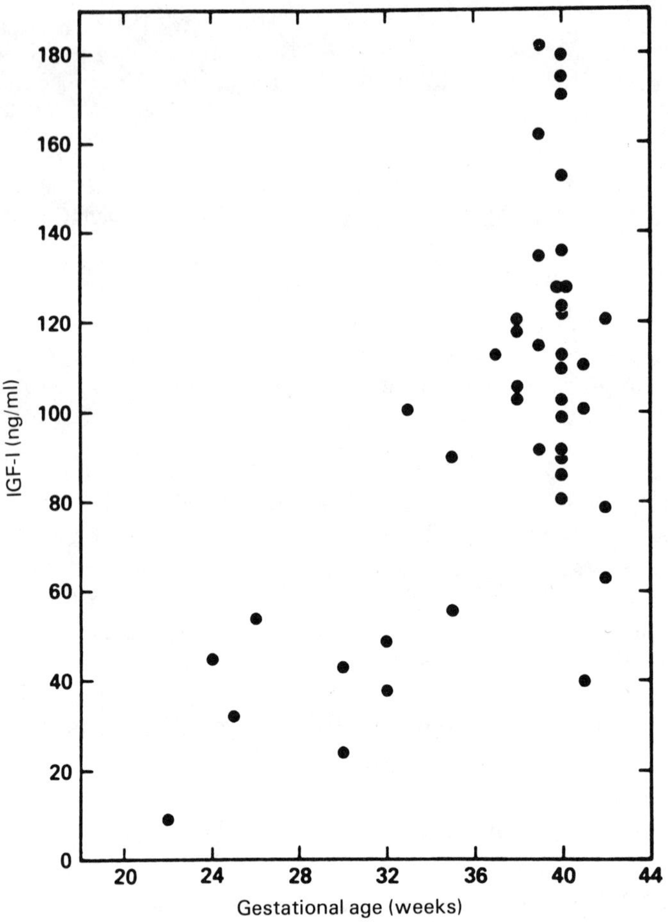

Figure 5 Plasma IGF-I levels by radioimmunoassay throughout gestation. Reproduced from Bennett *et al.*,[35] with permission.

mol.wt. of 40,000. This is in contrast to the findings for infants of 30 weeks or more or in postnatal blood, where most of the SMC was associated with a binding protein of mol.wt. 150,000.[37]

We examined the size of the IGF binding proteins in the serum of a 30-week gestation infant (birth weight = 940 g, one of twins). In agreement with the report of D'Ercole *et al.*,[37] when radiolabeled MSA was incubated with serum and then gel-filtered on Sephadex G-200, most of the bound tracer eluted approximately at the elution volume of albumin (Fig. 7A). The same result was obtained when MSA tracer binding was measured in individual fractions after gel filtration (Fig. 7A). Pools of Sephadex G-200 fractions were then gel-filtered on Sephadex G-50 in 1 M acetic acid to dissociate and separate IGF from binding proteins. The IGF generated by acid gel filtration was

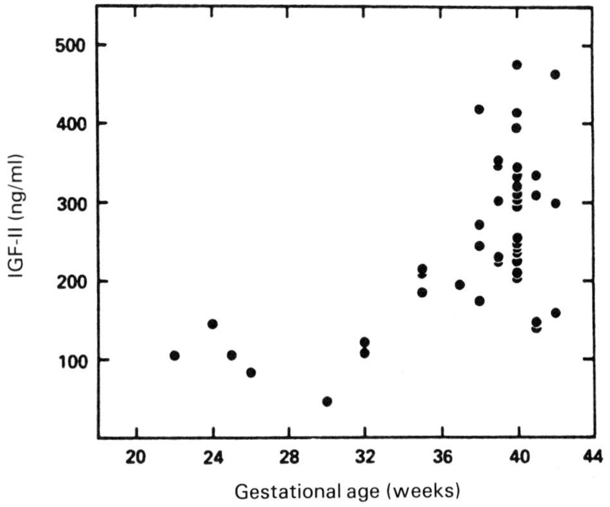

Figure 6 Plasma IGF-II levels by radioimmunoassay throughout gestation. Reproduced from Bennett *et al.*,[35] with permission.

measured by a competitive binding assay and was also found in the albumin region of the column (Fig. 7B). Binding capacity was determined by Scatchard analysis of MSA binding to stripped binding protein generated by the Sephadex G-50 acid gel filtration and was also found in the albumin region of the column (Fig. 7C). Importantly, no binding protein was found in the γ-globulin region of the column where the bulk of IGF binding protein is found in postnatal human serum.[38] Together with the results of D'Ercole *et al.*,[37] these findings indicate that IGF is found associated with a 40,000-mol.wt. binding protein in early to midgestation human serum. It is not clear whether this fetal binding protein is identical to a minor binding protein of similar size found in postnatal human serum.[38]

Recently Drop and Kortleve[39] have described a radioimmunoassay using antiserum raised against a 40,000-mol.wt. binding protein purified from midterm amniotic fluid. High concentrations of this binding protein were measured in midgestation cord blood with this radioimmunoassay suggesting that the 40,000-mol.wt. binding protein in midgestation human fetal blood may be identical or closely related to the binding protein found in amniotic fluid.

2.5. Clinical Correlations

2.5.1. Blood Levels versus Birth Weight

There have been a number of reports of a positive correlation between birth weight and IGF levels in normal infants by a variety of assays.[25,31,35,40] In addition, small-for-gestation infants have been found to have low levels of IGF by bioassay[41] and SMC radioimmunoassay.[5] With the exception of reports by Gluckman and Brinsmead[25] and D'Ercole and Underwood,[5] there has not been a positive correlation between IGF levels and birth length. Thus, the

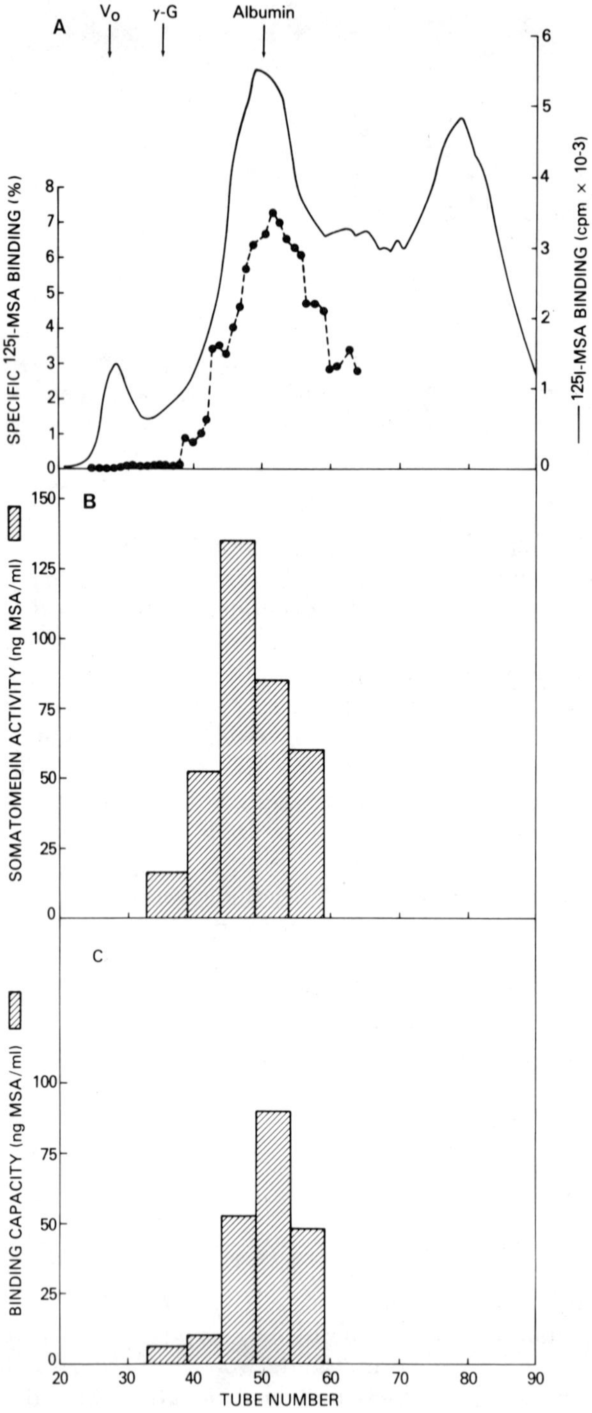

Figure 7 Circulating size of somatomedin activity in human cord serum from a 30-week gestation newborn. (A) One milliliter of cord serum was incubated with [^{125}I]MSA and then was chromatographed on Sephadex G-200, 0.05 M NH_4HCO_3, pH 8. Radioactivity was measured on individual fractions (——). One-hundred-microliter aliquots of the column fractions were incubated in duplicate with [^{125}I]MSA ± unlabeled MSA to measure specific binding activity (●– – –●) in each fraction; bound was separated from free using the charcoal method. (B and C) Pools were made of the column fractions in the top panel. These pools were gel-filtered on Sephadex G-50, 1 M acetic acid. Somatomedin activity was measured by a competitive protein binding assay, using partially purified rat serum binding protein, and is expressed as ng MSA/ml based on the volume of serum applied to the column (B). Scatchard analysis was performed on MSA binding to the stripped binding protein and is expressed as ng MSA bound/ml based on the volume of serum applied to the column (C). (We thank Dr. Ilene Fennoy for supplying the blood sample.)

question has been raised whether IGF levels in fetal blood simply reflect nutritional status rather than being causally related to fetal size.

2.5.2. *Syndrome of High Growth Hormone and Low IGF-I/SMC (Laron)*

In the syndrome described by Laron *et al.*,[42] serum growth hormone levels are elevated and IGF-I levels are very low. Of the original 22 patients, birth length was known in 10, and in five of these, birth lengths were 2 SD below the mean for the respective community. If IGF-I are correspondingly low *in utero*, these clinical observations may point to a role for IGF-I in fetal growth.

2.5.3. *Beckwith–Wiedemann Syndrome*

Ashton and Aynsley-Green[43] reported an elevated level of SM activity (2.8 IU/ml) by a chondrocyte bioassay in cord blood of an infant with Beckwith–Wiedemann syndrome (exomphalos, macroglossia, gigantism syndrome) compared to controls (0.72 ± 0.34).

2.5.4. *Transient Neonatal Diabetes Mellitus*

Blethen *et al.*[44] reported serial measurements of IGF-I (radioimmunoassay) and IGF-II (radioreceptor assay) levels in the plasma of an infant with transient neonatal diabetes mellitus. Cord blood IGF-I levels were low normal (0.21 IU/ml) while IGF-II levels were normal. With the onset of symptoms, IGF-I declined to 0.08 IU/ml and rose to 1.6 IU/ml during a period of rapid catchup growth. These findings raise the possibility that in addition to insulin deficiency, low IGF-I levels may contribute to the intrauterine growth retardation in this syndrome.

2.5.5. *Leprechaunism*

In this syndrome with severe intrauterine growth retardation, characteristic facies, and insulin resistance, the focus of attention has been on the insulin resistance as an explanation for the intrauterine growth retardation. However, cultured fibroblasts from one of these patients which show decreased insulin binding also exhibit decreased binding of IGF-I to its receptor.[45] These findings point to the possibility that the intrauterine growth retardation in this syndrome may be due, at least in part, to resistance to IGF as well as insulin.

3. Conclusions

There is a large body of indirect evidence which points to a role for IGF's in human fetal growth.

IGF receptors have been found in membrane preparations of placenta, liver, and brain as early as the first trimester. Presumably, the interaction of

IGF's with these receptors constitutes an early step in the biological response to these growth factors. Indeed, purified IGF's have been demonstrated to stimulate cell multiplication of human fetal fibroblasts in culture and stimulate [^{3}H]thymidine incorporation into DNA in human fetal chondrocytes.

The finding that human fetal fibroblasts produce IGF's raises the possibility that multiple human fetal tissues might produce IGF locally, as has been described in the mouse by D'Ercole *et al.*[46] It will be of interest to determine whether or not these human fibroblast cultures are producing IGF's distinct from IGF-I and IGF-II.

There are several caveats in the interpretation of blood levels of IGF's in the fetus. First, if local production is operative *in vivo*, the importance of circulating concentrations is minimized. Second, the IGF binding protein may be an important determinant of the bioactivity of the circulating IGF. For example, in rat interstitial fluid (represented by lymph) a 40,000-mol.wt. binding protein predominates.[47] It is possible, therefore, that the circulating form of IGF in midgestation may have more ready access to the plasma membrane than the larger 150,000-mol.wt. binding protein of late gestation and postnatal life. Also, if only free IGF is biologically active, then the binding affinity of the fetal 40,000-mol.wt. species relative to the binding affinity of the postnatal 150,000-mol.wt. species becomes important in determining the biological effectiveness of a particular circulating concentration of total IGF. There is also the parameter of responsiveness to IGF of fetal versus adult tissues in determining the effectiveness of a circulating level of IGF.

The experimental results of Sara and her colleagues[29] which point to a distinct human embryonic somatomedin are provocative. These data also imply that the midgestation brain receptor is distinctive in its IGF binding properties. It will be of interest to learn more about the physical properties and chemistry of this receptor.

In the rat the transition from predominance of the 40,000-mol.wt. binding protein in the fetus to the 150,000-mol.wt. form in postnatal life parallels the transition from fetal MSA to IGF-I in the postnatal animal.[48] Is it possible that accompanying this binding protein transition which occurs earlier in the human fetus (26–32 weeks)[37] there is also a shift in relative amounts of circulating IGF's, in this case from a distinct human embryonic IGF to IGF-I and IGF-II?

References

1. D. E. Hill, Fetal endocrine pancreas, *Clin. Obstet. Gynecol.* **23:**837, 1980.
2. M. W. Brinsmead and G. C. Liggins, Somatomedins and other growth factors in fetal growth, *Rev. Perinatal Med.* **3:**207, 1979.
3. V. R. Sara and K. Hall, Somatomedins and the fetus, *Clin. Obstet. Gynecol.* **23:**765, 1980.
4. D. J. Hill and R. D. G. Milner, Somatomedins and fetal growth, in: *The Fetus and Independent Life* (Ciba Foundation Symposium 86), Pitman, London, 1981, pp. 124–151.
5. A. J. D'Ercole and L. E. Underwood, Growth factors in fetal growth and development, in: *Fetal*

Endocrinology (M. J. Novy and J. A. Resko, eds.), Academic Press, New York, 1981, pp. 155–182.
6. P. D. Gluckman and G. C. Liggins, The regulation of fetal growth, in: *Fetal Physiology and Medicine: The Basis of Perinatology* (R. W. Beard and P. W. Nathanielsz, eds.), W. B. Saunders, London, 1984, pp. 511–557.
7. R. N. Marshall, L. E. Underwood, S. J. Voina, D. B. Foushee, and J. J. Van Wyk, Characterization of the insulin and somatomedin-C receptors in human placental cell membranes, *J. Clin. Endocrinol. Metab.* **39:**283, 1974.
8. B. Bhaumick, R. M. Bala, and M. D. Hollenberg, Somatomedin receptor of human placenta: solubilization, photolabeling, partial purification, and comparison with insulin receptor, *Proc. Natl. Acad. Sci. USA* **78:**4279, 1981.
9. S. D. Chernausek, S. Jacobs, and J. J. Van Wyk, Structural similarities between human receptors for somatomedin-C and insulin: Analysis by affinity labeling, *Biochemistry* **20:**7345, 1981.
10. W. H. Daughaday, I. K. Mariz, and B. Trivedi, A preferential binding site for insulin-like growth factor II in human and rat placental membranes, *J. Clin. Endocrinol. Metab.* **53:**282, 1981.
11. J. Massague and M. P. Czech, The subunit structure of two distinct receptors for insulin-like growth factors I and II and their relationship to the insulin receptor, *J. Biol. Chem.* **257:**5038, 1982.
12. C. L. Deal, H. J. Guyda, W. H. Lai, and B. I. Posner, Ontogeny of growth factor receptors in the human placenta, *Pediatr. Res.* **16:**820, 1982.
13. V. R. Sara, K. Hall, M. Misaki, L. Fryklund, N. Christensen, and L. Wetterberg, Ontogenesis of somatomedin and insulin receptors in the human fetus, *J. Clin. Invest.* **71:**1084, 1983.
14. M. M. Rechler, J. Zapf, S. P. Nissley, E. R. Froesch, A. C. Moses, J. M. Podskalny, E. E. Schilling, and R. E. Humbel, Interactions of insulinlike growth factors I and II and multiplication-stimulating activity with receptors and serum carrier proteins, *Endocrinology* **107:**1451, 1980.
15. M. Kasuga, E. Van Obberghen, S. P. Nissley, and M. M. Rechler, Structure of the insulin-like growth factor receptor in chick embryo fibroblasts, *Proc. Natl. Acad. Sci. USA* **79:**1864, 1982.
16. R. Rosenfeld, A. V. Thorsson, and R. L. Hintz, Increased somatomedin receptor sites in newborn circulating mononuclear cells, *J. Clin. Endocrinol. Metab.* **48:**456, 1979.
17. I. K. Ashton and A. E. Pocock, Action of multiplication-stimulating activity on [^{3}H]thymidine incorporation in rabit and human fetal chondrocytes *in vitro*, *J. Endocrinology* **99:**93, 1983.
18. E. R. Weidman and R. M. Bala, Direct mitogenic effects of human somatomedin on human embryonic lung fibroblasts, *Biochem. Biophys. Res. Commun.* **92:**577, 1980.
19. D. R. Clemmons and J. J. Van Wyk, Somatomedin-C and platelet-derived growth factor stimulate human fibroblast replication, *J. Cell. Physiol.* **106:**361, 1981.
20. P. R. Atkison, E. R. Weidman, B. Bhaumick, and R. M. Bala, Release of somatomedin-like activity by cultured WI-38 human fibroblasts, *Endocrinology* **106:**2006, 1980.
21. P. R. Atkison and R. M. Bala, Partial characterization of a mitogenic factor with somatomedin-like activity produced by cultured WI-38 human fibroblasts, *J. Cell. Physiol.* **107:**317, 1981.
22. L. Tato, M. V. L. Du Caju, C. Prevot, and R. Rappaport, Early variations of plasma somatomedin activity in the newborn, *J. Clin. Endocrinol. Metab.* **40:**534, 1975.
23. G. Giordano, E. Foppiani, F. Minuto, and D. Perroni, Growth hormone and somatomedin behavior in the newborn, *Acta Endocrinol.* **81:**449, 1976.
24. R. L. Hintz, J. M. Seeds, and R. E. Johnsonbaugh, Somatomedin and growth hormone in the newborn, *Am. J. Dis. Child.* **131:**1249, 1977.
25. P. D. Gluckman and M. W. Brinsmead, Somatomedin in cord blood: Relationship to gestation age and birth size, *J. Clin. Endocrinol. Metab.* **43:**1378, 1976.
26. J. J. Van Wyk, M. E. Svoboda, and L. E. Underwood, Evidence from radioligand assays that somatomedin-C and insulin-like growth factor-I are similar to each other and different from other somatomedins. *J. Clin. Endocrinol. Metab.* **50:**206, 1980.
27. A. J. D'Ercole, D. B. Foushee, and L. E. Underwood, Somatomedin-C receptor ontogeny and levels in porcine fetal and human cord serum, *J. Clin. Endocrinol. Metab.* **43:**1069, 1976.
28. K. Hall, G. Enberg, M. Ritzen, H. Svan, L. Fryklund, and K. Takano, Somatomedin A levels

in sera from healthy children and from children with growth hormone deficiency or delayed puberty, *Acta Endocrinol.* **94:**155, 1980.

29. V. R. Sara, K. Hall, C. H. Rodeck, and L. Wetterberg, Human embryonic somatomedin, *Proc. Natl. Acad. Sci. USA* **78:**3175, 1981.
30. W. H. Daughaday, B. Trivedi, and M. Kapadia, Measurement of insulin-like growth factor II by a specific radioreceptor assay in sera of normal individuals, patients with abnormal growth hormone secretion, and patients with tumor-associated hypoglycemia, *J. Clin. Endocrinol. Metab.* **53:**289, 1981.
31. P. D. Gluckman, J. J. Johnson-Barrett, H. H. Butler, B. W. Edgar, and T. R. Gunn, Studies of insulin-like growth factor-I and -II by specific radioligand assays in umbilical cord blood, *Clinical Endocrinology* **19:**405, 1983.
32. R. W. Furlanetto, L. E. Underwood, J. J. Van Wyk, and A. J. D'Ercole, Estimation of somatomedin-C levels in normals and patients with pituitary disease by radioimmunoassay, *J. Clin. Invest.* **60:**648, 1977.
33. R. M. Bala, J. Lopatka, A. Leung, E. McCoy, and R. G. McArthur, Serum immunoreactive somatomedin levels in normal adults, pregnant women at term, children at various ages and children with constitutional delayed growth, *J. Clin. Endocrinol. Metab.* **52:**508, 1981.
34. J. Zapf, H. Walter, and E. R. Froesch, Radioimmunological determination of insulin-like growth factors I and II in normal subjects and in patients with growth disorders and extrapancreatic tumor hypoglycemia, *J. Clin. Invest.* **68:**1321, 1981.
35. A. Bennett, D. M. Wilson, F. Liu, R. Nagashima, R. G. Rosenfeld, and R. L. Hintz, Levels of insulin-like growth factors I and II in human cord blood, *J. Clin. Endocrinol. Metab.* **57:**609, 1983.
36. A. C. Moses, S. P. Nissley, P. A. Short, M. M. Rechler, R. M. White, A. B. Knight, and O. Z. Higa, Increased levels of multiplication stimulating activity, an insulin-like growth factor in fetal rat serum, *Proc. Natl. Acad. Sci. USA* **77:**3649, 1980.
37. A. J. D'Ercole, D. F. Willson, and L. E. Underwood, Changes in the circulating form of serum somatomedin C during fetal life, *J. Clin. Endocrinol. Metab.* **51:**674, 1980.
38. R. M. White, S. P. Nissley, A. C. Moses, M. M. Rechler, and R. E. Johnsonbaugh, The growth hormone dependence of a somatomedin binding protein in human serum, *J. Clin. Endocrinol. Metab.* **53:**49, 1981.
39. S. L. S. Drop and D. J. Kortleve, Radioimmunoassay activity of a somatomedin binding protein, isolated from preterm amniotic fluid in pre- and postnatal serum, *Program of the Annual Meeting of the Endocrine Society,* 1983, abstr. 781.
40. I. K. Ashton and J. Vesey, Somatomedin activity in human cord plasma and relationship to birth size, insulin, growth hormone, and prolactin, *Early Human Dev.* **2**(2)**:**115, 1978.
41. T. P. Foley, Jr., R. DePhilip, A. Perricelli, and A. Miller, Low somatomedin activity in cord serum from infants with intrauterine growth retardation, *J. Pediatr.* **96:**605, 1980.
42. Z. Laron, A. Pertzelan, and M, Karp, Pituitary dwarfism with high serum levels of growth hormone, *Isr. J. Med. Sci.* **4:**883, 1968.
43. I. K. Ashton and A. Aynsley-Green, Plasma somatomedin activity in an infant with Beckwith-Wiedemann syndrome, *Early Human Dev.* **1**(4)**:**357, 1978.
44. S. L. Blethen, N. H. White, J. V. Santiago, and W. H. Daughaday, Plasma somatomedins, endogenous insulin secretion, and growth in transient neonatal diabetes mellitus, *J. Clin. Endocrinol. Metab.* **52:**144, 1981.
45. E. E. Van Obberghen-Schilling, M. M. Rechler, J. A. Romanus, A. B. Knight, S. P. Nissley, and R. E. Humbel, Receptors for insulin-like growth factor I are defective in fibroblasts cultured from a patient with leprechaunism, *J. Clin. Invest.* **68:**1356, 1981.
46. A. J. D'Ercole, G. T. Applewhite, and L. E. Underwood, Evidence that somatomedin is synthesized in multiple tissues in the mouse fetus, *Dev. Biol.* **75:**315, 1980.
47. A. C. Moses, S. P. Nissley, J. Passamani, R. M. White, and M. M. Rechler, Further characterization of growth hormone-dependent somatomedin-binding proteins in rat serum and demonstration of somatomedin binding protein produced by rat liver cells in culture, *Endocrinology* **104:**536, 1979.
48. R. M. White, S. P. Nissley, P. A. Short, M. M. Rechler, and I. Fennoy, Developmental pattern of a serum binding protein for MSA in the rat, *J. Clin. Invest.* **69:**1239, 1982.

49

Somatomedin C in the Ovine Fetus and Neonate

D. M. Styne, G. Van Vliet, A. M. Rudolph, J. Kitterman, H. Iwamoto, S. L. Kaplan, and M. M. Grumbach

1. Introduction

The chronically catheterized sheep model offers a useful approach to the study of fetal and neonatal neuroendocrine and pituitary maturation.(1) The ontogeny of somatomedin (SM) concentrations in the normal ovine fetus and after surgical or pharmacological manipulations has been investigated by bioassay,(2) radioreceptor assay,(3,4) and, more recently, radioimmunoassay.(5,6) We describe our studies of the ontogeny of somatomedin C/insulinlike growth factors (SMC/IGF-I), the site of secretion of SMC/IGF-I, and the pituitary hormones that could be involved in the control of production of SMC/IGF-I in the sheep fetus and neonatal lamb.

2. Methods

2.1. Assay

A specific SMC/IGF-I radioimmunoassay of acid–ethanol-extracted sheep plasma was used as recently described.(5) Antibody to SMC/IGF-I was generated by Dr. Van Wyk and Dr. Underwood and supplied by the National Hormone and Pituitary Program (NHPP). Either purified SMC/IGF-I kindly provided by Dr. R. Humbel(7) or synthetic SMC/IGF-I supplied by Dr. C. H. Li(8) was labeled

D. M. Styne • Department of Pediatrics, University of California–Davis, Davis, California 95616. ***G. Van Vliet, A. M. Rudolph, J. Kitterman, H. Iwamoto, S. L. Kaplan, and M. M. Grumbach*** • Department of Pediatrics, University of California–San Francisco, San Francisco, California 94143.

by the lactoperoxidase method and used as tracer. Lamb plasma was used as standard and assigned a value of 1 Unit/ml; this corresponded to 500 ng of purified SMC/IGF-I (Humbel). The plasma samples and standard were acid–ethanol-extracted and neutralized as described[9] to remove somatomedin binding proteins. The mean recovery was 83.5%.

Lamb plasma was used as standard in the radioimmunoassay and assigned a value of 1 U/ml. The assay sensitivity was 0.03 U/tube. Intra- and interassay coefficients of variation were 10% and 15%, respectively. The concentration of SMC/IGF-I was 0.04 U/ml in plasma from a hypophysectomized ewe, kindly supplied by Dr. L. Underwood, and 1.0 U/ml in plasma from an acromegalic ewe, provided by Dr. S. Davis.

2.2. *Site of Secretion*

Arteriovenous differences of SMC/IGF-I across the lungs, liver, kidney, and major vessels were determined during chronic catheterization at 135 days of gestational age to assess the site(s) of secretion of SMC/IGF-I in the fetal lamb. Catheterization of the portal vein, hepatic vein, umbilical artery, and inferior vena cava was carried out in five fetuses,[10] the femoral artery and renal vein in eight fetuses, the pulmonary artery and vein, inferior vena cava, and femoral artery in three fetuses, and the coronary artery and coronary sinus in one fetus. Catheters were maintained patent by the use of dilute heparin solutions, and samples were taken at least 72 hr after surgery and only used if the fetus maintained a normal pH (≥7.3) and PO_2 (≥20 mm Torr).

2.3. *Regulation of Secretion of SMC/IGF-I in the Fetus*

The time course of disappearance of plasma SMC/IGF-I after fetal cryohypophysectomy performed by the method of Brinsmead and Liggins[4] in four chronically catheterized fetuses of 105 days' gestational age was determined over 4 weeks. The *in vivo* criteria of a complete hypophysectomy were (1) no detectable plasma prolactin (PRL) or ovine growth hormone (oGH), (2) the absence of a rise in PRL after thyrotropin-releasing factor stimulation, and (3) the absence of a rise in oGH after a challenge with growth hormone-releasing factor (GRF). Confirmation of completeness of hypophysectomy was established by autopsy.

To determine the effect of exogenous human growth hormone (hGH) on the secretion of SMC/IGF-I from the liver, hGH was administered to each of two 135-day-old chronically catheterized hepatic preparation fetuses. hGH was given as a 1.5-mg bolus followed by a 22-hr infusion of 500 μg/hr into the portal vein. Changes in SMC/IGF-I following the peripheral i.v. administration of 10 μg/kg GRF to chronically catheterized ovine fetuses were also assessed.[11]

2.4. *In the Lamb*

Pituitary hGH from the NHPP, ovine PRL and ovine GH from Dr. Li were infused into the portal vein of 5 chronically catheterized lambs between 60–

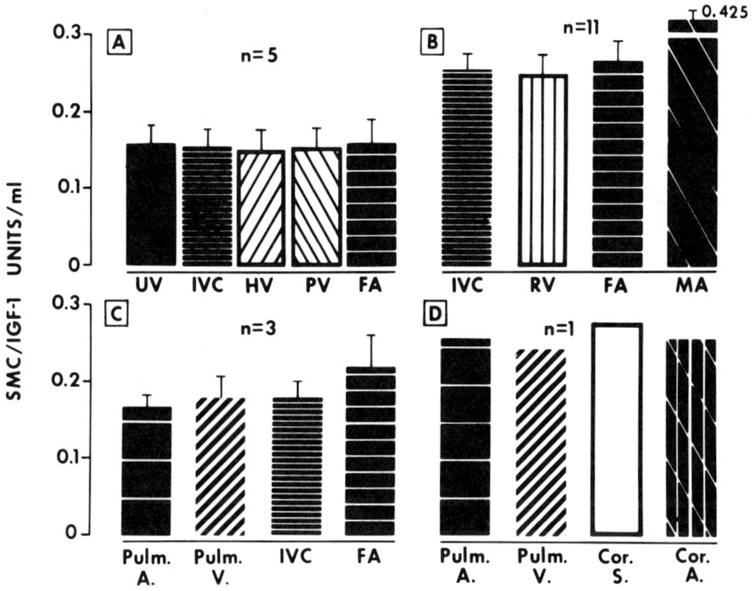

Figure 1 Mean ± SD plasma SMC/IGF-I in chronically catheterized vessels of the 135-day fetal lamb. UV, umbilical vein; IVC, inferior vena cava; HV, hepatic vein; PV, portal vein; FA, femoral artery; RV, renal vein; Pulm. A., pulmonary artery; Pulm. V., pulmonary vein; Cor. S., coronary sinus; Cor. A., coronary artery; MA, maternal artery. There was no significant difference between the SMC concentrations in any fetal vessels.

100 days postnatally (all were given as an initial 500 μg bolus followed by an infusion of 500 μg/hr for 4 hr). Plasma samples were obtained from the brachial artery and hepatic vein at 0, 2, 4, 6, and 24 hr for measurement of SMC/IGF-I.

To assess the effect of chronic hGH administration on plasma SMC/IGF-I, hGH (1 mg) was given IM daily for 10 days to two 50–60 day old lambs with chronic hepatic vessel catheters. Samples were obtained from hepatic vein and brachial artery for SMC/IGF-I daily and at 24 hr following the last injection of hGH.

2.5. Diurnal Variation of SMC/IGF-I in Postnatal Lamb

Plasma samples were obtained in postnatal lambs (1–4 weeks of age) from a peripheral vein every 30 min for 2 hr between 9–11 am and between 5–7 pm during the same 24-hr period.

3. Results

3.1. Studies in the Chronically Catheterized Ovine Fetus

There were no significant differences in the levels of SMC/IGF-I in samples obtained from the portal vein, inferior vena cava or umbilical artery. Comparable

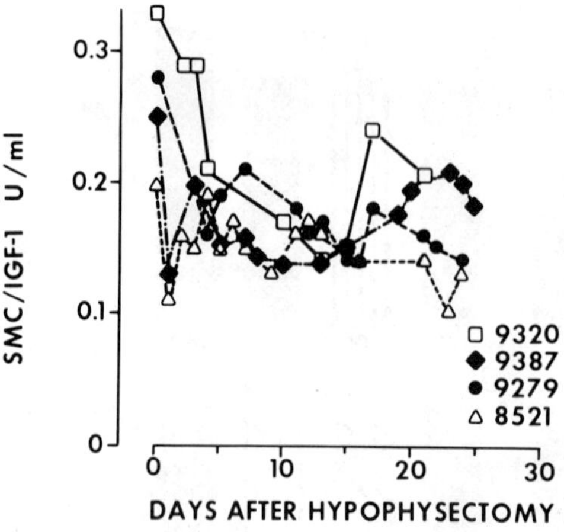

Figure 2 Plasma SMC/IGF-I in the chronically catheterized sheep fetus during a 4-week period after cryohypophysectomy (according to ref. 4). At the time of the procedure at 0 day, the fetuses were 105 days of gestational age.

levels of SMC/IGF-I were detected in the femoral artery, renal vein, coronary artery and vein, and pulmonary artery and vein (Fig. 1).

In the chronically catheterized fetal sheep, plasma SMC/IGF-I was lower before 100 days of gestation with a mean of 0.14 U/ml ± 0.02 (SD) than between 101 days and 144 days when the mean was 0.22 U/ml ± 0.01, a significant difference ($p < 0.005$) with a rise of 60%. Longitudinal data obtained from six chronically catheterized fetuses confirmed the constancy of SMC/IGF-I during the latter third of gestation. As early as the first postnatal day, plasma SMC/IGF-I rose, and the postnatal mean was 0.93 U/ml ± 0.07, a significant difference ($p < 0.0001$) versus late-gestation fetuses.

Following cryohypophysectomy, there was an initial fall in plasma SMC/IGF-I to 40–50% of preoperative values by 1–3 days after surgery. There was a gradual rise to 75% of preoperative levels by 25–28 days following cryohypohysectomy in the fetuses (Fig. 2).

Infusion of hGH into the portal vein of the fetuses over 22 hr achieved plasma levels of 450 ng/ml of hGH in the hepatic vein. There was no significant difference in plasma SMC/IGF-I concentrations in the right hepatic vein compared to levels in the vena cava or femoral or umbilical artery before, during, or following infusion of hGH (Fig. 3).

Following intravenous administration of 10 μg/kg GRF-44, there was a striking rise of plasma concentration of oGH within 3–5 min in the fetus. No significant increase in SMC/IGF-I was detected in samples obtained 120 min after i.v. GRF-44.

3.2. *Studies of SMC/IGF-I in the Lamb*

No diurnal variation of SMC/IGF-I was demonstrable in postnatal lambs. However, fluctuations in the concentration of SMC/IGF-I of 30% were detected during the 2-hr period of sampling in two 14-day-old lambs during the morning and the evening.

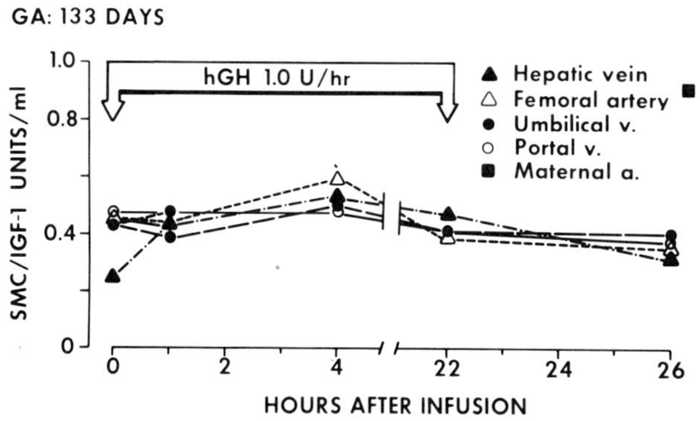

Figure 3 Plasma SMC/IGF-I in an ovine fetus of 135 days' gestational age during an infusion of human growth hormone (hGH) into the portal vein. hGH was given as a 3-IU (1.5 mg) bolus and then infused at 1 IU (0.5 mg) per hr for 2 hr.

A plasma SMC/IGF-I concentration gradient was not detected in samples obtained from the portal and hepatic vein in the lamb (Fig. 4). In our initial studies, administration of saline into the portal vein was associated with a 20–30% variation in the concentration of SMC/IGF-I in the hepatic vein and brachial artery during 6 hr of study. A 4-hr infusion of hGH (2.5 mg or 0.05

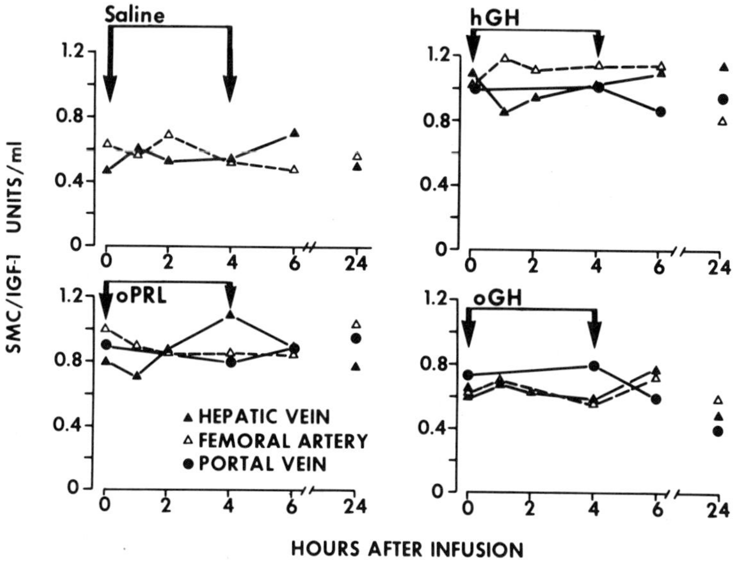

Figure 4 Plasma SMC/IGF-I in the chronically catheterized hepatic vein, femoral artery, and portal vein of a 60-day postnatal lamb during infusions into the portal vein. The 4 hr of infusion are denoted by the arrows. Human growth hormone (hGH), ovine prolactin (oPRL), and ovine growth hormone (oGH) were given as 0.5-mg boluses followed by 0.5 mg/hr infusions.

to 0.08 mg/kg) of hGH into the portal vein was carried out in five lambs. The concentration of hGH in the hepatic vein exceeded 100 ng/ml during the infusion. However, no SMC/IGF-I concentration gradient was found across the liver. In five lambs, ovine prolactin (2.5 mg) was infused over a 4-hr period in the portal vein and in three lambs, 2.5 mg of ovine growth hormone was infused over a 4-hr period. There was no significant change in SMC/IGF-I concentrations in the hepatic vein or brachial artery after infusion of hGH, oGH, or oPRL.

Chronic administration of hGH (1.0 mg, 0.025–0.035 mg/kg) given intramuscularly daily for 10 days to two lambs did not induce a significant rise in the concentration of SMC/IGF-I in the hepatic vein or brachial artery.

4. Discussion

We found low concentrations of plasma SMC/IGF-I between 78–98 days of gestation with a modest rise after 101 days of gestation. In contrast, Handwerger et al.[6] noted an eight fold rise in SMC/IGF-I in fetal lamb sera measured by RIA after acid treatment and neutralization. Methodologic differences in extraction techniques may be responsible for the conflicting results. In our prior report using a placental radioreceptor assay for SMA which detected both IGF-I and -II, a similar rise in fetal levels was seen after 100 days of gestation, similar to our present study.[3] Further, Brinsmead and Liggins detected a consistent rise of somatomedin throughout gestation using a placental radioreceptor assay with MSA as ligand which would measure IGF-I and IGF-II.[4] It is likely that the higher values of SM by radioreceptor assay than by IGF-I radioimmunoassay are due to elevated levels of IGF-II like SM rather than SMC/IGF-I. Gluckman[12] has reported measurements of IGF-II throughout gestation in the ovine model and demonstrated higher values in the fetus than the postnatal lamb. In all studies, there is a sharp rise in SMC/IGF-I during the neonatal period.

In the postnatal animal, but not in the fetus, pituitary GH is a major regulator for SM synthesis and release. Brinsmead and Liggins[4] found no decrease in fetal SM levels assayed by radioreceptor and bioassay methods following hypophysectomy. Falconer *et al.*[2] reported a 60% decrease in bioassayable SM in the hypophysectomized ovine fetus. We have detected a modest initial decrease in fetal SM levels within 3 days after hypophysectomy. Subsequently, the SM levels rise to values not significantly different than those of normal fetuses.

Based on *in vitro* perfusion studies, it has been proposed that the liver is a major site of SM synthesis postnatally.[13,14] Based on *in vitro* studies in the fetal mouse, the lung appears to be a major source of SM, with the liver and kidney as additional significant sites of somatomedin release.[15]

In our chronically catheterized model we found no arteriovenous difference of SMC/IGF-I across the liver, kidney, lung, heart, or placenta in the fetal lamb. Administration of a high dose of hGH in the portal vein of the chronically catheterized ovine fetus failed to induce an acute or delayed (24 hr) rise in SM levels.

Recently, Gluckman *et al.*[16] have demonstrated reduced fetal liver receptor binding for GH. A decreased number of GH receptors in the fetus at sites that postnatally are involved in the generation of SMC provides a possible explanation for the failure of GH to induce SM synthesis. Alternatively, if the disappearance rate or metabolic clearance of protein bound SMC/IGF-I is prolonged in the fetus, this may obscure or suppress hepatic production of additional SM. The doses of GH we used in the fetus were quite large, and delivered directly to the liver, but still less than the microgram doses used in the *in vitro* studies with rat liver.[13,14]

Some *in vitro* studies used a recirculating system in which small amounts of SM secreted in each circulation would be added to the sum of previously produced SM, thereby integrating the total secretion per unit of time. In the intact fetus, dilution of secreted SM into the entire circulatory volume may impede the detection of the release of small amounts of SM. An additional factor is the episodic variation in SMC/IGF-I, particularly noted in the postnatal lamb. Baxter[17] reported similar findings in the rat based on measurements of MSA. This variability may obscure small but significant changes in SMC/IGF-I.

Peripheral intravenous administration of growth hormone in dogs induces to a 30% rise in bioassayable SM in portal, hepatic, and renal vein by 2 hr.[18] A rise in bioassayable SM has been reported by 30–180 min in isolated rat liver following perfusion with growth hormone. There is precedence for GH exerting other rapid metabolic effects in sheep as peripheral administration of oGH (0.13–0.25 mg/kg) raises free fatty acid levels in adult sheep within 2 hr.[19]

Of interest is our observation that i.m. administration of hGH, 0.025–0.035 mg/kg for 10 days to two-month-old lambs did not induce a significant rise in levels of SMC/IGF-I in the hepatic vein. Maturation of responsiveness of SMC/IGF-I release to GH may not occur until later in the lamb.

In conclusion,

(1) SMC/IGF-I is not significantly growth hormone dependent in the fetal sheep as evidenced by the results of hypophysectomy or hGH infusions.

(2) Our data suggest that neither liver, lung, kidney, placenta, or heart are the predominant source of circulating SMC/IGF-I in the ovine fetus, but SMC/IGF-I may arise from multiple tissues. However, our studies would not have been sensitive to small arterio–venous differences of SMC/IGF-I. Alternatively, if SMC/IGF-I has a long halflife, it may reduce our ability to detect a small amount of active secretion.

(3) SMC/IGF-I is not GH- (either hGH or oGH) or prolactin-dependent in the neonatal lamb.

It would seem that IGF-II or similar peptides are the important SM's in the fetus and not SMC. Postnatally there is a transition period during which SMC rises and becomes progressively more GH-dependent, whereas the concentrations of IGF-II and possible other IGF-II-like peptides decrease from their high fetal levels.

References

1. P. D. Gluckman, M. M. Grumbach, and S. L. Kaplan, The neuroendocrine regulation and function of growth hormone and prolactin in the mammalian fetus, *Endocrinol. Rev.* **2**(4)**:**363, 1981.
2. J. Falconer, J. M. Forbes, I. C. Hart, J. S. Robinson, and G. D. Thorburn, Somatomedin-like activity in plasma after foetal hypophysectomy or nephrectomy and in experimental intrauterine growth retardation in sheep, *J. Endocrinol.* **83:**119, 1979.
3. P. D. Gluckman, K. Uthne, D. M. Styne, S. L. Kaplan, A. M. Rudolph, and M. M. Grumbach, Hormone ontogeny in the ovine fetus. IV. Serum somatomedin activity in the fetal and neonatal lamb and pregnant ewe: Correlation with maternal and fetal growth hormone, prolactin, and chorionic somatomammotropin, *Pediatr. Res.* **14:**194, 1979.
4. M. W. Brinsmead and G. C. Liggins, Serum somatomedin activity after hypophysectomy and during parturition in fetal lambs, *Endocrinology* **105**(1)**:**297, 1979.
5. G. Van Vliet, D. M. Styne, S. L. Kaplan, and M. M. Grumbach, Hormone ontogeny in the ovine fetus. XVI. Plasma immunoreactive somatomedin C/insulin-like growth factor I in the fetal and neonatal lamb in the pregnant ewe, *Endocrinology* **113**(5)**:**1716, 1983.
6. S. Handwerger, A. J. D'Ercole, L. E. Underwood, P. Gluckman, and G. C. Liggins, Serum somatomedin C concentrations in the fetal sheep increase markedly during gestation, *J. Endocrinol.* **96:**269, 1983.
7. E. Rinderknecht and R. E. Humbel, The amino acid sequence of human insulin-like growth factor I and its structural homology with proinsulin, *J. Biol. Chem.* **253:**2769, 1978.
8. C. H. Li, D. Yamashiro, D. Gospodarowicz, S. L. Kaplan, and G. Van Vliet, Total synthesis of insulin-like growth factor I (somatomedin C), *Biochemistry* **80:**2216, 1983.
9. W. H. Daughaday, I. K. Mariz, and S. L. Blethen, Inhibition of access of bound somatomedin to membrane receptor and immunobinding sites: A comparison of radioreceptor and radioimmunoassay of somatomedin in native and acid-ethanol-extracted serum, *J. Clin. Endocrinol. Metab.* **51:**781, 1980.
10. J. Bristow, A. M. Rudolph, and J. Itskovitz, A preparation for studying liver blood flow, oxygen consumption, and metabolism in the fetal lamb in utero, *J. Dev. Physiol.* **3:**255, 1981.
11. D. M. Styne, L. Cuttler, and M. M. Grumbach, Human pancreatic growth hormone releasing factor 1-44-NH2 (hpGRF) stimulates ovine growth hormone (oGH) secretion in the fetal and neonatal lamb, 65th Annual Meeting of the Endocrinology Society, San Antonio, 1983, abstr. 616.
12. P. D. Gluckman and J. H. Butler, Parturition-related changes in insulin-like growth factors-I and -II in the perinatal lamb, *J. Endocrinol.* **99:**223, 1983.
13. P. McConaghey and C. B. Sledge, Production of "sulphation factor" by the perfused liver, *Nature* **225:**1249, 1970.
14. L. S. Phillips, A. C. Herington, I. E. Karl, and W. H. Daughaday, Comparison of somatomedin activity in perfusates of normal and hypophysectomized rat livers with and without added growth hormone, *Endocrinology* **98:**606, 1976.
15. A. J. D'Ercole, G. T. Applewhite, and L. E. Underwood, Evidence that somatomedin is synthesized by multiple tissues in the fetus, *Dev. Biol.* **75:**315, 1980.
16. P. D. Gluckman, J. H. Butler, and T. B. Elliott, The ontogeny of somatotropic binding sites in ovine hepatic membranes, *Endocrinology* **112**(5)**:**1607, 1983.
17. R. C. Baxter, Z. Zaltsman, and J. R. Oliver, Pulsatility of immunoreactive somatomedin-C in chronically cannulated rats, *Endocrinology* **113**(2)**:**729, 1983.
18. R. Schimpff, M. Donnadieu, and M. Duval, Serum somatomedin activity measured as sulphation factor in peripheral, hepatic and renal veins in normal mongrel dogs: Early effects of intravenous injection of growth hormone, *Acta Endocrinol.* **93:**155, 1980.
19. J. M. Bassett and A. L. C. Wallace, Short-term effects of ovine growth hormone on plasma glucose, free fatty acids and ketones in sheep, *Metabolism* **15**(10)**:**933, 1966.

Contributors

Angela M. Acquaviva, Section on Growth and Development, Molecular, Cellular, and Nutritional Endocrinology Branch, National Institute of Arthritis, Diabetes, and Digestive and Kidney Diseases, National Institutes of Health, Bethesda, Maryland 20205

Eli Y. Adashi, Department of Obstetrics and Gynecology, University of Maryland, Baltimore, Maryland 21201

Kerstin Albertsson-Wikland, Department of Physiology, University of Göteborg, S-400 33 Göteborg, Sweden

G. P. August, Department of Pediatrics, Children's Hospital, National Medical Center, Washington, D.C., 20010

Sameul D. Balk, Department of Pathology, New England Deaconess Hospital, Boston, Massachusetts 02116

D. R. Bangham, National Institute for Biological Standards and Control, London NW3 6RB, England

J. J. Bell, Department of Pediatrics, College of Physicians and Surgeons, Columbia University, New York, New York 10032

Carol J. Berry, Department of Medicine and Clinical Research Facility, Emory University School of Medicine, Atlanta, Georgia 30322

Thomas A. Bewley, Laboratory of Molecular Endocrinology, University of California, San Francisco, California 94143

Juergen R. Bierich, Department of Pediatrics, University of Tubingen, 74 Tubingen, Federal Republic of Germany

Louise Bilezikjian, The Clayton Foundation Laboratories for Peptide Biology, The Salk Institute, La Jolla, California 92037

S. L. Blethen, Department of Pediatrics, Washington University School of Medicine, Saint Louis, Missouri 63110

Robert M. Blizzard, Department of Pediatrics, University of Virginia School of Medicine, Charlottesville, Virginia 22901

P. Bohlen, Neuroendocrinology Laboratories, Salk Institute, San Diego, California 92037

Johnny Brandt, KabiVitrum AB, Research and Development, Departments of Biochemistry and Analytical Chemistry, S-112 87 Stockholm, Sweden

P. Brazeau, Laboratoire de neuroendocrinologie expérimentale, Hôpital Notre-Dame, Montréal, Quebec, H21 4K8 Canada

A. Briskin, Neuroendocrinology Laboratories, Salk Institute, San Diego, California 92037

D. R. Brown, Department of Pediatrics, University of Minnesota, Minneapolis, Minnesota 55455

C. Bruno Bruni, Section on Growth and Development, Molecular, Cellular, and Nutritional Endocrinology Branch, National Institute of Arthritis, Diabetes, and Digestive and Kidney Diseases, National Institutes of Health, Bethesda, Maryland 20205

Pierre Chatelain, INSERM U 34, Hôpital Debrousse, 69005 Lyon, France

Rajender K. Chawla, Department of Medicine and Clinical Research Facility, Emory University School of Medicine, Atlanta, Georgia 30322

David R. Clemmons, Department of Pediatrics, University of North Carolina, School of Medicine, Chapel Hill, North Carolina 27514

Vittorio Coiro, Department of Physiology, University of Massachusetts Medical School, Worcester, Massachusetts 01605

The Committee on Growth Hormone Usage (S. L. Kaplan, G. P. August, S. A. Kaplan, M. H. MacGillivray, W. V. Moore, J. S. Parks, A. F. Parlow, A. W. Root, T. P. Foley, A. E. Wilhelmi, and L. E. Underwood, A. J. Johanson), The National Hormone and Pituitary Program, University of Maryland School of Medicine, Baltimore, Maryland 21201

Michael P. Czech, Department of Biochemistry, University of Massachusetts Medical School, Worcester, Massachusetts 01605

William H. Daughaday, Metabolism Division, Washington University School of Medicine, St. Louis, Missouri 63110

Heather J. Dean, Departments of Physiology and Pediatrics, University of Manitoba, Winnipeg, Manitoba, Canada R3E OW3

A. Joseph D'Ercole, Department of Pediatrics, University of North Carolina, School of Medicine, Chapel Hill, North Carolina 27514

David B. Donner, Memorial Sloan-Kettering Cancer Center and The Sloan-Kettering Division, Graduate School of Medical Sciences, Cornell University, New York, New York 10021

Andrea Dorato, Memorial Sloan-Kettering Cancer Center and The Sloan-Kettering Division, Graduate School of Medical Sciences, Cornell University, New York, New York 10021

Staffan Edén, Department of Physiology, University of Göteborg, S-400 33 Göteborg, Sweden

Catherine A. Egli, Department of Pediatrics, University of California–San Francisco, San Francisco, California 94143

F. Esch, Neuroendocrinology Laboratories, Salk Institute, San Diego, California 92037

Rina Eshet, Institute of Pediatric and Adolescent Endocrinology, Beilinson Medical Center, Petah Tikva 49 100, Israel, and Sackler School of Medicine, Tel Aviv University, Tel Aviv, Israel

William S. Evans, Department of Pediatrics, University of Virginia School of Medicine, Charlottesville, Virginia 22901

Angel Ferrandez, Hospital Infantil, E-S0009, Zaragosa, Spain

Timothy J. Fields, Department of Life Sciences, Indiana State University, Terre Haute, Indiana 47809

T. P. Foley, Department of Pediatrics, University of Pittsburgh, Pittsburgh, Pennsylvania 15213

S. Douglas Frasier, Department of Pediatrics and Physiology/Biophysics, University of Southern California School of Medicine and the Los Angeles County–University of Southern California Medical Center, Los Angeles, California 90033

Ulf Friberg, Department of Histology, Karolinska Institutet, S-10401 Stockholm, Sweden

Henry G. Friesen, Departments of Physiology and Pediatrics, University of Manitoba, Winnipeg, Manitoba, Canada R3E OW3

Luciano G. Frigeri, The Whittier Institute for Diabetes and Endocrinology, La Jolla, California 92037

Herwig Frisch, Department of Endocrinology, University-Kinderklinik Stien, A-1090 Wien IX-Vienna, Austria

Lawrence A. Frohman, Division of Endocrinology and Metabolism, Department of Internal Medicine, University of Cincinnati College of Medicine, Cincinnati, Ohio 45267

Linda Fryklund, KabiVitrum AB, Research and Development, Departments of Biochemistry and Analytical Chemistry, S-112 87 Stockholm, Sweden

Richard Furlanetto, Children's Hospital of Philadelphia, Philadelphia, Pennsylvania 19104

Lynne A. Gaynes, Endocrine Section, Metabolism Branch, National Cancer Institute, National Institutes of Health, Bethesda, Maryland 20205

Michael D. Geier, Genentech, Inc., South San Francisco, California 94080

Marie C. Gelato, Developmental Endocrinology Branch, National Institutes of Health, Bethesda, Maryland 20205

G. Yancey Gillespie, Department of Surgery and Cancer Research Center, University of North Carolina, School of Medicine, Chapel Hill, North Carolina 27514

James H. Goldman, Endocrine Division, Department of Medicine, Tufts University School of Medicine, New England Medical Center, Inc., Boston, Massachusetts 02111

H. Maurice Goodman, Department of Physiology, University of Massachusetts Medical School, Worcester, Massachusetts 01605

Genevieve Grichting, Department of Physiology, University of Massachusetts Medical School, Worcester, Massachusetts 01605

Melvin M. Grumbach, Department of Pediatrics, University of California–San Francisco, San Francisco, California 94143

R. Guillemin, Neuroendocrinology Laboratories, Salk Institute, San Diego, California 92037

Magnus Hagerman, KabiVitrum AB, Research and Development, Departments of Biochemistry and Analytical Chemistry, S-112 87 Stockholm, Sweden

David A. Harris, Department of Pediatrics, University of California–San Francisco, San Francisco, California 94143

Berthold P. Hauffa, Departments of Pediatrics, University of California–San Francisco, San Francisco, California 94143

William P. Heath, Department of Medicine and Clinical Research Facility, Emory University School of Medicine, Atlanta, Georgia 30322

Raymond L. Hintz, Department of Pediatrics, Stanford University, Stanford, California 94305

N. J. Hopwood, Department of Pediatrics, University of Michigan, Ann Arbor, Michigan 48109

James P. Hughes, Department of Life Sciences, Indiana State University, Terre Haute, Indiana 47809

J. Anthony Hulse, Department of Pediatrics, University of California–San Francisco, San Francisco, California 94143

René E. Humbel, Biochemisches Institut, Universität Zürich, Switzerland

Chung Ho Hung, Section of Pediatric Endocrinology, Department of Pediatrics, University of Kansas Medical Center, Kansas City, Kansas 66103

Ruth Illig, Universitatas Kinderklinik, CH-8032 Zurich, Switzerland

Olle G. P. Isaksson, Department of Physiology, University of Göteborg, S-400 33 Göteborg, Sweden

H. Iwamoto, Department of Pediatrics, University of California–San Francisco, San Francisco, California 94143

John-Olov Jansson, Department of Physiology, University of Göteborg, S-400 33 Göteborg, Sweden

Ann J. Johanson, Department of Pediatrics, University of Virginia School of Medicine, Charlottesville, Virginia 22901

Andrew J. S. Jones, Genentech, Inc., South San Francisco, California 94080

Susan Joyce, Department of Psychology, University of Virginia, Charlottesville, Virginia 22901

Donald L. Kaiser, Department of Pediatrics, University of Virginia School of Medicine, Charlottesville, Virginia 22901

Selna L. Kaplan, Department of Pediatrics, University of California–San Francisco, San Francisco, California 94143

Susan Kasper, Departments of Physiology and Pediatrics, University of Manitoba, Winnipeg, Manitoba, Canada R3E OW3

R. T. Kirkland, Department of Pediatrics, University of Texas, Houston, Texas 77030

J. Kitterman, Department of Pediatrics, University of California–San Francisco, San Francisco, California 94143

Jack L. Kostyo, Department of Physiology, University of Michigan Medical School, Ann Arbor, Michigan 48109

Karen Kover, Section of Pediatric Endocrinology, Department of Pediatrics, University of Kansas Medical Center, Kansas City, Kansas 66103

A. Avinoam Kowarski, Department of Pediatrics, University of Maryland School of Medicine, Baltimore, Maryland 21201

Michael H. Kutner, Department of Biometry and Clinical Research Facility, Emory University School of Medicine, Atlanta, Georgia 30322

Zvi Laron, Institute of Pediatric and Adolescent Endocrinology, Beilinson Medical Center, Petah Tikva 49 100, Israel, and Sackler School of Medicine, Tel Aviv University, Tel Aviv, Israel

Urban J. Lewis, The Whittier Institute for Diabetes and Endocrinology, La Jolla, California 92037

Choh Hao Li, Laboratory of Molecular Endocrinology, University of California, San Francisco, California 94143

Samuel Libber, Department of Pediatrics, Johns Hopkins University School of Medicine, Baltimore, Maryland 21205

N. Ling, Neuroendocrinology Laboratories, Salk Institute, San Diego, California 92037

Kenneth E. Lipson, Memorial Sloan-Kettering Cancer Center and The Sloan-Kettering Division, Graduate School of Medical Sciences, Cornell University, New York, New York 10021

Robert H. Lustig, Department of Pediatrics, University of California–San Francisco, San Francisco, California 94143

Margaret H. MacGillivray, Department of Pediatrics, State University of New York at Buffalo School of Medicine and The Children's Hospital of Buffalo, Buffalo, New York 14222

Kjell Madsen, Department of Histology, Karolinska Institutet, S-10401 Stockholm, Sweden

Joseph B. Martin, Neuroendocrinology Laboratory, Department of Neurology, Massachusetts General Hospital, Harvard Medical School, Boston, Massachusetts 02114

Synthia H. Mellon, Metabolic Research Unit, Department of Medicine, University of California, San Francisco, California 94143

George R. Merriam, Developmental Endocrinology Branch, National Institutes of Health, Bethesda, Maryland 20205

Claude J. Migeon, Department of Pediatrics, Johns Hopkins University School of Medicine, Baltimore, Maryland 21205

Tina Mitchell, Department of Psychology, University of Virginia, Charlottesville, Virginia 22901

Mark E. Molitch, Center for Endocrinology, Metabolism, and Nutrition and Department of Medicine, Northwestern University Medical School, Chicago, Illinois 60611

Wayne V. Moore, Section of Pediatric Endocrinology, Department of Pediatrics, University of Kansas Medical Center, Kansas City, Kansas 66103

Cristina Mottola, Department of Biochemistry, University of Massachusetts Medical School, Worcester, Massachusetts 01605

S. Peter Nissley, Endocrine Section, Metabolism Branch, National Cancer Institute, National Institutes of Health, Bethesda, Maryland 20205

John V. O'Connor, Genentech, Inc., South San Francisco, California 94080

Yoshitomo Oka, Department of Biochemistry, University of Massachusetts Medical School, Worcester, Massachusetts 01605

Kenneth C. Olson, Genentech, Inc., South San Francisco, California 94080

M. R. Pandian, Nichols Institute, San Juan Capistrano, California 92675

John S. Parks, Department of Pediatrics, Emory University School of Medicine, Atlanta, Georgia 30322

Bohdan Pavlu, KabiVitrum AB, Research and Development, Departments of Biochemistry and Analytical Chemistry, S-112 87 Stockholm, Sweden

Marilyn Perrin, The Clayton Foundation Laboratories for Peptide Biology, The Salk Institute, La Jolla, California 92037

Athalia Pertzelan, Institute of Pediatric and Adolescent Endocrinology, Beilinson Medical Center, Petah Tikva 49 100, Israel, and Sackler School of Medicine, Tel Aviv University, Tel Aviv, Israel

John A. Phillips III, Division of Genetics, Vanderbilt University School of Medicine, Nashville, Tennessee 37232

L. S. Phillips, Division of Endocrinology and Metabolism, Department of Medicine, Emory University School of Medicine, Atlanta, Georgia 30303

W. Jackson Pledger, Department of Pharmacology and Cancer Research Center, University of North Carolina, School of Medicine, Chapel Hill, North Carolina 27514

Leslie P. Plotnick, Department of Pediatrics, Johns Hopkins University School of Medicine, Baltimore, Maryland 21205

Michael A. Preece, Department of Growth and Development, Institute of Child Health, University of London, London WC1N 1EH, England

Salvatore Raiti, The National Hormone and Pituitary Program, University of Maryland School of Medicine, Baltimore, Maryland 21201

Matthew M. Rechler, Section on Growth and Development, Molecular, Cellular, and Nutritional Endocrinology Branch, National Institute of Arthritis, Diabetes, and Digestive and Kidney Diseases, National Institutes of Health, Bethesda, Maryland 20205

Seymour Reichlin, Endocrine Division, Department of Medicine, Tufts University School of Medicine, New England Medical Center, Inc., Boston, Massachusetts 02111

Jean Rivier, The Clayton Foundation Laboratories for Peptide Biology, The Salk Institute, La Jolla, California 92037

Alan D. Rogol, Department of Pediatrics, University of Virginia School of Medicine, Charlottesville, Virginia 22901

Joyce A. Romanus, Section on Growth and Development, Molecular, Cellular, and Nutritional Endocrinology Branch, National Institute of Arthritis, Diabetes, and Digestive and Kidney Diseases, National Institutes of Health, Bethesda, Maryland 20205

Stephen M. Rosenthal, Department of Pediatrics, University of California–San Francisco, San Francisco, California 94143

Michael J. Ross, Genentech, Inc., South San Francisco, California 94080

Daniel Rudman, Department of Medicine and Clinical Research Facility, Emory University School of Medicine, Atlanta, Georgia 30322

A. M. Rudolph, Department of Pediatrics, University of California–San Francisco, San Francisco, California 94143

William E. Russell, Department of Pediatrics, University of North Carolina, School of Medicine, Chapel Hill, North Carolina 27514

Libbie J. Russo, Section of Pediatric Endocrinology, Department of Pediatrics, University of Kansas Medical Center, Kansas City, Kansas 66103

Geneviève Sassolas, Centre de Médecine Nucléaire, Hôpital Neuro-Cardiologique, 69394 Lyon Cedex 3, France

Paul Sawchenko, Developmental Neurobiology Laboratory, The Salk Institute, La Jolla, California 92037

Jeanette M. Schepper, Department of Life Sciences, Indiana State University, Terre Haute, Indiana 47809

Elizabeth A. Schriock, Department of Pediatrics, University of California–San Francisco, San Francisco, California 94143

Hans Seifert, The Clayton Foundation Laboratories for Peptide Biology, The Salk Institute, La Jolla, California 92037

Morton B. Sigel, The Whittier Institute for Diabetes and Endocrinology, La Jolla, California 92037

Aviva Silbergeld, Institute of Pediatric and Adolescent Endocrinology, Beilinson Medical Center, Petah Tikva 49 100, Israel, and Sackler School of Medicine, Tel Aviv University, Tel Aviv, Israel

Bo Skoog, KabiVitrum AB, Research and Development, Departments of Biochemistry and Analytical Chemistry, S-112 87 Stockholm, Sweden

Eric P. Smith, Department of Pediatrics, University of North Carolina, School of Medicine, Chapel Hill, North Carolina 27514

Joachim Spiess, The Clayton Foundation Laboratories for Peptide Biology, The Salk Institute, La Jolla, California 92037

D. M. Styne, Department of Pediatrics, University of California–Davis, Davis, California 95616

Marjorie E. Svoboda, Department of Pediatrics, University of North Carolina, School of Medicine, Chapel Hill, North Carolina 27514

Larry Swanson, Developmental Neurobiology Laboratory, The Salk Institute, La Jolla, California 92037

Marta Szabo, Division of Endocrinology and Metabolism, Michael Reese Hospital, Chicago, Illinois 60616

J. M. Tanner, Department of Growth and Development, Institute of Child Health, University of London, London WC1N 1EH, England

Jeffrey E. Terrell, Section on Growth and Development, Molecular, Cellular, and Nutritional Endocrinology Branch, National Institute of Arthritis, Diabetes, and Digestive and Kidney Diseases, National Institutes of Health, Bethesda, Maryland 20205

Jennifer L. Thominet, Division of Endocrinology and Metabolism, Department of Internal Medicine, University of Cincinnati College of Medicine, Cincinnati, Ohio 45267

Michael O. Thorner, Department of Pediatrics, University of Virginia School of Medicine, Charlottesville, Virginia 22901

Robert A. Tolman, Endocrinology Research Program, Division of Diabetes, Endocrinology and Metabolic Diseases, National Institute of Arthritis, Diabetes and Digestive and Kidney Diseases, National Institutes of Health, Bethesda, Maryland 20205

Gene F. Tutwiler, McNeil Pharmaceutical, Spring House, Pennsylvania 08540

Louis E. Underwood, Department of Pediatrics, University of North Carolina, School of Medicine, Chapel Hill, North Carolina 27514

T. G. Unterman, Center for Endocrinology, Metabolism and Nutrition, Northwestern University Medical School, Chicago, Illinois 60611

Wylie Vale, The Clayton Foundation Laboratories for Peptide Biology, The Salk Institute, La Jolla, California 92037

Mary Lee Vance, Department of Pediatrics, University of Virginia School of Medicine, Charlottesville, Virginia 22901

Willard P. Vanderlaan, The Whittier Institute for Diabetes and Endocrinology, La Jolla, California 92037

G. Van Vliet, Department of Pediatrics, University of California–San Francisco, San Francisco, California 94143

Judson J. Van Wyk, Department of Pediatrics, University of North Carolina, School of Medicine, Chapel Hill, North Carolina 27514

Joan Vaughan, The Clayton Foundation Laboratories for Peptide Biology, The Salk Institute, La Jolla, California 92037

W. B. Wehrenberg, Neuroendocrinology Laboratories, Salk Institute, San Diego, California 92037

Lisa L. Wheeler, Department of Life Sciences, Indiana State University, Terre Haute, Indiana 47809

Robert M. White, Howard University Cancer Center, Departments of Oncology and Medicine, Howard University School of Medicine, Washington, D.C. 20059

Harvey J. Whitfield, Section on Growth and Development, Molecular, Cellular, and Nutritional Endocrinology Branch, National Institute of Arthritis, Diabetes, and Digestive and Kidney Diseases, National Insitutes of Health, Bethesda, Maryland 20205

Anders Wichman, KabiVitrum AB, Research and Development, Departments of Biochemistry and Analytical Chemistry, S-112 87 Stockholm, Sweden

Kazuyo Yamada, Memorial Sloan-Kettering Cancer Center and The Sloan-Kettering Division, Graduate School of Medical Sciences, Cornell University, New York, New York 10021

Yvonne W.-H. Yang, Section on Growth and Development, Molecular, Cellular, and Nutritional Endocrinology Branch, National Institute of Arthritis, Diabetes, and Digestive and Kidney Diseases, National Institutes of Health, Bethesda, Maryland 20205

Kin-Tak Yu, Department of Biochemistry, University of Massachusetts Medical School, Worcester, Massachusetts 01605

Klaus Zuppinger, Inselspital, CH-3010 Bern, Switzerland

Index